AFLATOXINS – BIOCHEMISTRY AND MOLECULAR BIOLOGY

Edited by **Ramón Gerardo Guevara-González**

Aflatoxins – Biochemistry and Molecular Biology
Edited by Ramón Gerardo Guevara-González

Published by InTech

Janeza Trdine 9, 51000 Rijeka, Croatia

Publishing Process Manager Petra Nenadic

Technical Editor Teodora Smiljanic

Cover Designer InTech Design Team

Additional hard copies can be obtained from orders@intechopen.com

Aflatoxins – Biochemistry and Molecular Biology,
Edited by Ramón Gerardo Guevara-González
p. cm.
ISBN 978-953-307-395-8

For EU product safety concerns:
IN TECH d.o.o., Prolaz Marije Krucifikse Kozulić 3, 51000 Rijeka, Croatia,
info@intechopen.com or visit our website at intechopen.com.

Meet the editor

Dr. Ramón Gerardo Guevara-Gonzalez, CA is a professor member of Biosystems engineering group, Faculty of Engineering, Universidad Autonoma de Queretaro, Mexico. He has published 69 articles related to areas of molecular biology and biochemistry of plants and microorganisms. He is a member of the national researchers systems (SNI) from Mexico. He has a recognition by the International Scientific Information in 2000, as the author of an article considered classic. He has worked aspects of biochemistry and molecular biology of aflatoxins especially antimutagenic effects of natural extracts.

Contents

Preface

Aflatoxins are produced by Aspergillus flavi group species and are thought to be one of the most cancerous natural substances known. Economically and biologically the most important fungal species able to produce the aflatoxins are Aspergillus flavus and Aspergillus parasiticus. The biosynthesis of aflatoxins, as all secondary metabolites, is strongly dependent on growth conditions such as substrate composition or physical factors like pH, water activity, temperature or modified atmospheres. Depending on the particular combination of external growth parameters the biosynthesis of aflatoxin can either be completely inhibited, albeit normal growth is still possible or the biosynthesis pathway can be fully activated. Knowledge about these relationships enables an assessment of which parameter combinations can control aflatoxin biosynthesis or which are conducive to phenotypic aflatoxin production. All these aspects are fascinating and strongly growing in knowledge based on the work of excellent researchers as those invited to write every chapter presented in this book. Finally, this book is an attempt to provide a wide and current approach of top studies in aflatoxins biochemistry and molecular biology, as well as some general aspects to researchers interested in this field.

Dr. Ramon Gerardo Guevara-Gonzalez
Professor
Biosystems Engineering Group
School of Engineering
Queretaro Autonomous University
Queretaro, Qro, Mexico

Part 1

Biosynthesis and Biotransformation

1

Biotransformation of Aflatoxin B1 and Its Relationship with the Differential Toxicological Response to Aflatoxin in Commercial Poultry Species

Gonzalo J. Diaz and Hansen W. Murcia
Facultad de Medicina Veterinaria y de Zootecnia
Universidad Nacional de Colombia, Bogotá D.C.,
Colombia

1. Introduction

Aflatoxin B1 (AFB1) is a highly toxic compound (LD_{50} = 1-50 mg/kg) for most animal species, although it is extremely toxic (LD50 < 1 mg/kg) for some highly susceptible species such as pigs, dogs, cats, rainbow trouts, and ducklings. The toxic effects of AFB1 are both dose and time dependent and two distinct forms of aflatoxicosis, namely acute and chronic, can be distinguished depending on the level and length of time of aflatoxin exposure. In many species acute poisoning is characterized by an acute hepatotoxic disease that manifests itself with depression, anorexia, icterus, and hemorrhages. Histologic hepatic lesions include periportal necrosis associated with bile duct proliferation and oval cell hyperplasia. Chronic aflatoxicosis resulting from regular low-level dietary intake of aflatoxins causes unspecific signs such as reduced weight gain, reduced feed intake, and reduced feed conversion in pigs and poultry, and reduced milk yield in cows. Another effect of chronic exposure is aflatoxin-induced hepatocellular carcinoma, bile duct hyperplasia and hepatic steatosis (fatty liver). However, these effects are species-specific and not all animals exposed to aflatoxin develop liver cancer. For example, the only poultry species that develops hepatocellular carcinoma after AFB1 exposure is the duck.

Differences in the susceptibility to acute and chronic AFB1 toxicosis have been observed among animals of different species. Animals having the highest sensitivity are the duckling, piglet, rabbit, dog and cat, while chickens, mice, hamsters, and chinchillas are relatively resistant. Further, mature animals are generally more resistant to AFB1 than young ones and females are more resistant than males. In general, in commercial poultry species, intake of feed contaminated with AFB1 results primarily in liver damage (the target organ of AFB1 is the liver), associated with immunosuppression, poor performance, and even mortality when the dietary levels are high enough. However, there is wide variability in specific species sensitivity to AFB1 and the susceptibility ranges from ducklings > turkey poults > goslings > pheasant chicks > quail chicks > chicks (Leeson et al., 1995). Even though there is still no clear explanation for this differential sensitivity, differences in susceptibility could be due to differences in AFB1 biotransformation pathways among species. The aim of the present chapter is to review the current knowledge on AFB1 biotransformation, with emphasis on commercial poultry species, and to correlate this information with the *in vivo* susceptibility to AFB1 in these species.

2. Biotransformation of aflatoxin B1

In general, the metabolism or biotransformation of xenobiotics (chemicals foreign to the organism) is a process aimed at converting the original molecules into more hydrophilic compounds readily excretable in the urine (by the kidney) or in the bile (by the liver). It has traditionally been conceptualized that this process occurs in two phases known as Phase I and Phase II, although some authors argue that this classification is no longer tenable and should be eliminated (Josephy et al., 2005). Phase I metabolism consists mainly of enzyme-mediated hydrolysis, reduction and oxidation reactions, while Phase II metabolism involves conjugation reactions of the original compound or the compound modified by a previous Phase I reaction. The current state of knowledge on the metabolism of AFB1 in different avian and mammalian species is summarized in Figure 1. As Figure 1 shows, a wide array of metabolites can be

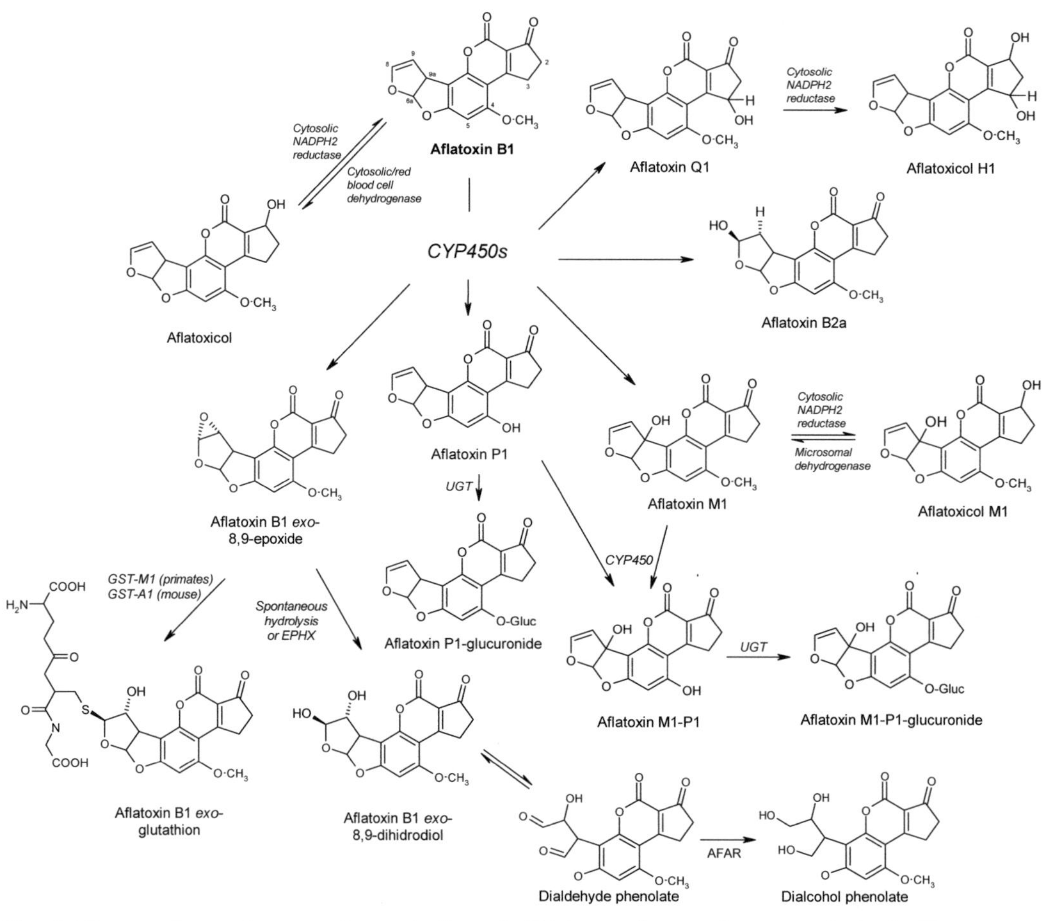

Fig. 1. Biotransformation reactions of aflatoxin B1 in poultry and mammals, including humans. The main CYP450s involved in these reactions are CYP1A1, CYP1A2, CYP2A6 and CYP3A4. Not all reactions occur in a single species.

produced directly from AFB1 (by oxidation and reduction reactions) or indirectly by further biotransformation of the metabolites formed. However, not all of these reactions occur in a single species and, in fact, only a few of them have been reported in poultry. Most AFB1 Phase I reactions are oxidations catalyzed by cytochrome P450 (CYP450) enzymes, but one reaction is catalyzed by a cytosolic reductase, corresponding to the reduction of AFB1 to aflatoxicol (AFL). Phase II reactions are limited to conjugation of the metabolite AFB1-*exo*-8,9-epoxide (AFBO) with glutathione (GSH, γ-glutamyl-cysteinyl-glycine), and conjugation of aflatoxins P1 and M1-P1 with glucuronic acid. Conjugation of AFBO with GSH is a nucleophilic trapping process catalyzed by specific glutathione transferase (GST) enzymes. The AFBO may also be hydrolyzed by an epoxide hydrolase (EPHX) to form AFB1-*exo*-8,9-dihydrodiol, although this reaction may also occur spontaneously. The dihydrodiol is in equilibrium with the dialdehyde phenolate form, which can be reduced by AFB1 aldehyde reductase (AFAR), an enzyme that catalyzes the NADPH-dependent reduction of the dialdehyde to dialcohol phenolate (Guengerich et al., 2001).

The translocation of xenobiotics across cell membranes by specific proteins known as transporters has been termed by some as "Phase III" metabolism. However, this process does not involve any modification of the xenobiotic structure and therefore it cannot be termed metabolism. This process, however, may have important implications on the toxic effect of a xenobiotic, particularly if the specific transporter involved in the translocation of the compound is not expressed normally, presents a genetic abnormality or becomes saturated. One transporter that has been identified as responsible for the translocation of a mycotoxin from the sinusoidal hepatic space into the hepatocyte is OATP (organic anion transporter polypeptide), which transports ochratoxin A (Diaz, 2000). However, no transporters for AFB1 have yet been described.

2.1 Phase I metabolism of aflatoxin B1

As mentioned before, the Phase I metabolism of AFB1 is carried out mainly by members of the CYP450 superfamily of enzymes. Their name comes from the absorption maximum at 450 nm when the reduced form complexes with carbon monoxide (Omura & Sato, 1964). CYP450s are membrane bound enzymes that can be isolated in the so-called microsomal fraction which is formed from endoplasmic reticulum when the cell is homogenized and fractionated by differential ultracentrifugation; microsomal vesicles are mainly fragments of the endoplasmic reticulum in which most of the enzyme activity is retained. The highest concentration of CYP450s involved in xenobiotic biotransformation is found in the endoplasmic reticulum of hepatocytes but CYP450s are present in virtually every tissue. CYP450s are classified into families identified by a number (e.g., 1, 2, 3, and 4), subfamilies identified by a letter (e.g., 2A, 2B, 2D, and 2E), and individual members identified by another number (e.g. CYP2A6, CYP2E1). Collectively, CYP450 enzymes participate in a variety of oxidative reactions with lipophilic xenobiotics and endogenous substrates including hydroxylation of an aliphatic or aromatic carbon, epoxidation of a double bond, heteroatom (*S*-, *N*- and *I*-) oxygenation and *N*-hydroxylation, heteroatom (*O*-, *S*-, and *N*-) dealkylation, oxidative group transfer, cleavage of esters, and dehydrogenation (Parkinson & Ogilvie, 2008). In regards to AFB1, CYP450s can hydroxylate, hydrate, *O*-demethylate, and epoxidate the molecule.

2.1.1 Hydroxylation and hydration of aflatoxin B1

CYP450s can produce at least three monohydroxylated metabolites from AFB1, namely aflatoxins M1 (AFM1), Q1 (AFQ1), and B_{2a} (AFB_{2a}) (Fig. 1). AFM1 was first isolated from the milk of cows and rats fed AFB1-contaminated peanut meal and it was initially termed "milk toxin" (de Iongh et al., 1964). It was later discovered that AFM1 is not a metabolite exclusive of mammals and, in fact, it is produced by crude or isolated microsomal liver preparations from many non-mammalian species. For example, AFM1 was found in most tissues of chickens receiving a diet containing 2,057 ppb AFB1 for 35 days (Chen et al., 1984); the highest level was found in the liver and kidneys, which relates to the important role of these organs in the biotransformation and elimination of xenobiotics, respectively.

AFQ1 results from the 3α-hydroxylation of AFB1 and it was first discovered as a major metabolite of AFB1 from monkey liver microsomal incubations (Masri et al., 1974). The predominant enzyme responsible for AFQ1 formation in human liver microsomes is CYP3A4 (Raney et al., 1992b) and AFQ1 is considered to be a major metabolite of AFB1 in humans and monkeys *in vitro* (Hsieh et al., 1974). Although AFQ1 has been detected as a minor metabolite of chicken and duck microsomal preparations (Leeson et al., 1995) it is considered to be a significant detoxication pathway of AFB1 (Raney et al., 1992b). In fact, AFQ1 is about 18 times less toxic for chicken embryos than AFB1 and it is not mutagenic in the *Salmonella typhimurium* TA 1538 test (Hsieh et al., 1974).

The hydration of the vinyl ether double bond (C8-C9) of AFB1 produces the 8-hydroxy derivative or hemiacetal, also known as AFB_{2a}. This metabolite was discovered in 1966 and, interestingly, it can be produced enzymatically (by both higher organisms and microbial metabolism), by photochemical degradation of AFB1, and by the treatment of AFB1 with acid (Lillehoj & Ciegler, 1969). The formation of the hemiacetal is difficult to assess *in vitro* because of strong protein binding, which probably involves the formation of Schiff bases with free amino groups (Patterson & Roberts, 1972). The ability of certain species to metabolize AFB1 into its hemiacetal at higher rates than others constitutes an important aspect of the resistance to the toxin, since the toxicity of AFB_{2a} is much lower than that of the parent compound. For instance, AFB_{2a} has been shown to be not toxic to chicken embryos at levels 100 times the LD_{50} of AFB1 (Leeson et al., 1995), and the administration of 1.2 mg of AFB_{2a} to one-day-old ducklings does not produce the adverse effects caused by the same dose of AFB1 (Lillehoj & Ciegler, 1969).

It has been generally considered that the monohydroxylated metabolites of AFB1 are "detoxified" forms of the toxin, which is probably the case for aflatoxins B_{2a} and Q1; however, AFM1 cannot be considered a detoxication product of AFB1. AFM1 is cytotoxic and carcinogenic in several experimental models and in ducklings its acute toxicity is similar to that of AFB1 (12 and 16 μg/duckling for AFB1 and AFM1, respectively). Also in ducklings, both AFB1 and AFM1 induce similar liver lesions; however, AFB1 induces only mild degenerative changes in the renal convoluted tubules whereas AFM1 causes both degenerative changes and necrosis of the tubules (Purchase, 1967).

2.1.2 *O*-Demethylation of aflatoxin B1

Another CYP450-mediated reaction of rat, mouse, guinea pig and rabbit livers is the 4-*O*-demethylation of AFB1. The phenolic product formed was initially isolated from monkey urine (Dalezios et al., 1971) and named aflatoxin P1 (the P comes from the word primate). AFP1 can be hydroxylated at the 9a position to form 4,9a-dihydroxyaflatoxin B1 (AFM1-P1, see Fig. 1), although this compound can also originate from AFM1 (Eaton et al., 1988). AFP1 is generally considered a detoxication product, mainly because it is efficiently conjugated with glucuronic acid (Holeski et al., 1987). There is no evidence that AFP1 or its 9a-hydroxy derivative are produced by any avian species (Leeson et al., 1995).

2.1.3 Epoxidation of aflatoxin B1

Another metabolic pathway of the vinyl ether double bond present in the AFB1 furofuran ring is its epoxidation. The resultant product, AFB1-*exo*-8,9-epoxide (AFBO), is an unstable, highly reactive compound, with a half-life of about one second in neutral aqueous buffer (Johnson et al., 1996), that exerts its toxic effects by binding with cellular components, particularly protein, DNA and RNA nucleophilic sites. AFBO is considered to be the active form responsible for the carcinogenicity and mutagenicity of AFB1 (Guengerich et al., 1998). The *endo*-8,9-epoxide of AFB1 can also be formed by rat and human microsomes (Raney et al., 1992a), but this form of the epoxide is not reactive. Once AFBO is formed it may be hydrolyzed, either catalytically or spontaneously, to form AFB1-8,9-dihydrodiol (AFB1-dhd) or it may be trapped with GSH. If AFB1-dhd is formed it may suffer a base-catalyzed furofuran ring opening to a dialdehyde (AFB1 α-hydroxydialdehyde), which is able to bind to lysine residues in proteins. The enzyme AFAR (see section 2) can protect against the dialdehyde by catalyzing its reduction to a dialcohol which is excreted in the urine either as the dialcohol itself or as a monoalcohol (Guengerich et al., 2001). AFAR activity, however, does not correlate with *in vivo* sensitivity to AFB1 in selected mammalian models (hamster, mouse, rat and pig) as it was demonstrated by Tulayakul et al. (2005). AFAR has been evidenced by immunoblot in the liver of turkeys (Klein et al., 2002) but its activity has not been investigated in this or any other avian species.

2.1.4 Reduction of aflatoxin B1

The C1 carbonyl group present in the cyclopentanone function of AFB1 can be reduced to a hydroxy group to form the corresponding cyclopentol AFL (Fig. 1). This reaction is not catalyzed by microsomal enzymes but by a cytosolic NADPH-dependent enzyme that in the case of the chicken has an estimated molecular weight of 46.5 KDa and is inhibited by the 17-ketosteroids androsterone, dehydroisoandrosterone and estrone (Chen et al., 1981). Formation of AFL was first reported in chicken, duck, turkey and rabbit liver cytosol (Patterson & Roberts, 1971), and it also occurs in quail (Lozano & Diaz, 2006). However, little or no activity has been observed in guinea pig, mouse or rat liver cytosol (Patterson & Roberts, 1971). AFL can be oxidized back to AFB1 by liver cytosol (Patterson & Roberts, 1972) and by red blood cells from several species (Kumagai et al., 1983). For this reason, AFL is considered to be a "storage" form of AFB1. The ratio of AFB1 reductase activity to AFL dehydrogenase activity *in vitro* has been observed to be higher in species that are extremely sensitive to acute aflatoxicosis (Wong & Hsieh, 1978), but the significance of this finding in poultry species remains to be determined. AFL cannot be considered a detoxified product of AFB1 since it is carcinogenic and mutagenic, it is acutely toxic to rabbits and it is correlated with susceptibility to AFB1 in some species (Kumagai et al., 1983). Further, AFL has the ability of inducing DNA adduct formation because the double bond between C-8 and C-9 is still present in this metabolite (Loveland et al., 1987). Conjugation of AFL with either glucuronic acid or sulfate would potentially be a true detoxication reaction because this step would prevent AFL from being reconverted to AFB1.

2.1.5 Reduction of aflatoxin B1 metabolites

The hydroxylated metabolites AFM1 and AFQ1 can also undergo the cytosolic reduction of the C1 carbonyl group in a reaction analogous to the reduction of AFB1 to AFL. The reduced metabolites of AFM1 and AFQ1 have been named aflatoxicol M1 (Salhab et al., 1977; Loveland et al., 1983) and aflatoxicol H1 (Salhab & Hsieh, 1975), respectively. Aflatoxicol H1 is a major metabolite of AFB1 produced by human and rhesus monkey livers *in vitro* (Salhab & Hsieh, 1975). Aflatoxicol M1 can also be produced from AFL and it can be oxidized back to AFM1 by a carbon monoxide-insensitive dehydrogenase activity associated with human liver microsomes (Salhab et al., 1977).

2.2 Phase II metabolism of aflatoxin B1

The most studied Phase II biotransformation reaction of any AFB1 metabolite is the nucleophilic trapping process in which GSH reacts with the electrophilic metabolite AFBO. Conjugation of AFBO with GSH is catalyzed by glutathione transferases (GST, 2.5.1.18), a superfamily of enzymes responsible for a wide range of reactions in which the GSH thiolate anion participates as a nucleophile. These intracellular proteins are found in most aerobic eukaryotes and prokaryotes, and protect cells against chemically-induced toxicity and stress by catalyzing the conjugation of the thiol group of GSH and an electrophilic moiety in the substrate. GSTs are considered the single most important family of enzymes involved in the metabolism of alkylating compounds and are present in most tissues, with high concentrations in the liver, intestine, kidney, testis, adrenal, and lung (Josephy & Mannervik, 2006). The soluble GSTs are subdivided into classes based on sequence similarities, a classification system analogous to that of the CYP450s. The classes are designated by the names of the Greek letters: Alpha, Mu, Pi, and so on, abbreviated in Roman capitals: A, M, P, etc. Within the class, proteins are numbered using Arabic numerals (e.g. GST A1, GST A2, etc.) and specific members are identified by the two monomeric units comprising the enzyme (e.g. GST A1-1, GST A2-2, GST M1-1, etc.). The microsomal GSTs (MGSTs) and its related membrane-bound proteins are structurally different from the soluble GSTs, forming a separate superfamily known as MAPEG (membrane-associated proteins in eicosanoid and GSH metabolism). MGSTs are not involved in the metabolism of AFB1 metabolites.
Another conjugation reaction reported for AFB1 metabolites is the conjugation of AFP1 and its 9a-hydroxy metabolite (aflatoxin M1-P1) with glucuronic acid. This conjugation has only been reported in rats and mice (Holeski et al., 1987; Eaton et al., 1988) and leads to the synthesis of detoxified products. Conjugation with glucuronic acid is catalyzed by enzymes known as UPD-glucuronosyltransferases (UGTs, Josephy & Mannevick, 2006), but the specific UGT involved in the conjugation of AFP1 and AFM1-P1 has not been described yet.

3. Biotransformation of aflatoxin B1 in poultry and its relationship with *in vivo* sensitivity

The role of poultry in mycotoxin research in general and aflatoxin research in particular is historically highly relevant since aflatoxins were discovered after a toxic Brazilian peanut meal caused the death of more than 100,000 turkeys of different ages (4-16 weeks) in England during the summer of 1960 (Blount, 1961). This mycotoxicosis outbreak was the first one ever reported for any animal species and for any mycotoxin. Initially only turkeys were affected but later ducklings and pheasants were also killed by the same misterious "X disease". Interestingly, no chickens were reported to have died from this new disease. Research conducted with poultry after the discovery of aflatoxins (reviewed by Leeson et al., 1995) has clearly shown that the *Gallus sp.* (which includes the modern commercial meat-type chickens and laying hens) is extremely resistant to aflatoxins while other commercial poultry species are highly sensitive. For instance, whereas ducklings and turkey poults exhibit 100% mortality at dietary levels of 1 ppm (Muller et al., 1970), chicks can tolerate 3 ppm in the diet without showing any observable adverse effects (Diaz & Sugahara, 1995). Interestingly, chickens are not only highly resistant to the adverse effects of AFB1 but some studies have reported a modest enhancement in the body weight of chickens exposed to dietary aflatoxins, a finding that has been characterized as an hormetic-type dose-response relationship (Diaz et al., 2008). At the molecular level, at least four mechanisms of action could potentially play a role in the resistance to AFB1: low formation of the putative reactive metabolite (AFBO) and/or AFL, high detoxication of the AFBO and/or AFL formed, intestinal biotransformation of AFB1 before it can reach the liver ("first-pass action"), and increased AFB1 (or toxic metabolites) efflux from the cells. It is important to note that translocation of xenobiotics and their metabolites from the hepatocytes (efflux) mediated by specific basolateral and canalicular transporters (Diaz, 2000) -a process sometimes referred to as Phase III metabolism-, has not been investigated for AFB1 in any species. However, both Phase I and Phase II metabolism appear to have a profound effect on the differential *in vivo* response to AFB1 in commercial poultry species. The formation of AFBO (by CYP450s) and AFL (by a cytosolic reductase) as well as the scarce information available about detoxication of AFBO through nucleophilic trapping with GSH in poultry will be discussed below.

3.1 Phase I metabolism of aflatoxin B1 in commercial poultry species

Research conducted by our group (Lozano & Diaz, 2006) showed that the microsomal and cytosolic biotransformation of AFB1 in chickens, quail, ducks and turkeys results in the formation of two major metabolites: AFBO (microsomes) and AFL (cytosol). The relative *in vivo* sensitivity to AFB1 in these species corresponds to ducks > turkeys > quail > chicken, and the aim of this work was to try to correlate the toxicological biochemical findings with the reported *in vivo* sensitivity. Using liver microsomal incubations it was demonstrated that turkeys produce the highest amount of AFBO (detected either as AFB1-dhd or AFB1-GSH) while chickens produce the least; duck and quails produce intermediate amounts (Fig. 2). AFB1 consumption (rate of AFB1 disappearance from the microsomal incubations) was also highest in turkeys, lowest in chickens and intermediate in quail and ducks. Interestingly, these two variables (AFBO production and AFB1 consumption) were highly correlated in the four species evaluated (Fig. 2).

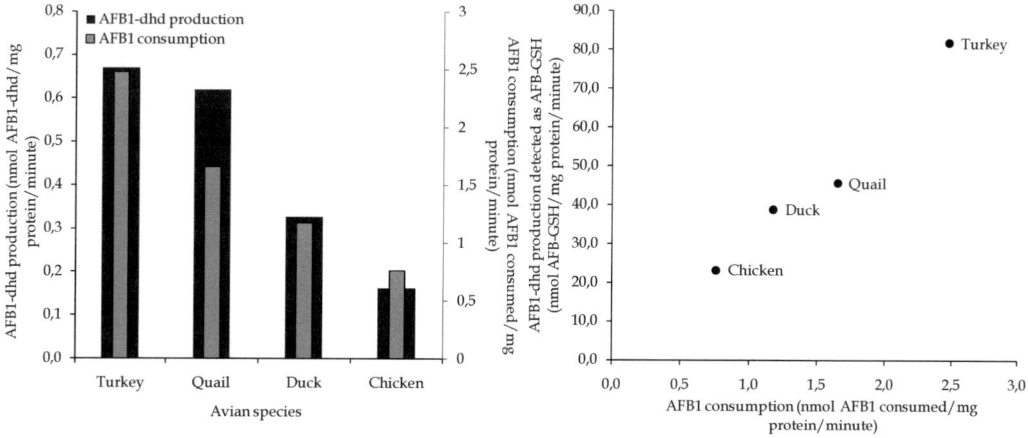

Fig. 2. AFBO production (measured as AFB1-dhd) and AFB1 consumption in turkey, quail, duck and chicken microsomal incubations (left) and relationship between AFBO formation (measured as AFB1-GSH) and AFB1 consumption (right).

Both biotransformation variables (AFBO formation and AFB1 disappearance) correlate well with the *in vivo* sensitivity observed for turkeys, quail and chickens (turkeys being highly sensitive, chickens being the most resistant and quail having intermediate sensitivity). However, other factor(s) besides AFBO formation and AFB1 consumption must play a role in the extraordinary high sensitivity of ducks to AFB1 because these biochemical variables did not correlate with the *in vivo* sensitivity for this particular species [ducks exhibit the highest *in vivo* sensitivity to AFB1 among these poultry species, not turkeys, as Rawal et al. (2010a) affirm].

The cytosolic metabolism of AFB1 in the same four poultry species shows a different trend compared with the microsomal metabolism (Fig. 3). Turkeys are again the largest producers of the cytosolic metabolite AFL but are followed by ducks, chickens and quail (instead of quail, ducks and chickens as it is observed for microsomal metabolism). As discussed before (see section 2.1.4), AFL is a toxic metabolite of AFB1 and it cannot be considered a detoxication product; therefore, it would be expected that sensitive species produce more AFL than resistant ones. However, no correlation between AFL production and *in vivo* sensitivity was observed. For instance, quail produced the lowest amount of AFL and it exhibits intermediate sensitivity to AFB1, while ducks, which are the most sensitive species, produced much less AFL than turkeys. AFB1 consumption by cytosol (rate of AFB1 disappearance from cytosolic incubations) was highest for the chicken, followed by turkeys, ducks and quail and there was no correlation between AFL formation and AFB1 consumption (Fig. 3). Further, as it was observed for AFL formation, there was no correlation between AFB1 disappearance from cytosol and *in vivo* sensitivity to AFB1. Investigation of the potential conjugation reactions of AFL might clarify the role of AFL formation on the *in vivo* sensitivity to AFB1 in poultry. It is possible that the high resistance of chickens to AFB1 might be due to an efficient reduction of AFB1 to AFL followed by conjugation and elimination of the AFL conjugate. Interestingly, it has been demonstrated that chick liver possesses much higher AFB1 reductase activity than duckling or rat liver (Chen et al., 1981).

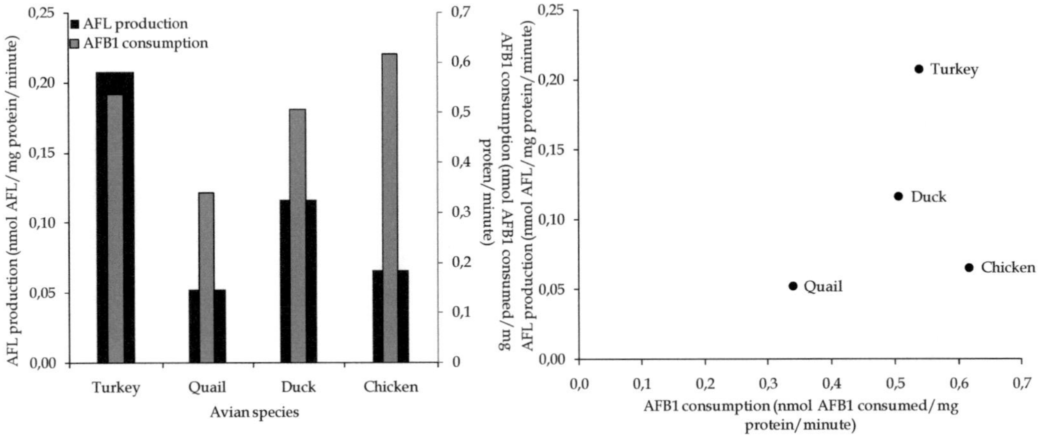

Fig. 3. AFL production and AFB1 consumption in turkey, quail, duck and chicken cytosolic incubations (left) and relationship between AFL formation and AFB1 consumption (right).

Even though the studies of Klein et al. (2000) in turkeys, and Lozano & Diaz (2006) in turkeys, chickens, ducks and quail had clearly demonstrated that hepatic microsomes from poultry were capable of bioactivating AFB1 into AFBO, there was only scarce information on the specific CYP450 enzymes responsible for this biotransformation reaction and it was limited to turkeys (Klein et al., 2000; Yip & Coulombe, 2006). In contrast, in humans, at least three CYP450s had been identified as responsible for AFB1 bioactivation to AFBO (CYP1A2, CYP2A6 and CYP3A4) (Omiecinski et al., 1999; Hasler et al., 1999), and there was evidence that the CYP3A4 human enzyme was the most efficient (Guengerich & Shimada, 1998). In view of this lack of information a series of studies were conducted by our group (Diaz et al., 2010a, 2010b, 2010c) in order to investigate which specific avian CYP450 orthologs were responsible for the bioactivation of AFB1 into AFBO. These studies were conducted by using specific human CYP450 inhibitors (α-naphthoflavone for CYP1A1/2, furafylline for CYP1A2, 8-methoxypsoralen for CYP2A6 and troleandomycin for CYP3A4), by correlating AFBO formation with human prototype substrate activity (ethoxyresorufin *O*-deethylation for CYP1A1/2, methoxyresorufin *O*-deethylation for CYP1A2, coumarin 7-hydroxylation for CYP2A6 and nifedipine oxidation for CYP3A4) and by investigating the presence of ortholog proteins in avian liver by immunoblot using antibodies specific against human CYP1A1, CYP1A2, CYP2A6 and CYP3A4. These series of studies revealed that the avian CYP2A6 ortholog is the main CYP450 enzyme responsible for the bioactivation of AFB1 into its epoxide form in all poultry species investigated. Evidences for this conclusion include the fact that AFBO production was inhibited by the CYP2A6 inhibitor 8-methoxypsoralen and that a significant correlation existed between coumarin 7-hydroxylation and AFB1 epoxidation activity in all species studied (Table 1). The finding of a protein by immunoblot using rabbit anti-human CYP450 polyclonal antibodies directed against the human CYP2A6 enzyme confirmed the existence of an immunoreactive protein in all birds studied (the putative CYP2A6 avian ortholog). These studies demonstrated for the first time the existence of the CYP2A6 human ortholog in avian species and they were the first reporting the role of this enzyme in AFB1 bioactivation in avian liver.

Poultry Species	7-Ethoxyresorufin-*O*-demethylation (CYP1A1/2)	7-Methoxyresorufin-*O*-demethylation (CYP1A2)	Coumarin 7-hydroxilation (CYP2A6)	Nifedipine oxidation (CYP3A4)
Turkey	0.32	-0.76	**0.90**	0.73
Quail	-0.09	0.21	**0.78**	0.07
Duck	**0.81**	**0.82**	**0.68**	**0.88**
Chicken	0.25	0.46	**0.83**	-0.24

Table 1. Pearson correlation coefficients for aflatoxin B1 epoxidation vs. prototype substrate activities of selected human CYP450 enzymes. Correlations in bold numbers are statistically significant ($P \leq 0.01$).

In turkeys, quail and chickens the CYP1A1 ortholog seems to have a minor role in AFB1 bioactivation, while in ducks there are evidences that AFB1 bioactivation is carried out not only by the CYP2A6 and CYP1A1 orthologs but also by the CYP3A4 and CYP1A2. The fact that four CYP450 enzymes are involved in AFB1 bioactivation in ducks could partially explain the high sensitivity of this species to AFB1. In turkey liver, AFB1 was reported to be activated to AFBO by a CYP 1A ortholog (Klein et al., 2000) that later was identified as the turkey CYP1A5 on the basis of its 94.7% sequence identity to the CYP1A5 from chicken liver (Yip & Coulombe, 2006). This enzyme was suggested to correspond to the human ortholog CYP1A2 (Yip & Coulombe, 2006). However, using human prototype substrates and inhibitors, Diaz et al. (2010a) found evidence for AFB1 bioactivation by CYP1A1 but not by CYP1A2 in turkey liver microsomes. Interestingly, the turkey CYP1A5 has a high amino acid sequence homology not only with the human CYP1A2 (62%) but also with the human CYP1A1 (61%) as reported by the UniProtKB database (http://www.uniprot.org) and the National Center for Biotechnology Information database (http://www.ncbi.nlm.nih.gov). It is possible that the turkey CYP1A5 enzyme cloned by Yip & Coulombe (2006) may in fact correspond to the human CYP1A1 ortholog or, even more interesting, to both the CYP1A1 and 1A2 human orthologs. Murcia et al. (2011) found a very high correlation between EROD (CYP1A1/2) and MROD (CYP1A2) activities in turkey liver microsomes ($r=0.88$, $P<0.01$) a finding that suggests that CYP1A1 and CYP1A2 activities in turkey liver are catalyzed by the same enzyme (i.e., the avian CYP1A5). The role of CYP1A5 turkey activity on the bioactivation of AFB1 in turkeys is further supported by the work of Guarisco et al. (2008) who found that dietary supplementation of the antioxidant butylated hydroxytoluene (BHT) partially protected against the adverse effects of AFB1, an effect that was accompanied by a reduction in EROD and MROD activities in the liver.

In regards to CYP3A4, Klein et al. (2000) found that this enzyme plays a minor role in the bioactivation of AFB1 in turkeys. This finding, however, could not be substantiated by Diaz et al. (2010a) who found no correlation between nifedipine oxidation (an indicator of CYP3A4 activity) and AFBO formation, and no effect on AFBO formation when the prototype inhibitor of human CYP3A4 activity troleandomycin was used. Induction of CYP3A4 activity by BHT in turkeys (as evidenced by increased nifedipine oxidation) was correlated with decreased *in vivo* adverse effects of AFB1 (Guarisco et al., 2008), which further supports the notion that CYP3A4 is not involved in AFB1 bioactivation in turkeys. This finding is of interest since CYP3A4 has been shown to be an activator of aflatoxins B1 and G1 in humans and other species (Parkinson & Ogilvie, 2008); however, in humans, CYP3A enzymes can form the AFBO only at relatively high substrate concentrations (Ramsdell et al., 1991). In contrast with turkeys, however, CYP3A4 does appear to play a role on AFB1 bioactivation in ducks (Diaz et al., 2010b). Duck microsomes show a high correlation between nifedipine oxidation and AFB1 epoxidation (Table 1) but the use of the specific human CYP3A4 inhibitor troleandomycin did not reduce AFBO production (Diaz et al., 2010b). A recent study reports the cloning of a turkey CYP3A37 expressed in *E. coli* able to biotransform AFB1 into AFQ1 (and to a lesser extent to AFBO) with an amino acid sequence homology of 76% compared with the human CYP3A4 (Rawal et al., 2010b). In this study the use of the inhibitors erythromycin (specific for human CYP3A1/4) and 17α-ethynylestradiol (specific for human CYP3A4) completely inhibited the production of AFBO. The results of the studies conducted with the CYP3A4 turkey ortholog indicate that the turkey enzyme is not sensitive to the CYP3A4 human inhibitor troleandomycin but that it is sensitive to erythromycin and 17α-ethynylestradiol. If this lack of sensitivity to troleandomycin also applies for the duck CYP3A4 ortholog, this could explain the results of Diaz et al. (2010b) previously described. In regards to the findings of Rawal et al. (2010b), it is important to note that the fact that a cloned gene expressed in a heterologous system (e.g.

E. coli) biotransforms AFB1 does not necessarily mean that this is a reflection of the situation in a biological system. Heterologously expressed enzymes typically exhibit a much different behavior than native ones. For instance, the enzyme affinity for nifedipine oxidation activity in turkey liver microsomes is much higher than that of the heterologously expressed turkey CYP3A37 (K_M values of 21 and 98 μM, respectively) (Murcia et al., 2011; Rawal et al., 2010b). Both *in vivo* studies and *in vitro* hepatic microsomal metabolism suggest that the turkey ortholog of the human CYP3A4 is most likely not involved in AFB1 bioactivation.

Large interspecies differences in enzyme kinetics and enzymatic constants for AFB1 epoxidation also exist among poultry species (Diaz et al., 2010a, 2010b, 2010c). Non-linear regression of these variables showed that turkey enzymes have the highest affinity and highest biotransformation rate of AFB1, as evidenced by the lowest K_M and highest $Vmax$ values compared with quail, duck and chicken enzymes (Fig. 4). This finding correlates well with the high *in vivo* sensitivity of turkeys to AFB1. In contrast, chicken enzymes showed the lowest affinity (highest K_M) and lowest biotransformation rate (lowest $Vmax$), findings that also correlate well with the high resistance of chickens to AFB1. Quail, a species with intermediate sensitivity to AFB1, also exhibited intermediate values for enzyme affinity and catalytic rate. In ducks, however, the enzymatic parameters of AFB1 biotransformation could not explain their high sensitivity to AFB1 since they had the second lowest catalytic rate ($Vmax$) and the third lowest enzyme affinity (K_M) for AFB1 (it was expected that ducks had the highest $Vmax$ and lowest K_M).

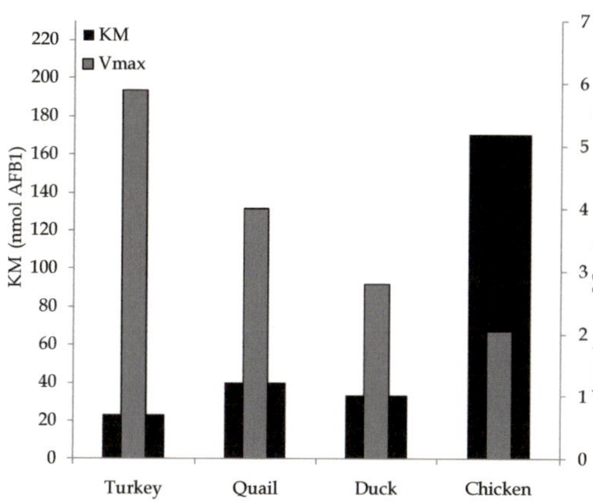

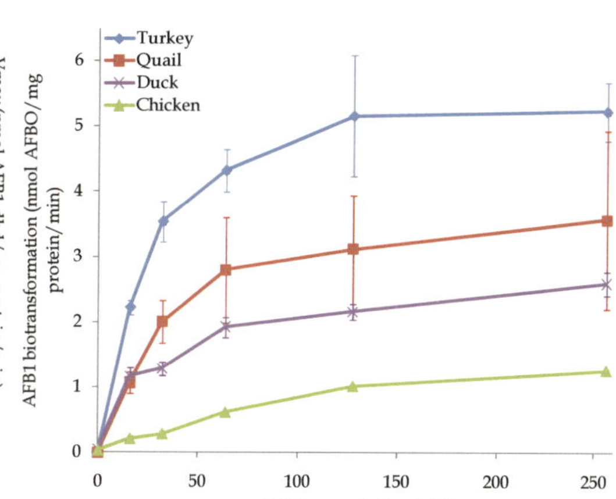

Fig. 4. Enzymatic constants K_M and $Vmax$ (left) and enzyme kinetics (right) of aflatoxin B1 epoxidation activity in liver microsomes of four poultry species.

Information for some CYP450 enzymes in turkey, chicken and quail can be found in the databases mentioned before (i.e. UniProtKB and GeneBank). Sequences for turkey CYP1A5 and CYP3A37, chicken CYP1A1, CYP1A4, CYP1A5 and CYP3A80, and Japanese quail CYP1A1, CYP1A4 and CYP1A5 have been reported. Surprisingly, however, there are no sequences reported for CYP2A6 despite the biochemical evidence for its existence in birds. As expected, a comparison of the human and avian CYP450 enzymes reveals a higher similarity among avian orthologs compared to human orthologs. Differences in protein structure between avian and human CYP450 enzymes could explain the differential response of the avian CYP450 orthologs to the human prototype substrate and inhibitors, which, nevertheless, are still useful tools in the investigation of CYP450 enzymes in birds.

3.2 Trapping and conjugation of aflatoxin B1 metabolites in commercial poultry species

Even though the ability to bioactivate AFB1 into AFBO is critical in the toxicological response to AFB1, in some species it is the ability to trap AFBO with GSH which ultimately determines the degree of AFB1-induced liver damage. For instance, both rats and mice exhibit high bioactivation rates of AFB1; however, mice are resistant to the hepatic carcinogenic effects of AFB1 while rats develop hepatocellular carcinoma. The reason for this differential response lies in the constitutive expression of high levels of an Alpha-class GST that catalyzes the trapping of AFBO in the mouse that is only expressed at low levels in the rat (Esaki & Kumagai, 2002). In fact, the induction of this enzyme in the rat leads to resistance to the development of hepatic carcinoma. Interestingly, in non-human primates it is Mu-class GSTs the ones responsible for AFBO trapping with GSH (Wang et al., 2000).

Turkeys are the only poultry species in which the role of GST-mediated trapping of AFBO with GSH has been investigated (Klein et al., 2000; 2002). At least six Alpha-class GSTs have been isolated, amplified and fully characterized from turkey, which exhibit similarities in sequence with human Alpha-class GSTs ranging from 53% to 90% (Kim et al., 2010). However, no soluble GST activity towards microsomally activated AFB1 has been found in liver cytosol from one-month old male turkeys (Klein et al., 2000), a finding that was later confirmed in male turkeys 9, 41, and 65 days of age (Klein et al., 2002). GSTs from the liver of one-day-old chicks (Chang et al., 1990) and nucleotide sequences of Alpha-class (Liu et al., 1993), Mu-class (Liu & Tam, 1991), Theta-class (Hsiao et al., 1995) and Sigma-class (Thomson et al., 1998) GSTs from chicken liver have been characterized, but there are no reports for their role in AFBO trapping with GSH.

The role of other conjugation reactions on AFB1 metabolism in poultry is still uncertain. Liver UGT activity and sulphotransferase (SULT) activity have been reported in bobwhite quail (Maurice et al., 1991), and in chickens, ducks and geese (Bartlet & Kirinya, 1976). However, no research on glucuronic acid conjugation or sulfate conjugation of AFB1 metabolites has been conducted in any commercial poultry species.

4. Concluding remarks

Research conducted recently has shown that there are clear differences in oxidative/reductive AFB1 metabolism that could explain the differential responses to AFB1 observed *in vivo* among turkeys, quail and chickens, but not ducks. The existence of a clear toxicological biochemical pattern that explains AFB1 sensitivity in three out of four species may be related to their different phylogenetic origins: Turkeys, quail and chickens are phylogenetically close to each other (all belong to the order Galliformes, family Phasianidae), but distant from ducks (order Anseriformes, family Anatidae). It is also interesting to note that while CYP2A6 (and to a lesser extent CYP1A1) is the major enzyme responsible for AFB1 bioactivation in the Galliformes studied, four enzymes (CYP1A1, 1A2, 2A6 and 3A4 ortholog activities) appear to be responsible for AFB1 bioactivation in ducks.

In regards to conjugation reactions, it has been demonstrated that turkeys do not express the GSTs responsible for AFBO trapping. However, the role of AFBO trapping by GSH has not been investigated in other poultry species and no information on the possible conjugation reactions of AFL has been reported for any avian species, either. Another pathway of AFBO metabolism that has not been investigated in poultry is the formation of AFB1-dhd and dialcohol (Fig. 1). Formation of AFB1-dhd may occur either spontaneously or through the action of a microsomal epoxide hydrolase (EPHX) and the possible role of EPHX in AFB1 biotransformation in birds is still unknown. The alternative pathway for AFB1-dhd, that is, the formation of an aflatoxin dialcohol through the action of the cytosolic enzyme AFAR, has not been investigated either. This topic is important to investigate since the dialcohol does not bind with proteins and therefore constitutes a true detoxication product.

Extra-hepatic localization of enzymes responsible for the biotransformation of AFB1 may also play a role in the differential response to AFB1 in birds. For instance, in humans CYP3A4 is the major enzyme involved in AFB1 bioactivation (Ueng et al., 1995) and this enzyme is highly expressed not only in the liver but also in the gastrointestinal tract. This "first-pass" effect may affect the absorption of unaltered AFB1 and therefore its ability to reach its target organ in humans. Finally, the so-called Phase III metabolism (basolateral and canalicular transport of xenobiotics) has been shown to determine sensitivity or resistance to xenobiotics in several experimental models. For instance, collie dogs are extremely sensitive to ivermectin due to the low expression of the transporter protein MDR1 (Diaz, 2000). The role of the translocation of AFB1 and its metabolites on AFB1 sensitivity/resistance needs to be investigated.

5. Acknowledgment

Thanks are due to the International Foundation for Science (Stockholm, Sweden) for partial funding of our research on *in vitro* metabolism of aflatoxin B1, to all people that have collaborated with this work (particularly Sandra Milena Cepeda and Amparo Cortés) and to the National University of Colombia in Bogotá, our second home.

6. References

Bartlet, A.L. & Kirinya, L.M. (1976). Activities of mixed function oxidases, UDP-glucuronyl transferase and sulphate conjugation enzymes in Galliformes and Anseriformes. *Quarterly Journal of Experimental Physiology and Cognate Medical Sciences*, Vol.61, No.2, (April 1976), pp. 105-119, ISSN 0033-5541

Blount, W.P. (1961). Turkey "X" disease. *Turkeys*, (March-April 1961), pp. 52-61

Chang, L.H.; Chuang, L.F.; Tsai, C.P.; Tu, C.P.D. & Tam, M.F. (1990). Characterization of glutathione S-transferases from day-old chick livers. *Biochemistry*, Vol.29, No.3, (January 1990), pp. 744-750, ISSN 0006-2960

Chen, S.C.G.; Wei, R.D. & Hsieh, D.P.H. (1981). Purification and some properties of chicken-liver aflatoxin B1 reductase. *Food and Cosmetics Toxicology*, Vol.19, pp. 19-24, ISSN 0278-6915

Chen, C.; Pearson, A.M.; Coleman, T.H.; Gray, J.I.; Pestka, J.J. & Aust, S.D. (1984). Tissue deposition and clearance of aflatoxins from broiler chickens fed a contaminated diet. *Food and Chemical Toxicology*, Vol.22, No.6, (June 1984), pp. 447-451, ISSN 0278-6915

Dalezios, J.; Wogan, G.N. & Weinreb, S.M. (1971). Aflatoxin P-1: a new aflatoxin metabolite in monkeys. *Science,* Vol.171, No.3971, (February 1971), pp. 584-585, ISSN 0036-8075

De Iongh, H.; Vles, R.O. & Van Pelt, J.G. (1964). Milk of mammals fed an aflatoxin-containing diet. *Nature,* Vol.202 , (May 1964), pp. 466-467, ISSN 0028-0836

Diaz, G.J. & Sugahara, M. (1995). Individual and combined effects of aflatoxin and gizzerosine in broiler chickens. *British Poultry Science,* Vol.36, No.5, (December 1995), pp. 729-736, ISSN 0007-1668

Diaz, G.J. (2000). Basolateral and canalicular transport of xenobiotics in the hepatocyte : a review. *Cytotechnology,* Vol.34, No.3, (November 2000), pp. 225-236, ISSN 0920-9069

Diaz, G.J.; Calabrese, E. & Blain, R. (2008). Aflatoxicosis in chickens (Gallus gallus): An example of hormesis?. *Poultry Science,* Vol.87, No.4, (April 2008), pp. 727-732, ISSN 0032-5791

Diaz, G.J.; Murcia, H.W. & Cepeda, S.M. (2010a). Bioactivation of aflatoxin B1 by turkey liver microsomes: responsible P450 enzymes. *British Poultry Science,* Vol.51, No.6, (December 2010), pp. 828-837, ISSN 0007-1668

Diaz, G.J.; Murcia, H.W. & Cepeda, S.M. (2010b). Cytochrome P450 enzymes involved in the metabolism of aflatoxin B1 in chickens and quail. *Poultry Science,* Vol.89, No.11, (November 2010), pp. 2461-2469, ISSN 0032-5791

Diaz, G.J.; Murcia, H.W. & Cepeda, S.M. (2010c). The role of selected cytochrome P450 enzymes on the bioactivation of aflatoxin B1 by duck liver microsomes. *Avian Pathology,* Vol.39, No.4, (August 2010), pp. 279-285, ISSN 0307-9457

Eaton, D.L.; Monroe, D.H.; Bellamy, G. & Kalman, D.A. (1988). Identification of a novel metabolite of aflatoxin B1 produced *in vitro* and in vivo in rats and mice. *Chemical Research in Toxicology,* Vol.1, No.2, (March 1988), pp. 108-114, ISSN 0893-228X

Esaki, H. & Kumagai, S. (2002). Glutathione-S-transferase activity towards aflatoxin epoxide in livers of mastomys and other rodents. *Toxicon,* Vol.40, No.7, (July 2002), pp. 941-945, ISSN 0041-0101

Guarisco, J.A.; Hall, J.O. & Coulombe, R.A. (2008). Mechanisms of butylated hydroxytoluene chemoprevention of aflatoxicosis - inhibition of aflatoxin B1 metabolism. *Toxicology and Applied Pharmacology,* Vol.227, No.3, (March 2008), pp. 339-346, ISSN 0041-008X

Guengerich, F.P. & Shimada, T. (1998). Activation of procarcinogens by human cytochrome P450 Enzymes. *Mutation Research/Fundamental and Molecular Mechanismsof Mutagenesis,* Vol.400, No.1-2, (May 1998), pp. 201-213, ISSN 0027-5107

Guengerich, F.P.; Jonhson, W.W.; Shimada, T.; Ueng, Y.F.; Yamazaki, H. & Langouet, S. (1998). Activation and detoxication of aflatoxin B1. *Mutatation Research/Fundamental and Molecular Mechanisms of Mutagenesis,* Vol.402, No.1-2, (June 1998), pp. 121-128, ISSN 0027-5107

Guengerich, F.P.; Cai, H.; McMahon, M.; Hayes, J.D.; Sutter, T.R.; Groopman, J.D.; Deng, Z. & Harris, T.M. (2001). Reduction of aflatoxin B1 dialdehyde by rat and human aldo-keto reductases. *Chemical Research in Toxicology*, Vol.14, No.6, (June 2001), pp. 727-737, ISSN 0893-228X

Hasler, J.A.; Estabrook, R.; Murray, M.; Pikuleva, I.; Waterman, M.; Capdevila, J.; Holla, V.; Helvig, C.; Falck, J.R.; Farrell, G.; Kaminsky, L.S.; Spivack, S.D.; Boitier, E. & Beaune, P. (1999). Human cytochromes P450. *Molecular Aspects of Medicine*, Vol.20, No.1-2, (February-April 1999), pp. 1-137, ISSN 0098-2997

Holeski, C.J.; Eaton, D.L.; Monroe, D.H. & Bellamy, G.M. (1987). Effect of phenobarbital on the biliary excretion of aflatoxin P_1-glucuronide and aflatoxin B_1-S-glutathione in the rat. *Xenobiotica*, Vol.17, No.2, (February 1987), pp. 139-153, ISSN 0049-8254

Hsiao, C.D.; Martsen, E.O.; Lee, J.Y.; Tsai, S.P. & Ming, F.T. (1995). Amino acid sequencing, molecular cloning and modeling of the chick liver class-theta glutathione S-transferase CL1. *Biochemical Journal*, Vol.312, No.1, (November 1995), pp. 91-98, ISSN 0264-6021

Hsieh, D.P.H.; Salhab, A.S.; Wong, J.J. & Yang, S.L. (1974). Toxicity of aflatoxin Q1 as evaluated with the chicken embryo and bacterial auxotrophs. *Toxicology and Applied Pharmacology*, Vol.30, No.2, (November 1974), pp. 237-242, ISSN 0041-008X

Johnson, W.W.; Harris, T.M. & Guengerich, F.P. (1996). Kinetics and mechanism of hydrolysis of aflatoxin B1 exo-8,9-epoxide and rearrangement of the dihydrodiol. *Journal of the American Chemical Society*, Vol.118, No.35, (September 1996), pp. 8213-8220, ISSN 0002-7863

Josephy, P.D.; Guengerich, F.P. & Miners, J.O. (2005). "Phase I" and "Phase II" drug metabolism: Terminology that we should phase out? *Drug Metabolism Reviews*, Vol.37, No.4, (November 2005), pp. 575-580, ISSN 0360-2532

Josephy, P.D. & Mannervik, B. (2006). *Molecular Toxicology*, (2nd edition), Oxford University Press, ISBN 978-0-19-517620-9, New York, USA

Kim, J.E.; Bauer, M.M.; Mendoza, K.M.; Reed, K.M. & Coulombe, R.A. (2010). Comparative genomics identifies new Alpha class genes within the avian glutathione S-transferase gene cluster. *Gene*, Vol.452, No.2, (March 2010), pp. 45-53, ISSN 0378-1119

Klein, P.J.; Buckner, R.; Kelly, J. & Coulombe, R.A. (2000). Biochemical basis for the extreme sensitivity of turkeys to aflatoxin B1. *Toxicology and Applied Pharmacology*, Vol.165, No.1, (May 2000), pp. 45-52, ISSN 0041-008X

Klein, P.J.; Van Vleet, T.R.; Hall, J.O. & Coulombe, R.A. (2002). Biochemical factors underlying the age-related sensitivity of turkeys to aflatoxin B1. *Comparative Biochemistry and Physiology Part C*, Vol.132, No.2, (June 2002), pp. 193-201, ISSN 1532-0456

Kumagai, S.; Nakano, N. & Aibara, K. (1983). Interactions of aflatoxin B1 and blood components of various species *in vitro*: Interconversion of aflatoxin B1 and aflatoxicol in the blood. *Toxicology and Applied Pharmacology*, Vol.67, No.2, (February 1983), pp. 292-301, ISSN 0041-008X

Leeson, S.; Diaz, G.J. & Summers, J.D. (1995). *Poultry metabolic disorders and mycotoxins.* University Books, ISBN 0-9695600-1-1, Guelph, Canada

Lillehoj, E.B. & Ciegler, A. (1969). Biological activity of aflatoxin B_{2a}. *Applied Microbiology*, Vol.17, No.4, (April 1969), pp. 516-519, ISSN 0099-2240

Liu, L.F. & Tam, M.F. (1991). Nucleotide sequence of class μ glutathione S-transferase from chicken liver. *Biochimica et Biophysica Acta (BBA) – Gene Structure and Expression*, Vol.1090, No.3, (November 1991), pp. 343-344, ISSN 0167-4781

Liu, L.F.; Wu, S.H. & Tam, M.F. (1993). Nucleotide sequence of class-α glutathione S-transferases from chicken liver. *Biochimica et Biophysica Acta (BBA) - Gene Structure and Expression*, Vol.1216, No.2, (November 1993), pp. 332-334, ISSN 0167-4781

Loveland, P.M.; Coulombe, R.A.; Libbey, L.M.; Pawlowski, N.E.; Sinnhuber, R.O.; Nixon, J.E. & Bailey, G.S. (1983). Identification and mutagenicity of aflatoxicol-M1 produced by metabolism of aflatoxin B1 and aflatoxicol by liver fractions from rainbow trout (*Salmo gairdneri*) fed β-naphthoflavone. *Food and Chemical Toxicology*, Vol.21, No.5, (October 1983), pp. 557-562, ISSN 0278-6915

Loveland, P.M.; Wilcox, J.S.; Pawolowski, N.E. & Bailey, G.S. (1987). Metabolism and DNA binding of aflatoxicol and aflatoxin B1 *in vivo* and in isolated hepatocytes from rainbow trout (*Salmo gairdneri*). *Carcinogenesis*, Vol.8, No.8, (August 1978), pp. 1065-1070, ISSN 0143-3334

Lozano, M.C. & Diaz, G.J. (2006). Microsomal and cytosolic biotransformation of aflatoxin B1 in four poultry species. *British Poultry Science*, Vol.47, No.6, (December 2006), pp. 734-741, ISSN 0007-1668

Masri, M.S.; Haddon, W.F.; Lundin, R.E. & Hsieh, D.P.H. (1974). Aflatoxin Q1. A newly identified major metabolite of aflatoxin B1 in monkey liver. *Journal of Agricultural and Food Chemistry*, Vol.22, No.3, (May-June 1974), pp. 512-515, ISSN 0021-8561

Maurice, D.V.; Lightsey, S.F.; Kuo-Tung, H.; Rhoades, J.F. (1991). Comparison of glutathione S-transferase activity in the rat and birds: tissue distribution and rhythmicity in chicken (*Gallus domesticus*) liver. *Comparative Biochemistry and Physiology B*, Vol.100, No.3, (March 1991), pp. 471-474, ISSN 1096-4959

Muller, R.D.; Carlson, C.W.; Semeniuk, G. & Harshfield, G.S. (1970). The response of chicks, ducklings, goslings, pheasants and poults to graded levels of aflatoxin. *Poultry Science*, Vol.49, No.5, (September 1970), pp. 1346-1350, ISSN 0032-5791

Murcia, H.W.; Cepeda, S.M. & Diaz, G.J. (2011). Enzymatic activity in turkey, duck, quail and chicken liver microsomes against four prototype cytochrome P450 human substrates and aflatoxin B1 (submitted).

Omiecinski, C.J.; Remmel, R.P. & Hosagrahara, V.P. (1999). Concise review of the cytochrome P450s and their roles in toxicology. *Toxicological Sciences*, Vol.48, No.2, (April 1999), pp. 151-156, ISSN 1096-6080

Omura, T. & Sato, R. (1964). The carbon monoxide-binding pigment of liver microsomes. I. Evidence for its hemoprotein nature. *Journal of Biological Chemistry*, Vol.239, No.7, (July 1964), pp. 2370-2378, ISSN 0021-9258

Parkinson, A. & Ogilvie, B.W. (2008). Biotransformation of xenobiotics, In: *Casarett & Doull's Toxicology – The basic science of poisons*, Klaassen, C.D. (Ed.), pp. 161-304, McGraw Hill Medical, ISBN 978-0-07147051-3, New York, USA

Patterson, D.S.P. & Roberts, B.A. (1971). The *in vitro* reduction of aflatoxins B1 and B2 by soluble avian liver enzymes. *Food and Cosmetics Toxicology*, Vol.9, No.6, (November 1971), pp. 829-837, ISSN 0278-6915

Patterson, D.S.P. & Roberts. B.A. (1972). Aflatoxin metabolism in duck-liver homogenates: The relative importance of reversible cyclopentenone reduction and hemiacetal formation. *Food and Cosmetics Toxicology*, Vol.10, No.4, (July 1972), pp. 501-512, ISSN 0278-6915

Purchase, I.F.H. (1967). Acute toxicity of aflatoxins M1 and M2 in one-day-old ducklings. *Food and Cosmetics Toxicology*, Vol.5, No.1, (January 1967), pp. 339-342, ISSN 0278-6915

Ramsdell, H.S.; Parkinson, A.; Eddy, A.C. & Eaton, D.L. (1991). Bioactivation of aflatoxin B1 by human liver microsomes: Role of cytochrome P450 IIIA enzymes. *Toxicology and Applied Pharmacology*, Vol.108, No.3, (May 1991), pp. 436-447, ISSN 0041-008X

Raney, K.D.; Coles, B.; Guengerich, F.P. & Harris, T.M. (1992a). The endo-8,9-epoxide of aflatoxin B1: A new metabolite. *Chemical Research in Toxicology*, Vol.5, No.3, (May 1992), pp. 333-335, ISSN 0893-228X

Raney, K.D.; Shimada, T.; Kim, D.H.; Groopman, J.D.; Harris, T.H. & Guengerich, F.P. (1992b). Oxidation of aflatoxins and sterigmatocystin by human liver microsomes: Significance of aflatoxin Q1 as a detoxication product of aflatoxin B1. *Chemical Research in Toxicology*, Vol.5, No.2, (March 1992), pp. 202-210, ISSN 0893-228X

Rawal, S.; Kim, J.E. & Coulombe, R.A. (2010a). Aflatoxin B1 in poultry: toxicology, metabolism and prevention. *Research in Veterinary Science*, Vol.89, No.3, (December 2010), pp. 325-331, ISSN 0034-5288

Rawal, S.; Yip, S.S.M. & Coulombe, R.A. (2010b). Cloning, expression and functional characterization of cytochrome P450 3A37 from turkey liver with high aflatoxin B1 epoxidation activity. *Chemical Research in Toxicology*, Vol.23, No.8, (August 2010), pp. 1322-1329, ISSN 0893-228X

Salhab, A.S. & Hsieh, D.P. (1975). Aflatoxicol H1: a major metabolite of aflatoxin B1 produced by human and rhesus monkeys *in vitro*. *Research Communications in Chemical Pathology and Pharmacology*, Vol.10, No.3, (March 1975), pp. 419-429, ISSN 0034-5164

Salhab, A.S.; Abramson, F.P.; Geelhoed, G.W. & Edwards, G.S. (1977). Aflatoxicol M1, a new metabolite of aflatoxicol. *Xenobiotica*, Vol.7, No.7, (July 1977), pp. 401-408, ISSN 0049-8254

Thomson, A.M.; Meyer, D.J. & Hayes, J.D. (1998). Sequence, catalytic properties and expression of chicken glutathione-dependent prostaglandin D_2 synthase, a novel class sigma glutathione S-transferase. *Biochemical Journal*, Vol.333, No.2, (July 1998), pp. 317-325, ISSN 0264-6021

Tulayakul, P.; Sakuda, S.; Dong, K.S & Kumagai, S. (2005). Comparative activities of glutathione-*S*-transferase and dialdehyde reductase toward aflatoxin B1 in livers of experimental and farm animals. *Toxicon*, Vol.46, No.2 (August 2010), pp. 204-209, ISSN 0041-0101

Ueng, Y.F.; Shimada, T.; Yamazaki, H. & Guengerich, F.P. (1995). Oxidation of aflatoxin B1 by bacterial recombinant human cytochrome P450 enzymes. *Chemical Research in Toxicology*, Vol.8, No.2, (March 1995), pp. 218-225, ISSN 0893-228X

Wang, C.; Bammler, T.K.; Guo, Y.; Kelly, E.J. & Eaton D.L. (2000). Mu-class GSTs are responsible for aflatoxin B1-8,9-epoxide-conjugating activity in the nonhuman primate *Macaca fascicularis* liver. *Toxicological Sciences*, Vol.56, No.1, (July 2000), pp. 26-36, ISSN 1096-6080

Wong, Z.A. & Hsieh, D.P. (1978) Aflatoxicol: major aflatoxin B1 metabolite in rat plasma. *Science*, Vol.200, No.4339, (April 1978), pp. 325-327, ISSN 0036-8075

Yip, S.S.M. & Coulombe, R.A. (2006). Molecular cloning and expression of a novel cytochrome P450 from turkey liver with aflatoxin B1 oxidizing activity. *Chemical Research in Toxicology*, Vol.19, No.1, (January 2006), pp. 30-37, ISSN 0893-228X

2

Control of Aflatoxin Biosynthesis in *Aspergilli*

Kenneth C. Ehrlich, Perng-Kuang Chang, Jiujiang Yu,
Jeffrey W. Cary and Deepak Bhatnagar
Southern Regional Research Center, New Orleans, Louisiana
USA

1. Introduction

Aflatoxins (AFs) are bisfurans that are polyketide-derived, toxic, and carcinogenic secondary metabolites produced by *Aspergillus flavus* on corn, peanuts, cottonseed, and tree nuts (Payne & Yu, 2010). While biosynthesis of these toxins has been extensively studied *in vitro*, much less is known about what causes the fungi to produce AFs under certain environmental conditions and only on certain plants. It is not yet known why wheat, soybean, and sorghum are resistant to AF contamination in the field whereas, under laboratory conditions *A. flavus* is able to colonize these plant tissues and produce AFs (Cleveland et al., 2009).

AF biosynthesis is a complicated process involving many levels of transcriptional and post-transcriptional control (Abbas et al., 2009; Chanda et al., 2009; Georgianna & Payne, 2009; Schmidt-Heydt et al., 2009). In this review we provide an overview of what is currently known about how environmental and nutritional factors stimulate or inhibit AF production. Environmental and nutritional signals interact directly with cell surface receptors or transport proteins. Usually the interaction sets up a cascade of cellular responses including activation of heat shock proteins or other chaperonin-like messengers as well as protein kinases or other protein modifying enzymes. These, in turn, activate transcription factors residing in the cytoplasm to allow them to cross the nuclear membrane boundary or, in some cases, to activate a DNA-binding protein already in the nucleus. The DNA-binding protein then has to find the correct location in the targeted gene's promoter and the region of chromatin containing the gene has to be in a transcriptionally competent (euchromatic) as opposed to an incompetent heterochromatic condensed state. Finding the correct location requires transcription-activating factors (TAFs) which either bind to chromatin and recruit the transcription factor to the DNA or bind to the transcription factor and guide it to the correct location. The stability of the transcription factor in the cell is another level of control of the transcriptional process. This stability is mediated by modification (ubiquitination or neddylation) (Busch et al., 2003). Recently, AF biosynthesis was shown to occur in dedicated peroxisomal vesicles (Roze et al., 2011). The ability to form and stabilize such vesicles could be critical to the coordination of the biosynthetic steps leading to AF formation. All of these processes are illustrated in Figure 1 and are discussed in more detail below.

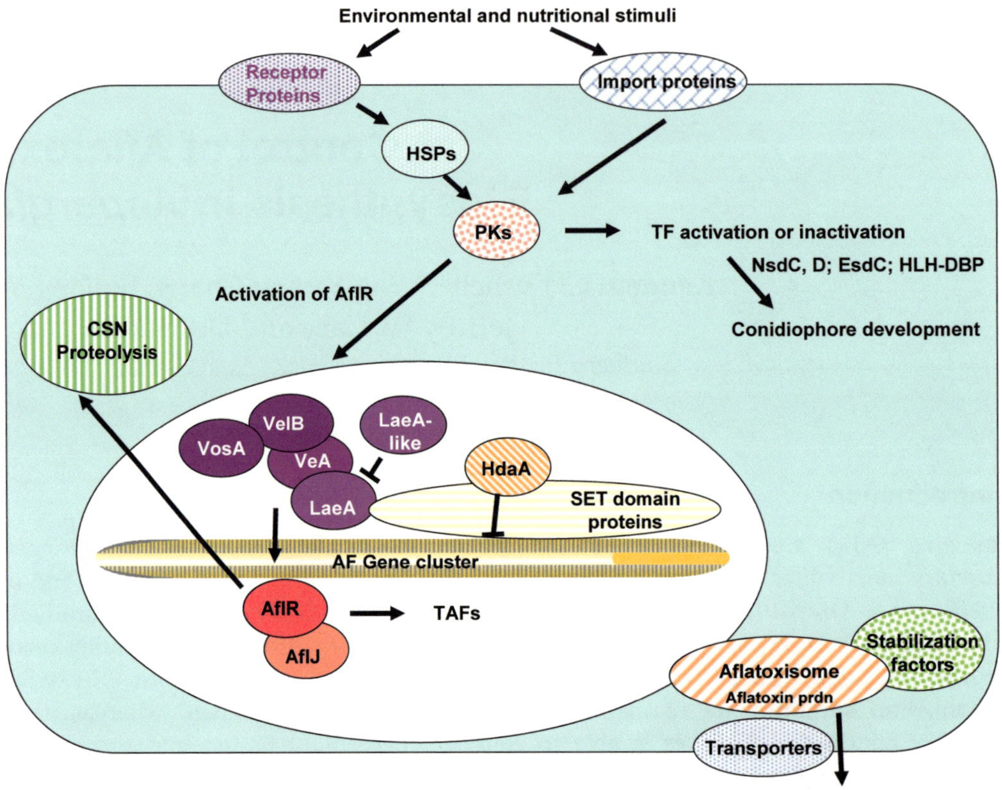

Fig. 1. Model showing factors involved in regulation of AF production. The cell is shown as a square and the nucleus as a large oval. Abbreviations: HSP-heat shock proteins, PK-protein kinase, CSN-COP9 signalosome, HLH_DBP-helix-loop-helix DNA-binding protein, TF-transcription factor, TAF-transcription activating factor, HdaA-histone deacetylase.

2. Transcriptional control of aflatoxin biosynthesis

2.1 The role of the pathway-specific transcription factor, AflR

Production of AF requires the coordinated transcription of about 30 clustered genes (Yu et al., 2004b). The genes for the 30 biosynthetic proteins are clustered within a 70-kb region of chromosome 3 (Fig. 2). In *A. flavus* the gene cluster is located 20 to 80 kb from the telomere depending on the *A. flavus* strain. The gene, *aflR,* encodes the sequence-specific Cys_6Zn_2 DNA-binding protein, AflR, which is responsible for transcriptional activation of most, if not all, AF structural genes (Chang et al., 1995; Ehrlich et al., 1998; Ehrlich et al., 1999b; Cary et al., 2000). AflR, as a typical Gal4-type transcription factor, has an N-terminal DNA binding domain and a C-terminal activation domain (Fig. 3). The DNA-binding domain recognizes the partially palindromic 11 bp double-strand motif, TCGSWNNSCGR (top strand only is shown, Fig. 3), in promoter regions of AF biosynthesis genes.

The strongest binding, based on electrophoretic mobility shift assays, is to sequences with a perfect 8 bp palindrome in an 11-bp motif, $TCG^{G}/_{C}NNN^{C}/_{G}CGA$. Footprinting studies showed that the preferred binding occurred to sequences in which the palindrome is flanked on the 5′-end by additional thymine or adenine residues. The binding motifs for

AflR almost always are within 200 bp of the gene's translational start site. In the *pksA* promoter region an additional AflR-binding site was found to be the AflR binding site for a short gene *hypC* (formerly *hypB1),* which encodes an enzyme necessary for oxidation of the AF precursor metabolite, norsolorinic acid anthrone.

Fig. 2. The AF cluster for *Aspergillus flavus, A. parasiticus and A. nomius*. Lettering on lower line is the current nomenclature for the genes using the *afl* designation(Yu et al., 2004a).

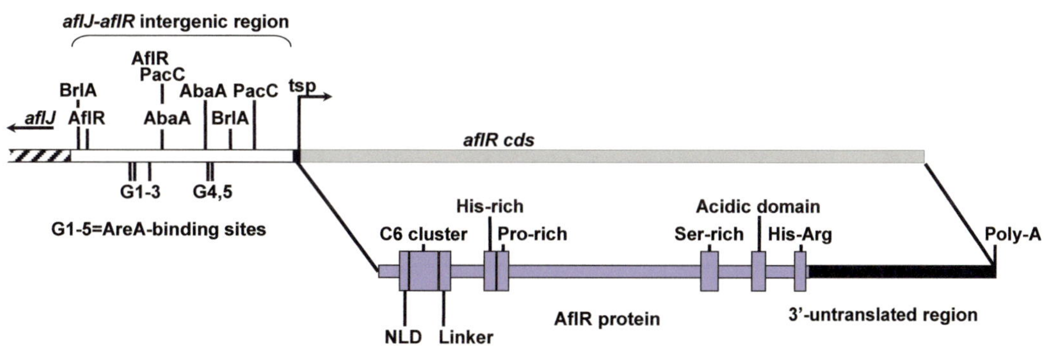

Fig. 3. Schematic depiction of the *aflR* gene. Consensus sequence elements for DNA-binding proteins are indicated on the *aflJ-aflR* intergenic region with their approximate positions. However, the *aflR* gene in each of the AF-producing species has fewer sites than those shown. Abbreviations: NL-nuclear localization; tsp-translational start point

In AflR a nuclear localization domain (RRARK) precedes the C6 cluster domain (CTSCASSKVRCTKEKPACARCIERGLAC) (Ehrlich et al., 1998). In many Gal4 type transcription factors the nuclear localization signal is within, not separate from, the C6 cluster as shown in Fig. 3). Furthermore many related C6 factors lack the underlined basic amino acids on the C-terminal side of the motif. A Blast search of *A. flavus* genome in the *Aspergillus* Comparative Database http://www.broadinstitute.org/annotation/genome/aspergillus_group/MultiHome.html with this sequence found only six other proteins with an E value=0.007 or lower (AFL2G_06146; AFL2G_11313; AFL2G_00473; AFL2G_02725; AFL2G_04045; AFL2G_08639). Since there are over 178 C6 transcription factor proteins in the database for *A. flavus* (Table 1), six is a low number for such proteins with this "AflR" type of DNA recognition domain. The amino acids immediately C-terminal to the C6-cluster, the "linker region", are presumed to determine DNA-binding specificity. This linker region (QYMVS**KR**MG**R**NP**R**) lacks basic amino acids at the N-terminal end of the region unlike the linker regions for many similar transcription factors, but possesses four basic amino acids in the C-terminal half of the motif. This set of amino acids may be a signature amino acid sequence that allows contact with the 11-bp $TCGN_5CGA$ DNA binding site recognized by AflR. A Blast search with this region gave only two close matches:

AFL2G_11313 and AFL2G_06146. These may also bind to a TCGN$_x$CGA DNA motif. The C-terminal activation domain for AflR, like those of other C6-type transcription factors in fungi, has a high number (8/28) of acidic amino acids and (5/28) basic amino acids and is otherwise not distinctive. The C-terminal 38 amino acid region (residues 408–444) also has runs of His and Arg, and acidic amino acids (**HH**PASPFSLLGFSGLEANL**RHR**LRAVSSDIIDYL**HR**E) and these are also part of the transcription factor activation domain (Xie et al., 2000).

Some other features of the AflR protein may be important for its stability and its ability to interact with other proteins (Ehrlich et al., 2003). The AflR protein sequences of several different AF-producing Aspergilli (*A. flavus* AF70, *A. pseudotamarii,* and *A. parasiticus* contained a histidine-rich motif (HAHTQAHTHAHSH, aa 103-113 in *A. flavus*) on the carboxy-terminal side of the DNA-recognition domain. The length of this motif varies in size for the different species. Garnier plot analysis predicted that this region has a coiled rather than a helical or beta-sheet configuration. Such regions may be involved in pH-controlled protein-protein interactions unique to AflR. Similar His-rich regions are found in a number of eukaryotic transcription factors (Janknect et al., 1991). Differences in length of these repeats among AflRs could be important in modulating AflR's activity at different pHs.

In all AF-producing *Aspergillus* species, a proline-rich region was adjacent to this site on the C-terminal side. AflR in *A. nomius* isolates had 11-12 proline residues while *A. flavus* and *A. parasiticus* isolates had eight. A key distinguishing feature shared by *A. nomius* and *A. bombycis* isolates is a serine-rich sequence, NSSDSSGSSRSSSSSSNSP, approximately 100 amino acids from the AflR C-terminus and immediately preceding a conserved domain rich in acidic amino acids. Seven serine residues were present in the homologous region from *A. pseudotamarii* and four in *A. flavus* and *A. parasiticus*. The serine-rich region in AflRs from *A. nomius* and *A. bombycis* is a distinct PEST (proline, glutamine, serine, and threonine-rich region) sequence that may be a target for ubiquitin-mediated proteolysis (Rechsteiner, 1988). No comparable region exists in AflRs from other species. PEST scores for this region (RSSSSSSNSPTTVSEER) were +18 and +12 in *A. nomius* and *A. bombycis*, respectively, which are comparable to scores of +10 to +13 for known PEST domains in transcription factors (Suske, 1999). *A. nomius* isolates have a second PEST domain (HPPPPPQSDQPPH, PEST score = +15). Most proteins with PEST sequences are regulatory molecules that require fast turnover to avoid improper function. The PEST sequences in *A. nomius* and *A. bombycis* either may reflect regulatory mechanisms different from other AF-producing taxa or may be non-functional remnants of an ancestral mechanism to regulate cellular levels of AflR at the post-transcriptional level.

Analysis of the promoter region of AflR (Figure 3; the figure shows the whole *aflJ-aflR* intergenic region) revealed several possible regulatory elements (Ehrlich et al., 1999a). A region from -93 to -123 (CATTTAGGCCTAAGTGCGA<u>**GGCAACGA**</u>AAAG) upstream of the translational start site is important for promoter activity. A partial AflR-binding site is present in this region (underlined) and may allow *aflR* expression to be self-regulated. The *aflR* promoter lacks a detectable TATA-box or CCAAT-box binding domains found in the promoter regions of many eukaryotic genes. The gene has been shown to be expressed even under conditions not conducive to AF biosynthesis. Therefore, low levels of AflR may be present in the cell, but after induction the levels increase and are able to activate the other genes in the gene cluster. When *A. parasiticus* is transformed with a plasmid containing *aflR* expressed under the control of the glycerolphosphate dehydrogense "housekeeping" promoter (*gpdA*), the fungi accumulate additional colored AF precursor metabolites, in particular, norsolorinic acid, indicating that higher levels of AflR cause increased expression of the early genes needed for production of the biosynthetic enzymes.

2.2 AflJ, a putative transcriptional co-activator of AF biosynthesis

The gene bidirectionally transcribed from *aflR* also is necessary for AF production (Meyers et al., 1998; Du et al., 2007). *aflJ* and *aflR* share a 737 bp intergenic region (Fig. 3). Knockout mutants of *aflJ* are unable to produce AF or its precursors (Meyers et al., 1998). Although AflJ has no recognized regions in the protein corresponding to enzymatic or regulatory domain, it has three putative membrane-spanning helices and a microbodies targeting signal. A microbody is a cytoplasmic organelle of a more or less globular shape that contains degradative enzymes bound within a single membrane. Microbody types include peroxisomes, glyoxisomes, glycosomes and Woronin bodies. A BlastP search of the GenBank database revealed some proteins having a methyltransferase domain (PFAM00891) but with certain regions missing (Fig. 4). PFAM00891 is a member of the superfamily cl10454 and includes O-methyltransferases that utilize S-adenosylmethionine as the substrate. In spite of the homology to known methyltransferases, it is unlikely that AflJ functions in this way. All of the AF-producing species of *Aspergillus*, the sterigmatocystin (ST)-producers, such as *A. nidulans*, and the dothistromin-producer, *Mycospaerella pini*, contain similar genes encoding an aflJ homolog. A tBlastN search of the *Aspergillus flavus* genome with the AflJ protein sequence revealed five hits (AFL2G_11558, E=-22; AFL2G_11312, E=-20; AFL2G_11323, E=-20; AFL2G_11580, E =-10; AFL2G_11922, E=-9). AFL2G_11323 has an OMT domain and AFL2G_11312 is a polyketide synthase with a methyltransferase domain.

There is still controversy about why Δ*aflJ* mutants fail to make AFs. In the original paper describing *aflJ* it was found that disruption of *aflJ* in *A. flavus* resulted in a failure to convert exogenously added pathway intermediates norsolorinic acid, sterigmatocystin, and *O*-methylsterigmatocystin to AF, indicating that these biosynthesis proteins were not made or were not active. The disrupted strain accumulated *pksA*, *nor1*, *ver1*, and *omtA* transcripts under conditions conducive to AF biosynthesis, but transcript levels for the early genes, *pksA* and *nor-1*, were significantly lower than in the parental strain (Du et al., 2007). Therefore, disruption of *aflJ* did not affect transcription of these genes. Although it was possible that Δ*aflJ* mutants failed to properly process RNA transcripts from AF cluster genes; recent studies found that the transcripts were processed normally (Du et al., 2007).

AflJ was shown to bind to AflR at a region within AflR's activation domain (Chang, 2003). Substitution of Arg429 and Arg431 in AflR with Leu residues abolished the binding. Deletions in targeted regions of AflJ, also prevented observable binding. AflJ stimulated the accumulation of AF and precursor metabolites in cultures when the transformant also contained a functional *aflR* gene. Transformants containing an extra copy of *aflR* but lacking an extra copy of *aflJ* had a reduced level of expression of *aflR* compared to transformants containing a second copy of both genes (Chang et al., 1995). Based on these results AflJ was classified as an AflR coactivator. From recent studies of Du, et al., *aflJ* was not found to be necessary for transcriptional activation of later genes (*ver1* and *omtA*) nor for *aflR*, but did upregulate expression of the polyketide synthase gene, *pksA* and *nor1*, two early genes necessary for the beginning steps in AF biosynthesis (Du et al., 2007). AflJ's transcription was regulated by AflR in *A. flavus* and *A. parasiticus*. An AflR-binding site near the translational start site of *aflJ* (Fig. 2) may mediate this activation. One AF-producing species (*A. nomius*) lacks an AflR-binding site in this region and therefore transcription of *aflJ* may not be regulated by AflR in this species (Ehrlich et al., 2003).

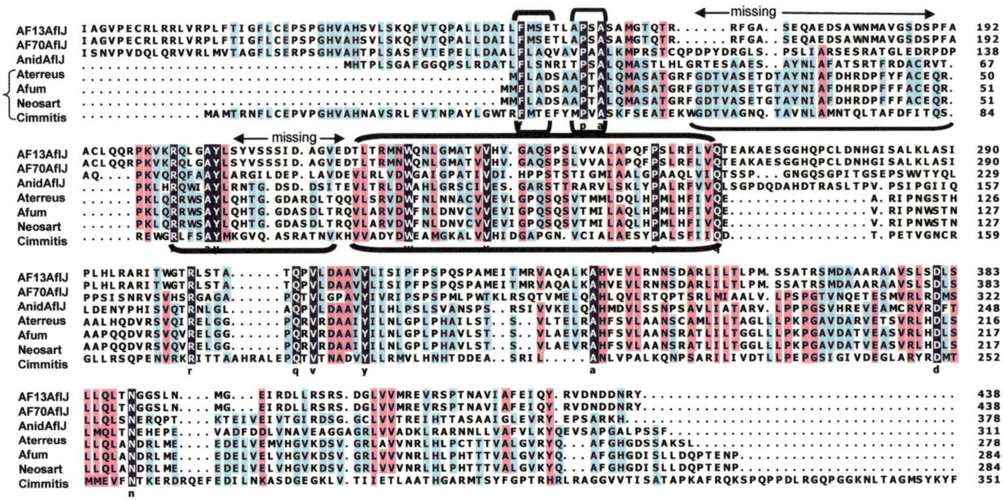

Fig. 4. Comparison of *A. flavus* AflJ sequence to those of non-aflatoxin producing species reveals partial homology to methyltransferase-2 family proteins. The bracketed proteins have *O*-methyltransferase domains. Regions bracketed at the bottom are the predicted methyltransferase domains.

One hypothesis to explain AflJ's role in transcription regulation is that, by binding to AflR, it allows AflR to interact with chromatin remodeling proteins such as LaeA (see below) (Ehrlich et al., 2011). Such interactions have been postulated to be necessary for opening up the chromatin region in which the AF cluster is located. Another possibility, that fits better with the likelihood that AflJ is a membrane-bound protein, as evidenced by the presence of transmembrane helices and its microbodies targeting signal, is that AflJ may be required for transmembrane transport of intermediates through intercellular compartments and, thereby, helps coordinate the localization of biosynthesis enzymes to a specialized organelle in the cell. The localization of AF biosynthesis has been postulated to occur in a dedicated peroxisomal vesicle, termed an aflatoxisome (Chanda et al., 2009). These two hypotheses will be discussed in more detail later in the paper.

2.3 Role of global transcriptional regulators

In concert with all eukaryotes, fungi make use of many different types of transcription factors to regulate cellular processes. Most of the important types that are relevant to control of development and secondary metabolite production are listed in Table 1. Transcription factors, such as AflR, which activate single genes or genes only in a related biosynthetic pathway generally act downstream of signaling cascades and can be activated in response to biological or environmental stimuli. These transcription factors are to be distinguished from globally acting factors which control the transcription of multiple sets of genes, sometimes in unrelated or more distantly related metabolic pathways. For fungi nutritional stimuli such as the carbon or nitrogen source as well as environmental stimuli such as temperature and pH are especially important for control of AF biosynthesis (O'Brian et al., 2007; Wilkinson et al., 2007). These nutrients activate global transcription factors such as CreA (needed for control of expression of sugar utilization genes) and AreA (needed for control of expression of nitrate utilization genes) (Reverberi et al., 2010). PacC is a global transcription factor involved in pH regulation of transcription (Arst & Penalva, 2003). When AF is produced under conditions that stimulate the development of asexual reproductive structures (the conidiospore or sclerotium), the global transcription factors that are needed for activation of genes involved in formation of such structures (BrlA, AbaA, NsdC, NsdD, LreA, LreB) also can affect expression of genes in AF biosynthesis (Georgianna & Payne, 2009; Schmidt-Heydt et al., 2009). The globally acting transcription factors involved in AF synthesis recognize short GC-rich sequences in the promoters of key genes in the biosynthesis cluster. For AreA the consensus GC-rich recognition sequence is HGATAR, for CreA it is SYGGGG, for PacC it is GCCARG, for AbaA it is CATTCY, and for BrlA it is MRAGGGR. Usually more than one consensus domain is needed for strong transcriptional regulation (Gomez et al., 2003), but if the globally acting factor is close to the pathway specific transcription factor in the promoter it may be effective in activating transcription. Disruption of this type of transcription factor gene usually has a large effect on ability to utilize nutrients, morphology, or growth at certain pHs depending on the factor, but rarely results in complete loss of expression of the AF biosynthesis genes. However they may be strongly downregulated if the factor is missing. In the *aflJ-aflR* intergenic region (Figure 2), depending on the species, there are as many as five HGATAR sites, zero to two BrlA sites, one or more PacC sites, several AbaA sites, and one possible CreA site (not shown in Fig. 2).

Types of transcription factors	Approx. No. in *A. flavus* genome	PFAM designation	Examples
Cys2His2 zinc finger (C2H2)	40	PF00096	BrlA, NsdC, PacC
C6 transcription factor (Cys6Zn2)	178	PF00105	AflR, AlcR, NirA
Helix-loop-helix (HLH)	10	PF00010	DevR, PalcA
Helix-turn-helix	2	PF00046	AdaA, Rpc82
Basic leucine zipper (bZip)	17	PF00170	AtfA, NapA, CpcA
Homeodomain (C4HC3) ring finger)	1	PF00319	PHD
Winged helix	33	PF08279	Hpa3, GlcD, Sin3
GATA factors	6	PF00320	NsdD,AreA, LreA
ATTS factor	1	PF01285	AbaA

Table 1. Types of transcription factors in *Aspergillus flavus*. *A. flavus* genome size is 36.8 Mb (8 chromosomes) with 13,487 predicted genes.

2.4 Involvement of transcriptional coactivators in AF biosynthesis

Besides globally acting transcription factors, which act in concert with a pathway-specific transcription factor to activate gene expression, proteins other than AflJ, are required for AF biosynthesis and, like AflJ, probably function as coactivators (Lewis & Reinberg, 2003; Daniel & Grant, 2007). Some of these proteins have been mentioned briefly above. The most important of these for AF biosynthesis are listed in Table 2.

Factor	Putative conserved domains	Putative role	GenBank Accession #
VeA	Fibronectin attachment pfam 07174	Mediates AF activation by binding to LaeA	AFLA_066460
VosA	Topoisomerase related; pfam09770	Spore viability, possible transcription factor; similar motifs to VeA and VelB	AFLA_026900
VelB	Nucleoside diphosphatase	Similar role to that of VeA, forms heterodimer with VosA and LaeA	AFLA_081490
LaeA	SAM-depdt Methyltransferase	Global regulator of transcription of secondary metabolite genes; presumed to be involved in specific chromatin remodeling by methylating histones.	AFLA_033290
LaeA-like	SAM-depdt Methyltransferase	Modulates effect of LaeA	AFLA_035950
AflJ	Microbodies signal	Binds to AflR; possible co-activator	
Bre2 (CclA)	SPRY	Chromatin remodeling; histone lysine N-methyltransferase	AFLA_089250
HdaA	Histone deacetylase	Histone deacetylase; controls heterochromatin formation	AFLA_025220
DmtA	Cytosine DNA methyltransferase	Putative DNA cytosine methyltransferase, required for sexual development	AFLA_056340
FluG	glutamine synthetase	Regulates BrlA activity for conidiophore development	AFLA_101920
FlbA	Pleckstrin, RGS	Conidiophore development	AFLA_134030
FlbC	C2H2 transcription factor	Regulates BrlA activity for conidiophore development	AFLA_137320
FlbD	MYB family DNA binding	Regulates BrlA activity for conidiophore development	AFLA_080170
MedA	Transcriptional regulator	Medusa genes, necessary for correct metulae development; proper temporal expression of BrlA	AFLA_136410
RcoA (TupA)	WD repeat	Effects growth and sexual, asexual development & secondary metabolism; transcriptional repressor	AFLA_054810

Table 2. Transcription-activating factors (TAFs) involved in aflatoxin biosynthesis

2.4.1 Role of *velvet* complex genes

The *velvet* family of genes, so named because mutants have a velvet-like colony appearance, is involved in regulation of conidophore developmental and concomitantly secondary metabolite production (Calvo, 2008). The encoded proteins act upstream of BrlA, whose role is the regulation of transcription of genes needed for spore formation. *A. nidulans* has both a sexual and asexual stage (Note: *Aspergillus* species thought to only be able to reproduce asexually have now been shown to be capable of sexual reproduction as well). When grown in the light, the fungus reproduces asexually whereas in the dark it reproduces sexually. The *velvet* genes mediate the response to light and inhibit asexual development when the fungus is grown in the dark. Light activates a series of receptors including the red light receptor,

FphA, and the blue light receptors, LreA and LreB, as well as the near UV light receptor, CryA. When activated FphA, LreA and LreB form a complex with VeA, and presumably affect its activity or allow it to enter the nucleus. CryA acts differently and affects *veA* expression (Bayram et al., 2008a). Wild-type strains of *A. nidulans* display light-dependent conidiation; strains bearing a mutation (*veA1*) conidiate vigorously regardless of illumination conditions (Mooney & Yager, 1990). VeA likely acts as a negative regulator of asexual development. The majority of *A. nidulans* strains used are derived from the *veA1* mutant because of its increased conidiation. Whereas the *veA1* mutation still permits ST formation in *A. nidulans,* deletion mutants of *veA* in *A. nidulans* or in *A. flavus* are unable to produce ST or AFs, respectively. VeA and VelB with LaeA (see below) form a tripartite complex. VelB and VosA also form a complex, at least in *A. nidulans.* Light-dependent conidiation in *A. nidulans* is mediated via VeA nuclear translocation (Stinnett et al., 2007) together with interaction of VeA with the phytochrome photoreceptor FphA (Purschwitz et al., 2009). *A. flavus* conidiates well regardless of the presence or absence of light.

Fig. 5. Alignment of *Velvet* family proteins showing the two most conserved regions in the proteins.

The *velvet* family consists of four proteins with two regions of homology (Figure 5). These proteins are VeA, a 574 amino acid protein with a nuclear localization domain and a possible transmembrane domain, VelB, a 361 amino acid protein lacking both of these domains, but containing a cluster of Asp residues at its C-terminus, VelC a 434 amino acid protein of unknown function, and VosA, a 449 amino acid protein with conserved domains suggestive of its being a DNA-binding or chromatin-binding protein. A conserved domain search in GenBank suggests that these proteins have domains typically found in proteins that affect chromatin formation, in particular a region called Asc-1, identified in VelC and VosA. ASC-1 homology or ASCH domains are a beta-barrel domain found in proteins that interact with RNA and could mediate the interaction between a transcription factor and the basal transcriptional machinery (Iyer et al., 2006). VosA, (viability of spores) was identified as a protein essential for asexual spore maturation (Ni & Yu, 2007). It may have a novel winged helix transcription activation domain near its C-terminus. It forms a complex with VelB (Sarikaya Bayram et al., 2010). VosA is required for trehalose biogenesis (Ni & Yu, 2007). Trehalose is a compound that helps to protect the spore from stress. Spores, for their long-term survival require high amounts of trehalose. It has been suggested that VosA may primarily control the activity of genes involved in the late process of sporulation, including trehalose biogenesis. VelB and VeA bind to LaeA ((Bayram et al., 2008b), see below). The competition for binding to VosA and LaeA makes VelB a partner in regulating the expression of genes involved in asexual sporulation and in secondary metabolite formation. The VeA-VelB heterodimeric complex of *A. nidulans* presumably forms in the cytoplasm and serves as the major pathway for VelB's entry into the nucleus. VelB and VosA predominantly interact in the nucleus. The same interactions of the *Velvet* genes presumably apply to the role of these proteins in transcription control of AF biosynthesis and sporogenesis in *A. flavus*.

2.4.2 Role of LaeA

Although there is much understanding from the literature on yeast of how Cys_6Zn_2 transcription factors function to activate gene expression, there is much less known about what activates the expression of these transcription factors. In a search for proteins that affect *aflR* expression in *A. nidulans*, a species that accumulates sterigmatocysin (ST), an AF precursor, a gene called *laeA* (Loss of AflR Expression), was isolated in which null mutants are unable to express AflR and lose the ability to make ST as well as other metabolites (Bok & Keller, 2004). *laeA* was predicted to encode a 375 amino acid protein with a S-adenosylmethionine-dependent methyltransferase domain typical of histone methyltansferases and argine methyltransferases. LaeA, however, lacked the SET and double loop domains typically found in such proteins and lacked a canonical nuclear localization signal, even though it was shown to reside in the nucleus. *laeA* expression, in *A. nidulans* was found to be downregulated by AflR, possibly because of AflR-binding sites in its promoter. No AflR-binding sites are present in the promoter of the *A. flavus* ortholog. Therefore, such regulation may be species-specific. The methyltransferase domain was shown to be required for LaeA's function. *laeA* in *A. nidulans* was also shown to be negatively regulated by protein kinase A and RasA, two signal transduction proteins shown to be involved in regulation of secondary metabolite gene activity and conidial development (see below). *laeA* null mutants showed little difference in spore production compared to the wild type, suggesting that the primary role of LaeA is to regulate expression of secondary metabolite gene clusters. It was proposed that LaeA may function as a unique, fungal secondary metabolite-specific regulator of chromatin organization necessary for activation of the genes in such clusters, including *aflR*.

In *ΔlaeA* mutants of *A. nidulans* silencing of *aflR* expression was found to be a consequence of the *aflR* being inside the cluster. When *aflR* was expressed in a locus outside the cluster, ST was produced even in the *ΔlaeA* mutants (Bok et al., 2006). Furthermore when a gene not associated with ST production was placed in the cluster, in the absence of functional LaeA, it was silenced. These results further suggested that chromosomal activity was mediated for secondary metabolite cluster genes by LaeA, and supported the hypothesis that LaeA plays a role in chromatin modification. In *A. parasiticus ΔlaeA* mutants, expression of *aflR* and other AF biosynthesis genes was also undetectable (Kale et al., 2007). Overexpression of *laeA* in strains having a functional copy of *aflR* increased *aflR*'s expression as well as the production of AFs. Furthermore, expression of *veA* was much lower in *ΔlaeA* mutants. Unlike *A. nidulans*, *A. flavus ΔlaeA* mutants showed decreased amounts of conidiation compared to the wild type and a complete absence of sclerotial production. In *ΔlaeA* mutants of *A. nidulans*, ST production was detected when a gene involved in heterochromatin maintenance encoding a histone deacetylase (*hdaA*), was disrupted. This result supported the role of LaeA in chromatin remodeling at the site of secondary metabolite gene clusters (Shwab et al., 2007). The coordinated involvement of both VeA and LaeA in secondary metabolite production was established when it was found that these two factors interact with each other and with VelB in the *velvet* family of light responsive factors. (Bayram et al., 2008b). VelB interacts with the N-terminus of VeA, whereas LaeA interacts with the C-terminus of VeA.

It is assumed that VeA, with its functional nuclear localization domain, facilitates nuclear transport of VelB and LaeA, which lack such domains. Once in the nucleus they function as coactivators to specifically target genes at the secondary metabolite locus, presumably by either aiding in the chromatin remodeling preceding AflR binding to the promoters of cluster genes or by directly interacting with AflR or other pathway-specific DNA-binding proteins involved in secondary metabolite gene cluster activation. Since overexpression of AflR is able to relieve the block in gene expression caused by loss of LaeA, it is possible that AflR, alone or in conjunction with globally acting factors, initiates the chromatin remodeling necessary for expression of genes in the biosynthesis gene cluster.

2.5 Regulation by chromatin conformation

In eukaryotes, DNA with the help of histones forms compact particles, the nucleosomes, where about 147 DNA base pairs make two turns around the histone protein octamer. DNA within nucleosomes is inaccessible to many transcription factors. Some transcription factors, so-called pioneering factors are still able to bind to their DNA recognition sites on the nucleosomal DNA. For most other transcription factors, the nucleosome must be opened up by specific chromatin remodeling proteins (Clapier & Cairns, 2009). Alternatively, the nucleosome can be partially unwrapped by destabilizing factors to allow temporal access to the transcription factor binding site. In many cases a transcription factor needs to compete for binding to its DNA binding site with other transcription factors and histone- or non-histone chromatin proteins. Pairs of transcription factors and other proteins can play antagonistic roles (activator versus repressor) in the regulation of the same gene.

Many of the known gene clusters involved in secondary metabolite formation are located in subtelomeric chromosomal regions in which the chromatin is normally inactive (heterochromatic) due to di- or trimethylation of lysine-9 or lysine-27 of histone-3 (H3K9Me3; H3K27Me3) (Palmer & Keller, 2010). Trimethylation of lysine 4 in histone 3 is associated with active chromatin as is histone lysine acetylation. Generally histone deacetylases (HDAC's) are involved in the formation of heterochromatin whereas histone methyltransferases are associated with either formation of euchromatin (active) or heterochromatin depending on the position of methylation (Jenuwein & Allis, 2001). The proposed function of LaeA is to loosen the chromatin specifically at certain gene clusters, such as the AF cluster, to make it accessible to global and pathway specific DNA binding transcription factors (Strauss & Reyes-Dominguez, 2010). The reason subtelomeric chromosomal regions are associated mainly with inactive chromatin is they are often repositories for foreign DNA, including transposons and retrotransposons (Shaaban et al., 2010). It is in the interest of the cell to prevent activity of the foreign DNA because it could lead to undesirable mutations of genes (by transposon insertion) critical for function of the organism.

In *Aspergillus* species the specific chromatin remodeling factors besides LaeA, are ClrD, a H3K9 methyltransferase, Bre2 (CclA), a H3K4 methyltransferase (part of the COMPASS complex, a complex of proteins in yeast responsible for H3K4 methylation), and HdaA, a histone deacetylase (Palmer et al., 2008; Reyes-Dominguez et al., 2008). Another protein,

HepA, encodes a heterochromatin protein (HP1), one of a family of proteins that bind di- or trimethylated H3K9, leading to gene silencing and heterochromatin formation (Reyes-Dominguez et al., 2010). Deletion mutants of these proteins have generally been associated with increased production of ST in *A. nidulans* except that deletion of *laeA* is associated with the loss of secondary metabolite production. In agreement with these gene disruption results, treatment of *A. nidulans* with the HDAC inhibitors, valproic acid or trichostatin A, increased metabolite production. The role of LaeA in the chromatin remodeling is unclear since it is still uncertain whether or not LaeA acts as a methyltransferase. One possibility is that it, along with associated transcription factors, act as a bridge connecting the basal transcriptional elements (the equivalent of a TATA binding protein) to the pathway specific transcription factor (AflR). Such a complex would resemble the SAGA (Spt-Ada-Gcn5 acetyltranferase) complex in yeast (Traven et al., 2006). Such complexes not only stabilize the transcription machinery, like a Mediator complex in yeast, but also actively recruit the RNA polymerase II to the transcription initiation site (Biddick et al., 2008). A schematic of these proposed interactions is shown in Figure 6.

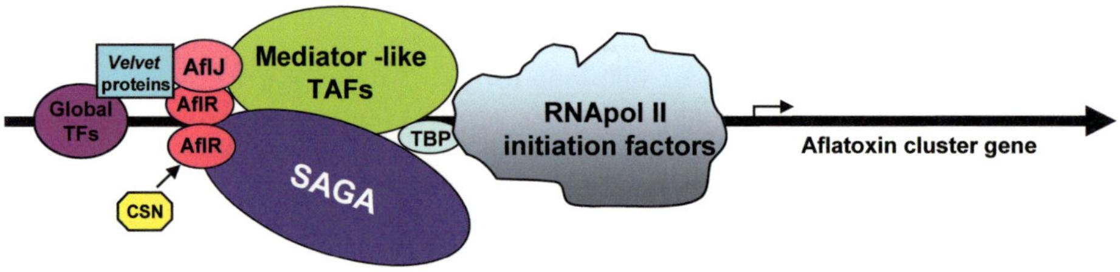

Fig. 6. Model of interactions of the aflatoxin cluster transcription regulatory factors with other proteins involved in the basal transcription machinery. Roles of SAGA and Mediator are described in a review by Traven, et al. (Traven et al., 2006).

In higher eukaryotes DNA methylation has been shown to affect gene expression (Clark & Melki, 2002). Because the level of DNA methylation in *Aspergillus* species is very low, it was considered unlikely to be involved in control of gene activity (Gowher et al., 2001). The loss of AF production and the induction of a *fluffy* phenotype in *A. parasiticus* clones treated with 5-azacytosine, a DNA methyltransferase inhibitor, as well as the presence in the genome of a DNA methyltransferase with predicted cytosine methylase function, suggested that DNA methylation may indeed play an unrecognized role in development and secondary metabolism (Tamame et al., 1983; Lee et al., 2008). DNA methyltransferases may function as chromatin remodeling agents in the absence of cytosine methylation. In support of this hypothesis, disruption of the gene for the DNA methyltransferase, *dmtA*, in *A. nidulans* yielded mutants unable to undergo sexual development (Lee et al., 2008). When gene expression by the nonaflatoxigenic *A. parasiticus* clones obtained by either 5-azacytosine treatment or by serial mycelial transfer (Chang et al., 2007) was compared to gene expression by the parental strain, the most highly downregulated genes were genes predicted to be involved in cell receptor recognition, signaling or transport (Wilkinson, et al., unpublished results). Surprisingly, the genes in the AF cluster were still transcribed at normal levels.

3. Signaling factors involved in AF biosynthesis

Transcription factors may be activated (or deactivated) through a signal-sensing domain by ligand binding, post-synthetic modifications, or proteolysis. The domain responsible for binding to a hormone or an elicitor is called the ligand binding domain (LBD) (Gomperts et al., 2009). This domain participates in several activities including hormone binding, homo- and/or heterodimerization, formation of a complex with a heat-shock protein, and transcriptional activation and repression. The binding of the ligand may affect the conformation and concomitantly the transcription factor's ability to activate gene expression. Ligand binding and postsynthetic modifications, such as phosphorylation, myristylation, acetylation, also can influence where a transcription factor is located within a cell and whether the transcription factor is in its active state and is capable of binding to DNA, coregulatory proteins, or to other transcription factors, possibly as homo- or hetero-dimers.

3.1 Cell surface receptors –environmental signaling

A large number of organic compounds either produced by plants or derived from other sources, such as antimicrobial agents, pesticides, and herbicides have been shown to either induce or inhibit AF production by various mechanisms, some still only poorly understood. An extensive list of inhibitors has been given in an early review (Zaika & Buchanan, 1987). The list of inhibitors compiled today is considerable longer (Holmes et al., 2008). Plant metabolites, such as gallic acid, neem leaf and fruit extracts, certain spice oils , anthocyanins, and related flavonoids have been shown to be inhibitory to fungal toxin production. Low-molecular-weight aldehydes, ketones, and alcohols from cotton leaf, corn, and soybean are also inhibitory. These plant metabolites probably affect the activity or biosynthesis of key developmental regulatory proteins. The plant oxylipins, 13-hydroperoxylinoleic acid and 9-hydroperoxy linoleic acid, affect development and toxin synthesis, likely through conversion of these molecules into psi factor (Tsitsigiannis et al., 2005). Most of these inhibitors directly impact the cellular signaling and regulatory networks necessary for activation of genes required for growth, development and AF production.

3.2 G protein signaling and protein kinases

Development and secondary metabolism in fungi occurs in response to nutrient availability, light, oxidative stimuli, osmotic conditions or salt concentrations. These environmental signals are usually sensed at the cell surface and the signal transmitted to the nucleus by a series of secondary messengers such as diacylglycerol, inositol phosphates, divalent calcium, cyclicAMP or cyclic GMP, and chaperonins (heat shock factors that rapidly transmit stress signals). The secondary messenger transduces the signal from the membrane to cellular targets to mediate morphological changes necessary for reproduction or survival and production of secondary metabolites. Transduction of the activating or inhibiting signal depends on G-protein signaling.

A G-protein-mediated signaling cascade in *A. nidulans* was found to regulate both asexual sporulation and sterigmatocystin (ST) production (Yu & Keller, 2005; Brodhagen & Keller, 2006). FadA, a subunit of a heterotrimeric G-protein involved in developmental regulation, promotes vegetative growth and represses both sexual/asexual development and ST production in *A. nidulans*, as well as AF production in *A. flavus* and *A. parasiticus*. (Calvo et al., 2002). Inhibition of FadA involves FlbA, a regulator of G-protein signaling (RGS). FlbA acts as a positive regulator of ST production (Hicks et al., 1997). FlbA appears to act upstream of AflR because AflR is unable to activate ST or AF biosynthesis in a Δ*flbA* mutant (Shimizu et al., 2003).

The cAMP-dependent protein kinase PkaA functions downstream of FadA in this same signaling pathway. PkaA possibly prevents ST production in *A. nidulans* by inhibiting *aflR* expression and as well as inactivating AflR (Shimizu & Keller, 2001; Shimizu et al., 2003). FluG is another protein in the *fluffy* protein family that is involved in control of asexual sporulation and ST production (Lee & Adams, 1994b; Lee & Adams, 1994a; Lee & Adams, 1996). The signal produced by FluG activates ST biosynthesis indirectly by activating *flbA* which represses FadA signaling (Seo et al., 2003; Seo et al., 2006). Another signaling protein involved in ST production is the GTP-binding protein RasA. RasA inhibits ST biosynthesis presumably by repression of *aflR* expression, although the molecular details of this repression are still unknown (Shimizu et al., 2003). Regulation of AflR by RasA may be at the post-transcriptional level and may be mediated by PkaA phosphorylation (Shimizu et al., 2003). A number of other protein kinases have been implicated in transduction of the environmental signals including the TOR kinase and MAP kinases. The TOR kinase, characterized in yeast as responsible for regulating nutrient mediated growth signaling, has recently been implicated in the biosynthesis of a fungal secondary metabolite (Teichert et al., 2006).

Nutrient regulation of AF and ST biosynthesis may be mediated by cAMP levels. cAMP has been shown to influence secondary metabolism in a number of species (Roze et al., 2004). Glucose and simple sugars in the growth medium act as inducers of AF biosynthesis, possibly by regulating levels of cAMP, and possibly by induction of the global transcription factor, CreA. Hormone-like signaling molecules, known as oxylipins, also contribute to regulation of secondary metabolites (Tsitsigiannis & Keller, 2006). These oxygenated lipid molecules mediate the balance of sexual to asexual spore production in aspergilli and are produced by fatty acid oxygenases encoded by ppo genes (Tsitsigiannis et al., 2004).

4. Specialized vesicles may be required for AF biosynthesis

AF and most secondary metabolites are hydrophobic organic compounds. Compartmentalization of hydrophobic substrates has been seen as a common feature of secondary metabolite production (Sirikantaramas et al., 2009). Recently, evidence has been obtained that the enzymes involved in AF biosynthesis are organized into a specialized peroxisomal vesicle where different oxidative steps occur after formation of the polyketide (Chanda et al., 2009). Development of aflatoxisomes was enhanced by conditions known to induce AF biosynthesis. Formation of the vesicle was inversely correlated with downregulation of a gene (*avaA*) required for vacuole formation (Ohsumi et al., 2002). Coordination of vesicle development was shown to be, in part, mediated by VeA and possibly by other proteins of the *Velvet* family. The organization and stability of such an organelle, obviously, is important for production of AF even if all of the biosynthetic genes are correctly transcribed, processed, and translated into proteins. The role of dedicated vesicles in the formation of AF could explain why in some cases, where the genes for AF biosynthesis are expressed and presumably the enzymes are made, no AF or precursor metabolites are formed. An example of this possibility is given below.

Previous research showed that 5-AC treatment and serial mycelial transfer in the absence of conidiation produced mutants with altered conidiophore development (a fluffy phenotype) (Tamame et al., 1983; Kale et al., 1994; Kale et al., 1996). Although these mutants were unable to produce AFs or precursor metabolites, as in some previous studies (Kale et al., 2007) they showed normal expression of most AF biosynthetic genes as well as normal expression of the aforementioned *laeA* and *veA*. A comparison of gene expression between these non-aflatoxigenic mutants and the parental strain by microarray analysis revealed that the most highly dysregulated genes were predicted to encode proteins targeted to cell membranes.

Among these are genes predicted to encode a protein kinase, an opsin homolog [a protein that may bind a photoreactive chromophore (Shichida & Matsuyama, 2009)], an integrin-repeat protein, and a calcium-binding protein. Since mycelial growth is unaffected in these mutants it is possible that the treatments affect proper formation of the vesicles necessary for AF biogenesis. Furthermore, other genes predicted to encode membrane-bound proteins are proteins involved in transport or exocytosis: a GPI (glycosylphosphatidylinositol)-anchored protein a GABA permease, a MFS transporter. Downregulation of expression of genes encoding membrane-bound proteins described above could explain the loss of normal development of asexual structures required for proper conidial formation and this phenotypic change in the mutant cells could prevent the formation of the vesicle needed for AF biogenesis (Chanda et al., 2009).

5. Protein turnover and its effect on AF biosynthesis

Another form of regulation that takes place post-translationally is control of transcription factor abundance by targeted degradation. Obviously if a transcription factor critical for a particular function is targeted for degradation, it no longer would be available for regulation of expression. As alluded to previously, some of the AflR proteins made by some *Aspergillus* species have PEST domains that may make these proteins destined for ubiquitin-mediated degradation (Rechsteiner, 1988). Ubiquitination is controlled by specialized ubiquitin ligases that reside in organelles called proteasomes. Another organelle related to the proteasome is called the COP9 Signalosome (CSN) (Busch et al., 2003). This multiprotein complex can both stabilize or destabilize other proteins by attaching or detaching a small protein (Nedd8) in a process called neddylation or deneddylation at the protein's ubiquitination site. The CSN complex also contains kinases that affect the activity of other regulatory factors.

In *A. nidulans* the COP9 signalosome was found to be a key regulator of light-dependent signaling and asexual and sexual development (Busch et al., 2003; He et al., 2005). The *Aspergillus* COP9 signalosome may control the abundance of the transcription factors that regulate these processes. Mutation of genes encoding *csnD* and *csnE* (components of CSN) (Busch et al., 2003) affects normal development in *A. nidulans* and pigmentation. In these mutants the abnormal mycelial pigmentation suggests that CSN regulates processes in both fungal development and secondary metabolism. When LaeA was used as the bait in a yeast two-hybrid assay with an *A. parasiticus* cDNA expression library as the prey, among the proteins binding to LaeA was COP9 signalosome complex subunit 5 (XM_001211499) (K. Ehrlich and B. M. Mack, unpublished data). This further suggests that the activity of LaeA may be modulated by specific interactions with CSN.

6. Conclusions

Expression of the genes in the AF biosynthesis cluster is mainly controlled by the pathway specific Cys_6Zn_2 DNA binding protein, AflR. While AflR appears to be necessary for the activation, a number of coactivators are important for fine-tuning of the timing of AflR's activity. These proteins, AflJ, LaeA, VeA, VelB and VosA, may form a complex in the nucleus to not only position AflR at the AF cluster genes but also alter the chromatin conformation in this locus in order to allow AflR and global transcriptional regulatory proteins to make contact with the basal transcription machinery. They may do this concomitantly with AflR binding or act to recruit AflR to the cluster. AflR expression is induced by simple sugars and inhibited by certain organic acids and aldehydes. Globally acting DNA-binding proteins are involved in transmitting the nutritional or environmental signal to the activation of the AF gene cluster. A signaling cascade involving cAMP-dependent protein kinase A plays a role in the activation. The regulatory models shown in Figures 1 and 6 resemble that for expression of developmentally regulated genes in yeast and other fungi.

7. References

Abbas, H.K., Wilkinson, J.R., Zablotowicz, R.M., Accinelli, C., Abel, C.A., Bruns, H.A. & Weaver, M.A. (2009). Ecology of *Aspergillus flavus*, regulation of aflatoxin production, and management strategies to reduce aflatoxin contamination of corn, *Toxin Reviews* 28, 142-153.

Arst, H.N. & Penalva, M.A. (2003). pH regulation in *Aspergillus* and parallels with higher eukaryotic regulatory systems, *Trends Genet* 19, 224-231.

Bayram, O., Biesemann, C., Krappmann, S., Galland, P. & Braus, G.H. (2008a). More than a repair enzyme: *Aspergillus nidulans* photolyase-like CryA is a regulator of sexual development, *Mol Biol Cell* 19, 3254-3262.

Bayram, O., Krappmann, S., Ni, M., Bok, J.W., Helmstaedt, K., Valerius, O., Braus-Stromeyer, S., Kwon, N.J., Keller, N.P., Yu, J.H. & Braus, G.H. (2008b). VelB/VeA/LaeA complex coordinates light signal with fungal development and secondary metabolism, *Science* 320, 1504-1506.

Biddick, R.K., Law, G.L. & Young, E.T. (2008). Adr1 and Cat8 mediate coactivator recruitment and chromatin remodeling at glucose-regulated genes, *PLoS One* 3, e1436.

Bok, J.W. & Keller, N.P. (2004). LaeA, a regulator of secondary metabolism in *Aspergillus* spp., *Eukaryot. Cell.* 3, 527-535.

Bok, J.W., Noordermeer, D., Kale, S.P. & Keller, N.P. (2006). Secondary metabolic gene cluster silencing in *Aspergillus nidulans, Mol. Microbiol.* 61, 1636-1645.

Brodhagen, M. & Keller, N.P. (2006). Signaling pathways connecting mycotoxin production and sporulation, *Mol Plant Pathology* 7, 285-301.

Busch, S., Eckert, S.E., Krappmann, S. & Braus, G.H. (2003). The COP9 signalosome is an essential regulator of development in the filamentous fungus *Aspergillus nidulans, Mol Microbiol* 49, 717-730.

Calvo, A.M. (2008). The VeA regulatory system and its role in morphological and chemical development in fungi, *Fungal Genet Biol* 45, 1053-1061.

Calvo, A.M., Wilson, R.A., Bok, J.W. & Keller, N.P. (2002). Relationship between secondary metabolism and fungal development, *Microbiol Mol Biol Rev* 66, 447-459.

Cary, J.W., Ehrlich, K.C., Wright, M., Chang, P.-K. & Bhatnagar, D. (2000). Generation of *aflR* disruption mutants of *A. parasiticus, Appl. Microbiol. Biotechnol.* 53, 680-684.

Chanda, A., Roze, L.V., Kang, S., Artymovich, K.A., Hicks, G.R., Raikhel, N.V., Calvo, A.M. & Linz, J.E. (2009). A key role for vesicles in fungal secondary metabolism, *Proc Natl Acad Sci U S A* 106, 19533-19538.

Chang, P.-K., Ehrlich, K.C., Yu, J., Bhatnagar, D. & Cleveland, T.E. (1995). Increased expression of *Aspergillus parasiticus aflR*, encoding a sequence-specific DNA-binding protein, relieves nitrate inhibition of aflatoxin biosynthesis, *Appl. Environ. Microbiol.* 61, 2372-2377.

Chang, P.K. (2003). The *Aspergillus parasiticus* protein AFLJ interacts with the aflatoxin pathway-specific regulator AFLR, *Mol. Genet. Genomics* 268, 711-719.

Chang, P.K., Wilkinson, J.R., Horn, B.W., Yu, J., Bhatnagar, D. & Cleveland, T.E. (2007). Genes differentially expressed by *Aspergillus flavus* strains after loss of aflatoxin production by serial transfers, *Appl Microbiol Biotechnol* 77, 917-925.

Clapier, C.R. & Cairns, B.R. (2009). The biology of chromatin remodeling complexes, *Annu Rev Biochem* 78, 273-304.

Clark, S.J. & Melki, J. (2002). DNA methylation and gene silencing in cancer: which is the guilty party?, *Oncogene* 21, 5380-5387.

Cleveland, T.E., Carter-Wientjes, C.H., De Lucca, A.J. & Boue, S.M. (2009). Effect of soybean volatile compounds on *Aspergillus flavus* growth and aflatoxin production, *J Food Sci* 74, H83-87.

Daniel, J.A. & Grant, P.A. (2007). Multi-tasking on chromatin with the SAGA coactivator complexes, *Mutat Res* 618, 135-148.

Du, W., Obrian, G.R. & Payne, G.A. (2007). Function and regulation of *aflJ* in the accumulation of aflatoxin early pathway intermediate in *Aspergillus flavus*, *Food Addit Contam* 24, 1043-1050.

Ehrlich, K.C., Cary, J.W. & Montalbano, B.G. (1999a). Characterization of the promoter for the gene encoding the aflatoxin biosynthetic pathway regulatory protein AFLR, *Biochim. Biophys. Acta.* 1444, 412-417.

Ehrlich, K.C., Mack, B.M., Cary, J.W., Bhatnagar, D. & Kale, S.P. (2011). A hypothesis to explain how LaeA specifically regulates certain secondary metabolite biosynthesis gene clusters, *World Mycotoxin Journal* 4, 53-58.

Ehrlich, K.C., Montalbano, B.G., Bhatnagar, D. & Cleveland, T.E. (1998). Alteration of different domains in AFLR affects aflatoxin pathway metabolism in *Aspergillus parasiticus* transformants, *Fungal Genet. Biol.* 23, 279-287.

Ehrlich, K.C., Montalbano, B.G. & Cary, J.W. (1999b). Binding of the C6-zinc cluster protein, AFLR, to the promoters of aflatoxin pathway biosynthesis genes in *Aspergillus parasiticus*, *Gene* 230, 249-257.

Ehrlich, K.C., Montalbano, B.G. & Cotty, P.J. (2003). Sequence comparison of *aflR* from different *Aspergillus* species provides evidence for variability in regulation of aflatoxin production, *Fungal Genet. Biol.* 38, 63-74.

Georgianna, D.R. & Payne, G.A. (2009). Genetic regulation of aflatoxin biosynthesis: from gene to genome, *Fungal Genet Biol* 46, 113-125.

Gomez, D., Garcia, I., Scazzocchio, C. & Cubero, B. (2003). Multiple GATA sites: protein binding and physiological relevance for the regulation of the proline transporter gene of *Aspergillus nidulans*, *Mol. Microbiol.* 50, 277-289.

Gomperts, B.D., Tatham, P.E.R. & Kramer, L.M. Signal transduction, *2nd* ed. Academic Press, San Diego, 2009.

Gowher, H., Ehrlich, K.C. & Jeltsch, A. (2001). DNA from *Aspergillus flavus* contains 5-methylcytosine, *FEMS Microbiol. Lett.* 205, 151-155.

He, Q., Cheng, P., He, Q. & Liu, Y. (2005). The COP9 signalosome regulates the *Neurospora* circadian clock by controlling the stability of the SCFFWD-1 complex, *Genes Dev* 19, 1518-1531.

Hicks, J.K., Yu, J.H., Keller, N.P. & Adams, T.H. (1997). *Aspergillus* sporulation and mycotoxin production both require inactivation of the FadA G alpha protein-dependent signaling pathway, *EMBO J.* 16, 4916-4923.

Holmes, R.A., Boston, R.S. & Payne, G.A. (2008). Diverse inhibitors of aflatoxin biosynthesis, *Appl Microbiol Biotechnol* 78, 559-572.

Iyer, L.M., Burroughs, A.M. & Aravind, L. (2006). The ASCH superfamily: novel domains with a fold related to the PUA domain and a potential role in RNA metabolism, *Bioinformatics* 22, 257-263.

Janknect, R., Sander, C. & Pongs, O. (1991). (HX)n repeats: a pH-controlled protein-protein interaction motif of eukaryotic transcription factors?, *FEBS Lett.* 295, 1-2.

Jenuwein, T. & Allis, C.D. (2001). Translating the histone code, *Science* 293, 1074-1080.

Kale, S.P., Bhatnagar, D. & Bennett, J.W. (1994). Isolation and characterization of morphological variants of *Aspergillus parasiticus* deficient in secondary metabolism production, *Mycol. Res.* 98, 645-652.

Kale, S.P., Cary, J.W., Bhatnagar, D. & Bennett, J.W. (1996). Characterization of experimentally induced, nonaflatoxigenic variant strains of *Aspergillus parasiticus*, *Appl. Environ. Microbiol.* 62, 3399-3404.

Kale, S.P., Cary, J.W., Hollis, N., Wilkinson, J.R., Bhatnagar, D., Yu, J., Cleveland, T.E. & Bennett, J.W. (2007). Analysis of aflatoxin regulatory factors in serial-transfer-induced non-aflatoxigenic *Aspergillus parasiticus*, *Food Addit. Contam.* 24, 1061-1069.

Lee, B.N. & Adams, T.H. (1994a). The *Aspergillus nidulans fluG* gene is required for production of an extracellular developmental signal and is related to prokaryotic glutamine synthetase, *Genes Dev* 8, 641-651.

Lee, B.N. & Adams, T.H. (1994b). Overexpression of *flbA*, an early regulator of *Aspergillus* asexual sporulation, leads to activation of *brlA* and premature initiation of development, *Mol. Microbiol.* 14, 323-334.

Lee, B.N. & Adams, T.H. (1996). FluG and *flbA* function interdependently to initiate conidiophore development in *Aspergillus nidulans* through *brlA* beta activation, *EMBO J.* 15, 299-309.

Lee, D.W., Freitag, M., Selker, E.U. & Aramayo, R. (2008). A cytosine methyltransferase homologue is essential for sexual development in *Aspergillus nidulans*, *PLoS One* 3, e2531.

Lewis, B.A. & Reinberg, D. (2003). The mediator coactivator complex: functional and physical roles in transcriptional regulation, *J Cell Sci* 116, 3667-3675.

Meyers, D.M., Obrian, G., Du, W.L., Bhatnagar, D. & Payne, G.A. (1998). Characterization of *aflJ*, a gene required for conversion of pathway intermediates to aflatoxin, *Appl. Environ. Microbiol.* 64, 3713-3717.

Mooney, J.L. & Yager, L.N. (1990). Light is required for conidiation in *Aspergillus nidulans*, *Genes Dev.* 4, 1473-1482.

Ni, M. & Yu, J.H. (2007). A novel regulator couples sporogenesis and trehalose biogenesis in *Aspergillus nidulans*, *PLoS One* 2, e970. doi:910.1371/journal.pone.0000970.

O'Brian, G.R., Georgianna, D.R., Wilkinson, J.R., Yu, J., Abbas, H.K., Bhatnagar, D., Cleveland, T.E., Nierman, W. & Payne, G.A. (2007). The effect of elevated temperature on gene transcription and aflatoxin biosynthesis, *Mycologia* 99, 232-239.

Ohsumi, K., Arioka, M., Nakajima, H. & Kitamoto, K. (2002). Cloning and characterization of a gene (*avaA*) from *Aspergillus nidulans* encoding a small GTPase involved in vacuolar biogenesis, *Gene* 291, 77-84.

Palmer, J.M. & Keller, N.P. (2010). Secondary metabolism in fungi: does chromosomal location matter?, *Curr Opin Microbiol* 13, 431-436.

Palmer, J.M., Perrin, R.M., Dagenais, T.R. & Keller, N.P. (2008). H3K9 methylation regulates growth and development in *Aspergillus fumigatus*, *Eukaryot Cell* 7, 2052-2060.

Payne, G.A. & Yu, J. (2010). Ecology, development and gene regulation in *Aspergillus flavus*. In: Masayuki, M. & Gomi, K. (Eds.), *Aspergillus: Molecular Biology and Genomics*. Caister Academic Press, Norfolk, UK, pp. 157-171.

Purschwitz, J., Muller, S. & Fischer, R. (2009). Mapping the interaction sites of *Aspergillus nidulans* phytochrome FphA with the global regulator VeA and the White Collar protein LreB, *Molecular Genetics and Genomics* 281, 35-42.

Rechsteiner, M. (1988). Regulation of enzyme levels by proteolysis: the role of PEST regions, *Adv. Enzyme Regulation* 27, 135-151.

Reverberi, M., Ricelli, A., Zjalic, S., Fabbri, A.A. & Fanelli, C. (2010). Natural functions of mycotoxins and control of their biosynthesis in fungi, *Applied Microbiology and Biotechnology* 87, 899-911.

Reyes-Dominguez, Y., Bok, J.W., Berger, H., Shwab, E.K., Basheer, A., Gallmetzer, A., Scazzocchio, C., Keller, N. & Strauss, J. (2010). Heterochromatic marks are associated with the repression of secondary metabolism clusters in *Aspergillus nidulans*, *Mol Microbiol* 76, 1376-1386.

Reyes-Dominguez, Y., Narendja, F., Berger, H., Gallmetzer, A., Fernandez-Martin, R., Garcia, I., Scazzocchio, C. & Strauss, J. (2008). Nucleosome positioning and histone H3 acetylation are independent processes in the *Aspergillus nidulans prnD-prnB* bidirectional promoter, *Eukaryot Cell* 7, 656-663.

Roze, L.V., Beaudry, R.M., Keller, N.P. & Linz, J.E. (2004). Regulation of aflatoxin synthesis by FadA/cAMP/protein kinase A signaling in *Aspergillus parasiticus*, *Mycopathologia* 158, 219-232.

Roze, L.V., Chanda, A. & Linz, J.E. (2011). Compartmentalization and molecular traffic in secondary metabolism: A new understanding of established cellular processes, *Fungal Genet Biol* 48, 35-48.

Sarikaya Bayram, O., Bayram, O., Valerius, O., Park, H.S., Irniger, S., Gerke, J., Ni, M., Han, K.H., Yu, J.H. & Braus, G.H. (2010). LaeA control of velvet family regulatory proteins for light-dependent development and fungal cell-type specificity, *PLoS Genet* 6, e1001226.

Schmidt-Heydt, M., Abdel-Hadi, A., Magan, N. & Geisen, R. (2009). Complex regulation of the aflatoxin biosynthesis gene cluster of *Aspergillus flavus* in relation to various combinations of water activity and temperature, *Int J Food Microbiol* 135, 231-237.

Seo, J.A., Guan, Y. & Yu, J.H. (2003). Suppressor mutations bypass the requirement of *fluG* for asexual sporulation and sterigmatocystin production in *Aspergillus nidulans*., *Genetics* 165, 1083-1093.

Seo, J.A., Guan, Y. & Yu, J.H. (2006). FluG-dependent asexual development in *Aspergillus nidulans* occurs via derepression, *Genetics* 172, 1535-1544.

Shaaban, M., Palmer, J.M., El-Naggar, W.A., El-Sokkary, M.A., Habib, E.E. & Keller, N.P. (2010). Involvement of transposon-like elements in penicillin gene cluster regulation, *Fungal Genetics and Biology* 47, 423-432.

Shichida, Y. & Matsuyama, T. (2009). Evolution of opsins and phototransduction, *Philos Trans R Soc Lond B Biol Sci* 364, 2881-2895.

Shimizu, K., Hicks, J.K., Huang, T.P. & Keller, N.P. (2003). Pka, Ras and RGS protein interactions regulate activity of AflR, a Zn(II)2Cys6 transcription factor in *Aspergillus nidulans, Genetics.* 165, 1095-1104.

Shimizu, K. & Keller, N.P. (2001). Genetic involvement of a cAMP-dependent protein kinase in a G protein signaling pathway regulating morphological and chemical transitions in *Aspergillus nidulans, Genetics.* 157, 591-600.

Sirikantaramas, S., Yamazaki, M. & Saito, K. (2009). A survival strategy: the coevolution of the camptothecin biosynthetic pathway and self-resistance mechanism, *Phytochemistry* 70, 1894-1898.

Stinnett, S.M., Espeso, E.A., Cobeno, L., Araujo-Bazan, L. & Calvo, A.M. (2007). *Aspergillus nidulans* VeA subcellular localization is dependent on the importin A carrier and on light, *Mol. Microbiol.* 63, 242-255.

Strauss, J. & Reyes-Dominguez, Y. (2010). Regulation of secondary metabolism by chromatin structure and epigenetic codes, *Fungal Genet Biol* 48, 62-69.

Tamame, M., Antequera, F., Villanueva, J.R. & Santos, T. (1983). High-frequency conversion to a "fluffy" developmental phenotype in *Aspergillus* spp. by 5-azacytidine treatment: evidence for involvement of a single nuclear gene, *Mol Cell Biol* 3, 2287-2297.

Teichert, S., Wottawa, M., Schonig, B. & Tudzynski, B. (2006). Role of the *Fusarium fujikuroi* TOR kinase in nitrogen regulation and secondary metabolism, *Eukaryot Cell* 5, 1807-1819.

Traven, A., Jelicic, B. & Sopta, M. (2006). Yeast Gal4: a transcriptional paradigm revisited, *EMBO Reports* 7, 496-499.

Tsitsigiannis, D., Kowieski, T.M., Zarnowski, R. & Keller, N.P. (2004). Endogenous lipogenic regulators of spore balance in *Aspergillus nidulans, Eukaryotic Cell* 3, 1398-1411.

Tsitsigiannis, D.I. & Keller, N.P. (2006). Oxylipins act as determinants of natural product biosynthesis and seed colonization in *Aspergillus nidulans, Mol Microbiol* 59, 882-892.

Tsitsigiannis, D.I., Kowieski, T.M., Zarnowski, R. & Keller, N.P. (2005). Three putative oxylipin biosynthetic genes integrate sexual and asexual development in *Aspergillus nidulans, Microbiology-Sgm* 151, 1809-1821.

Wilkinson, J.R., Yu, J., Abbas, H.K., Scheffler, B.E., Kim, H.S., Nierman, W.C., Bhatnagar, D. & Cleveland, T.E. (2007). Aflatoxin formation and gene expression in response to carbon source media shift in *Aspergillus parasiticus, Food Addit Contam.* 24, 1051-1060.

Xie, Y., Denison, C., Yang, S.H., Fancy, D.A. & Kodadek, T. (2000). Biochemical characterization of the TATA-binding protein-Gal4 activation domain complex, *J. Biol. Chem.* 275, 31914-31920.

Yu, J., Bhatnagar, D. & Cleveland, T.E. (2004a). Completed sequence of the aflatoxin pathway gene cluster in *Aspergillus parasiticus, FEBS Lett.* 564, 126-130.

Yu, J., Chang, P.K., Ehrlich, K.C., Cary, J.W., Bhatnagar, D., Cleveland, T.E., Payne, G.A., Linz, J.E., Woloshuk, C.P. & Bennett, J.W. (2004b). Clustered pathway genes in aflatoxin biosynthesis, *Appl Environ Microbiol* 70, 1253-1262.

Yu, J.H. & Keller, N. (2005). Regulation of secondary metabolism in filamentous fungi, *Annual Review of Phytopathology* 43, 437-458.

Zaika, L.L. & Buchanan, R.L. (1987). Review of compounds affecting the synthesis or bioregulation of aflatoxins, *J. Food Protection* 50, 691-708.

3

Aflatoxin Biosynthetic Pathway and Pathway Genes

Jiujiang Yu and Kenneth C. Ehrlich
USDA/ARS, Southern Regional Research Center, New Orleans, Louisiana USA

1. Introduction

Among the over 185 known species within the genus *Aspergillus, Aspergillus flavus* is the most economically important because it produces the toxic and carcinogenic aflatoxins. Its non-aflatoxigenic relative *A. oryzae* is used extensively for food fermentations (Jelinek et al., 1989). It is one of the most abundant soil-borne molds on earth. *A. flavus* fungus is a saprobe mold that is capable of surviving on many organic nutrient sources like plant debris, tree leaves, decaying wood, animal fodder, cotton, compost piles, dead insects and animal carcasses, stored grains, and even immunocompromised humans and animals (Klich, 1998). It has the ability to survive temperatures ranging from 12°C to 48°C, but the optimal growth temperature ranges from 28°C to 37°C. Its ability to grow at relatively high temperatures contributes to its pathogenicity toward humans and other warm blooded animals. For most of its lifecycle, the fungus exists in the form of mycelium or asexual spores known as conidia. Under adverse conditions such as lack of adequated nutrients or water, the fungal mycelium will transform to resistant structures called sclerotia which can survive extremely harsh environmental conditions. The fungus overwinters either as spores, sclerotia, or as mycelium in debris. When conditions become favorable the sclerotia germinate directly to produce new colonies or conidiophores with conidia (Bennett et al., 1986; Cotty, 1988; Chang et al., 2002).

Aflatoxins were first identified as the cause of a severe animal poisoning incident in England in 1960 called the Turkey X disease (Allcroft et al., 1961; Lancaster et al., 1961). *A. flavus* produces aflatoxin B_1 and B_2 whereas *A. parasiticus*, produces aflatoxins B_1, B_2, G_1, and G_2. These four major aflatoxins are named based on their blue (B) or green (G) fluorescence under ultraviolet light, and their relative mobility by thin-layer chromatography on silica gel. Aflatoxin M_1 is a hydroxylated derivative metabolized from aflatoxin B_1 by cows and secreted in milk (Van Egmond, 1989). In addition to aflatoxins B_1 and B_2, *A. flavus* also produces many other mycotoxins such as cyclopiazonic acid, kojic acid, beta-nitropropionic acid, aspertoxin, aflatrem and aspergillic acid (Goto et al., 1996).

The disease caused by ingestion of aflatoxins in contaminated food or feed is called aflatoxicosis. Acute aflatoxicosis occurs when aflatoxins are consumed at moderate to high levels. Depending on the level and duration of exposure, aflatoxins possess both hepatotoxic and carcinogenic properties. Symptoms in humans include vomiting, abdominal pain, alteration in digestion, limb and pulmonary edema, convulsions, rapid progressive jaundice, swollen liver, high fever, coma, and death. The predominant damage is to the liver (Scholl & Groopman, 1995); (Fung & Clark, 2004; Lewis et al., 2005), but acute damage to the kidneys and heart have been found (Richard & Payne, 2003). In liver aflatoxins irreversibly bind to protein and DNA to form adducts such as aflatoxin B_1-lysine in albumin and a guanyl-N7 adduct in DNA (Skipper & Tannenbaum, 1990). Disruption of the proteins and DNA bases in hepatocytes causes the toxicity (Tandon et al., 1978; Azziz-Baumgartner et al., 2005). Major outbreaks of acute aflatoxicosis from contaminated food in humans were reported in developing countries (Centers for Disease Control and Prevention, 2004; Lewis et al., 2005). For example, in western India in 1974, 108 persons died among 397 people affected with aflatoxin poisoning in more than 150 villages (Krishnamachari et al., 1975). A more recent incident of aflatoxin poisoning occurred in Kenya in July 2004 leading to the death of 125 people among 317 reported with illness due to consumption of aflatoxin contaminated maize (corn) (Centers for Disease Control and Prevention, 2004; Lewis et al., 2005). Acute toxicosis is not the only concern. World health authorities warn that low doses and long term dietary exposure to aflatoxins is also a major risk as chronic exposure can lead to hepatocellular carcinoma (Bressac et al., 1991; Hsu et al., 1991; Wogan, 1992; Fung & Clark, 2004).

Among the four major types of aflatoxins, aflatoxin B_1 is the most toxic and the most potent carcinogen in humans and animals including nonhuman primates, birds, fish, and rodents. Chronic exposure can result in suppressed immune response, malnutrition, proliferation of the bile duct, centrilobular necrosis and fatty infiltration of the liver, hepatic lesions, and even hepatomas. In animal models, aflatoxin B_1 is modified into a more toxic and carcinogenic by-product during detoxification by a cytochrome P450 monooxygenase in liver (Ngindu et al., 1982; Hsieh, 1989; Eaton & Gallagher, 1994; Lewis et al., 2005). The epoxide form of aflatoxin binds to guanine residues in DNA, forms guanyl-N7 adducts, and induces mutations. One mutation, a G to T transversion (Baertschi et al., 1989; Bressac et al., 1991) at the third base of codon 249, a mutation hot spot of the p53 tumor suppressor gene, is generally believed to be the mechanism for initiating hepatocarcinoma formation (Busby & Wogan, 1981; Hsu et al., 1991; Ozturk, 1991; Coursaget et al., 1993). The p53 gene encodes a transcription factor involved in cell cycle regulation. It is commonly mutated in human liver cancers (Groopman et al., 1994). Aflatoxin B_1 is also a potential immunosuppressive agent (Raisuddin et al., 1993). Chronic low level exposure of growing vertebrates to aflatoxins may enhance their susceptibility to infection and tumorigenesis (Raisuddin et al., 1993). AFB_1 also affects other organs and tissues, such as the lungs and the entire respiratory system (Kelly et al., 1997). Human hepatocarcinomas are also associated with hepatitis B virus (HBV) and C virus (HCV) infections (Peers et al., 1987; Hsieh, 1989; Wild et al., 1992). Together with aflatoxins these viruses significantly increased the risk of hepatoma in hepatitis patients (Chen et al., 1996a; Chen et al., 1996b; McGlynn et al., 2003; Arsura & Cavin, 2005). In developing countries, many children are exposed to aflatoxin before birth (Turner et al., 2007), while nursing (Polychronaki et al., 2007) and after weaning (Gong et al., 2004). An association of hepatocellular carcinoma and dietary exposure with aflatoxins has been established from patients living in high-risk areas of China, Kenya, Mozambique, Phillippines, Swaziland, Thailand, Transkei of South Africa (Lancaster et al., 1961; Zuckerman et al., 1967; Wong et al., 1977; Hsieh et al., 1985; Zhu et al., 1987; Huang & Hsieh, 1988; Wilson, 1989; Wogan, 1992; Eaton & Gallagher, 1994; Lewis et al., 2005).

Aspergillus flavus can grow in immunocompromised warm blooded mammals and can cause invasive and non-invasive aspergillosis in humans and animals (Denning et al., 1991; Denning, 1998; Mori et al., 1998; Denning et al., 2003). *A. flavus* is the second leading cause of aspergillosis slightly behind *A. fumigatus*. The incidence of aspergillosis caused by Aspergilli is rising due to the increase of immunocompromised patients in the population due to HIV infection (Denning, 1998; Nierman et al., 2005; Ronning et al., 2005).

A. flavus is a weak and opportunistic plant pathogen, affecting many agricultural crops such as maize (corn), cotton, groundnuts (peanuts), as well as tree nuts such as Brazil nuts, pecans, pistachio nuts, and walnuts. Preharvest contamination of these crops with aflatoxins is common. *A. flavus* also causes the spoilage of post harvest grains during storage. Because *A. flavus* lacks host specificity (St Leger et al., 2000) and can attack seeds of both monocots and dicots, and seeds produced both above ground (corn) as well as below the ground (peanuts). Under weather conditions favorable for its growth, *A. flavus* can cause ear rot on maize, resulting in significant economic losses to farmers (Robens, 2001; Richard & Payne, 2003; Robens & Cardwell, 2005).

2. Economic significance

Due to the toxic and carcinogenic properties of aflatoxins, only extremely low levels of aflatoxins in foods and feeds is allowed (Council for Agricultural Science and Technology, 2003; Fung & Clark, 2004). The International Agency for Research on Cancer (IARC) has designated aflatoxin as a human liver carcinogen (Van Egmond, 1989; van Egmond & Jonker, 2005; van Egmond et al., 2007). To minimize potential exposure to aflatoxins, maximum levels of aflatoxins in many commodities have been set at levels below 20 ppb by most countries (Van Egmond, 1989; van Egmond & Jonker, 2005; van Egmond et al., 2007). Regulatory guidelines of the U.S. Food and Drug Administration (FDA) specifically prevent the sale of commodities if contamination by aflatoxins exceeds 20 ppb total aflatoxins for interstate commerce of food and feedstuff and 0.5 ppb aflatoxin M_1 in milk. The European Commission has set the limits on groundnuts subject to further processing at 15 ppb for total aflatoxins and 8 ppb for aflatoxin B_1, and for nuts and dried fruits subject to further processing at 10 ppb for total aflatoxins and 5 ppb for aflatoxin B_1. The aflatoxin standards for cereals, dried fruits, and nuts intended for direct human consumption are even more stringent, and the limit for total aflatoxins is 4 ppb and 2 ppb for aflatoxin B_1 (van Egmond & Jonker, 2005).

Aflatoxin contamination of agricultural commodities poses a potential risk to livestock and human health (Lancaster et al., 1961; Bennett & Lee, 1979; Bennett, 1987; Jelinek et al., 1989; Cleveland & Bhatnagar, 1992; Eaton & Groopman, 1994; Hall & Wild, 1994; Bhatnagar et al., 2002; Bennett & Klich, 2003; Richard & Payne, 2003). It is not only a serious food safety concern, but it has significant economic implications for the agricultural industry worldwide because of restrictions limiting the trade of contaminated crops.

Since its discovery, extensive efforts have been made and expense incurred worldwide to monitor aflatoxin occurrence and to develop control strategies (Bennett, 1970; Bennett & Goldblatt, 1973; Bennett et al., 1976b; Papa, 1976; Papa, 1979; Papa, 1984). The hallmark discovery of a color mutant that accumulates the brick-red pigment, norsolorinic acid (NOR), in *A. parasiticus* marked a milestone in the understanding the chemistry of aflatoxin biosynthesis (Bennett et al., 1971; Bennett et al., 1976a; Bennett, 1979; Bennett et al., 1983). Since NOR is the earliest and the first stable aflatoxin precursor in the aflatoxin biosynthetic pathway (Hsieh et al., 1976; Dutton, 1988; Bennett et al., 1997), this discovery led to the identification of other key aflatoxin intermediates and established the primary metabolites in the aflatoxin pathway. It provided the opportunity to isolate the first aflatoxin pathway gene that encodes a reductase for the conversion from NOR to eventually aflatoxins (Hsieh & Mateles, 1970; Hsieh et al., 1973; Hsieh et al., 1976; Chang et al., 1992)Dutton, 1982 #357;Dutton, 1985 #313}. After the cloning of several important aflatoxin pathway genes, a 75 kb aflatoxin pathway gene cluster was established in *A. parasiticus* and *A. flavus* (Yu et al., 1995a). Discovery of the cluster promoted renewed interest in understanding aflatoxin biosynthesis by scientists all over the world. Significant progress has been made in elucidating the biosynthetic pathway, the pathway intermediates, genes, corresponding enzymes, and regulatory mechanisms (Bennett & Lee, 1979; Bennett et al., 1981; Cleveland et al., 1987; Bennett & Papa, 1988; Bhatnagar et al., 1992; Chang et al., 1993; Keller et al., 1993; Chang et al., 1995a; Chang et al., 1999a; Ehrlich et al., 1999a; Bennett & Klich, 2003; Chang, 2004; Yu et al., 2004c; Crawford et al., 2008a; Ehrlich, 2009; Ehrlich & Yu, 2009). At least 27 enzymatic steps have been characterized or proposed to be involved in bioconversion of aflatoxin intermediates to aflatoxins (Ehrlich, 2009). In this chapter, we focus on the aflatoxin biosynthetic pathway and the function of aflatoxin cluster genes. For detailed historical information on the aflatoxin pathway genes and gene cluster discovery, please refer to previous reviews (Yabe & Nakajima, 2004; Yu et al., 2004a; Yu et al., 2004c; Yu et al., 2011).

3. Aflatoxin biosynthetic pathway and genes involved

Attempts to decipher the aflatoxin biosynthetic pathway began shortly after the determination of the structure of these toxins (Goldblatt, 1969; Singh & Hsieh, 1977; FAO, 1995). The discovery of a colored mutant in *A. parasiticus* that accumulates norsolorinic acid (NOR) (Bennett et al., 1971; Lee et al., 1971; Bennett et al., 1997) paved the road for the establishment of aflatoxin biosynthetic pathway. With the rapid gene cloning and enzyme characterization, the enzymatic steps for biosynthesis of the 15 structurally defined aflatoxin pathway intermediates have been identified (Cleveland & Bhatnagar, 1987; Bhatnagar et al., 1989; Cleveland & Bhatnagar, 1990; Bhatnagar & Cleveland, 1991; Cleveland & Bhatnagar, 1991; Bhatnagar et al., 1992; Trail et al., 1995a; Yu et al., 1995a; Minto & Townsend, 1997; Townsend, 1997; Yu et al., 1997; Yu et al., 1998; Keller et al., 2000; Yu et al., 2004b; Crawford et al., 2008b). There are estimated to be 27 enzymatic steps in the aflatoxin biosynthesis (Ehrlich, 2009). As many as 30 genes are potentially involved in aflatoxin biosynthesis (Figure 1). The genes and corresponding enzymes have been extensively studied (Bennett et al., 1971; Lee et al., 1971; Yabe & Nakajima, 2004; Yu et al., 2004a; Yu et al., 2004c). In *A. flavus* and *A. parasiticus* the aflatoxin pathway genes are clustered within a 75-kb region of the fungal genome on chromosome III roughly 80 kb away from telomere (Wilson, 1989; Trail et al., 1995a; Trail et al., 1995b; Yu et al., 1995a; Townsend, 1997; Yu et al., 2004a; Yu et al., 2004c; Chang et al., 2005).

3.1 Acetate to norsolorinic acid (NOR)

Norsolorinic acid (NOR) was confirmed to be the first stable aflatoxin precursor (Bennett et al., 1971; Bennett, 1981; Bennett et al., 1983). A hexanoyl starter unit is the initial substrate for aflatoxin formation (Hsieh & Mateles, 1970). Two fatty acid synthases (FAS) and a polyketide synthase (NR-PKS, PksA) are involved in the synthesis of the polyketide from a hexanoyl starter unit. Seven iterative, malonyl-derived ketide extensions are required to produce norsolorinic acid anthrone (noranthrone) (Wilson, 1989; Trail et al., 1995a; Trail et al., 1995b; Brown et al., 1996a; Brown et al., 1996b; Watanabe et al., 1996; Watanabe & Townsend, 2002; Yabe & Nakajima, 2004; Crawford et al., 2006; Crawford et al., 2008a;

Crawford et al., 2008b). Mahanti et al. (Mahanti et al., 1996) cloned, by genetic complementation, a 7.5-kb large transcript which is required for NOR formation in a blocked *A. parasiticus* mutant. Its protein has high degree of similarity (67%) and identity (48%) to the beta-subunit of FASs (FAS1) of *Saccharomyces cerevisiae* and *Yarrowia lipolytica*. Metabolite feeding and gene disruption experiments further confirmed that *uvm8* encodes a subunit of a novel fatty acid synthase (FAS) directly involved in the backbone formation of the polyketide precursor of NOR during aflatoxin biosynthesis, therefore, on the basis of its function, the *uvm8* gene was renamed *fas-1A*. In the revised naming scheme, the *fas-1A* gene was renamed as *fas-1*, it encodes fatty acid synthase-1 in the aflatoxin biosynthetic pathway gene cluster (Figure 1). Another large transcript (*fas-2A*) which encodes an alpha-subunit of fatty acid synthase in the aflatoxin gene cluster was reported (Mahanti et al., 1996). The gene *fas-1A* and *fas-2A* were renamed *fas-1* and *fas-2*. They encode two fatty acid synthases (FASα and FASβ) (Payne, 1998). In *A. nidulans* the involvement of FASs in sterigmatocystin (ST) biosynthesis was also confirmed and were named *stcJ* and *stcK* in the ST cluster (Brown et al., 1996a; Brown et al., 1996b). The biochemical evidence for the role of a fatty acid synthase and a polyketide synthase (PKS) in the biosynthesis of aflatoxin was demonstrated (Watanabe & Townsend, 1996). Further details on the early stage of aflatoxin biosynthesis involving fatty acid synthases and polyketide synthases were reported (Watanabe et al., 1996; Hitchman et al., 2001; Watanabe & Townsend, 2002; Crawford et al., 2006). The N-acetylcysteamine thioester of hexanoic acid was incorporated into NOR in a *fas-1* disrupted transformant. A polyketide synthase gene (*pksA*) in *A. parasiticus* was demonstrated by gene disruption to be required for aflatoxin biosynthesis (Chang et al., 1995a). The predicted amino acid sequences of these PKSs contain the typical four conserved domains commonly found in other known PKS proteins: β-ketoacyl synthase (KS), acyltransferase (AT), acyl carrier protein (ACP), and thioesterase (TE) (Chang et al., 1995a). Townsend's group has dissected the functional domains of the PKS for aflatoxin biosynthesis (Crawford et al., 2008a; Crawford et al., 2008b; Crawford et al., 2008c). These include domains for the starter unit acyl transferase (SAT) which recognizes hexanoyl CoA and the N-acetylcysteamine thioester of hexanoic acid, the acyl carrier protein (ACP) , ketosynthase (KS), malonyl-CoA:ACP transacylase (MAT), product template (PT) allowing the iterative steps in forming the polyketide, and a thioesterase/Claisen-like cyclase (TE/CLC) (Crawford et al., 2008a). The predicted product converted by PksA is noranthrone. The conversion of noranthrone to NOR, the first stable intermediate in the pathway (Bennett et al., 1971; Lee et al., 1971; Papa, 1979; Bennett et al., 1981; Papa, 1982; Bennett et al., 1994; Bennett et al., 1997), is poorly defined, but it has been proposed to be catalyzed by a noranthrone oxidase, a monooxygenase, or to occur spontaneously (Dutton, 1988). Sequence analysis and enzymatic studies supports the contention that the *hypC* (a gene in the intergenic region of *pksA* and *nor-1*) gene product is the required noranthrone oxidase involved in the catalysis of the orxidation of norsolorinic acid anthrone to NOR (Ehrlich, 2009) The *fas-1*, *fas-2*, and *pksA* genes were renamed as *aflA*, *aflB* and *aflC* respectively (Wilson, 1989; Yu et al., 2004a; Yu et al., 2004c) (Figure 1). The *aflA*, *aflB* and *aflC* gene homologues in *A. nidulans* are *stcJ*, *stcK*, and *stcA*, respectively (Brown et al., 1996b).

3.2 Norsolorinic acid (NOR) to averantin (AVN)

The first stable AF intermediate was identified as NOR produced in *A. parasiticus* uv-generated disruption mutants (Bennett et al., 1971; Lee et al., 1971; Detroy et al., 1973;

Bennett et al., 1981) and in *A. flavus* (Papa, 1979; Papa, 1982) The NOR-accumulating mutants are leaky mutants whose aflatoxin biosynthesis is not completely blocked. By genetic complementation, the gene, *aflD* (*nor-1*), encoding a reductase was cloned (Chang et al., 1992). A recombinant Nor-1 protein expressed in *E. coli* catalyzed the reduction of NOR. Therefore, *aflD* (*nor-1*) encodes the ketoreductase needed for the conversion of the 1'-keto group in NOR to the 1'-hydroxyl group of AVN (Zhou & Linz, 1999). Disruption of the *aflD* (*nor-1*) gene also confirmed its involvement in conversion of NOR to AVN in aflatoxin biosynthesis (Trail et al., 1994). The *aflD* (*nor-1*) homologous gene in *A. nidulans* is *stcE* (Brown et al., 1996b). Genes homologous to *aflD* (*nor-1*), in the AF cluster, such as *aflE* (*norA*) and *aflF* (*norB*) are predicted to encode short chain aryl alcohol dehydrogenases. These proteins may also be able to catalyze the reduction of NOR to AVN depending on the reductive environment of the cell and may explain the leakiness of the *nor-1* mutation if they are able to complement Nor-1's function (Cary et al., 1996),

3.3 Averantin (AVN) to 5'-hydroxyaverantin (HAVN)

Radioisotope incorporation experiments established the earliest evidence for the conversion of AVN to HAVN (Bennett et al., 1980; McCormick et al., 1987). In these studies, three enzymatic steps can account for the conversion of NOR to averufin (AVF) (Yabe et al., 1991a). They are (i) NOR to AVN catalyzed by a reductase, (ii) NOR to HAVN catalyzed by a monooxygenase, and (iii) HAVN to AVF catalyzed by a second dehydrogenase. It was also proposed that the oxidation reactions are reversible and that NADPH was the preferred cofactor (Yabe et al., 1991b). The gene previously named *ord-1* encoding a P-450 monooxygenase was cloned and disrupted (Yu et al., 1997). Substrate feeding studies of the *ord*-1 mutant confirmed that HAVN is the intermediate in the conversion of AVN to AVF. The *ord-1* gene, which has a high degree of sequence similarity to *A. nidulans stcF* (Brown et al., 1996b), was renamed *aflG* (*avnA*).

3.4 5'-Hydroxyaverantin (HAVN) to oxoaverantin (OAVN), and averufin (AVF)

Numerous studies have established averufin as one of the key intermediates in aflatoxin formation (Lee et al., 1971; Hsieh, 1973; Lin & Hsieh, 1973; Lin et al., 1973; Fitzell et al., 1975; Singh & Hsieh, 1977; Keller et al., 2000). Several intermediates were reported to be involved in the conversion from AVN to AVF (Lin & Hsieh, 1973; Bhatnagar et al., 1992). One of these averufanin (AVNN), based on later studies was considered a shunt metabolite and not a genuine aflatoxin intermediate (Sakuno et al., 2003; Yabe & Nakajima, 2004). Chang et al. (Lee et al., 1971; Lin & Hsieh, 1973; Chang et al., 2000) characterized the cluster gene *aflH* (*adhA*) in *A. parasiticus* which encodes an alcohol dehydrogenase. It was showed that *adhA* deletion mutants accumulated predominantly HAVN and after prolonged growth the mutants were able to produce small amounts of AVNN consistant with AVNN being a shunt metabolite. Thus, HAVN might be converted directly to AVF or indirectly to AVF by an additional cytosolic enzyme. Sakuno et al. (Sakuno et al., 2003) characterized two cytosolic enzymes and a new aflatoxin intermediate named 5'-oxoaverantin (OAVN) as an intermediate between HAVN and AVF. The enzyme for the conversion from HAVN to OAVN is encoded by the *aflH* (*adhA*) gene. The *adhA* gene deletion mutant is leaky indicating that additional enzyme(s) or gene(s) may be involved in the conversion from OAVN to AVF. The enzymatic steps for aflatoxin biosynthesis and the possible involvement of additional enzymes have also been described (Townsend, 1997; Ehrlich, 2009; Ehrlich et al., 2010). Woloshuk and Payne (Woloshuk & Payne, 1994) identified an alcohol dehydrogenase gene, *adh1*, in *A. flavus*, expressed concurrently with aflatoxin pathway genes. No further report is made on the role of *A. flavus adh1* gene in aflatoxin synthesis. The *aflH* (*adhA*) gene in *A. flavus* and the *adhA* gene in *A. parasiticus* share no significant homology at either the DNA or the amino acid level.

3.5 Averufin (AVF) to versiconal hemiacetal acetate (VHA)

VHA was identified as an aflatoxin precursor formed by oxidation of AVF (Fitzell et al., 1977). The conversion of AVF to VHA involves the cytochrome P450 monooxidase, CypX, and another gene, *aflI* (*avfA*). Although *aflI* is required for the conversion, its oxidative role is unclear (Yu et al., 2000b). *A. nidulans* also has an *aflI* gene homolog (*stcO*) (Brown et al., 1996b; Yu et al., 2000b). Complementation of an averufin-accumulating mutant, *A. parasiticus* SRRC 165, with the *aflI* gene of *A. flavus* restored the strain's ability to convert AVF to VHA and to produce aflatoxins (Yu et al., 2000b). It is likely that the *aflI* (*avfA*) encoded protein along with CypX gene product is involved in the ring-closure step in the formation of hydroxyversicolorone. It is possible that the *avfA* gene product is assocated with the P450 monooxygenase to carry out the conversion as no additional intermediates other that AVF result from the disruption of either gene.

3.6 Versiconal hemiacetal acetate (VHA) to versiconal (VHOH, also abbreviated as VAL)

Several research groups have demonstrated that an esterase is involvement in the conversion of VHA to VHOH (VAL) (Schroeder et al., 1974; Yao & Hsieh, 1974; Bennett et al., 1976b; Fitzell et al., 1977; Hsieh et al., 1989; Yabe et al., 1991a; Yabe et al., 1991b; Kusumoto & Hsieh, 1996). The esterase was purified in *A. parasiticus* (Hsieh et al., 1989; Kusumoto & Hsieh, 1996). An esterase gene, *aflJ* (*estA*), in the aflatoxin gene cluster was identified (Yu et al., 2002). The homologous gene in the *A. nidulans* ST biosynthetic gene cluster is *stcI*. In the *A. parasiticus aflJ* (*estA*) deletion mutants, the accumulated metabolites were mainly VHA and versicolorin A (VERA) (Chang et al., 2004). A small amount of versiconol acetate (VOAc) and other downstream aflatoxin intermediates, including VHOH and versicolorin B also accumulated. A metabolic grid containing VHA, VOAc, VHOH, and versiconol (VOH) was previously described and it was suggested that the reactions from VHA to VHOH and from VOAc to VOH are catalyzed by the same esterase (Yabe et al., 1991a). Later, another metabolic grid containing versicolorone (VONE), VOAc, and VHA was identified (Yabe et al., 2003). Indeed, it has now been proven that the *estA*-encoded esterase catalyzes the conversion of both VHA to VHOH and VOAc to VOH during aflatoxin biosynthesis (Chang et al., 2004).

3.7 Versiconal (VHOH) to versicolorin B (VER B)

The enzymatic evidence that VHOH is converted toVERB by a cyclase was first provided by Lin and Anderson (Lin & Anderson, 1992). This enzyme was identified as versicolorin B synthase and was studied intensively by Townsend's laboratory (Zuckerman et al., 1967; Hsieh, 1973; McGuire et al., 1996; Silva et al., 1996; Silva & Townsend, 1997). The gene was cloned and named *vbs* (Zuckerman et al., 1967; Hsieh, 1973; Silva et al., 1996). The expected cyclase activity was demonstrated by the expressed recombinant protein of the *vbs* gene (Silva et al., 1996; Silva & Townsend, 1997). The VHOH cyclase (Lin & Anderson, 1992) and

VER B synthase (McGuire et al., 1996) were independently isolated from *A. parasiticus*. The enzyme catalyzes the side chain cyclodehydration of racemic VHA to VER B. This is another key step in aflatoxin formation since it closes the bisfuran ring of aflatoxin, the moiety ultimately responsible for aflatoxin's toxicity andcarcinogenicity. . The *vbs* gene was renamed *aflK* (*vbs*) a (Yu et al., 2004c). The homologous gene in the *A. nidulans* ST biosynthetic gene cluster is *stcN*.

3.8 Versicolorin B (VER B) to versicolorin A (VER A)

VER B is a critical branch point leading to the formation of either AFB_1/AFG_1 or AFB_2/AFG_2. Similar to AFB_2/AFG_2, VER B contains a tetrahydrobisfuran ring and, like AFB_1/AFG_1, VERA contains a dihydrobisfuran ring. The conversion of VER B to VER A requires desaturation of the bisfuran ring of VER B by an unstable microsomal enzyme that requires NADPH (Yabe et al., 1993). Disruption of *stcL* in *A. nidulans* (Kelkar et al., 1997) abolished ST synthesis and resulted in the accumulation of VER B. The *stcL* gene encodes a cytochrome P-450 monooxygenase. The homologue, *aflL* (*verB*), is present in the aflatoxin gene cluster of *A. parasiticus* and *A. flavus* strains. Cultural conditions appear to markedly affect the activity of VER B desaturase and thereby, the final ratio of AFB_1 to AFB_2 and AFG_1 to AFG_2 (Yabe & Nakajima, 2004).

3.9 Versicolorin A (VER A) to demethylsterigmatocystin (DMST) and versicolorin B (VER B) to demethyldihydrosterigmatocystin (DMDHST)

The formation of DMST and the biochemical conversion steps from VERA to DMST (and VerB to DHDMST) have been described in great detail (Henry & Townsend, 2005). The *aflM* (*ver-1*) gene (Skory et al., 1992), cloned by genetic complementation of VER A-accumulating *A. parasiticus* CS10, was shown to be responsible for the conversion of VER A to an intermediate that has not been isolated. The *aflM* (*ver-1*) gene was predicted to encode a ketoreductase, similar Nor-1. The *ver-1* homologue, *stcU*, (previously named *verA*) was identified in *A. nidulans* (Keller et al., 1994). Double mutation of *stcU* and *stcL* resulted in accumulation of only VER A (Keller et al., 1994). The *stcS* gene (previously named *verB*), another cytochrome P-450 monooxygenase gene, was also identified and studies showed that it is also involved in the conversion of VER A to an intermediate in the formation of DMST (possibly the first intermediate, which is then acted upon by Ver-1). Disruption of *stcS* resulted in the accumulation of VER A as did disruption of Ver-1 (Keller et al., 1995). Thus, both *stcU* and *stcS* are required for the conversion of VER A to DMST. The *stcS* homologue in *A. parasiticus*, named *aflN* (*verA*), has also been identified (Yu et al., 2004a; Yu et al., 2004c). A third enzyme is required for the conversion: hypA (aflY). This gene is predicted to encode a Baeyer-Villiger monooxygenase. Disruption of this gene also led to accumulation of VERA suggesting that, like VER-1, it acts as part of an enzyme complex without allowing the formation of an intermediate. A fourth enzyme, OrdB has also been implicated in the conversion, and like AvfA, its homolog, may be a helper protein for the monooxygenase, CypX.

3.10 Demethylsterigmatocystin (DMST) to sterigmatocystin (ST) and demethyldihydrosterigmatocystin (DMDHST) to dihydrosterigmatocystin (DHST)

Enzyme purification studies revealed that two *O*-methyltransferases, I and II, are involved in aflatoxin biosynthesis (Yabe et al., 1989). *O*-methyltransferase I catalyzes the transfer of the methyl from S-adenosylmethionine (SAM) to the hydroxyls of DMST and DHDMST to produce ST and DHST, respectively. This 43-kDa enzyme was purified from *A. parasiticus* and characterized (Yabe et al., 1998; Yabe et al., 1999). The corresponding gene, *dmtA*, was isolated from *A. parasiticus* based on a partial amino acid sequence of the purified enzyme (Motomura et al., 1999). Yu et al. (Yu et al., 2000b) concurrently isolated the same gene but named it *aflO* (*omtB*) (for *O*-methyltransferase B) from *A. parasiticus, A. flavus* and *A. sojae*. The predicted *dmtA*-encoded protein contains a consensus SAM-binding motif (Motomura et al., 1999). The *aflO* (*omtB*) homolog in *A. nidulans* was identified as *stcP*. This gene is required for the conversion of DMST to ST in *A. nidulans* as shown by gene disruption (Kelkar et al., 1996).

3.11 Sterigmatocystin (ST) to *O*-methylsterigmatocystin (OMST) and demethylsterigmatocystin (DMST) to dihydro-*O*-methylsterigmatocystin (DHOMST)

The gene for *O*-methyltransferase required for the conversion of ST to OMST and DHST to DHOMST was first cloned (Yu et al., 1993) from *A. parasiticus* by reverse genetics using antibodies raised against the purified *A. parasiticus O*-methyltransferase A (Keller et al., 1993). This gene was initially named *omt-1*, then *omtA* and finally renamed *aflP* (*omtA*) (Yu et al., 1993). The recombinant enzyme was expressed in *E. coli* and its activity to convert ST to OMST was demonstrated by substrate feeding studies (Yu et al., 1993). *O*-methyltransferase A has strict substrate-specificity and cannot methylate DMST or DHDMST. Thus, the *O*-methyltransferases A encoded by *aflP* (*omtA*) is the enzyme responsible for the conversion of ST to OMST and DMST to DHOMST. The genomic DNA sequence of this gene (*omtA*) was cloned from *A. parasiticus* and *A. flavus* (Yu et al., 1995b). This *aflP* (*omtA*) gene homologue was also detected in other aflatoxigenic and non-aflatoxigenic *Aspergillus* species (Klich et al., 1995). The absence of the *aflP* orthologue in *A. nidulans* is the reason that *A. nidulans* produces ST as the end product instead of aflatoxins.

3.12 *O*-methylsterigmatocystin (OMST) to aflatoxin B_1 (AFB_1) and aflatoxin G_1 (AFG_1) and demethyldihydrosterigmatocystin (DMDHST) to aflatoxin B_2 (AFB_2) and aflatoxin G_2 (AFG_2)

The relationship between B-group and G-group aflatoxin formation was proposed based on feeding experiments (Yabe et al., 1988). A P-450 monooxygenase gene in *A. flavus* named *ord-1* was shown to be necessory for this reaction (Prieto et al., 1996; Prieto & Woloshuk, 1997). This P-450 monooxygenase gene, *aflQ* (*ordA*), was cloned in *A. parasiticus* and demonstrated in a yeast system that it is involved in the conversion of OMST to AFB_1/AFG_1, and DHOMST to AFB_2/AFG_2 (Yu et al., 1998). Whether *aflQ* (*ordA*) gene product, OrdA, catalyzes two successive monooxygenase reactions in the later steps of aflatoxin biosynthesis is not clear. Studies (Yu et al., 1998) suggested that additional enzyme(s) is required for the synthesis of G-group aflatoxins. After the cloning and characterization of the *cypA* gene, it is clear that *cypA* encoded a cytochrome P450 monooxygenase for the formation of G-group aflatoxins (Ehrlich et al., 2004). Most recently, the *nadA* gene, which was shown, by gene profiling studies using microarray, to be a member of the aflatoxin gene cluster (Price et al., 2006; Yu et al., 2011) rather than belonging to the adjoining sugar utilization cluster as originally proposed (Yu et al., 2000a), was found to play a role in AFG_1/AFG_2 formation. Yabe's group recently disrupted the *nadA* gene and reported that NadA is a cytosolic enzyme for the conversion from a new aflatoxin intermediate named NADA, which is between OMST and AFG_1, to

AFG_1 (Cai et al., 2008). The *aflE* (*norA*) gene was initially believed to be involved in the conversion of NOR due to certain degree of sequence similarity to the *aflD* (*nor-1*) gene (Cary et al., 1996). However, recent studies support the hypothesis that the *aflE* (*norA*) is involved in the final two steps in AFB_1 formation (Ehrlich, 2009). In the same report, the transcript, *hypB*, a homolog of *hypC*, may be involved in one of the oxidation steps in the conversion of OMST to aflatoxins. *A. flavus* produces only AFB_1 and AFB_2, whereas *A. parasiticus* produces all four major aflatoxins, AFB_1, AFB_2, AFG_1, and AFG_2. Coincidentally, only the G-group aflatoxin producer, *A. parasiticus*, has intact *nadA* and *norB* genes. Preliminary data suggests that *norB* encodes another enzyme predominantly involved in AFG_1/AFG_2 formation (Ehrlich et al., 2008).

4. Regulation of aflatoxin biosynthesis

The aflatoxin pathway genes are found to be clustered in the genome of *A. flavus* and *A. parasiticus* (Yu et al., 1995a; Woloshuk & Prieto, 1998; Yu et al., 2004a; Yu et al., 2004c). These genes are expressed concurrently except for the regulatory gene *aflR*. In this gene cluster, a positive-acting regulatory gene, *aflR*, is located in the middle of the gene cluster. Adjacent to *aflR* a divergently transcribed gene, *aflS* (*aflJ*), was also found to be involved in the regulation of transcription (Meyers et al., 1998; Chang, 2004). Other physically unrelated genes, such as *laeA* and *veA*, also have been shown to exhibit a "global" regulatory role on aflatoxin biosynthesis (Kato et al., 2003; Bok & Keller, 2004; Calvo et al., 2004; Perrin et al., 2007).

4.1 Regulation by *aflR*

The *aflR* gene, encoding a 47 kDa sequence-specific zinc-finger DNA-binding protein is required for transcriptional activation of most, if not all, the structural genes of the aflatoxin gene cluster (Chang et al., 1993; Payne et al., 1993; Woloshuk et al., 1994; Chang et al., 1995b; Yu et al., 1996a; Yu et al., 1996b; Flaherty & Payne, 1997; Ehrlich et al., 1998; Chang et al., 1999a; Chang et al., 1999b). Like other Gal4-type regulatory proteins that bind to palindromic sequences, functional AflR probably binds as a dimer. It binds to the palindromic sequence 5'-TCGN5CGR-3' in the promoter regions of the structural genes (Ehrlich et al., 1999a; Ehrlich et al., 1999b). The AflR-binding motifs are found to be located from -80 to -600 bp, with the majority at the -100 to -200 bp, relative to the translation start site. AflR binds, in some cases, to a deviated sequence rather than the typical motif such as in the case of *aflG* (*avnA*). When there is more than one binding motif, only one of them is the preferred binding site such as in the case of *aflC* (*pksA*) (Ehrlich et al., 1999a; Ehrlich et al., 1999b). The more upstream motif is found to belong to another gene for turning on the expression of *hypC* (Ehrlich, unpublished observation). Deletion of *aflR* in *A. parasiticus* abolishes the expression of other aflatoxin pathway genes (Cary et al., 2000). Overexpression of *aflR* in *A. flavus* up-regulates aflatoxin pathway gene transcription and aflatoxin accumulation (Flaherty & Payne, 1997) in a fashion similar to that reported for *A. parasiticus* (Chang et al., 1995b). These results demonstrate that AflR is specifically involved in the regulation of aflatoxin biosynthesis. Indeed, all 23 upregulated genes, identified by transcription profiling using DNA microarray assays comparing wild-type and *aflR*-deleted *A. parasiticus* strains, have the consensus AflR binding motif in their promoter regions (Meyers et al., 1998; Price et al., 2006; Wilkinson et al., 2007a; Wilkinson et al., 2007b).

4.2 Regulation by *aflS* (*aflJ*)

The *aflS* (*aflJ*) gene, although not demonstrating significant homology with any other encoded proteins found in databases, is necessary for aflatoxin formation. In the *A. parasiticus aflR* transformants, the production of aflatoxin pathway intermediates was significantly enhanced in transformants that contained an additional *aflR* plus *aflS* (Chang et al., 1995b). Quantitative PCR showed that in the *aflS* knockout mutants, the lack of *aflS* transcript is associated with 5- to 20-fold reduction of expression of some aflatoxin pathway genes such as *aflC* (*pksA*), *aflD* (*nor-1*), *aflM* (*ver-1*), and *aflP* (*omtA*). The mutants lost the ability to synthesize aflatoxin intermediates and no aflatoxins were produced (Meyers et al., 1998). However, deletion of *aflS* (*aflJ*) did not have a discernible effect on *aflR* transcription, and vice versa. Du et al. (Du et al., 2007) showed that overexpression of *A. flavus aflS* (*aflJ*) did not result in elevated transcription of *aflM* (*ver-1*), *aflP* (*omtA*), or *aflR*, but it appears to have some effect on *aflC* (*pksA*), *aflD* (*nor-1*), *aflA* (*fas-1*), and *aflB* (*fas-2*) (Du et al., 2007), which are required for the biosynthesis of the early aflatoxin pathway intermediate, averantin. The mechanism(s) by which *aflS* modulates transcription of these pathway genes in concert with *aflR* is under investigation by gene profiling analysis using microarray technology.

4.3 Regulation by *laeA*

The novel global regulatory gene, *laeA* (for **l**ack of ***afl*R** expression), was first identified from *A. nidulans* (Bok & Keller, 2004). This gene is well conserved in fungi as shown by its presence in the genomes of all fungi so far sequenced. LaeA is a nuclear protein which contains an S-adenosylmethionine (SAM) binding motif and activates transcription of several other secondary metabolism gene clusters in addition to the AF cluster. Examples include the sterigmatocystin and penicillin clusters in *A. nidulans*, the gliotoxin cluster in *A. fumigatus*, and aflatoxin cluster in *A. flavus* (Bok & Keller, 2004; Bouhired et al., 2007). It also regulates genes required for virulence of *A. fumigatus* (Sugui et al., 2007). Perrin et al. (Perrin et al., 2007) carried out a whole-genome comparison of the transcriptional profiles of wild-type and *laeA*-deleted *A. fumigatus* strains and found that LaeA positively controls the expression of 20% to 40% of major classes of secondary metabolite biosynthesis genes. It also regulates some genes not associated with secondary metabolite clusters. Similar results were confirmed in gene expression profiling in *A. flavus* using microarrays to study the genetic mechanism of sclerotia formation (Yu, personal communication). The exact mechanism of how LaeA regulates secondary metabolism gene clusters is not yet known. Interestingly, when an unrelated gene such as *argB* was placed within the boundary of the ST gene cluster, it was co-regulated with other genes in the cluster. But, when a gene in the cluster, such as *aflR* was placed elsewhere in the genome, its regulation was not affected by LaeA (Bok et al., 2006). One proposed regulatory mechanism is that LaeA differentially methylates histone protein and it alters the chromatin structure for gene expression. Unlike the mentioned signaling factors, the primary role of LaeA is to regulate metabolic gene clusters, not sporulation, because *laeA*-deleted strains produced wild-type levels of conidia (Bok & Keller, 2004). Most recent analyses of nonaflatoxigenic *A. parasiticus sec-* (for secondary metabolism negative) variants generated through serial transfer of mycelia of the *sec+* parents show that *laeA* was expressed in both *sec+* and *sec-* strains (Kale et al., 2007). This result suggests that LaeA only exerts its effect on aflatoxin biosynthesis at a certain level and is independent of other regulatory pathways that are involved in fungal development.

4.4 Regulation by *veA*

The *veA* gene in *A. nidulans* (Mooney & Yager, 1990) is a gene initially found to be crucial for light-dependent conidiation. The light dependence is abolished by a mutation (*veA1*) which allows conidiation of *A. nidulans* to occur in the dark. A comparison of the light effect on sterigmatocystin production by *A. nidulans veA+* and *veA1* strains showed that both strains produced sterigmatocystin but the highest amount was produced by the *veA+* strain grown in darkness. However, *veA*-deleted *A. flavus* and *A. parasiticus* strains completely lost the ability to produce aflatoxin regardless of the illumination conditions (Duran et al., 2007; Stinnett et al., 2007). Under normal growth conditions, some *A. flavus* and all *A. parasiticus* strains produce conidia in both dark and light conditions. Stinnett et al. (Stinnett et al., 2007) showed that VeA contains a bipartite nuclear localization signal (NLS) motif and its migration to the nucleus is light-dependent and requires the importin α carrier protein. In the dark VeA is located mainly in the nucleus; under light it is located both in cytoplasm and nucleus. VeA has no recognizable DNA-binding seuqences and likely exerts its effect on sterigmatosyctin and aflatoxin production through protein-protein interactions with other regulatory factors. Post-translational modifications such as phosphylation and dephosphorylation may modulate its activity. Lack of VeA production in the *veA*-deleted *A. flavus* and *A. parasiticus* strains consequently abolishes aflatoxin production because a threshold concentration of nuclear VeA might be necessary to initiate aflatoxin biosynthesis.

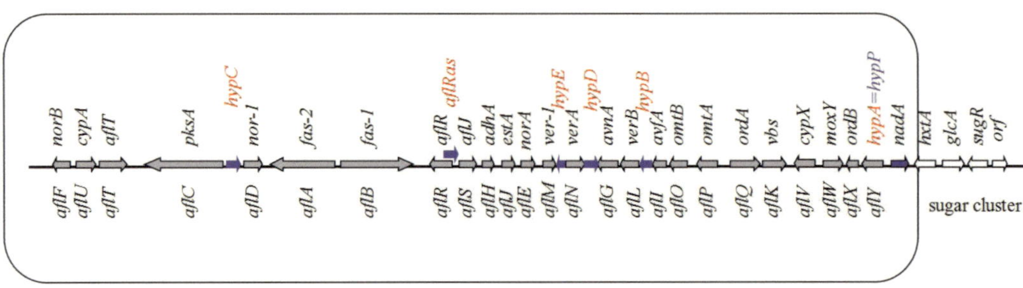

Fig. 1. Aflatoxin pathway gene cluster in *A. flavus*. This figure shows the order and location of the 30 aflatoxin pathway genes plus an *aflR* antisense gene clustered together in about 80 kb DNA region. The old gene names are labeled on top of the line and the new gene names sysmatically renamed according to gene convention are labeled below the line (Yu et al., 2004c). The transcripts of *hypA*, *hypB*, *hypC*, *hypD*, *hypE* and *aflRas* are identified through *Aspergillus flavus* EST. Arrows indicate the direction of gene transcription.

5. Conclusions

Aflatoxins are toxic and carcinogenic secondary metabolites produced mainly by *Aspergillus flavus* and *Aspergillus parasiticus* that contaminate preharvest crops and post harvest grains. Scientists worldwide have extensively studied biosynthesis of aflatoxins for more than 50 years. Aflatoxin biosynthesis is a complex process involving many intermediates and enzymes. Regulation of aflatoxin gene expression occurs at multiple levels and by multiple regulatory components. There are genetic factors, biotic and abiotic elements that affect aflatoxin formation. Recent studies have shed more light on the functions of the enzymes involved in each of the steps of aflatoxin biosynthesis, the genes encoding those enzymes, and the regulatory mechanisms of aflatoxin formation. Better understanding of the mechanisms of aflatoxin biosynthesis helps to identify natural inhibitors of fungal growth aflatoxin formation, and eventually will allow design of effective strategies to can reduce or eliminate aflatoxin contamination of food and feed commodities.

6. References

Allcroft, R., Carnaghan, R.B.A., Sargeant, K. & O'Kelly, J. (1961). A toxic factor in Brazilian groundnut meal, *Vet. Rec.* 73, 428-429.

Arsura, M. & Cavin, L.G. (2005). Nuclear factor-kappaB and liver carcinogenesis, *Cancer Lett.* 229, 157-169.

Azziz-Baumgartner, E., Lindblade, K., Gieseker, K., Rogers, H.S., Kieszak, S., Njapau, H., Schleicher, R., McCoy, L.F., Misore, A., DeCock, K., Rubin, C. & Slutsker, L. (2005). Case-control study of an acute aflatoxicosis outbreak, Kenya, 2004, *Aflatoxin Investigative Group. Environ. Health Perspect.* 113, 1779-1783.

Baertschi, S.W., Raney, K.D., Shimada, T., Harris, T.M. & Guengerich, F.P. (1989). Comparison rates of enzymatic oxidation of aflatoxin B1, aflatoxin G1, and sterigmatocystin and activities of the epoxides in forming guanyl-N adducts and inducing different genetic responses, *Chem. Res. Toxicol.* 2, 114-122.

Bennett, J.W. (1970). Microbiological aspects of the aflatoxin problem, *Amer Ass Feed Microscop Offic Proc* 118-131.

Bennett, J.W. (1979). Aflatoxins and anthraquinones from diploids of *Aspergillus parasiticus, J. Gen. Microbiol.* 113, 127-136.

Bennett, J.W. (1981). Loss of norsolorinic acid and aflatoxin production by a mutant of *Aspergillus parasiticus, J. Gen. Microbiol.* 124, 429- 432.

Bennett, J.W. (1987). Mycotoxins, mycotoxicoses, mycotoxicology and mycopathologia, *Mycopathologia* 100, 3-5.

Bennett, J.W., Bhatnagar, D. & Chang, P.K. (1994). The molecular genetics of aflatoxin biosynthesis, *FEMS Symp.* 51-58.

Bennett, J.W., Chang, P.K. & Bhatnagar, D. (1997). One gene to whole pathway: the role of norsolorinic acid in aflatoxin research, *Adv. Appl. Microbiol.* 45, 1-15.

Bennett, J.W. & Goldblatt, L.A. (1973). The isolation of mutants of *Aspergillus flavus* and *A. parasiticus* with altered aflatoxin producing ability, *Sabouraudia* 11, 235-241.

Bennett, J.W. & Klich, M. (2003). Mycotoxins, *Clin. Microbiol. Rev.* 16, 497-516.

Bennett, J.W., Kronberg, F. & Gougis, G. (1976a). Pigmented isolates from anthraquinone-producing mutants of *Aspergillus parasiticus, American Society for Microbiology* 76, 6.

Bennett, J.W., Kronberg, F.G., Goodman, L.A. & Seltman, M.A. (1983). Isolation of an anthraquinoe-accumulating mutant of *Aspergillus parasiticus* and partial characterization by dry column chromatography, *Mycologia* 75, 202-208.

Bennett, J.W. & Lee, L.S. (1979). Mycotoxins - their biosynthesis in fungi: aflatoxins and other bisfuranoids, *J Food Protection* 42, 805-809.

Bennett, J.W., Lee, L.S. & Cucullu, A.F. (1976b). Effect of dichlorvos on aflatoxin and versicolorin A production in *Aspergillus parasiticus, Bot. Gaz.* 137, 318-324.

Bennett, J.W., Lee, L.S., Shoss, S.M. & Boudreaux, G.H. (1980). Identification of averantin as an aflatoxin B1 precursor: placement in the biosynthetic pathway, *Appl. Environ. Microbiol.* 39, 835-839.

Bennett, J.W., Lee, L.S. & Vinnett, C.H. (1971). The correlation of aflatoxin and norsolorinic acid production, *J. American Oil Chemists Society* 48, 368-370.

Bennett, J.W., Leong, P.M., Kruger, S.J. & Keyes, D. (1986). Sclerotial and low aflatoxigenic morphological variants from haploid and diploid *Aspergillus parasiticus Experientia* 42, 848-851.

Bennett, J.W. & Papa, K.E. (1988). The aflatoxigenic *Aspergillus* spp., *Adv. Plant Pathol.* 6, 265-279.

Bennett, J.W., Silverstein, R.B. & Kruger, S.J. (1981). Isolation and characterization of two nonaflatoxigenic classes of morphological variants of *Aspergillus parasiticus, J. American Oil Chemists Society* 952.

Bhatnagar, D., Brown, R., Ehrlich, K. & Cleveland, T.E. (2002). Mycotoxins Contaminating Cereal Grain Crops: Their Occurrence and Toxicity. In: Khachatourians, G.G. & Arora, D.K. (Eds.), *Applied Mycology and Biotechnology*. Elsevier B.V., New York, pp. 171-196.

Bhatnagar, D. & Cleveland, T.E. (1991). Aflatoxin biosynthesis: developments in chemistry, biochemistry, and genetics, *Res. Bull. Iowa State Univ. Agric. Home Econ. Exp. Stn.* 391-405.

Bhatnagar, D., Cleveland, T.E. & Lillehoj, E.B. (1989). Enzymes in aflatoxin B1 biosynthesis: strategies for identifying pertinent genes, *Mycopathologia* 107, 75-83.

Bhatnagar, D., Ehrlich, K.C. & Cleveland, T.E. (1992). Oxidation- reduction reactions in biosynthesis of secondary metabolites. In: Bhatnagar, D., Lillehoj, E.B. & Arora, D.K. (Eds.), *Mycotoxins in Ecological Systems*. Marcel Dekker, New York, pp. 255-285.

Bok, J.-W. & Keller, N.P. (2004). LaeA, a regulator of secondary metabolism in *Aspergillus* spp., *Eukaryot. Cell* 3, 527-535.

Bok, J.W., Noordermeer, D., Kale, S.P. & Keller, N.P. (2006). Secondary metabolic gene cluster silencing in *Aspergillus nidulans, Mol. Microbiol.* 61, 1636-1645.

Bouhired, S., Weber, M., Kempf-Sontag, A., Keller, N.P. & Hoffmeister, D. (2007). Accurate prediction of the *Aspergillus nidulans* terrequinone gene cluster boundaries using the transcriptional regulator LaeA, *Fungal Genet. Biol.* 44, 1134-1145.

Bressac, B., Kew, M., Wands, J. & Ozturk, M. (1991). Selective G to T mutations of p53 gene in hepatocellular carcinoma from southern Africa, *Nature* 350, 429-431.

Brown, D.W., Adams, T.H. & Keller, N.P. (1996a). *Aspergillus* has distinct fatty acid synthases for primary and secondary metabolism, *Proc. Natl. Acad. Sci. U. S. A.* 93, 14873-14877.

Brown, D.W., Yu, J.H., Kelkar, H.S., Fernandes, M., Nesbitt, T.C., Keller, N.P., Adams, T.H. & Leonard, T.J. (1996b). Twenty-five coregulated transcripts define a sterigmatocystin gene cluster in *Aspergillus nidulans, Proc. Natl. Acad. Sci. U. S. A.* 93, 1418-1422.

Busby, W.F., Jr. & Wogan, G.N. (1981). Aflatoxins. In: Shank, R.C. (Ed.), *Mycotoxins and N-nitrosocompounds: Environmental Risks*. CRC Press, Boca Raton, pp. 3-45.

Cai, J., Zeng, H., Shima, Y., Hatabayashi, H., Nakagawa, H., Ito, Y., Adachi, Y., Nakajima, H. & Yabe, K. (2008). Involvement of the *nadA* gene in formation of G-group aflatoxins in *Aspergillus parasiticus, Fungal Genet Biol* 45, 1081-1093.

Calvo, A.M., Bok, J.-W., Brooks, W. & Keller, N.P. (2004). VeA is required for toxin and sclerotial production in *Aspergillus parasiticus, Appl. Environ. Microbiol.* 70, 4733-4739.

Cary, J.W., Ehrlich, K.C., Wright, M., Chang, P.K. & Bhatnagar, D. (2000). Generation of aflR disruption mutants of *Aspergillus parasiticus, Appl. Microbiol. Biotechnol.* 53, 680-684.

Cary, J.W., Wright, M., Bhatnagar, D., Lee, R. & Chu, F.S. (1996). Molecular characterization of an *Aspergillus parasiticus* dehydrogenase gene, *norA*, located on the aflatoxin biosynthesis gene cluster, *Appl. Environ. Microbiol.* 62, 360-366.

Centers for Disease Control and Prevention (2004). Outbreak of aflatoxin poisoning - eastern and central province, Kenya, *MMWR Morb Mortal Weekly Report* 53, 790-792.

Chang, P.K. (2004). Lack of interaction between AFLR and AFLJ contributes to nonaflatoxigenicity of *Aspergillus sojae, J. Biotechnol.* 107, 245-253.

Chang, P.K., Bennett, J.W. & Cotty, P.J. (2002). Association of aflatoxin biosynthesis and sclerotial development in *Aspergillus parasiticus, Mycopathologia* 153, 41-48.

Chang, P.K., Cary, J.W., Bhatnagar, D., Cleveland, T.E., Bennett, J.W., Linz, J.E., Woloshuk, C.P. & Payne, G.A. (1993). Cloning of the *Aspergillus parasiticus apa-2* gene associated with the regulation of aflatoxin biosynthesis, *Appl. Environ. Microbiol.* 59, 3273-3279.

Chang, P.K., Cary, J.W., Yu, J., Bhatnagar, D. & Cleveland, T.E. (1995a). The *Aspergillus parasiticus* polyketide synthase gene *pksA*, a homolog of *Aspergillus nidulans wA*, is required for aflatoxin B1 biosynthesis, *Mol. Gen. Genet.* 248, 270-277.

Chang, P.K., Ehrlich, K.C., Yu, J., Bhatnagar, D. & Cleveland, T.E. (1995b). Increased expression of *Aspergillus parasiticus aflR*, encoding a sequence-specific DNA-binding protein, relieves nitrate inhibition of aflatoxin biosynthesis, *Appl. Environ. Microbiol.* 61, 2372-2377.

Chang, P.K., Horn, B.W. & Dorner, J.W. (2005). Sequence breakpoints in the aflatoxin biosynthesis gene cluster and flanking regions in nonaflatoxigenic *Aspergillus flavus* isolates, *Fungal Genet. Biol.* 42, 914-923.

Chang, P.K., Skory, C.D. & Linz, J.E. (1992). Cloning of a gene associated with aflatoxin B1 biosynthesis in *Aspergillus parasiticus, Curr. Genet.* 21, 231-233.

Chang, P.K., Yabe, K. & Yu, J. (2004). The *Aspergillus parasiticus estA*- encoded esterase converts versiconal hemiacetal acetate to versiconal and versiconol acetate to versiconol in aflatoxin biosynthesis, *Appl. Environ. Microbiol.* 70, 3593-3599.

Chang, P.K., Yu, J., Bhatnagar, D. & Cleveland, T.E. (1999a). The carboxy-terminal portion of the aflatoxin pathway regulatory protein AFLR of *Aspergillus parasiticus* activates *GAL1::lacZ* gene expression in *Saccharomyces cerevisiae, Appl. Environ. Microbiol.* 65, 2508-2512.

Chang, P.K., Yu, J., Bhatnagar, D. & Cleveland, T.E. (1999b). Repressor- AFLR interaction modulates aflatoxin biosynthesis in *Aspergillus parasiticus, Mycopathologia* 147, 105-112.

Chang, P.K., Yu, J., Ehrlich, K.C., Boue, S.M., Montalbano, B.G., Bhatnagar, D. & Cleveland, T.E. (2000). *adhA* in *Aspergillus parasiticus* is involved in conversion of 5'-hydroxyaverantin to averufin, *Appl. Environ. Microbiol.* 66, 4715-4719.

Chen, C.J., Wang, L.Y., Lu, S.N., Wu, M.H., You, S.L., Zhang, Y.J., Wang, L.W. & Santella, R.M. (1996a). Elevated aflatoxin exposure and increased risk of hepatocellular carcinoma, *Hepatology* 24, 38- 42.

Chen, C.J., Yu, M.W., Liaw, Y.F., Wang, L.W., Chiamprasert, S., Matin, F., Hirvonen, A., Bell, D.A. & Santella, R.M. (1996b). Chronic hepatitis B carriers with null genotypes of glutathione S transferase MI and TI polymorphisms who are exposed to aflatoxin are at increased risk of hepatocellular carcinoma, *Am. J. Human Genet.* 59, 128-134.

Cleveland, T.E. & Bhatnagar, D. (1987). Individual reaction requirements of two enzyme activities, isolated from Aspergillus parasiticus, which together catalyze conversion of sterigmatocystin to aflatoxin B1, *Can. J. Microbiol.* 33, 1108-1112.

Cleveland, T.E. & Bhatnagar, D. (1990). Evidence for de novo synthesis of an aflatoxin pathway methyltransferase near the cessation of active growth and the onset of aflatoxin biosynthesis in *Aspergillus parasiticus* mycelia, *Can. J. Microbiol.* 36, 1-5.

Cleveland, T.E. & Bhatnagar, D. (1991). Molecular regulation of aflatoxin biosynthesis. In: Bray, G.A. & Ryan, D.H. (Eds.), *Proc Symp Mycotoxins Cancer Health*, pp. 270-287.

Cleveland, T.E. & Bhatnagar, D. (1992). Molecular strategies for reducing aflatoxin levels in crops before harvest. In: Bhatnagar, D. & Cleveland, T.E. (Eds.), *Molecular Approaches to Improving Food Quality and Safety*. Van Nostrand Reinhold, New York, pp. 205- 228.

Cleveland, T.E., Lax, A.R., Lee, L.S. & Bhatnagar, D. (1987). Appearance of enzyme activities catalyzing conversion of sterigmatocystin to aflatoxin B1 in late-growth-phase Aspergillus parasiticus cultures, *Appl. Environ. Microbiol.* 53, 1711-1713.

Cotty, P. (1988). Aflatoxin and sclerotial production by *Aspergillus flavus*: influence of pH, *Phytopathol.* 78, 1250-1253.

Council for Agricultural Science and Technology (2003), Mycotoxins: Risk in plant, animal, and human systems, Ames, IA.

Coursaget, P., Depril, N., Chabaud, M., Nandi, R., Mayelo, V., LeCann, P. & Yvonnet, B. (1993). High prevalence of mutations at codon 249 of the p53 gene in hyptocellular carcinomas from Senegal, *Br. J. Cancer* 67, 1395-1397.

Crawford, J.M., Dancy, B.C., Hill, E.A., Udwary, D.W. & Townsend, C.A. (2006). Identification of a starter unit acyl-carrier protein transacylase domain in an iterative type I polyketide synthase, *Proc. Natl. Acad. Sci. U. S. A.* 103, 16728-16733.

Crawford, J.M., Thomas, P.M., Scheerer, J.R., Vagstad, A.L., Kelleher, N.L. & Townsend, C.A. (2008a). Deconstruction of iterative multidomain polyketide synthase function, *Science* 320, 243-246.

Crawford, J.M., Vagstad, A.L., Ehrlich, K.C. & Townsend, C.A. (2008b). Starter unit specificity directs genome mining of polyketide synthase pathways in fungi, *Bioorg. Chem.* 36, 16-22.

Crawford, J.M., Vagstad, A.L., Whitworth, K.P., Ehrlich, K.C. & Townsend, C.A. (2008c). Synthetic strategy of nonreducing Iterative polyketide synthases and the origin of the classical "Starter-Unit Effect", *Chembiochem*.

Denning, D.W. (1998). Invasive aspergillosis, *Clin. Infect. Dis.* 26, 781-805.

Denning, D.W., Follansbee, S.E., Scolaro, M., Norris, S., Edelstein, H. & Stevens, D.A. (1991). Pulmonary aspergillosis in the acquired immunodeficiency syndrome, *N. Engl. J. Med.* 324, 654-662.

Denning, D.W., Riniotis, K., Dobrashian, R. & Sambatakou, H. (2003). Chronic cavitary and fibrosing pulmonary and pleural aspergillosis: Case series, proposed nomenclature and review, *Clin. Infect. Dis.* 37, S265-280.

Detroy, R.W., Freer, S. & Ciegler, A. (1973). Aflatoxin and anthraquinone biosynthesis by nitrosoguanidine-derived mutants of *Aspergillus parasiticus, Can. J. Microbiol.* 19, 1373-1378.

Du, W., Obrian, G.R. & Payne, G.A. (2007). Function and regulation of *aflJ* in the accumulation of aflatoxin early pathway intermediate in *Aspergillus flavus, J. Food Addit. Contam.* 24, 1043-1050.

Duran, R.M., Cary, J.W. & Calvo, A.M. (2007). Production of cyclopiazonic acid, aflatrem, and aflatoxin by *Aspergillus flavus* is regulated by *veA*, a gene necessary for sclerotial formation, *Appl. Microbiol. Biotechnol.* 73, 1158-1168.

Dutton, M.F. (1988). Enzymes and aflatoxin biosynthesis, *Microbiol. Rev.* 52, 274-295. Eaton, D. & Gallagher, E. (1994). Mechanisms of aflatoxin carcinogenesis, *Annu. Rev. Pharmacol. Toxicol.* 34, 135-172.

Eaton, D.L. & Groopman, J.D. The toxicology of aflatoxins: human health, veterinary, and agricultural significance. Academic Press, San Diego, CA, 1994. Ehrlich, K.C. (2009). Predicted roles of uncharacterized clustered genes in aflatoxin biosynthesis, *Toxins* 1, 37-58.

Ehrlich, K.C., Cary, J.W. & Montalbano, B.G. (1999a). Characterization of the promoter for the gene encoding the aflatoxin biosynthetic pathway regulatory protein AFLR, *Biochim. Biophys. Acta.* 1444, 412-417.

Ehrlich, K.C., Chang, P.-K., Scharfenstein, J.S.L., Cary, J.W., Crawford, J.M. & Townsend, C.A. (2010). Absence of the aflatoxin biosynthesis gene, *norA*, allows accumulation of deoxyaflatoxin B1 in *Aspergillus flavus* cultures, *FEMS Microbiol. Lett.* 305, 65-70.

Ehrlich, K.C., Chang, P.K., Yu, J. & Cotty, P.J. (2004). Aflatoxin biosynthesis cluster gene *cypA* is required for G aflatoxin formation, *Appl. Environ. Microbiol.* 70, 6518-6524.

Ehrlich, K.C., Montalbano, B.G., Bhatnagar, D. & Cleveland, T.E. (1998). Alteration of different domains in AFLR affects aflatoxin pathway metabolism in *Aspergillus parasiticus* transformants, *Fungal Genet. Biol.* 23, 279-287.

Ehrlich, K.C., Montalbano, B.G. & Cary, J.W. (1999b). Binding of the C6- zinc cluster protein, AFLR, to the promoters of aflatoxin pathway biosynthesis genes in *Aspergillus parasiticus, Gene* 230, 249-257.

Ehrlich, K.C., Scharfenstein, J.S.L., Montalbano, B.G. & Chang, P.-K. (2008). Are the genes nadA and norB involved in formation of aflatoxin G1, *International Journal of Molecular Sciences* 9, 1717- 1729.

Ehrlich, K.C. & Yu, J. (2009). Aflatoxin-like gene clusters and how they evolved. In: Varma, A.K. & Rai, M.K. (Eds.), *Mycotoxins in Food, Feed, and Bioweapons*. Springer Verlag, Heidelberg, Dordrecht, London, New York, pp. 65-76.

FAO (1995), Worldwide regulations for mycotoxins 1995, A Compendium. Italy, Rome.

Fitzell, D.L., Hsieh, D.P., Yao, R.C. & La Mar, G.N. (1975). Biosynthesis of averufin from acetate by *Aspergillus parasiticus, J. Agric. Food Chem.* 23, 442-444.

Fitzell, D.L., Singh, R., Hsieh, D.P. & Motell, E.L. (1977). Nuclear magnetic resonance identification of versiconal hemiacetal acetate as an intermediate in aflatoxin biosynthesis, *J. Agric. Food. Chem.* 25, 1193-1197.

Flaherty, J.E. & Payne, G.A. (1997). Overexpression of *aflR* leads to upregulation of pathway gene expression and increased aflatoxin production in *Aspergillus flavus*, *Appl. Environ. Microbiol.* 63, 3995-4000.

Fung, F. & Clark, R.F. (2004). Health effects of mycotoxins: a toxicological overview, *J. Toxicol. Clin. Toxicol.* 42, 217-234. Goldblatt, L.A. (1969). Aflatoxin-scientific background, control and implications. Acaderitic Press, New York.

Gong, Y., Hounsa, A., Egal, S., Turner, P.C., Sutcliffe, A.E., Hall, A.J., Cardwell, K. & Wild, C.P. (2004). Postweaning exposure to aflatoxin results in impaired child growth: a longitudinal study in Benin, West Africa, *Environ Health Perspect* 112, 1134-1138.

Goto, T., Wicklow, D.T. & Ito, Y. (1996). Aflatoxin and cyclopiazonic acid production by a sclerotium-producing *Aspergillus tamarii* strain, *Appl. Environ. Microbiol.* 62, 4036-4038.

Groopman, J.D., Wogan, G.N., Roebuck, B.D. & Kensler, T.W. (1994). Molecular biomarkers for aflatoxins and their application to human cancer prevention, *Cancer Res.* 54, 190-191.

Hall, A.J. & Wild, C.P. (1994). Epidemiology of aflatoxin-related disease. In: Eaton, D.L. & Groopman, J.D. (Eds.), *The Toxicology of Aflatoxins: Human Health, Veterinary, and Agricultural Significance*. Academic Press, San Diego, pp. 233-258.

Henry, K.M. & Townsend, C.A. (2005). Ordering the reductive and cytochrome P450 oxidative steps in demethylsterigmatocystin formation yields general insights into the biosynthesis of aflatoxin and related fungal metabolites, *J. Am. Chem. Soc.* 127, 3724-3733.

Hitchman, T.S., Schmidt, E.W., Trail, F., Rarick, M.D., Linz, J.E. & Townsend, C.A. (2001). Hexanoate synthase, a specialized type I fatty acid synthase in aflatoxin B1 biosynthesis, *Bioorg. Chem.* 29, 293-307.

Hsieh, D.P. (1973). Inhibition of aflatoxin biosynthesis of dichlorvos, *J. Agric. Food Chem.* 21, 468-470.

Hsieh, D.P., Cullen, J.M., Hsieh, L.S., Shao, Y. & Ruebner, B.H. (1985). Cancer risks posed by aflatoxin M1, *Princess Takamatsu Symp.* 16, 57-65.

Hsieh, D.P., Lin, M.T. & Yao, R.C. (1973). Conversion of sterigmatocystin to aflatoxin B1 by *Aspergillus parasiticus*, *Biochem. Biophys. Res. Commun.* 52, 992-997.

Hsieh, D.P., Lin, M.T., Yao, R.C. & Singh, R. (1976). Biosynthesis of aflatoxin. Conversion of norsolorinic acid and other hypothetical intermediates into aflatoxin B1, *J. Agric. Food Chem.* 24, 1170-1174.

Hsieh, D.P. & Mateles, R.I. (1970). The relative contribution of acetate and glucose to aflatoxin biosynthesis, *Biochem. Biophys. Acta.* 208, 482- 486.

Hsieh, D.P., Wan, C.C. & Billington, J.A. (1989). A versiconal hemiacetal acetate converting enzyme in aflatoxin biosynthesis, *Mycopathologia* 107, 121-126.

Hsieh, D.P.H. (1989). Potential human health hazards of mycotoxins. In: Natori, S., Hashimoto, H. & Ueno, Y. (Eds.), *Mycotoxins and Phycotoxins*. Elsevier, Amsterdam, pp. 69-80.

Hsu, I.C., Metcalf, R.A., Sun, T., Welsh, J.A., Wang, N.J. & Harris, C.C. (1991). Mutational hotspot in the p53 gene in human hepatocellular carcinomas, *Nature* 350, 427-428.

Huang, J.H. & Hsieh, D.P. (1988). Comparative study of aflatoxins M1 and B1 production in solid-state and shaking liquid cultures, *Proc. Natl. Sci. Counc. Repub. China* 12, 34-42.

Jelinek, C.F., Pohland, A.E. & Wood, G.E. (1989). Worldwide occurrence of mycotoxins in food and feeds--an update, *J. Assoc. Off. Anal. Chem.* 72, 223-230.

Kale, S.P., Cary, J.W., Hollis, N., Wilkinson, J.R., Bhatnagar, D., Yu, J., Cleveland, T.E. & Bennett, J.W. (2007). Analysis of aflatoxin regulatory factors in serial transfer-induced non-aflatoxigenic *Aspergillus parasiticus, J. Food Addit. Contam.* 24, 1061-1069.

Kato, N., Brooks, W. & Calvo, A.M. (2003). The expression of sterigmatocystin and penicillin genes in *Aspergillus nidulans* is controlled by *veA*, a gene required for sexual development, *Eukaryot. Cell* 2, 1178-1186.

Kelkar, H.S., Keller, N.P. & Adams, T.H. (1996). *Aspergillus nidulans stcP* encodes an *O*-methyltransferase that is required for sterigmatocystin biosynthesis, *Appl. Environ. Microbiol.* 62, 4296- 4298.

Kelkar, H.S., Skloss, T.W., Haw, J.F., Keller, N.P. & Adams, T.H. (1997). *Aspergillus nidulans stcL* encodes a putative cytochrome P- 450 monooxygenase required for bisfuran desaturation during aflatoxin/sterigmatocystin biosynthesis, *J. Biol. Chem.* 272, 1589- 1594.

Keller, N.P., Dischinger, H.C., Jr., Bhatnagar, D., Cleveland, T.E. & Ullah, A.H. (1993). Purification of a 40-kilodalton methyltransferase active in the aflatoxin biosynthetic pathway, *Appl. Environ. Microbiol.* 59, 479-484.

Keller, N.P., Kantz, N.J. & Adams, T.H. (1994). *Aspergillus nidulans verA* is required for production of the mycotoxin sterigmatocystin, *Appl. Environ. Microbiol.* 60, 1444-1450.

Keller, N.P., Segnar, S., Bhatnagar, D. & Adams, T.H. (1995). *stcS*, a putative P-450 monooxygenase, is required for the conversion of versicolorin A to sterigmatocystin in *Aspergillus nidulans, Appl. Environ. Microbiol.* 61, 3628-3632.

Keller, N.P., Watanabe, C.M., Kelkar, H.S., Adams, T.H. & Townsend, C.A. (2000). Requirement of monooxygenase-mediated steps for sterigmatocystin biosynthesis by *Aspergillus nidulans, Appl. Environ. Microbiol.* 66, 359-362.

Kelly, J.D., Eaton, D.L., Guengerich, F.P. & Coulombe, R.A., Jr. (1997). Aflatoxin B1 activation in human lung, *Toxicol. Appl. Pharmacol.* 144, 88-95.

Klich, M.A. (1998). Soil fungi of some low-altitude desert cotton fields and ability of their extracts to inhibit *Aspergillus flavus, Mycopathologia* 142, 97-100.

Klich, M.A., Yu, J., Chang, P.K., Mullaney, E.J., Bhatnagar, D. & Cleveland, T.E. (1995). Hybridization of genes involved in aflatoxin biosynthesis to DNA of aflatoxigenic and non-aflatoxigenic aspergilli, *Appl. Microbiol. Biotechnol.* 44, 439-443.

Krishnamachari, K.A., Bhat, R.V., Nagarajan, V. & Tilak, T.B. (1975). Hepatitis due to aflatoxicosis: an outbreak of hepatitis in parts of western India, *Lancet.* 1, 1061-1063.

Kusumoto, K.I. & Hsieh, D.P.H. (1996). Purification and characterization of the esterases involved in aflatoxin biosynthesis in *Aspergillus parasiticus, Can. J. Microbiol.* 42, 804-810.

Lancaster, M.D., Jenkins, F.P. & Philip, J.M. (1961). Toxicity associated with certain samples of ground nuts, *Nature* 192, 1095-1096.

Lee, L.S., Bennett, J.W., Goldblatt, L.A. & Lundin, R.E. (1971). Norsolorinic acid from a mutant strain of *Aspergillus parasiticus, J. American Oil Chemists Society* 48, 93-94.

Lewis, L., Onsongo, M., Njapau, H., Schurz-Rogers, H., Luber, G., Kieszak, S., Nyamongo, J., Backer, L., Dahiye, A.M., Misore, A., DeCock, K. & Rubin, C. (2005). Aflatoxin contamination of commercial maize products during an outbreak of acute aflatoxicosis in eastern and central Kenya, *Environ Health Perspect* 113, 1763- 1767.

Lin, B.K. & Anderson, J.A. (1992). Purification and properties of versiconal cyclase from *Aspergillus parasiticus, Arch. Biochem. Biophys.* 293, 67-70.

Lin, M.T. & Hsieh, D.P. (1973). Averufin in the biosynthesis of aflatoxin B, *J. Am Chem. Soc.* 95, 1668-1669.

Lin, M.T., Hsieh, D.P., Yao, R.C. & Donkersloot, J.A. (1973). Conversion of averufin into aflatoxins by *Aspergillus parasiticus, Biochemistry* 12, 5167-5171.

Mahanti, N., Bhatnagar, D., Cary, J.W., Joubran, J. & Linz, J.E. (1996). Structure and function of *fas-1A*, a gene encoding a putative fatty acid synthetase directly involved in aflatoxin biosynthesis in *Aspergillus parasiticus, Appl. Environ. Microbiol.* 62, 191-195.

McCormick, S.P., Bhatnagar, D. & Lee, L.S. (1987). Averufanin is an aflatoxin B1 precursor between averantin and averufin in the biosynthetic pathway, *Appl. Environ. Microbiol.* 53, 14-16.

McGlynn, K.A., Hunter, K., LeVoyer, T., Roush, J., Wise, P., Michielli, R.A., Shen, F.M., Evans, A.A., London, W.T. & Buetow, K.H. (2003). Susceptibility to aflatoxin B1-related primary hepatocellular carcinoma in mice and humans, *Cancer Res.* 63, 4594-4601.

McGuire, S.M., Silva, J.C., Casillas, E.G. & Townsend, C.A. (1996). Purification and characterization of versicolorin B synthase from *Aspergillus parasiticus.* Catalysis of the stereodifferentiating cyclization in aflatoxin biosynthesis essential to DNA interaction, *Biochemistry* 35, 11470-11486.

Meyers, D.M., Obrian, G., Du, W.L., Bhatnagar, D. & Payne, G.A. (1998). Characterization of *aflJ*, a gene required for conversion of pathway intermediates to aflatoxin, *Appl. Environ. Microbiol.* 64, 3713-3717.

Minto, R.E. & Townsend, C.A. (1997). Enzymology and molecular biology of aflatoxin biosynthesis, *Chem. Rev.* 97, 2537-2556.

Mooney, J.L. & Yager, L.N. (1990). Light is required for conidiation in *Aspergillus nidulans, Genes Dev.* 4, 1473-1482.

Mori, T., Matsumura, M., Yamada, K., Irie, S., Oshimi, K., Suda, K., Oguri, T. & Ichinoe, M. (1998). Systemic aspergillosis caused by an aflatoxin-producing strain of *Aspergillus flavus, Med. Mycol.* 36, 107-112.

Motomura, M., Chihaya, N., Shinozawa, T., Hamasaki, T. & Yabe, K. (1999). Cloning and characterization of the *O*-methyltransferase I gene (*dmtA*) from *Aspergillus parasiticus* associated with the conversions of demethylsterigmatocystin to sterigmatocystin and dihydrodemethylsterigmatocystin to dihydrosterigmatocystin in aflatoxin biosynthesis, *Appl. Environ. Microbiol.* 65, 4987-4994.

Ngindu, A., Johnson, B.K., Kenya, P.R., Ngira, J.A., Ocheng, D.M., Nandwa, H., Omondi, T.N., Jansen, A.J., Ngare, W., Kaviti, J.N., Gatei, D. & Siongok, T.A. (1982). Outbreak of acute hepatitis caused by aflatoxin poisoning in Kenya., *Lancet.* 1, 1346-1348.

Nierman, W.C., Pain, A., Anderson, M.J., Wortman, J.R., Kim, H.S., Arroyo, J., Berriman, M., Abe, K., Archer, D.B., Bermejo, C., Bennett, J., Bowyer, P., Chen, D., Collins, M.,

Coulsen, R., Davies, R., Dyer, P.S., Farman, M., Fedorova, N., Fedorova, N., Feldblyum, T.V., Fischer, R., Fosker, N., Fraser, A., Garcia, J.L., Garcia, M.J., Goble, A., Goldman, G.H., Gomi, K., Griffith-Jones, S., Gwilliam, R., Haas, B., Haas, H., Harris, D., Horiuchi, H., Huang, J., Humphray, S., Jimenez, J., Keller, N., Khouri, H., Kitamoto, K., Kobayashi, T., Konzack, S., Kulkarni, R., Kumagai, T., Lafon, A., Latge, J.P., Li, W., Lord, A., Lu, C., Majoros, W.H., May, G.S., Miller, B.L., Mohamoud, Y., Molina, M., Monod, M., Mouyna, I., Mulligan, S., Murphy, L., O'Neil, S., Paulsen, I., Penalva, M.A., Pertea, M., Price, C., Pritchard, B.L., Quail, M.A., Rabbinowitsch, E., Rawlins, N., Rajandream, M.A., Reichard, U., Renauld, H., Robson, G.D., Rodriguez de Cordoba, S., Rodriguez-Pena, J.M., Ronning, C.M., Rutter, S., Salzberg, S.L., Sanchez, M., Sanchez-Ferrero, J.C., Saunders, D., Seeger, K., Squares, R., Squares, S., Takeuchi, M., Tekaia, F., Turner, G., Vazquez de Aldana, C.R., Weidman, J., White, O., Woodward, J., Yu, J.H., Fraser, C., Galagan, J.E., Asai, K., Machida, M., Hall, N., Barrell, B. & Denning, D.W. (2005). Genomic sequence of the pathogenic and allergenic filamentous fungus *Aspergillus fumigatus*, *Nature* 438, 1151-1156.

Ozturk, M. (1991). p53 mutation in hepatocellular carcinoma after aflatoxin exposure, *Lancet.* 338, 1356-1359.

Papa, G. (1982). Norsolorinic acid mutant of *Aspergillus flavus*, *J. Gen. Microbiol.* 128, 1345-1348.

Papa, K.E. (1976). Linkage groups in *Aspergillus flavus*, *Mycologia* 68, 159-165.

Papa, K.E. (1979). Genetics of *Aspergillus flavus*: complementation and mapping of aflatoxin mutants, *Genet. Res.* 34, 1-9.

Papa, K.E. (1984). Genetics of *Aspergillus flavus*: linkage of aflatoxin mutants, *Can. J. Microbiol.* 30, 68-73.

Payne, G.A. (1998). Process of contamination by aflatoxin-producing fungi and their impacts on crops. In: Sinha, K.K. & Bhatnagar, D. (Eds.), *Mycotoxins in Agriculture and Food Safety*. Marcel Dekker, New York, pp. 279-306.

Payne, G.A., Nystrom, G.J., Bhatnagar, D., Cleveland, T.E. & Woloshuk, C.P. (1993). Cloning of the *afl-2* gene involved in aflatoxin biosynthesis from *Aspergillus flavus*, *Appl. Environ. Microbiol.* 59, 156-162.

Peers, F., Bosch, X., Kaldor, J., Linsell, A. & Pluijmen, M. (1987). Aflatoxin exposure, hepatitis B virus infection and liver cancer in Swaziland, *Int. J .Cancer* 39, 545-553.

Perrin, R.M., Fedorova, N.D., Bok, J.W., Cramer, R.A., Wortman, J.R., Kim, H.S., Nierman, W.C. & Keller, N.P. (2007). Transcriptional regulation of chemical diversity in *Aspergillus fumigatus* by LaeA, *PLoS Pathog.* 3, e50.

Polychronaki, N., West, R.M., Turner, P.C., Amra, H., Abdel-Wahhab, M., Mykkänen, H. & El-Nezami, H. (2007). A longitudinal assessment of aflatoxin M1 excretion in breast milk of selected Egyptian mothers, *Food Chem Toxicol* 45, 1210-1215.

Price, M.S., Yu, J., Nierman, W.C., Kim, H.S., Pritchard, B., Jacobus, C.A., Bhatnagar, D., Cleveland, T.E. & Payne, G.A. (2006). The aflatoxin pathway regulator AflR induces gene transcription inside and outside of the aflatoxin biosynthetic cluster, *FEMS Microbiol. Lett.* 255, 275-279.

Prieto, R. & Woloshuk, C.P. (1997). *ord1*, an oxidoreductase gene responsible for conversion of *O*- methylsterigmatocystin to aflatoxin in *Aspergillus flavus*, *Appl. Environ. Microbiol.* 63, 1661-1666.

Prieto, R., Yousibova, G.L. & Woloshuk, C.P. (1996). Identification of aflatoxin biosynthesis genes by genetic complementation in an *Aspergillus flavus* mutant lacking the aflatoxin gene cluster, *Appl. Environ. Microbiol.* 62, 3567-3571.

Raisuddin, S., Singh, K.P., Zaidi, S.I., Paul, B.N. & Ray, P.K. (1993). Immunosuppressive effects of aflatoxin in growing rats, *Mycopathologia* 124, 189-194.

Richard, J.L. & Payne, G.A. (2003), Mycotoxins: Risks in Plant, Animal and Human Systems, Council for Agricultural Science and Technology (CAST), Ames, IA. Robens, J.F. (2001). The costs of mycotoxin management to the USA: Management of aflatoxins in the United States, *APSnet Feature* *http://www.apsnet.org/online/feature/mycotoxin/top.html* 2-8.

Robens, J.F. & Cardwell, K. (2005). The cost of mycotoxin management in the United States. In: Abbas, H.K. (Ed.), *Aflatoxin and Food Safety*. CRC Press, Boca Raton, FL, pp. 1-12.

Ronning, C.M., Fedorova, N.D., Bowyer, P., Coulson, R., Goldman, G., Kim, H.S., Turner, G., Wortman, J.R., Yu, J., Anderson, M.J., Denning, D.W. & Nierman, W.C. (2005). Genomics of *Aspergillus fumigatus*, *Rev. Iberoam Micol. (RIAM)* 22, 223-228.

Sakuno, E., Yabe, K. & Nakajima, H. (2003). Involvement of two cytosolic enzymes and a novel intermediate, 5'-oxoaverantin, in the pathway from 5'-hydroxyaverantin to averufin in aflatoxin biosynthesis, *Appl. Environ. Microbiol.* 69, 6418-6426.

Scholl, P. & Groopman, J.D. (1995). Epidermiology of human exposures and its relationship to liver cancer. In: Eklund, M., Richard, J.L. & Mise, K. (Eds.), *Molecular Approaches to Food Safety: Issues Involving Toxic Microorganisms*. Alaken, Inc., Fort Collins, Co., pp. 169-182.

Schroeder, H.W., Cole, R.J., Grigsby, R.D. & Hein, H., Jr. (1974). Inhibition of aflatoxin production and tentative identification of an aflatoxin intermediate "versiconal acetate" from treatment with dichlorvos, *Appl. Microbiol.* 27, 394-399.

Silva, J.C., Minto, R.E., Barry, C.E., 3rd, Holland, K.A. & Townsend, C.A. (1996). Isolation and characterization of the versicolorin B synthase gene from *Aspergillus parasiticus*. Expansion of the aflatoxin B1 biosynthetic gene cluster, *J. Biol. Chem.* 271, 13600-13608.

Silva, J.C. & Townsend, C.A. (1997). Heterologous expression, isolation, and characterization of versicolorin B synthase from *Aspergillus parasiticus*. A key enzyme in the aflatoxin B1 biosynthetic pathway, *J. Biol. Chem.* 272, 804-813.

Singh, R. & Hsieh, D.P. (1977). Aflatoxin biosynthetic pathway: elucidation by using blocked mutants of *Aspergillus parasiticus*, *Arch. Biochem. Biophys.* 178, 285-292. Skipper, P.L. & Tannenbaum, S.R. (1990). Protein adducts in the molecular dosimetry of chemical carcinogens, *Carcinogenesis* 11, 507-518.

Skory, C.D., Chang, P.K., Cary, J. & Linz, J.E. (1992). Isolation and characterization of a gene from *Aspergillus parasiticus* associated with the conversion of versicolorin A to sterigmatocystin in aflatoxin biosynthesis, *Appl. Environ. Microbiol.* 58, 3527-3537.

St Leger, R.J., Screen, S.E. & Shams-Pirzadeh, B. (2000). Lack of host specialization in *Aspergillus flavus*, *Appl. Environ. Microbiol.* 66, 320-324.

Stinnett, S.M., Espeso, E.A., Cobeno, L., Araujo-Bazan, L. & Calvo, A.M. (2007). *Aspergillus nidulans* VeA subcellular localization is dependent on the importin alpha carrier and on light, *Mol. Microbiol.* 63, 242-255.

Sugui, J.A., Pardo, J., Chang, Y.C., Zarember, K.A., Nardone, G., Galvez, E.M., Mullbacher, A., Gallin, J.I., Simon, M.M. & Kwon- Chung, K.J. (2007). Gliotoxin is a virulence factor of *Aspergillus fumigatus*: *gliP* deletion attenuates virulence in mice immunosuppressed with hydrocortisone, *Eukaryot. Cell* 6, 1562- 1569.

Tandon, H.D., Tandon, B.N. & Ramalingaswami, V. (1978). Epidemic of toxic hepatitis in India of possible mycotoxic origin, *Arch. Pathol. Lab. Med.* 102, 372-376.

Townsend, C.A. (1997). Progress towards a biosynthetic rationale of the aflatoxin pathway, *Pure Appl. Chem.* 58, 227-238.

Trail, F., Chang, P.-K., Cary, J. & Linz, J.E. (1994). Structural and functional analysis of the *nor-1* gene involved in the biosynthesis of aflatoxins by *Aspergillus parasiticus*, *Appl. Environ. Microbiol.* 60, 4078-4085.

Trail, F., Mahanti, N. & Linz, J. (1995a). Molecular biology of aflatoxin biosynthesis, *Microbiol.* 141, 755-765.

Trail, F., Mahanti, N., Rarick, M., Mehigh, R., Liang, S.H., Zhou, R. & Linz, J.E. (1995b). Physical and transcriptional map of an aflatoxin gene cluster in *Aspergillus parasiticus* and functional disruption of a gene involved early in the aflatoxin pathway, *Appl. Environ. Microbiol.* 61, 2665-2673.

Turner, P.C., Collinson, A.C., Cheung, Y.B., Gong, Y., Hall, A.J., Prentice, A.M. & Wild, C.P. (2007). Aflatoxin exposure in utero causes growth faltering in Gambian infants, *Int J Epidemiol* 36, 1119-1125.

van Egmond, H.P. (1989). Current situation on regulations for mycotoxins. Overview of tolerances and status of standard methods of sampling and analysis, *J. Food Addit. Contam.* 6, 139-188.

van Egmond, H.P. & Jonker, M.A. (2005). Worldwide regulations on aflatoxins. In: Abbas, H.K. (Ed.). CRC Press, Taylor and Francis Group, Boca Raton, FL, pp. 77-93.

van Egmond, H.P., Schothorst, R.C. & Jonker, M.A. (2007). Regulations relating to mycotoxins in food: Perspectives in a global and European context, *Anal Bioanal Chem* 389, 147-157.

Watanabe, C.M. & Townsend, C.A. (1996). Incorporation of molecular oxygen in aflatoxin B1 biosynthesis, *J. Org. Chem.* 61, 1990-1993.

Watanabe, C.M. & Townsend, C.A. (2002). Initial characterization of a type I fatty acid synthase and polyketide synthase multienzyme complex NorS in the biosynthesis of aflatoxin B1, *Chem. Biol.* 9, 981-988.

Watanabe, C.M., Wilson, D., Linz, J.E. & Townsend, C.A. (1996). Demonstration of the catalytic roles and evidence for the physical association of type I fatty acid synthases and a polyketide synthase in the biosynthesis of aflatoxin B1, *Chem. Biol.* 3, 463-469.

Wild, C.P., Shrestha, S.M., Anwar, W.A. & Montesano, R. (1992). Field studies of aflatoxin exposure, metabolism and induction of genetic alterations in relation to HBV infection and hepatocellular carcinoma in The Gambia and Thailand, *Toxicol. Lett.* 64-65, 455-461.

Wilkinson, J.R., Yu, J., Abbas, H.K., Scheffler, B.E., Kim, H.S., Nierman, W.C., Bhatnagar, D. & Cleveland, T.E. (2007a). Aflatoxin formation and gene expression in response to carbon source media shift in *Aspergillus parasiticus*, *J. Food Addit. Contam.* 24, 1051-1060.

Wilkinson, J.R., Yu, J., Bland, J.M., Nierman, W.C., Bhatnagar, D. & Cleveland, T.E. (2007b). Amino acid supplementation reveals differential regulation of aflatoxin biosynthesis in *Aspergillus flavus* NRRL 3357 and *Aspergillus parasiticus* SRRC 143, *Appl. Microbiol. Biotechnol.* 74, 1308-1319.

Wilson, D.M. (1989). Analytical methods for aflatoxins in corn and peanuts, *Arch. Environ. Contam. Toxicol.* 18, 308-314. Wogan, G.N. (1992). Aflatoxins as risk factors for hepatocellular carcinoma in humans, *Cancer Res.* 52, 2114-2118.

Woloshuk, C.P., Foutz, K.R., Brewer, J.F., Bhatnagar, D., Cleveland, T.E. & Payne, G.A. (1994). Molecular characterization of *aflR*, a regulatory locus for aflatoxin biosynthesis, *Appl. Environ. Microbiol.* 60, 2408-2414.

Woloshuk, C.P. & Payne, G.A. (1994). The alcohol dehydrogenase gene *adh1* is induced in *Aspergillus flavus* grown on medium conducive to aflatoxin biosynthesis, *Appl. Environ. Microbiol.* 60, 670-676.

Woloshuk, C.P. & Prieto, R. (1998). Genetic organization and function of the aflatoxin B1 biosynthetic genes, *FEMS Microbiol. Lett.* 160, 169- 176.

Wong, J.J., Singh, R. & Hsieh, D.P. (1977). Mutagenicity of fungal metabolites related to aflatoxin biosynthesis, *Mutat. Res.* 44, 447- 450.

Yabe, K., Ando, Y. & Hamasaki, T. (1988). Biosynthetic relationship among aflatoxins B1, B2, G1, and G2, *Appl. Environ. Microbiol.* 54, 2101-2106.

Yabe, K., Ando, Y. & Hamasaki, T. (1991a). A metabolic grid among versiconal hemiacetal acetate, versiconol acetate, versiconol and versiconal during aflatoxin biosynthesis, *J. Gen. Microbiol.* 137, 2469-2475.

Yabe, K., Ando, Y., Hashimoto, J. & Hamasaki, T. (1989). Two distinct *O*-methyltransferases in aflatoxin biosynthesis, *Appl. Environ. Microbiol.* 55, 2172-2177.

Yabe, K., Chihaya, N., Hamamatsu, S., Sakuno, E., Hamasaki, T., Nakajima, H. & Bennett, J.W. (2003). Enzymatic conversion of averufin to hydroxyversicolorone and elucidation of a novel metabolic grid involved in aflatoxin biosynthesis, *Appl. Environ. Microbiol.* 69, 66-73.

Yabe, K., Matsushima, K., Koyama, T. & Hamasaki, T. (1998). Purification and characterization of O-methyltransferase I involved in conversion of demethylsterigmatocystin to sterigmatocystin and of dihydrodemethylsterigmatocystin to dihydrosterigmatocystin during aflatoxin biosynthesis, *Appl. Environ. Microbiol.* 64, 166-171.

Yabe, K., Matsuyama, Y., Ando, Y., Nakajima, H. & Hamasaki, T. (1993). Stereochemistry during aflatoxin biosynthesis: conversion of norsolorinic acid to averufin, *Appl. Environ. Microbiol.* 59, 2486- 2492.

Yabe, K. & Nakajima, H. (2004). Enzyme reactions and genes in aflatoxin biosynthesis, *Appl. Microbiol. Biotechnol.* 64, 745-755.

Yabe, K., Nakamura, M. & Hamasaki, T. (1999). Enzymatic formation of G-group aflatoxins and biosynthetic relationship between G- and B- group aflatoxins, *Appl. Environ. Microbiol.* 65, 3867-3872.

Yabe, K., Nakamura, Y., Nakajima, H., Ando, Y. & Hamasaki, T. (1991b). Enzymatic conversion of norsolorinic acid to averufin in aflatoxin biosynthesis, *Appl. Environ. Microbiol.* 57, 1340-1345.

Yao, R.C. & Hsieh, D.P. (1974). Step of dichlorvos inhibition in the pathway of aflatoxin biosynthesis, *Appl. Microbiol.* 28, 52-57.

Yu, J., Bhatnagar, D. & Cleveland, T.E. (2004a). Completed sequence of aflatoxin pathway gene cluster in *Aspergillus parasiticus, FEBS Lett.* 564, 126-130.

Yu, J., Bhatnagar, D. & Cleveland, T.E. (2004b). Genetics and biochemistry of aflatoxin formation and genomics approach for eliminating aflatoxin contamination. In: Romeo, J.T. (Ed.), *Recent Advances in Phytochemistry* Elsevier, New York, pp. 224-242.

Yu, J., Cary, J.W., Bhatnagar, D., Cleveland, T.E., Keller, N.P. & Chu, F.S. (1993). Cloning and characterization of a cDNA from *Aspergillus parasiticus* encoding an O-methyltransferase involved in aflatoxin biosynthesis, *Appl. Environ. Microbiol.* 59, 3564-3571.

Yu, J., Chang, P.K., Bhatnagar, D. & Cleveland, T.E. (2000a). Cloning of a sugar utilization gene cluster in *Aspergillus parasiticus, Biochim. Biophys. Acta.* 1493, 211-214.

Yu, J., Chang, P.K., Bhatnagar, D. & Cleveland, T.E. (2002). Cloning and functional expression of an esterase gene in *Aspergillus parasitcus, Mycopathologia* 156, 227-234.

Yu, J., Chang, P.K., Cary, J.W., Bhatnagar, D. & Cleveland, T.E. (1997). *avnA*, a gene encoding a cytochrome P-450 monooxygenase, is involved in the conversion of averantin to averufin in aflatoxin biosynthesis in *Aspergillus parasiticus, Appl. Environ. Microbiol.* 63, 1349-1356.

Yu, J., Chang, P.K., Cary, J.W., Wright, M., Bhatnagar, D., Cleveland, T.E., Payne, G.A. & Linz, J.E. (1995a). Comparative mapping of aflatoxin pathway gene clusters in *Aspergillus parasiticus* and *Aspergillus flavus, Appl. Environ. Microbiol.* 61, 2365-2371.

Yu, J., Chang, P.K., Ehrlich, K.C., Cary, J.W., Bhatnagar, D., Cleveland, T.E., Payne, G.A., Linz, J.E., Woloshuk, C.P. & Bennett, J.W. (2004c). Clustered pathway genes in aflatoxin biosynthesis, *Appl. Environ. Microbiol.* 70, 1253-1262.

Yu, J., Chang, P.K., Ehrlich, K.C., Cary, J.W., Montalbano, B., Dyer, J.M., Bhatnagar, D. & Cleveland, T.E. (1998). Characterization of the critical amino acids of an *Aspergillus parasiticus* cytochrome P- 450 monooxygenase encoded by *ordA* that is involved in the biosynthesis of aflatoxins B1, G1, B2, and G2, *Appl. Environ. Microbiol.* 64, 4834-4841.

Yu, J., Chang, P.K., Payne, G.A., Cary, J.W., Bhatnagar, D. & Cleveland, T.E. (1995b). Comparison of the *omtA* genes encoding *O*-methyltransferases involved in aflatoxin biosynthesis from *Aspergillus parasiticus* and *A. flavus, Gene* 163, 121-125.

Yu, J., Fedorova, F., Montalbano, B.G., Bhatnagar, D., Cleveland, T.E., Bennett, J.W. & Nierman, W.C. (2011). Mycotoxin gene expression in *Aspergillus flavus* in response to temperature as detected by RNA- Sequencing *FEMS Microbiol. Lett.* submitted.

Yu, J., Woloshuk, C.P., Bhatnagar, D. & Cleveland, T.E. (2000b). Cloning and characterization of *avfA* and *omtB* genes involved in aflatoxin biosynthesis in three *Aspergillus* species, *Gene* 248, 157- 167.

Yu, J.H., Butchko, R.A., Fernandes, M., Keller, N.P., Leonard, T.J. & Adams, T.H. (1996a). Conservation of structure and function of the aflatoxin regulatory gene *aflR* from *Aspergillus nidulans* and *A. flavus*, *Curr. Genet.* 29, 549-555.

Yu, J.H., Wieser, J. & Adams, T.H. (1996b). The *Aspergillus* FlbA RGS domain protein antagonizes G protein signaling to block proliferation and allow development, *EMBO J.* 15, 5184-5190.

Zhou, R. & Linz, J.E. (1999). Enzymatic function of the nor-1 protein in aflatoxin biosynthesis in Aspergillus parasiticus, *Appl. Environ. Microbiol.* 65, 5639-5641.

Zhu, J.Q., Zhang, L.S., Hu, X., Xiao, Y., Chen, J.S., Xu, Y.C., Fremy, J. & Chu, F.S. (1987). Correlation of dietary aflatoxin B1 levels with excretion of aflatoxin M1 in human urine, *Cancer Res.* 47, 1848-1852.

Zuckerman, A.J., Rees, K.R., Inman, D. & Petts, V. (1967). Site of action of aflatoxin on human liver cells in culture, *Nature* 214, 814-815.

4

Conserved Regulatory Mechanisms Controlling Aflatoxin and Sterigmatocystin Biosynthesis

Ana M. Calvo and Sourabh Dhingra
Northern Illinois University, USA

1. Introduction

Filamentous fungi produce a wide array of compounds known as secondary metabolites or natural products (i.e. Adrio and Demain, 2003; Reverberi et al 2010; Brakhage and Schroeckh 2011). The exact function of these compounds is unknown but it is postulated that they provide fungi with an advantage in their ecological niche (Keller et al., 2005; Georgianna and Payne 2009). The broad group of natural products includes compounds such as antibiotics, pigments and also mycotoxins (Fig.1). Among mycotoxins, aflatoxins (AF), specifically AF B1, are the most potent carcinogenic natural compounds known, and they are mainly produced by the opportunistic plant pathogens *Aspergillus flavus* and *Aspergillus parasiticus* (Squire 1989; Sweeny and Dobson 1999; Payne and Brown, 1998). Other AF-producers among *Aspergilli* include *Aspergillus nomius, Aspergillus ochraceoroseus, Aspergillus bombycis* and *Aspergillus pseudotamari* (Cary and Ehrlich, 2006). *Aspergillus flavus* and *Aspergillus parasiticus* have the ability to colonize oil seed crops of agricultural importance, such as corn, cotton, peanuts, sorghum and tree nuts. Ingestion of aflatoxin contaminated food can cause hepatocellular carcinoma, immunotoxicity, and teratogenic effects (Dvorackova and Kusak, 1990; Trail et al., 1995; Wogan et al., 1992). Various developed countries have strict regulation on the amount of AF allowed in food commodities. Contaminated crops above the permitted limit have to be destroyed, resulting in economic losses. The annual estimated loss due to AF contamination is attributed at approximately $270 million in the USA alone (Richard and Payne, 2003).

It is known that AF and a related mycotoxin called sterigmatocystin (ST) (Fig.2) are synthesized through the same conserved biosynthetic pathway, in which ST is the penultimate precursor. ST is produced by several *Aspergillus* species, including *Aspergillus nidulans*, one of the most characterized eukaryotic systems that has been used as a model organism for more than 60 years (Pontecorvo et al., 1953). The elevated number of characterized *A. nidulans* genes and mutant strains makes this model fungus ideal for genetic and molecular studies. There is a physical and genetic map of the eight chromosomes in *A. nidulans.* The whole *A. nidulans* genome has been sequenced and annotated, and has been compared with other *Aspergillus* genomes (David et al., 2008; Galagan et al., 2005). Structural and signaling pathway genes controlling ST production in *A. nidulans* are also found in AF-producer *Aspergillus* spp. (Hicks et al., 1997; reviewed by

Calvo et al., 2002; Calvo, 2008). For these reasons this model system is especially productive in the study of the AF/ST gene clusters and regulatory pathways directing mycotoxin production. In this review we will focus on these common regulatory mechanisms governing AF and ST biosynthesis in *A. flavus* and *A. parasiticus,* and in *A. nidulans.*

2. AF/ST gene clusters

ST/AF gene clusters remain some of the best characterized mycotoxin gene clusters. Both encode enzymes participating in 29 metabolic steps along with two regulatory proteins, AflR and AflS. (Cary et al., 2009). Both gene clusters extend to approximately 70 kb in the genome (Brown et al., 1996; Cary and Ehrlich, 2006; Ehrlich et al., 2005; Cary et al., 2009). The order and direction of genes in AF and ST clusters is conserved for all genes except four in AF pathway and three in ST which are inverted. It has been hypothesized that these differences might have been the result of gene reorganization from an ST-producing ancestor by recombination and duplications of near-telomeric regions where these clusters are found (Cary et al., 2009; Carbone et al., 2007). Based on phylogenetic analysis, the clusters may be 450 million years old (Galagan et al., 2005).
The enzymatic reactions and chemistry involved in the AF/ST biosynthetic pathway have already been extensively covered in previous reviews (Bhatnagar et al., 2003; Bennett and Klich, 2003; Hicks et al., 2002; Huffman et al., 2010; Keller et al., 2005; Minto and Townsend, 1997; Payne and Brown, 1998; Yabe and Nakajima, 2004), and will not be addressed in this review. Beyond the genus *Aspergillus,* other fungi present semi-conserved pathways, for example *Dothistroma septosporum,* which produces a metabolite that resembles versicolorin B, a precursor of both ST and AF. This suggests further cluster conservation across fungal genera (Bradshaw and Zhang, 2006).

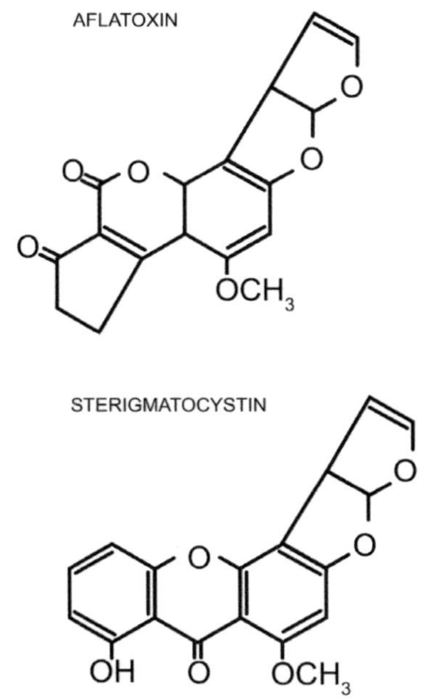

Fig. 1. Chemical structures of the mycotoxins aflatoxin and sterigmatocystin.

3. AflR and AflS

In addition to the structural genes, two regulatory genes are also included in these clusters, *aflR* and *aflS* (formerly known as *aflJ*). The gene products AflR and AflS, are specific regulators for AF/ST gene cluster activation and for concomitant AF/ST production. AflR, encodes a transcription factor with a cysteine rich Gal4 type bi-nuclear zinc finger cluster (Chang et al., 1995; Yu et al., 1996). AflR has been shown to control most of the steps of the ST/AF biosynthetic pathway. AflR binds DNA at the consensus motif 5'TCGN5CGA3' in the promoter of biosynthetic genes (Fernandes et al., 1998). More than one AflR binding site is common in promoters of AF genes (Cary et al., 2009; Cary et al., 2000). Despite the differences between *A. nidulans* and *A. flavus* AflR, the function is conserved (Yu et al., 1996). Deletion *aflR* (Δ*aflR*) strain in *A. flavus* and *A. nidulans* fails to produce AF and ST respectively. Additionally, heterologously expressed *A. flavus aflR* in the deletion *A. nidulans* Δ*aflR* strain is able to activate the ST gene cluster, supporting functional conservation (Yu et al., 1996). In *A. parasiticus*, a microarray comparison of Δ*aflR* and wild-type strains revealed eighteen genes involved in AF pathway to be differentially expressed (Price et al., 2006). Constitutive expression of *aflR* was shown to increase the transcript of AF cluster genes along with 50-fold greater aflatoxin production (Flaherty and Payne 1997).

In addition to AflR, AflS also plays a role in the regulation of AF biosynthesis (Meyers et al., 1998; Chang et al., 1993; 2002; Chang, 2003). *aflS* lies next to *aflR* in the AF cluster. The gene products AflR and AflS were found to interact in *A. parasiticus* (Chang, 2003). Deletion of *aflS* has been shown to decrease expression of several biosynthetic genes in the cluster (Meyers et al., 1998), but does not alter *aflR* expression (Chang, 2003). The Δ*aflS* mutant still produces reduced but detectable levels of AF in *A. parasiticus* (Meyers et al., 1998). Interestingly, overexpression of *aflS*, particularly in combination with overexpression of *aflR* synergically increased expression of biosynthetic genes and AF production (Chang, 2003). *Aspergillus sojae*, is unable to produce AF in spite possessing the AF gene cluster in its genome; its *aflR* gene has a mutation that results in a truncated protein, which fails to interact with AflS (Chang, 2004).

Our genomic sequence analysis indicates that *aflS* is also present in *A. nidulans*, and that it is located next to *aflR* in the ST gene cluster (Calvo et al., unpublished data). As in the case of *aflS* in the AF cluster, *A. nidulans aflS* transcriptional direction is opposite with respect to that of *aflR*, sharing an intergenic region of 599 bp.

4. Global regulation

4.1 VeA and VeA-interacting proteins

Global regulatory mechanisms that control different cell functions, including morphogenesis and secondary metabolism tend to be conserved across fungal species. The *velvet* gene or *veA* is an example of global regulation. VeA is unique to fungi and is highly conserved in Ascomycetes (Myung et al., 2011). This regulator controls the morphological balance between sexual and asexual development (Kim et al., 2002; Li et al., 2006; Calvo, 2008). Interestingly, our laboratory showed for the first time that VeA also controls the production of numerous secondary metabolites, including ST and AF production in *A. nidulans*, and *A. flavus* and *A. parasiticus* (Kato et al., 2003; Calvo et al., 2004; Duran et al., 2007; Carry et al., 2007). In *A. nidulans* and later in *A. flavus* it has been demonstrated that *veA* is necessary for *aflR* transcription, and therefore controls the expression of the ST/AF clustered genes (Kato et al., 2003; Duran et al., 2007).

The effect of VeA on secondary metabolism is broad. In *A. nidulans* the production of other metabolites, including penicillin, is also affected by the absence of *veA* (Kato et al., 2003). In *A. flavus*, the biosynthesis of other mycotoxins is also *veA*-dependent, such as the case of cyclopiazonic acid and aflatrem, where we showed that expression of aflatrem genes requires a *veA* wild-type allele (Duran et al., 2007).

VeA is also conserved across fungal genera, for example the *veA* homolog FvVE1 in *Fusarium verticilliodies* (Li et al., 2006; Myung et al., 2009), *AcveA* in *Acremonium chrysogenum* (Dreyer et al., 2007) and *ve-1* in *Neurospora crassa* (Bayram et al., 2008). Studies in these organisms also link *veA* to secondary metabolism. The *veA* regulatory system has been extensively reviewed (Calvo, 2008). VeA is transported to the nucleus (Stinnett et al., 2007) where it forms interactions with other regulatory proteins that also influence morphogenesis, and secondary metabolism, including production of mycotoxin (Purschwitz et al., 2008; Calvo, 2008; Bayram et al., 2008; Bayram et al., 2010).

In *A. nidulans* it has been shown that VeA interacts with a protein called LaeA (Bayram et al., 2008). *laeA* is also necessary for ST/AF production. Unlike *veA*, which has been shown to control *A. nidulans* asexual/sexual morphological development, *laeA* has a mild effect on morphogenesis, influencing Hülle cell numbers (Bayram et al., 2010). However, in *A. flavus* both *veA* and *laeA* have an important role in morphogenesis, being necessary for sclerotial production (Duran et al., 2007; Duran et al., 2009; Kale et al., 2008). LaeA is a putative methyl transferase, with an S-adenosylmethionine binding domain (Bok and Keller, 2004). In *A. flavus* it has been shown that *laeA* is a negative regulator of *veA* transcription (Kale et al., 2008). In addition, increased expression of *veA* (in a strain with multiple *veA* copies) results in decreased *laeA* expression (Amaike and Keller, 2009), suggesting a mutual negative transcriptional regulatory feedback control.

In addition to transcriptional regulation between *veA* and *laeA*, recent studies also suggest posttranslational regulation (Bayram et al., 2010); a new form of the VeA protein with higher molecular weight has been detected in Δ*laeA* strains, indicating that in the wild type, *laeA* prevents a modification of the VeA protein (Bayram et al., 2010).

Another protein from the *velvet* family, VelB, has been shown to interact with VeA. The Δ*velB* strain produces a reduced and delayed but still detectable amount of ST. VelB protein also interacts with VosA, a positive regulator of sporogenesis (Ni and Yu 2007; Bayram et al., 2010). Homologs of *velB* and *vosA* are present in *A. flavus* genome (data not shown). Other VeA-interacting proteins, such as the light-response mediating proteins FphA, LreA and LreB (Purschwitz et al., 2008), also influence morphological and metabolic changes in response to environmental stimuli affecting mycotoxin production (see section below corresponding to the effect of light).

4.2 G protein signalling

Several cellular functions, including fungal growth, morphogenesis and secondary metabolism are governed by G-protein signalling pathways. For example, the *A. nidulans fadA* gene encodes an α subunit of an heterotrimeric G protein complex where FadA interacts with Gβ (SfaD) and Gγ (GpgA) subunits (Rosen et al., 1999; Seo et al., 2005). Mutations in FadA blocking the intrinsic GTPase activity results in a permanently active stage of this protein (Hicks et at., 1997). It is likely that activation of FadA upregulates adenylyl cyclase and cAMP-dependent kinase PKA. Activation of FadA or overexpression of PKA results in an increase in vegetative growth and reduction of asexual development,as well as reduction in ST production (Hicks et al., 1997; Shimizu and Keller, 2000). It has been shown that overexpression of PKA represses *aflR* expression (Shimizu and Keller, 2000). In addition to transcriptional regulation of ST genes, PKA also regulates ST production at posttranslational level. PKA negatively regulates the localization of AflR protein in the nucleus by phosphorylation (Shimizu et al., 2003). In *A. parasiticus* PKA also negatively regulates AF production (Roze et al., 2004). Additionally, expression of the FadA constitutive active form also resulted in a decrease of AF intermediates and a reduction in conidiation in *A. parasiticus* (Hicks et al., 1997). These findings strongly indicate a conservation of the FadA-PKA signaling pathway in regulating ST and AF production.

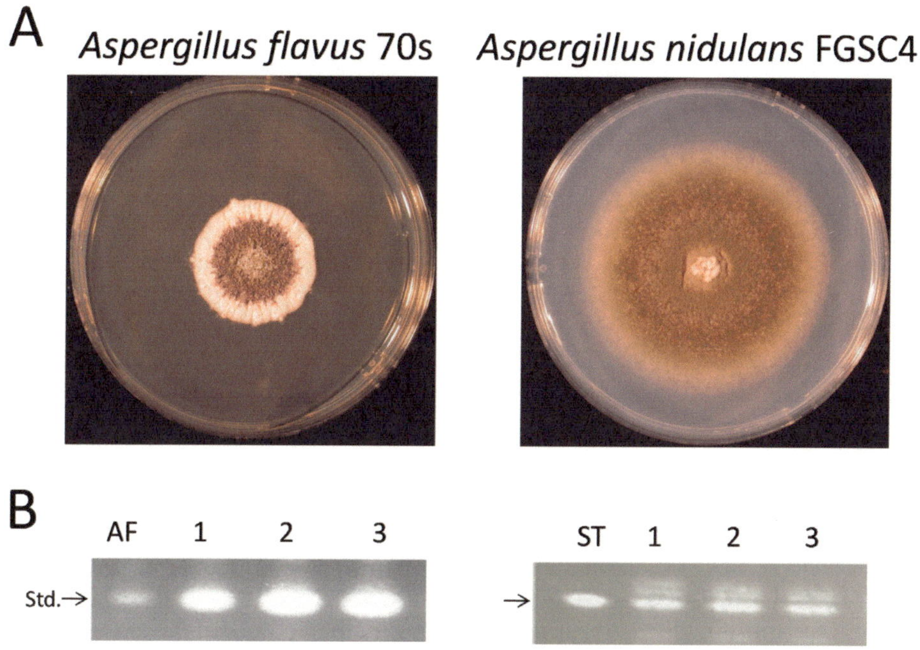

Fig. 2. *Aspergillus flavus* (70S) and *Aspergillus nidulans* (FGSC 4) wild type strains.
A) Point-inoculated cultures incubated for five days. B) Thin layer chromatography analysis of the mycotoxins aflatoxin and sterigmatocystin. The chemical analysis was carried out as previously described (Kato et al., 2003; Duran et al., 2007). Std. = Standard, AF = Aflatoxin, ST = Sterigmatocystin. 1-3 represent replicates.

In the wild type, FadA is regulated by FluG, which activates transcription of *flbA*, encoding a GTPase activating protein that negatively regulates FadA (Lee and Adams, 1994; 1996; Seo et al., 2003. Both, *fluG* and *flbA* are necessary for ST production (Calvo et al., 2002). Homologs are also present in the *A. flavus* genome (data not shown).
Further studies showed of role of the FadA-interacting proteins, SfaD and GpgA subunits, and PhnA, a positive regulator of Gβγ, as positive regulators of *aflR* and consequently ST biosynthesis (Seo and Yu, 2006).

4.3 MpkB signalling

In eukaryotes **M**itogen **A**ctivated **P**rotein (MAP) kinases play a crucial role in signal transduction to the nucleus to control differentiation and cell growth (Banuett, 1998; Bradwell, 2006; Brown et al., 1996). For instance, in *Saccharomyces cerevisiae*, the MAP kinase *FUS3* is known to regulate mating (Doi et al., 1994; Zhan et al., 1997). In *A. nidulans* and other filamentous fungi, homologs of *FUS3* have been characterized (i.e. Atoui et al., 2008; Chen et al., 2004; Cousin et al., 2006; Di Petro et al., 2001; Jenczmionka et al., 2003; Li D. et al., 2005; Mey et al., 2002; Moriwaki et al., 2007; Rauyaree et al., 2005; Ruiz-Roldman et al., 2001; Takano et al., 2000; Xu et al., 1996;). Our laboratory has shown that the *A. nidulans FUS3* homolog, *mpkB*, controls sexual development and secondary metabolism (Paoleti et al., 2007; Atoui et al., 2008). *mpkB* affects expression of *laeA*, a global regulator of secondary metabolism (See Section Global Regulation). The *A. nidulans* Δ*mpkB* strain presents decreased transcript levels of *aflR* and of ST structural genes, consequently the Δ*mpkB* mutant failed to produce ST. Production of other secondary metabolites such as the antibiotic penicillin and the anti-tumoral compound terrequinone A was also affected (Atoui et al., 2008). Since *FUS3* homologs are also present in other fungal species, its role may be conserved in the AF-producing fungi.

4.4 Metabolic channeling

Chanda et al. (2009) also showed the importance of aflatoxigenic vesicles called aflatoxisomes in aflatoxin biosynthesis, storage and transport in *A. parasiticus*. High vesicle numbers were directly linked to AF-inducing growth conditions and inversely linked to lower expression of *vb1* (homolog of *AvaA* in *A. nidulans*) and *vps16* (another protein of Class C Vps tethering complex blocked by Sortin3). The wild type strain of *A. parasiticus* showed decreased transcript levels of *vb1* and *vps16*, whereas transcripts levels were constant in the Δ*veA* strain. This suggests a direct role for VeA in aflatoxisome production by down-regulating *vb1* and *vps16*.

4.5 Chromosome remodeling

Secondary metabolism gene clusters are often located in the chromosomal near-telomeric regions (Nierman et al., 2005; Rehmeyer et al., 2006), associated with epigenetic regulation. In 2004, Roze et al. identified CRE1 protein binding sites involved in the cAMP response. CRE1 protein recruits a histone acetyl transferase to the promoters of AF genes, resulting in acetylation of histones, specifically H4, favoring transcriptional activation (Shahbazian and Grunstein, 2007; Spencer and Davie, 1999; Daniel et al., 1998). Roze et al., (2007) showed the effect of histone H4 acetylation on transcript level of "early", "middle" and "late" genes of the AF cluster that leads to aflatoxin in *A. parasiticus*.The proposed model involves the activation of cAMP/PKA pathway by a decrease of glucose sensed by G-protein couple receptors. Activation of PKA would lead to CRE1 protein binding and recruiting of the histone acetyl transferase. This is followed by histone acetylation and AF gene expression, enhanced by AflR binding to the accessible promoter regions.

The effect of histone acetylation on mycotoxin production was further supported by *A. nidulans* studies, where deletion of the histone deacetylase gene, *hdaA,* lead to an increase in ST production by activation of transcription of ST genes. Deletion of *hdhA* also increased the production of other secondary metabolites such as penicillin (Shwab et al., 2007).

5. Environmental factors

5.1 Light

The effect of light on production of AF in *A. flavus* and *A. parasiticus* has been previously described (Joffe and Lisker, 1969; Bennett and Dunn et al., 1981). In these studies light-mediated regulation of mycotoxin production was conditioned by the incubation temperature. Light regulates morphological development as well as ST production in *A. nidulans*. A strain with a *veA* wild-type allele (*veA*+) develops asexually forming conidiophores in the light, while in the dark this fungus preferentially forms fruiting bodies called cleistothecia. *A. nidulans* produces more ST in the dark than in the light when cultured on glucose minimum medium. In the dark VeA is efficiently transported to the nucleus by the importin-α KapA (Stinnett et al., 2007), leading to ST biosynthesis. Interestingly, *A. nidulans veA1* mutants, with a truncated VeA protein missing the first 36 amino acids, produce less toxin compared to the *veA*+ strains (Stinnett et al., 2007). The truncation at the VeA N-terminal in *veA1* strains results in alterations of the binding with KapA and VeA transport (Stinnett et al., 2007; Araujo-Bazan et al., 2009).

Differential VeA localization was observed using blue and red light. Blue light had a similar effect to that with white light, decreasing VeA transport to nucleus, however, efficient nuclear localization was observed under red light, a result similar to that in dark conditions (Stinnett et al., 2007). This coincides with greater ST production under red light as compared to blue light. A red phytochrome-like protein, FphA, interacts directly with VeA in the nucleus (Purschwitz et al., 2008). Also, LreA and LreB, orthologs of the *N. crassa* blue collar protein CW1 and CW2 respectively, interact with VeA indirectly via FphA. Under light FphA negatively regulates VeA nuclear transport (Purschwitz et al., 2008). FphA and LreA/B proteins have opposite effects regulating secondary metabolism. Δ*fphA* strain of *A. nidulans* produced more ST than the wild type, whereas a reduction of toxin is observed in the Δ*lreA* and Δ*lreB* mutants. Homologs of *fphA*, *lreB* and *lreA* are present in the *A. flavus* genome, suggesting that the corresponding gene products could also be involved in AF regulation (data not shown).

5.2 Carbon source

It has been shown that AF is produced from glucose derived acetyl-CoA (Buchanan and Lewis, 1984, Shantha and Murthy, 1981). Simple sugar like glucose, sucrose, fructose or sorbitol enhances AF production, while complex carbon sources like peptone, galactose, xylulose, mannitol and lactose are not conducive to AF biosynthesis (Calvo et al., 2002 and references therein). A microarray analysis conducted on *A. parasiticus* showed that AF genes were differentially expressed when comparing expression levels in YE (medium with low sugar) and YES (medium with sucrose) (Wilkinson et al., 2007). A 10-fold increase in AF was observed within 48 hours in YES medium.

Interestingly, in a recent *A. nidulans* study we showed that the concentration of glucose is predominant to the effect of light on ST production (Atoui et al., 2010). Higher levels of ST were produced in the dark at low glucose concentration (1%) with respect to those produced in light cultures, however, more ST was observed in cultures exposed to light when higher amounts of glucose were added (i.e. 2%). These results also correlate with differential *aflR* expression levels in each case (Atoui et al., 2010). These findings indicate that the response to different environmental factors, such as light and carbon source on secondary metabolism occur integratively leading to an adaptation to a complex environment.

5.3 Nitrogen source

Nitrogen metabolism affects mycotoxin production. However the effect of nitrogen varies depending on the nitrogen source utilized. Whereas organic nitrogen sources induce AF production, nitrate as the sole nitrogen source in the medium is non-conducive to AF production (Georgianna and Payne, 2009). This differs from ST production in *A. nidulans*, where ST levels increase in medium with nitrate and are reduced when ammonia is used (Feng and Leonard, 1998).

In *A. parasiticus*, Chang et al.(2000) reported a major nitrogen utilization regulatory factor called AreA. This protein has been previously well characterized in *A. nidulans* where it has been shown to be a member of the GATA family of transcription factors and mediates nitrogen metabolite repression (Wilson and Arst, 1998). Overexpression of *aflR* in *A. parasiticus* overcomes the nitrate inhibition of AF production suggesting that AreA regulatory effect on mycotoxin biosynthesis could be linked to *aflR* regulation (Chang et al., 1995). Later it was demonstrated that AreA is able to bind to the intergenic region between *aflR* and *aflJ*, where several GATA binding sites are located (Chang et al., 2000).

Analysis of the effect of nitrate on AF biosynthesis and *aflR* and *alfS* expression in different *A. flavus* strains showed variability. This could be due variation in the number of AreA binding sites present in the intergenic region between *aflR-aflJ* (Ehrlich and Cotty, 2002). In addition, Marzluf (1997) suggests that nitrate utilization requiring induction and expression of several enzymes for reduction of nitrate to ammonium, could cause a delay in nitrogen utilization for AF biosynthesis. This may also be viewed as inhibition of biosynthesis (Ehrlich and Cotty, 2002).

5.4 Temperature

The effect of temperature on ST and AF biosynthesis is not conserved. While AF is produced between 25-30°C, ST is produced at a higher temperature, 37°C. Surprisingly, Schmidt-Heydt et al. (2009) showed that *aflS* (*aflJ*) expression levels were high at 37°C, however only low amounts of AF were detected. In addition, a microarray analysis performed by OBrian et al. (2007), showed 144 genes to be differentially expressed when *A. flavus* was grown at 28°C compared to cultures at 37°C. Out of these, 103 genes were upregulated at 28°C, 25% of which are involved in secondary metabolism. In this case, not only *aflJ*(*aflS*) but also *aflR* expression levels were similar at both 28°C and 37°C.

A SILAC study by Georgianna et al. (2008) on *A. flavus* revealed differences in protein levels at 28°C and at 37°C, while RNA transcripts did not change. This could be attributed to posttranslational regulation, where enzymes might not be present or as functional at the higher temperatures. This could also be attributed to the fact that in this study the authors only analyzed the cytosolic proteins, therefore concentration levels might vary due to different subcellular location or possible membrane interactions. Liu and Chu (1998) reported differences in AflR protein levels at 29°C when compared to those at 37°C using anti-AflR in *A. parasiticus*. AflR levels were 4 times lower at 37°C when compared to protein levels at 28°C. The difference in temperature-dependent AF production is most likely not related to phosphorylation of AflR by PKA, since the AflR subcellular localization does not change at 28°C or 37°C (Georgiana and Payne, 2009).

5.5 pH

It has been reported that *Aspergillus* species produce more AF/ST at acidic pH as compared to alkaline pH (Keller et al., 1997), which in *A. nidulans* correlates with an increase in *stcU* expression in acidic environment. This may be attributed in part to the role of alkaline transcription factor PacC, a zinc finger transcription factor that binds to the consensus sequence 5'-GCCAAG-3' (Tilburn et al., 1995). Under acidic condition PacC is in its inactive form whereas under alkaline conditions, PacC is proteolytically cleaved to gain its active form. In its active form PacC activates expression of alkali expressed genes and represses acidic expressed genes by binding to the consensus sequence in the promoter region (Tilburn et al., 1995). PacC binding sites have been reported in the promoter of *aflR*, and putative binding sites have also been found on various genes in the cluster of ST/AF (Ehrlich et al., 1999; Keller et al., 1997). Putative binding sites have also been found in the promoter region of *veA* in different *Aspergilli* (Calvo et al., unpublished data), which, as mentioned above, is necessary for ST/AF production (Kato et al., 2003; Calvo et al., 2004; Duran et al., 2007).

Interestingly, PacC has antagonistic roles in controlling penicillin and mycotoxin production (Espeso et al., 1993). PacC-dependent penicillin production and expression of *ipnA*, gene encoding the isopenicillin N synthetase, increased under alkaline conditions when compared to acidic conditions.

In contrast with the reports mentioned above, a recent study by Delgado-Virgen and Guzman-de-Pena (2009) showed increased ST production in alkaline pH with respect to the levels produced at acidic pH. According to this report, *aflR* transcript levels are elevated in the "alkalinity mimicking" pacCc14 mutant. This opposite effect with respect to previous reports could be due to differences in culture conditions utilized in these two studies (for example Keller et al. used complete minimal medium while Delgado-Virgen and Guzman-de-Pena used Käfer minimal medium).

5.6 Plant metabolites

Fungal morphogenesis and secondary metabolism can be affected by plant-based compounds (Holmes et al., 2008). For example, volatile aldehydes, jasmonic acid and methyl jasmonate from the plant lipoxygenase pathway have been shown to reduce AF production (i.e. Zeringue and McCormick, 1990, Wright et al., 2000; Norton, 1999; Bhatnagar and MaCormick, 1988; Goodrich-Tanrikulu et al., 1995). Precursors of the lipoxygenase pathway called oxylipins, a family of oxidized polyenoic fatty acids, are found in different kingdoms, including plantae, fungi and monera (Tsitsgiannis and Keller, 2006). They might contribute to a communication established across kingdoms (Tsitsgiannis et al., 2004). In *A. nidulans,* oxylipin signal molecules called psi factors (Champe et al., 1987), are similar to 13S-hydroperoxy linoleic acid and 9S-hydroperoxy linolenic acid commonly found in plants (Calvo et al., 1999; Calvo et al., 2001; Tsitsigiannis et al., 2004). Both 9S-hydroperoxy linoleic acid and 13-hydroperoxy linolenic acid have been shown to differentially affect mycotoxin production; while the former has a stimulatory effect, the latter repressed its synthesis (Burow et al., 1997). Also, these fatty acids affect the balance between sclerotia/cleistothecial formation and conidiation in *A. flavus*, *A. parasiticus* and *A. nidulans* (Calvo et al., 1999). It is possible that these compounds of plant origin could mimic or interfere with the fungal psi factor regulatory mechanism. In *A. nidulans*, several oxylipin genes, *ppoA*, *ppoB* and *ppoC,* are responsible for synthesis of linoleic- and linolenic-derived psi factors. ST analysis of oxylipin deletion strains reveal that Δ*ppoB* and Δ*ppoC* strains produced more toxin than the control strain. On the other hand, Δ*ppoA* strain produced less toxin than the wild type. Δ*ppoA*, Δ*ppoC* and Δ*ppoA*, Δ*ppoB*, Δ*ppoC* double and triple mutants trains did not produce any detectable ST (Tsitsgiannis and Keller, 2006). The findings described indicate a direct correlation between presence of oxylipins and mycotoxin production.

Other compounds such as ethylene and CO_2 have been described to reduce AF production (Cary et al., 2009, and reference herein), while other molecules such as 2-ethyl-1-hexanol stimulate mycotoxin biosynthesis or affect it in a dose-dependent manner, such as the case of 2-buten-1-ol (Roze et al., 2007). Interestingly, ethylene, 2-ethyl-1-hexanol and 2-ethyl-1-hexanol have also been found to be produced by *Aspergillus* species, such as *A. parasiticus* and *A. nidulans,*where they could play a role as signal molecules (Roze et al., 2004; Roze et al., 2007).

5.7 Nitric oxide

Other gaseous molecules have been described as signaling molecules in higher eukaryotes. Plamer et al. (1987) and Ignarro et al. (1987) showed that the endothelium-dependent relaxing factor (EDRF) was indeed nitric oxide (NO), produced from L-Arginine (Palmer et al., 1988). Hausladen and Gow et al. (1998) identified a flabohemoglobin 'denitrosolase' in *E. coli* able to metabolize NO to nitrate. Thus flabohemoglobins were showed to decrease nitrosative stress by detoxifying NO. At low levels, NO is an important signaling molecule controlling numerous cell functions in higher eukaryotes (Wendehenne et al., 2001, Romanov et al., 2008; Roselli et al., 1998; Masuda et al., 2001; Kim et al., 2007; Lamattina et al., 2003). However,until recently there were no studies of NO as a signaling molecule in fungi. In *A. nidulans* two flabohemoglobins, FhbA and FhbB, have been reported (Schinko et al., 2010). A recent study in our laboratory (Baidya et al., 2011) showed a link between NO and morphological development and ST production in *A. nidulans*. This was the first report of the effect of NO in morphogenesis and secondary metabolism in the fungal kingdom (Baidya et al., 2011). Deletion of *fhbA* results in an increase of sexual development and decrease of ST biosynthesis (Baidya et al., 2011). The reduction of ST was linked to a decrease of *aflR* expression. Homologs of *fhbA* and *fhbB* are present in the *A. flavus* genome (unpublished data). The implications of this study suggest a broader possible regulatory effect of NO that might also impact AF production as well as the production of other fungal metabolites.

6. Conclusion

AF, commonly produced by *A. flavus* and *A. parasiticus*, is a potent mutagenic and carcinogenic natural compound that constitutes a health threat worldwide. To set the basis for the development of efficient control strategies to prevent or reduce the negative impact associated with AF contamination it is essential to uncover the regulatory networks that lead to mycotoxin biosynthesis in these fungi. In this review we have shown that studies in the model filamentous fungus *A. nidulans* have facilitated rapid progress in this field, revealing genetic mechanisms governing the production of the mycotoxin ST that are also found conserved in the regulation and production of AF by *A. flavus* or *A. parasiticus*. This, combined with the availability of new technologies leading to the genome sequencing of *A. nidulans, A. flavus* and other filamentous fungi, has further expanded our knowledge on AF/ST regulation and biosynthesis.

The elucidation of the genetic mechanisms that operate in response to different environmental factors is of particular importance, since some of these responses are adaptations that allow these fungi to colonize plants leading to AF contamination of crops. Part of these response mechanisms to environmental cues involves the participation of proteins and formation of protein interactions that are unique to fungi, for example VeA and

VeA-interacting proteins, initially characterized in *A. nidulans*. For these reasons these protein complexes constitute potential targets for control strategies to reduce the production of AF and possibly mycotoxin production in other fungi.

Although there are some disparities in the regulation of ST and AF, most of the control mechanisms governing the synthesis of these mycotoxins are conserved. *A. nidulans* is an excellent model organism to elucidate the complexity of these regulatory networks and provide insight in remediating the impact of AF contamination.

7. References

Adrio, J. L., and A. L. Demain. 2003. Fungal biotechnology. *International Microbiology* 6 (3):191-199.

Amaike, S., and N. P. Keller. 2009. Distinct Roles for VeA and LaeA in Development and Pathogenesis of Aspergillus flavus. *Eukaryotic Cell* 8 (7):1051-1060.

Andrianopoulos, A., S. Kourambas, J. A. Sharp, M. A. Davis, and M. J. Hynes. 1998. Characterization of the Aspergillus nidulans nmrA gene involved in nitrogen metabolite repression. *Journal of Bacteriology* 180 (7):1973-1977.

Araujo-Bazan, L., S. Dhingra, J. Chu, J. Fernandez-Martinez, A. M. Calvo, and E. A. Espeso. 2009. Importin alpha is an essential nuclear import carrier adaptor required for proper sexual and asexual development and secondary metabolism in Aspergillus nidulans. *Fungal Genetics and Biology* 46 (6-7):506-515.

Arst, H. N., and D. J. Cove. 1973. Nitrogen metabolite repression in Aspergillus-nidulans. *Molecular & General Genetics* 126 (2):111-141.

Atoui, A., D. P. Bao, N. Kaur, W. S. Grayburn, and A. M. Calvo. 2008. Aspergillus nidulans natural product biosynthesis is regulated by mpkB, a putative pheromone response mitogen-activated protein kinase. *Applied and Environmental Microbiology* 74 (11):3596-3600.

Atoui, A., C. Kastner, C. M. Larey, R. Thokala, O. Etxebeste, E. A. Espeso, R. Fischer, and A. M. Calvo. Cross-talk between light and glucose regulation controls toxin production and morphogenesis in Aspergillus nidulans. *Fungal Genetics and Biology* 47 (12):962-972.

Baidya, S., J. W. Cary, W. S. Grayburn, and A. M. Calvo. 2011. Role of nitric oxide and flavohemoglobin homologous genes in *Aspergillus nidulans* sexual development and mycotoxin production. *Applied Environmental Microbiology,* June 3 [Epub ahead of print]

Banuett, F. 1998. Signalling in the yeasts: An informational cascade with links to the filamentous fungi. *Microbiology and Molecular Biology Reviews* 62 (2):249-+.

Bardwell, L. 2006. Mechanisms of MAPK signalling specificity. *Biochemical Society Transactions* 34:837-841.

Bayram, O., S. Krappmann, M. Ni, J. W. Bok, K. Helmstaedt, O. Valerius, S. Braus-Stromeyer, N. J. Kwon, N. P. Keller, J. H. Yu, and G. H. Braus. 2008. VelB/VeA/LaeA complex coordinates light signal with fungal development and secondary metabolism. *Science* 320 (5882):1504-1506.

Bayram, O., S. Krappmann, S. Seiler, N. Vogt, and G. H. Braus. 2008. Neurospora crassa ve-1 affects asexual conidiation. *Fungal Genetics and Biology* 45 (2):127-138.

Bayram, O. S., O. Bayram, O. Valerius, H. S. Park, S. Irniger, J. Gerke, M. Ni, K. H. Han, J. H. Yu, and G. H. Braus. LaeA Control of Velvet Family Regulatory Proteins for Light-Dependent Development and Fungal Cell-Type Specificity. *Plos Genetics* 6 (12).

Bennett, J. W., J. J. Dunn, and C. I. Goldsman. 1981. Influence of white-light on production of aflatoxins and anthraquinones in Aspergillus-parasiticus. *Applied and Environmental Microbiology* 41 (2):488-491.

Bennett, J. W., and M. Klich. 2003. Mycotoxins. *Clinical Microbiology Reviews* 16 (3):497-+.

Bhatnagar, D., K. C. Ehrlich, and T. E. Cleveland. 2003. Molecular genetic analysis and regulation of aflatoxin biosynthesis. *Applied Microbiology and Biotechnology* 61 (2):83-93.

Bhatnagar, D., and S. P. McCormick. 1988. The inhibitory effect of neem (azadirachta-indica) leaf extracts on aflatoxin synthesis in Aspergillus-parasiticus. *Journal of the American Oil Chemists Society* 65 (7):1166-1168.

Bok, J. W., and N. P. Keller. 2004. LaeA, a regulator of secondary metabolism in Aspergillus spp. *Eukaryotic Cell* 3 (2):527-535.

Bradshaw, R. E., and S. G. Zhang. 2006. Biosynthesis of dothistromin. *Mycopathologia* 162 (3):201-213.

Brakhage, A. A., and V. Schroeckh. Fungal secondary metabolites - Strategies to activate silent gene clusters. *Fungal Genetics and Biology* 48 (1):15-22.

Brown, D. W., J. H. Yu, H. S. Kelkar, M. Fernandes, T. C. Nesbitt, N. P. Keller, T. H. Adams, and T. J. Leonard. 1996. Twenty-five coregulated transcripts define a sterigmatocystin gene cluster in Aspergillus nidulans. *Proceedings of the National Academy of Sciences of the United States of America* 93 (4):1418-1422.

Brown, M. P., C. S. Brown-Jenco, and G. A. Payne. 1999. Genetic and molecular analysis of aflatoxin biosynthesis. *Fungal Genetics and Biology* 26 (2):81-98.

Buchanan, R. L., and D. F. Lewis. 1984. Regulation of aflatoxin biosynthesis - effect of glucose on activities of various glycolytic-enzymes. *Applied and Environmental Microbiology* 48 (2):306-310.

Burow, G. B., T. C. Nesbitt, J. Dunlap, and N. P. Keller. 1997. Seed lipoxygenase products modulate Aspergillus mycotoxin biosynthesis. *Molecular Plant-Microbe Interactions* 10 (3):380-387.

Calvo, A. M. 2008. The VeA regulatory system and its role in morphological and chemical development in fungi. *Fungal Genetics and Biology* 45 (7):1053-1061.

Calvo, A. M., J. Bok, W. Brooks, and N. P. Keller. 2004. veA is required for toxin and sclerotial production in Aspergillus parasiticus. *Applied and Environmental Microbiology* 70 (8):4733-4739.

Calvo, A. M., H. W. Gardner, and N. P. Keller. 2001. Genetic connection between fatty acid metabolism and sporulation in Aspergillus nidulans. *Journal of Biological Chemistry* 276 (28):25766-25774.

Calvo, A. M., L. L. Hinze, H. W. Gardner, and N. P. Keller. 1999. Sporogenic effect of polyunsaturated fatty acids on development of Aspergillus spp. *Applied and Environmental Microbiology* 65 (8):3668-3673.

Calvo, A. M., R. A. Wilson, J. W. Bok, and N. P. Keller. 2002. Relationship between secondary metabolism and fungal development. *Microbiology and Molecular Biology Reviews* 66 (3):447-+.

Carbone, I., J. H. Ramirez-Prado, J. L. Jakobek, and B. W. Horn. 2007. Gene duplication, modularity and adaptation in the evolution of the aflatoxin gene cluster. *Bmc Evolutionary Biology* 7.

Cary, J. W., and K. C. Ehrlich. 2006. Aflatoxigenicity in Aspergillus: molecular genetics, phylogenetic relationships and evolutionary implications. *Mycopathologia* 162 (3):167-177.

Cary, J. W., B. G. Montalbano, and K. C. Ehrlich. 2000. Promoter elements involved in the expression of the Aspergillus parasiticus aflatoxin biosynthesis pathway gene avnA. *Biochimica Et Biophysica Acta-Gene Structure and Expression* 1491 (1-3):7-12.

Cary, J. W., G. R. Obrian, D. M. Nielsen, W. Nierman, P. Harris-Coward, J. Y. D. Bhatnagar, T. E. Cleveland, G. A. Payne, and A. M. Calvo. 2007. Elucidation of veA-dependent genes associated with aflatoxin and sclerotial production in Aspergillus flavus by functional genomics. *Applied Microbiology and Biotechnology* 76 (5):1107-1118.

Cary, J. W., L. Szerszen, and A. M. Calvo. 2009. Regulation of Aspergillus flavus Aflatoxin Biosynthesis and Development. In *Mycotoxin Prevention and Control in Agriculture.*

Champe, S. P., P. Rao, and A. Chang. 1987. An endogenous inducer of sexual development in Aspergillus-nidulans. *Journal of General Microbiology* 133:1383-1387.

Chanda, A., L. V. Roze, S. Kang, K. A. Artymovich, G. R. Hicks, N. V. Raikhel, A. M. Calvo, and J. E. Linz. 2009. A key role for vesicles in fungal secondary metabolism. *Proceedings of the National Academy of Sciences of the United States of America* 106 (46):19533-19538.

Chang, P. K. 2003. The Aspergillus parasiticus protein AFLJ interacts with the aflatoxin pathway-specific regulator AFLR. *Molecular Genetics and Genomics* 268 (6):711-719.

Chang, P. K. 2004. Lack of interaction between AFLR and AFLJ contributes to nonaflatoxigenicity of Aspergillus sojae. *Journal of Biotechnology* 107 (3):245-253.

Chang, P. K., J. W. Bennett, and P. J. Cotty. 2002. Association of aflatoxin biosynthesis and sclerotial development in Aspergillus parasiticus. *Mycopathologia* 153 (1):41-48.

Chang, P. K., J. W. Cary, D. Bhatnagar, T. E. Cleveland, J. W. Bennett, J. E. Linz, C. P. Woloshuk, and G. A. Payne. 1993. Cloning of the Aspergillus-parasiticus apa-2 gene associated with the regulation of aflatoxin biosynthesis. *Applied and Environmental Microbiology* 59 (10):3273-3279.

Chang, P. K., K. C. Ehrlich, J. J. Yu, D. Bhatnagar, and T. E. Cleveland. 1995. Increased expression of Aspergillus-parasiticus aflR, encoding a sequence-specific dna-binding protein, relieves nitrate inhibition of aflatoxin biosynthesis. *Applied and Environmental Microbiology* 61 (6):2372-2377.

Chang, P. K., K. Matsushima, T. Takahashi, J. J. Yu, K. Abe, D. Bhatnagar, G. F. Yuan, Y. Koyama, and T. E. Cleveland. 2007. Understanding nonaflatoxigenicity of Aspergillus sojae: a windfall of aflatoxin biosynthesis research. *Applied Microbiology and Biotechnology* 76:977-984.

Chang, P. K., J. J. Yu, D. Bhatnagar, and T. E. Cleveland. 2000. Characterization of the Aspergillus parasiticus major nitrogen regulatory gene, areA. *Biochimica Et Biophysica Acta-Gene Structure and Expression* 1491 (1-3):263-266.

Chen, C. B., A. Harel, R. Gorovoits, O. Yarden, and M. B. Dickman. 2004. MAPK regulation of sclerotial development in Sclerotinia sclerotiorum is linked with pH and cAMP sensing. *Molecular Plant-Microbe Interactions* 17 (4):404-413.

Chiang, Y. M., B. R. Oakley, N. P. Keller, and C. C. C. Wang. Unraveling polyketide synthesis in members of the genus Aspergillus. *Applied Microbiology and Biotechnology* 86 (6):1719-1736.

Cousin, A., R. Mehrabi, M. Guilleroux, M. Dufresne, T. Van der Lee, C. Waalwijk, T. Langin, and G. H. J. Kema. 2006. The MAP kinase-encoding gene MgFus3 of the non-appressorium phytopathogen Mycosphaerella graminicola is required for penetration and in vitro pycnidia formation. *Molecular Plant Pathology* 7 (4):269-278.

Daniel, P. B., W. H. Walker, and J. F. Habener. 1998. Cyclic AMP signaling and gene regulation. *Annual Review of Nutrition* 18:353-383.

David, H., I. S. Ozcelik, G. Hofmann, and J. Nielsen. 2008. Analysis of Aspergillus nidulans metabolism at the genome-scale. *Bmc Genomics* 9.

Davis, N. D., U. L. Diener, and Agnihotr.Vp. 1967. Production of aflatoxins b1 and g1 in chemically defined medium. *Mycopathologia Et Mycologia Applicata* 31 (3-4):251-&.

Delgado-Virgen, F., and D. Guzman-de-Pena. 2009. Mechanism of Sterigmatocystin Biosynthesis Regulation by pH in Aspergillus Nidulans. *Brazilian Journal of Microbiology* 40 (4):933-942.

Di Pietro, A., F. I. Garcia-Maceira, E. Meglecz, and M. I. G. Roncero. 2001. A MAP kinase of the vascular wilt fungus Fusarium oxysporum is essential for root penetration and pathogenesis. *Molecular Microbiology* 39 (5):1140-1152.

Doi, K., A. Gartner, G. Ammerer, B. Errede, H. Shinkawa, K. Sugimoto, and K. Matsumoto. 1994. Msg5, a novel protein phosphatase promotes adaptation to pheromone response in Saccharomyces-cerevisiae. *Embo Journal* 13 (1):61-70.

Dreyer, J., H. Eichhorn, E. Friedlin, H. Kurnsteiner, and U. Kuck. 2007. A homologue of the Aspergillus velvet gene regulates both cephalosporin C biosynthesis and hyphal fragmentation in Acremonium chrysogenum. *Applied and Environmental Microbiology* 73 (10):3412-3422.

Duran, R., J. W. Cary, and A. M. Calvo. 2009. The role of veA on Aspergillus flavus infection of peanuts, corn and cotton. *Open Mycology Journal* 3:27-36.

Duran, R. M., J. W. Cary, and A. M. Calvo. 2007. Production of cyclopiazonic acid, aflatrem, and aflatoxin by Aspergillus flavus is regulated by veA, a gene necessary for sclerotial formation. *Applied Microbiology and Biotechnology* 73:1158-1168.

Dvorackova, I., and V. Kusak. 1990. Hepatocellular carcinoma a 28-year necropsy review. *Journal of Environmental Pathology Toxicology and Oncology* 10 (4-5):220-224.

Ehrlich, K. C., J. W. Cary, and B. G. Montalbano. 1999. Characterization of the promoter for the gene encoding the aflatoxin biosynthetic pathway regulatory protein AFLR. *Biochimica Et Biophysica Acta-Gene Structure and Expression* 1444 (3):412-417.

Ehrlich, K. C., and P. J. Cotty. 2002. Variability in nitrogen regulation of aflatoxin production by Aspergillus flavus strains. *Applied Microbiology and Biotechnology* 60 (1-2):174-178.

Ehrlich, K. C., J. Yu, and P. J. Cotty. 2005. Aflatoxin biosynthesis gene clusters and flanking regions. *Journal of Applied Microbiology* 99 (3):518-527.

Espeso, E. A., J. Tilburn, H. N. Arst, and M. A. Penalva. 1993. pH regulation is a major determinant in expression of a fungal penicillin biosynthetic gene. *Embo Journal* 12 (10):3947-3956.

Feng, G. H., and T. J. Leonard. 1998. Culture conditions control expression of the genes for aflatoxin and sterigmatocystin biosynthesis in Aspergillus parasiticus and A-nidulans. *Applied and Environmental Microbiology* 64 (6):2275-2277.

Fernandes, M., N. P. Keller, and T. H. Adams. 1998. Sequence-specific binding by Aspergillus nidulans AflR, a C-6 zinc cluster protein regulating mycotoxin biosynthesis. *Molecular Microbiology* 28 (6):1355-1365.

Flaherty, J. E., and G. A. Payne. 1997. Overexpression of aflR leads to upregulation of pathway gene transcription and increased aflatoxin production in Aspergillus flavus. *Applied and Environmental Microbiology* 63 (10):3995-4000.

Galagan, J. E., S. E. Calvo, C. Cuomo, L. J. Ma, J. R. Wortman, S. Batzoglou, S. I. Lee, M. Basturkmen, C. C. Spevak, J. Clutterbuck, V. Kapitonov, J. Jurka, C. Scazzocchio, M. Farman, J. Butler, S. Purcell, S. Harris, G. H. Braus, O. Draht, S. Busch, C. D'Enfert, C. Bouchier, G. H. Goldman, D. Bell-Pedersen, S. Griffiths-Jones, J. H. Doonan, J. Yu, K. Vienken, A. Pain, M. Freitag, E. U. Selker, D. B. Archer, M. A. Penalva, B. R. Oakley, M. Momany, T. Tanaka, T. Kumagai, K. Asai, M. Machida, W. C. Nierman, D. W. Denning, M. Caddick, M. Hynes, M. Paoletti, R. Fischer, B. Miller, P. Dyer, M. S. Sachs, S. A. Osmani, and B. W. Birren. 2005. Sequencing of Aspergillus nidulans and comparative analysis with A. fumigatus and A. oryzae. *Nature* 438 (7071):1105-1115.

Georgianna, D. R., A. M. Hawkridge, D. C. Muddiman, and G. A. Payne. 2008. Temperature-dependent regulation of proteins in Aspergillus flavus: Whole organism stable isotope labeling by amino acids. *Journal of Proteome Research* 7 (7):2973-2979.

Georgianna, D. R., and G. A. Payne. 2009. Genetic regulation of aflatoxin biosynthesis: From gene to genome. *Fungal Genetics and Biology* 46 (2):113-125.

Gidijala, L., Jakw Kiel, R. A. L. Bovenberg, I. J. Van Der Klei, and M. A. Van Den Berg. Biosynthesis of active pharmaceuticals: beta-lactam biosynthesis in filamentous fungi. In *Biotechnology and Genetic Engineering Reviews, Vol 27*. Loughborough: Nottingham University Press.

Goodrich-Tanrikulu, M., K. Howe, A. Stafford, and M. A. Nelson. 1998. Changes in fatty acid composition of Neurospora crassa accompany sexual development and ascospore germination. *Microbiology-Sgm* 144:1713-1720.

Han, K. H., J. A. Seo, and J. H. Yu. 2004. A putative G protein-coupled receptor negatively controls sexual development in Aspergillus nidulans. *Molecular Microbiology* 51 (5):1333-1345.

Hausladen, A., A. J. Gow, and J. S. Stamler. 1998. Nitrosative stress: Metabolic pathway involving the flavohemoglobin. *Proceedings of the National Academy of Sciences of the United States of America* 95 (24):14100-14105.

Hicks, Julie K., Kiminori Shimizu, and Nancy P. Keller. 2002. Genetics and biosynthesis of aflatoxins and sterigmatocystin. *The Mycota: A comprehensive treatise on fungi as experimental systems for basic and applied research. Agricultural applications*:55-69.

Hicks, J. K., J. H. Yu, N. P. Keller, and T. H. Adams. 1997. Aspergillus sporulation and mycotoxin production both require inactivation of the FadA G alpha protein-dependent signaling pathway. *Embo Journal* 16 (16):4916-4923.

Holmes, R. A., R. S. Boston, and G. A. Payne. 2008. Diverse inhibitors of aflatoxin biosynthesis. *Applied Microbiology and Biotechnology* 78 (4):559-572.

Hsieh, D. P. H., and Mateles. 1970. Relative contribution of acetate and glucose to aflatoxin biosynthesis. *Biochimica Et Biophysica Acta* 208 (3):482-&.

Huffman, J., R. Gerber, and L. C. Du. Recent Advancements in the Biosynthetic Mechanisms for Polyketide-Derived Mycotoxins. *Biopolymers* 93 (9):764-776.

Hynes, M. J. 1975. Studies on role of area gene in regulation of nitrogen catabolism in Aspergillus-nidulans. *Australian Journal of Biological Sciences* 28 (3):301-313.

Ignarro, L. J., G. M. Buga, K. S. Wood, R. E. Byrns, and G. Chaudhuri. 1987. Endothelium-derived relaxing factor produced and released from artery and vein is nitric-oxide. *Proceedings of the National Academy of Sciences of the United States of America* 84 (24):9265-9269.

Jenczmionka, N. J., F. J. Maier, A. P. Losch, and W. Schafer. 2003. Mating, conidiation and pathogenicity of Fusarium graminearum, the main causal agent of the head-blight disease of wheat, are regulated by the MAP kinase gpmk1. *Current Genetics* 43 (2):87-95.

Joffe, A. Z., and N. Lisker. 1969. Effects of light, temperature, and ph value on aflatoxin production in vitro. *Applied Microbiology* 18 (3):517-&.

Kachholz, T., and A. L. Demain. 1983. Nitrate repression of averufin and aflatoxin biosynthesis. *Journal of Natural Products* 46 (4):499-506.

Kale, S. P., L. Milde, M. K. Trapp, J. C. Frisvad, N. P. Keller, and J. W. Bok. 2008. Requirement of LaeA for secondary metabolism and sclerotial production in Aspergillus flavus. *Fungal Genetics and Biology* 45 (10):1422-1429.

Kato, N., W. Brooks, and A. M. Calvo. 2003. The expression of sterigmatocystin and penicillin genes in Aspergillus nidulans is controlled by veA, a gene required for sexual development. *Eukaryotic Cell* 2 (6):1178-1186.

Keller, N. P., C. Nesbitt, B. Sarr, T. D. Phillips, and G. B. Burow. 1997. pH regulation of sterigmatocystin and aflatoxin biosynthesis in Aspergillus spp. *Phytopathology* 87 (6):643-648.

Keller, N. P., G. Turner, and J. W. Bennett. 2005. Fungal secondary metabolism - From biochemistry to genomics. *Nature Reviews Microbiology* 3 (12):937-947.

Kim, H. C., J. S. Byun, T. K. Lee, C. W. Jeong, M. Ahn, and T. Shin. 2007. Expression of nitric oxide synthase isoforms in the testes of pigs. *Anatomia Histologia Embryologia-Journal of Veterinary Medicine Series C* 36 (2):135-138.

Kim, H. S., K. Y. Han, K. J. Kim, D. M. Han, K. Y. Jahng, and K. S. Chae. 2002. The veA gene activates sexual development in Aspergillus nidulans. *Fungal Genetics and Biology* 37 (1):72-80.

Lafon, A., K. H. Han, J. A. Seo, J. H. Yu, and C. d'Enfert. 2006. G-protein and cAMP-mediated signaling in aspergilli: A genomic perspective. *Fungal Genetics and Biology* 43 (7):490-502.

Lamattina, L., C. Garcia-Mata, M. Graziano, and G. Pagnussat. 2003. Nitric oxide: The versatility of an extensive signal molecule. *Annual Review of Plant Biology* 54:109-136.

Lamb, H. K., J. S. Ren, A. Park, C. Johnson, K. Leslie, S. Cocklin, P. Thompson, C. Mee, A. Cooper, D. K. Stammers, and A. R. Hawkins. 2004. Modulation of the ligand binding properties of the transcription repressor NmrA by GATA-containing DNA and site-directed mutagenesis. *Protein Science* 13 (12):3127-3138.

Lee, B. N., and T. H. Adams. 1994. Overexpression of flbA, an early regulator of aspergillus asexual sporulation, leads to activation of brlA and premature initiation of development. *Molecular Microbiology* 14 (2):323-334.

Lee, B. N., and T. H. Adams. 1996. fluG and flbA function interdependently to initiate conidiophore development in Aspergillus nidulans through brlA beta activation. *Embo Journal* 15 (2):299-309.

Li, D., P. Bobrowicz, H. H. Wilkinson, and D. J. Ebbole. 2005. A mitogen-activated protein kinase pathway essential for mating and contributing to vegetative growth in Neurospora crassa. *Genetics* 170 (3):1091-1104.

Li, S. J., K. Myung, D. Guse, B. Donkin, R. H. Proctor, W. S. Grayburn, and A. M. Calvo. 2006. FvVE1 regulates filamentous growth, the ratio of microconidia to macroconidia and cell wall formation in Fusarium verticillioides. *Molecular Microbiology* 62 (5):1418-1432.

Liu, B. H., and F. S. Chu. 1998. Regulation of aflR and its product, AflR, associated with aflatoxin biosynthesis. *Applied and Environmental Microbiology* 64 (10):3718-3723.

Marzluf, G. A. 1997. Genetic regulation of nitrogen metabolism in the fungi. *Microbiology and Molecular Biology Reviews* 61 (1):17-&.

Masuda, M., T. Kubota, and T. Aso. 2001. Effects of nitric oxide on steroidogenesis in porcine granulosa cells during different stages of follicular development. *European Journal of Endocrinology* 144 (3):303-308.

Mey, G., K. Held, J. Scheffer, K. B. Tenberge, and P. Tudzynski. 2002. CPMK2, an SLT2-homologous mitogen-activated protein (MAP) kinase, is essential for pathogenesis of Claviceps purpurea on rye: evidence for a second conserved pathogenesis-related MAP kinase cascade in phytopathogenic fungi. *Molecular Microbiology* 46 (2):305-318.

Meyers, D. M., G. Obrian, W. L. Du, D. Bhatnagar, and G. A. Payne. 1998. Characterization of aflJ, a gene required for conversion of pathway intermediates to aflatoxin. *Applied and Environmental Microbiology* 64 (10):3713-3717.

Minto, R. E., and C. A. Townsend. 1997. Enzymology and molecular biology of aflatoxin biosynthesis. *Chemical Reviews* 97 (7):2537-2555.

Monahan, B. J., M. C. Askin, M. J. Hynes, and M. A. Davis. 2006. Differential expression of Aspergillus nidulans ammonium permease genes is regulated by GATA transcription factor AreA. *Eukaryotic Cell* 5 (2):226-237.

Moriwaki, A., J. Kihara, C. Mori, and S. Arase. 2007. A MAP kinase gene, BMK1, is required for conidiation and pathogenicity in the rice leaf spot pathogen Bipolaris oryzae. *Microbiological Research* 162 (2):108-114.

Morozov, I. Y., M. Galbis-Martinez, M. G. Jones, and M. X. Caddick. 2001. Characterization of nitrogen metabolite signalling in Aspergillus via the regulated degradation of areA mRNA. *Molecular Microbiology* 42 (1):269-277.

Morozov, I. Y., M. G. Martinez, M. G. Jones, and M. X. Caddick. 2000. A defined sequence within the 3 ' UTR of the areA transcript is sufficient to mediate nitrogen metabolite signalling via accelerated deadenylation. *Molecular Microbiology* 37 (5):1248-1257.

Myung, K., S. J. Li, R. A. E. Butchko, M. Busman, R. H. Proctor, H. K. Abbas, and A. M. Calvo. 2009. FvVE1 Regulates Biosynthesis of the Mycotoxins Fumonisins and Fusarins in Fusarium verticillioides. *Journal of Agricultural and Food Chemistry* 57 (11):5089-5094.

Myung, K., N. C. Zitomer, M. Duvall, A. E. Glenn, R. T. Riley and A. M. Calvo. The conserved global regulator VeA is necessary for symptom production and mycotoxin synthesis in maize seedlings by *Fusarium verticillioides. Plant Pathology.* article published online: 21 JUL 2011. DOI: 10.1111/j.1365-3059.2011.02504.x"

Ni, M., and J. H. Yu. 2007. A Novel Regulator Couples Sporogenesis and Trehalose Biogenesis in Aspergillus nidulans. *Plos One* 2 (10).

Nielsen, K. F., J. M. Mogensen, M. Johansen, T. O. Larsen, and J. C. Frisvad. 2009. Review of secondary metabolites and mycotoxins from the Aspergillus niger group. *Analytical and Bioanalytical Chemistry* 395 (5):1225-1242.

Nierman, W. C., A. Pain, M. J. Anderson, J. R. Wortman, H. S. Kim, J. Arroyo, M. Berriman, K. Abe, D. B. Archer, C. Bermejo, J. Bennett, P. Bowyer, D. Chen, M. Collins, R. Coulsen, R. Davies, P. S. Dyer, M. Farman, N. Fedorova, N. Fedorova, T. V. Feldblyum, R. Fischer, N. Fosker, A. Fraser, J. L. Garcia, M. J. Garcia, A. Goble, G. H. Goldman, K. Gomi, S. Griffith-Jones, R. Gwilliam, B. Haas, H. Haas, D. Harris, H. Horiuchi, J. Huang, S. Humphray, J. Jimenez, N. Keller, H. Khouri, K. Kitamoto, T. Kobayashi, S. Konzack, R. Kulkarni, T. Kumagai, A. Lafton, J. P. Latge, W. X. Li, A. Lord, W. H. Majoros, G. S. May, B. L. Miller, Y. Mohamoud, M. Molina, M. Monod, I. Mouyna, S. Mulligan, L. Murphy, S. O'Neil, I. Paulsen, M. A. Penalva, M. Pertea, C. Price, B. L. Pritchard, M. A. Quail, E. Rabbinowitsch, N. Rawlins, M. A. Rajandream, U. Reichard, H. Renauld, G. D. Robson, S. R. de Cordoba, J. M. Rodriguez-Pena, C. M. Ronning, S. Rutter, S. L. Salzberg, M. Sanchez, J. C. Sanchez-Ferrero, D. Saunders, K. Seeger, R. Squares, S. Squares, M. Takeuchi, F. Tekaia, G. Turner, C. R. V. de Aldana, J. Weidman, O. White, J. Woodward, J. H. Yu, C. Fraser, J. E. Galagan, K. Asai, M. Machida, N. Hall, B. Barrell, and D. W. Denning. 2005. Genomic sequence of the pathogenic and allergenic filamentous fungus Aspergillus fumigatus. *Nature* 438 (7071):1151-1156.

Norton, R. A. 1999. Inhibition of aflatoxin B-1 biosynthesis in Aspergillus flavus by anthocyanidins and related flavonoids. *Journal of Agricultural and Food Chemistry* 47 (3):1230-1235.

Obrian, G. R., D. R. Georgianna, J. R. Wilkinson, J. Yu, H. K. Abbas, D. Bhatnagar, T. E. Cleveland, W. Nierman, and G. A. Payne. 2007. The effect of elevated temperature on gene transcription and aflatoxin biosynthesis. *Mycologia* 99 (2):232-239.

Palmer, R. M. J., D. S. Ashton, and S. Moncada. 1988. Vascular endothelial-cells synthesize nitric-oxide from l-arginine. *Nature* 333 (6174):664-666.

Palmer, R. M. J., A. G. Ferrige, and S. Moncada. 1987. Nitric-oxide release accounts for the biological-activity of endothelium-derived relaxing factor. *Nature* 327 (6122):524-526.

Paoletti, M., F. A. Seymour, M. J. C. Alcocer, N. Kaur, A. M. Caivo, D. B. Archer, and P. S. Dyer. 2007. Mating type and the genetic basis of self-fertility in the model fungus Aspergillus nidulans. *Current Biology* 17 (16):1384-1389.

Payne, G. A., and M. P. Brown. 1998. Genetics and physiology of aflatoxin biosynthesis. *Annual Review of Phytopathology* 36:329-362.

Perrin, R. M., N. D. Fedorova, J. W. Bok, R. A. Cramer, J. R. Wortman, H. S. Kim, W. C. Nierman, and N. P. Keller. 2007. Transcriptional regulation of chemical diversity in Aspergillus fumigatus by LaeA. *Plos Pathogens* 3 (4):508-517.

Pontecorvo, G., J. A. Roper, L. M. Hemmons, K. D. Macdonald, and A. W. J. Bufton. 1953. The genetics of Aspergillus-nidulans. *Advances in Genetics Incorporating Molecular Genetic Medicine* 5:141-238.

Price, M. S., J. J. Yu, W. C. Nierman, H. S. Kim, B. Pritchard, C. A. Jacobus, D. Bhatnagar, T. E. Cleveland, and G. A. Payne. 2006. The aflatoxin pathway regulator AflR induces gene transcription inside and outside of the aflatoxin biosynthetic cluster. *Fems Microbiology Letters* 255 (2):275-279.

Purschwitz, J., S. Mueller, C. Kastner, M. Schoser, H. Haas, E. A. Espeso, A. Atoui, A. M. Calvo, and R. Fischer. 2008. Functional and physical interaction of blue- and red-light sensors in Aspergillus nidulans. *Current Biology* 18 (4):255-259.

Rauyaree, P., M. D. Ospina-Giraldo, S. Kang, R. G. Bhat, K. V. Subbarao, S. J. Grant, and K. F. Dobinson. 2005. Mutations in VMK1, a mitogen-activated protein kinase gene, affect microsclerotia formation and pathogenicity in Verticillium dahliae. *Current Genetics* 48 (2):109-116.

Rehmeyer, C., W. X. Li, M. Kusaba, Y. S. Kim, D. Brown, C. Staben, R. Dean, and M. Farman. 2006. Organization of chromosome ends in the rice blast fungus, Magnaporthe oryzae. *Nucleic Acids Research* 34 (17):4685-4701.

Reverberi, M., A. Ricelli, S. Zjalic, A. A. Fabbri, and C. Fanelli. Natural functions of mycotoxins and control of their biosynthesis in fungi. *Applied Microbiology and Biotechnology* 87 (3):899-911.

Richard, J.L., and G.A. Payne. 2003. Mycotoxins: Risks in Plant and Animal Systems. *Council for Agricultural Science and Technology.*

Romanov, G. A., S. N. Lomin, N. Y. Rakova, A. Heyl, and T. Schmulling. 2008. Does NO play a role in cytokinin signal transduction? *Febs Letters* 582 (6):874-880.

Rosen, S., J. H. Yu, and T. H. Adams. 1999. The Aspergillus nidulans sfaD gene encodes a G protein beta subunit that is required for normal growth and repression of sporulation. *Embo Journal* 18 (20):5592-5600.

Rosselli, M., P. J. Keller, and R. K. Dubey. 1998. Role of nitric oxide in the biology, physiology and pathophysiology of reproduction. *Human Reproduction Update* 4 (1):3-24.

Roze, L. V., A. E. Arthur, S. Y. Hong, A. Chanda, and J. E. Linz. 2007. The initiation and pattern of spread of histone H4 acetylation parallel the order of transcriptional activation of genes in the aflatoxin cluster. *Molecular Microbiology* 66 (3):713-726.

Roze, L. V., R. M. Beaudry, A. E. Arthur, A. M. Calvo, and J. E. Linz. 2007. Aspergillus volatiles regulate aflatoxin synthesis and asexual sporulation in Aspergillus parasiticus. *Applied and Environmental Microbiology* 73 (22):7268-7276.

Roze, L. V., R. M. Beaudry, N. P. Keller, and J. E. Linz. 2004. Regulation of aflatoxin synthesis by FadA/cAMP/protein kinase A signaling in Aspergillus parasiticus. *Mycopathologia* 158 (2):219-232.

Roze, L. V., A. M. Calvo, A. Gunterus, R. Beaudry, M. Kall, and J. E. Linz. 2004. Ethylene modulates development and toxin biosynthesis in Aspergillus possibly via an ethylene sensor-mediated signaling pathway. *Journal of Food Protection* 67 (3):438-447.

Roze, L. V., M. J. Miller, M. Rarick, N. Mahanti, and J. E. Linz. 2004. A novel cAMP-response element, CRE1, modulates expression of nor-1 in Aspergillus parasiticus. *Journal of Biological Chemistry* 279 (26):27428-27439.

Ruiz-Roldan, M. C., F. J. Maier, and W. Schafer. 2001. PTK1, a mitogen-activated-protein kinase gene, is required for conidiation, appressorium formation, and pathogenicity of Pyrenophora teres on barley. *Molecular Plant-Microbe Interactions* 14 (2):116-125.

Scazzocchio, C. 2000. The fungal GATA factors. *Current Opinion in Microbiology* 3 (2):126-131.

Schinko, T., H. Berger, W. Lee, A. Gallmetzer, K. Pirker, R. Pachlinger, I. Buchner, T. Reichenauer, U. Guldener, and J. Strauss. Transcriptome analysis of nitrate assimilation in Aspergillus nidulans reveals connections to nitric oxide metabolism. *Molecular Microbiology* 78 (3):720-738.

Schmidt-Heydt, M., A. Abdel-Hadi, N. Magan, and R. Geisen. 2009. Complex regulation of the aflatoxin biosynthesis gene cluster of Aspergillus flavus in relation to various combinations of water activity and temperature. *International Journal of Food Microbiology* 135 (3):231-237.

Seo, J. A., Y. J. Guan, and J. H. Yu. 2003. Suppressor mutations bypass the requirement of fluG for asexual sporulation and sterigmatocystin production in Aspergillus nidulans. *Genetics* 165 (3):1083-1093.

Seo, J. A., K. H. Han, and J. H. Yu. 2005. Multiple roles of a heterotrimeric G-protein gamma-subunit in governing growth and development of Aspergillus nidulans. *Genetics* 171 (1):81-89.

Seo, J. A., and J. H. Yu. 2006. The phosducin-like protein PhnA is required for G beta gamma-mediated signaling for vegetative growth, developmental control, and toxin biosynthesis in Aspergillus nidulans. *Eukaryotic Cell* 5 (2):400-410.

Shahbazian, M. D., and M. Grunstein. 2007. Functions of site-specific histone acerylation and deacetylation. *Annual Review of Biochemistry* 76:75-100.

Shantha, T., and V. S. Murthy. 1981. Influence of tricarboxylic-acid cycle intermediates and related metabolites on the biosynthesis of aflatoxin by resting cells of Aspergillus-flavus. *Applied and Environmental Microbiology* 42 (5):758-761.

Shimizu, K., J. K. Hicks, T. P. Huang, and N. P. Keller. 2003. Pka, Ras and RGS protein interactions regulate activity of aflR, a Zn(II)2Cys6 transcription factor in Aspergillus nidulans. *Genetics* 165 (3):1095-1104.

Shimizu, K., and N. P. Keller. 2001. Genetic involvement of a cAMP-dependent protein kinase in a g protein signaling pathway regulating morphological and chemical transitions in Aspergillus nidulans. *Genetics* 157 (2):591-600.

Shwab, E. K., J. W. Bok, M. Tribus, J. Galehr, S. Graessle, and N. P. Keller. 2007. Histone deacetylase activity regulates chemical diversity in Aspergillus. *Eukaryotic Cell* 6 (9):1656-1664.

Smith, C. A., D. Robertson, B. Yates, D. M. Nielsen, D. Brown, R. A. Dean, and G. A. Payne. 2008. The effect of temperature on Natural Antisense Transcript (NAT) expression in Aspergillus flavus. *Current Genetics* 54 (5):241-269.

Spencer, V. A., and J. R. Davie. 1999. Role of covalent modifications of histones in regulating gene expression. *Gene* 240 (1):1-12.

Squire, R. A. 1981. Ranking animal carcinogens - a proposed regulatory approach. *Science* 214 (4523):877-880.

Stinnett, S. M., E. A. Espeso, L. Cobeno, L. Araujo-Bazan, and A. M. Calvo. 2007. Aspergillus nidulans VeA subcellular localization is dependent on the importin alpha carrier and on light. *Molecular Microbiology* 63 (1):242-255.

Sweeney, M. J., and A. D. W. Dobson. 1999. Molecular biology of mycotoxin biosynthesis. *Fems Microbiology Letters* 175 (2):149-163.

Takano, Y., T. Kikuchi, Y. Kubo, J. E. Hamer, K. Mise, and I. Furusawa. 2000. The Colletotrichum lagenarium MAP kinase gene CMK1 regulates diverse aspects of fungal pathogenesis. *Molecular Plant-Microbe Interactions* 13 (4):374-383.

Tilburn, J., S. Sarkar, D. A. Widdick, E. A. Espeso, M. Orejas, J. Mungroo, M. A. Penalva, and H. N. Arst. 1995. The Aspergillus PacC zinc-finger transcription factor mediates regulation of both acid-expressed and alkaline-expressed genes by ambient pH. *Embo Journal* 14 (4):779-790.

Trail, F., N. Mahanti, and J. Linz. 1995. Molecular-biology of aflatoxin biosynthesis. *Microbiology-Uk* 141:755-765.

Tribus, M., J. Galehr, P. Trojer, G. Brosch, P. Loidl, F. Marx, H. Haas, and S. Graessle. 2005. HdaA, a major class 2 histone deacetylase of Aspergillus nidulans, affects growth under conditions of oxidative stress. *Eukaryotic Cell* 4 (10):1736-1745.

Tsitsigiannis, D., T. M. Kowieski, R. Zarnowski, and N. P. Keller. 2004. Endogenous lipogenic regulators of spore balance in Aspergillus nidulans. *Eukaryotic Cell* 3 (6):1398-1411.

Tsitsigiannis, D. I., and N. P. Keller. 2006. Oxylipins act as determinants of natural product biosynthesis and seed colonization in Aspergillus nidulans. *Molecular Microbiology* 59 (3):882-892.

Wendehenne, D., A. Pugin, D. F. Klessig, and J. Durner. 2001. Nitric oxide: comparative synthesis and signaling in animal and plant cells. *Trends in Plant Science* 6 (4):177-183.

Wilkinson, J. R., J. Yu, H. K. Abbas, B. E. Scheffler, H. S. Kim, W. C. Nierman, D. Bhatnagar, and T. E. Cleveland. 2007. Aflatoxin formation and gene expression in response to carbon source media shift in Aspergillus parasiticus. *Food Additives and Contaminants* 24 (10):1051-1060.

Wilson, R. A., and H. N. Arst. 1998. Mutational analysis of AREA, a transcriptional activator mediating nitrogen metabolite repression in Aspergillus nidulans and a member of the "streetwise" GATA family of transcription factors. *Microbiology and Molecular Biology Reviews* 62 (3):586-+.

Wogan, G. N. 1992. Aflatoxins as risk-factors for hepatocellular-carcinoma in humans. *Cancer Research* 52 (7):S2114-S2118.

Wright, M. S., D. M. Greene-McDowelle, H. J. Zeringue, D. Bhatnagar, and T. E. Cleveland. 2000. Effects of volatile aldehydes from Aspergillus-resistant varieties of corn on Aspergillus parasiticus growth and aflatoxin biosynthesis. *Toxicon* 38 (9):1215-1223.

Xu, J. R., and J. E. Hamer. 1996. MAP kinase and cAMP signaling regulate infection structure formation and pathogenic growth in the rice blast fungus Magnaporthe grisea. *Genes & Development* 10 (21):2696-2706.

Yu, J. H., R. A. E. Butchko, M. Fernandes, N. P. Keller, T. J. Leonard, and T. H. Adams. 1996. Conservation of structure and function of the aflatoxin regulatory gene aflR from Aspergillus nidulans and A-flavus. *Current Genetics* 29 (6):549-555.

Yu, J. J., P. K. Chang, K. C. Ehrlich, J. W. Cary, D. Bhatnagar, T. E. Cleveland, G. A. Payne, J. E. Linz, C. P. Woloshuk, and J. W. Bennett. 2004. Clustered pathway genes in aflatoxin biosynthesis. *Applied and Environmental Microbiology* 70 (3):1253-1262.

Zeringue, H. J., and S. P. McCormick. 1990. Aflatoxin production in cultures of Aspergillus-flavus incubated in atmospheres containing selected cotton leaf-derived volatiles. *Toxicon* 28 (4):445-448.

Zhan, X. L., R. J. Deschenes, and K. L. Guan. 1997. Differential regulation of FUS3 MAP kinase by tyrosine-specific phosphatases PTP2/PTP3 and dual-specificity phosphatase MSG5 in Saccharomyces cerevisiae. *Genes & Development* 11 (13):1690-1702.

Part 2

Biomarkers and Breeding

5

Identification of Gene Markers in Aflatoxin-Resistant Maize Germplasm for Marker-Assisted Breeding

Robert L. Brown[1], Abebe Menkir[2], Zhi-Yuan Chen[3], Meng Luo[3] and Deepak Bhatnagar[1]
[1]USDA-ARS-SRRC, New Orleans, LA
[2]International Institute of Tropical Agriculture, Ibadan,
[3]Louisiana State University Agricultural Center, Baton, Rouge, LA
[1,3]USA
[2]Nigeria

1. Introduction

Aflatoxins, the toxic and highly carcinogenic secondary metabolites of *Aspergillus flavus, A. parasiticus* are the most widely investigated of all mycotoxins due to their role in establishing the significance of mycotoxins in animal diseases, and to the regulation of their presence in food (Brown et al., 1998; Dorner et al., 1999). Aflatoxins pose serious health hazards to humans and domestic animals, because they frequently contaminate agricultural commodities (CAST, 1979; Diener et al., 1987). Presently, numerous countries have established or proposed regulations for controlling aflatoxins in food and feeds (Haumann, 1995); the US Food and Drug Administration (FDA) has limits of 20 ppb, total aflatoxins, on interstate commerce of food and feed, and 0.5 ppb of aflatoxin M_1 on the sale of milk. However, many countries, especially in the developing world, experience contamination of domestic-grown commodities at alarmingly greater levels than does the U.S. Evidence of this was demonstrated in a study that revealed a strong association between exposure to aflatoxin and both stunting (a reflection of chronic malnutrition) and being underweight (a reflection of acute malnutrition) in West African children (Gong et al., 2002). Also, a 2004 outbreak of acute aflatoxicosis in Kenya, due to ingestion of contaminated maize, resulted in 125 deaths (Probst et al., 2007).

Recognition of the need to control aflatoxin contamination of food and feed grains has elicited various approaches from researchers to eliminate this toxin from maize and other susceptible crops. The approach to enhance host resistance through conventional or molecular breeding gained renewed attention following the discovery of natural resistance to *A. flavus* infection and aflatoxin production in maize (Gardner et al., 1987; King & Scott, 1982; Widstrom et al., 1987; Scott & Zummo, 1988; Campbell & White, 1995; Brown et al., 1995, 1999). During the past two decades, maize genotypes with natural preharvest resistance to aflatoxin production have been identified through field screening (Scott & Zummo, 1988; Campbell & White, 1995; Warner et al., 1992). However, there is always a need to continually identify and utilize additional sources of maize genotypes with aflatoxin-resistance.

An important contribution to the identification/investigation of kernel aflatoxin-resistance has been the development of a rapid laboratory screening assay. The kernel screening assay (KSA), was developed and used to study resistance to aflatoxin production in GT-MAS:gk kernels (Brown et al., 1993, 1995]. The KSA is designed to address the fact that aflatoxin buildup occurs in mature and not developing kernels. Although, other agronomic factors (e.g. husk tightness) are known to affect genetic resistance to aflatoxin accumulation in the field, the KSA measures seed-based resistance. The seed, of course, is the primary target of aflatoxigenic fungi, and is the edible portion of the crop. Therefore, seed-based resistance represents the core objective of maize host resistance. Towards this aim, the KSA has demonstrated proficiency in separating susceptible from resistant seed [Brown et al., 1993, 1995). This assay has several advantages, as compared to traditional field screening techniques (Brown et al., 1995): 1) it can be performed and repeated several times throughout the year and outside of the growing season; 2) it requires few kernels; 3) it can detect/identify different kernel resistance mechanisms; 4) it can dispute or confirm field evaluations (identify escapes); and 5) correlations between laboratory findings and inoculations in the field have been demonstrated. The KSA can, therefore, be a valuable complement to standard breeding practices for preliminary evaluation of germplasm. However, field trials are necessary for the final confirmation of resistance.

One drawback to using the known resistant maize lines to develop commercial lines is their poor agronomic quality (Brown et al., 1999). To overcome this, markers need to be identified to facilitate the incorporation of aflatoxin-resistance into lines with commercially-acceptable genetic backgrounds. The expression of maize kernel proteins has been implicated in kernel resistance to *A. flavus* infection/aflatoxin production (Cordero et al., 1992, 1994; Guo, et al., 1996; Huang et al., 1997). Using reverse genetics to identify genes that are associated with aflatoxin-resistance may lead to the discovery of breeding markers. These protein/gene markers could be used to transfer resistance to good genetic backgrounds while excluding undesirable traits. The purpose of this review is to highlight the discovery of resistance-associated proteins (RAPs) and their potential as breeding markers.

2. Discovery of Resistance-Associated Proteins (RAPs)

The development of the KSA by Brown *et al.* (Brown et al., 1995) facilitated the verification of maize kernel resistance under laboratory conditions in a short time. This accelerated the discovery of knowledge surrounding host resistance mechanisms. Using this assay, Brown *et al.* (Brown et al., 1993) discovered the existence of subpericarp resistance in maize kernels and that the expression of this resistance requires a live embryo, the latter indicating a potential role for kernel proteins in resistance. Guo *et al.* (1996) found that imbibition of kernels, before inoculation with *A. flavus*, significantly increased aflatoxin-resistance of susceptible maize genotypes. Further investigation revealed that susceptible genotypes were able to induce antifungal proteins upon fungal infection (Guo et al., 1996), suggesting that susceptible lines have the ability to induce an active defense mechanism after fungal infection. The usefulness of the KSA as an investigative tool is aided by the fact that KSA results correlate well with field results (Brown et al., 1995) and that aflatoxin buildup occurs after kernel maturity, a developmental phase where constitutive factors required for kernel resistance are highlighted by the KSA (Brown et al., 1995). Agronomic factors contributing to resistance have to be evaluated during field trials.

Examination of kernel proteins of several maize genotypes revealed differences between genotypes resistant or susceptible to aflatoxin contamination (Guo et al., 1997, 1998). Imbibed susceptible kernels contained increased levels of germination-induced ribosome inactivating protein (RIP) and zeamatin; both proteins have demonstrated growth-inhibitory activity *in vitro* against *A. flavus* (Guo et al., 1998). In another study, two kernel proteins were identified from a resistant maize inbred line (Tex6), which may contribute to resistance to aflatoxin contamination (Huang et al., 1997). When a commercial maize hybrid was inoculated with toxgenic and atoxigenic strains of *A. flavus* at milk stage, one chitinase and one β-1,3-glucanase isoform were detected in maturing infected kernels, while another isoform was detected in maturing uninfected kernels (Ji et al., 2000). Lozovaya et al., 1998) reported that the presence of *A. flavus* caused an increase in β-1,3-glucanase activity in callus tissues from a resistant genotype, but not from a susceptible one. A more rapid and stronger induction of the PR-1 and PR-5 genes in maize leaves has also been observed in an incompatible interaction when compared to a compatible interaction upon pathogen infection (Morris et al., 1998). A 14 kDa trypsin inhibitor protein (TI) was found to express at high levels in resistant lines but at low levels or is missing in susceptible ones (Chen et al., 1998). This protein demonstrated antifungal activity against *A. flavus* and several other pathogenic fungi (Chen et al., 1999a), possibly through inhibition of fungal α-amylase activity and production (Chen et al., 1999b). This could limit the availability of simple sugars needed for fungal growth and aflatoxin production (Woloshuk et al., 1997).

The above-studies indicate an important role for kernel proteins in disease resistance. Further investigation, supporting earlier work by Guo (1996), found that both constitutive and inducible proteins are required for kernel resistance to *A. flavus* infection and aflatoxin production (Chen et al., 2001). In fact, one major difference between resistant and susceptible genotypes is that resistant lines constitutively produce higher levels of antifungal proteins compared to susceptibles. Therefore, research on resistance genes/proteins has focused heavily on the identification of constitutively-produced kernel resistance-associated proteins or RAPs.

2.1 Using comparative proteomics for RAP discovery

To assist in the further identification of RAPs, proteomics approaches have been employed. This increased protein resolution and detection sensitivity by 10 to 20 fold over conventional approaches and, thus, enhanced ability to identify more constitutively-expressed RAPs. Kernel proteins from several resistant and susceptible genotypes were compared using large format 2-D gel electrophoresis. A number of protein spots, either unique or 5-fold up-regulated in resistant lines, were detected, isolated from preparative 2-D gels and identified using ESI-MS/MS after in-gel digestion with trypsin (Chen et al., 2002, 2007a). These proteins can be grouped into three categories based on their peptide sequence homology: (1) storage proteins, such as globulins (GLB1, GLB2), and late embryogenesis abundant proteins (LEA3, LEA14); (2) stress-responsive proteins, such as aldose reductase (ALD), glyoxalase I (GLX I) and heat shock proteins, and (3) antifungal proteins, including TI. In total, 21 proteins upregulated in resistant versus susceptible lines have been identified using comparative proteomics (Table 1).

Antifungals	Stress-related	Storage
Zeamatin	Aldose reductase (ALD)	Globulin I
Trypsin Inhibitor 14 kD	Cold-regulated (ZmCORp)	Globulin II
Trypsin inhibitor 10 kD	Water stress inducible (WSI)	Cupin domain (Zmcup)
Ribosome inactivating (RIP)	Anionic peroxidase	Late embryogenesis (LEAIII)
B-1,3,-glucanase	Small heat shock protein	LEA 14
PR 10	Glyoxalase (GLX I)	
PR 10.1	Peroxiredoxin (PER1)	

Table 1. Resistance-associated proteins (RAPs) identified by proteomics[1,2]

No investigation has been conducted to determine the possible direct involvement of stress-related proteins in host fungal resistance. However, increased temperatures and drought, which often occur together, are major factors associated with aflatoxin contamination of maize kernels (Payne, 1998). Unique or higher levels of hydrophilic storage or stress-related proteins, such as the aforementioned, may put resistant lines at an advantage for the ability to synthesize proteins and defend against pathogens while under stress. Further studies including physiological and biochemical characterization, genetic mapping, plant transformation using RAP genes, RNAi gene silencing experiments and marker-assisted breeding should clarify the roles of stress-related RAPs in kernel resistance (Brown et al., 2003).

To conduct the above-described comparative proteomics studies, composite profiles for resistance and for susceptibility were developed from 2 D gels of several resistant or susceptible maize lines. This was done to homogenize nonresistance-related differences among lines within each group, and, therefore, facilitate the identification of resistance-related proteins. In using the composite gel approach, only those proteins that were five-fold upregulated in resistant versus susceptible lines were studied to minimize the chance of identifying proteins unrelated to host resistance.

An advancement in the aforementioned 2-D approach is the use of difference gel electrophoresis or DIGE in RAP discovery (Figure 1) (Luo et al., 2010). This represents an advancement because it eliminates side-by-side gel comparisons and therefore, gel-to-gel variability. Protein samples extracted from frozen embryo or endosperm are run on the same gel with an internal reference using CyDye DIGE fluorescent dyes for labeling (Minden et al., 2009). Treated samples are labeled by Cy5, control by Cy3 and the internal standard by Cy2. The internal standard consists of pooled protein from controls and treated samples. For IEF, Cy2-, Cy3-, and Cy5- labeled samples are mixed with an appropriate buffer and applied to an Immobiline DryStrip (pH 3-10). After the second dimension is completed, the CyDye-labeled gels are visualized using a fluorescent Image Reader. Other advantages of DIGE are that gel staining is unnecessary, quantifying protein spots using DIGE is easier and more accurate than the standard 2 D approach, and identified proteins of interest can be recovered from the same 2-D gel and used directly for downstream MS/MS analysis for mass finger printing or *de novo* peptide sequencing.

---

[1] Table is adaptation and updated version of table from Cary et al., 2009 reference.

[2] Each protein was up-regulated in aflatoxin-resistant versus –susceptible lines.

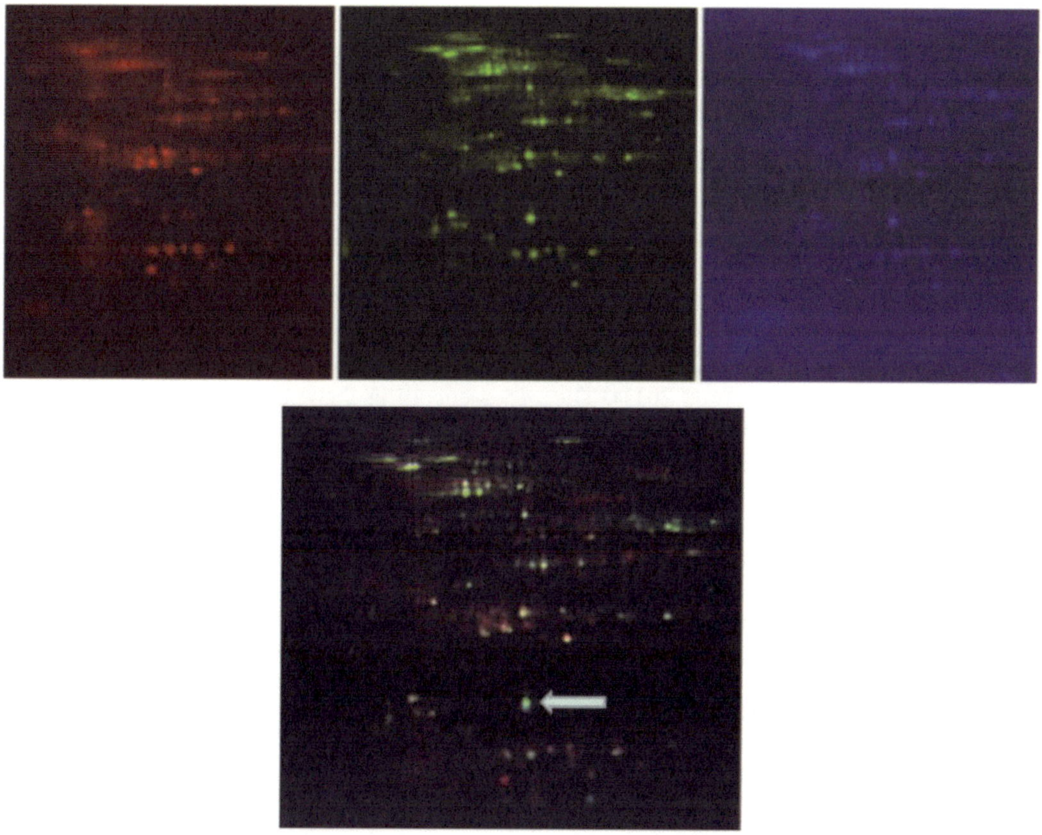

Fig. 1. DIGE experiment involving two closely-related maize lines varying in aflatoxin accumulation. **R**= resistant; **S**= susceptible. Red image: probe labeled by Cy3 = **S**. Green image: probe labeled by Cy5 = **R**; Blue image: probe labeled by Cy2 = reference. Bottom image: Images overlapped highlighting Heat Shock Protein 17.2 (arrow), 6.5 fold more abundant in **R** sample vs. **S** sample.

2.1.1 Employing closely-related breeding lines to enhance RAP discovery

Recently, the screening of progeny generated through a collaborative breeding program between IITA-Nigeria (International Institute of Tropical Agriculture) and the Southern Regional Research Center of USDA-ARS in New Orleans (SRRC) facilitated the identification of closely-related lines from the same backcross differing significantly in aflatoxin accumulation, and proteome analysis of these lines has been conducted (Menkir, et al., 2006; Brown et al., 2001; Chen et al., 2005, 2011). Investigating maize lines sharing close genetic backgrounds should enhance the identification of RAPs without the confounding effects experienced with lines of diverse genetic backgrounds.

The IITA-SRRC collaboration has attempted to combine resistance traits of the U.S. resistant inbred lines with those of African lines, originally selected for resistance to ear rot diseases and for demonstrated potential aflatoxin-resistance (*via* KSA) (Menkir et al., 2006; Brown et al., 2001). Five elite tropical inbred lines from IITA adapted to the Savanna and mid-altitude ecological zones of West and Central Africa were crossed with four US resistant maize lines in Ibadan, Nigeria. The F1 crosses were backcrossed to their respective US inbred lines and self-pollinated thereafter. The resulting lines were selected through the S4 generation for resistance to foliar diseases and for desirable agronomic characteristics under conditions of severe natural infection in their respective areas of adaptation. Promising S5 lines were screened with the KSA and five pairs of closely-related lines were shown to be significantly different in aflatoxin resistance, while sharing as high as 97% genetic similarity (Chen et al., 2005, 2011). Using these lines in proteomic comparisons to identify RAPs has advantages: (1) gel comparisons and analyses become easier; and (2) protein differences between resistant and susceptible lines as low as twofold can be identified with confidence. In addition, the likelihood of identifying proteins that are directly involved in host resistance is increased.

In a preliminary proteomics comparison of constitutive proteins between those African closely-related lines, a new category of resistance-associated proteins (putative regulatory proteins) was identified, including a serine/threonine protein kinase and a translation initiation factor 5A (Chen et al., 2011). The genes encoding these two resistance associated regulatory proteins are being cloned and their potential role in host resistance to *A. flavus* infection and aflatoxin production will be further investigated.

Conducting proteomic analyses using lines from this program not only enhances chances of identifying genes important to resistance, but may have immediate practical value. The IITA-SRRC collaboration has recently registered and released six inbred lines with aflatoxin-resistance in good agronomic backgrounds, which also demonstrate good levels of resistance to southern maize blight and southern maize rust (Menkir et al., 2008). Resistance field trials for these lines on U.S. soil will be conducted; the ability to use resistance in these lines commercially will depend on having excellent markers, since seed companies desire insurance against the transfer of undesirable traits into their elite genetic backgrounds. The fact that this resistance is coming from good genetic backgrounds is also a safeguard against the transfer of undesirable traits.

2.1.2 Proteomic investigation of maize rachis and silk tissues

A study was conducted to investigate the proteome of rachis tissue, maternal tissue that supplies nutrients to kernels (Pechanova, 2006). An interesting finding in this study is that after infection by *A. flavus*, rachis tissue of aflatoxin-resistant genotypes did not up-regulate PR proteins as these were already high in controls where they had strongly and constitutively accumulated during maturation. However, rachis tissue of aflatoxin-susceptible lines did not accumulate PR proteins to such an extent during maturation, but increased them in response to fungal infection. Given the relationship of the rachis to kernels, these results support findings of Chen *et al.* (2001), who demonstrated that higher constitutive levels of proteins in resistant versus susceptible kernels was a primary factor that determined kernel genetic resistance to aflatoxin contamination.

Another study was conducted to identify proteins in maize silks that may be contributing to resistance against *A. flavus* infection/colonization [Peethambaran et al., 2010). Antifungal bioassays were performed using silk extracts from two aflatoxin-resistant and two-susceptible inbred lines. Silk extracts from resistant inbreds showed greater anti-fungal activity compared to susceptible inbreds. Comparative proteomic analysis of the two resistant and susceptible inbreds led to the identification of antifungal proteins including three chitinases that were differentially-expressed between resistant and susceptible lines. When tested for chitinase activity, silk proteins from extracts of resistant lines also showed significantly higher chitinase activity than that from susceptible lines. Differential expression of chitinases in maize resistant and susceptible inbred silks suggests that these proteins may contribute to resistance.

3. Characterization of RAPs

A literature review of the RAPs that have been identified indicates that storage and stress-related proteins may play important roles in enhancing stress tolerance of host plants. The expression of storage protein GLB1 and LEA3 has been reported to be stress-responsive and ABA-dependant (Thomann et al., 1992). Transgenic rice overexpressing a barley LEA3 protein HVA1 showed significantly increased tolerance to water deficit and salinity (Xu et al., 1996).

The role of GLX I (Table 2) in stress-tolerance was first highlighted in an earlier study using transgenic tobacco plants overexpressing a *Brassica juncea* glyoxalase I (Veena et al., 1999). The substrate for glyoxalase I, methylglyoxal, is a potent cytotoxic compound produced spontaneously in all organisms under physiological conditions from glycolysis and photosynthesis intermediates, glyceraldehydes-3-phosphate and dihydroxyacetone phosphate. Methylglyoxal is an aflatoxin inducer even at low concentrations; experimental evidence indicates that induction is through upregulation of aflatoxin biosynthetic pathway transcripts including the *AFLR* regulatory gene (Chen et al., 2004). Therefore, glyoxalase I may be directly affecting resistance by removing its aflatoxin-inducing substrate, methylglyoxal.

PER1, a 1-cys peroxiredoxin antioxidant identified in a proteomics investigation (Chen et al., 2007a), was demonstrated to be an abundant peroxidase (Table 2), and may play a role in the removal of reactive oxygen species. The PER1 protein overexpressed in *Escherichia coli* demonstrated peroxidase activity *in vitro*. It is possibly involved in removing reactive oxygen species produced when maize is growing under stress (Chen et al., 2007a).

Another RAP that has been characterized further is the pathogenesis-related protein 10 (PR10) (Table 2). It showed high homology to PR10 from rice (85.6% identical) and sorghum (81.4% identical). It also shares 51.9% identity to intracellular pathogenesis-related proteins from lily (AAF21625) and asparagus (CAA10720), and low homology to a RNase from ginseng [48]. The PR10 overexpressed in *E. coli* exhibited ribonucleolytic and antifungal activities. In addition, an increase in the antifungal activity against *A. flavus* growth was observed in the leaf extracts of transgenic tobacco plants expressing maize *PR10* gene compared to the control leaf extract (Chen et al., 2006). This evidence suggests that PR10 plays a role in kernel resistance by inhibiting fungal growth of *A. flavus*. Further, its expression during kernel development was induced in the resistant line GT-MAS:gk, but not in susceptible Mo17 in response to fungal inoculation (Chen et al., 2006). Recently, a new *PR10* homologue was identified from maize (*PR10.1*) (Xie et al., 2010). *PR10* was expressed at higher levels in all tissues compared to *PR10.1*, however, purified PR10.1 overexpressed in *E. coli* possessed 8-fold higher specific RNase activity than *PR10* (Xie et al., 2010). This homologue may also play a role in resistance.

Evidence supporting a role for *PR10* in host resistance is also accumulating in other plants. A barley *PR10* gene was found to be specifically induced in resistant cultivars upon infection by *Rhynchosporium secalis*, but not in near-isogenic susceptible plants (Steiner-Lange et al., 2003). In cowpea, a *PR10* homolog was specifically up-regulated in resistant epidermal cells inoculated with the rust fungus *Uromyces vignae* Barclay (Mould et al., 2003). A *PR10* transcript was also induced in rice during infection by *Magnaporthe grisea* (McGee et al., 2001).

To directly demonstrate whether selected RAPs play a key role in host resistance against *A. flavus* infection, an RNA interference (RNAi) vector to silence the expression of endogenous

RAP genes (such as *PR10, GLX I* and *TI*) in maize through genetic engineering was constructed (Chen et al., 2004b, 2010). The degree of silencing using RNAi constructs is greater than that obtained using either co-suppression or antisense constructs, especially when an intron is included (Wesley et al., 2001). Interference of double-stranded RNA with expression of specific genes has been widely described (Fire et al., 1998; Gura, 2000). Although the mechanism is still not well understood, RNAi provides an extremely powerful tool to study functions of unknown genes in many organisms. This posttranscriptional gene silencing (PTGS) is a sequence-specific RNA degradation process triggered by a dsRNA, which propagates systemically throughout the plant, leading to the degradation of homologous RNA encoded by endogenous genes, and transgenes.

Both particle bombardment and *Agrobacterium*-mediated transformation methods were used to introduce the RNAi vectors into immature maize embryos. The former was used to provide a quick assessment of the efficacy of the RNAi vector in gene silencing. The latter, which can produce transgenic materials with fewer copies of foreign genes and is easier to regenerate, was chosen for generating transgenic kernels for evaluation of changes in aflatoxin-resistance. It was demonstrated using callus clones from particle bombardment that *PR10* expression was reduced by an average of over 90% after the introduction of the RNAi vector Chen et al., 2010). The transgenic kernels also showed a significant increase in susceptibility to *A. flavus* infection and aflatoxin production. The data from this RNAi study clearly demonstrated a direct role for PR10 in maize host resistance to *A. flavus* infection and aflatoxin contamination (Chen et al., 2010).

RNAi vectors to silence other RAP genes, such as *GLX I* and *TI*, have also been constructed, and introduced into immature maize embryos through both bombardment and *Agrobacterium* infection (Chen et al., 2007b). It will be very interesting to see the effect of silencing the expression of these genes in the transgenic kernels on host resistance to *A. flavus* infection and aflatoxin production.

ZmCORp, a protein with a sequence similar to cold-regulated protein and identified in the above-proteomic studies, was shown to exhibit lectin-like hemagglutination activity against fungal conidia and sheep erythrocytes (Table 2) (Baker et al., 2009a). When tested against *A. flavus*, ZmCORp inhibited germination of conidia by 80% and decreased mycelial growth by 50%. Quantitative real-time RT-PCR revealed *ZmCORp* to be expressed 50% more in kernels of a resistant maize line versus a susceptible. ZmTIp, a 10 kDa trypsin inhibitor, had an impact on *A. flavus* growth, but not as great as the previously-mentioned 14 kDa TI (Baker et al., 2009b).

3.1 Mapping genes

Chromosome regions associated with resistance to *A. flavus* infection and inhibition of aflatoxin production in maize have been identified through Restriction Fragment Length Polymorphism (RFLP) analysis in three "resistant" lines (R001, LB31, and Tex6) in an Illinois breeding program, after mapping populations were developed using B73 and/or Mo17 elite inbreds as the "susceptible" parents (White et al., 1995, 1998). Chromosome regions associated with inhibition of aflatoxin in studies considering all 3 resistant lines demonstrated that there are some chromosome regions in common. Regions on chromosome arms 2L, 3L, 4S, and 8S may prove promising for improving resistance in commercial lines through marker assisted breeding (White et al., 1998). In some cases, chromosomal regions were associated with resistance to *Aspergillus* ear rot and not aflatoxin inhibition, and vice versa, whereas other chromosomal regions were found to be associated with both traits. This suggests that these two traits may be at least partially under separate genetic control.

QTL studies involving other populations have identified chromosome regions associated with low aflatoxin accumulation. In a study involving 2 populations from Tex6 x B73, conducted in 1996 and 1997, promising QTLs for low aflatoxin were detected in bins 3.05-6, 4.07-8, 5.01-2, 5.05-5, and 10.05-10.07 (Paul et al., 2003). Environment strongly influenced detection of QTLs for lower toxin in different years; QTLs for lower aflatoxin were attributed to both parental sources. In a study involving a cross between B73 and resistant inbred Oh516, QTL associated with reduced aflatoxin were identified on chromosomes 2, 3 and 7 (bins 2.01 to 2.03, 2.08, 3.08, and 7.06) (Busboom & White, 2004). QTLs contributing resistance to aflatoxin accumulation were also identified using a population created by B73 and resistant inbred Mp313E, on chromosome 4 of Mp313E (Brooks et al., 2005). This confirmed the findings of an earlier study involving Mp313E and susceptible Va35 (Davis et al., 2000). Another QTL in this study, which has similar effects to that on chromosome 4, was identified on chromosome 2 (Brooks et al., 2005). A recent study to identify aflatoxin-resistance QTL and linked markers for marker-assisted breeding was conducted using a population developed from Mp717, an aflatoxin-resistant maize inbred, and NC300, a susceptible inbred adapted to the southern US. QTL were identified on all chromosomes, except 4, 6, and 9; individual QTL accounted for up to 11% of phenotypic variance in aflatoxin accumulation (Warburton et al., 2009).

RAP Gene	**Activity vs. A. flavus**	**Resistance-related enzyme rx**	**Mapping Bin**	**Other**
Heat Shock a	nda	nda	1.03	
PR - 10	+[1]	Rnase	1.03	knockout=Suscept
TI-14 kDa	+high	Inhib. trypsin	2.06	Inhib. amylase
WSI	nda	nda	3.07	
Zeamatin	+low	Inhib. Trypsin	7.04	
Heat Shock b	nda	nda	8.01	
ZmCorp	+	Lectin	8.04	
GLXI	nda	Forms D-lactate	10.3	knockout
RIP	+	lytic	nda	
PER 1	nda	peroxidase	nda	
B-1,3 glucanase	+	glucanase	nda	

Table 2. Evidence supporting the candidacy of selected RAPs as breeding markers[1]

A number of RAP genes identified in the proteomics studies have been mapped to chromosomal location (Table 2) using the genetic sequence of B73 now available online

[1] + denotes presence of activity; nda denotes no data available.

(http://archive.maizesequence.org/index.html). Using the DNA sequence of the RAPs and blasting them against the B73 sequence allowed us to place each gene into a virtual bin, allowing us to pinpoint the chromosomal location to which each gene maps (Brown et al., 2010). The chromosomes involved include the above-mentioned chromosomes 1, 2, 3, 7, 8 and 10, some in bins closely located to those described above. This adds support to proteomic data and characterization results that suggest the involvement of 14 kDa TI, water stress inducible protein, zeamatin, one of the heat shock, cold-regulated, glyoxalase I and PR10 proteins in aflatoxin-resistance. From the above QTL investigations, it is observed that variation can exist in the chromosomal regions associated with *Aspergillus* ear rot and aflatoxin inhibition in different mapping populations. This suggests the presence of different genes for resistance in the different identified resistant germplasm. It will be important to map resistant lines investigated through proteomics or to obtain data from associative mapping panels regarding gene location.

4. Conclusion

Host resistance as a strategy for eliminating aflatoxin contamination of maize is closer to being a reality due to the identification of genotypes with natural resistance to aflatoxin accumulation and the development of new inbred lines through breeding. However, to exploit this resistance for the benefit of maize growers, markers have to be identified to facilitate the transfer of resistance to elite proprietary backgrounds that have commercial value. The identification of resistance-associated proteins goes a long way towards providing the novel markers that will be indispensible to any commercial breeding undertaking. Characterization studies including RNAi gene silencing and gene mapping are instrumental in building a case for the involvement of selected RAPs in kernel resistance to aflatoxin contamination.

Here, a listing of RAPs identified through comparative proteomics is presented along with evidence of the potential of selected RAPs as breeding markers. Investigations of RAPs, as discussed above, not only impact the development of commercially-useful resistant maize lines, but provide an expanding base of knowledge concerning nature's requirements for creating a durable resistance against the opportunistic pathogen, *A. flavus*. It remains to be determined, how the different categories of proteins, antifungal, stress-related, storage and others contribute to the total picture of resistance. Future investigations (e.g., proteomics and microarray analysis) may also impact aflatoxin-resistance through the discovery of RAPs down-regulated in resistant lines, RAPs induced upon fungal infection and also factors involved in the regulation of RAPs. These discoveries will not only contribute to the development of aflatoxin-resistant maize lines, they may aid other susceptible crops and assist in meeting the challenges of other mycotoxin-producing fungi, while enhancing our understanding of host plant interactions with fungi.

5. Acknowledgement

Research discussed in this review received support from the National Research Initiative Competitive Grants Program, USAID Linkage Program-IITA, Nigeria, USDA-ARS Office of International Research Programs (OIRP) -USAID Collaborative Support Program.

6. References

Baker, R., Brown, R.L., Cleveland, T.E., Chen, Z.-Y. & Fakhoury, A. (2009a). COR, a maize lectin-like protein with antifungal activity against *Aspergillus flavus*. *Journal of Food Protection,* Vol. 72, pp. 120–127

Baker, R., Brown, R.L., Cleveland, T.E., Chen, Z.-Y. & Fakhoury, A. (2009b). ZmTI, a maize trypsin inhibitor with limited activity against *Aspergillus flavus*. *Journal of Food Protection,* Vol. 72, pp. 85–188

Brooks, T.D., Williams, W.P., Windham, G.L., Wilcox, M.C. & Abbas, H. (2005). Quantitative trait loci contributing resistance to aflatoxin accumulation in maize inbred Mp313E. *Crop Science,* Vol. 45, 171–174

Brown, R.L., Cotty, P.J., Cleveland, T.E. & Widstrom, N.W. (1993). Living maize embryo influences accumulation of aflatoxin in maize kernels. *Journal of Food Protection,* Vol. 56, 967–971

Brown, R.L., Cleveland, T.E., Payne, G.A., Woloshuk, C.P., Campbell, K.W. & White, D.G. (1995). Determination of resistance to aflatoxin production in maize kernels and detection of fungal colonization using an *Aspergillus flavus* transformant expressing *Escherichia coli* β-glucuronidase. *Phytopathology,* Vol. 85, 983–989

Brown, R.L., Bhatnagar, D., Cleveland, T.E. & Cary, J.W. (1998). Recent advances in preventing mycotoxin contamination. In: *Mycotoxins in Agriculture and Food Safety;* K.K. Sinha, & D. Bhatnagar, (Eds), pp. 351-379, Marcel Dekker: New York, NY, USA

Brown, R.L., Chen, Z.-Y., Cleveland, T.E. & Russin, J.S. (1999). Advances in the development of host resistance to aflatoxin contamination by *Aspergillus flavus*. *Phytopathology,* Vol. 89, pp. 13–117

Brown, R.L., Chen, Z.-Y., Menkir, A., Cleveland, T.E., Cardwell, K., Kling, J. & White D.G. (2001). Resistance to aflatoxin accumulation in kernels of maize inbreds selected for ear rot resistance in West and Central Africa. *Journal of Food Protection,* Vol. 64, pp. 396–400

Brown, R.L., Chen, Z.-Y., Menkir, A. & Cleveland, T.E. (2003). Using biotechnology to enhance host resistance to aflatoxin contamination of maize—A Mini-review. *African Journal of Biotechnology,* Vol. 2, pp. 557–562

Brown, R.L., Chen, Z.-Y., Warburton, M., Luo, M., Menkir, A., Fakhoury, A. & Bhatnagar, D. (2010). Discovery and characterization of proteins associated with aflatoxin-resistance: evaluating their potential as breeding markers. *Toxins,* Vol. 2, pp. 919-933

Busboom, K.N. & White, D.G. (2004). Inheritance of resistance to aflatoxin production and Aspergillus ear rot of maize from the cross of inbreds B73 and Oh516. *Phytopathology,* Vol. 94, pp. 1107–1115

Campbell, K.W. & White, D.G. (1995). Evaluation of maize genotypes for resistance to aspergillus ear rot, kernel infection, and aflatoxin production. *Plant Disease,* Vol. 79, pp. 1039–1045

Cary, J.W., Rajasekaran, K., Yu, J., Brown, R.L., Bhatnagar, D. & Cleveland, T.E. (2009). Transgenic approaches for pre-harvest control of mycotoxin contamination in crop plants. *World Mycotoxin Journal,* Vol. 2, pp. 203–214

CAST. (1979). Aflatoxins and other mycotoxins: An agricultural perspective; CAST Report No. 80; Council for Agricultural Science and Technology: Ames, IA

Chen, Z.-Y., Brown, R.L., Lax, A.R., Guo, B.Z., Cleveland, T.E. & Russin, J.S. (1998). Resistance to *Aspergillus flavus* in maize kernels is associated with a 14 kDa protein. *Phytopathology,* Vol. 88, pp. 276–281

Chen, Z.-Y., Brown, R.L., Lax, A.R., Cleveland, T.E. & Russin, J.S. (1999a). Inhibition plant pathogenic fungi by a maize trypsin inhibitor over-expressed in *Escherichia coli. Applied and Environmental Microbiology,* Vol. 65, pp. 1320–1324

Chen, Z.-Y., Brown, R.L., Russin, J.S., Lax, A.R. & Cleveland, T.E. (1999b). A maize trypsin inhibitor with antifungal activity inhibits *Aspergillus flavus* α-amylase. *Phytopathology,* Vol. 89, 902–907

Chen, Z.Y., Brown, R.L., Cleveland, T.E., Damann, K.E. & Russin, J.S. (2001). Comparison of constitutive and inducible maize kernel proteins of genotypes resistant or susceptible to aflatoxin production. *Journal of Food Protection,* Vol. 64, pp. 1785–1792

Chen, Z.-Y., Brown, R.L., Damann, K.E. & Cleveland, T.E.(2002). Identification of unique or elevated levels of kernel proteins in aflatoxin-resistant maize genotypes through proteome analysis. *Phytopathology,* Vol. 92, pp. 1084–1094

Chen, Z.Y., Brown, R.L., Damann, K.E. & Cleveland, T.E.(2004). Identification of a maize kernel stress-related protein and its effect on aflatoxin accumulation. *Phytopathology,* Vol. 94, pp. 938–945

Chen, Z.Y., Brown, R.L., Cleveland, T.E. & Damann, K.E. (2004b). Investigating the roles of an aflatoxin resistance-associated protein in maize using RNAi. *Phytopathology,* Vol. 94, pp. S18

Chen, Z.Y., Brown, R.L., Menkir, A., Damann, K.E. & Cleveland, T.E. (2005). Proteome analysis of near isogenic maize lines differing in the level of resistance against *Aspergillus flavus* infection/aflatoxin production. *Phytopathology,* Vol. 95, pp. S19

Chen, Z.Y., Brown, R.L., Rajasekaran, K., Damann, K.E. & Cleveland, T.E. (2006). Evidence for involvement of a pathogenesis-related protein in maize resistance to *Aspergillus flavus* infection/aflatoxin production. *Phytopathology,* Vol. 96, pp. 87–95

Chen, Z.Y., Brown, R.L., Damann, K.E. & Cleveland, T.E. (2007a) Identification of maize kernel endosperm proteins associated with resistance to aflatoxin contamination by *Aspergillus flavus. Phytopathology,* Vol. 97, pp. 1094–1103

Chen, Z.-Y.[1], Brown, R.L.[2], Damann, K.E.[1] & Cleveland, T.E. [2] (2007b). [1]Louisiana State University, Baton Rouge, LA; [2]USDA-ARS-SRRC, New Orleans, LA. Unpublished work

Chen, Z.-Y., Brown, R.L., Damann, K.E. & Cleveland, T.E. (2010). PR10 expression in maize and its effect on host resistance against *Aspergillus flavus* infection/aflatoxin production. *Molecular Plant Pathology,* Vol. 11, pp. 69–81

Chen, Z.-Y., Brown, R. L., Menkir, A. & Cleveland, T.E. (2011) Identification of resistance-associated proteins in closely- related maize lines varying in aflatoxin accumulation. *Molecular Breeding,* Vol. 27, DOI 10. 1007/s11032-011-9597-3

Cordero, M.J., Raventós, D. & San Segundo, B. (1992). Induction of PR proteins in germinating maize seeds infected with the fungus *Fusarium moniliforme. Physiological. and Molecular Plant Pathology,* Vol. 41, pp. 189-200

Cordero, M.J., Raventos, D. & San Segundo, B. (1994). Differential expression and induction of chitinases and β-1,3-glucanases in response to fungal infection during germination of maize seeds. *Molecular and Plant Microbe Interactions,* Vol. 7, pp. 23-31

Davis, G.L., Windham, G.L. & Williams, W.P. (2000) QTL for aflatoxin reduction in maize. *Proceedings of the Maize Genetics Conference Abstracts,* Vol. 41, pp. P139, Coeur d'Alene, Idaho, March, 2000

Diener, U.L., Cole, R.J., Sanders, T.H., Payne, G.A. & Lee, L.S. (1987). Epidemiology of aflatoxin formation by *Aspergillus flavus. Annual Review of Phytopathology,* Vol. 25, pp. 249-270

Dorner, J.W., Cole, R.J. & Wicklow, D.T. (1999). Aflatoxin reduction in maize through field application of competitive fungi. *Journal of Food Protection,* Vol. 62, pp. 650-656

Fire, A., Xu, S., Montgomery, M.K., Kostas, S.A., Driver, S.E. & Mello C.C. (1998). Potent and specific genetic interference by double stranded RNA in *Caenorhabditis elegans. Nature,* Vol. 391, pp. 806-811

Gardner, C.A.C., Darrah, L.L., Zuber, M.S. & Wallin, J.R. (1987). Genetic control of aflatoxin production in maize. *Plant Disease,* Vol. 71, pp. 426-429

Gong, Y.Y., Cardwell, K., Hounsa, A., Egal, S., Turner, P.C., Hall, A.J. & Wild, C.P. (2002). Dietary aflatoxin exposure and impaired growth in young children from Benin and Togo: cross sectional study. *British Medical Journal,* Vol. 325, pp. 20-21

Guo, B.Z., Russin, J.S., Cleveland, T.E., Brown, R.L. & Widstrom, N.W. (1996). Resistance to aflatoxin contamination in maize as influenced by relative humidity and kernel germination. *Journal of Food Protection,* Vol. 59, pp. 276-281

Guo, B.Z., Chen, Z.-Y., Brown, R.L., Lax, A.R., Cleveland, T.E., Russin, J.S., Mehta, A.D., Selitrennikoff, C.P. & Widstrom, N.W. (1997). Germination induces accumulation of specific proteins and antifungal activities in maize kernels. *Phytopathology,* Vol. *87,* pp. 1174-1178

Guo, B.Z., Brown, R.L., Lax, A.R., Cleveland, T.E., Russin, J.S. & Widstrom, N.W. (1998). Protein profiles and antifungal activities of kernel extracts from maize genotypes resistant and susceptible to *Aspergillus flavus. Journal of Food. Protection,* Vol. 61, pp. 98-102

Gura, T. (2000). A silence that speaks volumes. *Nature,* Vol. 404, pp. 804-808, ISSN

Haumann, F. (1995). Eradicating mycotoxins in food and feeds. *Inform,* Vol. 6,pp. 248-256

Huang, Z., White, D.G. & Payne, G.A. (1997). Maize seed proteins inhibitory to *Aspergillus flavus* and aflatoxin biosynthesis. *Phytopathology,* Vol. *87,*pp. 622-627

Ji, C., Norton, R.A., Wicklow, D.T. & Dowd, P.E. (2000). Isoform patterns of chitinase and β-1,3-glucanase in maturing maize kernels (*Zea mays* L.) associated with *Aspergillus flavus* milk stage infection. *Journal of Agricultural and Food Chemistry,* Vol. 48, pp. 507–511

King, S.B. & Scott, G.E. (1982). Screening maize single crosses for resistance to preharvest infection of kernels by *Aspergillus flavus. Phytopathology,* Vol. *72*, pp. 942

Lozovaya, V.V., Waranyuwat, A. & Widholm, J.M. (1998). β-1,3-glucanase and resistance to *Aspergillus flavus* infection in maize. *Crop Science,* Vol. *38*, pp. 1255–1260

Luo, M., Brown, R.L., Chen, Z.-Y., Menkir, A., and Bhatnagar, D. (2010). Identification of gene marker sets for screening maize lines for resistance to aflatoxin contamination. *Proceedings of the 3rd International Conference of Plant Molecular Breeding*, pp. 74, Beijing China, Sept. 5-9, 2010.

McGee, J.D., Hamer, J.E. & Hodges, T.K. (2001). Characterization of a PR-10 pathogenesis-related gene family induced in rice during infection with *Magnaporthe grisea. Molecular and Plant Microbe Interactions,* Vol. *14*, pp. 877–886

Menkir, A., Brown, R.L., Bandyopadhyay, R.., Chen, Z.-Y. & Cleveland, T.E. (2006). A USA-Africa collaborative strategy for identifying, characterizing, and developing maize germplasm with resistance to aflatoxin contamination. *Mycopathologia,* Vol. 162, pp. 225–232

Menkir, A., Brown, R.L., Bandyopadhyay, R. & Cleveland, T.E. (2008). Registration of six tropical maize germplasm lines with resistance to aflatoxin contamination. *Journal of Plant Registrations* Vol. 2, pp. 246–250

Minden, J.S., Dowd, S.R., Meyer, H.E., & Stuhler, K. (2009). Difference gel electrophoresis. *Electrophoresis,* Vol. 30, pp. S156-S161

Morris, S.W., Vernooij, B., Titatarn, S., Starrett, M., Thomas, S., Wiltse, C.C. Frederiksen, R.A.., Bhandhufalck, A. & Hulbert, S. (1998). Induced resistance response in maize. *Molecular and Plant Microbe Interactions,* Vol. *11*, pp. 643–658

Mould, M.J., Xu, T., Barbara, M., Iscove, N.N. & Heath. M.C. (2003). cDNAs generated from individual epidermal cells reveal that differential gene expression predicting subsequent resistance or susceptibility to rust fungal infection occurs prior to the fungus entering the cell lumen. *Molecular and. Plant Microbe Interactions,* Vol. *16*, pp. 835–845

Paul, C., Naidoo, G., Forbes, A., Mikkilineni, V., White, D. & Rocheford, T. (2003) Quantitative trait loci for low aflatoxin production in two related maize populations. *Theoretical and Applied Genetics,* Vol. 07, pp. 263–270

Payne, G.A. (1998). Process of contamination by aflatoxin-producing fungi and their impact on crops. In: *Mycotoxins in Agriculture and Food Safety*; K.K. Sinha, & D. Bhatnagar, (Eds), pp. 279-306, Marcel Dekker, New York, NY, USA,

Pechanova, O. (2006). Proteomic analysis of maize rachis from inbred lines resistant and susceptible to *Aspergillus flavus*. Ph.D. Dissertation, Mississippi State University, USA, pp. 230.

Peethambaran, B., Hawkins, L., Windham, G.L., Williams, W.P. & Luthe, D.S. (2010). Antifungal activity of maize silk proteins and role of chitinases in *Aspergillus flavus* resistance. *Toxin Reviews,* Vol. 29, pp. 27–39

Probst, C., Njapau, H. & Cotty, P.J. (2007). Outbreak of an acute aflatoxicosis in Kenya in 2004: Identification of the causal agent. *Applied and Environmental Microbiology,* Vol. *73*, pp. 2762–2764

Scott, G.E. & Zummo, N. (1988). Sources of resistance in maize to kernel infection by *Aspergillus flavus* in the field. *Crop Science,* Vol. 28, pp. 505–507

Steiner-Lange, S., Fischer, A., Boettcher, A., Rouhara, I., Liedgens, H., Schmelzer, E. & Knogge, W. (2003). Differential defense reactions in leaf tissues of barley in response to infection by *Rhynchosporium secalis* and to treatment with a fungal avirulence gene product. *Molecular and Plant Microbe Interactions,* Vol. 16, pp. 893–902

Thomann, E.B., Sollinger, J., White, C. & Rivin, C.J. (1992). Accumulation of group 3 late embryogenesis abundant proteins in *Zea mays* embryos. *Plant Physiology,* Vol. 99, pp. 607–614

Veena, V., Reddy, S. & Sopory, S.K. (1999). Glyoxalase I from Brassica juncea: Molecular cloning, regulation and its over-expression confer tolerance in transgenic tobacco under stress. *Plant Journal,* Vol. 17, pp. 385–395

Warburton, M.L., Brooks, T.D., Krakowsky, M.D., Shan, X., Windham, G.L. & Williams, W.P. (2009). Identification and mapping of new sources of resistance to aflatoxin accumulation in maize. *Crop Science,* Vol. 49, pp. 1403–1408

Warner, S.A.J., Scott, R. & Draper, J. (1992). Characterization of a wound-induced transcript from the monocot asparagus that shares similarity with a class of intracellular pathogenesis-related (PR) proteins. *Plant Molecular Biology,* Vol. 19, pp. 555–561

Wesley, S.V., Helliwell, C.A., Smith, N.A., Wang, M.B., Rouse, D.T., Liu, Q., Gooding, P.S. Singh, S.P., Abbott, D., Stoutjesdijk, P.A., Robinson, S.P., Gleave, A.P., Green, A.G. & Waterhouse, P.M. (2001). Construct design for efficient, effective and high-throughput gene silencing in plants. *Plant Journal,* Vol. 27, pp. 581–590

White, D.G., Rocheford, T.R., Kaufman, B. & Hamblin, A.M. (1995). Chromosome regions associated with resistance to *Aspergillus flavus* and inhibition of aflatoxin production in maize. *Proceedings of the USDA-ARS Aflatoxin Elimination Workshop,* pp. 8, Atlanta, GA, USA, October, 1995

White, D.G., Rocheford, T.R., Naidoo, G., Paul, C. & Hamblin, A.M. (1998). Inheritance of molecular markers associated with, and breeding for resistance to Aspergillus Ear Rot and aflatoxin production in maize using Tex6. *Proceedings of the USDA-ARS Aflatoxin Elimination Workshop,* pp. 4-6, St. Louis, MO, USA, October, 1998

Widstrom, N.W., McMillian, W.W. & Wilson, D.M. (1987). Segregation for resistance to aflatoxin contamination among seeds on an ear of hybrid maize. *Crop Science,* Vol. 27, pp. 961–963

Woloshuk, C.P., Cavaletto, J.R. & Cleveland, T.E. (1997). Inducers of aflatoxin biosynthesis from colonized maize kernels are generated by an amylase activity from *Aspergillus flavus*. *Phytopathology,* Vol. 87, pp. 164–169

Xie, Y.R., Chen, Z.-Y., Brown, R.L. & Bhatnagar, D. (2010). Expression and functional characterization of two pathogenesis-related protein10 genes from *Zea mays*. *Journal of. Plant Physiology*, Vol. 167, pp. 121–130

Xu, D., Duan, X., Wang, B., Hong, B., Ho, T.H.D. & Wu, R. (1996). Expression of a late embryogenesis abundant protein gene HVA1, from barley confers tolerance to water deficit and salt stress in transgenic rice. *Plant Physiology*, Vol. 110, pp. 249–257

6

Biomarkers of Aflatoxin Exposure and Its Relationship with the Hepatocellular Carcinoma

Alessandra Vincenzi Jager, Fernando Silva Ramalho,
Leandra Náira Zambelli and Carlos Augusto Fernandes Oliveira
Universidade de São Paulo
Brazil

1. Introduction

Mycotoxins are secondary metabolites produced by fungi that grow naturally in foodstuffs. They are able to generate a wide variety of toxic effects in vertebrates, including men (Coulombe, 1991). Toxigenic fungi may contaminate foodstuffs in the most different phases of production and processing, from cultivation to transport and storage. Mycotoxins show high chemical stability and may persist in the foodstuff even after fungi were removed by common manufacturing and packaging processes (Chu, 1991).

Diseases caused by mycotoxins are called mycotoxicoses. They are diffuse syndromes that cause lesions mainly in organs such as liver, kidneys, epithelial tissue (skin and mucous membranes) and central nervous system, depending on the type of the toxin. Two or more toxins may also occur simultaneously, leading to intensified toxic effects on the susceptible organism (Orsi et al., 2007).

Aflatoxins are mycotoxins produced by fungi in the genus *Aspergillus*, species *A. flavus*, *A. parasiticus* and *A. nomius* (Moss, 1998). These fungi are distributed worldwide, and their optimal growth conditions are relative humidity of 80-85% and temperature around 30ºC (Coulombe, 1991).

Nowadays, 18 similar compounds are called aflatoxins. However, the most important in medical terms are types B_1, B_2, G_1 e G_2 (Coulombe, 1991). Aflatoxin B_1 (AFB_1), besides being the most frequently found in plant substrates, has the greatest toxigenic power. Aflatoxins B_2 (AFB_2), G_1 (AFG_1) and G_2 (AFG_2) have about 50, 20 and 10% of AFB_1 toxigenic power, respectively (Leeson et al., 1995).

AFB_1 is a genotoxic compound, and is considered to be one of the most potent natural mutagens. Liver carcinogenesis is the most important effect of chronic aflatoxin exposure. This toxicity has been widely demonstrated - mainly in relation to AFB_1 - in many animal species, including fish, birds, rodents, carnivores and primates (Busby & Wogan, 1984). Based on available studies, the *International Agency for Research on Cancer* (IARC) concluded, in 1987, that there was enough evidence to classify AFB_1 in Group 1 - human carcinogen (Rothschild, 1992)

One of the most important aspects in risk analysis of chemical substances is to determine the degree of human exposure (World Health Organization [WHO], 2002), a particularly difficult task for contaminants present in foodstuffs. However, it is possible to indirectly estimate the degree of exposure based on data on consumption of contaminated foodstuffs, and on the average occurrence of the toxin. In this estimation, the degree of exposure is

measured in terms of probable daily intake (PDI) per unit of body weight, and is generally expressed in ng/kg of body weigh (BW) / day. In risk analysis, PDI is compared with tolerable daily intake (TDI) determined in toxicological studies. In spite of the genotoxic characteristic of this toxin, there is no consensus on tolerable daily intake of AFB_1.
Taking into account aflatoxin toxicity and the lack of an established TDI, several countries determined regulations on maximum aflatoxin levels allowed in foodstuffs. Table 1 summarizes some data of a report by the Food and Agriculture Organization of the United Nations (FAO, 2004). It may be noted that the European Community and the Mercosur standardized their regulations, although some countries kept some food items with additional country regulations. Foodstuffs characteristic of each country, frequency of consumption of these items and climate characteristics apparently influence maximum limits adopted in each region, although there is a consensus that these limits should comply with the ALARA (as lowest as reasonable accepted) criterion recommended by the FAO (2004).

2. AFB_1 biotransformation

Biotransformation is a process by which the body transforms foreign substances (xenobiotics) in new chemical compounds (metabolites), that is, a process in which the initial compound is modified to be eliminated by the biological system (Guenguerich, 1999). After oral ingestion, AFB_1 is efficiently absorbed and biotransformed before urinary and fecal excretion (Figure 1).
Absorbed AFB_1 and its metabolites are excreted in urine and feces. Breastfeeding mothers who consume contaminated foodstuffs may also shed aflatoxins metabolites in their milk. Studies in animals demonstrated that in normal conditions, 50% of AFB_1 oral dose is quickly absorbed in the duodenum and reach the liver by the portal system (Wilson et al., 1985). AFB_1 is concentrated in the liver and, in lesser amounts, in the kidneys. It may also be found in mesenteric venous blood as free AFB_1 or as water-soluble metabolites (Wogan et al., 1967).
Enzymes of the cytochrome P450 (CYP) family, CYP1A2, CYP3A4 and CYP2A6, are responsible for the biotransformation of absorbed aflatoxins (Essigmann et al., 1982). These enzymes convert AFB_1 into its carcinogenic form, AFB-8,9-epoxide, which bonds covalently to DNA and serum albumin, producing AFB_1-N^7-guanine and lysine adducts, respectively (Essigmann et al. 1977; Sabbioni et al. 1987). The bond between AFB_1 and DNA modifies the structure and biological activity of DNA, leading to the basic mutagenic and carcinogenic mechanisms of the toxin. Studies with rat livers showed that AFB_1-N^7-guanine adducts may be removed after they are formed, leaving apurinic sites in the DNA molecule (Hsieh et al., 1991). Vacant sites tend to be filled with adenine, causing a guanine to thymine transversion and generating a highly significant point of mutation (Aguillar et al., 1993).
Besides being epoxided, AFB_1 can be also oxidized into several other derivatives. The main hydroxylated metabolites are aflatoxin M_1 (AFM_1), aflatoxin Q_1 (AFQ_1), a demethylated metabolite, aflatoxin P_1 (AFP_1), and a reduced metabolite, aflatoxicol (Figure 1). AFM_1 may be activated to form AFM_1-8,9-epoxide, which binds to DNA and is excreted in urine as AFM_1-N^7-guanine (Egner et al, 2003). AFQ_1 and AFP_1 are not significantly oxidized by human microsomes, and are not considered to be genotoxic (Raney et. al, 1992). Metabolites AFM_1, AFQ_1 and AFP_1 are not good substrates for epoxidation, are less genotoxic than AFB_1, and consequently, are considered detoxification products. However, because of the high toxicity reported for AFM_1, researchers should be cautious when labeling this compound a "detoxification product" (Neal et al., 1998).

Country	Food Product	Mycotoxin	Limit (μg/kg)
United States	All foods except milk	AFB_1, AFB_2, AFG_1, AFG_2	20
	Milk	AFM_1	0.5
Canada	Nuts and nut products	AFB_1, AFB_2, AFG_1, AFG_2	15
Mercosur (Brazil, Argentina, Paraguay and Uruguay)	Peanuts, maize, and maize products Fluid milk Powdered milk	AFB_1, AFB_2, AFG_1, AFG_2 AFM_1	20 0.5 5
Bosnia & Herzegovina	Wheat, maize, rice and cereals Beans	AFB_1, AFG_1	1 5
China	Maize and maize products, peanut and peanut products, peanut oil, irradiated peanut	AFB_1	20
	Rice, irradiated rice, edible vegetable oil		10
	Soy bean sauce, grain paste, vinegar, other grains, beans, fermented foods, fermented bean products, starch products, fermented wine, red rice, butter cake, pastry biscuit and bread, food additive alpha-amylase, food additive glucoamylase preparation, salad oil		5
India	All food products	AFB_1, AFB_2, AFG_1, AFG_2 and AFM_1	30
European Union	Groundnuts, nuts and dried fruits, processed products intended for direct human consumption or as ingredients in foodstuffs	AFB_1 AFB_1, AFB_2, AFG_1, AFG_2	2 4
Australia and New Zealand	Peanuts and tree nuts	AFB_1, AFB_2, AFG_1, AFG_2	15
Chile	All foods	AFB_1, AFB_2, AFG_1, AFG_2	5
	Milk	AFM_1	0.05
Japan	All foods	AFB_1	10
Israel	Nuts, peanuts, maize flour, figs and their products, and other foods	AFB_1 AFB_1, AFB_2, AFG_1, AFG_2	5 15
	Milk and milk products	AFM_1	0.05
Italy	Infusion plants	AFB_1 AFB_1, AFB_2, AFG_1, AFG_2	5 10
	Baby food	AFM_1	0.01
Mexico	Cereals and cereal products Corn flour for tortillas	AFB_1, AFB_2, AFG_1, AFG_2	20 12

EU member states: Austria, Belgium, Denmark, Finland, France, Germany, Greece, Ireland, Italy, Luxembourg, the Netherlands, Portugal, Spain, Sweden, the United Kingdom

Table 1. Limits for mycotoxin contamination in food products destined for human consumption in different countries.

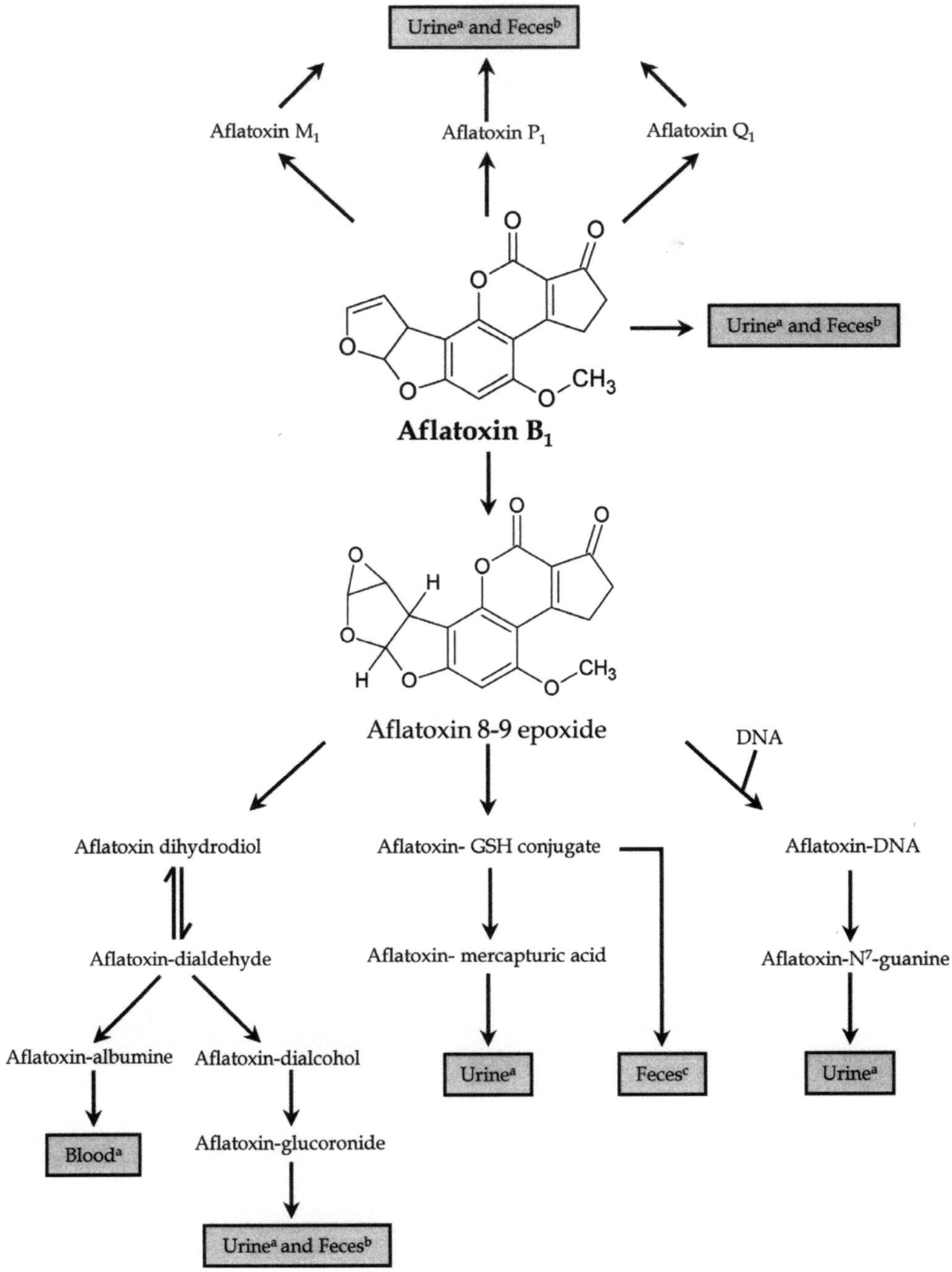

Source: Adapted from Mykkanen et al., 2005.

Fig. 1. Pathways of aflatoxin B_1 biotransformation and excretion in humans.
(a) Experimental and human evidendence of excretion of this metabolite; (b) Scarce or no evidence available; (c) only experimental evidence available (no data for humans).

3. Role of aflatoxins in the etiology of hepatocellular carcinoma

Hepatocellular carcinoma (HCC) represents more than 80% of primary malignant tumors of the liver, and it is the 7th to 9th most common type of cancer worldwide affecting men and women, respectively. About 315,000 new cases of HCC are reported annually, a total of 4.1% of all malignant tumors in the world population. Although it is a relatively uncommon tumor, HCCs are aggressive, and mortality rates reach significant values, with about 312,000 deaths a year, and maximal survival rates of 5% in 5 years. Occurrence of HCC is associated with some degree of chronic liver disease in 90% of the cases, and it is an important cause of death in cirrhosis patients. HCC incidence has been growing, and may be directly related to the frequency of hepatitis C virus infection and longer survival of cirrhosis patients (Yang & Roberts, 2010).

HCC incidence in Africa and southeastern China is far greater than in the rest of the world. Besides known risk factors of western countries such as viral hepatitis and alcohol consumption, these populations are exposed to aflatoxin. The toxin is ingested in contaminated and stored foodstuffs, such as peanuts, maize, soybeans and rice. The association between AFB_1 and HCC is based on the ability of the toxin to induce a specific mutation of gene *p53* (Bressac et al., 1991).

In Brazil, HCC is not included among the ten most common types of cancers, probably because of underreporting. Estimates show that there are 2 to 3,000 diagnoses of the disease every year, with a national incidence of 1:100,000 inhabitants/year. Incidence of this neoplasm is greater in the north, northeast and southeast than in the south of the country. The greatest frequency occurs in the states of Amazonas, Bahia and Espírito Santo. (Pimenta & Massabki, 2010). In São Paulo, incidence is a little greater than the mean of the country, affecting about 2:100.000 inhabitants/year. In terms of mortality, HCC is the 7th death cause and is responsible for 4% of the deaths by cancer in Brazil, annually. HCC incidence rate in Brazil is associated with advanced cirrhosis in 71.2% of the cases, as observed in the rest of the world. However, serology for viral hepatitis is negative in 42% of the cases HCC, even with regional discrepancies (Gonçalves et al., 1997). This difference may be related to the exposure to AFB_1. This relationship, however, was not analyzed in the whole country.

Apparently, HCC progress is not an occasional event. Hepatocarcinogenesis seems to be a multifactorial process in which extrinsic stimuli induce gene changes in mature hepatocytes, leading to successive proliferation and cell death cycles that culminate in the production of monoclonal populations. Several lines of evidence suggest that hepatocarcinogenesis may begin in preneoplastic lesions, such as regenerative macronodules and low or high grade dysplastic nodules. Accumulation of genetic changes and new mutations in preneoplastic lesions would probably cause HCC (Theise et al., 2002).

In molecular terms, many derangements observed in HCC may be preferentially attributed to cirrhosis and inflammatory activity, and others are inherent to dysplastic nodules and to HCC itself. In early stages of chronic hepatitis, there are significant changes in the expression of growth factors, proteases and metalloproteinases, besides somatic changes, reduced apoptosis and increased expression of oncogenes and transcriptional factors. In general, these changes become more prominent and complex as the lesion progresses to fibrosis, cirrhosis, dysplastic nodules and, finally, HCC (Coleman, 2003). No tumor-suppressor gene exclusively associated with HCC has been identified. However, all molecular changes accumulated by chronic hepatitis and cirrhosis in repeated aggression / regeneration cycles, directly contribute to hepatocarcinogenesis (Fausto & Weber, 1993). Aneuploidy is also a common event in hepatocarcinogenesis. HCC is characterized by a considerable loss of heterozygosity, and includes several chromosomes, such as 1p, 4q, 6q, 8p, 8q, 9p, 13q, 16p, 16q, and 17p. Mutations in several critical genes , such *p73, p53, Rb, APC, DLC-1* (deleted in liver cancer), *p16, GSTP1, PTEN, IGF-2, BRCA2, SOCS-1, Smad2 and Smad4, β-catenine, c-myc, and cyclin D1* were also identified (Fujimori et al., 1991; Tsuda et al., 1992).

Impaired control of cell cycle is an important event in carcinogenesis. The first observations involving carcinogenesis and cyclins were related to detection of the incorporation of Hepatitis B virus DNA to *cyclin A* gene in HCC (Wang et al. 1990), and to amplification of *cyclin D1* gene in some cell lineages of colon carcinoma (Leach et al., 1993). The *p16/cyclin D1/RB* pathway (retinoblastoma) may be considered the greatest cell cycle regulator. *RB* and *p16* act as tumor supressor genes, and *cyclin D1* as an oncogene (Weinberg, 1995; Ito et al., 1999). Aberrant expression of both cyclin-dependent kinases (CDK) and CDK-inhibitors has an important role in HCC development. High expression of *cyclin D1* in HCC is variable, ranging from 6 to 76% in different studies. Among positive regulators of cell cycle, changes in *cyclin D1, A* and *B1* expression compared with normal tissues have been associated with increased cell growth and development of neoplasms (Ito et al., 1999).

Analysis of aberrant expression of *cyclin D1*, its biological role and its relationship with mutations in *p53* in cases of HCC demonstrated that *cyclin D1* of was normally expressed in healthy livers, but it was highly reduced in 40% of the livers affected by HCC (Peng et al., 1998). Lower expression of *cyclin D1* RNAm was associated with larger and less differentiated tumors. Increased expression of *cyclin D1* was observed in only 5.6% of the cases. On the other hand, *cyclin E* shows increased expression in 56% of the HCC cases. Overexpression of *cyclin E* was associated with little differentiation and with invasiveness, but not with tumor volume. Thus, decreased expression of *cyclin D1* and increased expression of *cyclin E* are intimately associated with mutation in *p53*. Besides, overexpression of *cyclin E* and concomitant loss of *p53* function seem to contribute to HCC progression (Peng et al., 1998 ; Jung et al., 2001).

There are three important inhibitors of cell cycle progression in the Cip/Kip family: $p27^{KIP1}$, $p21^{WAF1}$, and $p57^{KIP2}$. The most comprehensively studied of these inhibitors, in terms of clinical significance in the evolution of human cancer, is $p27^{KIP1}$. Expression of $p27^{KIP1}$ is marked in non-proliferating cells, and it has important roles in the regulation of both quiescence and progression in G1 phase, by means of inhibition of cyclin / CDK complexes. Loss of $p27^{KIP1}$ acts together with mutations of several oncogenes and suppressor genes, stimulating tumor growth. Reduced production of the protein synthesized by $p27^{KIP1}$ is significantly involved in the stage and volume of primary tumors. Thus, $p27^{KIP1}$ has been described as a crucial negative regulator of HCC progression. Its increased expression is considered an independent variable in favorable prognosis of HCC (Ito et al., 2001; Fiorentino et al., 2000). Reduced $p21^{WAF1}$ expression is mainly related to mutation in gene *p53* in HCC, and also contributes to hepatocarcinogenesis. However, $p21^{WAF1}$ loss was not identified as an independent factor in HCC bad prognosis (Ito et al., 2001). Compared with healthy livers, expression of $p57^{KIP2}$ is significantly decreased in HCC lesions. This decreased expression of $p57^{KIP2}$ was associated with highly aggressive tumors, characterized by more advanced stages, little differentiation, larger size, portal invasion, intense cellular growth and low disease-free survival rates (Ito et al., 2001)

In terms of frequency, the most common molecular changes observed in HCC cases are *p53* (20-70%), *cyclin D* (11%), *p16Ink4* (0-50%), *Rb* (15%) and *β-catenin* (16-26%), from which only *p53* mutation was reported to be associated with hepatitis B virus gene interaction and exposure to aflatoxin (Ozturk, 1999).

Although mutations in *p53* pathway have an important role in HCC pathogenesis in cases of cirrhosis, changes in cell cycle regulator genes $p21^{waf1/cip1}$ and $p27^{Kip1}$ are more involved in

HCC cases unrelated to cirrhosis (Tretiakova et al., 2010). These cases could be, in part, attributed to aflatoxin. Therefore, other molecular changes in genes $p21^{waf1/cip1}$ and $p27^{Kip1}$ should also be assessed in individuals exposed to aflatoxin.

4. Biomarkers of aflatoxin exposure

Biomarkers measure cellular, biological or molecular changes in biological tissues, cells or fluids, providing information on disease or exposure to a given substance. As biomarkers are used to measure or indicate biological processes, detection of specific biomarkers may aid identification, diagnosis and treatment of individuals who are affected and at risk, but still asymptomatic. Development of biomarkers for environmental agents should be based on specific knowledge on metabolism, formation of by-products and general mechanism of action (Groopman & Kensler, 1999).

Biomarkers may be classified in four categories: internal dose, biologically effective dose, early biological response, and altered structure/function. (Figure 2). Internal dose is the amount of substance that is metabolized. Individual characteristics determine susceptibility to exposure, such as the ability to activate / detoxify carcinogens, ability to repair DNA changes, nutritional status and immunity, age, sex and socioeconomic status (WHO, 1993).

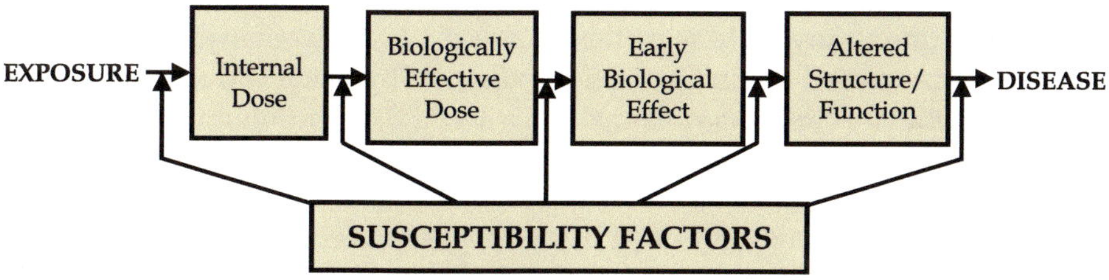

Source: Adapted from Groopman & Kensler (1999)

Fig. 2. Classification of biomarkers.

Biomarkers of exposure and effect for aflatoxins have been validated in comprehensive studies in animals and humans. Dose-response relationship between AFM_1 and AFB_1-N^7-guanine levels and incidence of liver tumors was first established in animals (Groopman et al., 1992b). Biomarkers were then evaluated in humans to determine sensitivity, specificity, accuracy and reliability parameters. Later validation in epidemiological studies evaluated intra and inter-subject variability, the relationship biomarker–external dose and the feasibility of using them in large population studies (Groopman et al., 1992a; Groopman et al., 1992c).

In a study carried out in Shanghai, People's Republic of China, 18,244 volunteers were followed up for three years. The analysis included individual interviews on eating habits, possible previous exposure to aflatoxins, and collection of urine samples (Ross et al., 1992; Qian et al., 1994). Cases and controls were compared to detect associations between aflatoxin markers, infection by Hepatitis B virus (HBV), and hepatocellular carcinoma. Data showed a 340% increase in the relative risk for HCC when aflatoxin biomarkers were detected in urine. Relative risk in individuals showing positive results for HBV was 730% greater. Subjects who showed aflatoxin markers in urine and were positive for HBV infection had a relative risk of developing HCC 5,900% greater. These data support the relationship between the two major causes of HCC: HBV infection and exposure to AFB_1. Besides, when individual metabolites were stratified for HCC incidence, the presence of AFB_1-N^7-guanine adduct led to 200 to 300% increase in the relative risk of developing HCC.

After the Shanghai study, a trial developed in Taiwan with about 15,000 volunteers also analyzed the relationship between HBV, exposure to AFB_1, and incidence of HCC; the results of this trial confirmed the findings of the previous study (Yu et al., 1997). Risk of developing HCC along with AFB_1 exposure was more pronounced among those individuals infected by HBV and with detectable levels of AFB_1-N^7-guanine in their urine.

5. Occurrence of aflatoxin biomarkers in biological fluids

In past decades, several studies reported the presence of aflatoxins, metabolites and biomarkers in urine (Table 2). Zhu et al. (1987) analyzed 252 urine samples from inhabitants of Guangxi province, People's Republic of China, and reported a correlation between total daily ingestion of AFB_1 and excretion of AFM_1. Between 1.2 and 2.2 of AFB_1 ingested daily was found in urine as AFM_1, in levels ranging from 0 to 3.2 ng/mL. In a later study, the same urine samples were analyzed again and levels of AFB_1-N^7-guanine adduct were also correlated with AFB_1 ingestion (Groopman et al., 1992c). Total amounts of AFB_1-N^7-guanine excreted in urine in a three-day period ranged from < 50 and 3250 ng and about 0.2% of AFB_1 ingested was excreted in urine as AFB_1-N^7-guanine. In the same study, levels of the metabolite AFP_1 did not show a significant statistical correlation between dietary exposure and excretion in urine, and the metabolite AFQ_1 was observed in few samples. Percentage of AFB_1 excreted in urine as any of these metabolites was 4.4% in women and 7.6% in men.

In another study, also carried out in Guangxi province, AFB_1-lysine adduct was determined in serum samples of 42 inhabitants and compared with AFB_1 ingestion and AFM_1 excretion in urine (Gan et al., 1988). Significant correlation coefficients were found between AFB_1-lysine levels in serum and AFM_1 in urine, and between AFB_1-lysine in serum and dietary exposure to AFB_1. It is estimated that 1.4 and 2.3% of AFB_1 ingested is covalently bound to albumin.

Qian et al. (1994) detected 55 cases of hepatocellular carcinoma. From these cases, urine samples of 50 individuals and 267 control samples were analyzed for levels of AFB_1-N^7-guanine, AFM_1, AFP_1 and AFB_1. The metabolite detected in the greatest concentration was AFP_1, (0.59-16.0 ng/mL), whereas AFM_1 ranged from 0.17-5.2 ng/mL, and 0.3 to 1.81 ng/mL for AFB_1-N^7-guanine adduct.

Wild et al. (1992) carried out a study with 20 individuals in Gambia, West Africa, and also confirmed the validity of AFB_1-lysine as a biomarker. Parallel evaluation of the same individuals by Groopman et al. (1992a) for AFB_1-N^7-guanine in urine, confirmed not only the correlation between this metabolite and AFB_1, ingestion, but also demonstrated the correlation between levels of AFB_1-lysine in serum and AFB_1-N^7-guanine in urine. AFG_1 was the most frequent metabolite observed in urine, as a consequence of the high concentration of aflatoxin found in the foodstuffs consumed by the individuals analyzed, compared with other studies in which the diet analyzed did not have AFG_1. Besides, metabolites AFQ_1 and AFP_1 were also determined, and AFM_1 was observed in some samples.

Levels of AFB_1-N^7-guanine adducts in urine (Groopman et al., 1992a; Groopman et al., 1992c) and AFB_1-lysine in blood (Gan et al., 1988) show the biological effective dose of aflatoxin to which the individual has been exposed. Concentration of AFB_1-N^7-guanine in urine shows exposure to AFB_1 in a 1 to 2-day period, whereas concentration of AFB_1-lysine in serum indicates 2 to 3-month exposure (Wild et al., 1992).

Urinary and fecal excretion of metabolites AFQ_1 e AFM_1 and urinary excretion of AFB_1-N^7-guanine were evaluated in 83 university students in China (Mykkanen et al., 2005). Mean fecal AFQ_1 concentration (137 ng/g, moist weight) was about 60 times greater than mean

AFM_1 concentration (2.3 ng/g, moist weight). In urine, mean AFQ_1 concentration was 10.4 ng/mL, and 0.04 ng/mL and 0.38 ng/mL for AFM_1 and AFB_1-N^7-guanine, respectively. The authors emphasized that, compared with other studies, differences in concentrations and frequencies of AFQ_1 and AFM_1 in their study may be attributed to differences in age and diet of the subjects. Participants of this study were young adults, 18-24 years of age, whereas in previous trials, individuals were 25 to 65 years old. Expression of CYP3A enzymes, which produce AFQ_1, decreases about 25-40% with age in animals and humans, and consumption of foodstuffs rich in flavonoids, such as green tea, may increase AFQ_1 formation by activation of these enzymes.

In Brazil, Scussel et al. (2006) evaluated the presence of AFB_1-lysine adduct in blood samples of 50 subjects in the city of Sao Paulo, in 1999. The adduct was detected in 62% of the samples, in a concentrations ranging from 0 - 57.3 pg AFB_1-lysine/ mg blood albumin. Mean concentration in positive samples was 14.9 pg/mg. Sixty-five urine samples from inhabitants of the city of Piracicaba, state of Sao Paulo, were analyzed for AFM_1 and 65% of them showed concentrations greater or equal to 1.8 pg/mL, with mean concentration of 5.96 pg/mL (Romero et al., 2010). Correlation between probable aflatoxin intake - estimated by means of questionnaires on the frequency of consumption - and AFM_1 levels in urine were not significantly correlated.

AFM_1 is also excreted in milk during lactation, and several studies demonstrated the presence of this metabolite in human milk. In the Arab Emirates, AFM_1 was detected in milk in concentrations ranging from 5 to 3400 pg/mL (Abdulrazzaq et al., 2003). In Australia, AFM_1 levels ranged from 28 to 1031 pg/mL, and in Thailand, from 39 to 1736 pg/mL (El-Nezami et al., 1995). In a study carried out in Gambia (Zarba et al., 1992), 0.09 to 0.43 % AFB_1 ingested in the diet was excreted in milk as AFM_1. In Brazil, this metabolite was studied in samples collected from human milk banks. From 50 samples analyzed, only one was contaminated by AFM_1 at a concentration of 0.024 ng/mL (Navas et al., 2005). In a recent study carried out with 160 lactating mothers in Iran, AFM_1 was detected in 157 samples, with concentrations ranging from 0.3 to 26.7 ng/kg (Sadeghi et al, 2009).

Aflatoxins were also detected in samples of umbilical cord blood, demonstrating they can cross the placenta, starting exposure to this carcinogen in the uterus (Wild et al., 1991; Turner et al., 2007).

Quantitative determination of several metabolites in complex matrices, such as serum and urine, requires specific and sensitive methods for a large number of samples. Particularly for AFB_1-lysine adduct in serum, methods may include radioimmunoassay (RIA; Gan et al., 1988), enzyme linked imunosorbent assay (ELISA; Wild et al., 1992), or purification with immunoaffinity columns followed by separation by high performance liquid chromatography (HPLC) and detection by fluorescence (Wild et al., 1992; Wang et al., 1996). As all these methods require antibodies for detection and/or purification, results will necessarily reflect the capacity, specificity and/or sensitivity of the antibody (Wang et al., 2001). Results obtained using ELISA, RIA and fluorescence were significantly different (Sheabar et al., 1993; Wild et al., 1990). ELISA is highly sensitive, but it is less specific and shows higher concentration of AFB_1-lysine due to the concomitant detection of adducts from reactions with other amino acids and ingestion of aflatoxins of similar structure, such as AFG_1. HPLC-fluorescence is specific for AFB_1-lysine, but it is not sensitive enough for epidemiological studies.

A recently developed method combines solid phase extraction and liquid chromatography-mass spectrometry (HPLC-MS/MS), showing high specificity and sensitivity (McCoy et al., 2005). The method uses a stable isotope internal standard to correct recovery and equipment variability. This method showed to be adequate for routine quantification of adducts in human serum (Scholl et al., 2006b).

Sample/ Country	Aflatoxin and metabolite	no. samples	% positive samples	Level [a] Mean	Level [a] Range	Ref.
Urine/ Brazil	AFM_1	69	65	5.96 pg/mL	1.8-39.9 pg/mL	Romero et al. (2010)
Urine/ Bosnia and Herzegovi-na	AFB_1					Aljicevic & Hamzic (2010)
	HCC patients	30	100	N.S.	0.05-0.26 µg/kg	
	Control group	30	100		0.05-0.15 µg/kg	
Urine/ China	AFM_1	145	54	NS	0.003-0.243 ng/mL	Sun et al.(1999)
Urine/ China	AFM_1	42	NS	NS	0.01-3.2 ng/mL	Zhu et al. (1987)
				NS	40-4800 ng/day	
Urine/ China	AFM_1	29	89	192 ng/day	0.9-3569 ng/day	Wang et al. (2001)
	AFB_1-N^7-G		41	407	64.9-1789	
	AFB_1-Merc		89	103	6.6-494	
	AFQ_1		26	92.2	77.3-137	
	AFP_1		30	664	80.4-3569	
Urine/ China	AFB_1-N^7-G (placebo)	39				El-Nezami et al (2006)
	Beginning		59.5	0.54 ng/mL	0.29-1.03 ng/mL	
	Week 3		64.3	0.63	0.34-1.16	
	Week 5		57.1	0.46	0.25-0.86	
	Final		54.2	0.45	0.24-0.83	
	AFB_1-N^7-G (intervention with probiotics[c])	44				
	Beginning		51.3	0.42 ng/mL	0.22-0.82 ng/mL	
	Week 3		43.6	0.27	0.15-0.47	
	Week 5		38.5	0.19	0.11-0.31	
	Final		61.5	0.45	0.26-0.79	
Urine/ China	**AFM_1** (intervention with GTP)	352	100			Tang et al. (2008)
	<u>Beginning</u>					
	Placebo			59.41 pg/mg	0.42-141.99 pg/mg	

Sample/ Country	Aflatoxin and metabolite	no. samples	% positive samples	Level [a] Mean	Level [a] Range	Ref.
				crea	crea	
	500 mg GTP			60.85	0.59-746.10	
	1000 mg GTP			40.12	0.52-308.27	
	<u>1st month</u>					
	Placebo			61.67	0.52-881.39	
	500 mg GTP			15.03	0.38-64.27	
	1000 mg GTP			20.06	0.77-51.50	
	<u>3rd month</u>					
	Placebo			78.66	0.24-1276.25	
	500 mg GTP			16.12	0.18-222.35	
	1000 mg GTP			25.95	0.12-338.85	
	AFB_1 - Merc (intervention with GTP)					
	<u>Beginning</u>					
	Placebo			8.67 pg/mg crea	0.43-41.15 pg/mg crea	
	500 mg GTP			10.31	0.38-50.77	
	1000 mg GTP			9.32	0.60-67.71	
	<u>1st month</u>					
	Placebo			9.95	0.09-57.92	
	500 mg GTP			79.53	1.57-362.47	
	1000 mg GTP			79.48	0.30-465.62	
	<u>3rd month</u>					
	Placebo			6.11	0.43-50.58	
	500 mg GTP			97.76	11.32-501.48	
	1000 mg GTP			96.60	18.20-560.30	
Urine/ China	Total aflatoxin (AFB_1-N^7-G, AFP_1. AFB_1. AFQ_1)	42	NS	NS NS	1.5-2.3 ng/mL 3300-6600 ng/day	Groopman et al (1992c)
Urine/ China	AFM_1	317	67	NS	0.17-5.2 ng/mL	Qian et al.(1994)
	AFB_1-N^7-G		49	NS	0.3-1.81	
	AFP_1		53	NS	0.59-16	
	AFB_1		71	NS	NE	

Sample/ Country	Aflatoxin and metabolite	no. samples	% positive samples	Level [a] Mean	Range	Ref.
Urine/ China	AFM_1 AFQ_1 AFB_1-N^7-G	83	83	0.04 ng/mL 10.4 0.38	0.01-0.33 ng/mL 3.4-23.3 0.0-2.15	Mykkanen et al. (2005)
Urine/ Egypt	AFB_1 AFM_1 AFG_1	20	30	NS NS 1.1[b]	< 1.5 ng/mL < 2.5 ng/mL	Al-Saadany (1993)
Urine/ Egypt	AFB_1	60	61	NS	0.01-0.15 ng/mL	Hatem et al. (2005)
Urine / Egypt (Children 1-2.5 years old)	AFB_1 AFB_2 AFG_1 AFG_2 AFM_1	50	38	189[b] pg/mL 1.4 76.6 2.2 5.5	- pg/mL 0.8-2.2 72.1-81.1 0.9-8.0 5.0-6.2	Polychonaki et al. (2008)
Guinea (Children 2-4 years old)	AFB_1 AFB_2 AFG_1 AFG_2 AFM_1	50	86	2682 pg/mL 5.7 709[b] 19.0 97.0	179-18000 pg/mL 0.6-43 - 1.4-199 8.0-801	
Urine/ Gambia	AFB_1-N^7-G (AFG_1. AFM_1. AFP_1. AFQ_1 also detected)	20	NS	NS	48.2-7099 ng/day	Groopman et al (1992b)
Urine/ United States	AFB_1-lys AFM_1	184	20.6 11.7	3.84 pg/mg alb 223.85 pg/mg crea	1.01-16.57 pg/mg alb 1.89-935.49 pg/mg alb	Johnson et al. (2010)
Serum/ Benin	AFB_1-alb	480	99	NS	5-1064 pg/mg alb	Gong et al. (2002)
Serum/ Brazil	AFB_1-alb	50	62	14.9 pg/mg alb	0-57.3 pg/mg alb	Scussel et al. (2006)
Serum/ Egypt	AFB_1	60	61	NS	0.04-0.69 ng/mL	Hatem et al. (2005)
Serum/ Egypt	AFB_1 AFM_1 AFM_2	20	55	NS NS 0.2[b] ng/mL	< 4.5 ng/mL <0.5 -	Al-Saadany (1993)

Sample/ Country	Aflatoxin and metabolite	no. samples	% positive samples	Level [a] Mean	Range	Ref.
Serum/ Gambia	AFB_1-alb	20	NS	44 pg/mg alb	NS	Wild et at. (1992)
Serum/ Gambia	AFB_1-alb	357	100		NS	Wild et at. (2000)
	Feb/Mar			83.2 pg/mg alb		
	July/Aug			34.9 pg/mg alb		
Serum/ China	AFB_1-alb	64	100	0.9972 pmol/mg alb	0.3325-2.2703 pmol/mg alb	Jiang et al.(2005)
Serum/ Gambia	AFB_1-alb	444	100	NS	2.2-459 pg/mg alb	Turner et al. (2000)
Serum/ Gambia	AFB_1-alb	117	100	29.3 pg/mg alb	2.2-254 pg/mg alb	Wild et at. (1993)
Serum/ Guinea	AFB_1-alb	600	95	NS	9.4-22 pg/mg alb	Sylla et al. (1999)
Serum/ Ghana	AFB_1-alb (pregnant women)	755	100	10.9 pg/mg alb	0.44-268.73 pg/mg alb	Shuaib et al. (2010)
Serum/ Ghana	AFB_1-alb	507	100	0.94 pmol/mg alb	0.1-4.4 pmol/mg alb	Tang et al. (2009)
Serum/ Gambia	AFB_1-alb					Turner et al. (2007)
	Mothers	119	100	38.9 pg/mg alb	23.3-64.1 pg/mg alb	
	Umbilical cord	99	48.5	2.5	2.5-7.9	
	Children (4 months)	118	11	2.5	2.5-2.5	
Serum/ China	AFB_1-alb	42	NS	NS	30-340 pg/mg alb	Gan et al. (1988)

Table 2. Aflatoxins and metabolites in human urine and serum. (a) The unit is expressed only in the first row; (b) Only one positive sample; (c) *Lactobacillus rhamnosus LC705* and *Propionibacterium freudenreichii* subsp. *shermanii* (1:1, m:m), 2 - 5 x 10^{10} colony forming units/d. NS, not specified; GTP, green tea polyphenols; alb, albumin; crea, creatinine.

Methods used to determine AFB_1-N^7-guanine adduct include immunoassays (Groopman et al., 1992a), HPLC with UV detection (Groopman et al., 1992c), or fluorescence (Wang et al., 1999; Mykkanen et al., 2005). Egner et al. (2006) described a method using HPLC-MS/MS in the analysis of AFB_1-N^7-guanine in urine, also based on the use of stable isotype internal standard. Precision and accuracy were far superior than previous procedures. Together with the analysis of AFB_1-lysin, determination of these two biomarkers in urine and serum samples is precise, accurate, specific and selective. Determination of residual aflatoxin and metabolites AFM_1, AFP_1 and AFQ_1 in urine has been carried out using HPLC-fluorescence (Tang et al., 2008; Polychronaki et al., 2008; Romero et al., 2010). However, HPLC-MS/MS has recently been used successfully to determine AFB_1, AFB_2, AFG_1, AFG_2, AFM_1 and AFP_1 in urine (Everley et al., 2007).

6. Concluding remarks

Current concepts derived from intensive research on biotransformation, mechanisms of toxicity and evidence of the role of aflatoxins in the etiology of human liver cancer were summarily presented in this chapter. AFB_1 exerts its effects after conversion to the reactive compound AFB_1-epoxide by means of cytochrome P450-dependent enzymes. This epoxide can form derivatives with cellular macromolecules, including proteins, RNA and DNA. Reaction with DNA occurs with guanines in codon 249 of tumor suppressor gene *p53*. Although mutations in *p53* pathway have an important role in HCC pathogenesis other molecular changes in genes $p21^{waf1/cip1}$ and $p27^{Kip1}$ should also be assessed in individuals exposed to aflatoxin.
Primary biotransformation of AFB_1 also produces hydroxylated and less toxic derivatives, such as aflatoxins Q_1 and P_1. Intra and interspecies differences in the pathways of activation/detoxification are directly related to the susceptibility of animals to aflatoxin effects. In humans, individual biomonitoring of AFB_1 metabolites such as AFB_1-N^7-guanine have demonstrated that aflatoxins constitute an important risk factor for hepatocellular carcinoma in highly exposed populations. Some of these studies also show synergism between aflatoxins and hepatitis B virus in the development of human HCC. Based on these concepts, and taking into account the frequent detection of aflatoxins in foodstuffs worldwide, further investigations are needed to assess the level of dietary exposure to these toxins and its impact on human health.

7. References

Abdulrazzaq, Y.M.; Osman, N.; Yousif, Z.M. & Al-Falahi, S. (2003). Aflatoxin M_1 in breast-milk of UAE women. *Annals of tropical paediatrics*, Vol.23, No.3, (September 2003), pp.173-179, ISSN 0272-4936

Aguillar, F.; Hussain, S.P. & Ceerutti, P. (1993). Aflatoxin B_1 induces the transversion of G-T in codon 249 of the p53 tumor suppressor gene in human hepatocytes. *Proceeding of the National Academic Sciences of the United States of America,* Vol.90, No.18, (September 1993), pp.8586-8590, ISSN 0027:8424

Aljicevic, M. & Hamzic, S. (2010). Aflatoxin in the urine of patients suffering from HCC and HBV. *Healthmed,* Vol.4, No.4, (January 2010), pp.852-856, ISSN 1840:2291

Al-Saadany, A.; El-Hennawy, S.; Amra, H.; Iman Abd El-Reheim & Ezzat, S. (1993). A study of aflatoxins and Kwashiorkor in Benha. *Benha Medical Journal,* Vol.10, No.3, (September 1993), pp.125-138, ISSN 1110-208X

Bressac, B.; Kew, M.; Wands, J. & Ozturk, M. (1991). Selective G-mutation to T-mutations of p53 gene in hepatocellular carcinoma in Southern Africa. *Nature,* Vol.350, No.6317, (April 1991), pp. 429-431, 1991, ISSN 0028-0836

Busby, W.F. & Wogan, G.N (1984). Aflatoxins. In: *Chemical carcinogens,* Searle, C.E., , pp. 945-1136, American Chemical Society, ISBN 0841208697 , Washington, D.C., USA

Chu, F.S. (1991). Mycotoxins: food contamination, mechanism, carcinogenic potential and preventive measures. *Mutatation Research,* Vol.259, No. 3-4, (March-April 1991), pp.291-306, ISSN 0921-8262

Coleman, W.B. (2003). Mechanisms of human hepatocarcinogenesis. *Current Molecular Medicine,* Vol. 3, No.6, (September 2003), pp. 573–588, 2003, ISSN 1566-5240

Coulombe, R.A. (1991). Aflatoxins. In: *Mycotoxins and phytoalexins,* Sharma, R.P. & Salunkhe, p.103-143, CRC Press, ISBN 0849388333, Boca Raton

Egner, P.A.; Groopman, J.D.; Wang, J.S.; Kensler, T.W. & Friesen, M.D. (2006). Quantification of Aflatoxin-B_1-N^7-Guanine in Human Urine by High-Performance Liquid Chromatography and Isotope Dilution Tandem Mass Spectrometry. *Chemical Research in Toxicology,* Vol.19, No.9, (September 2006), pp.1191-1195, ISSN 0893-228X

Egner, P.A.; Yu, X.; Johnson, J.K.; Nathasingh, C.K.; Groopman, J.D.; Kensler, T.W. & Roebuck, B.D. (2003). Identification of aflatoxin M_1-N^7-guanine in liver and urine of tree shrews and rats following administration of aflatoxin B_1. *Chemical Research in Toxicology,* Vol.16, No.9, (September 2003), pp.1174-1180, ISSN 0893-228X

El-Nezami, H.; Nicoletti, G.; Neal, G.E.; Donohue, D.C. & Ahokas, J.T. (1995). Aflatoxin M_1 in human breast milk samples from Victoria, Australia and Thailand. *Food and Chemical Toxicology,* Vol.33, No.3, (March 1995), pp.173-179, ISSN 0278-6915

El-Nezami, H.S.; Polychronaki, N.N.; Ma, J.; Zhu, H.; Ling, W.; Salminen, E.K.; Juvonen, R.O.; Salminen, S.J.; Poussa, T. & Mykkanen, M. (2006). Probiotic supplementation reduces a biomarker for increased risk of live cancer in young men from Southern China. *The American Journal of Clinical Nutrition,* Vol.83, No.5, (May 2006), pp.1199-1203, ISSN 0002-9165

Essigmann, J.M.; Croy, R..G; Nadzan, A.M.; Busby, W.F.; Reinhold, V.N.; Buchi, G. & Wogan, G.N. (1977). Structural identification of the major DNA adduct formed by aflatoxin B_1 in vitro. Proceedings of the National Academy of Sciences of the United States of America, Vol.74, No.5, pp.1870-1874, ISSN 0027-8424

Essigmann, J.M.; Croy, R.G.; Bennett, R.A. & Wogan, G.N. (1982). Metabolic activation of aflatoxin B_1: patterns of DNA adduct formation, removal, and excretion in relation to carcinogenesis. *Drug Metabolism Reviews,* Vol.13, No.4, pp. 581-602, ISSN 0360-2532

Everley, R.A.; Ciner, F.L.; Zhang, D.; Scholl, P.F.; Groopman, J.D. & Croley, T.R. (2007). Measurement of aflatoxin and aflatoxin metabolites in urine by liquid chromatography–tandem mass spectrometry. *Journal of Analytical Toxicology,* Vol.31, No.3, (April 2007), pp.150-156, ISSN 0146-4760

Fausto, N. & Webber, E.M. (1993). Control of liver growth. *Critical Reviews in Eukaryotic Gene Expression,* Vol.3, No.2 pp.117-135, ISSN 1045-4403

Fiorentino, M.; Altimari, A.; D'Errico, A.; Cukor, B.; Barozzi, C.; Loda, M. & Grigioni, W.F. (2000). Acquired expression of p27 is a favorable prognostic indicator in patients with hepatocellular carcinoma. *Clinical Cancer Research,* Vol.6, No.10, (October 2000), pp.3966-3972, ISSN 1078-0432

Food and Agriculture Organization of the United Nations - FAO. (2004). Worldwide regulations for mycotoxins in food and feed in 2003. FAO Food and Nutrition Paper No. 81. pp. 1728–3264, Food and Agriculture Organization of the United Nations, ISBN 9251051623, Rome, Italy.

Fujimori, M.; Tokino, T.; Hino, O.; Kitagawa, T.; Imamura, T.; Okamoto, E.; Mitsunobu, M.; Ishikawa, T.; Nakagama, H. & Harada, H. (1991). Allelotype study of primary hepatocellular carcinoma. *Cancer Research*, Vol.51, No.1, (January 1991), pp.89-93, ISSN 0008-5472

Gan, L.S.; Skipper, P.L.; Peng, X.C.; Groopman, J.D.; Chen, J.S.; Wogan, G.N. & Tannenbaum, S.R. (1988). Serum albumin adducts in the molecular epidemiology of aflatoxin carcinogenesis: correlation with aflatoxin B_1 intake and urinary excretion of aflatoxin M_1. *Carcinogenesis*, Vol.9, No.7, (July 1988), pp.1323-1325, ISSN 0143-3334

Gonçalves, C.S.; Pereira, F.E.L. & Gayotto, L.C.C. (1997). Hepatocellular carcinoma in Brazil: report of a national survey (Florianopolis, SC, 1995). *Revista do Instituto de Medicina Tropical de São Paulo*, Vol.39, No.3, (May-June 1997), pp.165-70, 1997, ISSN 0036-4665

Gong, Y.Y.; Cardwell, K.; Hounsa, A.; Egal, S.; Turner, P.c.; Hall, A.J. & Wild, C.P. (2002). Dietary aflatoxin exposure and impaired growth in young children from Benin and Togo: cross sectional study. *British Medical Journal*, Vol.325, No.7354, (July 2002), pp.20-21, ISSN 0959-535X

Groopman, J.D.; Hall, A.J.; Whittle, H.; Hudson, G.J.; Wogan, G.N.; Montesano, R. & Wild, C.P. (1992a). Molecular dosimetry of aflatoxin-N^7-guanine in human urine obtained in The Gambia, West Africa. *Cancer epidemiology, biomarkers & prevention*, Vol. 1, No.3, (March-April 1992), pp.221-227, ISSN 1055-9965

Groopman, J.D.; Hasler, J.A.; Trudel, L.J.; Pikul, A.; Donahue, P.R. & Wogan, G.N. (1992b). Molecular dosimetry in rat urine of aflatoxin-N^7-guanine and other aflatoxin metabolites by multiple monoclonal antibody affinity chromatography and immunoaffinity/high performance liquid chromatography. *Cancer Research*, Vol.52, No.2, (January 1992), pp.267-274, ISSN 0008-5472

Groopman, J.D. & Kensler, T.W. (1999). The light at the end of the tunnel for chemical-specific biomarkers: daylight or headlight? *Carcinogenesis*, Vol.20, No.1, (January 1999), pp.1-11, ISSN 0143-3334

Groopman, J.D.; Zhu, J.Q.; Donahue, P.R.; Pikul, A.; Zhang, L.S.; Chen, J.S. & Wogan, G.N. (1992c). Molecular dosimetry of urinary aflatoxin-DNA adducts in people living in Guangxi Autonomous Region, People's Republic of China. *Cancer Research*, Vol.52. No.1, (January 1992), pp.45-52, ISSN 00088-5472

Guengerich, F.P. (1999). Cytochrome P-450 3A4: regulation and role in drug metabolism. *Annual Review of Pharmacology Toxicology*, Vol.39, pp.1-17, ISSN 0362-1642

Hatem, N.L.; Hassab, H.M.; Abd Al-Rahman, E.M.; El-Deeb, S.A. & El-Sayed Ahmed, R.L. (2005). Prevalence of aflatoxins in blood and urine of Egyptian infants with protein-energy malnutrition. *Food & Nutrition Bulletin*, Vol.26, No.1, pp.49-56, ISSN 0379-5721

Hsieh, D.P.H & Atkinson, D.N. Bisfuranoid mycotoxins: their genotoxicity and carcinogenicity. (1991). *Advances in Experimental Medicine and Biology*, Vol.283, pp.525-532, ISSN 0306437376

International Agency for Research on Cancer - IARC. (1993). Some Naturally Occurring Substances: Food Items and Constituents, Heterocyclicaromatic Amines and Mycotoxins. In: *IARC monographs on the evaluation of carcinogenic risks to humans*. pp. 245–395, International Agency for Research on Cancer, ISBN 928321256-8, Lyon, France

Ito, Y.; Matsuura, N.; Sakon, M.; Miyoshi, E.; Noda, K.; Takeda, T.; Umeshita, K.; Nagano, H.; Nakamori, S.; Dono, K.; Tsujimoto, M,; Nakahara, M.; Nakao, K.; Taniguchi, N. & Monden, M. (1999). Expression and prognostic roles of the G1-S modulators in hepatocellular carcinoma: P27 independently predicts the recurrence. *Hepatology*, Vol.30, No.1, (July 1999), pp.90-99, ISSN 0270-9139

Ito, Y.; Takeda, T.; Sakon, M.; Tsujimoto, M.; Monden, M.; Matsuura, N. (2001). Expression of p57/Kip2 protein in hepatocellular carcinoma. *Oncology*, Vol.61, No.3, pp.221-225, ISSN 0030-2414

Jiang, Y.; Jolly, P.E.; Ellis, W.O.; Wang, J.S.; Phillips, T.D. & Williams, J.H. (2005). Aflatoxin B_1 albumin adduct levels and cellular immune status in Ghanaians. *International Immunology*, Vol.17, No.6, (June 2005), pp.807-814, ISSN 0953-8178

Johnson, N.M.; Qian, G.; Xu, L.; Tietze, D.; Marroquin-Cardona, A.; Robinson, A.; Rodriguez, M.; Kaufman, L.; Cunningham, K.; Wittmer, J.; Guerra, F.; Donnelly, K.C; Williams, J.H.; Wang, J.-S. & Phillips, T.D. (2010). Aflatoxin and PAH exposure biomarkers in a U.S. population with a high incidence of hepatocellular carcinoma. *Science of the Total Environment*, Vol.408, No.23, (November 2010), pp.6027-6031, ISSN 0048-9697

Jung, Y.J.; Lee, K.H.; Choi, D.W.; Han, C.J.; Jeong, S.H.; Kim, K.C.; Oh, J.W.; Park, T.K. & Kim, C.M. (2001). Reciprocal expressions of cyclin E and cyclin D1 in hepatocellular carcinoma. *Cancer Letters*, Vol.168, No.1, (July 2001), pp.57-63, ISSN 0304-3835

Leach, F.S.; Elledge, S.J.; Sherr, C.J.; Willson, J.K.; Markowitz, S.; Kinzler, K.W. & Vogelstein, B. (1993). Amplification of cyclin genes in colorectal carcinomas. *Cancer Research*, Vol.53, No.9, (May 1993), pp.1986-1989, ISSN 00088-5472

Leeson, S.; Diaz, G.J. & Summers, J.D. (1995). *Poultry metabolic disorders and mycotoxins*. Guelph, University Books, ISBN 096956001, Ontario, Canada

McCoy, L., Scholl, P. F., Schleicher, R., Groopman, J. D., Powers, C. & Pfeiffer, C. M. (2005). Analysis of aflatoxin B_1-lysine adducts in serum by isotope dilution liquid chromatography tandem mass spectrometry. *Rapid communications in mass spectrometry*, Vol.19, No.16, pp.2203-2210, ISSN 0951-4198

Moss, M.O. Recent studies of mycotoxins. (1998). *Journal of Applied Microbiology*, Vol.84, Supplement, pp.62S-76S, ISSN 1364- 5072

Mykkanen, H.; Zhu, H.; Salminen, E.; Juvonen, R.O.; Ling, W.; Ma, J.; Polychronaki, N.; Kemlainen, H.; Mykkanen, O.; Salminen, S. & El-Nezami, H. (2005). Fecal and urinary excretion of aflatoxin B_1 metabolites (AFQ_1, AFM_1 and AFB-N^7-guanine) in young Chinese males. *International Journal of Cancer*, Vol.115, No.6, (July 2005), pp.879-84, ISSN 0020-7136

Navas, A.; Sabino, M. & Rodriguez-Amaya, D. B. Aflatoxin M_1 and ochratoxin A in a human milk bank in the city of São Paulo, Brazil. (2005). *Food Additives and Contaminants*, Vol.22, No.5, (May 2005), pp. 57-462, ISSN 0265-203X

Neal, G.E.; Eaton, D.L.; Judah, D.J. & Verma, A. Metabolism and toxicity of aflatoxins M_1 and B_1 in human-derived in vitro systems. (1998). *Toxicology and Applied Pharmacology*, Vol.151, No.1, (July 1998), pp.152-158, ISSN 0041-008X

Ozturk, M. (1999). Genetic aspects of hepatocellular carcinogenesis. *Seminars in Liver Disease*, Vol.19, No.3, pp.235-242, ISSN 0272-8087

Peng, S.Y.; Chou, S.P. & Hsu, H.C. (1998). Association of downregulation of cyclin D1 and of overexpression of cyclin E with p53 mutation, high tumor grade and poor prognosis in hepatocellular carcinoma. *Journal of Hepatology*, Vol.29, No.2, (August 1998), pp.281-289, ISSN 0168-8278

Pimenta, J.R. & Massabki, P.S. (2010). Carcinoma hepatocelular: um panorama clínico. *Revista da Sociedade Brasileira de Clínica Médica*, Vol.8, pp.59-67, ISSN 1679-1010

Polychronaki, N.; Wild, C.P.; Mykkanen, H.; Amra, H.; Abdel-Wahhab, M.; Sylla, A.; Diallo, M.; El-Nezami, H. & Turner, P.C. (2008). Urinary biomarkers of aflatoxin exposure in young children from Egypt and Guinea. *Food and Chemical Toxicology*, Vol.46, No.2, (February 2008), pp.519-526, ISSN 0278-6915

Orsi, R.; Oliveira, C.A.F.; Dilkin, P.; Xavier, J; Direito, G.; Correa, B. (2007). Effects of oral administration of aflatoxin B1 and fumonisin B1 in rabbits (Oryctolagus cuniculus). *Chemico-Biological Interactions*, Vol. 170, No.3 (December 2007), pp. 201-208, ISSN 0009-2797

Qian, G.S.; Ross, R.K.; Yu, M.C.; Yuan, J.M.; Gao, Y.T.; Henderson, B.E.; Wogan, G.N. & Groopman, J.D. (1994). A follow-up study of urinary markers of aflatoxin exposure and liver cancer risk in Shanghai, People's Republic of China. *Cancer epidemiology, biomarkers & prevention*, Vol.3, No.1, (January February 1994), pp.3-10, ISSN 1055-9965

Raney, K.D.; Shimada, T.; Kim, D.H.; Groopman, J.D.; Harris, T.M. & Guengerich, F.P. (1992). Oxidation of aflatoxins and sterigmatocystin by human liver microsomes: significance of aflatoxin Q_1 as a detoxication product of aflatoxin B_1. *Chemical Research in Toxicology*, Vol.5, No.2, (March-April 1992), pp.202-10, ISSN 0893-228X

Romero, A.C.; Ferrreira, T.R.B.; Dias, C.T.S.; Calori-Domingues, M.A. & Gloria, E.M. (2010). Occurrence of AFM_1 in urine samples of a Brazilian population and association with food consumption. *Food Control*, Vol.21, No. 4, (April 2010), pp.554-558, ISSN 0956-7135

Ross, R.K.; Yuan, J.M.; Yu, M.C.; Wogan, G.N.; Qian, G.S.; Tu, J.T.; Groopman, J.D.; Gao, Y,T. & Henderson, B.E. (1992). Urinary aflatoxin biomarkers and risk of hepatocellular carcinoma. *Lancet*, Vol.339, No.8799, (April 1992), pp.943-946, ISSN 0140-6736

Sabbioni, G.; Skipper, P.L.; Buchi, G. & Tannenbaum, S.R. (1987). Isolation and characterization of the major serum albumin adduct formed by aflatoxin B_1 in vivo in rats. *Carcinogenesis*, Vol.8, No.6, (June 1987), pp.819-824, ISSN 0143-3334

Sadeghi, N.; Oveisi, M.R.; Jannat, B.; Hajimahmoodi, M.; Bonyani, H. & Jannat, F. (2009). Incidence of aflatoxin M_1 in human breast milk in Tehran, Iran. *Food Control*, Vol.20, No.1, (January 2009), pp.75-78, ISSN 0956-7135

Scholl, P.F.; Turner, P.C.; Sutcliffe, A.E.; Sylla, A.; Diallo, M.S.; Frisen, M.D.; Groopman, J.D. & Wild, C.P. (2006). Quantitative comparison of aflatoxin B_1 serum albumin adducts in humans by isotope dilution mass spectrometry and ELISA. *Cancer epidemiology, biomarkers & prevention*, Vol.15, No.4, (April 2006), pp.823-826, ISSN 1055-9965

Scussel, V.M.; Haas, P.; Gong, Y. Y.; Turner, C. P. & Wild, C. P. (2006). Study of aflatoxin exposure in a Brazilian population using an aflatoxin-albumin biomarker. *11th International IUPAC Symposium on Mycotoxins and Phycotoxins. Proceedings*, pp.197-202, ISBN 978-90-8686-007-4

Sheabar, F.Z.; Groopman, J.D.; Qian, G.-S. & Wogan, G.N. (1993). Quantitative Analysis of Aflatoxin Albumin adducts. *Carcinogenesis*, Vol.14, No.6, (June 1993), pp. 1203-1208, ISSN 0143-3334

Shuaib, F.M.B.; Jolly, P.E.; Ehiri, J.E.; Yatich, N.; Jiang, Y.; Funkhouser, E.; Person, S.D.; Wilson, C.; Ellis, W.O.; Wang, J-S. & Williams, J.H. (2010). Association between birth outcomes and aflatoxin B_1 biomarker blood levels in pregnant women in Kumasi, Ghana. *Tropical Medicine and International Health*, Vol. 15, No.2, (February 2010), pp.160-167, ISSN 1360-2276

Sun, Z.; Lu, P.; Gail, M.H.; Pee D.; Zhang, Q.; Ming, L..; Wang, J.; Wu, Y.; Liu, G. & Zhu, Y. (1999). Increased risk of hepatocellular carcinoma in male hepatitis B surface antigen carriers with chronic hepatitis who have detectable urinary aflatoxin metabolite M_1. *Hepatology*, Vol.30, No.2, (August 2010), pp.379-383, ISSN 0270-9139

Sylla, A.; Diallo, M.S.; Castegnaro, J. & Wild, C.P. (1999). Interactions between hepatitis B virus infection and exposure to aflatoxins in the development of hepatocellular carcinoma: a molecular epidemiology approach. *Mutatation Research*, Vol.428, No.1-2, (July 1999), pp.187-196, ISSN 0027-5107

Tang, L.; Meng, T.; Xu, L.; Luo, H.; Huang, T.; Yu, J.; Zhang, L.; Gao, W.; Cox, S.B. & Wang, J-S. (2008). Modulation of aflatoxin biomarkers in human blood and urine by green tea polyphenols intervention. *Carcinogenesis*, Vol.29, No.2 , (February 2008), pp.411-417, ISSN 0143-3334

Tang, L.; Xu, L.; Afriyie-Gyawu, E.; Liu, W.; Wang, P.; Tang Y.; Wang, Z.; Huebner, H.J.; Ankrah, N.-A.; Ofori-Adjei, D.;. Williams, J.H.; Wang, J.-S. & Phillips, T.D. (2009). Aflatoxin-albumin adducts and correlation with decreased serum levels of vitamins A and E in an adult Ghanaian population. *Food additives and contaminants*, Vol.26, No.1, pp.108-118, ISSN 0265-203X

Theise N.D.; Park Y.N. & Kojiro M. (2002). Dysplastic nodules and hepatocarcinogenesis. *Clinics in Liver Disease*, Vol.6, No.2, (May 2002), pp.497-512, ISSN 1089-3261

Tretiakova, M.S.; Shabani-Rad, M.T.; Guggisberg, K.; Hart, J.; Anders, R.A. & Gao, Z.H. (2010). Genomic and immunophenotypical differences between hepatocellular carcinoma with and without cirrhosis. *Histopathology*, Vol.56, No.6, (May 2010), pp.683-693, ISSN 0309-0167

Tsuda, H.; Oda, T.; Sakamoto, M. & Hirohashi, S. (1992). Different pattern of chromosomal allele loss in multiple hepatocellular carcinomas as evidence of their multifocal origin. *Cancer Research*. Vol.52, No.6, (March 1992), pp.1504-1509, ISSN 0008-5472

Turner, P.C.; Collison, A.C.; Cheung, Y.B.; Gong, Y.Y.; Hall, A.J.; Prentice, A.M. & Wild, C.P. (2007). Aflatoxin exposure in utero causes growth faltering in Gambia infants. *International Journal of Epidemiology*, Vol.36, No.5, (October 2007), pp.1119-1125, ISSN 0300-5771

Turner, P.C.; Mendy, M.; Whittle, H.; Fortuin, M.; Hall, A.J. & Wild, C.P. (2000). Hepatitis B infection and aflatoxin biomarker levels in Gambian children. *Tropical Medicine & International Health*, Vol.5, No.12, (December 2000), pp.837-841, ISSN 1360-2276

Wang, J.; Chenivesse, X.; Henglein, B. & Brechot, C. (1990). Hepatitis B virus integration in a cyclin A gene in a hepatocellular carcinoma. *Nature*, Vol.343, No.6258 (February 1990), pp.555-557, ISSN 0028-0836

Wang, J.S.; Huang, T.; Su, J.; Liang, F.; Wei, Z.; Liang, Y.; Luo, H.; Kuang, S.Y.; Qian, G.S.; Sun, G.; He, X.; Kensler, T.W. & Groopman, J.D. (2001). Hepatocellular carcinoma and aflatoxin exposure in Zhuqing Village, Fusui County, People's Republic of China. *Cancer epidemiology, biomarkers & prevention*, Vol.10, No.2, (February 2001), pp.143-146, ISSN 1055-9965

Wang, J.-S.; Shen, X.; He, X.; Zhu, Y.-R.; Zhang, B. C.; Wang, J.-B.; Qian, G.-S.; Kuang, S.-Y.; Zarba, A.; Egner, P. A.; Jacobson, L. P.; Munoz, A.; Helzlsouer, K. J.; Groopman, J. D. & Kensler, T. W. (1999). Protective alterations in phase 1 and 2 metabolism of aflatoxin B_1 by oltipraz in residents of Qidong, People's Republic of China. *Journal of the national cancer institute*, Vol.91, No.4, (February 1999), pp.347-354, ISSN 0027-8874

Wang, L.Y.; Hatch, M,; Chen, C.J.; Levin, B.; You, S.L.; Lu, S.N.; Wu, M.H.; Wu, W.P.; Wang, L.W.; Wang, Q.; Huang, G.T.; Yang, P.M.; Lee, H.S. & Santella, R.M. (1996). Aflatoxin exposure and risk of hepatocellular carcinoma in Taiwan. *International Journal of cancer*, Vol.67, No.5, (September 1996), pp.620-625, ISSN 0020-7136

Weinberg, R.A. (1995). The retinoblastoma protein and cell cycle control. *Cell*, Vol.81, No.3 (May 1997), pp.323-330, ISSN 0092-8674

Wild, C.P.; Fortuin, M.; Donato, F.; Whittle, H.C.; Hall, A.J.; Wolf, C.R. & Montesano, R. (1993). Aflatoxin, liver enzymes, and hepatitis B virus infection in Gambian children. *Cancer epidemiology, biomarkers & prevention*, Vol.2, No.6, (November-December 1993), pp.555-61, ISSN 1055-9965

Wild, C.P.; Hudson, G.J.; Sabbioni, G.; Chapot, B.; Hall, A.J.; Wogan, G.N.; Whittle, H.; Montesano, R. & Groopman, J.D. (1992). Dietary intake of aflatoxins and the level of albumin-bound aflatoxin in peripheral blood in The Gambia, West Africa. *Cancer epidemiology, biomarkers & prevention*, Vol.1, No.3, (March-April 1992), pp.229-234, ISSN 1055-9965

Wild, C.P.; Jiang, Y.-Z.; Sabbioni, G.; Chapot, B. & Montesano, R. (1990). Evaluation of methods for quantification of aflatoxin-albumin adducts and their application to human exposure assessment. *Cancer Research*, Vol.50, (January 1990), pp.245-250 ISSN 0008-5472

Wild, C.P.; Rasheed, F.N.; Jawla, M.F.; Hall, A.J.; Jansen, L.A. & Montesano, R. (1991). In-utero exposure to aflatoxin in West Africa. *The Lancet*, Vol. 337, No. 8756, (June 1991), pp.1602, ISSN 0140-6736

Wild, C.P.; Yin, F.; Turner, P.C.; Chemin, I.; Chapot, B.; Mendy, M.; Whittle, H.; Kirk, G.D. & Hall, A.J. (2000). Environmental and genetic determinants of aflatoxin-albumin adducts in the Gambia. *International Journal of cancer*, Vol.86, No. 1, (April 2000), pp.1-7, pp. ISSN 0020-7136

Wilson R.; Ziprin, R.; Ragsdale, S. & Busbee, D. (1985). Uptake and vascular transport of ingested aflatoxin. *Toxicology Letters*, Vol.29, No.2-3, (December 1985), pp.169-176, ISSN 0378-4274

Wogan, G.N.; Edward, G.S. & Shank, R.C. (1967). Excretion and tissue distribution of radioactivity from aflatoxin B_1-C14 in rats. *Cancer Research*, Vol. 27, No.10P1, 1729-1736, ISSN 0008-5472

World Health Organization (WHO). (1993). *Biomarkers and risk assessment: concepts and principles*, World Health Organization, ISBN 9241571551, Geneva

World Health Organization (WHO). (2002). *Evaluation of certain mycotoxins in food*, World Health Organization, ISBN 9241209062, Geneva

Yang, J.D. & Roberts, L.R. (2010). Hepatocellular carcinoma: A global view. *Nature Review Natural Gastroenterology & Hepatology*. Vol.7, No. 8, (August 2010), pp. 448-458, ISSN 1759-5045

Yu, M.W.; Lien, J.P.; Chiu, Y.H.; Santella, R.M.; Liaw, Y.F. & Chen, C.J. (1997). Effect of aflatoxin metabolism and DNA adduct formation on hepatocellular carcinoma among chronic hepatitis B carriers in Taiwan. *Journal of Hepatology*, Vol.27, No.2, (August 1997), pp.320–330, ISSN 0168-8278

Zarba, A.; Wild, C.P.; Hall, A.J.; Montesano, R.; Hudson, G.J. & Groopman, J.D. (1992). Aflatoxin M_1 in human breast milk from The Gambia, West Africa, quantified by combined monoclonal antibody immunoaffinity chromatography and HPLC. *Carcinogenesis*, Vol.13, No.5, (May 1992), pp.891-894, ISSN 0143-3334

Zhu, J.Q.; Zhang, L.S.; Hu, X.; Xiao, Y.; Chen, J.S.; Xu, Y.C.; Fremy, J. & Chu, F.S. (1987). Correlation of dietary aflatoxin B_1 levels with excretion of aflatoxin M_1 in human urine. *Cancer Research*, Vol.47, No.7, (April 1987), pp.1848-1852, ISSN 0008-5472

7

The Use of Proteomics as a Novel Tool in Aflatoxin Research

E. Razzazi-Fazeli, M. Rizwan, C. Mayrhofer and K. Nöbauer
University of Veterinary Medicine Vienna, Austria

1. Introduction

The field of system biology among the disciplines, genomics, transcriptomics, proteomics and other "omics" technologies, has become one of the important scientific fields nowadays and will play a major role in investigating biological processes on a global level. Huge financial, technological and organisation efforts are needed to perform system biology based approaches. Progresses in molecular biology have led to the sequencing of the whole genome of several organisms, ranging from bacteria to various mammals including human. While the term "genomics" has already been used for a long time, new technologies such as transcriptomics, proteomics and metabolomics have been introduced during the last two decades and are beginning to expand rapidly. In general the "omics technologies" deal with hundreds or thousands of genes and/or their products such as mRNA, proteins or metabolites. Genomics is the study of genes, their variation and function by sequencing and mapping them. Similarly, transcriptomics deals with information at the messenger RNA (mRNA) level in an organism, tissue or cells at a given time by providing quantitative or semi-quantitative data. The transcriptome is however, not a straight copy of the genome, since the sequence of RNA molecules can be altered due to differential splicing and RNA editing. In contrast to the genome, which is static, the transcriptome is changing depending on environmental signals. Although, genomics and transcriptomics provide huge amount of information for understanding biological processes, the knowledge regarding the products of genes and transcripts remained underutilized. While, in last three decades predominantly genes and gene expression profiles were investigated and genomics has become the major side in biosciences, the analysis of proteins in a global approach was neglected mostly. However, since the 1990s also proteomics has gained increasing interest in the field of biosciences. The term proteomics was mentioned for the first time in Siena/Italy by Marc Wilkins in 1994 on occasion of the symposium on "2D Electrophoresis: from protein maps to genomes" (Wilkins 1997). Proteomics is defined as the qualitative and quantitative comparison of proteins in a cell or organism under defined conditions at a given time point. In other words proteomics is a term coined to comprise a field that attempts to understand the expression, function and regulation of the entire set of proteins encoded by an organism (Liebler 2002). In addition, since the proteins within a cell are the functioning units, their expression is strongly influenced by the environmental signals such as drugs, toxins, stress, age and other surrounding conditions. It is a complementary technology to genomic as well as transcriptomic research (Wilkins et al., 2007). Proteomics aims to deliver statistically significant quantitative as well as qualitative data, which have to be validated by different methods like western blotting and additionally verified using transcriptomics data. Different strategies have been used within this new discipline such as analyzing the whole proteome, determining the interactome, investigating the secretome and verifying glycoproteome as well as the phosphoproteome etc.

While Protein chemistry has long tradition, proteomics can be seen as the "child" of this discipline. However, there are major differences between traditional protein chemistry and its "child" proteomics: While protein chemistry is interested in only one or a set of proteins, its function and structure, proteomics deal with hundreds of proteins in complex mixtures and their discovery by database matching. Identification and quantification of proteins in a complex mixture is an analytical challenge. If comparing proteomics with genomics and transcriptomics fundamental discrepancy from methodological point of view can be distinguished. DNA and mRNA are physico-chemically homogenous molecules, whereas the researchers in case of proteins are confronted with a different situation. In the case of proteins neither amplification nor hybridization is possible. The proteins exhibit a fully different chemical behaviour like hydrophobicity and hydrophilicity as well as completely different concentration ranges of low and high abundant proteins ranging from millimol to atomol. Additionally, posttranslational modifications, proteolytic degradation and protein turnover lead to a much complex situation in the analysis of proteome compared to genome or transcriptome. Therefore selective and sensitive detection methods and a combination of different techniques are necessary to be able to perform reliable proteome analysis. As a consequence, the cost of instruments and the use of various methods are much higher. Another complementary OMICS technology is metabolomics, which deals with the separation, identification and quantitative determination of thousands of metabolites in biological samples. As a consequence the combination of all systems biology based data will lead to knowledge about the functional biology of whole-organism (Humphery-Smith & Hecker 2006). As other scientific fields proteomics has also found broad interest by mycotoxicologists and fungal biologist. Thus, in the last decades the field of fungal proteomics and related areas have been expanding rapidly (Fig. 1).

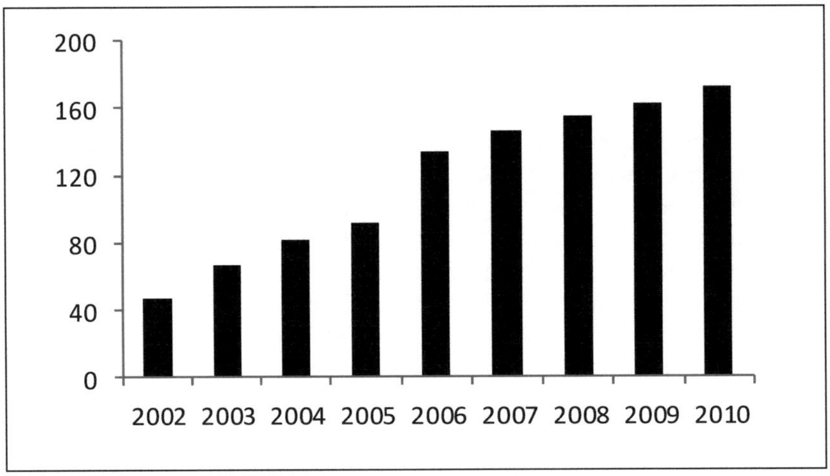

Fig. 1. The number of new articles (including reviews) that have appeared in each of the past eight years related to fungal proteomics

2. Methods in proteome analysis

The early development of proteomics goes back to the introduction of two dimensional gel electrophoresis (2-DE). The current methods of choice in proteomics are one- or two-dimensional polyacrylamide gel electrophoresis (1-DE or 2-DE) followed by nanoHPLC combined with electrospray mass spectrometry (LC-ESI-MS) or matrix-assisted laser-desorption ionization / time of flight mass spectrometry (MALDI-TOF). Accordingly, two strategies for the analysis of proteome can be distinguished, namely "gel-based" and "gel-free" proteomics. In "gel-based" proteomics polyacrylamide gel electrophoresis (PAGE) in either one or two dimensions are used. (Fig.2). Mainly 2-DE in combination with either MALDI-TOF-TOF or nanoHPLC-MS-MS is applied. Due to high resolution power of 2-DE hundreds of proteins can be separated in a single gel. Additionally, 1-DE in combination with nanoHPLC-MS/MS is frequently used in proteomics (Chandramouli & Qian 2009).

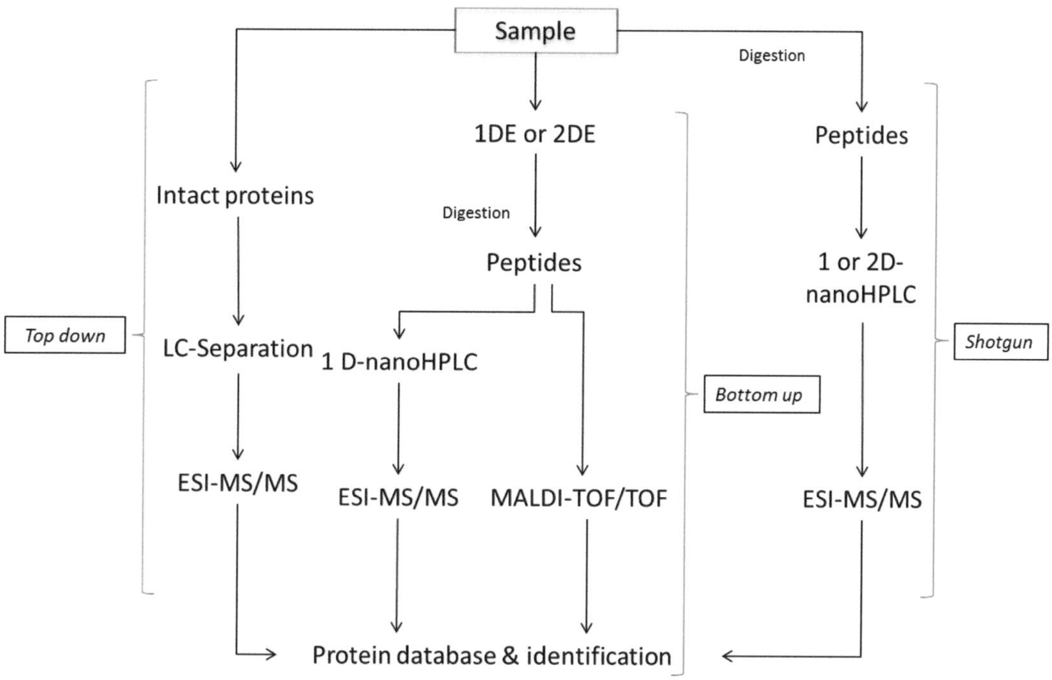

Fig. 2. Schematic illustration of strategies used in proteome analysis

In contrast to the "gel-based" approach, the "gel-free" approach uses chromatographic separation of internal peptides derived from digested proteins. Ion-exchange (IE) and reverse-phase liquid chromatography (RP-HPLC) are used in two dimensional approaches to achieve high resolution separation of peptides (2D-LC), which is also a versatile and powerful tool. However, a "gel-based" proteomic approach is much cheaper than a "gel-free" one if taking the instrument cost into consideration.

Despite the separation performance, the first step in a proteomics approach is the sample preparation, which is extremely crucial since the results are deeply influenced by this first step.

2.1 Sample preparation strategies in fungal proteomics

The aim of the sample preparation in proteomics is the effective extraction of all expressed proteins from an organism or tissue with the possible highest efficiency, by solubilizing them. Without appropriate extraction and solubilisation, further separation and analysis of proteins is definitely not possible and the proteomics approach will fail (Posch 2008). Especially in fungal proteomics, to our experiences, the appropriate sample preparation plays a key role in protein extraction from the cell. Filamentous fungi contain a rigid cell wall and the extraction of intracellular proteins is therefore difficult. The cell wall is involved in many important functions, such as physical protection, osmotic stability, selective permeability barrier, immobilized enzyme transport, cell to cell interactions, and morphogenesis. Additionally, the cell wall is involved in virulence, pathogenicity, antigenicity, immunomodulation and adhesion to host substrate in pathogen fungi (Chaffin et al., 1998). Most of the studies were reported on the cell wall of *Saccharomyces cerevisiae* (De Groot et al., 2005; Klis et al., 2006; Ruiz-Herrera et al., 2006), which is generally composed of glucans (with β-1,3 and β-1,6 linkage), chitin (N-acetylglucosamine polymers), and proteins. These proteins are often highly O- and/or N-mannosylated leading to an elevated complexity (Pitarch et al., 2008). Generally, the cell wall proteins are very difficult to analyze due to their low solubility, hydrophobic nature and low quantity.

Cell lysis is necessary in order to extract the intracellular proteins from filamentous fungi. There are numerous cell lysis protocols used for extraction of proteins, which include: grinding with liquid nitrogen using mortar and pestle (Hernandez-Macedo et al., 2002; Grinyer et al., 2005; Shimizu & Wariishi 2005; Shimizu et al., 2005; Fernandez-Acero et al., 2006; Kniemeyer et al., 2006; Yajima & Kav 2006), mechanical grinding using glass beads (Nandakumar & Marten 2002), chemical lysis (Riezman et al., 1983), and enzymatic lysis (Conzelmann et al., 1988). In a review by Nandakumar & Marten, different lysis methods were compared to extract intracellular proteins of *A. oryzae* for 1-DE and 2-DE (Nandakumar & Marten 2002). The authors tested four lysis cell protocols: (i) boiling in strong alkali, (ii) boiling in sodium dodecyl sulfate (SDS), (iii) chemical lysis in Y-PER® reagent, and (iv) mechanical lysis via rapid agitation with glass beads in a Mini-BeadBeater®. The authors reported that the "mechanical lysis via rapid agitation with glass beads" method seems to be most suitable for the protein extraction and showed good patterns on 1-DE and 2-DE gel electrophoresis.

Most of the protein extraction methods and lysis protocols applied to fungal proteomics were overtaken from plant proteomics with some modifications. Mainly detergents like SDS or 3-[(3-Cholamidopropyl) dimethylammonio]-1-propanesulfonate (CHAPS) as well as chaotropic agents (urea and thiourea) in combination with a reducing agents such as DTT are used. In order to prevent proteolytic degradation of proteins and thus, changes in the proteome pattern, protease activity has to be inhibited. Consequently it is necessary to add protease inhibitors to the lysis buffer. In general a combination of different protease inhibitors in form of a cocktail is recommended. Furthermore, the addition of carrier ampholytes improves the solubility of proteins (Westermeier et al., 2008).

Major proteins of *Aspergillus ochraceus* were extracted with the help of a mortar and pestle in liquid nitrogen and were separated on 1-DE as well as 2-DE (Rizwan et al., 2010). Similarly, intracellular proteins from *A. fumigatus* were extracted by mortar and pestle using liquid nitrogen followed by brief sonication (Carberry et al., 2006).

2.1.1 Protein precipitation

During the cell lysis other interfering substances like phospholipids and nucleic acids can be co-extracted and will be visualized in the acidic part of the gel (Westermeier et al., 2008).

Therefore the crude extract should be purified prior to 2-DE. A very common strategy is to perform precipitation to remove the contaminants after cell lysis. Proteins can be easily and effectively precipitated by ammonium sulfate, chloroform/methanol, trichloroacetic acid (TCA) etc. In the case of TCA precipitation, TCA has to be removed by washing the pellets with acetone in order to prevent problems with isoelectric focusing (IEF). Sodium hydroxide treatment for short time resulted in improved solubilization as studied by (Nandakumar et al., 2003). Another advantage of precipitation is that other interfering substances like detergents and salts can be removed as well (Lopez 2007). In addition, precipitation leads to irreversible inactivation of proteases (Westermeier et al., 2008). Following the precipitation the proteins have to be again dissolved by adding chaotropic agents like urea and thiourea as well as detergents. Especially the addition of zwitterionic detergents like CHAPS were reported to facilitate the protein resolubilization (Kniemeyer et al., 2006). Mechin et al., 2003 reported the most appropriate resolubilization mixture should consist of two detergents (CHAPS and SB3-10), two chaotropes (urea and thiourea), two reducing agents dithiothreitol (DTT) and tris(2-carboxyethyl)phosphine (TCEP) and two types of carrier ampholytes (pH 4-6.5 and pH 3-10 ranges). Despite advantages of precipitation in removing interfering substances, the procedure of precipitation and resolubilization rarely delivers 100% recovery of all proteins resulting into changes in protein pattern (Lopez 2007).

2.2 Two-dimensional gel electrophoresis in proteome expression analysis

Decades after introducing 2-DE, this technique is still the most important and basic research tool used in protein chemistry. Moreover, the establishment of this technique contributed significantly in the development and establishment of proteomics. 2-DE was first introduced independently by (Klose, 1975) and (O'Farrell, 1975). Thereby proteins are separated based on their isoelectric points (pI) in the first dimension and on their molecular masses in the second dimension. Proteins have zwitterionic character due to their acidic and basic groups resulting in different charges depending on the pH of their environment. The pI is the pH at which the net charge of the protein is zero and no migration of the molecules in an electric field is possible. The technique of separating proteins by their pI is called isoelectric focusing (IEF). There are two possibilities to perform IEF, either by using carrier ampholytes generated pH gradients or applying immobilized pH gradient (IPG) (Smith 2009). While in the original method carrier ampholytes in thin gel rods in glass or plastic tubes were used, nowadays mainly IPG strips are used in the first dimension. IPGs consist of a pH gradient in gels prepared by co-polymerization of acrylamide monomers with acrylamide derivates containing carboxylic groups (Lopez 2007). In comparison with the carrier ampholytes based method, the IPG strips are more stable, highly reproducible and allow the focusing of acidic as well as basic proteins (Görg & Weiss 2000; Westermeier & Barnes 2004). Furthermore, IPGs exhibit a large separation capacity and no gradient drift. Meanwhile IPGs in different type of pH gradients and strip lengths are commercially available. The use of narrow-range gradients (pH 3–6, 4–7, 5–8, and 7–10) in various lengths (7, 11, 17, 18, and 24 cm) will improve the resolution of protein separation. The prepared protein extracts can be then loaded by *in-gel* rehydration of IPGs or by cup loading procedure. Once the sample is loaded on the strips, the electric field is applied at 8000 or 10,000 V in a dedicated apparatus with an effective cooling system at 20°C temperature.

Following equilibrating with the anionic detergent SDS, the strips are loaded onto a SDS-PAGE apparatus. SDS denatures and binds to the proteins building a complex with a net negative charge on each protein. The basic principle of the separation is that negatively charged proteins will migrate through the polyacrylamide toward the anode upon the application of an electric field. By applying electric current the proteins are separated based on their size or stokes radii (Laemmli 1970). Dependent on the percentage of polyacrylamide the logarithm of the molecular weight is a linear function of the distance of migration of the proteins. Thus, by using known marker proteins the molecular mass of the proteins can be estimated. Mostly a homogeneous gel is used in second dimension since the proteins are separated in a first dimension (Westermeier et al., 2008). Fig. 3 shows a 2-D GE separation of whole protein extract from *A. ochraceus* NRRL 5175. The protein extract (350 µg) was subjected to IEF followed by SDS-PAGE and Coomassie blue staining. Numbers refer to protein identifications. As can be seen in Fig. 3 on the y-axis of the gel the molecular weight of the proteins appear and on the x-axis the pH gradient (pH 4-10).

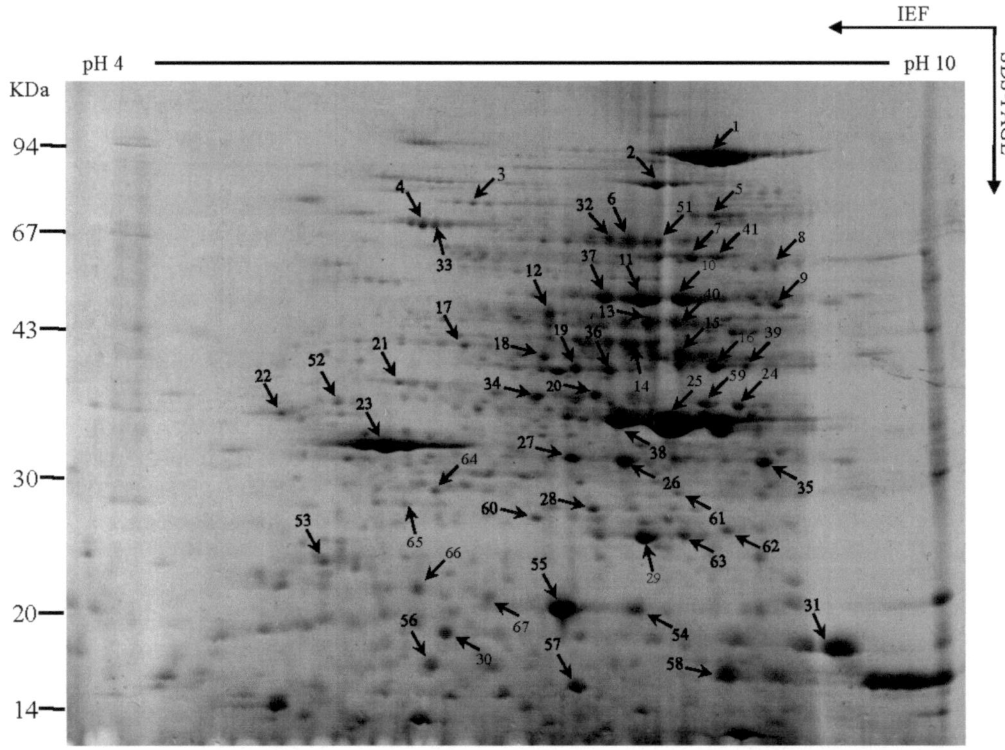

Fig. 3. Two dimensional Gel of the whole protein extract from *A. ochraceus* (Rizwan, 2010)

After the electrophoretic separation, the proteins can be detected by different staining methods that exhibit various sensitivities. There are many dyes commercially available like coomassie blue, colloidal coomassie blue and silver nitrate with sensitivities of 50 ng protein/spot, 10 ng protein/spot and 0.5 ng protein/spot respectively (Chevalier 2010). Additionally, there are fluorescent dyes such as Deep Purple, Sypro Ruby etc., which show sensitivities of about 1 ng protein/spot. In the case of fluorescent dyes an appropriate fluorescent scanner has to be used to visualize the spots. MS-compatibility of silver staining is reached by doing the staining without glutaraldehyde.

The introduction of the "Difference in Gel Electrophoresis (2D-DIGE)" technique (GE Healthcare) was a major progress in semi quantitative gel based proteomics (Ünlü et al., 1997). DIGE is a pre-electrophoretic multiplexed fluorescent staining method. Chemically modified cyanine dyes (CyDye) that have different excitation and emission wavelengths are used. The major advantage of 2D-DIGE is the ability to reduce, to a large extent, gel to gel variation. This is due to the fact that proteins of different proteome states (samples) migrate in the same gel and under same electrophoretic conditions. There are three different dyes Cy2, Cy3 and Cy5, which label the proteins by covalently binding to the lysine residues (minimal labeling). In minimal labeling only 400 pmol/L of dye is added to 50 µg of protein, which means that only 3 to 5% of the proteins will be labeled and 95–97% proteins remain unlabeled (Westermeier et al., 2008).

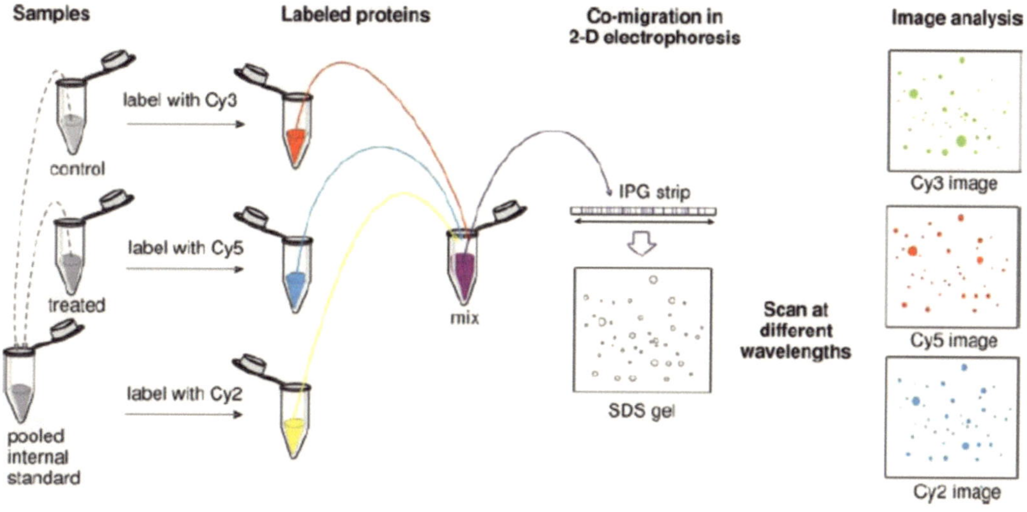

Fig. 4. The principle of 2D-DIGE from (Westermeier et al., 2008).

Mostly two samples (control and treated) in an experiment are labeled by Cy3 and Cy5. Additionally, aliquots of each sample are pooled and then labeled by Cy2 as internal standard (Fig 4, 5). In contrast to lysine minimal labeling, saturating labeling of cysteine residues by Cy3 and Cy5 can be performed, which is very sensitive (Sitek et al., 2005; Westermeier et al., 2008). After electrophoretic separation the gel can be visualized by a fluorescent scanner by exciting each of the dyes with a different wavelength. By using an appropriate software it is possible to obtain semi-quantitative comparison of different proteome patterns within the same gel. Another advantage of 2D-DIGE is its wide linear dynamic range and the possibility to visualize the spots by silver or Coomassie blue staining afterwards.

A drawback is however, that only two different fluorescent reagents are commercially offered for "complete DIGE" and the costs of the reagents are rather high-priced for larger proteomics projects. Furthermore, limitations in loading capacity, quantitative reproducibility, difficulties in handling, and interfacing problems to mass spectrometry limit the analysis depth and comprehensiveness of the gel-based proteomics studies.

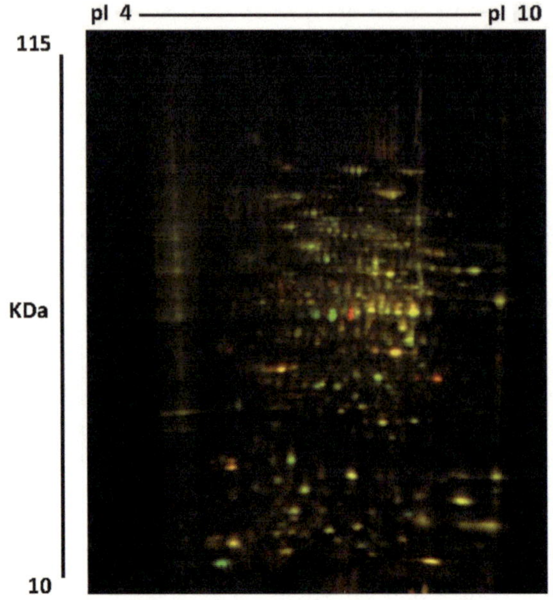

Fig. 5. 2D-DIGE of proteins from *A. ochraceus* NRRL 5221 cultivated in malt extract (ME) and yeast extract (YES) broth. Green (Cy3): 5221 cultivated in ME, and red (Cy5): 5221 cultivated in YES (25µg of protein extract was loaded). Yellow represent those proteins expressed in both cultures (Performed by Ingrid Miller, with friendly permission), (Rizwan, 2010)

Overall, 2-DE was and is still the method with the highest resolution for separating complex mixture of proteins. One of the advantages of 2-DE is the possibility to extract the separated proteins and/or its internal peptides from the gel for MS-based identification. For biochemical characterization of proteins after 2-DE it is possible to detect them by performing western blotting. Beside the high separation capacity large amount of protein can be loaded onto the gel.

In fact, the introduction of IPGs has tremendously facilitated the 2-DE technique. On the basis of IPGs and their commercial availability, 2-DE became more user friendly. In addition, sample loading capacity in 2-DE is satisfactory, however not unlimited. As an example the loading capacity of narrow-range IPGs is much higher than those of broad-range IPGs (Chevalier 2010). Nevertheless, 2-DE shows limitations and advantages like any other method. The detection of low-abundance proteins as well as high molecular weight proteins are major problems in 2-DE. Another drawback of 2-DE is the weakness of this technique when separating very alkaline proteins like ribosomal proteins or histones (Lopez 2007; Chevalier 2010). Additionally, there are also some innate limitations in reproducibility and dynamic range as well as the limitation of detecting post-translational modifications.

Especially in the case of membrane proteins 2-DE has difficulties to resolve and separate proteins due to their hydrophobicity. Their poor solubility leads to "smearing" effects within the IPG strips (Liebler 2002). In these cases a different 2-DE strategy was performed by using 16-benzyldimethyl-n-hexadecylammonium chloride (16-BAC) a cationic detergent in the first dimension followed by a SDS-PAGE separation in the second dimension (Zahedi et al., 2007).

Additionally, gel-free proteomic methods, such as the multidimensional protein identification technology are also becoming popular in proteomics. However, there is up to now no technology that matches 2-DE in its ability for routine parallel expression profiling of large sets of complex protein mixtures. 2-DE delivers a map of intact proteins, which reflects changes in protein expression level, isoforms, or post-translational modifications. Therefore 2-DE using IPGs combined with protein identification by mass spectrometry is at present the workhorse for the majority of current proteome projects.

2.3 Mass spectrometry

Mass spectrometry has revolutionized the analysis of biomolecules from the beginning of the 1990s. Without mass spectrometry (MS) the identification and characterization of proteins, glycans and many other biomolecules in a global approach would not be possible. The mass spectrometer consists of an interface, an ion source, mass analyzer and detector. After ionization of the molecules in the source the ions are separated based on their mass to charge ratio (*m/z*) in the mass analyzer and will be subsequently detected. The so called gas phase ions are accelerated in a field-free space, magnetic or electromagnetic fields towards the detector. Different mass analyzers such as time of flight (TOF), magnetic sectors, quadrupoles, quadrupole ion traps and linear ion traps are commonly used (Wilkins et al., 2007; Faull et al., 2008). Other high resolution mass analyzers such as orbitrap and Fourier transform ion-cyclotron resonance (FTICR) are also used in proteomic studies. Hybrid instruments such as a combination of quadrupole/quadrupole and quadrupole/TOF or linear ion trap/orbitrap analyzer are applied routinely (Fig. 6).

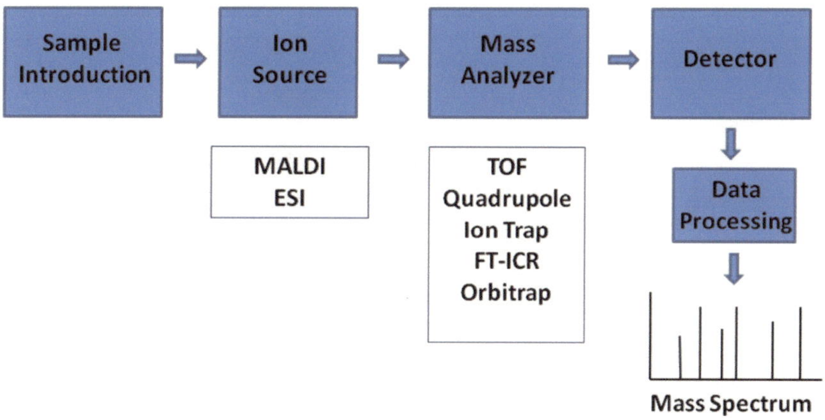

Fig. 6. Block diagram of a mass spectrometer in proteomic approach (modified from (Faull et al., 2008)

Currently, the most commonly used ionization techniques for protein identification are MALDI and ESI in combination with tandem mass spectrometry (MS/MS). These are soft ionization methods which allow ionization of biomolecules without destroying them. The development of desorption ionization methods such as MALDI and the introduction of ESI in combination with HPLC have promoted mass spectrometry to an essential tool in proteomics (Karas et al., 1987; Karas 1988; Hillenkamp et al., 1990). With the help of the so called "biological mass spectrometry" it is possible not only to identify the proteins but also to determine the amino acid sequence and characterize the post-translational modifications of proteins. The development of both soft ionization methods was honored with the Nobel Prize for chemistry in the year 2002.

2.3.1 Matrix Assisted Laser Desorption Ionization (MALDI)

In MALDI, gas-phase ions are produced from large, non-volatile and thermally unstable compounds such as proteins. The ionization of biomolecules is facilitated by using appropriate matrices. The sample is mixed with these mostly low molecular weight compounds, for example α-cyano-4-hydroxy cinnamic acid or 2,5-dihydoxybenzoic acid, which have a strong absorption at the laser wavelength. The matrix along with the sample is then spotted on a target plate or mixed directly on the target. The matrix imparts a key role by strongly absorbing the laser light energy and causing, indirectly, the vaporization of the analyte when the laser pulses of known wavelength hit the crystals inside the source of the mass spectrometer (Fig. 7). The matrix also serves as a proton donor in positive ionization mode and receptor in negative ionization mode to ionize the analyte (Hillenkamp & Peter-Katalini 2007). The ionization depends critically on the matrix–analyte combination, not on the number of acidic or basic groups of the analyte (Beavis et al., 1990). The ions produced are then accelerated towards the analyzer by applying an electric field. In most cases time of flight (TOF) is used as the mass analyzer for MALDI.

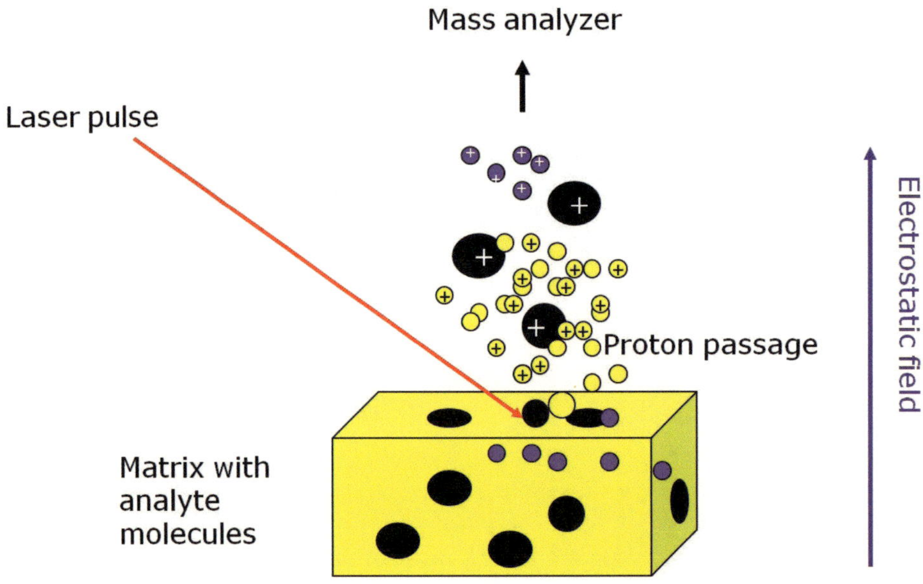

Fig. 7. Schematic illustration of the MALDI ionization. Modified from (Lottspeich & Zorbas 2006)

2.3.2 Electrospray Ionization (ESI)

The basic principle of ESI involves the introduction of a continuous stream of liquid through a capillary to the ion source at atmospheric pressure (Cech & Enke 2001). An electric field is obtained by applying a voltage between the capillary tip and the counter-electrode, i.e. the mass spectrometer entrance. Droplets are generated and vaporized continuously towards the entrance. Droplet fission will occur when the repulsion between the charges on the surface becomes too high, producing new smaller droplets, as shown in Fig. 8 (Wilkins et al., 2007; Westermeier et al., 2008). The ESI interface can be coupled to HPLC, capillary or nano-HPLC (flow rates of 20 nl/min-200 µl/min) as well as to capillary electrophoresis. Generally, multiply charged molecular ions are generated in ESI reducing the mass of proteins of interest. Therefore large proteins can also be analyzed by ESI in mass spectrometers with limited dynamic range.

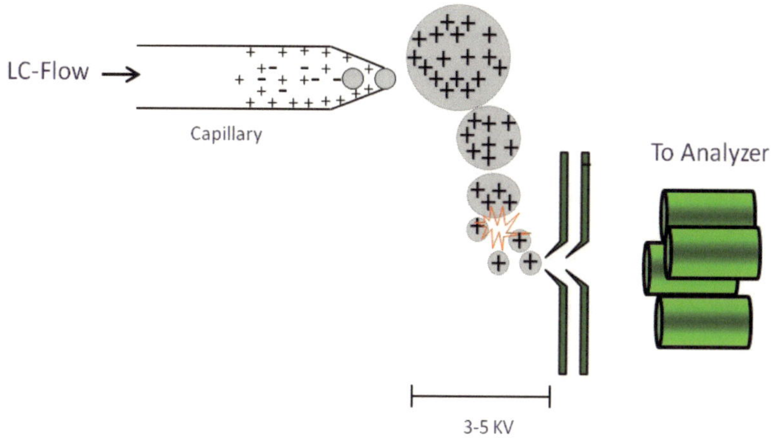

Fig. 8. Principle of ionization in ESI interface

In the last decade, nano electrospray has become the method of choice, when performing HPLC-MS/MS for separating internal peptides and identifying them. Nano electrospray was first introduced by (Wilm & Mann, 1996) and has found its main application field in proteomics. The use of nano flow has several advantages over the conventional LC. The sensitivity is much higher since the column I.D. and flow rate are reduced to 20–100 µm and 50–600 nl/min respectively. Typically 300 nl/min for a 75 µm I.D. column are used in separation of peptides. Additionally the nano flow rate is highly compatible to MS since better ionization efficiency is achieved with reduced flows resulting in an improvement of mass spectrometric sensitivity. In contrast to 2-DE, the on-line nano-HPLC-MS can be easily automated. Recently, using dedicated nano ultra-performance LC (UPLC), columns with smaller particles (1–3 µm) can be used by pressures around 700 bar. Nowadays, the use of the reversed-phase nanoHPLC-ESI-MS/MS has become state of the art for the separation of complex peptide mixtures and identification of unknown proteins. Performing nano-HPLC, an appropriate nano spray source has to be exploited.

If complex protein mixtures (shotgun proteomics) have to be analyzed by nano-HPLC-MS/MS, a better peak resolution in HPLC needs to be achieved. In this case the combination of several chromatographic techniques is necessary. Two-dimensional nanoHPLC is performed using strong cation exchange chromatography in the first dimension and a reversed-phase chromatography in the second dimension (2D-LC). This separation strategy of peptides has become the key component of "gel free" proteomics, which is also called Multidimensional Protein Identification Technology (MudPIT). Nevertheless, 2D-LC is a highly sophisticated technique, which needs well trained human resources. For complex peptide mixtures the number of protein identifications can be increased by approximately 25%, when using two dimensional nano-HPLC (Westermeier et al., 2008).

2.3.3 Identification of proteins

There are three different MS based methods used for global protein identification - "shotgun", "bottom-up" and "top-down"-proteomics (Fig. 2). In "shotgun"-proteomics the whole cell lysate or tissue is digested and the internal peptides are then separated by high resolution one or two dimensional nanoHPLC coupled with ESI-MS/MS. In contrast, in "bottom-up"-proteomics the proteins are separated prior to proteolytic digestion and the internal peptides of proteins are used to search genomic databases for protein identification. In "top-down"-proteomics, the biological material will not undergo a proteolytic digestion (Wehr, 2006). Herein, the intact proteins are directly separated by HPLC and then CID is performed in order to obtain fragments of the proteins, which deliver sequence information. The intact masses of proteins together with MS/MS spectra are then used for identifying and characterizing proteins. In the "top-down"-approach special tailored software is needed. Additionally, high resolution separation methods combined with high resolution mass spectrometers are necessary to obtain reliable results in "top-down"-proteomics. The "bottom-up" approach is the most widely used strategy followed by "shotgun"-proteomics. The "top-down"-proteomics is employed much less than the other approaches.

2.3.3.1 Sample pre-treatment prior to mass spectrometry

Following separation by 1-DE or 2-DE and visualization of the bands/spots the proteins are either manually excised or picked by robots. Destaining prior to further steps is recommended. The spots/bands are washed and the proteins are reduced followed by alkylation. The gel pieces are dried using acetonitrile and the proteins are *in-gel* digested using proteases, mostly trypsin (Fig. 9). The resulting peptides are extracted from the gel and then analyzed by using off-line methods such as MALDI-TOF or are further separated by a chromatographic method such as nanoHPLC coupled with ESI-MS/MS. In case of MALDI analysis the peptide solution needs to be desalted using a miniaturized reverse phase chromatography. For shotgun proteomics the cell lysate is digested in solution including a reduction and alkylation step prior to the proteolytic digest.

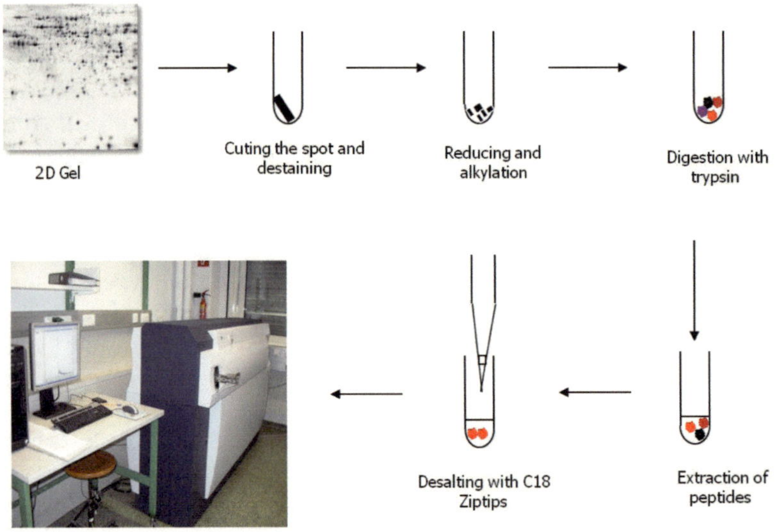

Fig. 9. Workflow and treatment of the gel spots prior to MS analysis.

2.3.3.2 Protein identification by MALDI-TOF/TOF

For protein identification, peptide mass fingerprint (PMF) and tandem mass spectra have to be obtained. PMF is based on the determination of internal peptide masses, which are specific for each protein due to its sequence (Wilkins et al., 2007). It is a rapid and efficient strategy for identifying proteins which nevertheless shows also a number of limitations due to uncertainties. As a consequence MALDI-TOF/TOF instruments have been developed which are able to perform MS/MS. Nowadays it is very easy and fast to analyze post source decay (PSD) ions in order to evaluate the PMF identification results using this type of instruments. It is also possible to obtain sequence information or to perform *de novo* peptide sequencing by collision-induced dissociation (CID).

2.3.3.3 Protein identification by on line LC-ESI-MS/MS

In tandem mass spectrometry parent ions (ionized peptides) will be selected in a first analyzer (MS1) followed by CID with a neutral gas. The fragmented daughter ions will then be separated in a second analyzer (MS2) and detected afterwards (Fig. 11). In the case of peptides the CID will result in specific fragmentation of the polypeptide backbone (Westermeier et al., 2008). The singly charged peptide fragments deliver valuable sequence and structural information. An alternative to PMF is the identification of proteins based on "Peptide Fragment Fingerprinting" (PFF) by nanoHPLC-ESI-MS/MS. In this approach MS/MS spectra are taken into account (Wilkins et al., 2007). Also many post translational modifications like phosphorylation as well as glycosylation of proteins can be studied. In an ion trap the CID fragmentation can be performed in the same trap but hybrid instruments need to have a collision cell.

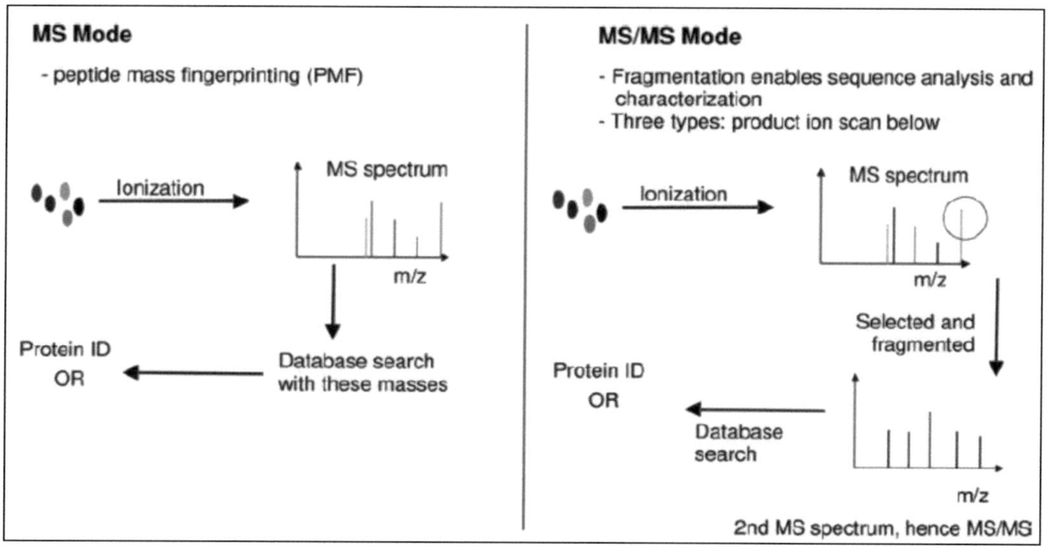

Fig. 10. Schematic illustration of protein identification using MS and MS/MS mode from (Westermeier et al., 2008).

2.3.3.4 Database search

Peptide mass fingerprint or sequences obtained from tandem mass spectra will be matched to virtual protein sequence information in various databases such as Swiss-Prot, which contain information of annotated protein sequences. Search engines like Mascot or Sequest are used (http://www.matrixscience.com , http://fields.scripps.edu/sequest/). However, in the case of some fungi there are only few proteins available in protein databases so *de novo* peptide sequencing needs to be performed. With the obtained information protein homology searches such as MS homology, SPIDER, MS BLAST etc. can be carried out (Rizwan et al., 2010).

In the last decade a number of instrumental progresses have led to enormous improvements in biological mass spectrometry. More sensitive instruments with higher resolution and mass accuracy have been introduced. Nevertheless, technical and biological variances should be taken into account, while performing proteomic techniques.

3. Proteomic approaches in aflatoxin research

Contaminations of agricultural commodities with aflatoxins are not only a serious health risk to humans and animals but also result in economic losses. Therefore, strategies were taken into account to reduce the post- and pre-harvest aflatoxin contamination. There are many factors like nutritional and/or environmental signals, which influence aflatoxin biosynthesis. It has been shown that also the production of aflatoxins is highly dependent not only on Aspergillus species but also on host plants (Bhatnagar et al., 2008). In addition, drought stress and pre-harvest aflatoxin contamination were studied by analyzing defense and stress-related proteins in plant tissue. In some studies *Aspergillus flavus* infection of corn was associated with the expression of stress-related proteins and antifungal proteins. There are some review papers dealing with the application of proteomics in studying fungal biology (Kim et al., 2008). Additionally, Bhatnagar et al. have reviewed the potential of OMICS technologies including genomics, proteomics, metabolomics for solving aflatoxin contamination problem (Bhatnagar et al., 2008). In a further mass spectrometry-based proteomics study fungal cell wall glycoproteins have been analysed (Yin et al., 2008). Nutritional factors as well as environmental influences play a significant role in effecting

Year	*Aspergillus* species	No. of separated peptides	No. of identified peptides	Reference
2001	*A. fumigatus*	19	9	(Bruneau et al., 2001)
2002	*A. nidulans*	200	5	(Melin et al., 2002)
2004	*A. flavus*	110	22	(Medina et al., 2004)
2005	*A. flavus*	>50	51	(Medina et al., 2005)
	A. nidulans	NA	2	(Ström et al., 2005)
2006	*A. fumigatus*	51	26	(Asif et al., 2006)
	A. oryzae	>110	29	(Oda et al., 2006)
	A. fumigatus	>800	37	(Kniemeyer et al., 2006)
	A. fumigatus	180	54	(Carberry et al., 2006)
2007	*A. nidulans*	927	30	(Kim et al., 2007 b)
2009	*A. fumigatus*	-	-	(Kniemeyer et al., 2009)
	A. fumigatus	>700	78	(Zhang et al., 2009)
2010	*A. ochraceus*	>500	26	(Rizwan et al., 2010)

Table 1. Timetable of proteomics studies in Aspergillus species

aflatoxin biosynthesis. One of these factors is drought stress, which is an important factor in preharvest contamination of agricultural commodities with aflatoxins. The conventional breeding, genetic and system biology approaches are valuable tools in improving crop resistance to drought stress resulting on a better aflatoxin management and control strategies. Other approaches have applied proteomics for studying *A. flavus* secreted proteins, effect of environmental signals such as temperature on regulation of proteins in *A. flavus* and studying hepatocarcinogenicity of aflatoxins. In this part we discuss the potential of proteomics to investigate the aflatoxin problematic.

3.1 Proteomics in studying host resistance mechanism to *Aspergillus flavus*

One of the approaches in the field of aflatoxin research with regard to proteomics was to study the aflatoxin resistance proteins in host plants such as corn. The investigation on proteins associated with host resistance has been shown to be a possible strategy for controlling aflatoxin contamination of plants (Brown et al., 2010). An early paper was published by Chen and his co-workers analysing proteins of different corn genotypes, which were found to be resistant or susceptible to *Aspergillus flavus* (Chen et al., 1998). The authors have identified a protein with molecular weight of 14-kDa in relatively high concentrations in kernels of resistant corn genotypes. The protein showed 100% homology of N-terminal sequence to a corn trypsin inhibitor described in a previous work (Swartz et al., 1977). In susceptible hybrids the mentioned 14-kDa protein was not present or only in low concentrations. After the protein extraction and purification by an affinity column using type III trypsin, 1-DE was performed, showing high levels of a 14-kDa protein band in resistant genotypes. The authors suggested that the 14-kDa protein may be useful as a selectable marker for resistance to *A. flavus*.

In a further study the same authors (Chen et al., 2002) performed a proteomics approach to identify potential markers in maize resistant to *A. flavus*. The authors used 2-DE to discover variations in kernel protein expression between genotypes resistant and susceptible. Eleven protein spots, that were upregulated or only present in the resistant maize, were subjected to ESI-MS/MS and conventional Edman degradation for peptide sequencing. From each spot 2 to 9 peptides were sequenced to positively identify the proteins. Based on peptide sequence homology, spots were identified as globulin 1 and 2, which can be categorized as storage proteins. Furthermore, late embryogenesis abundant proteins (LEA) associated with drought were identified. Additionally, heat, water, and osmotic-stress related proteins (aldose reductase, WSI18 and HSP16.9) could be identified.

As *A. flavus* infects crops prior to harvest, the most explored strategy is the investigation of pre-harvest host resistance. One of the methods contributing to the development of host resistance is a kernel screening assay, which shows some advantages compared to traditional breeding methods. Brown et al., 2006 have reviewed efforts and methods dealing with resistance in corn. Using proteome analysis and subtractive approaches several proteins associated with resistance could be identified and characterised. Another review was published by Chen et al. dealing with the identification of aflatoxin resistance-related proteins and genes by using proteomics and RNAi gene silencing (Chen et al., 2009).

The same group of authors has studied the negative correlation between the expression of pathogenesis-related protein (PR10) and the kernel resistance against *A. flavus* infection (Chen et al., 2010). Using an RNAi gene silencing vector in maize embryos the expression of PR10 was reduced in transgenic kernels. Changes in fungal colonization were detected and aflatoxin production was significantly increased. Interestingly, a proteomic approach was conducted in order to proof the RNAi-silenced kernels. The authors could observe a significant reduction in PR10 production on protein level in RNAi-silenced lines.

In a previously published paper Brown et al., 2010 have reviewed different approaches to identify and characterize proteins associated with aflatoxin resistance in kernel. This review emphasizes that the use of proteomics enabled the identification of a number of aflatoxin resistance associated proteins. Some of them were mapped to chromosomal locations. Additionally, genomic approaches were applied and by QTL (quantitative trait locus) mapping some chromosomal regions, which are associated with low aflatoxin accumulation were found (Brown et al., 2010). The investigations on aflatoxin resistance associated proteins and the regulation of their expression will lead to development of aflatoxins resistant plants. In the future the resistance associated proteins can be used as novel breeding markers.

In a further study Chen et al., 2007 have identified proteins in maize kernel endosperm, which are responsible for resistance to aflatoxin contamination. In resistant lines ten protein spots have been found to be more than two-fold upregulated in comparison to susceptible lines. The following proteins were identified by combination of 2-DE and ESI-MS/MS: globulin-2 protein, late embryogenesis abundant proteins (LEA3 and LEA14), a stress-related peroxiredoxin antioxidant (PER1), heat-shock proteins (HSP17.2), a cold-regulated protein (COR), and an antifungal trypsin-inhibitor protein (TI) (Chen et al., 2007).

Another factor affecting the agricultural commodities is the drought stress. In fact the preharvest contamination of maize, peanuts and other products with aflatoxins has been observed to be higher especially in the drought years, having devastating economical loses. In a recently published review (Guo at al., 2008) have investigated the potential of genetics, genomics and proteomics in understanding the relationship between drought stress and preharvest aflatoxin contamination in agricultural products. In this paper, factors affecting aflatoxin production, host resistance and the tools used to study the responses to drought stress are reviewed. Furthermore, the application of corn proteomics and the relationship to host resistance are discussed. In addition, the function and expression of storage and stress-related proteins, which may enhance the stress tolerance of host plants, were highlighted. Different proteomic approaches revealed that resistant lines have elevated levels of stress-related proteins, antifungal and storage proteins in comparison to susceptible lines (Chen et al., 2002; Guo et al., 2008).

Using proteomic tools a number of proteins could be separated by 2-DE and were further identified using mass spectrometry. Different categories of proteins were found, like resistance associated proteins which were divided into 3 groups: stress-responsive proteins, storage proteins and antifungal proteins indicating that storage and stress-responsive proteins may play an important role in enhancing stress-tolerance of host plant (Chen et al., 2002; Bhatnagar et al., 2008).

Based on research done in this field, proteomic approaches can help scientists to understand mechanisms involved in host resistance as well as stress. The achieved results will lead to new strategies for improving plant resistance against fungal contamination. The discovery of storage and stress related proteins as biomarkers of plant tissues, with regard to aflatoxins, will help breeders to find appropriate strategies to improve plant resistance and stress tolerance of host plants.

3.2 Fungal proteomic approach studying secreted proteins in *Aspergillus flavus*

Filamentous fungi normally secrete a broad spectrum of structural proteins and enzymes into their environment that play amongst others an important role in the nutrition of the mycelia. This fact is used in the biotechnology industry for the production of enzymes needed in food, feed and pharmaceutical industry. The analysis of secreted proteins of *Aspergillus flavus* was performed in order to investigate enzymes and other proteins in this fungus. Additionally, analysis of the fungal extracellular proteome can lead to a better understanding of the pathogenicity of these organisms. In general, proteomics of filamentous fungi is still a relatively new approach. There are only some papers dealing with secretome analysis of filamentous fungi. Most of them have focused on purification and characterisation of single proteins. However there are very little investigations on the global analysis of fungal extracellular proteome.

One of the first papers applying a proteomic approach was published by Medina et al. The authors analyzed rutin-induced secreted proteins of *A. flavus* (Medina et al., 2004). In this study, proteomic analysis to identify the extracellular enzymes of *A. flavus,* grown on the flavonoid rutin as the only source of carbohydrates, were used. 1-DE and 2-DE were used to separate the secreted proteins. The protein spots were analysed by MALDI-TOF MS, where thereby 15 rutin-induced and 7 non-induced proteins could be identified. However, more than 90 protein spots had no positive matches in the database and remained unidentified. There is just very little genome sequence information available for *A. flavus,* making the protein identification difficult.

In a further study Medina et al. have analyzed the secreted proteins from *A. flavus* (Medina et al., 2005), which was grown in rutin-containing, glucose-containing and dextrose-containing media. Culture broths from all three media were collected after 3, 7, 10 and 14 days. After sample preparation, the proteins were separated by 1-DE and 2-DE. Differentially secreted proteins were then further analysed by a nano-HPLC system combined with an ion trap MS. Under three different substrate conditions, 51 unique proteins could be identified. 27% of them were proteins with known function involved in carbohydrate metabolism and 22% were proteins involved in proteolysis and peptidolysis. There were also two identified proteins in the category of redox enzymes and two proteins in the category of electron/proton transport proteins. Due to the lack of available genomic sequences and protein sequences in databases, most of the identifications were performed by using homology BLAST searching. The protein identification will be easier as soon as more genome sequences are available.

Recently a review was published by (Bouws et al., 2008) dealing with different aspects, including the relevance of fungal secretome analysis in biotechnology with regards to secreted enzymes. This indicates that the study of secreted proteins will become an important field of fungal proteomics in the future.

3.3 Quantitative proteomics using stable isotope labelling amino acids in cell culture (SILAC) as a new strategy in studying *Aspergillus flavus* biology

The biosynthesis of mycotoxins in fungi has been shown to be highly dependent on their environment. It is due to the regulation of enzymes involved in biosynthesis pathways of these toxins. In an interesting study, changes in response to environmental stimuli were investigated in fungus *A. flavus* (Georgianna et al., 2008). In this approach temperature-dependent regulation of proteins in *A. flavus* were investigated by using stable isotope labelling of amino acids in the growing culture.

Stable isotope labelling amino acids in cell culture (SILAC) is normally used for relative quantification of protein levels. Mostly the amino acids arginine or lysine are labelled by ^{13}C or ^{15}N isotopes. As the proteolytic enzyme trypsin cleavages proteins on these two amino acids sides, at least some peptides will be labelled N-terminal. In this study authors have performed a modified SILAC procedure using $^{13}C_6$-arginine for labelling. The effect of conducive (28°C) and non-conducive (37°C) temperatures for aflatoxin biosynthesis was studied. The cultures in two different approaches were compared: at 28 °C for $^{13}C_6$-arginine versus to 37°C with $^{12}C_6$-arginine and 28 °C for $^{13}C_6$-arginine versus 28 °C for $^{12}C_6$-arginine.

Fungal mycelium was harvested and proteins were extracted. After the sample preparation, 1-DE was performed to separate proteins. Each lane was divided into forty bands, which were cut from the gel. Proteins were *in-gel* digested with trypsin and the internal peptides were extracted from the gel. The resulting peptides were separated by nano-HPLC and were analysed using a high resolution mass spectrometer, a linear ion trap combined with a Fourier transform ion cyclotron resonance mass spectrometer (LTQ-FTICR-MS). At 28 °C about 18 proteins were found to be upregulated whereas at 37°C 31 proteins were higher expressed. In this study a satisfactorily labelling efficiency of 78 ± 6.6% was achieved and the fungal organism accumulated the labelled arginine very fast (Georgianna et al., 2008).

Additionally, the authors performed a transcriptomics approach with the same fungal cultures in order to verify the proteomics results and studied the effect of temperature on gene transcription. RNA quantity was determined by RNA microarray and the correlation of gene transcription with protein expression data was studied. The protein expression was observed to have only a moderate correlation between transcript and protein levels for the same gene. At 37°C the changes in enzyme expression involved in aflatoxin biosynthesis were dependent on the repression of the transcription of aflatoxin pathway genes (Georgianna et al., 2008).

In a further approach the same group of authors have performed "top-down"-proteomics combined with SILAC (Collier et al., 2008). Two cultures of *A. flavus*, one grown in a standard medium with $^{13}C_6$-arginine, the other in medium with $^{12}C_6$-arginine, at 28 °C for 24 h, were prepared. Fungal mycelium was then separated by filtration and lysis was performed by grinding with liquid nitrogen. The lysates of both cultures were combined in a 1:1 mixture. The mixture of intact proteins was then analysed using a nano reversed-phase HPLC coupled to a LTQ-FTICR mass spectrometer. In total 1318 intact proteins or fragments were detected corresponding to 659 SILAC pairs of which 22 proteins could be identified. The authors have reported that an incorrect quantification was observed in the case of proteins with greater numbers of arginine, while proteins with fewer arginine had consistent quantification. The labelling efficiency was similar to previous study published by the same authors, being 75-85%.

These initial approaches could clearly demonstrate the potential of quantitative proteomics using SILAC in studying the effects of environmental signals on fungal cells. It could be shown that changes in protein patterns are really measurable. However, additional investigations are needed in order to study the influence of different growth conditions, not only temperature, but also light, agricultural commodity or nutrition as well as antifungal agents on cell biochemistry.

3.4 Proteomics in studying aflatoxin induced hepatocarcinogenesis

Aflatoxins have been shown to be potent carcinogens and liver toxins inducing hepatocellular carcinoma. The consumption of high amounts of aflatoxin over a long period are associated with liver cirrhosis and/or primary liver carcinomas (Seow et al., 2001; Reiter et al., 2009). Especially aflatoxin B1 is a potent carcinogenic agent which undergoes metabolic activation by cytochrom P450 isoenzymes. The products are the carcinogen aflatoxin B1-8,9-exo-peroxide and the DNA adduct N7-guanin- aflatoxin B1 (Li et al., 2008).
A recent paper investigates the carcinogenic effect of aflatoxins on mammalian cells by analysing the different protein pattern of liver biopsies from animals exposed to aflatoxin B1. The liver proteomes before and after induced hepatocarcinogenesis were analysed in order to identify proteins responsible for hepatocellular carcinoma (Li et al., 2008).
In this study hepatocellular carcinoma was induced in adult tree shrews (*Tupaia belangeri chinensis*). Liver tissue of control and treated animals as well as human liver tissues with and without hepatocellular carcinoma were analysed in the same way. The proteomic approach included the separation of proteins by 2-DE followed by MALDI-TOF/TOF mass spectrometric identification of internal peptides. About 1200 spots on every gel could be detected. Differentially expressed proteins were analysed and those proteins (123 spots), with more than two-fold altered expression levels, could be identified. The results revealed that the expression of peroxiredoxin II was upregulated in hepatocellular carcinoma tissue both in tree shrew and in human samples. The authors confirmed the proteomics results by RT-PCR and Western blot. Peroxiredoxins are antioxidant proteins, which protect the cell against oxidative stress (Li et al., 2008).
Such findings contribute to better understanding the mechanism of hepatocarcinogenesis, when investigating the key proteins responsible for tumorigenesis.

4. Conclusion

The application of "OMICS" technologies especially proteomics in mycotoxin research and mycotoxicology is still at the beginning if compared to biomedical approaches. In the recent years there is however an enormous increase of publications in this field. In mycotoxicology, the study of effects of environmental signals on the expression of proteins will become one of the major issues, since mycotoxin production has been shown to be dependent on the growth and cultivation conditions. The complexicity of proteomics is quite higher than those of genomics due to the fact that the proteome changes according to environmental signals and other factors influencing the cell. Additionally, proteins are processed within the cell by post translational modification, ubiquitination and proteolytic degradation. Proteomic approaches require very expensive equipments and highly trained scientists. Many developments, especially in the field of mass spectrometry, will help to overcome problems regarding sensitivity and mass accuracy as well as mass resolution. It will also contribute to better and reliable protein identifications as well as quantification in proteomics. A further challenge is the systematic and efficient analysis of vast data, which needs suitable bioinformatic tools.
In future the combination of major OMICS technologies like genomics, transcriptomics, proteomics and metabolomics will significantly accelerate the understanding of fungal cell life, its secondary metabolite machinery and cellular responses to its environment. As a consequence the achieved knowledge will help to prevent and/or reduce mycotoxin contamination in agricultural commodities.

5. Acknowledgement

The authors want to thank Ingrid Miller for performing 2D-DIGE as well as Monika Hideghetyova and Hossein Razzazi-Fazeli for technical assistance during preparation of this manuscript.

6. References

Asif, A. R., M. Oellerich, et al. (2006). Proteome of conidial surface associated proteins of Aspergillus fumigatus reflecting potential vaccine candidates and allergens. *J Proteome Res* Vol.5, No.4, pp.954-962

Beavis, R., B. Chait, et al. (1990). Matrix-assisted lasser desorption/ionization. *Physics* Vol.66, pp.269

Bhatnagar, D., K. Rajasekaran, et al. (2008). The 'omics' tools: genomics, proteomics, metabolomics and their potential for solving the aflatoxin contamination problem. *World Mycotoxin Journal* Vol.1, No.1, pp.3-12

Bouws, H., A. Wattenberg, et al. (2008). Fungal secretomes—Nature's toolbox for white biotechnology. *Appl Microbiol Biotechnol* Vol.80, No.3, pp.381-388

Brown, R., Z. Chen, et al. (2006). Proteomics to identify resistance factors in corn-a review. *Mycotoxin Research* Vol.22, No.1, pp.22-26

Brown, R. L., Z.-Y. Chen, et al. (2010). Discovery and Characterization of Proteins Associated with Aflatoxin-Resistance: Evaluating Their Potential as Breeding Markers. *Toxins* Vol.2, No.4, pp.919-933

Bruneau, J. M., T. Magnin, et al. (2001). Proteome analysis of Aspergillus fumigatus identifies glycosylphosphatidylinositol-anchored proteins associated to the cell wall biosynthesis. *Electrophoresis* Vol.22, No.13, pp.2812-2823

Carberry, S., C. M. Neville, et al. (2006). Analysis of major intracellular proteins of Aspergillus fumigatus by MALDI mass spectrometry: identification and characterisation of an elongation factor 1B protein with glutathione transferase activity. *Biochem Biophys Res Commun* Vol.341, No.4, pp.1096-1104

Cech, N. B. and C. G. Enke (2001). Practical implications of some recent studies in electrospray ionization fundamentals. *Mass Spectrom Rev* Vol.20, No.6, pp.362-387

Chaffin, W. L., J. L. Lopez-Ribot, et al. (1998). Cell wall and secreted proteins of Candida albicans: Identification, function, and expression. *Microbiol Mol Biol Rev* Vol.62, No.1, pp.130-180

Chandramouli, K. and P. Y. Qian (2009). Proteomics: Challenges, Techniques and Possibilities to Overcome Biological Sample Complexity. *Human Genomics and Proteomics: HGP* Vol.2009,

Chen, Z.-Y., R. L. Brown, et al. (2002). Identification of Unique or Elevated Levels of Kernel Proteins in Aflatoxin-Resistant Maize Genotypes Through Proteome Analysis. *Phytopathology* Vol.92, No.10, pp.1084-1094

Chen, Z.-Y., R. L. Brown, et al. (2009). Identifying Aflatoxin Resistance-related Proteins/Genes through Proteomics and RNAi Gene Silencing. *Peanut Science* Vol.36, No.1, pp.35-41

Chen, Z.-Y., R. L. Brown, et al. (1998). Resistance to Aspergillus flavus in Corn Kernels Is Associated with a 14-kDa Protein. *Phytopathology* Vol.88, No.4, pp.276-281

Chen, Z. Y., R. Brown, et al. (2007). Identification of maize kernel endosperm proteins associated with resistance to aflatoxin contamination by Aspergillus flavus. *Phytopathology* Vol.97, No.9, pp.1094-1103

Chen, Z. Y., R. L. Brown, et al. (2010). PR10 expression in maize and its effect on host resistance against Aspergillus flavus infection and aflatoxin production. *Mol Plant Pathol* Vol.11, No.1, pp.69-81

Chevalier, F. (2010). Review Highlights on the capacities of" Gel-based" proteomics. *Proteome Science*, Vol.8, No.23

Collier, T. S., A. M. Hawkridge, et al. (2008). Top-down identification and quantification of stable isotope labeled proteins from Aspergillus flavus using online nano-flow reversed-phase liquid chromatography coupled to a LTQ-FTICR mass spectrometer. *Analytical Chemistry* Vol.80, No.13, pp.4994-5001

Conzelmann, A., H. Riezman, et al. (1988). A major 125-kd membrane glycoprotein of Saccharomyces cerevisiae is attached to the lipid bilayer through an inositol-containing phospholipid. *EMBO J* Vol.7, No.7, pp.2233-2240

De Groot, P. W., A. F. Ram, et al. (2005). Features and functions of covalently linked proteins in fungal cell walls. *Fungal Genet Biol* Vol.42, No.8, pp.657-675

Faull, K. F., A. N. Dooley, et al. (2009). An Introduction to the Basic Principles and Concepts of Mass Spectrometry. In Whitelegge J. P. (Edited), *Comprehensive Analytical Chemistry* Vol.52, pp.2-44, Elsevier, Amsterdam

Fernandez-Acero, F. J., I. Jorge, et al. (2006). Two-dimensional electrophoresis protein profile of the phytopathogenic fungus Botrytis cinerea. *Proteomics* Vol.6 Suppl 1, pp.S88-96

Georgianna, D. R., A. M. Hawkridge, et al. (2008). Temperature-dependent regulation of proteins in Aspergillus flavus: whole organism stable isotope labeling by amino acids. *Journal of Proteome Research* Vol.7, No.7, pp.2973-2979

Görg, A. and W. Weiss (2000). *Proteome research: two-dimensional gel electrophoresis and identification methods*. Springer verlag.

Grinyer, J., S. Hunt, et al. (2005). Proteomic response of the biological control fungus Trichoderma atroviride to growth on the cell walls of Rhizoctonia solani. *Curr Genet* Vol.47, No.6, pp.381-388

Guo, B., Z. Y. Chen, et al. (2008). Drought stress and preharvest aflatoxin contamination in agricultural commodity: genetics, genomics and proteomics. *J Integr Plant Biol* Vol.50, No.10, pp.1281-1291

Hernandez-Macedo, M. L., A. Ferraz, et al. (2002). Iron-regulated proteins in Phanerochaete chrysosporium and Lentinula edodes: differential analysis by sodium dodecyl sulfate polyacrylamide gel electrophoresis and two-dimensional polyacrylamide gel electrophoresis profiles. *Electrophoresis* Vol.23, No.4, pp.655-661

Hillenkamp, F., M. Karas, et al. (1990). Matrix assisted UV-laser desorption/ionization: A new approach to mass spectrometry of large biomolecules. *Biological Mass Spectrometry, Burlingame et al.,(eds.), Elsvier Science Pub., Amsterdam,* pp.49-60

Hillenkamp, F. and J. Peter-Katalini (2007). *MALDI MS: A practical guide to instrumentation, methods and applications*. Vch Verlagsgesellschaft Mbh.

Humphery-Smith, I. and M. Hecker (2006). *Microbial proteomics: functional biology of whole organisms*. Wiley-Interscience.

Karas, M., D. Bachmann, et al. (1987). Matrix-assisted ultraviolet laser desorption of non-volatile compounds. *International Journal of Mass Spectrometry and Ion Processes* Vol.78, pp.53-68

Karas, M. a. H., F (1988). Laser desorption ionization of proteins with molecular masses exceeding 10,000 Da. *Analytical Chemistry* Vol.60, No.20, pp.2299-2301

Kim, Y., M. Nandakumar, et al. (2008). The state of proteome profiling in the fungal genus Aspergillus. *Briefings in Functional Genomics & Proteomics* Vol.7, No.2, pp.87-94

Kim, Y., M. P. Nandakumar, et al. (2007). Proteome map of Aspergillus nidulans during osmoadaptation. *Fungal Genet Biol* Vol.44, No.9, pp.886-895

Klis, F. M., A. Boorsma, et al. (2006). Cell wall construction in Saccharomyces cerevisiae. *Yeast* Vol.23, No.3, pp.185-202

Klose, J. (1975). Protein mapping by combined isoelectric focusing and electrophoresis of mouse tissues. *Human Genetics* Vol.26, No.3, pp.231-243

Kniemeyer, O., F. Lessing, et al. (2009). Proteome analysis for pathogenicity and new diagnostic markers for *Aspergillus fumigatus*. *Medical Mycology* Vol.47, supp 1, pp.248 - 254

Kniemeyer, O., F. Lessing, et al. (2006). Optimisation of a 2-D gel electrophoresis protocol for the human-pathogenic fungus Aspergillus fumigatus. *Curr Genet* Vol.49, No.3, pp.178-189

Laemmli, U. K. (1970). Cleavage of structural proteins during the assembly of the head of bacteriophage T4. *Nature* Vol.227, pp.680-685

Li, Y., X. Qin, et al. (2008). Proteome analysis of aflatoxin B1 induced hepatocarcinogenesis in tree shrew (Tupaia belangeri chinensis) and functional identification of candidate protein peroxiredoxin II. *Proteomics* Vol.8, No.7, pp.1490-1501

Liebler, D. C. (2002). *Introduction to proteomics: tools for the new biology*. Humana Pr Inc.

Lopez, J. (2007). Two-dimensional electrophoresis in proteome expression analysis. *Journal of chromatography B* Vol.849, No.1-2, pp.190-202

Lottspeich, F. and H. Zorbas (2006). *Bioanalytik* Elsevier Spektrum Verlag, München.

Mechin, V., L. Consoli, et al. (2003). An efficient solubilization buffer for plant proteins focused in immobilized pH gradients. *Proteomics* Vol.3, No.7, pp.1299-1302

Medina, M. L., P. A. Haynes, et al. (2005). Analysis of secreted proteins from Aspergillus flavus. *Proteomics* Vol.5, No.12, pp.3153-3161

Medina, M. L., U. A. Kiernan, et al. (2004). Proteomic analysis of rutin-induced secreted proteins from Aspergillus flavus. *Fungal Genet Biol* Vol.41, No.3, pp.327-335

Melin, P., J. Schnurer, et al. (2002). Proteome analysis of *Aspergillus nidulans* reveals proteins associated with the response to the antibiotic concanamycin A, produced by Streptomyces species. *Mol Genet Genomics* Vol.267, No.6, pp.695-702

Nandakumar, M. P. and M. R. Marten (2002). Comparison of lysis methods and preparation protocols for one- and two-dimensional electrophoresis of Aspergillus oryzae intracellular proteins. *Electrophoresis* Vol.23, No.14, pp.2216-2222

Nandakumar, M. P., J. Shen, et al. (2003). Solubilization of Trichloroacetic Acid (TCA) Precipitated Microbial Proteins via NaOH for Two-Dimensional Electrophoresis. *J Proteome Res* Vol.2, No.1, pp.89-93

O'Farrell, P. H. (1975). High resolution two-dimensional electrophoresis of proteins. *Journal of Biological Chemistry* Vol.250, No.10, pp.4007

Oda, K., D. Kakizono, et al. (2006). Proteomic analysis of extracellular proteins from Aspergillus oryzae grown under submerged and solid-state culture conditions. *Appl Environ Microbiol* Vol.72, No.5, pp.3448-3457

Pitarch, A., C. Nombela, et al. (2008). Cell Wall Fractionation for Yeast and Fungal Proteomics: *Methods Mol Biol.* Vol. 425, pp.217-239

Posch, A. (2008). *2D Page: sample preparation and fractionation.* Humana Press.

Reiter, E., J. Zentek, et al. (2009). Review on sample preparation strategies and methods used for the analysis of aflatoxins in food and feed. *Molecular nutrition & food research* Vol.53, No.4, pp.508-524

Riezman, H., T. Hase, et al. (1983). Import of proteins into mitochondria: a 70 kilodalton outer membrane protein with a large carboxy-terminal deletion is still transported to the outer membrane. *EMBO J* Vol.2, No.12, pp.2161-2168

Rizwan, M., I. Miller, et al. (2010). Proteome analysis of *Aspergillus ochraceus*. *Mycotoxin Research* Vol.26, No.3, pp.171-180

Rizwan, M. (2010). Proteome profiling of *Aspergillus ochraceus* by 2-D gel electrophoresis and MALDI-TOF/TOF mass spectrometry, Inaugural-Dissertation, University of Veterinary Medicine Vienna

Ruiz-Herrera, J., M. V. Elorza, et al. (2006). Molecular organization of the cell wall of Candida albicans and its relation to pathogenicity. *FEMS Yeast Res* Vol.6, No.1, pp.14-29

Seow, T. K., R. C. M. Y. Liang, et al. (2001). Hepatocellular carcinoma: from bedside to proteomics. *Proteomics* Vol.1, No.10, pp.1249-1263

Shimizu, M. and H. Wariishi (2005). Development of a sample preparation method for fungal proteomics. *FEMS Microbiol Lett* Vol.247, No.1, pp.17-22

Shimizu, M., N. Yuda, et al. (2005). Metabolic regulation at the tricarboxylic acid and glyoxylate cycles of the lignin-degrading basidiomycete Phanerochaete chrysosporium against exogenous addition of vanillin. *Proteomics* Vol.5, No.15, pp.3919-3931

Sitek, B., J. Luttges, et al. (2005). Application of fluorescence difference gel electrophoresis saturation labelling for the analysis of microdissected precursor lesions of pancreatic ductal adenocarcinoma. *Proteomics* Vol.5, No.10, pp.2665-2679

Smith, R. (2009). Two-dimensional electrophoresis: an overview. *Methods Mol Bio.* Vol.519, pp.1-16

Ström, K., J. Schnurer, et al. (2005). Co-cultivation of antifungal Lactobacillus plantarum MiLAB 393 and Aspergillus nidulans, evaluation of effects on fungal growth and protein expression. *FEMS Microbiol Lett* Vol.246, No.1, pp.119-124

Swartz, M. J., H. L. Mitchell, et al. (1977). Isolation and characterization of trypsin inhibitor from opaque-2 corn seeds. *Journal of Biological Chemistry* Vol.252, No.22, pp.8105-8107

Unlü, M., M. E. Morgan, et al. (1997). Difference gel electrophoresis. A single gel method for detecting changes in protein extracts. *Electrophoresis* Vol.18, No.11, pp.2071-2077

Westermeier, R. and N. Barnes (2004). *Electrophoresis in practice.* WILEY-VCH Verlag GmbH & Co. KGaA, Weinheim.

Westermeier, R., T. Naven, et al. (2008). *Proteomics in practice: a guide to successful experimental design.* WILEY-VCH Verlag GmbH & Co. KGaA, Weinheim

Wilkins, M. R. (1997). *Proteome research: new frontiers in functional genomics.* Springer Verlag.

Wilkins, M. R., R. D. Appel, et al., Eds. (2007). *Proteome research: concepts, technology and application*, Springer Verlag.

Wehr T. (2006), Top-Down versus Bottom-Up Approaches in Proteomics, LCGC North America, Volume 24, Issue 9

Wilm, M. and M. Mann (1996). Analytical properties of the nanoelectrospray ion source. *Analytical Chemistry* Vol.68, No.1, pp.1-8

Yajima, W. and N. N. Kav (2006). The proteome of the phytopathogenic fungus *Sclerotinia sclerotiorum*. *Proteomics* Vol.6, No.22, pp.5995-6007

Yin, Q. Y., P. W. J. de Groot, et al. (2008). Mass spectrometry-based proteomics of fungal wall glycoproteins. *Trends in Microbiology* Vol.16, No.1, pp.20-26

Zahedi, R. P., J. Moebius, et al. (2007). Two-dimensional BAC/SDS-PAGE for membrane proteomics. *Subcellular Proteomics*, Vol.43, pp.13-20

Zhang, L., D. Feng, et al. (2009). Comparative proteomic analysis of an Aspergillus fumigatus mutant deficient in glucosidase I (AfCwh41). *Microbiology* Vol.155, No.7, pp.2157-2167

8

Genetic Resistance to Drought in Maize and Its Relationship in Aflatoxins Production

Ramón Gerardo Guevara-González et al.*
CA Ingeniería de Biosistemas. División de Estudios de Posgrado. Facultad de Ingeniería. Universidad Autónoma de Querétaro. C.U. Cerro de las Campanas, S/N, Colonia Las Campanas. C.P. 76010. Santiago de Querétaro, Querétaro, México

1. Introduction

Maize (Zea mays L.) is one of the most important crops in the world. It is the third most important food grain crop in the developing world and is estimated that the demand for maize in developing countries will grow by 50%, from 558 million tons in 1995 to 837 million tons in 2020. Much of this increased demand will be needed by domestic supply for developing countries, which will require intensifying production on existing agricultural land (Ribaut and Ragot, 2007).

Drought is one of the prime abiotic stress in crops in the world. Crop yield losses due to drought stress are considerable. Particularly in maize, as an example, drought is the major stress affecting productivity in Africa leading up to 70% or total crop loss (Muoma et al., 2010; Ashraf, 2010). Although a variety of approaches have been used to alleviate the problem of drought, plant breeding, either conventional breeding or genetic engineering, seems to be an efficient and economic means of tailoring crops to enable them to grow successfully in drought-prone environments (Ashraf, 2010).

In turn, aflatoxins are found to contaminate a wide variety of important agricultural products such as corn, peanuts, tree nuts and cottonseed especially under extreme heat and drought conditions (Payne, 1998; Chen et al., 2003). Aflatoxin contamination significantly reduces the value of grain as an animal feed and export commodity (Chen et al., 2002).

---

* Angela María Chapa-Oliver[1], Laura Mejía-Teniente[1], Irineo Torres-Pacheco[1], Moises Alejandro Vazquez-Cruz[1], Juan Jesús Cervantes-Landaverde[2], Ricardo Ernesto Preciado-Ortiz[3] and Ernesto Moreno- Martinez[4]
1 CA Ingeniería de Biosistemas. División de Estudios de Posgrado. Facultad de Ingeniería, Universidad Autónoma de Querétaro. C.U. Cerro de las Campanas, S/N, Colonia Las Campanas, C.P. 76010, Santiago de Querétaro, Querétaro, México,
2 Biotecnología, Facultad de Química. Universidad Autónoma de Querétaro. C.U. Cerro de las Campanas, S/N, Colonia Las Campanas. C.P. 76010. Santiago de Querétaro, Querétaro, México,
3 Campo Experimental Bajío. Instituto Nacional de Investigaciones Forestales Agrícolas y Pecuarias. Km 6.5 Carretera Celaya-San Miguel Allende, Celaya Gto. México,
4 UNIGRAS, FES-Cuautitlán, UNAM, Cuautitlán Izcalli, Edo. de México, Apdo. Postal 25. México.

Therefore; aflatoxins have emerged as a major concern in agriculture and health sectors because of their harmful effects on human and animal health as well as for ubiquitous presence of aflatoxigenic fungi in many agricultural commodities under field and storage conditions; It is a great problem throughout the world. Aflatoxins are toxic secondary metabolites produced by Aspergillus flavus and *A. parasiticus*. They are potent carcinogens and poses health hazards to humans and domestic animals (Tubajika and Damann, 2001).
The production of aflatoxin is induced by *Aspergillus* species under high temperatures and low relative humidity (Widstrom et al., 2003). Infection of maize by *Aspergillus flavus* with the subsequent aflatoxin accumulation, it is why represents a serious risk in maize growing under drought conditions (Diener, et al.,1987; Payne etal., 1998). Drought can reduce the ability of maize to resist the invation of *A. flavus,* because it negatively affects the expression of genes encoding associated.resistance.
Substantial effort has been made to identify corn genotypes that resist infection by *A. flavus.* Although the most desirable and effective control of A. flavus and aflatoxin contamination is through the development of genetically resistant maize genotypes, successful management of aflatoxin in the field will require host resistance combined with management strategies such as appropriate nitrogen fertilization, population densities, insect control, and irrigation (Tubajika and Damann, 2001). Field studies demonstrate that reduction of drought stress by irrigation reduces aflatoxin contamination in corn and peanut. Drought tolerant corn varieties were also found to produce significantly less aflatoxins in the field under drought conditions compared to aflatoxin-resistant controls; this suggests a possible association between drought tolerance and aflatoxin resistance in corn (Brown et al., 2004).
Gene expression studies of plants in response to biotic or abiotic stress also found that disease resistance-related genes could be regulated by abiotic stress and vice versa. Further examination of host plant and pathogen interactions revealed that plant responses to abiotic stress and pathogen infections were mediated through several common regulatory genes or factors. The presence of "cross-talk" between responses to abiotic stress and biotic stress provides new approaches for enhancing host resistance to biotic stress through the up regulation of key signal transduction factors. Recent efforts to identify molecular and genetic markers for corn kernel resistance as well as studies in host plant-pathogen interactions have suggested a correlation between stress tolerance and plant disease resistance (Brown et al., 2004).

2. Maiz drought tolerance

Abiotic stress presents a major challenge in the quest for sustainable food production as this may reduce the potential yields by 70% in crop plants. Coping with plant environmental stress is the foundation of sustainable agriculture. Stress is a phenomenon that limits crop productivity or destroys biomass. Stress can be also biotic, caused by insects and diseases, or abiotic, which may include drought, flooding, salinity, metal toxicity, mineral deficiency, adverse pH, adverse temperature, and air pollution (Borlaug and Dowswell, 2005).
Within abiotic stress, drought is regarded as the most damaging. Moreover, water limitation is one of the most important constraints for agriculture and recently, global warming may be worsening this situation in most agricultural regions (Gosal et al., 2009; Xoconostle-Cázares et al., 2010).

The complex nature of drought tolerance limits its management through conventional breeding methods (Gosal et al., 2009). For example, stress hydric in flower formation stage reduces the number of kernels per plant, whereas in the grain filling stage the size of the kernel is reduced (Zarco et al., 2005; Grant et al., 1989).

Conventional breeding and genetic engineering, are becoming an art through which crop varieties of high quality and yield are being developed. Breeding for any desired trait undoubtedly requires a significant amount of genetic variation at intra-specific, inter-specific or inter-generic levels. The contributions of plant breeding to food production at global level have been enormous during the 20th century. There has been most important plant breeding break-through for almost all commercially important crops including major ones such as maize, wheat, rice, cotton, among others. However, relatively little breeding work has been carried out on improving crops for drought tolerance. (Ashraf, 2010).

Through conventional breeding, genetic variability for drought tolerance among crops/crop cultivars or among sexually compatible plant species can be identified, and the genetic variation so identified can be introduced through different mating designs into cultivars/lines with good agronomic characteristics (Ashraf, 2010). Conventional breeders have made considerable strides in developing drought tolerant lines/cultivars of some important food crops. One of the breeding approach started in Mexico in the 1970s at the International Maize and Wheat Improvement Center (CIMMYT). This started with the intention of developing drought tolerant maize (Bänziger et al., 2004). In 1997, CIMMYT spanned its breeding program to southern Africa aimed at improving maize for the drought-hit areas. A number of maize hybrids developed by the CIMMYT scientists were found superior to all those developed by private enterprises. The maize hybrids were superior in terms of growth and grain yield under drought-prone environments (Bänziger et al., 2004).

In 2006, plant breeders at the Crops Research Institute (CRI) based at Kumasi, Ghana, have developed a highly drought tolerant maize cultivar 'Obatanpa GH' in collaboration with the International Institute of Tropical Agriculture (IITA), Ibadan, the CIMMYT, Mexico, and the Sasakawa Global 2000. Similarly, 16 early maturing maize inbred lines (from TZEI 1 to TZEI 16) resistant to a scrounging weed Striga hermonthica (Del.) Benth were produced by the IITA. All these lines were found to be highly resistant to water limited conditions (Ashraf, 2010).

Marker-assisted breeding (MAB) and transgenic approach are diverse biotechnologies. In MAB desirable genes can be tagged so they can be easily selected within the breeding population, whereas through the transgenic approach the desirable genes can be transferred from one species to another. A large number of genomic regions of a crop germplasm can be examined for their breeding value through MAB, which facilitates the breeder to pool genes of diverse origins (Vinh and Paterson, 2005).

3. Molecular markers for drought tolerance identified in maize

Given that the drought resistance or tolerance in maize is clearly a qualitative character (Zarco et al., 2005). Through MAB it is now possible to examine the usefulness of thousands of genomic regions of a crop germplasm under water limited regimes. By examining the breeding value of each of the genomic regions, the breeder can coalesce genes of multifarious origins in novel ways, which was not possible previously with conventional breeding tools and protocols (Ashraf, 2010).

Given that the drought resistance or tolerance in maize is clearly a quantitative character (Zarco et al., 2005). But, like tolerance to other abiotic stress, drought stress is controlled by many minor genes (polygenes) that have additive effects in their expression (Thi Lang and Chi Buu, 2008. The loci on chromosomes housing such types of genes are now referred to as quantitative trait loci (QTL) (Ashraf, 2010). In a QTL analysis, phenotypic evaluation is carried out in a large number of plants from a segregating population for a variety of genetic markers. Then, the whole population, or only a part of it, is genotyped. Finally, appropriate statistical analysis is performed to pinpoint the loci controlling a trait (Asins, 2002). Due to the intricacy of the abiotic stress tolerance and the problems encountered in phenotypic based selection, the QTL mapping has been considered as imperative to the use of DNA markers for improving stress tolerance (Ashraf, 2010).

Natural genetic variation of a crop can be exploited either via direct selection under stressful conditions (simulated or natural) or via mapping of QTL and subsequent marker-assisted selection (Ashraf et al., 2008). QTL mapping allows assessing the locations, numbers, magnitude of phenotypic effects, and pattern of gene action (Vinh and Paterson, 2005). The role of polygenes in controlling a trait has been widely assessed by traditional means, but the use of DNA markers and QTL mapping has made it convenient to dissect the complex traits (Ashraf, 2010).

Recent molecular biology tools have undoubtedly led to the development of DNA markers that have been effectively used to identify QTL a number of traits in different crops. Ashraf et al. (2008) have listed a variety of DNA markers such as RFLPs, RAPDs, CAPS, PCRindels, AFLPs, microsatellites (SSRs), SNPs, and DNA sequences being currently in use to examine the inheritance of stress tolerance. QTL mapping for the drought tolerance trait has been done in different crops, the most notable being maize, wheat, barley, cotton, sorghum, and rice (Sari-Gorla et al., 1999; Ashraf, 2010).

Associations between markers and traits were first reported in maize by Stuber and Moll (1972) using isozymes. The advent of abundant DNA-based molecular markers allowed the construction of genetic maps. In maize, a linkage analysis between the manifestation of some key characteristics like male and female flowering time, anthesis-silking interval, plant height, and molecular markers [RFLP, microsatellites (SSR) and AFLP] was carried out under different water regimes using a maize population consisting of 142 RILs derived from selfing the F1 population from a cross B73×H99. Linkage analysis showed that, the QTL identified for male flowering time and plant height were the same under well-watered and water-stressed conditions (Sari-Gorla et al., 1999).

A marker-assisted backcross (MABC) selection program for improving grain yield under water limited conditions in tropical maize was conducted at CIMMYT, Mexico, which involved the crossing of drought resistant line Ac7643 with a drought susceptible line CML247. Marker-based selection was carried out stepwise on all four generations (from BC1F1 to BC2F3). After the four consecutive MABC cycles, the 70 BC2F3 individuals exhibiting the closest allelic composition at target and non-target loci were bred with two CIMMYT testers (CML254 and CML274). Thirty genotypes were selected on the basis of their performance in terms of grain yield and some key agronomic traits. However, the best five MABC-derived hybrids produced yield about 50% more than that of control hybrids, but in contrast, under mild water stress, there was no difference between MABC-derived hybrids and the control plants. This confirms that the expression of genetic variation for drought tolerance mainly depends on the severity of drought stress (Ribaut and Ragot, 2007).

4. Correlations between drought and heat stress tolerance and resistance to Aspergillus infection in maize

Some studies have shown that heat stress plays an important role in the susceptibility of corn to aflatoxin contamination (Payne, 1998; Abbas et al., 2002). Under these conditions stress the plant can result in high aflatoxin levels. Plant stress facilitates greater colonization of corn kernel and infection by *A. flavus* in the field. In general it appears that there is a relationship between temperature-moisture and fungal infection and consequently aflatoxin contamination. It has been reported that when conditions of high temperature and drought occur together during the growing season, increases growth of the fungus and toxin production in grains (Payne, 1998). In this direction, there are results that indicate that irrigating corn fields to reduce drought stress also reduced fungal infection and aflatoxin production (Jones et al. 1981) or lower soil temperature reduces aflatoxin contamination in peanut (Hill et al., 1983).There are another results that reinforce this notion in peanuts (Cole et al., 1985; Dorner et al. 1989). Studies on the influence of irrigation and subsoiling on infection and aflatoxin production in corn, have suggested that water stress appears to be a major factor affecting aflatoxin contamination (Payne et al. 1986). Even it has been found in peanuts an inverse relationship between the amount of water supplied and fungal colonization with aflatoxin production (Wotton and Strange 1987).On the other hand, Tubajika and Damann, (2000) found that corn drought tolerant lines all had significantly lower levels of ear rot and aflatoxin contamination compared to the aflatoxin resistant. In summary, available information suggests a possible association between drought tolerance and aflatoxin resistance in corn. But it has been difficult to detect genetic markers or chromosomal regions associated with kernel resistance quantitative trait loci (QTL) because enormous influence of environmental factors on phenotypic expression of resistance (Davis et al ., 1999, Paul et al., 2003).

Chen et al., (2002) performed a proteomic approach to identify proteins whose level of expression was associated with kernel resistance against *A. flavus* infection and aflatoxin production. They compared resistant with susceptible genotypes using large format two-dimensional gel electrophoresis. Over a dozen proteins spots were identified and sequenced. These proteins were categorized as follows: storage proteins (globulin 1 and globulin 2), late embryogenesis abundant (LEA) proteins related to drought or desiccation (LEA3 and LEA14), water- or osmo-stress related proteins (WSI18 and aldose reductase), heat-stress related proteins (HSP16.9), and antifungal proteins which include a trypsin inhibitor (Chen et al., 2002). The majority of those proteins were stress-related proteins and highly hydrophilic storage proteins which suggest that kernel resistance may require high levels of these kinds of proteins (Brown et al., 2004).

5. Stress-related proteins

Recent studies have found higher levels of stress-related proteins and highly hydrophilic storage proteins in kernels of resistant genotypes compared with susceptible genotypes. This may enable resistant kernels to effectively induce an active defense response upon fungal attack, even under stress caused by heat or drought (Chen et al., 2004).

Chen et al. (2004) conducted an study conducted in which a GLX-I protein was identified on the basis of peptide sequence analysis. This protein was expressed at higher levels in resistant maize kernel embryos in contrast with the susceptible ones. Sequence homology comparisons indicate that maize GLX-I belongs to the long-type glyoxalase family (280 to 295 amino acids), which contains two highly homologous domains. GLX-I is present in many organisms, such as fungi, plants, and animals. It catalyzes the conversion of MG, a potent cytotoxic compound, to nontoxic D-lactate in the presence of glutathione and GLX-II (Johansen et al., 2000). MG is known to arrest growth and react with DNA and protein and increase sister chromatid exchanges (Payne, 1998). Recent studies found that, in addition to dehydrin and group 3 late embryogenesis abundant proteins, *glx-I* was induced during drought stress in *Sporobolus stapfianus*, and in response to salt and water stress in *B. juncea* and tomato, suggesting an important role for GLX-I in conferring tolerance to plants under those stress conditions (Payne, 1998; Chen et al., 2004). Levels of GLX-I activity were examined in dry, noninfected, and infected kernels of resistant and susceptible lines and it was found that resistant lines generally have higher constitutive levels of GLX-I activity than susceptible ones. In addition, the level of MG did not increase in resistant genotypes. This lack of increase could be due to the relatively higher levels of GLX-I activity observed in resistant infected kernels. An elevation in MG content in susceptible genotypes, combined with low GLX-I activities, could weaken the kernel's ability to defend against fungal infection (Chen et al., 2004).

6. Antifungal proteins

Plants are exposed to a large number of pathogenic fungi. Although they do not have an immune system, plants have evolved a variety of potent defense mechanisms, including the synthesis of low-molecular-weight compounds, proteins, and peptides that have antifungal activity (Selitrennikoff, 2001). These proteins appear to be involved in either constitutive or induced resistance to fungal attack. Several classes of antifungal proteins involve inhibition of the synthesis of the fungal cell wall or disrupt cell wall structure and/or function; others perturb fungal membrane structure, resulting in fungal cell lysis (Selitrennikoff, 2001).

Plants when exposed to pathogens such as fungi and viruses produce low-molecular-weight antimicrobial compounds called phytoalexins, antimicrobial peptides, and small proteins and up-regulate a number of antimicrobial proteins. These plant proteins, called pathogenesis-related (PR) proteins, have been classically divided into five groups, PR-1, -2, -3, -4, and -5, based on serological and amino acid sequence analyses (Van Loon, 1985). PR-1 proteins are accumulated to high levels after pathogen infection and are antifungal both in planta and in vitro (Niderman et al., 1995). PR-1 proteins have been found in rice, wheat, maize, tobacco, *Arabidopsis thaliana* barley, and many other plants (Selitrennikoff, 2001).

PR-2 proteins have (1,3) β-endoglucanase activity in vitro and have been grouped into three classes on the basis of amino acid sequence analysis. The antifungal activity of plant (1,3)β-glucanases is thought to occur by PR-2 proteins hydrolyzing the structural (1,3)β glucan present in the fungal cell wall, particularly at the hyphal apex of filamentous molds where glucan is most exposed, resulting in a cell wall that is weak. This weakened cell wall results in cell lysis and cell death (Belfa et al., 1996).

A number of enzymatic assays have shown PR-3 (chitinases) proteins to have in vitro chitinase activity. Chitinases have been isolated from fungi, plants, and bacteria and have potent antifungal activity against a wide variety of human and plant pathogens. PR-3 proteins are endochitinases that cleave cell wall chitin polymers in situ, resulting in a weakened cell wall and rendering fungal cells osmotically sensitive (Selitrennikoff, 2001).

PR-4 proteins are chitin-binding proteins PR-4 proteins have been isolated from potato, tobacco, barley, tomato, and many other plants. The antifungal activity of these kinds of proteins is likely the result of protein binding to nascent fungal cell wall β-chitin. By mechanisms not understood these results in disrupted cell polarity, with concomitant inhibition of growth (Bormann et al., 1999).

PR-5 proteins are known as TL proteins several TL proteins cause cell permeability changes in fungal cells with a cell wall but have no or little effect on protoplasts. For example, zeamatin (a TL protein from corn) caused very rapid cell lysis of *N. crassa*, even at 4°C; lysis occurred primarily at subapical regions (Roberts and Selitrennikoff 1990).

7. Biochemical pathways

Biotic or abiotic stress alone was able to induce the expression of genes involved in both biotic and abiotic stress responses. Fungal infection represents a unique kind of stress to host plants. In response to such stress, plants not only induce specific antifungal genes, but also up regulate general stress-related genes (Brown et al., 2004).

An increasing body of evidence suggests that a subset of plant responses to biotic and abiotic stress is shared, such as the generation of reactive oxygen species (ROS), the activation of mitogen-activated protein kinases (MAPKs), and hormone modulations (Brown et al., 2004). Plant–pathogen recognition causes the rapid activation of appropriate defenses. Some of the components in the signal transduction pathways have been identified and characterized.

One of the mayor consequences of drought stress is the loss of protoplasmic water leanding to the concentration of ions such as Cl- and NO3-. At high concentrations these ions effectively inhibit metabolic functions (Hartung et al., 1998). Also, the concentrations of protoplasmic metabolites and the loss of water from the cell lead to the formation of a glassy state. Under this condition, the chances of molecular interactions can cause protein denaturation and membrane fusion (Hoekestra et al., 2001). To maintain cell turgor and metabolic functions under drought stress, plants generate reactive oxygen species (ROS), including hydrogen peroxide (Inze and Montagu,1995). ROS production is recognized as a common event in plant response to biotic and abiotic stress. The mechanism of how ROS leads to downstream responses is still not clear, however, the requirement of specific MAPKs has been implicated (Kovtun et al., 2000). Accumulation of hydrogen peroxide, eventually, induce the expression of detoxification and stress protection proteins as Heat Shock Proteins (HSPs)(Kovtun et al., 2000). Some reports suggests that HSPs function as molecular chaperones which are involved in ATP-dependant protein unfolding or assembly/disassembly actions and prevent protein denaturation during stress (Pelham, 1986). Termotolerance have been associated to expression of HSPs in maize (Preczewski et al., 2000)

The mitogen activated protein kinase (MAPK) cascades are the major components downstream of receptors that transduce extracellular stimuli into intracellular responses (Zhang and Klessig, 2001). One of the mechanisms by which different stimuli converge onto one MAPK is believed to involve several unrelated kinases that function as MAPKKKs to initiate the MAPK cascade. Several plant kinases have been identified as MAPKKKs, including EDR1 and NPK1/ANPs (Zhang and Klessig, 2001)

A variety of plant hormones, including salicylic acid (SA), jasmonate (JA), ethylene, and abscisic acid, have been implicated in mediating responses to a wide range of biotic and

abiotic stress. The roles of these hormones are dependent upon the particular hostpathogen interaction (Diaz et al., 2002; Brown et al., 2004). The the effect of phytohormones is also regulated by other factors. For example, the MAPK kinase kinase, EDR1, negatively regulates SA-inducible defenses, whereas MAPK 4 appears to differentially regulate SA and JA signals. These findings also suggest that MAPK modulates cross-talk between different plant defense pathways (Hammond-Kosacky and Parkerz, 2003).
Everything discussed in this article, on the complex role of drought in relation to aflatoxin corn contamination, shows the current research situation and paths toward solving this important and serious public health problem

8. References

Abbas HK, Williams WP, Windham GL, Pringle HC III, Xie W, Shier WT (2002). Aflatoxin and fumonisin contamination of commercial corn (Zea mays) hybrids in Mississippi. J. Agric. Food Chem. 50:5246-5254.

Ashraf M, Athar HR, Harris PJC, Kwon TR (2008). Some prospective strategies for improving crop salt tolerance. Adv Agron 97:45-110.

Ashraf M (2010). Inducing drought tolerance in plants: Recent advances. Biotechnology Advances 28:169-183.

Bänziger M, Setimela PS, Hodson D, Vivek B (2004) Breeding for improved drought tolerance in maize adapted to southern Africa. Proceedings of the 4th International Crop Science Congress. Brisbane, Australia: Published on CDROM; 2004. 26 Sep - 1 Oct.

Beffa, R., and F. Meins, Jr. 1996. Pathogenesis-related function of plant β-1,3-glucanases investigated by antisense transformation—a review. Gene 179:97–103.

Borlaug, N.E., and C. R. Dowswell. 2005. Feeding a world of ten billion people: A 21st century challenge. In proc. of "In the wake of double helix: From the green revolution to the gene revolution" 27–31st May 2003 Bologna, Italy.

Bormann C, Baier D, Horr I, Raps C, Berger J, Jung G, Schwartz H (1999). Characterization of a novel, antifungal, chitin-binding protein Streptomyces tendae Tu901 that interferes with growth polarity. J. Bacteriol. 181:7421–7429.

Brown RL, Chen ZY, Cleveland TE (2009). Evidence for an association in corn between stress tolerance and resitance to Aspergillus flavus infection and aflatoxins contamination. Afr. J. Biotechnol. 12:693-699.

Chen ZY, Brown RL, Damann KE, Cleveland TE (2002). Identification of unique or elevated levels of kernel proteins in aflatoxin-resistant maize genotypes through proteome analysis. Phytopathology 92:1084-1094.

Chen ZY, Brown RL, Damann KE, Cleveland TE (2004). Identification of a Maize Kernel Stress-Related Protein and Its Effect on Aflatoxin Accumulation. Phytopathology 94:938-945.

Diaz J, ten Have A, van Kan JA (2002). The role of ethylene and wound signaling in resistance of tomato to Botrytis cinerea. Plant Physiol. 129:1341-1351.

Diener, U. L., Cole, R. J., Sanders, T. H., Payne, G. A., Lee, L. S., and Klich, M. A. 1987. Epidemiology of aflatoxin formation by Aspergillus flavus. Annu. Rev. Phytopathol. 25:249-270.

Gosal SS, Wani SH, Kang MS (2009). Biotechnology and Drought Tolerance. Journal of Crop Improvement 23:19-54.

Hammond-Kosacky KE, Parkerz JE (2003). Deciphering plant–pathogen communication: fresh perspectives for molecular resistance breeding. Curr. Opin. in Biotechnol. 14:177–193.

Hartung W, Schiller P, Karl -Josef D (1998). Physiology of poikilohydric plants. Prog. Bot. 59:299-327.

Hoekstra F, Golovina J, Buitinik (2001). Mechanisms of plant dessication tolerance. Trends plant Sci. 6: 431-438.

Inze D. and Montagu MV (1995). Oxidative stres in plants. Curr. Opin. Biotechnol., 6:153-158.

Johansen KS, Svendsen II, Rasmussen SK (2000). Purification and cloning of the two domain glyoxalase I from wheat bran. Plant Sci. 155:11-20.

Jones RK, Duncan HE, Hamilton PB (1981). Planting date, harvest date, and irrigation effects on infection and aflatoxin production by Aspergillus flavus in field corn. Phytopathology 71:810-816.

Kovtun YWL, Chiu G, Tena and J. Sheen (2000). Functional analysis of oxidative stress-activated mitogen-activated protein kinase cascade in plants. Proc. Nat. Acad. Sci. USA. 97: 2940-2945.

Moreno OJ, Kang MS (1999). Aflatoxins in maize: The problem and genetic solutions. Plant Breed.118:1-16.

Muoma J, Ombori O, Jesse M. (2010). Signal Transduction: One of the Current Molecular Approaches in the Management of Drought Stress in the Sub-Saharan Region. Biotech. 9: 469-476.

Niderman T, Genetet I, Buryere T, Gees R, Stintzi A, Legrand M, Fritig B, Mosinger E (1995). Pathogenesis-related PR-1 proteins are antifungal. Isolation and characterization of three 14-kilodalton proteins of tomato and of a basic PR-1 of tobacco with inhibitory activity against Phytophthora infestans. Plant Physiol. 108:17–27.

Payne GA, Cassel DK, Adkins CR (1986). Reduction of aflatoxin contamination in corn by irrigation and tillage. Phytopathology 76:679-684.

Payne GA (1998). Process of contamination by aflatoxin producing fungi and their impact on crops. In Sinha KK, Bhatnagar D (eds.), Mycotoxins in Agriculture and Food Safety. Marcel Dekker, New York, NY, pp. 279-306.

Ribaut JM, Ragot M. (2007) Marker-assisted selection to improve drought adaptation in maize: the backcross approach, perspectives, limitations, and alternatives. J Exp Bot 58:351–60.

Roberts W, Selitrennikoff CP (1990). Zeamatin, an antifungal protein from maize with membrane-permeabilizing activity. J. Gen. Microbiol. 136:1771–1778.

Sari-Gorla M, Krajewski P, Di Fonzo N, Villa M, Frova C (1999). Genetic analysis of drought tolerance in maize by molecular markers. II. Plant height and flowering. Theor Appl Genet 99:289–95.

Selitrennikoff CP (2001). Antifungal proteins. Applied and Enviromental Microbiology. 67:2883-2894.

Stuber CW, Moll RH (1972). Frequency changes of isozyme alleles in a selection experiment for grain yield in maize (Zea mays L.). Crop Science 12:337–340.

Thi Lang N, Chi Buu B (2008). Fine mapping for drought tolerance in rice (Oryza sativa L.). Omonrice 16:9-15.

Thomann EB, Sollinger J, White C, Rivin CJ (1992). Accumulation of group 3 late embryogenesis abundant proteins in Zea mays embryos. Plant Physiol. 99:607-614.

Tubajika KM, Damann KE (2001). Sources of Resistance to Aflatoxin Production in maize. J. Agric. Food Chem. 49:2652-2656.

Van Loon LC (1985). Pathogenesis-related proteins. Plant Mol. Biol. 116: 111–116.

Vinh NT, Paterson AH (2005). Genome mapping and its implication for stress resistance in plants. In: Ashraf M, Harris PJC, editors. Abiotic stresses: plant resistance through breeding and molecular approaches.

Widstrom NW, Guo BZ, Wilson DM (2003). Integration of crop management and genetics for control of preharvest aflatoxin contamination of corn. J. Toxicol.Toxin Rev., 22: 199-227.

Xoconostle-Cázares B, Ramírez-Ortega FA, Flores-Elenes L, Ruíz-Medrano R (2010). Drought Tolerance in Crop Plants. Am. J. Plant Physiol., 5:241-256.

Xu D, Duan X, Wang B, Hong B, Ho THD, Wu R (1996). Expression of a late embryogenesis abundant protein gene HVA1, from barley confers tolerance to water deficit and salt stress in transgenic rice. Plant Physiol. 110:249-257.

Zhang S, Klessig DF (2001). MAPK cascades in plant defense signaling. Trends Plant Sci.6:520-527.

Mahuku George. 2010. Varietal improvement for managing Aspergillus flavus and aflatoxin in maize International Mycotoxin Conference - MycoRed. 2010 Penang, Malaysia.

Zarco Perelló Emiliano, González Hernández Victor A., López Peral ta María Cristina y Salinas Moreno Yolanda. 2005. Physiological markers for drough tolerance in maize (Zea Mays L.). Agrociencia 39: 517-528.

Part 3

Pathogenesis and Toxicology

9

The Molecular Pathogenesis of Aflatoxin with Hepatitis B Virus-Infection in Hepatocellular Carcinoma

Hai-Xia Cao and Jian-Gao Fan
Department of Gastroenterology, Xinhua Hospital, Shanghai Jiaotong University School of Medicine, Shanghai, China

1. Introduction

Hepatocellular carcinoma (HCC) is the fifth most common cancer worldwide and represents the third cause of mortality among deaths from cancer (Semela & Heim, 2011; EI-Serag & Rudolph, 2007; Yuen et al, 2009; Lodato et al, 2006; Hainaut & Boyle, 2008). There are upwards of 600, 000 new cases each year in Asia and sub-Saharan Africa, where populations suffer both from a high prevalence of hepatitis B virus (HBV) infection and largely uncontrolled aflatoxin exposure in food, and more than 200, 000 HCC-related deaths annually in the People's Republic of China alone (Kew, 2002; Wang et al, 2002). Aflatoxins may play a causative role in 5%-28% of all global HCC cases (Liu & Wu, 2010).

Aflatoxin is a kind of mycotoxins produced by the mold Aspergillus flavus, which can be found in legumes, corns, soybeans, rice, milk, and cheese throughout their lives. Aflatoxin B1 (AFB1) is regarded as a class I carcinogen by the World Health Organization (Henry et al, 2001; López et al, 2002). Aflatoxin B1, the most commonly occurring and potent of the aflatoxins is associated with a specific AGG to AGT transversion mutation at codon 249 of the p53 gene in human HCC, providing mechanistic support to a causal link between exposure and disease (Goldman & Shields, 2003; Sugimura, 2000; Wild & Montesano, 2009). Prospective epidemiological studies have shown a more than multiplicative interaction between HBV and aflatoxins in terms of HCC risk. In this chapter, the available evidence for the mechanism of aflatoxin with HBV-infection in HCC will be reviewed.

2. Mechanisms of aflatoxins in HCC

Aflatoxins, first described in the early 1960s, are the most toxic and carcinogenic compounds among the known mycotoxins. They are produced by several Aspergillus species and consist of at least 16 structurally related furanocoumarins, of which AFB1, AFB2, AFG1, and AFG2 are the four most abundant aflatoxins. The designation B and G came from the Blue or Green fluorescence emitted by these compounds under long-wave ultraviolet light, respectively (McLean & Dutton, 1995). Human exposure to aflatoxins is mainly through consumption of contaminated staples, such as maize and pea-nuts. Contaminations are often the result of inappropriate storage of grain, which leads to the infestation by aflatoxin-producing fungi (Williams et al, 2004). Liver is the primary target organ for acute and chronic injury. Ingestion of high doses of aflatoxin over a short time period can result in acute aflatoxicosis, with symptoms such as hemorrhagic necrosis of the liver, bile duct proliferation, edema and lethargy (Williams et al, 2004). Chronic exposure to low or moderate doses of aflatoxin may lead to the development of HCC.

AFB1 is a potent human carcinogen (IARC, 1993), which occurs in the low or sub micrograms per kilogram range and is regulated by legislation in the European Union (EU) at 2 μg/kg in foods for direct human consumption (European Community, 2006). Several studies have evaluated an association between the risk for HCC and exposure to AFB1. A prospective case-control study from China mainland showed that individuals with the presence of urinary aflatoxin biomarkers had a significantly increased risk of HCC after adjusting for HBV surface antigen (HBsAg) seropositivity and cigarette smoking (Qian et al, 1994). These data were further supported by a community-based cohort study from Taiwan which found that elevated AFB1 exposure measured by detectable AFB1-albumin adducts was an independent risk factor for HCC after adjustment for other important confounding factors (odds risk : 5.5, 95% confidence interval : 1.2-24.5) (Chen et al, 1996a).

2.1 Metabolism of AFB1

The cytochrome P450 (CYP) enzymes are a superfamily of hemeproteins that are important in the oxidative, peroxidative and reductive metabolism of endogenous compounds and participate in the chemical carcinogenesis process (Gonzalez & Lee, 1996). AFB1 is activated by CYP enzymes, mainly CYP 1A2 and 3A4, to form four major metabolites: AFM1, AFQ1, AFB1-endo-8, 9-epoxide and AFB1-exo-8, 9-epoxide. AFM1 and AFQ1 are polarmolecules; AFB1-endo-8, 9-epoxide reacts poorly with DNA (McLean & Dutton, 1995; Guengerich et al, 1998; Iyer et al, 1994). The overall contribution of these enzymes to AFB1 metabolisms in vitro depends on the affinity of the enzyme but in vivo it also depends on expression levels in human liver where CYP 3A4 is predominant. Expression of CYP 1A1/2 and 3A4 in liver tissues of hepatocellular carcinoma cases and controls was detected and their relationship to HBV and AFB1-DNA adducts was also investigated (Pfeifer et al, 1993). For CYP 3A4, in contrast to control tissues, there was a significant association with AFB1-DNA adducts in tumor and adjacent non-tumor tissues in patients with HCC. Human beings show large interindividual variations in metabolism activities that lead to different susceptibilities to the genotoxic actions of carcinogens (Zhang et al, 2000).

A model using human liver epithelial cell lines stably expressing cytochrome P450 cDNA revealed that CYP 1A2 and CYP 3A4 both contribute to the formation of AFB1-induced p53 mutation whereas CYP 2A6 does not appear to play a significant role (Macé et al, 1997). In an in vitro study, inhibition of CYP 1A2 and CYP 3A4 by oltipraz, a drug which has been reported to inhibit AFB1 activation in human hepatocytes, was shown (Langouët et al, 1995).

AFB1 can also be metabolized by NADPH-dependent reductase into a carcinogenic metabolite aflatoxicol (AFL) (Salhab & Edwards, 1977; Wong & Hsieh, 1976). AFL acts as a reservoir of AFB1, prolonging its lifetime in body, as it can be reconverted to AFB1, which then can be further metabolized. Formation of AFL does not decrease the toxicity of AFB1 because it can also bind to DNA and is as potent carcinogen like AFB1 (Bailey et al, 1994).

Glutathione S-transferase (GST) is a family of conjugation enzymes involved in the metabolism of exogenous and endogenous lipophilic compounds for their excretion and detoxification. For AFB1, the detoxification pathway is via GST-mediated conjugation with reduced glutathione (GSH) to form AFB1 exo- and endo-epoxide GSH conjugates (Guengerich et al, 1998; Raney et al, 1992; Johnson WW, 1997a). Members of the GST family, such as GST-μ (GSTM1) and GST-θ (GSTT1), are important candidates for involvement in susceptibility to AFB-associated HCC, because they may regulate an individual's ability to metabolize the ultimate carcinogen of aflatoxin, the exo-epoxide (Johnson WW, 1997a). Epidemiological studies have suggested that genetic polymorphisms in AFB1 metabolizing enzymes are factors in individual susceptibility to aflatoxin-related HCC (McGlynn et al, 1995; Chen et al, 1996b). GSTM1 genotype can be categorized into two classes: the homozygous deletion genotype (GSTM1null genotype) and genotypes with one or two alleles present (GSTM1 non-null genotype); GSTT1 can also be deleted (Pemble et al, 1994; Rebbeck et al, 1999). Carriers of GSTM1 and GSTT1 homozygous null genotypes lack of the corresponding enzyme activities (Pemble et al, 1994). Chen et al (1996b) documented a biological gradient between serum AFB1-albumin adduct levels and HCC risk among chronic HBsAg carriers who had null GSTM1 and GSTT1 genotypes but not among those who had non-nullgenotypes in a Taiwan population. Wild et al (2000) reported an association between the GSTM1 null genotype and AFB1-albumin adducts in a Gambian population, although the association was restricted to people who were not infected with HBV. The effect of aflatoxin exposure on HCC risk was also more pronounced among chronic HBsAg carriers with the GSTT1 null genotype than those who were non-null (Sun et al, 2001). Based on the above studies conducted in different places and others not reviewed, whether or not there are interactions among AFB1, HBV infection and GSTs genotypes in the development of HCC is still controversial.

2.2 Damage to DNA

The main genotoxic product of AFB1 metabolites is AFB1-exo-8, 9-epoxide, an unstable metabolite which reacts with DNA, and forms adducts with a yield of 98% (Johnson & Guengerich, 1997). This reactivity is 1000-fold greater than that of its endo-isomer (Iyer et al, 1994). AFB1-exo-epoxide can intercalate between the bases in DNA, and can form covalent bonds with DNA by electrophilic attack on the N7 position in guanines, leading to the formation of the adduct 8, 9-dihy-dro-8-(N7-guanyl)-9-hydroxyaflatoxin B1 (AFB1-N7-Gua). This adduct is fairly unstable due to a positive charge on the imidazole ring. As a result, AFB1-N7-Gua adducts undergo processing reactions that include either depurination to form an apurinic (AP) site, or opening of the imidazole ring to form a more stable, AFB1–formamidopyrimidine (AFB1–FAPY) adduct (Smela et al, 2001). It has been proposed that AFB1 can also cause DNA damage through production of reactive oxygen species (Petitjean et al, 2007). AFB1-DNA bulky adducts, including AFB1-N7-Gua and AFB1–FAPY, are repaired mainly by Nucleotide Excision Repair (NER) (Bedard & Massey, 2006) but all AFB1-DNA adducts are not repaired with the same efficiency. The AP site is repaired by Base Excision Repair (BER) (Bedard & Massey, 2006). AFB1-induced DNA lesions that escape DNA repair can result in mutations (Smela et al, 2001). AFB1-N7 -Gua, AFB1-FAPY and AP sites are all candidate precursors of mutations, predominantly induce G-T transversions.

As this is the pathway of its toxicological effects, the two most extensively studied biomarkers are the urinary aflatoxin N-7-guanine adduct, which results from AFB1 reaction with DNA, and aflatoxin-lysine adduct, which is obtained by digestion of the aflatoxin–albumin adduct occurring in serum (Shephard, 2009). Early experimental studies around 1980 demonstrated that the major aflatoxin-nucleic acid adduct, AFB1-N7-Gua, was excreted exclusively in the urine of exposed rats (Bennett et al, 1981; Egner et al, 2006). The serum aflatoxin-albumin adduct was also examined as a biomarker of exposure because the longer half-life of albumin would be expected to integrate exposures over longer time periods, i.e., months instead of days. Studies in experimental models found that the formation of aflatoxin-DNA adducts in liver, urinary excretion of the aflatoxin-nucleic acid adduct, and formation of the serum albumin adduct were highly correlated events (Groopman, 1994).

2.3 AFB1 inducing p53 mutaion (G-T transversions)

AFB1 covalently binds to guanine and cytosine residues of DNA both in vivo and in vitro and forms AFB1-DNA adducts, it also forms RNA and protein adducts impairing DNA, RNA and ultimately protein synthesis (Santella et al, 1998; Meneghini & Schumacher, 1977; Amstad & Cerutti, 1983). The presence of AFB1-DNA adducts can contribute to genetic alterations involved in the development of HCC. In 1977, Lin et al (1977) reported that adduct formation by metabolically activated reactive intermediates with hepatocyte DNA could lead to mutations in the host genome. The p53 tumor suppressor gene is the most frequently mutated gene in human cancers. Two groups found at the same time, that mutations of the p53 gene on chromosome 17 are frequent in HCC and a point mutation at the third position of codon 249 resulting in a G:C to T:A transversion was common in HCC tissues which were collected in China and Africa (Hsu et al, 1991; Bressac et al, 1991).

This mutation is present in up to 50% of patients with HCC who are indigenous to geographic regions with high exposure to AFB1(Bressac et al, 1991; Coursaget et al, 1993; Hsu et al, 1991; Ming et al, 2002). In contrast, it is absent in patients with HCC from regions with low exposure to AFB1 as well as in cancers other than HCC(Aguilar et al, 1994; Challen et al, 1992). The mutant allele is defined as R249S and the mutant protein as p.R249S. R249S accounts for more than 90% of TP53 mutations found in HCC cases from regions with high aflatoxin exposure levels, including Qidong in China and several sub-Saharan African countries. This percentage drops to around 30–40% in Taiwan and Hong Kong, where aflatoxin exposure levels are considered to be moderate. Among HCCs from low exposure regions, such as the United States and Japan, R249S only accounts for less than 6% of TP53 mutations(Petitjean et al, 2007).

This hot spot mutation in HCC from regions with high levels of dietary aflatoxins links this genetic change to exposure to aflatoxins. In recent years, the p53 codon 249 mutation has also been detected in plasma or serum DNA of HCC patients(Kirk et al, 2000; Jackson et al, 2001; Kuang et al, 2005). This mutated DNA may serve as a biomarker of exposure to AFB1 and for detection of early HCC(Jackson et al, 2001).

2.4 P53 mutation in HCC

The crystal structure of the wild-type p53 core domain in complex with DNA reveals that the arginine at position 249 provides four essential bridges to other residues in the L2 and L3 loops of the DNA-binding surface, but does not make direct contacts with DNA(Cho et al, 1994). Replacement of the arginine by a serine thermodynamically destabilizes the core domain and impairs its DNA-binding capacity(Bullock et al, 2000). Nuclear Magnetic Resonance (NMR) showed that the general structure of p.R249S core domain was similar to that of wild-type p53, but that the mutant demonstrated local structural distortion around position 249 in the L3 loop, with increased flexibility of the beta-sandwich scaffold(Friedler et al, 2004). The presence of a serine at position 249 induces a reorientation of M243. In the wild-type core domain, M243 is exposed to solvent when the protein is not bound to DNA but is buried within at the interface between p53 oligomers when complexed with target DNA. In T-p53C-R249S, M243 displaces M246 from its buried location within a hydrophobic pocket of the zinc-binding region, leading to the formation of a short alpha-helix and a local conformational change that displaces the DNA-contact residue R248. This effect may explain the loss of the DNA-binding capacity of p.R249S.

p.R249S has lost DNA-binding and transactivation capacities towards most, if not all, promoters that contain p53 consensus binding sequence. In yeast assays, its residual activity towards p53-dependent promoters is of less than 20% of that of wild-type p53, similar to most other "hotspot" p53 mutants. At biological level, Ponchel and colleagues(1994) have reported an increase in colony formation but not of tumorigenicity in nude mice upon transfection of R249S in Hep3B (p53-null hepatoblastoma cells). Using non-immortalized human epithelial cells, Schleger and colleagues (1999) found that p.R249S expression increased the size but not the number of colonies in clonogenicity assays, but did not prolong the lifespan of the cells. Two studies have addressed the effects of transgenic expression of the murine homologous p.R246S in mouse liver. Yin and colleagues (1998) have reported enhanced cell cycle activity in the liver due to an increased entry into G1 phase. Thus, none of the above mechanisms provide a convincing functional explanation for the apparent selection of p.R249S in aflatoxin-induced HCC.

Tumor-derived p53 mutant proteins contribute to carcinogenesis through three overlapping mechanisms (Gouas et al, 2010): loss of wild-type p53 trans-activation function (loss of function); capacity to inhibit the activity of wild-type p53 (dominant-negative effect) and possible 'gain-of-function' effects, by which mutant proteins have acquired new, pro-oncogenic properties (Brosh & Rotter, 2009; Oren & Rotter, 2010). Several mechanisms underlying such gain-of-function effects have been described, including: transactivation of gene enhancing proliferation, inhibiting apoptosis or chemoresistance or increasing invasiveness, inflammation and angiogenesis (Brosh et al., 2009); interaction with various proteins, in particular TAp63 and TAp73, the products of two genes related to TP53 that exert differentiation and growth suppressive effects during development and morphogenesis (Bergamaschi et al, 2003; Gaiddon et al, 2001; Marin et al, 2000; Strano et al, 2002). It is not clear whether p.R249S exerts such gain-of-function effects which needs further studies.

2.5 Others changes

Moreover, it has been recently demonstrated that AFB1-albumin adducts in patients with HCC correlate significantly with the presence of plasma DNA hypermethylation and mutations in the p16 and p53 tumor suppressor genes (Zhang et al, 2006). AFB1-induced HCC in Fischer 344 rats showed activating mutations in codon 12 of K-ras but in human HCC, the incidence of point mutation of K-ras and N-ras oncogenes was low (Tsuda et al, 1989). In an in vitro study, AFB1 interfered with the molecular mechanisms of cell cycle regulation (Ricordy et al, 2002). Gursoy-Yuzuqullu (2011) showed that genotoxic doses of AFB1 induce an incomplete and inefficient checkpoint response in human cells. This defective response may contribute to the mutagenic and carcinogenic potencies of aflatoxins. AFB1 also induced mitotic recombination (Stettler & Sengstag, 2001) and minisatellite rearrangements (Kaplanski et al, 1997). Mitotic recombination and genetic instability may therefore be alternative mechanisms by which aflatoxin contributes to genetic alterations in HCC (Wild & Turner 2002).

Long and colleagues(2009) report that there were interactions between the genetic polymorphism of XPD codon 751 and AFB1-exposure years, and imply that this polymorphism may have functional significance in HCC induced by AFB1.While XPD protein, encoded by XPD gene, is a DNA-dependent ATPase/helicase that is associated with the TFIIH transcription-factor complex, and plays a role in NER pathway. During NER, XPD participates in the opening of the DNA helix to allow the excision of the DNA fragment containing the damaged base (Benhamou & Sarasin, 2002; Manuguerra et al, 2006). These results suggest that the genotypes of XPD with codon 751 Gln alleles may increase the risk of AFB1-related HCC and the NER pathway may play an important role in the mechanism of action of this genotoxin.

3. Interaction between AFB and HBV in HCC

The marked worldwide heterogeneity of HCC incidence in HBV-endemic regions might be related, at least in part, to aflatoxin exposure (Yu &Yuan 2004; Yuen et al, 2009; Lodato et al, 2006). It should be stressed that areas with high exposure to AFB1 are also characterized by a high prevalence of HBV infection. AFB1 is independent of the risk conferred by HBV, however concomitant exposure to both HBV and AFB1 markedly increases the risk of HCC. Ross et al (1992) formerly demonstrated a synergistic interaction between HBV and AFB1 in the development of liver cancer. Subsequently, Sun et al (1999) followed a cohort of Chinese men with chronic HBV for 10 years and found that the relative risk of HCC was significantly increased in subjects with detectable AFM1 levels. Aflatoxin exposure in association with HBV infection induces a 60-fold increase in risk of HCC(RR: 59.4, 95% CI: 16.6-212.0), while aflatoxin alone increases the risk fourfold.(Yu &Yuan 2004; Ming et al, 2002; Turner et al, 2002). In a case-control study in The Gambia, the codon 249ser mutation was examined in the plasma of HCC cases, cirrhosis patients and controls. The presence of both the codon 249ser mutation and HBV infection was associated with an OR = 399 (95% CI: 48.6–3270) (Kirk et al, 2004; Kirk et al, 2005). HBV transgenic mice overexpressing HBsAg in the liver showed more HCC than non-transgenic littermates when exposed to AFB1 (Sell & Heim, 1991).

Aflatoxin exposures multiplicatively increase the risk of liver cancer in people chronically infected with hepatitis B virus (HBV), which illustrates the deleterious impact that even low toxin levels in the diet can have on human health (Qian et al, 1994; Ross et al, 1992; Wang LY et al, 1996). Ghebranious and Sell (1998) have investigated the induction of HCC in p.R246S transgenic mice: they found that the presence of p.R246S enhanced aflatoxin-induced formation of liver tumor in conjunction with a HBsAg transgene.

Recently, a large-scale quantitative analysis of TP53 mutations in the serum of healthy individuals from The Gambia, a country with high aflatoxin exposure levels and endemic for HBV infection, has been performed. The study revealed the presence of R249S DNA in the serum or plasma of apparently healthy subjects, with a seasonal variation that recapitulated the known variation in exposure to aflatoxin. Moreover, the presence and mean concentrations of R249S DNA in the plasma or serum were significantly associated with HBV carriage, with 44% of carriers showing detectable levels of the mutation compared to only 24% of non-carriers. These results are consistent with the hypothesis that HBV infection may specifically contribute to the acquisition and/or the retention of R249S mutation in hepatocytes (Gouas et al, 2009).

However, the biology underlying this statistical interaction is not fully understood. Several lines of evidence suggest that HBV infection may enhance the mutagenic effects of AFB1. First, the presence of the virus may interfere with AFB1 metabolism. The expression of

CYP450 enzymes that metabolize AFB1 is increased in HBsAg-transgenic mice (Kirby et al, 1994). Moreover, Gambian children and adolescents chronically infected with HBV show higher concentrations of AFB1 adducts than uninfected individuals (Kew, 2003). Second, viral replication and chronic inflammation due to infection induces oxidative stress in hepatocytes, which may contribute to the mutation load at codon 249 (Hussain et al, 1994) .Third, HBV may also promote these mutations through indirect mechanisms, e.g. by inducing chronic inflammation which, in turn, increases hepatocyte turnover rate and the risk of acquisition of a mutation such as R249S. Another possible mechanism of interference between HBV and mutagenesis at codon 249 may occur at the level of cellular DNA repair. AFB1–DNA adducts, in particular, AFB1-N7-Gua and AFB1-FAPY, are repaired through the NER pathway. The viral antigen HBx can decrease NER efficiency, probably through physical interaction with wild-type p53 and/or with the components of the NER machinery, such as XPB and XPD DNA helicases (Jia et al, 1999). The inhibition of NER by HBx may therefore lead to the persistence of AFB1-DNA adducts, which increases the risk for mutations. However, this mechanism does not account for the specific retention of R249S, unless some form of sequence-specificity in preferential repair of DNA at other positions than the third base of codon 249 is concerned. The high rate of R249S mutation may also be the consequence of the fact that this mutant provides some kind of selective advantage to infected liver cells. This may lead to the clonal expansion of p.R249S-expressing hepatocytes and may increase the probability for further accumulation of other cancer-prone mutations. This property may be due to a special capacity of p.R249S to form complexes with selected viral antigens. It has been reported that the HBx protein physically associates with p53 and apparently blocks its normal function in vitro and in vivo (Feitelson et al, 1993; Wang XW et al, 1994). In agreement with this view, in mice carrying HBx transgene, the X protein may bind to p53 and induce its retention into the cytoplasm, thus functionally inactivating its function (Ueda et al, 1995; Lin Y et al, 1997). Whether this interaction is of relevance for mutant p53, and in particular for p.R249S, is not clear. Of note, those HBx transgenic mice have a higher levels of G: C–T:A transversions induced by AFB1 than AFB1-exposed wild-type mice (Madden et al, 2002).

There maybe exists other mechanisms, including the predisposition of HBV-infected hepatocytes to aflatoxin induced DNA damage; an increase in susceptibility to chronic HBV infection in aflatoxin-exposed individuals; and oxidative stress exacerbated by co-exposure to aflatoxins and chronic hepatitis infection. Some epidemiological studies have clearly shown that two factors of importance in determining the risk of HCC in the natural history of HBV infection are the age at primary infection and the presence of serum HBeAg or HBV DNA, biomarkers of active viral replication, in patients with chronic active hepatitis (Yang et al, 2002). In addition, hepatitis B virus X (HBx) protein affects various cellular functions relevant to cancer development, namely p53 and oxidative DNA damage (Hussain et al, 2007). This correlation may be indicative of AFB1 exposure itself inducing oxidative stress, HBV could predispose hepatocytes to the carcinogenic action of aflatoxins. For example, human liver epithelial cells, expressing wild-type p53 and transfected with HBx gene were more sensitive to the cytotoxic action of AFB1-8, 9-epoxide than were the parent cells (Sohn et al, 2000). The HBx expressing cells were also more prone to apoptosis and to induction of mutations at codon 249 of the p53 gene. Aflatoxin exposure may alter the effects of the hepatitis virus infection, perhaps affecting susceptibility to infection or viral replication. HepG2 cells transfected with re-circularised HBV and treated with AFB1 showed a 2-3-fold increase in HBsAg at 96 hours post-treatment (Banerjee et al, 2000). DNA damage can also increase viral DNA integration into the host genome (Dandri et al, 2002) and it is possible that AFB1 could exert this effect directly, or indirectly via the oxidative stress mentioned above. In addition, the recent study in Guangxi, China (Xu et al, 2010) found additive effects of HBV BCP mutations and high serum AFB1-lysine adduct level on the risk of developing HCC.

A recent study provided the first evidence that cirrhosis may play a contributory role in the pathogenesis of AFB1-induced HCC (Jiang et al, 2010). Kuniholm (2008) and co-workers used an ultrasound-based method to diagnose the presence of cirrhosis in 97 black Africans. A score of at least 7 out of a possible 11 points on the ultrasound-based scale was the criterion for the diagnosis of cirrhosis. This method has 77.8% sensitivity and 92.5% specificity in comparison with liver biopsy in identifying cirrhosis in HBV-infected patients (Lin DY et al, 1993; Hung et al, 2003). Three hundred and ninety seven individuals with no evidence of liver disease and a normal serum AFB1 concentration served as controls. Long-term exposure to AFB1 was assessed in the patients with cirrhosis and the controls on the basis of two observations: A history of lifetime groundnut (peanut) intake or the finding in the serum of a genetic marker of heavy exposure to AFB1, the 249 ser p53 mutation. An increased relative risk of cirrhosis development of 2.8 (95% confidence interval 1.1-7.7) was calculated using a history of life-time dietary intake of groundnuts as the criterion for significant exposure, and of 3.8 (95% confdence interval 1.5-9.6) using the finding of the 249 ser p53 mutation in serum as the criterion, allowing for the possible confounding effect of HBV and HCV infection in each instance.

4. Conclusion

Further understanding of the interaction of HBV infection, genetic variation and exposure to environmental chemical carcinogens will help to elucidate mechanisms of human hepatocarcinogenesis and develop more effective strategies for HCC prevention. At present, simple, low-technology, and inexpensive practices can result in a striking decrease in exposure to AFB1.

5. References

Aguilar F, Harris CC, Sun T, et al. (1994) Geographic variation of p53 mutational profle in nonmalignant human liver. Science, 1994 May 27; 264(5163): 1317-1319

Amstad P, Cerutti P.(1983) DNA binding of aflatoxin B1 by co-oxygenation in mouse embryo fibroblasts C3H/10T1/2. Biochem Biophys Res Commun, 1983 May 16; 112(3): 1034-1040

Bailey GS, Loveland PM, Pereira C, et al.(1994) Quantitative carcinogenesis and dosimetry in rainbow trout for aflatoxin B1 and aflatoxicol, two aflatoxins that form the same DNA adduct. Mutat Res.1994 Aug;313(1):25-38

Banerjee R., Caruccio L., Zhang Y.J, et al. (2000) Effects of carcinogen-induced transcription factors on the activation of hepatitis B virus expression in human hepatoblastoma HepG2 cells and its implication on hepatocellular carcinomas, .Hepatology 2000 Aug 32(2):367–374.

Bedard L.L., Massey T.E. (2006) Aflatoxin B1-induced DNA damage and its repair. Cancer Lett. 2006 Sep 28; 241(2):174–183.

Benhamou S, Sarasin A.(2002) ERCC2/XPD gene polymorphisms and cancer risk. Mutagenesis 2002 Nov; 17(6):463-469.

Bennett RA, Essigmann JM, Wogan GN. (1981). Excretion of an aflatoxin-guanine adduct in the urine of aflatoxin B1-treated rats. Cancer Res. 1981 Feb;41(2):650-654

Bergamaschi D, .Gasco M, Hiller L, et al. (2003) p53 polymorphism influences response in cancer chemotherapy via modulation of p73-dependent apoptosis. Cancer Cell, 2003 Apr;3(4):387-402.

Bressac B, Kew M, Wands J, et al.(1991) Selective G to T mutations of p53 gene in hepatocellular carcinoma from southern Africa. Nature. 1991 Apr 4; 350(6317): 429-431

Brosh R, Rotter V.(2009) When mutants gain new powers: news from the mutant p53 field. Nat. Rev. Cancer.2009 Oct;9(10), :701-713.

Bullock A.N, Henckel J., Fersht A.R.(2000) Quantitative analysis of residual folding and DNA binding in mutant p53 core domain: definition of mutant states for rescue in cancer therapy, Oncogene.2000 Mar 2; 19 (10):1245-1256.

Challen C, Lunec J, Warren W, et al.(1992) Analysis of the p53 tumor-suppressor gene in hepatocellular carcinomas from Britain. Hepatology 1992 Dec; 16(6): 1362-1366

Chen CJ, Wang LY, Lu SN, et al.(1996a) Elevated aflatoxin exposure and increased risk of hepatocellular carcinoma. Hepatology. 1996 Jul; 24(1): 38-42.

Chen CJ, Yu MW, Liaw YF, et al.(1996b). Chronic hepatitis B carriers with null genotypes of glutathione S-transferase M1 and T1 polymorphisms who are exposed to aflatoxin are at increased risk of hepatocellular carcinoma. Am J Hum Genet. 1996 Jul; 59(1): 128-134

Cho Y, Gorina S, Jeffrey P.D., et al.(1994) Crystal structure of a p53 tumor suppressor-DNA complex: understanding tumorigenic mutations. Science. 1994 Jul 15;265 (5170) :346-355.

Coursaget P, Depril N, Chabaud M, et al.(1993). High prevalence of mutations at codon 249 of the p53 gene in hepatocellular carcinomas from Senegal. Br J Cancer. 1993 Jun; 67(6): 1395-1397

Dandri M, Burda M.R, Bürkle A, et al.(2002) Increase in de novo HBV DNA integrations in response to oxidative DNA damage or inhibition of poly(ADP-ribosyl)ation. Hepatology.2002 Jan;35 (1): 217-223.

Egner PA, Groopman JD, Wang JS, et al.(2006). Quantification of aflatoxin-B1-N7-Guanine in human urine by high-performance liquid chromatography and isotope dilution tandem mass spectrometry. Chem Res Toxicol. 2006 Sep;19(9):1191-1195

EI-Serag HB, Rudolph KL.(2007).Hepatocellular carcinoma: epidemiology and molecular carcinogenesis. Gastroenterology. 2007 Jun; 132(7): 2557-2576.

European Commission Regulation (EC) No 1881/2006 of 19 December 2006. Off J Eur Union L364/5

Feitelson M.A., Zhu M., . Duan L.X, et al.(1993).Hepatitis B x antigen and p53 are associated in vitro and in liver tissues from patients with primary hepatocellular carcinoma. Oncogene.1993 May;8(5): 1109-1117.

Friedler A., DeDecker B.S, Freund S.M., et al.(2004) Structural distortion of p53 by the mutation R249S and its rescue by a designed peptide: implications for mutant conformation. J Mol Biol.2004 Feb 6; 336 (1): 187-196

Gaiddon C., Lokshin M., Ahn J., et al. (2001) A subset of tumor-derived mutant forms of p53 down-regulate p63 and p73 through a direct interaction with the p53 core domain. Mol Cell Biol. 2001 Mar; 21(5):1874-1887.

Ghebranious N., Sell S.(1998).The mouse equivalent of the human p53ser249 mutation p53ser246 enhances aflatoxin hepatocarcinogenesis in hepatitis B surface antigen transgenic and p53 heterozygous null mice. Hepatology.1998 Apr; 27 (4): 967–973.

Goldman R, Shields PG. (2003).Food mutagens. J Nutr. 2003 Mar;133 Suppl 3:965S–973S.

Gonzalez FJ, Lee YH.(1996) Constitutive expression of hepatic cytochrome P450 genes. FASEB J. 1996 Aug; 10(10): 1112-1117

Gouas D, Shi H, Hainaut P. (2009).The aflatoxin-induced TP53 mutation at codon 249 (R249S): biomarker of exposure, early detection and target for therapy. Cancer Lett. 2009 Dec 1;286(1):29-37

Gouas DA, Shi H, Hautefeuille AH, et al.(2010) Effects of the TP53 p.R249S mutant on proliferation and clonogenic properties in human hepatocellular carcinoma cell lines: interaction with hepatitis B virus X protein. Carcinogenesis.2010 Aug;31(8):1475-1482

Gursoy-Yuzuqullu O, Yuzugullu H, Yilmaz M, et al.(2011) Aflatoxin genotoxicity is associated with a defective DNA damage response by passing p53 activation. Liver Int.2011 Apr; 31(4):561-571

Groopman JD.1994. Molecular dosimetry methods for assessing human aflatoxin exposures. In The Toxicology of Aflatoxins: Human Health, Veterinary, and Agricultural Significance, ed. DL Eaton, JD Groopman, pp. 259–79. San Diego: Academic

Guengerich FP, Johnson WW, Shimada T, et al. (1998) Activation and detoxication of aflatoxin B1. Mutat Res.1998 Jun 18;402 (1-2):121–128.

Hainaut P, Boyle P. (2008).Curbing the liver cancer epidemic in Africa. Lancet. 2008 Feb 2; 371(9610): 367–368.

Henry S.H., Whitaker T., Rabbani I., et al. (2001) Aflatoxin M1: Joint ExpeCommittee on Food Additives (JECFA). Hamilton, Ontar Canada

Hsu IC, Metcalf RA, Sun T, et al.(1991) Mutational hotspot in the p53 gene in human hepatocellular carcinomas. Nature. 1991 Apr 4; 350(6317): 427-428

Hung CH, Lu SN, Wang JH, et al.(2003) Correlation between ultrasonographic and pathologic diagnoses of hepatitis B and C virus-related cirrhosis. J Gastroenterol. 2003; 38(2): 153-157

Hussain SP, Aguilar F, Amstad P, et al, Oxy-radical induced mutagenesis of hotspot codons 248 and 249 of the human p53 gene, Oncogene.1994 Aug; 9 (8): 2277–2281.

Hussain SP, Schwank J, Staib F, et al.(2007) TP53 mutations and hepatocellular carcinoma: insights into the etiology and pathogenesis of liver cancer. Oncogene.2007 Apr 2; 26 (15): 2166–2176

International Agency for Research on Cancer (1993) WHO IARC monographs on the evaluation of carcinogenic risks to humans. Some naturally occurring substances: food items and constituents, heterocyclic aromatic amines and mycotoxins.aflatoxins. IARC, Lyon 6:245–395

Iyer RS, Coles BF, Raney KD, et al.(1994), DNA adduction by the potent carcinogen aflatoxin B1: mechanistic studies. J Am Chem Soc.1994, 116 (5) :1603–1609.

Jackson PE, Qian GS, Friesen MD, et al.(2001) Specifc p53 mutations detected in plasma and tumors of hepatocellular carcinoma patients by electrospray ionization mass spectrometry. Cancer Res. 2001 Jan 1; 61(1): 33-35

Jia L, Wang XW, Harris CC. (1999) Hepatitis B virus X protein inhibits nucleotide excision repair. Int J Cancer. 1999 Mar 15;80 (6) :875–879

Jiang W, Wang XW, Unger T, et al.(2010). Cooperation of tumor-derived HBx mutants and p53-249(ser) mutant in regulating cell proliferation, anchorage-independent growth and aneuploidy in a telomerase-immortalized normal human hepatocyte-derived cell line. Int J Cancer. 2010 Sep1;127(5):1011-1020

Johnson WW, Ueng YF, Widersten M, et al.(1997) Conjugation of highly reactive aflatoxin B1 exo-8, 9-epoxide catalyzed by rat and human glutathione transferases: estimation of kinetic parameters. Biochemistry. 1997 Mar 18; 36(11): 3056-3060

Johnson WW, Guengerich FP. (1997) Reaction of aflatoxin B1 exo-8, 9-epoxide with DNA: kinetic analysis of covalent binding and DNA-induced hydrolysis. Proc Natl Acad Sci USA 1997 Jun 10;94 (12) :6121–6125.

Kaplanski C, Chisari FV, Wild CP.(1997) Minisatellite rearrangements are increased in liver tumours induced by transplacental afatoxin B1 treatment of hepatitis B virus transgenic mice, but not in spontaneously arising tumours. Carcinogenesis 1997 Apr; 18(4): 633-639

Kew MC. (2002) Epidemiology of hepatocellular carcinoma. Toxicology 2002 Dec 27;(181-182):35–38

Kew M.C. (2003) Synergistic interaction between aflatoxin B1 and hepatitis B virus in hepatocarcinogenesis. Liver Int. 2003 Dec;23 (6): 405–409.

Kirby GM, Chemin I, Montesano R, et al.(1994) Induction of specific cytochrome P450s involved in aflatoxin B1 metabolism in hepatitis B virus transgenic mice. Mol Carcinog.1994 Oct;11 (2) :74–80.

Kirk GD, Camus-Randon AM, Mendy M, et al. (2000) Ser-249 p53 mutations in plasma DNA of patients with hepatocellular carcinoma from The Gambia. J Natl Cancer Inst. 2000 Jan 19; 92(2): 148-153

Kirk GD, Lesi OA, Mendy M, et al.(2004) The Gambia liver cancer study: infection with hepatitis B and C and the risk of hepatocellular carcinoma in West Africa. Hepatology. 2004 Jan; 39 (1): 211–219.

Kirk GD, Lesi OA, Mendy M, et al.(2005) 249(ser)TP53 mutation in plasma DNA, hepatitis B viral infection, and risk of hepatocellular carcinoma. Oncogene. 2005 Sep 1; 24 (38): 5858–5867.

Kuang SY, Lekawanvijit S, Maneekarn N, et al.(2005) Hepatitis B 1762T/1764A mutations, hepatitis C infection, and codon 249 p53 mutations in hepatocellular carcinomas from Thailand. Cancer Epidemiol Biomarkers Prev. 2005 Feb; 14(2): 380-384

Kuniholm MH, Lesi OA, Mendy M, et al. (2008) Aflatoxin exposure and viral hepatitis in the etiology of liver cirrhosis in the Gambia, West Africa. Environ Health Perspect. 2008 Nov; 116(11): 1553-1557

Langouët S, Coles B, Morel F, et al. (1995) Inhibition of CYP1A2 and CYP3A4 by oltipraz results in reduction of aflatoxin B1 metabolism in human hepatocytes in primary culture. Cancer Res. 1995 Dec 1; 55(23): 5574-5579

Lin DY, Sheen IS, Chiu CT, et al.(1993) Ultrasonographic changes of early liver cirrhosis in chronic hepatitis B: a longitudinal study. J Clin Ultrasound. 1993 Jun; 21(5): 303-308

Lin JK, Miller JA, Miller EC. (1977) 2, 3-Dihydro-2-(guan-7-yl)-3-hydroxy-aflatoxin B1, a major acid hydrolysis product of aflatoxin B1-DNA or -ribosomal RNA adducts formed in hepatic microsome-mediated reactions and in rat liver in vivo. Cancer Res. 1977 Dec; 37(12): 4430-4438

Lin Y, Nomura T, Yamashita T, et al.(1997) The transactivation and p53-interacting functions of hepatitis B virus X protein are mutually interfering but distinct. Cancer Res. 1997 Nov 15;57 (22): 5137–5142.

Liu Y, Wu F.(2010) Global burden of aflatoxin-induced hepatocellular carcinoma: a risk assessment. Environ Health Perspect. 2010 Jun; 118(6):818–824.

Lodato F, Mazzella G, Festi D, et al.(2006).Hepatocellular carcinoma prevention: a worldwide emergence between the opulence of developed countries and the economic constraints of developing nations. World J. Gastroenterol. 2006 Dec 7; 12(45): 7239–7249.

Long XD, Ma Y, Zhou YF, et al. (2009) XPD codon 312 and 751 polymorphisms, and AFB1 exposure, and hepatocellular carcinoma risk. BMC Cancer.2009 Nov 17;9:400

López C, Ramos L, Bulacio L, et al. (2002) Aflatoxin B1 content in patients with hepatic diseases. Medicina.2002;62(4):313–316.

Macé K, Aguilar F, Wang JS, et al.(1997) Aflatoxin B1-induced DNA adduct formation and p53 mutations in CYP450-expressing human liver cell lines. Carcinogenesis. 1997 Jul; 18(7): 1291-1297

Madden CR, Finegold MJ, Slagle BL. (2002) Altered DNA mutation spectrum in aflatoxin b1-treated transgenic mice that express the hepatitis B virus x protein. J Virol. 2002 Nov;76 (22): 11770–11774.

Manuguerra M, Saletta F, Karagas MR, et al.(2006) XRCC3 and XPD/ERCC2 single nucleotide polymorphisms and the risk of cancer: a HuGE review.Am J Epidemio. 2006 Aug 15; 164(4):297-302

Marin MC, Jost CA, Brooks LA, et al. (2000) A common polymorphism acts as an intragenic modifier of mutant p53 behaviour. Nat Genet. 2000 May; 25(1): 47–54.

McGlynn KA, Rosvold EA, Lustbader ED, et al. (1995). Susceptibility to hepatocellular carcinoma is associated with gentic variation in the enzymatic detoxifcation of afatoxin B1. Proc Natl Acad Sci USA, 1995 Mar 14; 92(6): 2384-2387

McLean M, Dutton MF.(1995) Cellular interactions and metabolism of aflatoxin: an update. Pharmacol. Ther.1995 Feb; 65 (2): 163–192.

Meneghini R, Schumacher RI. (1977) Aflatoxin B1, a selective inhibitor of DNA synthesis in mammalian cells. Chem Biol Interact. 1977 Sep;18(3): 267-276

Ming L, Thorgeirsson SS, Gail MH, et al.(2002) Dominant role of hepatitis B virus and cofactor role aflatoxin in hepatocarcinogenesis in Qidong, China. Hepatology. 2002 Nov; 36(5): 1214–1220.

Oren M, Rotter V. (2010) Mutant p53 gain-of-function in Cancer. Cold Spring Harb Perspect Biol.2010 Feb; 2(2):a001107

Pemble S, Schiroeder KR, Spencer SR, et al.(1994) Human glutathione S-transferase theta (GSTT1): cDNA cloning and the characterization of a genetic polymorphism. Biochem J. 1994 May 15; 300(pt1): 271-276

Petitjean A, Mathe E, Kato S, et al.(2007) Impact of mutant p53 functional properties on TP53 mutation patterns and tumor phenotype: lessons from recent developments in the IARC TP53 database. Hum Mutat. 2007 Jun;28 (6):622–629.

Pfeifer AM, Cole KE, Smoot DT, et al.(1993) Simian virus 40 large tumor antigen-immortalized normal human liver epithelial cells express hepatocyte characteristics and metabolize chemical carcinogens. Proc Natl Acad Sci USA 1993 Jun 1; 90(11): 5123-5127

Ponchel F, Puisieux A, Tabone E, et al.(1994) Hepatocarcinoma-specific mutant p53-249ser induces mitotic activity but has no effect on transforming growth factor beta 1-mediated apoptosis. Cancer Res. 1994 Apr 15;54 (8):2064–2068.

Qian GS, Ross RK, Yu MC, et al.(1994). A follow-up study of urinary markers of aflatoxin exposure and liver cancer risk in Shanghai, People's Republic of China. Cancer Epidemiol Biomarkers Prev. 1994 Jan-Feb; 3(1): 3-10

Raney KD, Meyer DJ, Ketterer B, et al.(1992) Glutathione conjugation of aflatoxin B1 exo- and endo-epoxide by rat and human glutathione S-transferases. Chem Res Toxicol. 1992 Jul-Aug;5(4): 470-478

Rebbeck TR, Walker AH, Jaffe JM, et al.(1999) Glutathione S-transferase-mu (GSTM1) and -theta (GSTT1) genotypes in the etiology of prostate cancer. Cancer Epidemiol Biomarkers Prev. 1999 Apr; 8(4 Pt 1): 283-287

Ricordy R, Gensabella G, Cacci E, et al.(2002) Impairment of cell cycle progression by aflatoxin B1 in human cell lines. Mutagenesis 2002 May; 17(3): 241-249

Ross RK, Yuan JM, Yu MC, et al. (1992) Original article. Urinary aflatoxin biomarkers and risk of hepatocellular carcinoma. Lancet.1992 Apr 18; 339(8799):943–946

Salhab AS, Edwards GS.(1977) Production of aflatoxicol from aflatoxin B1 by postmitochondrial liver fractions.J Toxicol Environ Health.1977 Jan;2(3):583-587

Santella RM, Chen CJ, Zhang YJ, et al. (1998) Biological markers of aflatoxin B1 in hepatocellular cancer in Taiwan. In: Mendelsohn ML, Mohr LC and Peeters JP, editors. Biomarkers: medical and workplace applications. Washington, DC: Joseph Henry Press, 1998: 355-364

Schleger C, Becker R, Oesch F, et al.(1999) The human p53 gene mutated at position 249 per se is not sufficient to immortalize human liver cells. Hepatology. 1999 Mar;29 (3): 834–838

Sell S, Hunt JM., Dunsford HA, et al.(1991) Synergy between hepatitis B virus expression and chemical hepatocarcinogens in transgenic mice. Cancer Res. 1991 Feb 15;51 (4) :1278–1285.

Semela D, Heim M. (2011).Hepatocellular carcinoma. Ther Umsch. 2011 Apr;68(4):213-217.

Shephard GS.(2009) Aflatoxin analysis at the beginning of the twenty-first century.Anal Bioanal Chem.2009 Nov;395(5):1215-1224

Smela ME, Currier SS, Bailey EA, et al.(2001) The chemistry and biology of aflatoxin B(1): from mutational spectrometry to carcinogenesis. Carcinogenesis 2001 Apr;22 (4): 535–545

Sohn S, Jaitovitch-Groisman I, .Benlimame N, et al.(2000) Retroviral expression of the hepatitis B virus x gene promotes liver cell susceptibility to carcinogen-induced site specific mutagenesis. Mutat Res. 2000 Jun 30; 460 (1): 17–28.

Stettler PM, Sengstag C.(2001) Liver carcinogen aflatoxin B1 as an inducer of mitotic recombination in a human cell line. Mol Carcinog 2001 Jul; 31(3): 125-138

Strano S, Fontemaggi G, Costanzo A, et al. (2002) Physical interaction with human tumor-derived p53 mutants inhibits p63 activities. J Biol Chem.2002 May 24;277(21):18817–18826

Sugimura T.(2000) Nutrition and dietary carcinogens. Carcinogenesis.2000 Mar;21(3):387-395.

Sun CA, Wang LY, Chen CJ, et al.(2001) Genetic polymorphisms of glutathione S-transferases M1 and T1 associated with susceptibility to aflatoxin-related hepatocarcinogenesis among chronic hepatitis B carriers: a nested case-control study in Taiwan. Carcinogenesis. 2001 Aug; 22(8): 1289-1294

Sun Z, Lu P, Gail MH, et al.(1999) Increased risk of hepatocellular carcinoma in male hepatitis B surface antigen carriers with chronic hepatitis who have detectable urinary aflatoxin metabolite M1. Hepatology.1999 Aug;30(2):379-383

Tsuda H, Hirohashi S, Shimosato Y, et al.(1989) Low incidence of point mutation of c-Ki-ras and N-ras oncogenes in human hepatocellular carcinoma. Jpn J Cancer Res. 1989 Mar; 80(3): 196-199

Turner PC, Sylla A, Diallo MS, et al.(2002) The role of aflatoxins and hepatitis viruses in the etiopathogenesis of hepatocellular carcinoma: a basis for primary prevention in Guinea-Conakry, West Africa. J Gastroenterol Hepatol. 2002 Dec; 17 Suppl: S441–S448.

Ueda H, Ullrich SJ, Gangemi JD, et al.(1995) Functional inactivation but not structural mutation of p53 causes liver cancer. Nat. Genet. 1995 Jan;9 (1): 41–47.

Wang LY, Hatch M, Chen CJ, et al. (1996). Aflatoxin exposure and risk of hepatocellular carcinoma in Taiwan. Int J Cancer. 1996 Sep 4; 67(5):620–625

Wang XW, Forrester K, Yeh H, et al.(1994) Hepatitis B virus X protein inhibits p53 sequence-specific DNA binding, transcriptional activity, and association with transcription factor ERCC3. Proc Natl Acad Sci USA.1994 Mar 15; 91 (6): 2230–2234.

Wang XW, Hussain SP, Huo TI, et al.(2002) Molecular pathogenesis of human hepatocellular carcinoma. Toxicology. 2002 Dec 27;181-182:43–47

Wild CP, Yin F, Turner PC, et al.(2000) Environmental and genetic determinants of aflatoxin-albumin adducts in the Gambia. Int J Cancer. 2000 Apr 1; 86(1): 1-7

Wild CP, Turner PC.(2002) The toxicology of aflatoxins as a basis for public health decisions. Mutagenesis 2002 Nov; 17(6): 471-481.

Wild CP, Montesano R.(2009) A model of interaction: Aflatoxins and hepatitis viruses in liver cancer aetiology and prevention. Cancer Lett. 2009 Dec 1;286(1):22-28

Williams JH, Phillips TD, Jolly PE, et al. (2004) Human aflatoxicosis in developing countries: a review of toxicology, exposure, potential health consequences, and interventions. Am J Clin Nutr. 2004 Nov;80 (5): 1106–1122

Wong JJ, Hsieh DP.(1976) Mutagenicity of aflatoxins related to their metabolism and carcinogenic potential.Proc Natl Acad Sci USA.1976 Jul;73(7):2241-2244

Xu L, Qian G, Tang L, et al. (2010) Variations of hepatitis B virus and serum aflatoxin-lysine adduct on high risk of hepatocellular carcinoma in Southern Guangxi, China.J Hepatol.2010 Oct;53(4):671-676

Yang HI, Lu SN, Liaw YF, et al.(2002) Hepatitis B e antigen and the risk of hepatocellular carcinoma. N Engl J Med. 2002 Jul 18:347 (2002) 168–174.

Yin L, Ghebranious N, Chakraborty S, et al.(1998) Control of mouse hepatocyte proliferation and ploidy by p53 and p53ser246 mutation in vivo. Hepatology.1998 Jan; 27 (1) :73–80.

Yu MC, Yuan JM. (2004) Environmental factors and risk for hepatocellular carcinoma. Gastroenterology 2004 Nov; 127(5 Suppl 1): S72–S78

Yuen MF, Hou JL, Chutaputti A, et al.(2009). Hepatocellular carcinoma in Asia Pacific region. J Gastroenterol Hepatol.2009 Mar; 24(3): 346–353.

Zhang YJ, Chen SY, Tsai WY, et al. (2000) Expression of cytochrome P450 1A1/2 and 3A4 in liver tissues of hepatocellular carcinoma cases and controls from Taiwan and their relationship to hepatitis B virus and aflatoxin B1- and 4-aminobiphenyl-DNA adducts. Biomarkers. 2000 July 1;5(4):295-306

Zhang YJ, Rossner P Jr, Chen Y, et al. (2006)Aflatoxin B1 and polycyclic aromatic hydrocarbon adducts, p53 mutations and p16 methylation in liver tissue and plasma of hepatocellular carcinoma patients. Int J Cancer 2006 Sep1; 119(5): 985-991

10

A Comprehensive Review of Male Reproductive Toxic Effects of Aflatoxin

Mohammad A. Akbarsha[1,3], Faisal Kunnathodi[1] and Ali A. Alshatwi[2]
[1]*Department of Animal Science, Bharathidasan University, Tiruchirappalli,*
[2]*Department of Food Science and Nutrition,*
College of Food Science and Agriculture, King Saud University, Riyadh, KSA
[3]*King Saud University, Riyadh, KSA*
[1]*India*
[2,3]*Saudi Arabia*

1. Introduction

Aflatoxins, highly oxygenated, heterocyclic, difuranocoumarin compounds that could be present in human foods and animal feedstuffs, are an important group of mycotoxins produced by the fungi *Aspergillus flavus, A. parasiticus* and *A. nomius* (Diaz et al., 2008). Other species of Aspergillus such as *A. bombycis, A. ochraceoroseus* and *A. pseudotamari* may also produce aflatoxins (Bennett & Klich, 2003; Klich et al., 2000; Mishra & Das, 2003). On a worldwide scale, the aflatoxins are found in stored food commodities and oil seeds such as corn, peanuts, cottonseed, rice, wheat, oats, barley, sorghum, millet, sweet potatoes, potatoes, sesame, cacao beans, almonds, etc., which on consumption pose health hazards to animals, including aquaculture species of fish, and humans (Abdel-Wahab et al., 2008; Hussein & Brassel, 2001). Health effects occur in fish, companion animals, livestock, poultry and humans because aflatoxins are potent hepatotoxins, immunosuppressants, mutagens, carcinogens and teratogens. Public health concerns center on both primary poisoning from aflatoxins in commodities, food and feedstuffs, and relay poisoning from aflatoxins in milk (Coppock & Christian, 2007). There are four major natural aflatoxins (AFs), AFB1, AFB2, AFG1 and AFG2. The hierarchy of toxicity of different aflatoxins is in the order AFB1>AFG1>AFB2>AFG2. There are two additional metabolic products of aflatoxins B1 and B2, *viz.,* M1 and M2. More than 5 billion people in developing countries worldwide are at risk of chronic exposure to naturally occurring aflatoxins through contaminated foods (Shephard, 2003; Williams et al., 2004) and more so in the tropical regions, where the climatic conditions favour luxurious growth of *Aspergillus spp*, and people rely on commodities such as cereals, oilseeds, spices, tree nuts, milk, meat and dried fruits that are potentially contaminated by aflatoxins (Strosnider et al., 2006).

Symptoms of aflatoxicosis include feed refusal, decreased feed efficiency, stunted growth, decreased milk production and impaired reproductive efficiency (Diekman & Green, 1992; Oguz & Kurtoglu, 2000; Pier, 1992; Raju & Devegowda, 2000). Aflatoxins in general, and AFB1 in particular, can induce DNA damage, gene mutation, sister-chromatid exchanges and other chromosomal anomalies, which account for their genotoxic, teratogenic and carcinogenic properties (Batt et al., 1980; International Agency for Research on Cancer

[IARC], 1993; Ray-Chaudhuri et al., 1980). AFB1 can form adducts with DNA, RNA and protein, which form the major basis of the health risks (Sun et al., 2001; Williams et al., 2004). Epidemiological and experimental studies have implicated aflatoxins in male reproductive health, and the present review is an attempt to put together the knowledge in a comprehensive perspective.

2. Aflatoxins in sperm and semen

Aflatoxins or their metabolites can reach the testis (Bukovjan et al., 1992) and be present in the semen through this route (Ibeh et al., 1994; Picha et al., 1986; Uriah et al., 2001). Aflatoxins have been detected in boar sperm (Picha et al., 1986) and human semen (Ibeh et al., 1994). In a cross sectional study, Ibeh et al., (1994) found a relationship between aflatoxin levels in serum of infertile men compared to controls: 40% of semen from infertile men had aflatoxins and 50% of spermatozoa were abnormal, whereas 8% of semen from fertile individuals had aflatoxins and only 10-15% were abnormal. The concentrations of aflatoxins detected in the semen were consistently higher among infertile compared to the fertile men. This study was supported by experiments conducted in rats, and the results were in agreement with the observations in the human samples. Uriah et al., (2001) reported translocation of aflatoxin B1 in humans from blood to semen through the blood-testis barrier. In the boars, the highest AF residues in sperm were recorded in March to May and were related with aflatoxin concentration in the feed ration. The group of boars with fertility disorder had more AF in their sperm (up to 100 $pmol^{-1}$), lower sperm concentration, impaired survival of spermatozoa and a large proportion of abnormal spermatozoa (Picha et al.,1986). When ram epididymal sperm were put in different concentrations of aflatoxin, on one-hour post-incubation in control group 81.25% of sperm cells were alive of which 82.88% were motile. The lowest motility (15.93%) was observed in 62.5 ppb aflatoxin-exposed sperm. Sperm viability did not change significantly after 2nd and 3rd hr incubation but significantly decreased in 4th and 5th hr post- incubation. The results of the experiment showed that aflatoxin could decrease motility of sperm obtained from ejaculation or epididymis (Tajik et al., 2007). Ibeh et al., (2000) cultured oocytes for *in vitro* fertilization (IVF) in IVF medium containing AFB_1 and exposed to sperm cells. Epididymal sperm capacitated in IVF medium, with or without AFB_1, were exposed to oocytes. AFB_1 exposure significantly reduced the mean number of ova fertilized. Exposure of sperm to AF caused significant reduction in their motility.

3. Some classical experimental studies on testicular effects of AFs

One of the earliest reports indicating impairment of reproductive efficiency due to AF toxicity was that of Maryamma & Sivadas, (1975) who found that continuous feeding of a diet containing 0.7 ppm AF produced testicular degeneration in male goats. Subsequently, there have been other reports of AFB_1 causing delay in physiological and behavioural sexual maturation (Ottinger & Doerr, 1980) and also delayed testicular development in juvenile Japanese quail (Doerr & Ottinger, 1980). Sharlin et al., (1980) found decreased semen volumes and testes weights, and disruption of the germinal epithelium in mature male white Leghorn chicks. Another study conducted by Sharlin et al., (1981) to investigate the relative importance of ingestion of aflatoxin versus decreased feed consumption led to the conclusion that even though decreased feed consumption did not produce symptoms of aflatoxicosis, it had accounted for 60% of the effects of aflatoxin on reproduction. AFB1 toxicity leads to reduction in size and weight of testis, with mild testicular degeneration to complete disappearance of cellular components accompanied by interstitial cell proliferation and reduced estrogen concentration in rat (Gopal et al., 1980).

Ikegwuonu et al., (1980) observed some degeneration in the testis of aflatoxin treated rats, accounting for the loss of germ cells. Further, the authors provided biochemical insight into the toxicity, and postulated inhibition of testicular ribose 5′-phosphate, which in turn might lead to the impairment of testicular nucleic acid synthesis. It was postulated that prolonged intake of aflatoxin leads to the disturbance of the ensemble of transaminases, particularly GOT and GPT activities, which can adversely affect testicular protein synthesis resulting in decrease of the testicular weight. Aflatoxin impairs protein biosynthesis by forming adducts with DNA, RNA and protein, inhibits RNA synthesis, DNA-dependent RNA polymerase activity, and causes degranulation of endoplasmic reticulum (Cullen & Newberne, 1994; Groopman et al., 1996). Reduction in protein content has also been reported in the testis of aflatoxin-treated mice (Nair & Verma, 2000), which could be responsible for the reduced enzyme activities.

Piskac et al., (1982) showed that prolonged administration of aflatoxin to male rats and pigs resulted in different degrees of dystrophy leading to the destruction and atrophy of spermiogenic epithelium and oedema formation in the tissue. According to Hafez et al., (1982), aflatoxins affect sperm counts and morphology in buffalo bulls. The effect of dietary aflatoxin has also been reported to be clastogenic for meiotic chromosomes, and capable of inducing abnormalities in sperm head morphology and decreasing sperm count in mouse. In this case, the abnormal chromosomes were found to have both structural changes such as breaks, gaps, fragments, translocations, terminal associations as well as gross changes which include numerical changes, clumping and stickiness (Sinha & Prasad, 1990).

Feeding of adult roosters with AF-contaminated diet produced several toxic manifestations which included atrophy of the testes, decrease in the diameter of the seminiferous tubules, decrease in the height of seminiferous epithelium, thickening in intertubular area of the testes, and increase in the abundance of interstitial cells. In some cases, there was no spermatogenesis in the testis (Ortatatli et al., 2002). Desquamation of seminiferous epithelium and degeneration of the desquamated or necrotic cells have been reported (Jayakumar et al., 1988; Sharlin et al., 1980). Ortatatli et al., (2002) observed focal lymphoid cell infiltration in testes in adult roosters fed AF-contaminated diet, which has already been reported to occur in other organs such as liver, kidney and pancreas due to aflatoxicosis (Dafalla et al., 1987; Esapda et al., 1997; Kiran et al., 1998).

Evidence for impairment of Leydig cell function with a resultant drop in testosterone in testis preceding disruption of spermatogenesis in rats was provided by Egbunike et al., (1980, 1982). Recently, Abu El-Saad & Mahmoud, (2009) found decreased levels of FSH, LH and testosterone in AF-treated rat. However, no significant differences were observed in testosterone production and secretion by isolated testicular cells of control or aflatoxin treated male chickens when incubated *in vitro* with different concentrations of LH (Clarke & Ottinger, 1989). This could be due to the species differences in rate of exposure of aflatoxin (dietary *vs.* intraperitoneal) or potency of aflatoxin (aflatoxin mixture *vs.* purified aflatoxin B1). However, there was an unexpected suppression of testosterone production even in the presence of 1600ng/ml LH. The similar response of isolated testicular cells from both aflatoxin-treated and control males when exposed to varying amounts of LH indicated a lack of effect of dietary aflatoxin on the steroidogenic capacity of testicular cells *in vitro*.

4. Gross histopathological changes in the testis of AFB1 treated mouse

Aflatoxin B1 was tested for male reproductive toxic effects in our laboratory, and the observations were published in a series of papers. In addition to the already known ones, several newer manifestations were reported. One of our early studies (Faridha et al., 2006) aimed at finding gravimetric, histopathological and histometric changes in the testis of Swiss mouse in response to treatment of aflatoxin B1 (AFB1) in a chronic toxicity testing over different periods of time. AFB1, suspended in corn oil and ethanol, was administered through intra-peritoneal route to 90 day old Swiss mouse at a daily dose of 50ug/kg body weight for 7, 15, 35, 45 days. The testicles and seminal vesicles of the animals were subjected to histopathological analysis adopting paraffin/resin embedding and light microscopy. Computer-assisted histometric analysis of several parameters was also made. Gravimetric analysis of testicles and seminal vesicles revealed duration-dependent decrease in their respective weights (Table 1). In the mice treated for 15 days, the weight of testicles decreased significantly to 73%, in those treated for 35 days to 68% and in those treated for 45 days to 51%. Weight of the seminal vesicles also decreased to 76% in mice treated for 15 days, to 69% in those treated for 35 days and to 59% in those treated for 45 days Histopathological changes were observed in the testis of mice belonging to all the four experimental groups and the impact clearly reflected dependence on the duration of treatment. In general the trend was decrease in size of the seminiferous tubules (STs) (Table 2).

Duration of Treatment	**Weight of testicles (mg)**		**Weight of the seminal vesicles (mg)**	
	Control	**Experimental**	**Control**	**Experimental**
7 days	213±16	203±13 (95)	85±8	79±7 (93)
15 days	217±14	158±12*(73)	84±9	64±6*(76)
35 days	219±21	148±08*(68)	86±8	59±6*(69)
45 days	220±19	112±06*(51)	88±7	52±6*(59)

Table 1. Weight of the paired testicles and seminal vesicles of control and AFB1-treated mice (Mean ± SD). *p<0.01. Number in parenthesis, percentage of the control value.

Duration of Treatment	Perimeter (um)		Diameter (um)	
	Control	AFB1-treated	Control	AFB1-treated
7 days	444.69±10.64	389.92±16.34*	165.65±3.70	120.56±4.71*
15 days	453.00±8.12	332.11±15.09*	164.76±3.19	103.98±8.45*
35 days	462.39±5.34	309.10±19.92*	163.84±2.09	94.01±3.25*
45 days	464.03±2.91	232.53±11.62*	164.56±2.32	81.44±5.71*

Table 2. Perimeter and diameter of the seminiferous tubules of mice treated AFB1. Each value is mean ± SD of 25 measurements made at x400 with sections from the right testis of 5 animals. *p<0.001

Critical observation of the individual STs revealed almost total absence of elongating spermatids (Fig. 1A-D). Spermiated spermatozoa were invariably absent in the lumen. The height of the seminiferous epithelium (SE) either increased or decreased and, correspondingly, the lumen was either almost obliterated or increased. A duration-dependent appearance of uni- (Fig. 2A-D) and multinucleate (Fig. 3) giant cells was noticed. The SE of the mice treated for 15, 35 and 45 days possessed small to large vacuoles or empty spaces increasing in magnitude in relation to the duration of treatment. The vacuoles were empty or contained cell debris. Cell shrinkage and necrosis or pycnosis of the nuclei were also noticed. Giant cells were noticed in the epithelium as well as in the lumen, and they possessed vacuolated cytoplasm and pycnotic nuclei or nuclei with marginalized chromatin.

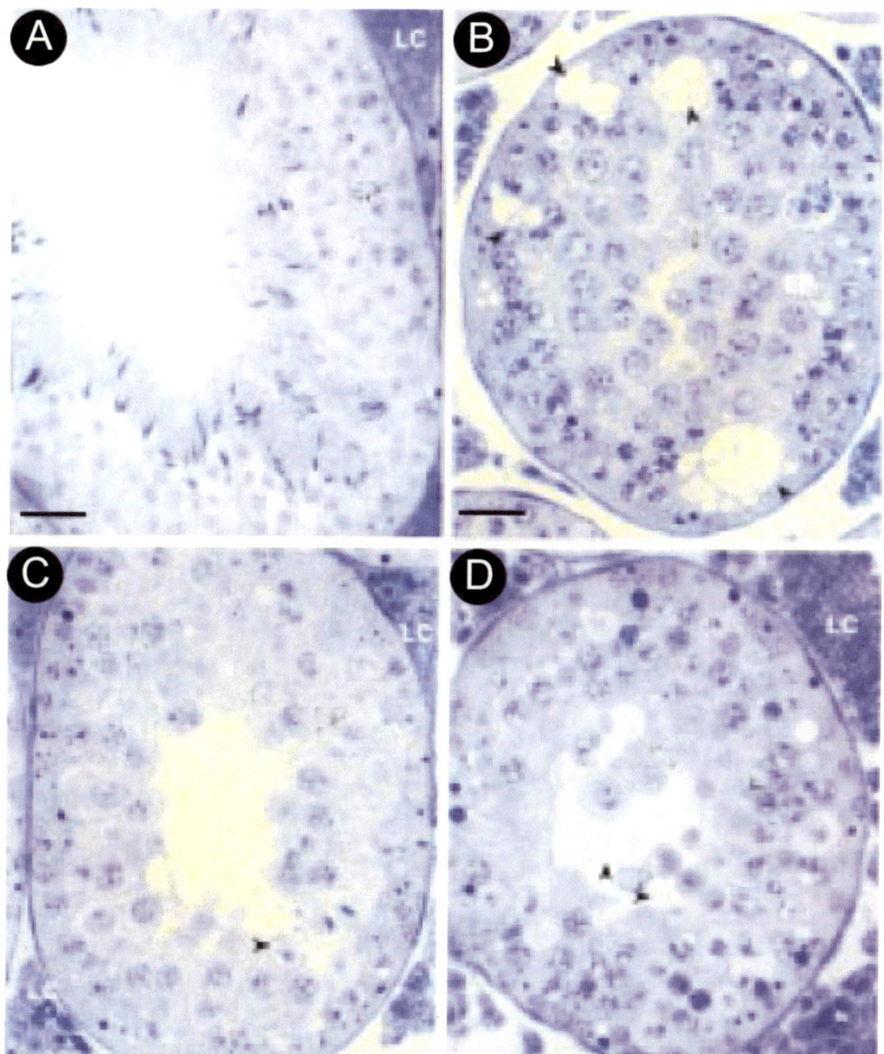

Fig. 1. A-D. Seminiferous tubule of control and treated mice; **A:** Control; **B:** Treated (7days). Note loss of intercalary germ cells (arrowheads). **C:** Treated (15 days). Note loss of germ cells (arrowheads) from the epithelium. The Leydig cells are densely granulated and/or vacuolated. **D:** Same as C, a different tubule. Note absence of elongating spermatids and presence of uninucleate giant cells towards the lumen (arrowhead). Semithin sections, TBO staining. Scale bar 18µm.

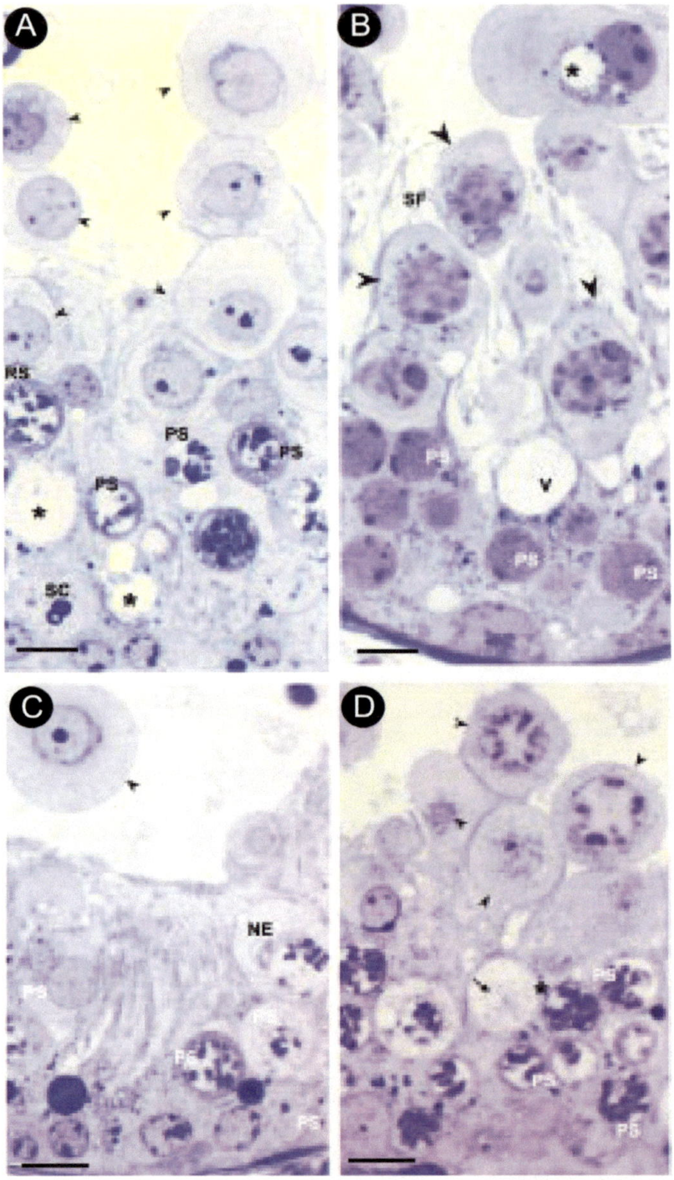

Fig. 2. A-D. Seminiferous epithelium of treated mice. **A:** Shows uninucleate giant cells (arrowheads) (which are spermatids) in the epithelium, damage to chromatin of pachytene spermatocytes, and loss of intercalary germ cells (asterisks). **B:** The uninucleate giant spermatid (arrowhead) is seen in the lumen of the seminiferous tubule. Necrosis of pachytene spermatocytes is also evident (asterisks). **C:** The UNGCs (arrowheads) are pachytene spermatocytes. Note doubling of size of the nucleus, compared to those which underlie them. The giant cells are in the process of being released, and one of them is vacuolated (asterisk). **D:** The giant cells (arrowheads) are in the process of being released into the lumen. In the area marked with asterisks, germ cells are totally lost. NE, necrosis; PS, pachytene spermatocytes; SC, Sertoli cell; SF, Sertoli cell fibrosis. Semithin sections, TBO staining. Scale bar, 4μm.

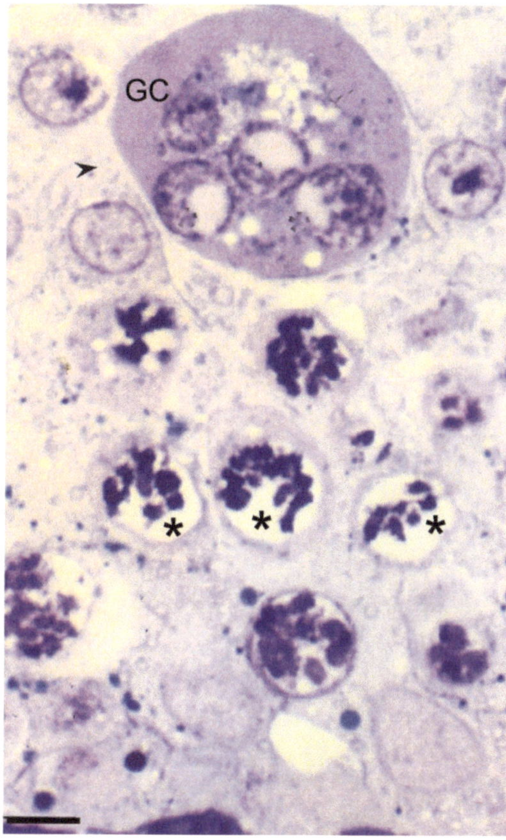

Fig. 3. Seminiferous epithelium of a treated mouse showing a multinucleate giant cell (GC). The MNGC, with nuclei containing marginalized chromatin, lies towards the lumen. Note spermatocytes arrested in M2 (asterisk). Note abnormality in the entire adluminal compartment. The basal compartment is intact. Semithin section, TBO staining. Scale bar 4μm.

Critical observation of the STs of AFB1-treated mice, particularly those in the 35 and 45 day treatment groups, revealed occurrence of pachytene spermatocytes or spermatids of size double that of the respective normal cells (Fig. 2A-D). Such cells are designated as uninucleate giant cells (UNGCs). They were present in the epithelium along the luminal profile (Figs. 1D, 2A-D), some projecting into the lumen but still adherent to the Sertoli cells or lying loose in the lumen. In several cases the UNGC possessed highly vacuolated cytoplasm, and the nucleus was altered in morphology.

Another observation made in several of the STs of the AFB1-treated mice belonging to 15, 35 and 45 day treatment groups was occurrence of multinucleate giant cells (MNGCs) or symplasts (diameter, 40-52 μm) (Fig. 3). Such cells possessed two to 16 nuclei. The nuclei were either intact or had marginalized chromatin. The cytoplasm indicated little to extensive vacuolation. One of the observations was appearance of large cells (diameter 20-30 μm) containing several micronuclei (Fig. 4). Such cells are designated as multiple (or meiotic) micronucleate giant cells (MMGCs). They were present in the epithelium as well as the lumen; when present in the epithelium, they were separated from the Sertoli cells to a great extent, indicating that they are released into the lumen and would result in the appearance of vacuoles in the epithelium. The micronuclei had the appearance of dot-like dense chromatoid bodies. In a few tubules UNGCs, MNGCs and MMGCs coexisted. Loss of germ cells in a few tubules was so acute that hardly any germ cell was present in the ad-luminal compartment, with the epithelium manifesting small to large vacuoles. In some of the tubules the Sertoli cells themselves, from above the level of the ectoplasmic specialization, *i.e.*, the tight junctions of the blood-testis barrier had broken away and such broken portions were carrying with them the pachytene spermatocytes, rendering the epithelium comparable to Sertoli cell-only syndrome, though careful observation revealed the presence of spermatogonia (Fig. 5). The immature germ cells thus lost from the STs could be traced to the rete testis.

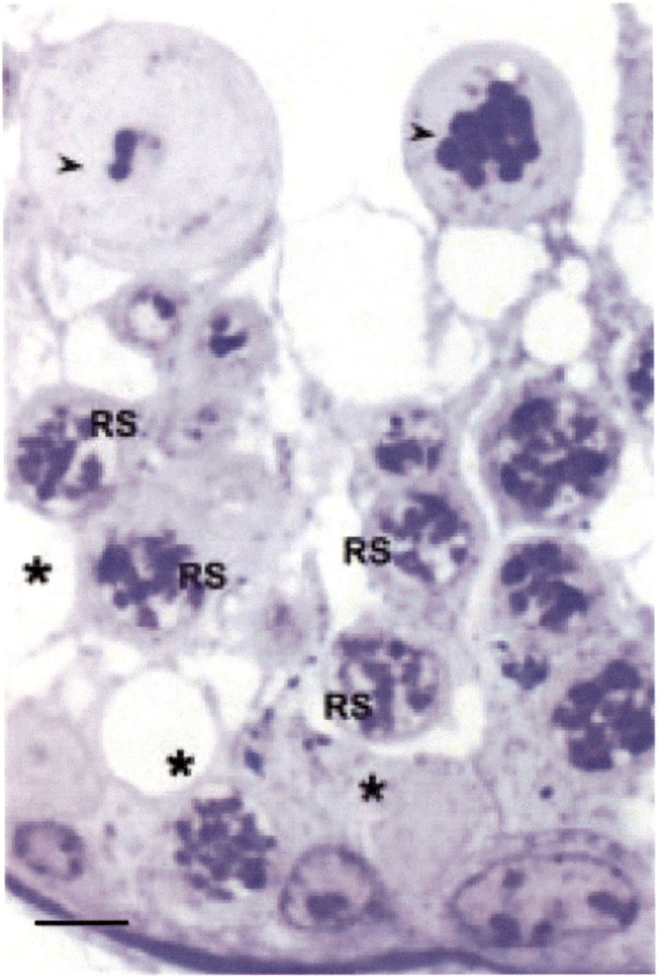

Fig. 4. In this seminiferous tubule, chromatin of round spermatids (RS) is damaged and some RS are missing (asterisks). Two cells arrested in M2, with the chromosome pairs constituting the micronuclei, are shown (arrowheads). Semithin section, TBO staining. Scale bar, 4µm.

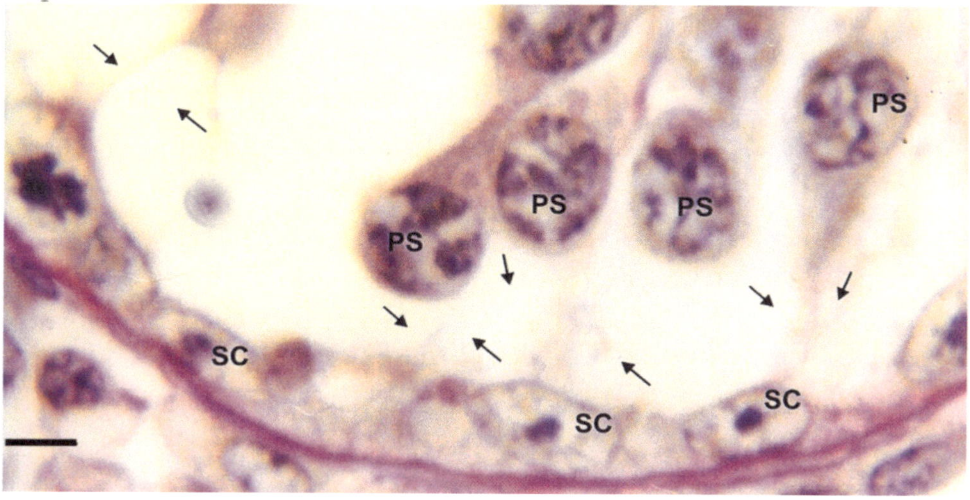

Fig. 5. Pachytene spermatocytes including portions of Sertoli cells are being lost (arrows). But the body of the Sertoli cell and the few basal compartment germ cells are intact. Paraffin section, hematoxylin and eosin staining. Scale bar, 4 µm.

5. Multinucleate giant cells (symplasts) and their origin

Whereas the seminiferous tubules of control mice did not contain any multinucleate giant spermatid or symplastic spermatid (hereinafter referred to as symplasts), the 50 sections of seminiferous tubules of the treated mice counted for symplasts, 28 had 1-17 symplasts (Faridha et al., 2007). The symplasts possessed two to several nuclei (Fig. 3), and the maximum number of nuclei in a symplast was 16. The symplasts, mostly spherical, measured a diameter of 12–20 µm as compared to 6–8 µm of the normal step1 round spermatids. The nuclei of both the normal round spermatids and symplasts measured the same diameter, 5-7 µm. Though the nuclei had normal appearance in a few symplasts, the chromatin was either marginalized (Fig. 6A) or fragmented (Fig. 6B) in the others. In the symplasts, which possessed nuclei with normal morphology, the cytoplasm was intact whereas in those which possessed nuclei with pathological manifestations, the cytoplasm was mostly vacuolated (Fig. 6B, C). The constituent spermatids of the symplasts progressed in spermiogenesis only up to step 8, as seen in the development of acrosome, which indicated that the cells did not progress beyond this step (Fig. 6D) and were released from the Sertoli cells.

The origin of symplasts was traced to the opening of cytoplasmic bridges connecting spermatids (Fig. 7). The bridge connecting normal spermatids measured 0.1–0.2 µm diameter and its lining had an electron-dense plaque extending to a short distance into the cells connected by the bridge. Towards the origin of symplasts, the perimeter of the bridge increased. In the constituent spermatids, the nuclear chromatin underwent marginalization, indicating apoptotic morphology. Subsequently, the cytoplasm of one of the constituent cells in the case of a prospective binucleate symplast or all the constituent cells excepting one in the case of a prospective multinucleate symplast was squeezed into the remaining cell. This resulted in one of the constituent cells becoming larger than the other(s), thus becoming cytoplasm-rich. This was followed by the entry of the nucleus/nuclei of the cytoplasm-poor cell(s) into the cytoplasm-rich cell. Even at this stage, the widened cytoplasmic bridge

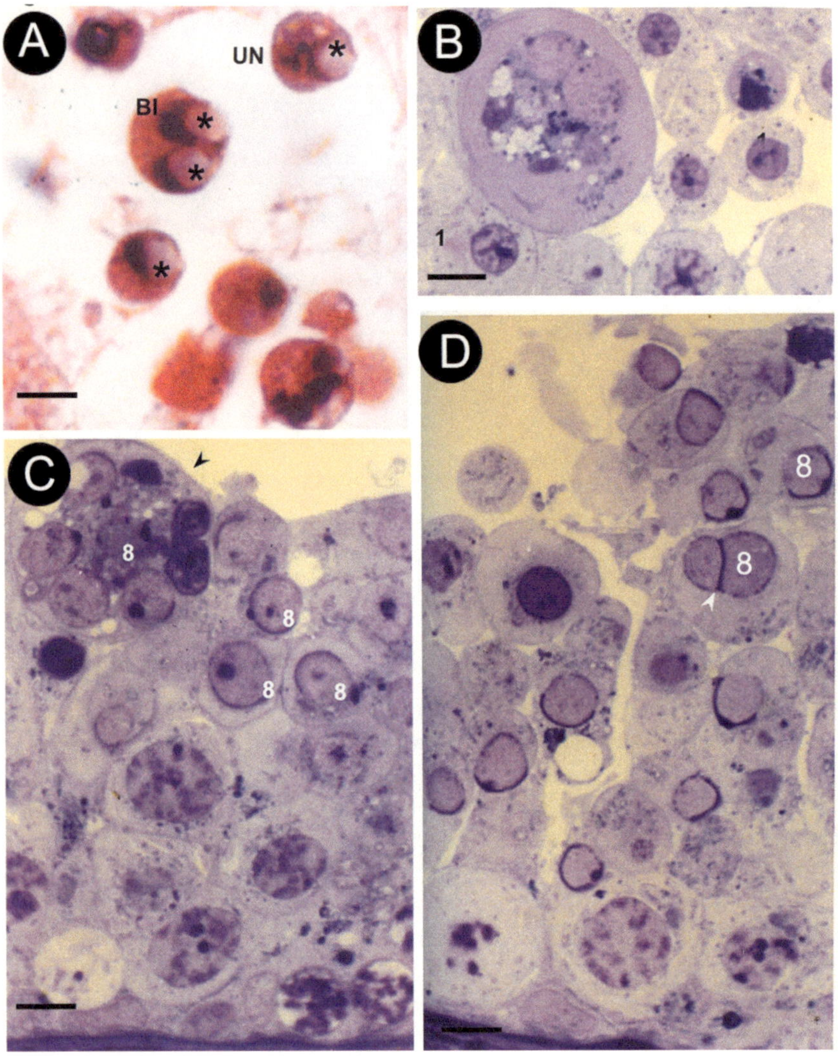

Fig. 6. Aspects of multinuceate giant cell of aflatoxin treated mice. A: A binucleate symplast (BI) with nuclei containing crescentic / marginalized chormatin (asterisks); and uniculceate giant cell (UN) also with nucleus containing marginalized chromatin (asterisks) and vacuolated cytoplasm. B: A multinucleate giant cell with the nuclei surrounded by compact cytoplasm. The normal round spermatids are in step 1 of spermiogenesis (1). C: A multinucleate giant cell (arrowhead) with pycnotic nuclei and vacuolated cytoplasm. The spermatids are in step 8 of spermiogenesis (8). D: Uninucleate and binucleate (arrowhead) giant spermatids at step 8 of spermiogenesis (8). A: Paraffin section, hematoxylin and eosin staining; B-D: Semithin section, TBO staining. Scale bar, 4 µm.

persisted, and did not collapse totally. Since the perimeter of the widened cytoplasmic bridge was not large enough for the nucleus/nuclei of the cytoplasm-poor cell(s) to pass through, it/they responded with change to a thimble shape. During this penultimate stage of origin of symplast, the cytoplasm of the cytoplasm-poor cell(s) was almost bereft of organelles whereas that of the cytoplasm-rich cell was not only rich in organelles but vacuoles too. It was only after or consequent upon the cytoplasm-rich cell having become bi- or multinucleate, the symplast was established to its final spherical shape, with no trace of the cytoplasmic bridge(s). Though loss of integrity of the intercellular bridges between male germ cell clones has been suggested as the mechanism underlying the generation of symplastic spermatids induced due to cytochalasin D (Russell et al., 1987) and *sys* insertional mutation (MacGregor et al., 1990), this report was the first to provide unambiguous evidence for opening of the cytoplasmic bridges to lead to the formation of multinucleate spermatids, and substantiates the mechanism proposed earlier (MacGregor et al., 1990; Russell et al., 1987).

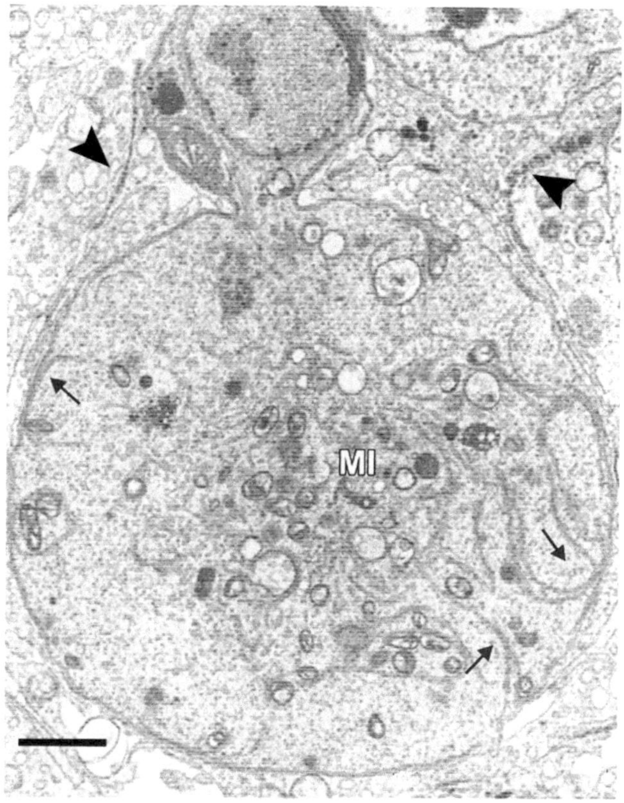

Fig. 7. Transmission electron micrograph showing symplast formation. Note the widened cytoplasmic bridge (arrowheads). A cell on top is cytoplasm poor and the one at bottom is cytoplasm rich. Prominant mitochondria (MI), endoplasmic reticulum (arrows). Scale bar, 1 μm.

Though the essential components of the vertebrate germ cell intercellular bridge have not been until now described, cytoskeletal proteins actin (Russell et al., 1987) and tubulin (MacGregor et al., 1990) have been demonstrated in the walls of the cytoplasmic bridges. Though both these proteins could be targets of agents that disrupt cytoplasmic bridges between spermatids, since cytochalasins, like AFs, are also of fungal origin, the target for AFB1 in the seminiferous epithelium could be actin microfilaments as has been proposed for cytochalasin D (Russell et al., 1987). Alternatively, AFB1 treatment would bring about oxidative damage to the cells (Abu El-Saad & Mahmoud, 2009) and the disruption of the cytoskeletal element in the cytoplasmic bridge would be a consequence of this damage (Lin et al., 2006).

6. Multiple/meiotic micronucleate giant cells (MMGCs) and their origin

The origin of MMGCs could be traced to meiotic metaphase I (M1) and metaphase II (M2) cells (Faisal et al., 2008a). It occurred due to failure of separation of chromosome bivalents in the case of M1 or failure of splitting of the centromere of replicated univalents in the case of M2, in both cases accompanied by or caused due to failure of the spindle apparatus. Delay in meiotic progression was indicated in the thorough asynchrony of the stages in the cycle of seminiferous epithelium. MMGCs invariably appeared detached from Sertoli cells. With the failure of spindle apparatus, the bivalents (in the case of M1) and the replicated univalents (in the case of M2) were arrested from progression towards completion of meiotic division (Fig. 8).

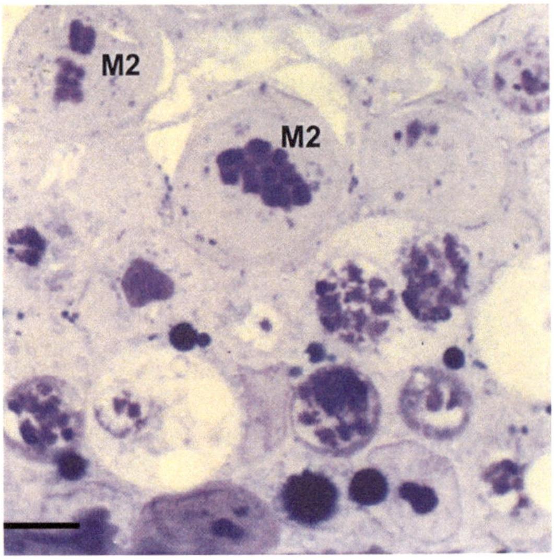

Fig. 8. Failure of meiotic chromosomes (M2) to move to the poles due to problem in spindle fibers, resulting in meiotic microculei. Semithin section, TBO staining. Scale bar, 4 µm.

Ultrastructural evidence for disruption of spindle fiber as the cause of micronucleation was also obtained. In the control mice, cells in early metaphase of second meiotic division had the chromosomes aligned in the equatorial plate and the spindle fibers appeared in the vicinity of the centrioles. Subsequently, the different chromosomes were closely aligned along the metaphasic plate, and the spindle fibers established connection with the centromeres. The separation of bivalents resulted in the univalents arriving at the poles, marking the telophase. In several of the M1 and M2 cells of AFB1-treated mice, not only the spindle fibers were absent, but the bivalents in the case of M1 and the replicated univalents in the case of M2 tended to disaggregate, each becoming a micronucleus. Each micronucleus was formed from a bivalent of M1 cell or a replicated univalent of M2 cell (Figures, in Faisal et al., 2008a).

Meiotic micronuclei are produced in the testicular germ cells by clastogenic or aneuploidogenic agents. Sinha & Prasad, (1990) provided evidence confirming the clastogenic property of AFB1. AFB1 is presumed to be aneuploidogenic also, and the concept of failure of spindle apparatus leads to the generation of meiotic micronuclei is further strengthened by the observation of intact kinetochore in the chromosomes. It was suggested that AFB1 affects assembly of tubulin into microtubules and/or brings about tubulin depolymerization, which would ultimately cause failure of pole-ward movement of the chromosomes.

7. Manifestations in the epididymis to AFB1 treatment

Little is known about the extent of the damaging effect of aflatoxins on the male reproductive tract, particularly the epididymis. Epididymis being the critical organ where in the spermatozoa arrive from the testis and undergo physiological maturation so as to become motile and fertilizable, any toxic manifestation here will explain why spermatozoa become morphologically abnormal and/or physiologically defective and unviable; alternatively, the epididymis would play a protective role so as to safeguard the spermatozoa. Agnes & Akbarsha, (2001) made a pioneering study on the effect of aflatoxin in mouse epididymis. Treatment of male mice with aflatoxin B1 through intra-peritoneal route, in a chronic toxicity testing, resulted in several histopathological changes in the epididymis. Light as well as transmission electron microscopic observations of the sections of epididymis of AFB1-treated mice revealed the presence of small or large vacuoles in the epithelial lining of all segments of the epididymis (Fig. 9). These vacuoles were enclosed in large pale epithelial cells which were quite different in organization from the other epididymal epithelial cell types (principal, clear, narrow, apical and basal cells, and intra-epithelial leucocytes). These cells were designated pale vacuolated epithelial cells (PVECs). The lumen of the vacuole contained spermatozoa and debris or an amorphous to dense PAS-positive material (Fig. 9), or all three materials. There were short microvilli extending from the cell into the vacuole. The vacuole appeared to arise as a result of the degeneration of a principal cell that led to fistula formation, during which the content of the ductal lumen and the principal cell fistula merged and spermatozoa from the ductal lumen entered into the fistula. The neighbouring intact principal cells bent over the degenerating principal cell, cutting off its continuity with the ductal lumen. The basal cell flanking the principal cell apparently developed into a PVEC and enclosed the disintegrating principal cell, including the spermatozoa that had entered it.

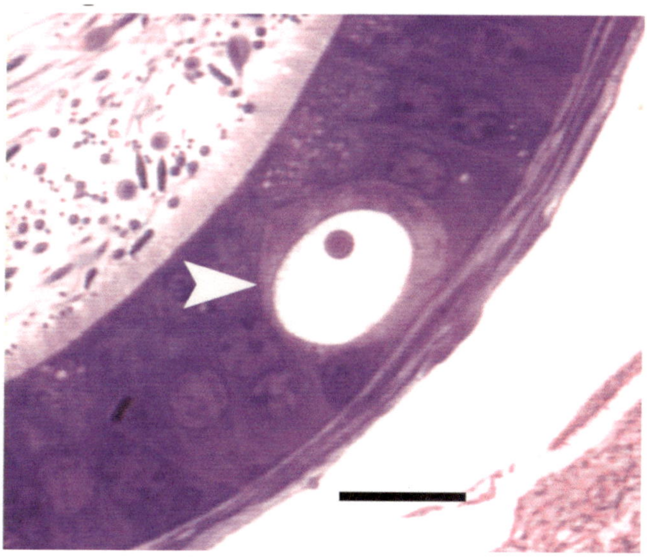

Fig. 9. Section of epididymal duct at caput of a treated mouse showing a pale vacuolated epithelial cell (arrowhead) with a large vacuole containing a dense PAS positive material. Paraffin section, PAS and hematoxylin staining. Scale bar, 4 µm.

Presumably, the PVEC acts upon the material enveloped, through digestion in the vacuole, followed by endocytotic uptake, lysosomal digestion and absorption. Hence, it was proposed that the PVEC develops from the basal cell as a protective device against the autoimmune response to spermatozoa in the context of pathological changes in the principal cells. Though the underlying mechanism of development of PVEC may be either due to androgen deprivation or direct toxicity of AFB1 to the epididymis, the onset of the development of PVEC is due to the pathological change in one or more of the principal cells.

Subsequently, Faisal et al., (2008b) reported the presence of epididymosomes in the AFB1 treated rats (Fig. 10). Epididymosomes, the apocrine secretions from the epithelium of epididymis, are found to be associated with a complex mixture of proteins and play a critical role in the transfer of proteins to sperm surface towards their post-testicular maturation (Frenette et al., 2006; Saez et al., 2003; Thimon et al., 2008). Two or more epididymal spermatozoa embedded in a dense matrix were observed. Such spermatozoa underwent disintegration to varying degrees (Fig. 11) starting with the outer membrane and then the mitochondrial sheath/fibrous sheath, microtubule doublets and ODFs, in that order. From the transmission electron micrographs, it was seen that when the lumen abounded with the defective spermatozoa, there was profuse discharge of epididymosomes. This was further strengthened by the observation of abundant matrix-entangled spermatozoa in the epididymal lumen (Fig. 11). Thus, it was suggested that the epididymosomes in this context are concerned with contributing the dense matrix and the enzymatic mechanism for degradation/dissolution of the defective spermatozoa, thereby excluding the normal sperm from the enzymatic degradation, which is an aspect of versatility of epididymis.

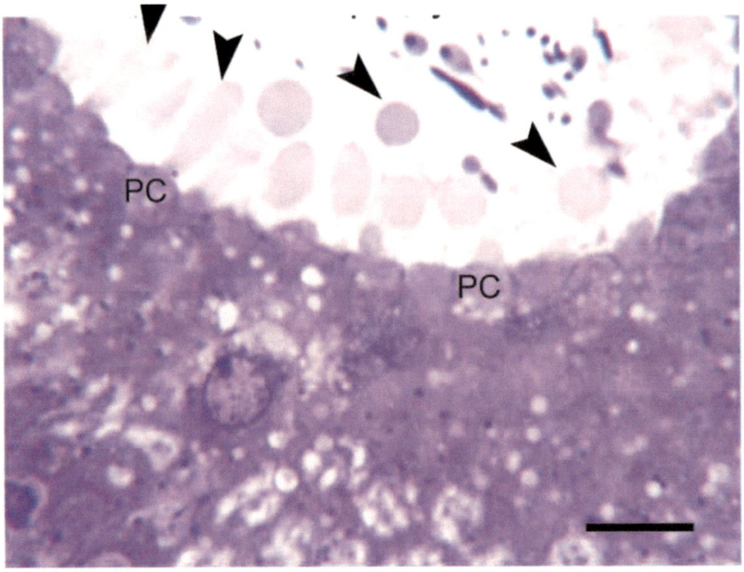

Fig. 10. Section of epdididymal duct at initial segment of a treated mouse showing release of epididymosomes (arrowheads) from the principal cells (PC). The lumen contains epididymosomes (arrowheads) and a few sperm. Semithin section, TBO staining. Scale bar, 4 µm.

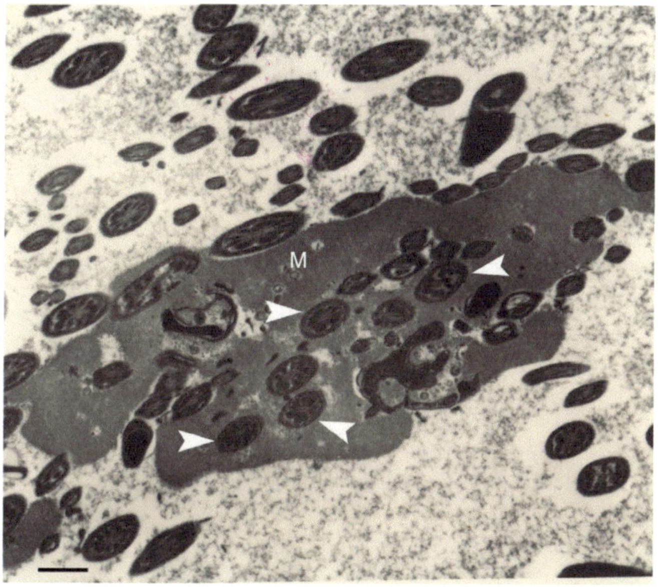

Fig. 11. A transmission electron micrograph showing corpus epididymidal spermatozoa (arrowheads) embedded in a dense matrix (M). Note the disintegration of spermatozoa to various degrees. The dense matrix is surrounded by normal epidiymal plasma, in which normal spermatozoa are found. Scale bar, 1 µm.

Epididymal epithelial cells are, by and large, terminally differentiated cells and do not usually divide unless in case of induction into mitosis. Under this background, we found in about 60% of AF treated mice and rats the principal cells of the initial segment of the epididymis were provoked into mitosis (Fig. 12A, B). Thus, AF could be a potent mitogenic agent, and potentially carcinogenic agent in respect of epididymis (Agnes, Faisal and Akbarsha, unpublished observation).

8. Manifestations in sperm

Agnes & Akbarsha, (2003) assessed the changes in the sperm. There was little change in the sperm concentration of mice treated AFB1 for 7 and 15 days, whereas in the mice treated for 35 and 45 days there was a drastic reduction in sperm concentration. In the mice treated for 35 days, this decreased to about 32% of the control and in those treated for 45 days it decreased to 19%. Sperm motility also displayed the same trend, again in a duration-dependent manner.
The percentage of sperm with abnormal morphology increased on AFB1 treatment in a manner dependent on the duration of the treatment. The various head abnormalities included head without the hook, unusual head shapes, vacuolation of the head and incomplete head. The major tail abnormality was bent or coiled tail. In each treatment group 20–40% of the mice had sperm head detached from the flagellum. Also, a considerably high percentage of sperm had sticky flagellum. Several sperm remained fused in varying numbers over short to long distances and several sperm were agglutinated. More recent studies also reported low sperm concentration, reduction in sperm motility, increased sperm abnormal morphologies and, additionally, decrease in the viability of spermatozoa of mice treated with aflatoxin (Abu El-Saad & Mahmoud, 2009; Mathuria & Verma, 2008).

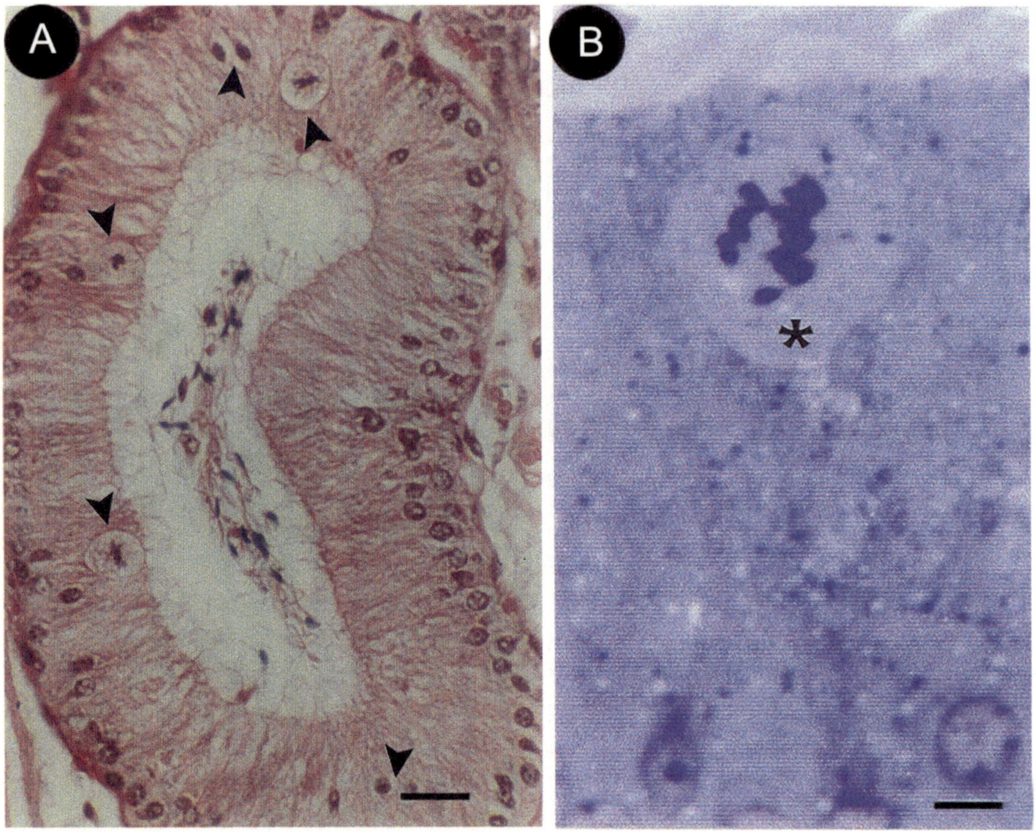

Fig. 12. Sections of the initial segment of the epididymis of treated mice. A: Many mitotic cells (arrowheads) are shown. B: One of the cells in mitosis clearly shown. A, paraffin section, PAS and hematoxylin staining. Scale bar, 10 µm; B, semithin section, TBO staining. Scale bar, 4 µm.

Another major observation was retention of predominant cytoplasmic droplet (CD) by the cauda epididymidal sperm of AFB1-treated mice (Fig. 13A). The quantitative assessment of retention of CD revealed that it increased in the duration-dependent manner of the treatment. In several such spermatozoa, large highly electron dense inclusions were found in the CD (Agnes & Akbarsha, 2003). Spermatozoa with two axonemes in a common cytoplasm were observed among which, in few cases, the axonemes contained the lamellar and vesicular elements of the CD (Fig. 13B).

Apart from the impact on the sperm count, another important factor to be accounted for determining fertility status in the male is the motility of the sperm. In this study, sperm motility was found to be impaired. The factors affecting motility are to be looked at among the endogenous and exogenous factors namely machinery for motility and contribution of the epididymis towards the physiological maturation of the sperm, respectively (Cooper et al., 1998). Considering these histopathological changes, it was speculated that AFB1 treatment through a direct effect on epididymis or indirectly through the Leydig cells, affects the epididymal function of physiological maturation of sperm, leading to an impairment of sperm motility. However, a direct effect of AFs on the epididymal sperm count can not be ruled out, since Ibeh et al., (1994) have shown AF to be present in the human semen, and Picha et al., (1986) reported high levels of AFB1 residues in the seminal plasma of boars.

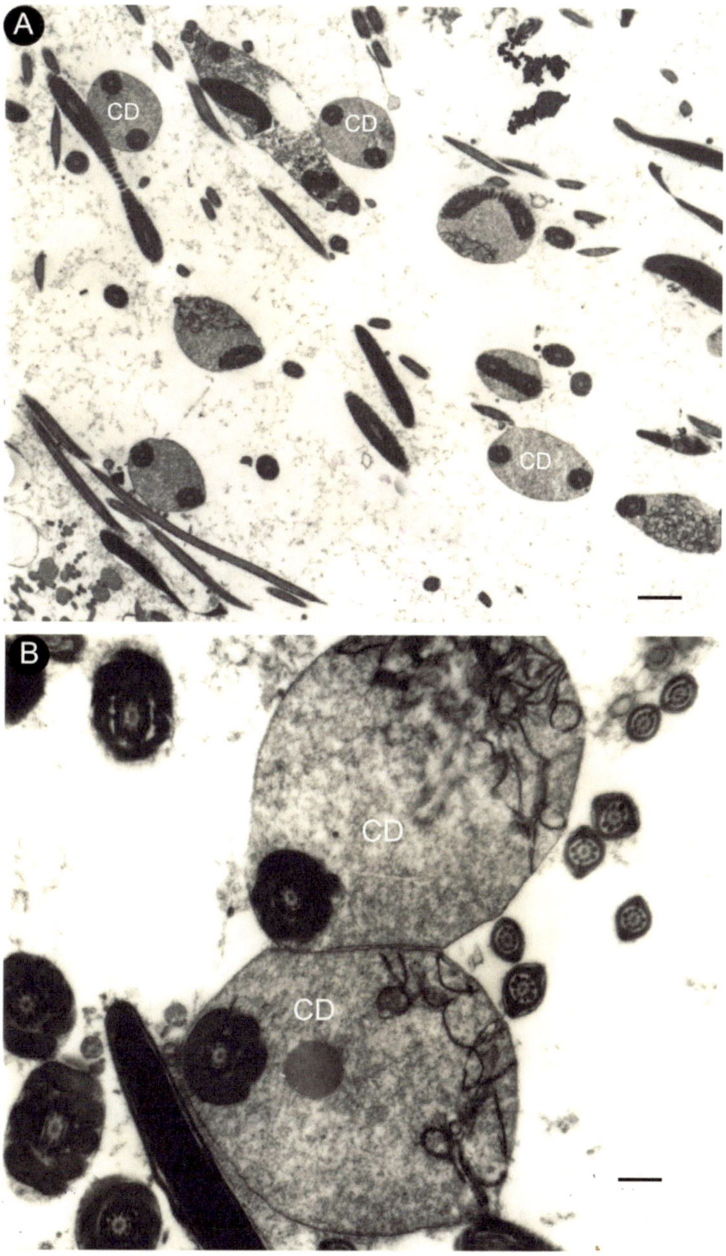

Fig. 13. Transmission electron micrographs showing cauda epididymidal spermatozoa retaining the cytoplasmic droplet (CD). A: Almost all the sperm retaining the cytoplasmic droplet. B: Two seprmatozoa retaining CD, which contains lamellar and vesicular elements, charateristic of CD. A: Scale bar, 2 μm; B: 0.2 μm.

Another intriguing observation was extrusion of one or more outer dense fibres (ODFs) along with the respective microtubule doublets of the axoneme at the midpiece -principal piece junction and/or connecting piece of rat sperm (Fig. 14) (Faisal et al., 2008b). The ODFs took either of two following courses. In one course the ODFs were disorganized and lost their connection with the axoneme. In the second course ODFs underwent slow disintegration such that in some sections there was no trace of ODFs on one side but those on the other side were intact. The affected spermatozoa did not exhibit forward progressive motility but a few exhibited sideways lashing of the flagellum at the very early stages but, subsequently, the lashing also stopped. There were also sperm mid-piece sections without any trace of plasma membrane, mitochondrial sheath and even axoneme in some cases, leaving only the ODFs intact. In many spermatozoa ODFs, in varying numbers, also disintegrated. These sections were not revealing identity as belonging to sperm mid-piece (Faisal et al, 2008b).

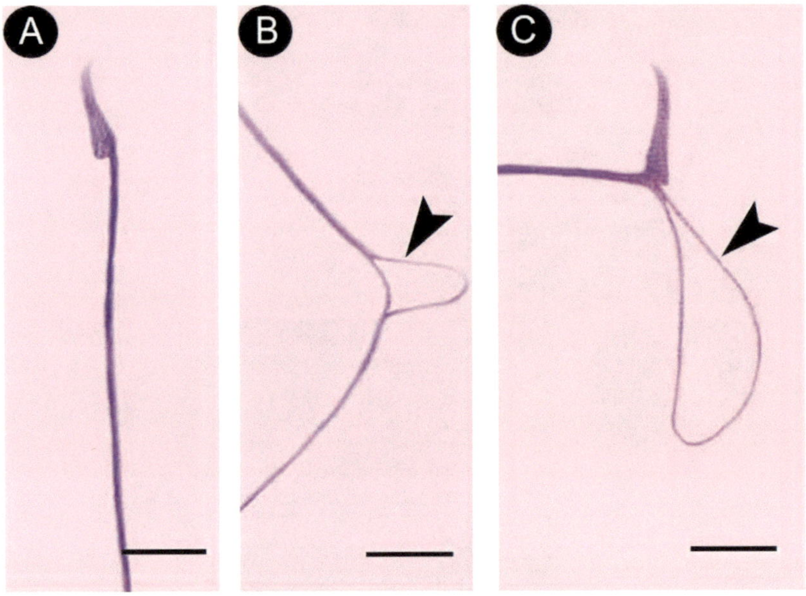

Fig. 14. Eosin and nigrossin stained spermatozoa of rat. A: Normal sperm, B: ODFs on one side are extruded (arrowhead) at the midpiece-principal piece junction, C: ODFs on one side extruded (arrowhead) at the connecting piece. Scale bar, 4 µm.

The manifestations in the principal piece were different from the above. Here, the fibrous sheath and the plasma membrane were lifted off from the ODFs - axoneme complex. With the fibrous sheath (FS) remaining intact, the ODFs and the axonemal doublets on one side or all around the circumference disintegrated, and in the latter case the principal piece in transverse section appeared as an empty vesicle. In some transverse sections of spermatozoa, the ODFs on one side were missing but such missing ODFs were found outside the fibrous sheath, i.e., between the fibrous sheath and the sperm plasma membrane.

9. Effect of AFB1–treatment on Leydig cells

In the mice treated AFB1, two trends were noticed. In the mice treated AFB1 for 7 days the Leydig cells underwent hypertrophy, and dark dense vesicles accumulated in the cytoplasm

(Fig. 15). In the mice treated for 15 and more days, there was a duration-dependent hyperplasia of the Leydig cells, distortion of shape of their nuclei and appearance in their cytoplasm of large vacuoles or dense granules. Histometric analysis of Leydig cells of AFB1-treated mice showed increase in the counts of Leydig cells per unit area and decrease in the Leydig cell nuclear diameter; the changes were dependent on the duration of treatment (Table 3) (Faridha et al., 2006).

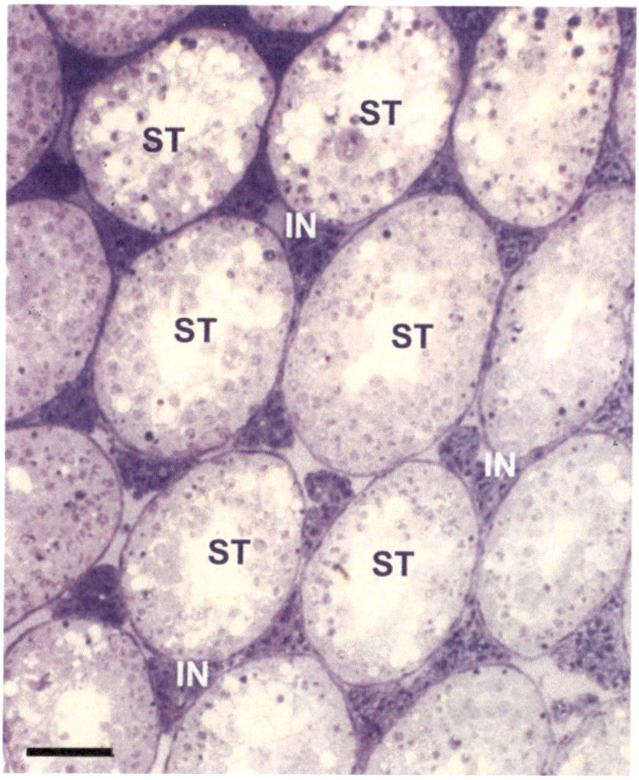

Fig. 15. Section of the testis of a treated mouse showing seminiferous tubules (ST), wtih various histopathological changes, and the interstitium (IN) showing Leydig cells which are dense, hypertrophied and densely vacuolated. Semithin section, TBO staining. Scale bar, 20 μm.

Duration of treatment	Counts per 10^3 um^2 area		Leydig cell perimeter (um)		Leydig cell nuclear diameter (um)	
	Control	Treated	Control	Treated	Control	Treated
7 days	20.93±1.32	14.15±2.86*	123.32±8.47	62.12±6.43*	5.52±0.63	4.68±0.43
15 days	19.98±1.36	20.93±3.19	118.86±9.66	54.41±5.83*	5.36±0.86	3.47±0.64*
35 days	20.32±1.86	28.86±2.68*	121.92±10.86	42.12±4.94*	5.62±0.81	3.16±0.67*
45 days	20.18±1.43	33.70±3.92*	124.86±10.32	31.68±4.66*	5.43±0.48	2.45±0.52*

Table 3. Leydig cell counts, perimeter and nuclear diameter of Leydig cells of AFB1-treated mice. Each value is mean ± SD of 25 measurements made at x400 with sections from the right testis of 5 animals. *p<0.001

10. Effect of AFB1 on fertility in the male

There was no change in the litter size of female mice mated with male mice treated AFB1 for 7 days. In the 15 day treatment group there was a significant decrease in the litter size, whereas in 35 and 45 day treatment groups the females mated with the treated males did not deliver a litter (Table 4) (Faridha et al., 2006). It was earlier reported that a number of young pups had abnormalities such as stumpy tail and blindness of one eye, and there was greater mortality of the pups (Agnes & Akbarsha, 2003).

Duration of AFB1 treatment	Litter size	
	Control	Experiment
7 days	9.8±1.46	9.8±1.79
15 days	9.8±1.48	2.2±1.48
35 days	9.9±1.72	Nil*
45 days	9.2±2.28	Nil*

Table 4. Results of fertility test of treatment group.

11. Conclusions

It has long been suspected, based on epidemiological studies on humans and animals, and experimental studies on fish, poultry, cattle, ram, boar, rat, mouse, etc., that dietary aflatoxins, on chronic exposure at small doses, could be causing disturbance to male reproductive mechanisms. In this background, a series of investigations were undertaken by the authors of this chapter and their students where in Swiss mouse and Wistar rat were treated with aflatoxin B1 through intra-peritoneal route, at a concentration of 20 μg per kg body weight per day (50 μg per kg bw in one study), in chronic male reproductive toxicity testing, for selected durations in relation to the duration of one spermatogenic cycle of the respective animals. The investigations led to the conclusion that aflatoxin B1 is severely toxic to male reproductive mechanisms. The manifestations include severe histopathological changes in the testis, affecting both spermatogenic and androgenic compartments. In the spermatogenic compartment the seminiferous epithelium is severely disrupted resulting in loss of germ cells to various degrees. This loss is preceded by hampering of division (mitotic as well as meiotic) of germ cells, resulting in uninucleate and symplastic giant cells. Meiotic micronucleate giant cells are also produced in large numbers. Tubulin of microtubules of the spindle apparatus appears to be the immediate target to aflatoxin in this case. The affected germ cells are prematurely released from the Sertoli cell. Thus spermatogenesis is severely hampered, resulting in decrease of sperm counts. Motility and viability of the spermatozoa are also impaired. Spermatozoa end up with a variety of abnormal morphologies. Leydig cells undergo hypertrophy and/or hyperplasia, and thorough cytoplasmic vacuolation, which indicate impairment of androgen secretion. The epididymis also undergoes histopathological changes, the most important of which is degeneration of principal cells of the epithelium, access of spermatozoa into these cells, and development of pale vacuolated epithelial cells to deal with such spermatozoa so as to circumvent an autoimmune response to the sperm antigens. Aflatoxin could also be mitogenic in the principal cells of initial segment of the epididymis, suggesting carcinogenic potential of aflatoxin in the epididymis. The fertility of the treated animals is highly compromised. Thus, chronic exposure of humans and animals to aflatoxins, which is possible through dietary contamination, particularly in the tropical climate of developing countries, can bring about deterioration of male reproductive health.

12. Acknowledgments

The authors are grateful to Dr. Agnes Victor Fernandez, Dr. A. Faridha, Dr. R. Girija, Ms. A. Radha, Mr. A. Riyasdeen, and Mr. Md Zeeshan for the technical help. The transmission electron microscopy facility of Christian Medical College and Hospital, Vellore, India, is gratefully acknowledged. The financial assistance through two major projects to Dr. M.A. Akbarsha, from the Department of Science and Technology, New Delhi, and the Senior Research Fellowship to Dr. Faisal Kunnathodi from Council for Scientific and Industrial Research, New Delhi, are acknowledged. Dr. M.A. Akbarsha and Dr. Ali A. Alshatwi thank King Saud University, Riyadh, for support in various forms.

13. References

Abdel-Wahab, M.; Mostafa, M.; Sabry, M.; el-Farrash, M. & Yousef, T. (2008). Aflatoxins as a risk factor for hepatocellular carcinoma in Egypt, Mansoura Gastroenterology Center study. *Hepato-gastroenterology*, Vol.55, No.86-87, (September-October 2008), pp. 1754-1759, ISSN 0172-6390

Abu El-Saad, A.S. & Mahmoud, H.M. (2009). Phytic Acid Exposure Alters AflatoxinB1-induced Reproductive and Oxidative Toxicity in Albino Rats (*Rattus norvegicus*). *Evidence Based Complementary and Alternative Medicine*, Vol.6, No.3, (September 2009), pp. 331-341, ISSN 1533-2101

Agnes, V.F. & Akbarsha, M.A. (2001). Pale vacuolated epithelial cells in epididymis aflatoxin-treated mice. *Reproduction*, Vol.122, No.4, (October 2001), pp. 629–641, ISSN 1470-1626

Agnes, V.F. & Akbarsha, M.A. (2003). Spermatotoxic effect of aflatoxin B(1) in albino mouse. *Food and Chemical Toxicology*, Vol.41, No.1, (January 2003), pp. 119–130, ISSN 0278-6915

Batt, T.R.; Hsueh, J.L.; Chen, H.H. & Huang, C.C. (1980). Sister chromatid exchanges and chromosome aberrations in V79 cells induced by aflatoxin B1, B2, G1 and G2 with or without metabolic activation. *Carcinogenesis*, Vol.1, No.9, (September 1980), pp. 759-763, ISSN 0143-3334

Bennet, J.W. & Klich, M. (2003). Mycotoxins. *Clinical Microbiology Reviews*, Vol.16, No.3, (July 2003), pp. 497-516, ISSN 0893-8512

Bukovjan, K.; Hallmannová, A.; Karpenko, A. & Prosek, J. (1992). Detection of aflatoxin B1 in tissues of free-living game animals (*Lepus europaeus, Phasisnus colchicus, Capreolus capreolus, Anas platyrhynchos*). *Zentralblatt für Veterinärmedizin. Reihe B*, Vol.39, No.9, (November 1992), pp. 695-708, ISSN 0514-7166

Clarke, R.N. & Ottinger, M.A. (1989). Factors related to decreased testosterone concentrations in the peripheral circulation of the maturing male chicken (*Gallus domesticus*) fed aflatoxin . *Animal Reproduction Science*, Vol.18, No.1-3, (February 1989), pp. 25-34, ISSN 0378-4320

Cooper, N.J.; Holland, M.K. & Breed, W.G. (1998). Extratesticular sperm maturation in the brush-tail possum, *Trichosurus vulpecula. Journal of Reproduction & Fertility Supplement*, Vol.53, pp. 221-226 ISSN 1470-1626

Coppock, R. W. & Christian, R.G. (2007). Aflatoxins, In: *Veterinary Toxicology: Basic and Clinical Principles,* R. Gupta, (Ed.), 117-133, ISBN 978-0-12-370467-2, Academic Press, USA

Cullen, J.M. & Newberne, P.M. (1994). Acute hepatotoxicity of aflatoxins, In: Eaton, D.L., Groopman, J.D. (Eds.), 3-26, Toxicology of Aflatoxins, ISBN 0-12-228255-8, Academic Press, San Diego

Dafalla, R.; Yagi, A.I. & Adam, S.E.I. (1987). Experimental aflatoxicosis in hybro-type chicks; sequential changes in growth and serum constituents and histopathological changes. *Veterinary and Human Toxicology*, Vol.29, No.3, (June 1987), pp. 222–225, ISSN 0145-6296

Diaz, G.J.; Calabrese, E. & Blain, R. (2008). Aflatoxicosis in chickens (*Gallus gallus*): an example of hormesis?. *Poultry Science*, Vol.87, No.4, (April 2008), pp. 727-732, ISSN 0032-5791

Diekman, M. A. & Green, M. L. (1992). Mycotoxins and reproduction in domestic livestock. *Journal of Animal Science,* Vol.70, No. 5, (May 1992), pp. 1615-1627, ISSN 0021-8812

Doerr, J.A. & Ottinger, M.A. (1980). Delayed reproductive development resulting from aflatoxicosis in juvenile Japanese quail. *Poultry Science*, Vol.59, No.9, (September 1980), pp. 1995-2001, ISSN 0032-5791

Egbunike, G.N.; Emerole, G.O.; Aire, T.A. & Ikegwuonu, F.I. (1980). FI Sperm production rates, sperm physiology and fertility in rats chronically treated with sub-lethal doses of aflatoxin B1. *Andrologia*, Vol.12, No.5,(September-October 1980),pp. 467-475, ISSN 0303-4569

Egbunike, G.N. (1982). Steroidogenic and spermatogenic potentials of the male rat after acute treatment with aflatoxin B1. *Andrologia*, Vol.14, No.5, (September-October 1982), pp. 440-446, ISSN 0303-4569

Espada, Y.; Gopegui, R.R.; Cuadradas, C. & Cabanes, F.J. (1997). Fumonisin mycotoxicosis in broiler plasma proteins and coagulation modifications. *Avian Diseases*, Vol.41, No. 1, (January–March 1997), pp. 73–79, ISSN 1933-5334

Faisal, K.; Faridha, A. & Akbarsha, M.A. (2008a). Induction of meiotic micronuclei in spermatocytes *in vivo* by aflatoxin B1: Light and transmission electron microscopic study in Swiss mouse. *Reproductive Toxicology*, Vol. 26, No.3-4, (November-December 2008), pp. 303-309, ISSN 0890-6238

Faisal, K.; Periasamy, V.S.; Sahabudeen, S.; Radha, A.; Anandhi, R. & Akbarsha, M.A, (2008b). Spermatotoxic effect of aflatoxin B1 in rat: extrusion of outer dense fibers and axonemal microtubule doublets of sperm flagellum. *Reproduction*, Vol.135, No.3, (March 2008), pp. 303-310, ISSN 1470-1626

Faridha, A.; Faisal, K. & Akbarsha, M.A. (2006). Duration-dependent histopathological and histometric changes in the testis of aflatoxin-treated mice. *Journal of Endocrinology and Reproduction*, Vol.10, No.2, (December 2006), pp. 117-133, ISSN 0971-913X

Faridha, A.; Faisal. K., & Akbarsha, M.A. (2007). Aflatoxin treatment brings about generation of multinucleate giant spermatids (symplasts) through opening of cytoplasmic bridges: light and transmission electron microscopic study in Swiss mouse. *Reproductive Toxicology*, Vol. 24, No. 3-4, (November-December 2007), pp. 403-408, ISSN 0890-6238.

Frenette, G.; Thabet, M. & Sullivan, R. (2006). Polyol pathway in human epididymis and semen. *Journal of Andrology*, Vol.27, No.2, (March-April 2006),pp. 233-239, ISSN 0196-3635

Gopal, T.; Oehme, F.W.; Liao, T.F. & Chen, C.L. (1980). Effects of intratesticular aflatoxin B1 on rat testes and blood estrogens. *Toxicology Letters*, Vol.5, No. 3-4, (March 1980), pp. 263-267, ISSN 0378-4274

Groopman, J.D.; Wang, J.S. & Scholl, P. (1996). Molecular biomarkers for aflatoxins: from adducts to gene mutations to human liver cancer. *Canadian Journal of Physiology and Pharmacology*, Vol.74, No.2, (February 1996), pp. 203-209, ISSN 0008-4212

Hafez, A.H.; Megalla, S.E. & Mahmed, A.A. (1982). Aflatoxin and aflatoxicosis. III. Effect of dietary aflatoxin on the morphology of buffalo bull spermatozoa. *Mycopathologia*, Vol.77, No.3, (March 1982), pp. 141-144, ISSN 0301-486X

Hussein, H.S. & Brasel, J.M. (2001). Toxicity, metabolism, and impact of mycotoxins on humans and animals. *Toxicology*, Vol.167, No.2, (October 2001), pp. 101-134, ISSN 0300-483X

IARC. (1993). In: *Food Items and Constituents*, Heterocyclic Aromatic Amines and Mycotoxins Lyon, France

Ibeh, I.N.; Saxena, D.K. & Uraih, N. (2000). Toxicity of aflatoxin: effects on spermatozoa, oocytes, and *in vitro* fertilization. *Journal of Environmental Pathology, Toxicology and Oncology*, Vol.19, No.4, pp. 357-361, ISSN 0731-8898

Ibeh, I.N.; Uraih, N. & Ogonar, J.I. (1994). Dietary exposure to aflatoxin in human male infertility in Benin City, Nigeria. *International Journal of Fertility and Menopausal Studies*, Vol.39, No.4 (July-August 1994), pp. 208-214, ISSN 1069-3130

Ikegwuonu, F.I.; Egbunike, G.N.; Emerole, G.O. & Aire, T.A. (1980). The effects of aflatoxin B_1 on some testicular and kidney enzyme activity in rat. *Toxicology*, Vol. 17, No.1, pp. 9-16, ISSN 0300-483X

Jayakumar, P.M.; Valsala, K.V. & Rajan , A. (1988). Experimental aflatoxicosis in the duck with special reference to pathology of the testis. *Kerala Journal of Veterinary Science*, Vol.19, No. 2, pp. 122–128, ISSN 0374-8774

Kiran, M.M.; Demet, O.; Ortatatli, M. & Oguz, H. (1998). The preventive effect of polyvinyl-polypyrrolidone on aflatoxicosis in broilers. *Avian Pathology*, Vol. 27, No.3, pp. 250–255, ISSN 0307-9457

Klich, M.A.; Mullaney, E.J.; Daly, C.B. & Cary, J.W. (2000). Molecular and physiological aspects of aflatoxin and sterigmatocystin biosynthesis by *Aspergillus tamarii* and *A. ochraceoroseus*. *Applied Microbiology and Biotechnology*, Vol.53, No.5, (May 2000), pp. 605-609, ISSN 0175-7598

Lin, W.C.; Liao, Y.C.; Liau, M.C.; Lii, C.K. & Sheen, L.Y. (2006). Inhibitory effect of CDAII, a urinary preparation, on aflatoxin B (1)-induced oxidative stress and DNA damage in primary cultured rat hepatocytes. *Food and Chemical Toxicology*, Vol.44, No.4, (April 2006), pp. 546–51, ISSN 0278-6915

MacGregor, G.R.; Russell, L.D.; Van Beek, M.E.; Hanten, G.R.; Kovac, M.J.; Kozak, C.A.; Meistrich, M.L. & Overbeek, P.A. (1990). Symplastic spermatids (sys): a recessive insertional mutation in mice causing a defect in spermatogenesis. *Proceedings of the National Academy of Sciences*, Vol.87, No.13, (July 1990), pp. 5016–5020, ISSN 0027-8424

Mathuria, N. & Verma, R.J. (2008). Curcumin ameliorates aflatoxin-induced toxicity in mice spermatozoa. *Fertility & Sterility*, Vol. 90, No.3, (September 2008), pp. 775–780, ISSN 0015-0282

Maryamma, K.I. & Sivadas, C.G. (1975). Aflatoxicosis in goats (an experimental study*). The Indian Veterinary Journal*, Vol.52, pp. 385-392, ISSN 0019 - 6479

Mishra, H. N. & Das, C. (2003). A review on biological control and metabolism of aflatoxin. *Critical Reviews In Food Science And Nutrition*, Vol.43, No.3, (January 2003), pp. 245-264, ISSN 1040-8398

Nair, A. & Verma, R.J. (2000). Effect of aflatoxin on testis of mouse and amelioration by vitamin E. *Indian Journal of Toxicology*, Vol. 7, No.2, (March 2000), pp. 109-116

Ortatatli, M.; Ciftci, M.K.; Tuzcu, M. & Kaya, A. (2002). The effects of aflatoxin on the reproductive system of roosters. *Research in Veterinary Science*, Vol.72, No.1, (February 2002), pp. 29-36, ISSN 0034-5288

Oguz, H. & Kurtoglu, V. (2000). Effect of clinoptilolite on fattening performance of broiler chickens during experimental aflatoxicosis. *British Poultry Science*, Vol. 41, No.4, (September 2000), pp. 512–517, ISSN 0007-1668

Ottinger, M. A. & Doerr, J. A. (1980). Early influence of aflatoxin upon sexual maturation in the male Japanese quail. *Poultry Science*, Vol. 59, No.8, (August 1980), pp. 1750–1754, ISSN 0032-5791

Picha, J.; Cerovský, J. & Pichová, D. (1986). Fluctuation in the concentration of sex steroids and aflatoxin B1 in the seminal plasma of boars and its relation to sperm production. *Veterinární Medicína*, Vol.31, No.6, (June 1986), pp. 347-357, ISSN 0375-8427

Pier, A.C. (1992). Major biological consequences of aflatoxicosis in animal production. *Journal of Animal Science*, Vol.70, No.12, (December 1992), pp. 3964-3967, ISSN 0021-8812

Piskac, A.; Drábek, J.; Halouzka, R. & Groch, L. (1982). The effect of long term administration of aflatoxins on the health status of male rats and pigs with respect to morphological changes in the testes. *Veterinární Medicína*, Vol.27, No.2, (February 1982), pp. 101-111, ISSN 0375-8427

Raju, M.V. & Devegowda, G. (2000). Influence of esterified-glucomannan on performance and organ morphology, serum biochemistry and haematology in broilers exposed to individual and combined mycotoxicosis (aflatoxin, ochratoxin and T-2 toxin). *British Poultry Science*, Vol.41, No.5, (December 2000), pp. 640-650, ISSN 0007-1668

Ray-Chaudhuri, R.; Kelley, S. & Iype, P.T. (1980). Induction of sister chromatid exchanges by carcinogens mediated through cultured rat liver epithelial cells. *Carcinogenesis*, Vol.1, No.9, (September 1980), pp. 779-786, ISSN 0143-3334

Russell, L.D.; Vogl, A.W. & Weber, J.E. (1987). Actin localization in male germ cell intercellular bridges and the disruption of selected bridges by cytochalasinD. *The*

American Journal of Anatomy. Vol.180, No.1, (September 1987), pp. 25–40, ISSN 0002-9106

Saez, F.; Frenette, G. & Sullivan, R. (2003). Epididymosomes and prostasomes: their roles in posttesticular maturation of the sperm cells. *Journal of Andrology*, Vol.24, No.2, (March-April 2003), pp. 149-154, ISSN 0196-3635

Sharlin, J.S.; Howarth, B. Jr. & Wyatt, R.D. (1980). Effect of dietary aflatoxin on reproductive performance of mature White Leghorn males. *Poultry Science*, Vol.59, No.6, (June 1980), pp. 1311-1315, ISSN 0105-6263

Sharlin, J.S.; Howarth, B. Jr.; Thompson, F.N. & Wyatt, R.D. (1981). Decreased reproductive potential and reduced feed consumption in mature white leghorn males fed aflatoxin. *Poultry Science*, Vol.60, No.12, (December 1981), pp. 2701- 2708, ISSN 0105-6263

Shephard, G.S. (2003). Aflatoxin and food safety: recent African perspectives. *Toxin Reviews*, Vol.22, No.2-3, (January 2003), pp. 267-286, ISSN 1556-9543

Sinha, S. P. & Prasad, V. (1990). Effect of dietary concentration of crude aflatoxin on meiotic chromosomes, sperm morphology and sperm count in mice, *Mus musculus*. *Proceedings of the Indian National Science Academy. Part B, Biological Sciences*, Vol.56, No.3, pp. 269–276, ISSN 0073-6600

Strosnider, H.; Azziz-Baumgartner,E.; Banziger, M.; Bhat, R.V.; Breiman, R.; Brune, M.N.; DeCock, K.; Dilley, A.; Groopman, J.; Hell, K.; Henry, S.H.; Jeffers, D.; Jolly, C.; Jolly, P.; Kibata, G.N.; Lewis, L.; Liu, X.; Luber, G.; McCoy, L.; Mensah, P.; Miraglia, M.; Misore, A.; Njapau, H.; Ong, C.N.; Onsongo, M.T.K.; Page, S.W.; Park, D.; Patel, M.; Phillips, T.; Pineiro, M.; Pronczuk, J.; Rogers, H.S.; Rubin, C.; Sabino, M.; Schaafsma, A.; Shephard, G.; Stroka, J.; Wild, C.; Williams, J.T. & Wilson, D. (2006). Workgroup report: public health strategies for reducing aflatoxin exposure in developing countries. *Environmental Health Perspectives*, Vol. 114, No.12, (December 2006), pp. 1898-1903, ISSN 0091-6765

Sun,C.A.; Wang, L.Y.; Chen, C.J.; Lu, S.L.; You, S.L.; Wang, L.W.; Wang, Q.; Wu, D.M. & Santella, R.M. (2001). Genetic polymorphisms of glutathione S-transferase M1 and T1 associated with susceptibility to aflatoxin-related hepatocarcinogenesis among chronic hepatitis B carriers: a nested case-control study in Taiwan. *Carcinogenesis*, Vol.22, No.8, (August 2001), pp. 1289–1294, ISSN 0143-3334

Tajik, P.; Mirshokraee, P. & Khosravi, A. (2007). Effects of different concentrations of aflatoxin B on ram epididymal and ejaculatory sperm viability and motility *in vitro*. *Pakistan Journal of Biological Sciences*, Vol.10, No.24, (December 2007), pp. 4500-4504, ISSN 1028-8880

Thimon, V.; Frenette, G.; Saez, F.; Thabet, M. & Sullivan, R. (2008). Protein composition of human epididymosomes collected during surgical vasectomy reversal: a proteomic and genomic approach. *Human Reproduction*, Vol.23, No.8, (August 2008), pp. 1698–1707, ISSN 0268-1161

Uriah, N.; Ibeh, I.N. & Oluwafemi, F. (2001). A study on the impact of aflatoxin on human reproduction. *African Journal of Reproductive Health*, Vol.5, No.1, (April 2001), pp. 106-110, ISSN 1118-4841

Williams, J.H.; Phillips, T.D.; Jolly, P.E.; Stiles, J.K.; Jolly, C.M. & Aggarwal, D. (2004). Human aflatoxicosis in developing countries: a review of toxicology, exposure, potential health consequences, and interventions. *The American Journal of Clinical Nutrition*, Vol.80, No.5, (November 2004), pp. 1106-1122, ISSN 0002-9165

11

Aflatoxicosis in Layer and Breeder Hens

Milad Manafi
Department of Animal Science, Malayer University, Malayer
Iran

1. Introduction

During the past few decades there has been a steady increase in global production of poultry meat and eggs. Although the high nutritive value of eggs and poultry meat has resulted in increasing demand, food quality and safety factors are becoming increasingly significant in determining market value of poultry products. Poultry production is one of the fastest growing sectors of Iranian agriculture. Egg production is increasing at the rate of 4-6 per cent per annum, while broiler production at the rate of 10-12 per cent.

At present, Iran is the largest producer of poultry meat in neighboring countries and ninth largest producer in the world, thanks to a 753% growth in meat production from 195 thousand ton in 1978 to 1468 1thousand ton in 2008-09. Broilers are the major source of meat supply in the country. About 270 million broilers are produced every year. Consequent to increased production, per capita consumption /availability has also increased from 7 eggs in 1961 to 42 eggs in 2008. The per capita consumption of poultry meat has increased from 5.4kg in 1978 to 21.8kg in 2008-09. This enormous growth and spurt in poultry production has put a tremendous pressure on proper feeding of poultry in order to sustain the poultry industry in Iran (Manafi, 2010).

As mycotoxins are one of the major factors suppressing poultry productivity and also product quality, control of their impact is critical (Oguz, 2011).

According to the United Nation's Food and Agriculture Organization (FAO), approximately 25% of world's grain supply is contaminated with mycotoxins. The greatest economic impact of mycotoxin contamination is felt by crop and poultry producers, as well as food and feed producers.

Contamination of poultry feeds with mycotoxins is one of the major problems associated with the feeding of poultry. Mycotoxins are the toxic metabolites synthesized by a certain naturally growing fungi on animal feed, feed ingredients and other agricultural crops. More than 350 mycotoxins have been identified so far in feedstuffs. Aflatoxin is the most commonly occurring mycotoxin in Iran. Aflatoxins are a group of secondary metabolites produced by a certain species of fungus of the genus *Aspergillus* (especially *A. flavus* and *A. parasiticus*). These fungi are capable of growing and contaminating the grains and cereals at any time before and after the harvest, during storage, transportation and processing of feed ingredients and the formulated feeds after processing. The spores of the fungi remain dormant but when the level of moisture is more than 12 per cent with a temperature of 25-35°C, with humidity of 80 per cent and adequate aeration initiate their growth. Mycotoxins have adverse effect on both health and productivity in almost all species of domestic animals including poultry. In general, mycotoxicosis results in reduced feed intake, diminished feed conversion, decrease in production and subsequently increased susceptibility to various infections depending upon the type of toxins ingested (Xue et al., 2010).

Mycotoxins	Fungi
Aspergillus toxins	
• **Aflatoxins B_1, B_2, G_1 and G_2**	*Aspergillus flavus* and *Aspergillus parasiticus*
• **Cyclopiazonic acid**	*Aspergillus flavus*
• **Ochratoxins**	*Aspergillus ochraceus*
• **Sterigmatocystin**	*Aspergillus versicolor*
Pencillium Toxins:	
• **Ochratoxins**	*Penicillium viridicatum*
• **Citrinin**	*Pencillium citrinum*
Fusarium Toxins	
• **T-2 Toxin, HT-2 Toxin, Diacetoxyscirpenol (DAS), Monoacetoxyscirpenol (MAS)**	*Fusarium tricinctum,* *Fusarium solani*
• **Deoxynivalenol (DON, vomitoxin)**	*Fusarium graminearum*
• **Zearalenone**	*Fusarium graminearum, Fusarium roseum*
• **Fumonisins B_1, B_2**	*Fusarium moniliforme* *Fusarium proliferatum*
• **Moniliformin**	*Fusarium moniliforme*
Ergot toxins	
• **Ergopeptines**	*Claviceps purpurea*
• **Ergovaline**	*Acremonium coenophialum*

Table 1. Significant mycotoxins and fungi (molds) in foods and feeds

Aflatoxin contamination of feedstuffs has been reported to be of a wide range from 1 to 900µg/kg in commonly used ingredients as well as mixed feed samples in developing countries (Mohanamba et al., 2007). Poultry industry suffers greater economic losses due to the greater susceptibility of the species in comparison with other animals to the toxin apart from continuing intermittent occurrences in feeds (Fraga et al., 2007; Thapa, 2008).

2. Incidence of aflatoxin

It is imperative that food contaminated with aflatoxin is considered unsafe for human and animal health. Aflatoxins occur over a wide variety of substrates of practical importance to poultry feeding (maize, groundnut/meal, cottonseed meal, sunflower extractions, rice, soya bean meal and compounded feeds.) Because of increasing awareness of the risk of aflatoxin contamination of foods and feeds, this has opened a new vista to conduct survey of feed stuffs which are commonly contaminated with aflatoxin. The details of the survey are presented in Table 2.

Author and year	Ingredient	Level (ppb)
Shetty et al. (1987)	Mixed poultry feed	30-1610
Jelinek et al. (1989)	Corn and corn products Peanuts	0.1-1970 0.2-5000
Devegowda et al. (1990)	Groundnut Maize Bajra	48-900 32-1000 12-15
Hegazy et al. (1991)	Poultry feed	1-2000
Devegowda & Arvind (1993)	Maize Ground nut cake Others	25-1002 45-1500 10-80
Jindal et al. (1993)	Poultry feeds	> 300 30-160
Dhavan & Choudary (1995)	Feed ingredients and mixed feeds	High concentrations
Sala & Ueno (1997)	Maize	20-100
Chandrasekharan (2000)	Maize	21to 500
Pandey et al. (2001)	Maize Wheat Groundnut extraction	948 285 225
Chandrasekharan et al. (2002)	Different feed samples	0 - 50
Wang et al. (2003)	Different feed samples	8.27
Manafi, (2007)	Poultry feeds	500
Manafi et al. (2009)	Mash Poultry feeds	450
Manafi et al. (2010)	Pelleted Poultry feeds	470

Table 2. The results of surveys conducted by various investigators on natural occurrence of aflatoxins in various feeds and feed stuffs

Results of the contamination monitoring program for mycotoxins from 1976 till date showing that, much of the monitored grain contained aflatoxin above 20ppb, higher than the regulatory limits in feeds of most countries (Jelinek et al., 1989).

2.1 Safe/permissible level of mycotoxins in poultry feeds

What is a safe level? Can a contaminated grain source be safely fed to other animals if not poultry? What will be the economic impact of a given level of contamination? These are some of the questions commonly asked by people involved in poultry and livestock farming.

Strictly speaking **there is no *safe level.*** With reference to mycotoxins the risk directly depends on the level of the major mycotoxins in the feed and also on the co-occurrence and levels of other mycotoixns.

In order to reduce the toxic and economical impact of mycotoxins, several countries regulate the levels of certain mycotoxins in foods and feeds. Worldwide food and feed legislation safeguards the health of consumers and the economic interests of animal producers and traders. Virtually all countries with fully developed market economies have regulations with the exception of some African countries.

Since the main consumers of poultry products are humans, it becomes relevant to also view the problem of mycotoxins residues in poultry products from a human health standpoint.

Mycotoxin	**Poultry production**	**Carry over in Meat and Eggs**
Aflatoxin B_1	+	Liver
Ochratoxin A	+	Hatching eggs
Cyclopiazonic acid	+	Meat and eggs
Deoxynivalenol	+	Hatching Eggs
Zearalenone	+	Eggs
T-2 toxin	-	-
Diacetoxyscripenol	-	-
Fusarochromanone (*Fusarium* toxin)	+	Hatching Eggs
Aurofusarin (pigment)	+	Eggs

Table 3. Occurrence of mycotoxin residues in poultry products

3. Toxicity and mode of action

Aflatoxin B_1 is found to be highly toxic (6.1mg/kg body weight) to chicken as compared to other Aflatoxins. Chronic aflatoxicosis resulting from regular low level dietary intake of aflatoxin caused reduced weight gain, decrease in feed intake and poor feed efficiency. The important biochemical effects of aflatoxin B_1 are inhibition of DNA replication and RNA synthesis (Kichou & Walser, 1994). Hsieh (1985) reported inhibition of elongation and/or termination of the translational process of protein synthesis, interference in successive steps in mitochondrial respiratory chain, alteration in immune response and exert carcinogenic, teratogenic and mutagenic effects by reacting with nucleophillic sites in macromolecular components. Further, it was stated that aflatoxin is accumulated in liver and the high content of microsomal cytochrome P-450 enzymes of hepatic cells favors the formation of DNA- aflatoxin adducts. Hence, liver is the major target organ for the aflatoxin toxicity. Among avian species, the most susceptible are ducks and turkeys followed by pheasants, chickens and quails (Diaz et al., 1995).

4. Aflatoxicosis

Aflatoxicosis caused by consumption of aflatoxins represents one of the most serious diseases to man, as well as poultry, livestock and other animals.

Aflatoxicosis in poultry is characterized by hemorrhages, anorexia, mortality, decreased feed efficiency and production, pathological changes in the liver, kidney and bile duct. The economic loss in the poultry industry due to aflatoxicosis is estimated to run upto millions of dollars (Raju et al., 2005).

5. Aflatoxicosis in commercial layers

The most prominent manifestations of experimental aflatoxicosis in layers are reduced egg production and egg weight, increased liver fat and alterations in some serum biochemical parameters.

Sims et al. (1970) fed *ad libitum* aflatoxin -contaminated diet having levels of 2.00 to 8.00ppm aflatoxin B_1 for 17 days and observed a significant reduction in egg production. Egg weight was not affected and also they could not detect any fluorescent metabolites in the eggs or liver of hens fed dietary aflatoxin.

Hamilton & Garlich (1971) fed the Single Comb White Leghorn hens with 1.25-200ppm dietary aflatoxin for three weeks and reported a dose related decrease in egg production and egg size, but shell thickness was not affected. The lipid content of liver was significantly increased in aflatoxin fed hens (5.00ppm) when compared with the control group.

Garlich et al. (1973) reported that the White Leghorn hens receiving 20.00ppm of aflatoxin in their diet for seven days did not adversely affect egg production but plasma calcium, protein, cholesterol and triglycerides were all decreased. In this study, delayed adverse effect of aflatoxin on egg production was observed. Once the hens were returned to a control diet for recovery, egg production began to decline significantly from the first day of the recovery period. Egg production reached to a minimum of 35 per cent, seven days later and then returned to the level of the control group, 19 days after the withdrawal of the contaminated diet. This delayed effect on egg production emphasizes the severe epidemiological problem of mycotoxins. Under field conditions, the feed causing the problem can be totally consumed before its adverse effects are noticed to undertake any therapeutic measure to solve the problem.

Huff et al. (1975) investigated the effect of graded levels of dietary aflatoxin up to 10.00ppm on layers. After four weeks, liver size and liver lipid content were increased, while egg production and egg size were decreased. Dry weight and lipid content of the yolk were not affected but yolk and plasma carotenoid concentrations were elevated.

McDaniel et al. (1979) reported that feeding of 200ppb aflatoxin in the diets did not significantly alter shell thickness of eggs obtained from layers. They concluded a trend with the known phenomenon of inverse relationship between age of bird and egg shell thickness.

Boulton et al. (1981) recorded a significant reduction in HI titers in layer breeders at 500ppb levels of aflatoxin.

Iqbal et al. (1983) fed the White Leghorn layers up to 5.00ppm dietary aflatoxin for three periods each consisting of 28 days. They reported that feeding 1.00ppm level of aflatoxin resulted in a significant reduction in hen day egg production and 2.00ppm level onwards feed efficiency was adversely affected. Congested and haemorrhagic livers, enlarged spleens, and immature ova with congestion were commonly seen. However, none of the levels affected feed consumption, body weight, egg weight, shell percentage, Haugh unit scores and serum protein levels. According to Dalvi & McGowan (1984), chronic aflatoxin toxicity in birds was characterized by drop in egg production. Washburn et al. (1985) reported that dietary aflatoxin at 5.00ppm fed for three weeks had no detrimental effect on shell strength but egg weight was significantly reduced.

Johri & Sadagopan (1989) reported a significant reduction in hen day egg production of laying quails when fed with 0.50 or 0.75ppm aflatoxin. Johri et al. (1990) studied the effect of low levels of dietary aflatoxin (0.00-0.75ppm) in Japanese quail fed toxic diet for 100 d and reported that egg production, protein utilization and body weight were adversely affected by 0.50 and 0.75ppm, whereas feed consumption and hatchability of fertile eggs were adversely affected by 0.30ppm. At 0.75ppm level, fertility of eggs and serum total protein decreased and serum glutamic pyruvic transaminase (ALT) increased.

Aflatoxin when added at 0 and 10ppm, with tryptophan to a layer ration, showed significant reduction in egg production percentage (Rogers et al., 1991). Rao &Joshi (1993) included

1.25, 2.50, 5 and 10ppm aflatoxin B_1 in layer rations for four weeks and found decreased egg production in birds receiving 5 and 10ppm of aflatoxin B_1.

Fernandez et al. (1994) reported a significant reduction in egg production and oral lesions in layer chicken treated with 120ppb onwards for varying periods.

Azzam & Gabal, (1998) reported a significant reduction in egg production of commercial layers fed with high levels of aflatoxin for six weeks.

Kubena et al. (1999) studied the effect of diets containing 50 or 100mg/kg moniliformin fed to White Leghorn laying hens for 420 d and observed that egg production was reduced by approximately 50 per cent by the end of the second 28-d laying period. Egg weights were reduced by the 100mg/kg toxin. The hens in toxin-treated group also had significantly lower body weight than the other treatments. Mortality was minimal except in hens fed with 100mg toxin/kg diet.

Mukhopadhy et al. (2000) have also reported a significant reduction in egg production in commercial layers exposed to 500ppb aflatoxin given for 90 days.

Ginzberg et al. (2000) reported that the yolk color in the group fed on 5 per cent of *Spirulina* algae was 2.4 times darker compared to the control laying hens.

Nimruz (2002) found that yolk color index of layers was significantly improved by the addition of *Spirulina* in feed. He concluded that Zeaxanthin content in the yolk tended to increase significantly with the dosage of *Spirulina*.

Kim et al. (2003) found reduction in serum calcium, phosphorous and ALT and increase in gamma glutamyl transferase (GGT) levels in laying hens by dietary levels of 500ppb of aflatoxin given from week 67 in laying hens.

Chowdhury & Smith (2004) reported decrease in feed efficiency when layers fed *Fusarium* mycotoxins contaminated diets compared with control groups.

Ogido et al. (2004) reported an increase in feed consumption and decrease in egg weight in Japanese quails fed with combination of 50ppb of aflatoxin B_1 and 10ppm of fumonisin B_1 for 140d.

Verma et al. (2004) reported decrease in hen day egg production, egg weight, feed consumption, shape index, albumen index and Haugh unit due to feeding 1ppm of aflatoxinB_1 for 42 d to White Leghorn hens aged 42 weeks.

Svetlana Grigorova (2005) reported that when adding 2 per cent and 10 per cent of dry biomass from fresh water algae of Chlorella genus in the combined forages for laying hens, the yolk pigmentation became significantly more intensive by 2.5 units by the Roche's scale.

Ninety-six laying hens fed with 2.50ppm of aflatoxin B_1 for four weeks by Zaghini et al. (2005) showed decrease in egg weight, egg shell weight and increased protein percentage in albumen. They reported that aflatoxin influenced color parameters, which was attributed to interference of aflatoxin B_1 with lipid metabolism and pigmentary substances deposition in yolk. Further, no aflatoxin B_1 or aflatoxin M_1 residues were found in eggs of the experimental groups.

Pandey & Chauhan (2007) reported that feeding of aflatoxin B_1 at the dose rate of 2.50, 3.13, 3.91mg/kg to the White Leghorn layers from first week to 40 weeks of age did not affect the body weight but resulted in decreased feed consumption, reduction in both egg production and egg weight at 3.91mg/kg level and caused 11-47 per cent dose-dependent mortality. They also reported that feeding aflatoxin B_1 at the dose rate of 2.50, 3.19 and 3.91mg/kg to the White Leghorn layers resulted in paleness of breast muscles, discolored livers, enlarged and pale kidney. Enlarged hearts and lungs were noticed at 3.13 and 3.19mg/kg levels. However, there were no changes in the intestine and spleen at all levels, but the Bursa of

Fabricius was oedematous and enlarged at 3.91mg/kg level. Lymphoid depletion and lymphocytolysis and reticuloendothelial cell hyperplasia in the spleen were also observed in all the toxin fed groups.

Denli et al. (2008) reported a reduced daily feed consumption, egg mass, and serum triglyceride concentrations, while increase in the relative liver weight, the serum activity of alkaline phosphatase, and the serum concentration of uric acid in twenty-eight Hisex Brown laying hens of 47 weeks of age fed with ochratoxin A for 3 weeks when compared those fed with the control diet.

Thapa (2008) reported a significant reduction in egg production of layers fed with varying levels of aflatoxin for three periods.

6. Aflatoxicosis in breeders

When aflatoxin (20.00ppm) was incorporated into feed of mature broiler breeder males for four weeks, no alteration in spermatozoa counts, semen volume, or semen DNA, RNA or protein content was recorded (Briggs et al., 1974).

Howarth & Wyatt (1976) fed broiler breeder hens 5 and 10ppm of aflatoxin in their diet for four weeks and reported no reduction in fertility, whereas hatchability of fertile eggs declined significantly from 95.00 per cent in the control to 68.90 and 48.50 per cent, respectively in 5 and 10ppm aflatoxin fed groups. Egg production decreased significantly during weeks three and four after initiation of toxin feeding in hens fed with 10 and 5ppm aflatoxin, respectively. They also observed enlarged fatty and friable liver and enlarged spleens by feeding aflatoxin at the dose levels of 0.00, 5.00 and 10.00ppm. Further, they did not observe any latent effect of aflatoxin or its metabolites on the performance of the surviving chicks hatched from broiler breeder hens, fed with 0.00, 5.00 and 100µg/kg of aflatoxin for four weeks.

Sharlin et al. (1981) reported decreased semen volume and testes weight and disruption of the germinal epithelium in mature White Leghorn males fed with 20.00ppm aflatoxin for five weeks. They also noticed decrease in feed intake and body weight. However, there was no effect on per cent fertile eggs or per cent hatchability of fertile eggs from hens artificially inseminated with spermatozoa from the treated males.

When laying hens and mature cocks were fed diets containing 8.10ppm aflatoxin B_1 or 1.60ppm aflatoxin G_1 for three weeks, egg production ceased. Histological examination of the ovaries showed follicular atresia. On the contrary, no testicular lesions were seen in the males (Hafez et al., 1982).

Jayakumar et al. (1988) fed aflatoxin B1 at rate of 25µg/duck, daily for three months and noticed reduced fertility and hatchability. Khan et al. (1989) injected 26.00, 81.00 and 216.00 ng/egg of aflatoxin B_1 and reported that lethal dose was 216.00 ng/egg and it caused mortality of chick embryo by the fourth day of incubation.

Tiwari et al. (1989) compared the hatchability of chicks hatched from aflatoxin containing eggs and concluded that it was low in comparison to chicks hatched from aflatoxin free eggs. Further, they studied the post-hatch performance of chicks hatched from aflatoxin containing eggs and observed lower weight gains and impaired defense system in chicks fed on normal diet.

In a study by Abdelhamid & Dorra (1990) where the maternal diet contained 100.00ppb of aflatoxin, citrinin or patulin for six weeks, the chicks had significantly higher weight than the control.

Rao (1990) observed a drastic deterioration in semen quality of breeder cocks fed with 1.000ppm aflatoxin. The traits affected were semen volume, semen concentration, motility and abnormalities.

Stephen et al. (1991) reported a significant drop in egg production in layer chicken fed with 5.00 and 10.00ppm aflatoxin for three weeks.

Nelson-oritiz & Qureshi (1992) assessed single dose exposure of six day-old embryos to 0.100, 0.500 and 1.00µl of aflatoxin B_1 and concluded that rate of mortality of the embryos was dose related. Chick embryos, administered different levels of aflatoxin or ochratoxin on the chorio-allontoic membrane showed decreased weight and length. Further, abnormalities like everted viscera, exposed brain, crossed beak, underdeveloped eyes and head and twisted limbs were observed.

Bergsjo et al. (1993) reported chick developmental anomalies when laying hens were fed diets containing 4.90mg of DON/kg of feed for 10 weeks.

Diaz & Sugahara (1995) reported that birds fed aflatoxin at 0.66 or 3.00µg/kg diet did not show any adverse effect on chick performance.

Muthiah (1996) conducted an experiment to study the effect of graded dietary levels of aflatoxin B_1 (0.00, 0.50, 1.00 and 1.50ppm) on the reproductive performance of layer breeders. He reported that the sperm motility and concentration were not affected while the percentage of sperm abnormality increased when aflatoxin B_1 was included in the diet of breeder cocks. The feed consumption was significantly decreased and egg production declined in proportion to the level of aflatoxin B_1 incorporation in the diets. There was no effect on fertility but hatchability was affected. The chicks hatched from breeder hens, received graded levels of aflatoxin in their diets did not show any effects on body weight, weight gain, mortality and feed consumption during the 0-8 weeks post-hatch performance period.

Cotter & Weinner (1997) reported lowered hatchability in broiler breeder hens fed with four levels viz., 0.00, 308.00, 610.00 and 1834.00ppb of aflatoxin.

Brake et al. (1999) conducted an experiment by feeding diets with different levels of diacetoxyscirpenol (DAS) (ranging from 0.00 to 20.00mg/kg) to broiler breeders between 67 to 69 wk of age. They observed no effect on egg production, when DAS was fed upto the level of 5.00ppm. Furher, they have demonstrated that feeding diets contaminated with 10.00 and 20.00mg of DON per kg of feed decreased the fertility in broiler breeder males, though there was no difference in the volume of semen produced.

Brake et al. (2000) reported that there were dose-related decreases in body weight and feed consumption indicating feed refusal, as well as dose-related increases in the extent of mouth lesions of broiler breeders fed with 0.00, 5.00, 10.00, or 20.00mg DAS/kg diet from 24 to 25 wk of age.

Stanley et al. (2004) reported that feeding aflatoxin at the rate of 3mg/kg to 35 week's old Cobb broiler breeder hens for three weeks significantly reduced serum total protein, albumin, calcium and phosphorus levels.

Sypecka et al. (2004), reported that only trace amounts of *Fusarium* mycotoxins are transferred into the eggs of laying hens, which are unlikely to be of significance with respect to embryonic mortality.

Yegani et al. (2006) reported no effect in feed consumption, body weight, and egg production. However, increase in early embryonic mortality (1 to 7d) in eggs from birds fed contaminated grains with deoxynivalenol (12.60mg/kg of feed) was observed in broiler breeder hens. They also reported that the ratio of chick weight to egg weight was not affected. Weight gains of chicks fed a standard broiler starter diet at 7, 14, and 21 d of age were also not significantly affected by previous dietary treatments for the dam. Feeding of contaminated diets did not affect semen volume, sperm concentration, viability, and motility. There was no effect of diet on the relative weights of liver, spleen, kidney and testes.

Histological evidence of adverse effects of aflatoxin on the germinal epithelium of the testes was reported in immature chickens dosed with 200μg of aflatoxin/day/chick for 35 days (Mohan et al., 2008).

Manafi et al. (2009) reported the aflatoxin fed at the levels of 300, 400 and 500ppb for three periods, each with duration of three weeks to broiler breeders from 28 to 36 weeks of age. Inclusion of 500ppb aflatoxin in the diet significantly ($P \leq 0.05$) affected feed consumption, feed efficiency, egg production, egg weight, fertility, hatchability, embryonic mortality, GGT and ALT levels in the serum, organ lesion scores (liver, kidney, proventriculus and gizzard), relative weights of heart as well as liver, antibody titers against Newcastle and Infectious bursal diseases when compared to that of control. The results indicated no significant ($P \geq 0.05$) effect of aflatoxin on body weight, shell thickness, Haugh unit, residue in eggs, sperm count, per cent live sperm, yolk color index and relative weight of spleen when compared to that of control.

Serum alkaline phosphatase levels were significantly higher, serum alkaline aminotransferase ($P=0.068$) and gamma-glutamyltransferase ($P=0.067$) levels tended to increase ($P<0.05$) in 58-wk-old Ross 308 broiler breeders fed with 100 μg aflatoxin-contaminated diet than those of hens fed the uncontaminated diet (Matur et al., 2010).

7. Aflatoxin residue in eggs

Although the concentration of mycotoxins and their metabolites are generally much lower in eggs than in animal feeds and are not likely to cause acute intoxications in humans. However, the residues of carcinogenic mycotoxins such as aflatoxin B_1 and M_1, (aflatoxin M_1 is a polar metabolite of aflatoxin B_1) and ochratoxin A, when present in animal products are a threat to human health and must be monitored. The limit for aflatoxin B_1 in complete feeds is 0.02mg/kg.

Trucsksess et al. (1983) were able to detect aflatoxin B_1 and M_1 residues in eggs of hens fed contaminated feed. After 7 days of withdrawal only trace amounts remained in eggs. According to Wolzak et al. (1985) clearance of aflatoxin occurs faster from the albumen than from the yolk.

Aflatoxin and some of their metabolites can be carried over from feed to eggs in ration ranging from 5,000:1 to 66,200:1 and even to 125,000:1, whereas in other trials no measurable residual aflatoxin B_1 or its metabolites were found in eggs (Oliveira et al., 2002).

Zaghini et al. (2005) reported that no traces of aflatoxin B_1 or aflatoxin M_1 residues were found in eggs of layer hens supplemented with diet containing 2.50ppm aflatoxin B_1.

In another study, Salwa & Anwer (2009) reported no traces of aflatoxin in the eggs of layers fed with 25.00, 50.00 and 100ppb of aflatoxin in their diet for 60 days.

8. Counteraction of aflatoxicosis

The infestation of agricultural products, intended for human and animal consumption with toxigenic fungi that are capable of producing highly toxic metabolites has been a worldwide problem. Increased efforts are being undertaken to develop cost effective and safe procedures and products to effectively deal with the decontamination and remediation of mycotoxin contamination in feedstuffs. The available approaches were reviewed by Trenholm et al. (1996) and Devegowda et al. (2003).

The methods aimed at preventing or reducing the level of mycotoxin contamination were classified as preventive or curative. The following approaches were recommended:

1. Prevention of the initial growth of moulds and subsequent production of mycotoxin.
2. Detection of mycotoxin in feed and selective removal of contaminated portions.
3. Inactivation or destruction of the toxin by physical, chemical and biological means.
4. Utilization of mycotoxin resistant genetic resources.

8.1 Physical methods

According to Park & Liang (1993) Mycotoxins (aflatoxin, ochratoxin, T-2 toxin and citrinin) are highly soluble in organic solvents. Their extraction from the feed stuffs using several solvents or mixture of solvents has been proved to be highly effective.

Scott (1989) opined that thermal treatment appears to have little effect on the toxin content as mycotoxins are heat resistant. Irradiation of feed stuffs may reduce the toxin content considerably. Exposure of contaminated feed ingredients to sunlight may also prove to be effective. These methods have little practical applicability.

The utilization of mycotoxin-binding adsorbents, which do not get absorbed from the GIT and instead bind physically with mycotoxin, is the most applied physical method of protecting animals against the harmful effects of mycotoxin contaminated feed and has gained considerable attention in recent times. The efficiency of the adsorption depends on the chemical structure of both the adsorbent and the mycotoxin. Before applying this technique for routine use, it is essential to establish that the adsorbent does not remove essential nutrients from the diet (Manafi et al., 2009).

Clays like hydrated sodium calcium aluminosilicates, activated carbon, bentonite, clays and special polymers are made of two or more mineral-oxide layers. These layers are stacked parallel units of silica and alumina sheets. The silica form tetrahedral sheets and the alumina forms octahedral sheets. Some of these clay particles have the ability to absorb moisture and will expand while others do not. The difference is due to clay chemistry and the elements (cations) that are components of the layers.

8.2 Herbal methods

Application of some herbal extracts of plant origin like turmeric (*Curcuma longa*), garlic (*Allium sativum*) and asafetida (*Ferula asafetida*) have shown to counteract aflatoxicosis in animals and poultry through their antioxidant activity.

Several herbal products contain antioxidant substances capable of scavenging free radicals and enhancing antioxidant enzymes. Nyandieka et al. (1990) reported that use of ethanolic extract of *Cassia senna,* (herb) as laxative inhibited the mutagenic effects of aflatoxin B_1. Feeding of the extract of *Azadirachta indica* prevented metabolic activation of aflatoxin B_1 to its epoxide derivative. Hepatic antioxidant status of rats was enhanced by feeding of a phenolic-lignin enriched extract of the fruit *Schisandra chinensis* and provided hepato protection against aflatoxin B_1.

Oxidative changes (increased peroxides, reduced antioxidant enzyme activity) in liver and kidney due to aflatoxin B_1 were reversed in rats by feeding a root/rhyzome extract of *Picrorhiza kurroa* (Picroliv) and a seed extract of *Silybum marianum* (Silymarin) (Weiss, 2002). Similarly, Rosamarinic acid, a phenolic component of *Boragnaceae* species of plants (sage,

basil, and mint) reduced free radical oxygen formation, and inhibited protein/DNA synthesis as well as apoptosis of human hepatoma cells caused by aflatoxin B_1 and ochratoxin A (Manafi et al., 2009).
According to Gowda & Ledoux (2008) ellagic acid, a phenolic compound of strawberries and grapes showed anticarcinogenic activity and inhibited aflatoxin B_1 mutagenicity. S-methyl methane thiosulfonate present in cabbage and onion suppressed chromosomal aberrations due to aflatoxin B_1 in rat bone marrow cells. Diterpenes, cafestol and kahweol present in green and roasted coffee beans prevented the covalent binding of aflatoxin B_1 to DNA by modulation of the carcinogenesis enzyme system.
Iqbal et al. (1983) observed chemoprotective effect of piperine (1-piperoyl piperidine), an alkaloid of pepper against aflatoxin by inhibiting cytochrome P 450 bioactivation of aflatoxin B_1. The protective effect of chlorophylline (a derivative of the green pigment chlorophyll) against aflatoxin B_1 was also observed. The toxic effects of aflatoxin in chicken was reversed by the administration of an alcoholic extract of African nut meg. The carbonyl functional groups of the curcuminoids are thought to be responsible for their antimutagenic and anticarcinogenic action. Further, strong inhibitory effect of curcumin on superoxide anion generation was noticed.

8.3 Use of enzymes

The enzymes are believed to break the functional atomic group of the mycotoxin molecule and thereby render them nontoxic (Kumar et al., 1993). Enzymes viz., carboxyesterase present in the microsomal fraction of the liver, esterase and epoxidase are being tried for their practical applicability in the field conditions (Pasteiner, 1997).

8.4 Nutritional manipulations

Increasing the crude protein content and supplementation of additional levels of riboflavin, pyridoxine, folic acid and choline showed protective effect against aflatoxicosis (Ehrich et al., 1986). Anti-oxidants like BHT and β-napthoflavone, vitamin C and vitamin E offer protection against aflatoxin induced genotoxicity in *in vitro* studies (Johri et al., 1990).
Increase in dietary protein levels and supplementation of L-phenylalanine was revealed to be effective against aflatoxicosis and ochratoxicosis. Devegowda et al. (1998) reported that the supplementation of the diet with selenium and methionine partially alleviated the adverse effects of aflatoxin respectively.

8.5 Nutritional manipulations

Increasing the crude protein content and supplementation of additional levels of riboflavin, pyridoxine, folic acid and choline showed protective effect against aflatoxicosis (Ehrich et al., 1986). Anti-oxidants like BHT and β-napthoflavone, vitamin C and vitamin E offer protection against aflatoxin induced genotoxicity in *in vitro* studies (Johri et al., 1990).
Increase in dietary protein levels and supplementation of L-phenylalanine was revealed to be effective against aflatoxicosis and ochratoxicosis. Devegowda et al. (1998) reported that the supplementation of the diet with selenium and methionine partially alleviated the adverse effects of aflatoxin respectively.

8.6 Biological methods

With the awareness of potential harmful effects of chemicals used for counteracting mycotoxins and the cost involvement with their usage has prompted the scientists to look for alternative methods which are applicable and safe. A rapid explosion in the field of feed industry through the biotechnological methods has opened a new possibility of degradation of mycotoxins by microorganisms. Several yeasts, moulds and bacterial strains posses the ability either to destroy or transform mycotoxins successfully (Phillips et al., 1988).

8.6.1 Bacterial degradation

Acid producing bacteria's such as *Lactobacillus plantarum* and *Lactobacillus acidophilus* were found to detoxify aflatoxin in maize. Rumen bacteria were found to degrade ochratoxin A (OA), T-2 toxin and zearalenone (ZEA) (Linderfelser & Ceigler, 1970).
He et al. (1992) detoxified moldy maize diet containing 5.00ppm vomitoxin, microbially through incubation with the contents of large intestine of chickens having a detoxified vomitoxin of 2.10ppm.

8.6.2 Protozoan degradation

Tetrahymena pyriformis at a dose rate of $22x10^6$ cells detoxified aflatoxin B_1, by converting it into its hydroxyl products to an extent of 5 per cent in 24 hours and 67 per cent in 48 hours (Robertson et al., 1970). Intact rumen fluid containing various protozoa was reported to metabolize T-2 toxin and ochratoxin while no effect on aflatoxin was noted (Kiessling et al., 1984).

8.6.3 Fungal degradation

Some of the species of fungi have been found to detoxify aflatoxin. An intracellular substance was found to be responsible for *A. flavus* and *A. parasiticus* to degrade the formed toxins in a culture when their mycelium was subjected to fragmentation. The peroxidase enzymes produced by the fungal mycelium, which can catalyze hydrogen peroxide into free radicals, reacts with aflatoxin (Dvorak, 1989).

8.6.4 Degradation by yeast

Yeasts are being primarily used as growth promoters in poultry and animal feeds. Besides their beneficial effects on feed utilization and rich concentration of many vitamins, certain species and strains of yeasts have been observed to detoxify mycotoxins through its degradation (Cooney, 1980).
In the early 1990s, a commercially yeast culture preparation of *Saccharomyces cerevisiae*[1026], which was earlier noted as digestive aid and a growth promoter, was found to improve hatchability (McDaniel, 1991) and broiler body weights (Stanley et al., 1993). Further investigations lead to the establishment of yeast culture preparation's ability to adsorb aflatoxins in poultry feeds (Devegowda et al., 1995).
Supplementation of live cells of *Saccharmyces cerevisiae*[1026] was found to be beneficial in counteracting the adverse effect of several mycotoxins (Stanley et al., 1993). It has also improved serum total protein and HI titer against Newcastle disease in aflatoxin fed broilers (Devegowda et al., 1996). In an another study, inactivated yeast at 0.2 per cent in the diet was found to alleviate the growth depression effects of aflatoxin up to 200ppb level (Devegowda et al., 1998).
Mycotoxin binding ability of Maannanoligosaccharides has been demonstrated in various *in vitro* trials (Devegowda et al., 1998) and *in vivo* trials (Raju & Devegowda, 2000; Swamy et al., 2004).

9. Conclusion

The addition of aflatoxin in layer and breeder hens could lead to aflatoxicosis which can adversely affect the performance and reproductive efficacy of the birds. Aflatoxicosis in poultry is characterized by hemorrhages, anorexia, mortality, decreased feed efficiency and production, pathological changes in the liver, kidney and bile duct. The economic loss in the poultry industry due to aflatoxicosis is estimated to run up to millions of dollars. In commercial Layers, the most prominent manifestations of experimental aflatoxicosis are reduced egg production and egg weight, increased liver fat and alterations in some serum biochemical parameters.In case of breeder hens, many of parameters like changes in body weight, feed consumption, feed conversion ratio, egg production, egg weight, shell thickness, Haugh unit score, yolk color index, fertility, hatchability and embryonic mortality and some of biochemical and immunological parameters like serum levels of GGT and ALT, visceral organ weight and organ lesions and serum antibody titers for ND and IBD could be altered due to aflatoxicosis. This chapter briefly reviews the impact of aflatoxicosis in commercial layer and breeder hens.

10. References

Abdelhamid, A.M. & Dorra, T.M. (1990). Study on effects of feeding laying hens on separate mycotoxins (aflatoxin, patulin, or citrinin) contaminated diets on the egg quality and tissue constituents. *Arch. Tierernahr.*, 40: 305-315.

Azzam, A.H. & Gabal, M.A. (1998). Aflatoxin and immunity in layer hens. *Avian Path.*, 27: 570-577.

Bergsjo, B., Herstad, O. & Nafstad, I. (1993). Effects of feeding deoxynivalenol-contaminated oats on reproductive performance in White Leghorn hens. *Br. Poult. Sci.*, 34: 147–159.

boulton, S.L., dick, J.W. & hughes, B.L. (1981). Effects of dietary aflatoxin and ammonia - inactivated aflatoxin on Newcastle disease antibody titers in layer - breeders. *Avian Dis.*, 26: 1-6.

Brake, J., Hamilton, P.B. & Kittrell, R.S. (1999). Effect of the trichothecene mycotoxin diacetoxyscirpenol on fertility and hatchability of broiler breeders. *Poult. Sci.*, 78: 1690–1694.

Brake, J., Hamilton, P.B. & Kittrell. (2000). Effects of the trichothecene mycotoxin diacetoxyscirpenol on feed consumption, body weight and oral lesions of broiler breeders. *Poult. Sci.*, 79; 856-863.

Briggs, D.M., Wyatt, R.D. & Hamilton, P.B. (1974). The effect of dietary aflatoxin on semen characteristics of mature broiler breeder males. *Poult. Sci.*, 53: 2115-2119.

Chandrashekharan, D. (2000). Mycotoxins: A threat to livestock and public health, Personal Communication.

Chandrashekharan, D., Sudaram, T.K. & Natrajan, A. (2002). Mycotoxin scenario in India and control strategies. IV Biennial Conference and Exhibition, Kolkata, pp. 34-41.

Chowdhury, S.R. & SMITH, T.K., 2004. Effects of feeding blends of grains naturally contaminated with *Fusarium* mycotoxins on performance and metabolism of laying hens. *Poult. Sci.*, 83: 1849–1856.

Cooney, D.D. (1980). Activated charcoal: antidotal and other medical uses. Marcel Dekker Inc., New York, USA.

Cotter, P.F. & Weinner, J. (1997). Bio-MOS modulates the phyto haemagglutinin wattle reaction in commercial pullets. *Proc. 18th Annual Meeting of Southern Poultry Science Society, Atlanta, GA, USA,* S36 (Abst.).

Dalvi, R.R. & McGowan, C. (1984). Experimented induction of chronic aflatoxicosis in chickens by purified charcoal, phenobarbital and reduced glutathione. *Poult. Sci.,* 63: 485-491.

Denli, M., Blandon, J.C., Guynot, M.E., Salado, S. & Perez, J.F. (2008). Efficacy of a new ochratoxin-binding agent (OcraTox) to counteract the deleterious effects of ochratoxin A in laying hens. *Poult. Sci.,* 87: 2266-2272

Devegowda, G., Barmase, B.S. & Devurkar, U. (1990). Surveillance of aflatoxin contamination in feed ingredients and feeds in and around Bangalore during 1988-1989. Proc. *XIII Annual Conf. of Indian Poult. Sci. Assoc.,* Bombay, 20-22.

Devegowda, G. & Arvind, B.I.R. (1993). Survey of aflatoxin B_1 contamination of feedstuffs. In: *Proc. 6th Anim. Nutr. Res. Workers Conf., Bhubaneswar,* pp. 194.

Devegowda, G., Aravind, B.I.R. & Morton, M.G. (1996). *Saccharmyces cerevisiae* and mannanoligosaccharides to counteract aflatoxicosis in broilers. In: *Proc. Australian Poult. Sci. Symp., Sydney, Australia,* 8: 103-106.

Devegowda, G., Raju, M.V.L.N. & Swamy, H.V.L.N. (1998). Mycotoxins: Novel solutions for their counteraction. *feedstuffs.* 70: (50): 12-16.

Devegowda, G., Aravind, K.L., Murthy, T.N.K. & Reddy, B.N. (2003). Mycotoxins: Economic risk and control-The biological solution. In: *Proc. of 3rd international poultry show and seminar organized by WPSA,* Dhaka, Bangladesh. pp: 94-98.

Dhavan, A.S. & Choudary, M.R. (1995). Incidence of aflatoxins in animal feed stuffs: A decade's scenario in India. *J.A.O.A.C.* 78: 693-698.

Diaz, G.J. & Sugahara, M. (1995). Individual and combined effects of aflatoxin and gizzerosine in broiler chickens. *Br. Poult. Sci.,* 36: 729-736.

Dvorak, M. (1989). Ability of bentonite and natural zeolite to adsorb aflatoxin from liquid media. *Vet. Med. (Praha).* 34: 733-741.

Ehrich, M., Driscoll, C. & Larsen, C. (1986). Ability of ethoxiquin and butylated hydroxytoluene to counteract deleterious effects of dietary aflatoxin in chicks. *Avian Dis.,* 30: 802-807.

Fernandez, A., Verde, M.T., Gascon, M., Ramos, J., Gomez, J., Luco, D.F. & Chavez, G. (1994). Variations of clinical biochemical parameters of laying hens and broiler chickens fed aflatoxin containing feed. *Avian Path.,* 23: 37-47.

Garlich, J.D., Tung, H.T. & Hamilton, O.B. (1973). The effects of short term feeding of aflatoxin on egg production and some plasma constituents of the laying hen. *Poult. Sci.,* 52: 2206-2211.

Ginzberg A., Cohen, M., Sod-Moriah, U.A., Shany, S., Rosenshtrauch, A. & Arad, S.M. (2000). Chickens fed with biomass of the red microalga Porphyridium sp. Have reduced blood cholesterol level and modified fatty acid composition in egg yolk. *J. Appl. Phyc.,* 12: 325-330.

Gowda, N.K.S. & Ledoux, D.R. (2008). Use of antioxidants in amelioration of mycotoxin toxicity: a review. *Anim. Nutr. Food Tech.,* 8: 1-11.

Hafez, A.H., Megalla, S.E., Abdel Fattah, H.M. & Kamel, Y.Y. (1982). Aflatoxin and aflatoxicosis. Effect of aflatoxin on ovaries and testicles in mature domestic fowls. *Mycopathologia,* 77: 137-139.

Hamilton, P.B. & Garlich, J.D. (1971). Aflatoxin as a possible cause of fatty liver syndrome in laying hens. *Poult. Sci.,* 50: 800-804.

He, P., Young, L.G. & Forsberg, C. (1992). Microbial transformation of deoxynivalenol (vomitoxin). *App. Environ. Microbiol.,* 58: 3857-3863.

Hegazy, S.M., Azzam, A. & Gabal, M.A. (1991). Interaction of naturally occurring aflatoxins in poultry feed and immunization against fowl cholera. *Poult. Sci.,* 70: 2425-2428.

Howarth, B. JR. & Wyatt, R.D. (1976). Effect of dietary aflatoxin on fertility, hatchability and progeny performance of broiler breeder hens. *App. Envi. Micro.,* 31: 680-684.

Hsieh, D.P.H. (1985). An assessment of liver cancer risk posed by aflatoxin B_1 in western world. In: *Trichothecenes and other mycotoxins.* 11: 369-383.

Huff, W.E., Wyatt, R.D. & Hamilton, P.B. (1975). Effects of dietary aflatoxin on certain egg yolk parameters. *Poult. Sci.,* 54: 2014-2018.

Iqbal, Q.K., Rao, P.V. & Reddy, S.J. (1983). Dose-response relationship of experimentally induced aflatoxicosis in commercial layers. *Indian J. Anim. Sci.,* 53: 1277-1280.

Jayakumar, P.M., Valsala, K.V. & Rajan, A. (1988). Experimental aflatoxicosis in the duck with special reference to pathology of the testes. *J. Vet. Sci.,* 19: 122-128.

Jelinek, C.F., Ponland, A.E. & Wood, G.E. (1989). Worldwide occurrence of mycotoxins in foods and feeds - An update. *J.A.O.A.C.,* 72: 223-230.

Jindal , N., Mahipal, S.K. & Mahajan, N.K. (1993). Effect of hydrated sodium calcium aluminosilicate on prevention of aflatoxicosis in broilers. *Indian J. Anim. Sci.,* 63: 649-652.

Johri, T.S. & Sadagopan, V.R. (1989). Aflatoxin occurrence in feed stuffs and its effect on poultry production. *J. Toxical. Toxin Rev.,* 8:281-287.

Johri, T.S., Agrawal, R. & Sadagopan, V.R. (1990). Effects of low dietary levels of aflatoxin on laying quails (Coturnix coturnix japonica) and their response to dietary modifications. *Indian J. Anim. Sci.,* 60: 355-359.

Khan, B.A., Hussain, S.S. & Ahmad, M.A. (1989). Toxicity of aflatoxin B_1 to chick embryo. *Pak. J. Scientific and Industrial Res.,* 32: 353-354.

Kichou, F. & Walser, M.M. (1994). Effects of aflatoxin B_1 on chicken chondrocytes in culture. *Avian Dis.,* 38: 11-15.

Kiessling, K.H., Petersson, H., Sandholm, K. & Olsen, M. (1984). Metabolism of aflatoxin, ochratoxin, zearalenone and three trichothecenes by intact rumen fluid, rumen protozoa and rumen bacteria. *Appl. Env. Microbiol.,* 47: 1070-1073.

Kim, J.G., Lee, L.W., Kim, P.G., Roh, W.S. & Shintani, H. (2003). Reduction of aflatoxins by Korean soybean paste and its effect on cytotoxicity and reproductive toxicity. Part 3. Inhibitory effects of Korean soybean paste (Doenjang) on aflatoxin toxicity in laying hens and aflatoxin accumulation in their eggs. *J. Food Protec.,* 66: 866-873.

Kubena, L.F., Harvey, R.B., Buckley, S.A. Bailey, R.H. & Rottinghaus, G.E. (1999). Effects of long-term feeding of diets containing moniliformin, supplied by *Fusarium fujikuroi* culture material, and fumonisin, supplied by *Fusarium moniliforme* culture material, to laying hens. *Poult. Sci.,* 78: 1499-1505.

Kumar, A.A., Chanda, S.K., Rao, A.T., Biscoi, P.C. & Mishra, P.K. (1993). Clinico-pathological changes in experimental aflatoxicosis in quails. *Indian J. Poult. Sci.,* 28:150-153.

Linderfelser, L.A. & Ciegler, A. (1970). Studies on aflatoxin detoxification in shelled corn by ensiling. *J. Agri. Food Chem.,* 18: 640-643.

Manafi, M., Umakantha, B. & Pirany, N. (2007). Efficacy of graded levels of high grade Bentonite (Ashafeed) to lessen individual and combined toxicities of T-2 toxin and Aflatoxin. In: *Proc. 16th World Vet. Poult. Conf., Beijing, China.* pp. 304-305.

Manafi, M. (2009). Comparative Efficacy of Bentonite, *Spirulina platensis* and Glucomannan Mycotoxin Binders on Aflatoxicosis in Broiler Breeders and Carry Over Effects on Progeny Performance. Ph.D. thesis submitted to Karnataka Veterinary, Animal and Fisheries Sciences University, Bangalore, India.

Manafi, M. (2010). Effect of AFlatoxicosis in Hatching Eggs Quality and Immune Status in Broiler Breeders. In: *World Poult. Sci. J. (Spp.),* 66: 704-705.

Matur, E., Ergul, E., Akyazi, I., Eraslan, E. & Ciraklit, Z. T. (2010). The effects *Saccharomyces cerevisiae* extract on the weight of some organs, liver and pancreatic digestive enzyme activity in breeder hens fed diets contaminated with aflatoxins. *Poult. Sci.,* 89:2213-2220.

McDaniel, G., Ronald, D.A. & Coleman, M.A. (1979). The effect of egg shell quality on hatchability and embryonic mortality. *Poult. Sci.,* 58:10-13.

Mohan, K., Jayakumar, K., Narayana Swamy, H.D. Shridhar, N. B. Bayer M.D. & Manafi, M. (2008). Hepatotoxicity of Acetaminophen in Chickens. Journal of Veterinary Pharmacology and Toxicology. 7(1-2): 48-49.

Mohanamba, T., Rao, M. R. & Habibi, S.M.M. (2007). Aflatoxin contamination in animal feeds. *Indian Vet. J.,* 84: 416.

Mukhopadhy, H.K., Paul, W.M. Dorairajan, N. & Chandran, N.D.J. (2000). Drop in egg production due to aflatoxin B_1 contamination in feed. *Int. J. Anim. Sci.,* 15: 123-126.

Muthiah, J. (1996). Studies on the effect of aflatoxin B_1 on reproduction performance of egg type breeders and their amelioration. Ph.D. thesis submitted to Tamil Nadu Veterinary and Animal Science University, Madras, India.

Nelson-Oritz, D.L. & Qureshi, M.A. (1992). Effects of AFB_1 embryonic exposure on chicken mononuclear phagocytic cell functions. *Dev. Corp. Immunal.,* 16: 187-196.

Nimruz R. (2002). A new natural source of carotenoids for poultry. *Inter. Feeding,* 41:50-51

Nyandieka, H.S., Wakhis, J. & Kilonzo, M.M. (1990). Association of reduction of AF B_1–induced liver tumors by antioxidants with increased activity of microsomal enzymes. *Indian J. Med. Res.,* 92: 332-336.

Ogido, R, Oliveira, C., Ledoux, D., Rottinghaus, G., Correa, B., Butkeraitis, T., Reis, P., Goncales, E. & Albuquerque, R. (2004). Effects of prolonged administration of aflatoxin B_1 and fumonisin B_1 in laying Japanese quail. *Poult. Sci.,* 83: 1953-1958.

Oguz, H. (2011). A review from experimental trials on detoxification of aflatoxin in poultry feed. *Eurasian J. Vet. Sci.,* 27 (1): 1-12.

Oliveira, C.A.F., Rosmaninho, J.F., Butkeraitis, P., Correa, B., Reis, T.A., Guerra, J.L., Albuquerque, S.R. & Moro, M.E.G. (2002). Effect of low levels of dietary aflatoxin B_1 on laying of Japanese quail. *Poult. Sci.,* 81: 976-980.

Pandey, I., Chauhan, S.S. & Singh, K.S. (2001). Studies on the infestation of aflatoxin B_1 in poultry feeds in the Tarai region of U.P., *Indian J. Poult. Sci.,* 36: 221-223.

Pandey, I. & Chauhan, S.S. (2007). Studies on production performance and toxin residues in tissues and eggs of layer chickens fed on diets with various concentrations of aflatoxin B_1. *Br. Poult. Sci.,* 48: 713-723.

Pasteiner, S. (1997). Copying with mycotoxin contaminated feed stuffs. *Feed Int.,* 18: 12-16.

Park, D.L. & Liang, B. (1993). Perspectives on aflatoxin control for human food and animal feeds. *Trends Food Sci. Tech.*, 4: 334-342.

Phillips, T.D., Kubena, L.F., Harvey, R.B., Taylor & Heidelbaugh, N.D. (1988). Hydrated sodium calcium aluminosilicate: a high affinity sorbent for aflatoxin. *Poult. Sci.*, 67: 243-247.

Raju, M.V.L.N. & Devegowda, G. (2000). Influence of esterified glucomannan on performance and organic morphology, serum biochemistry and hematology in broilers exposed to individual and combined mycotoxicosis (aflatoxin, ochratoxin and T-2 toxin). *Br. Poult. Sci.*, 41: 640-650.

Raju, M.V.L.N., Rama RAO, S.V., Radhika, K. & Chawak, M.M. (2005). Dietary supplementation of *Spirulina* and its effects on broiler chicken exposed to aflatoxicosis, *Indian J. Poult. Sci.*, 40: 36-40.

Rao, R.E. (1990). *CFTRI Manual Lecture*, pp.12.

Rao, V.N & Joshi, H.C. (1993). Effect of certain drugs on acutely induced aflatoxicosis in chicken. *Indian Vet. J.*, 70: 344-347.

Robertson, J.A., Tennisson, D.J. & Boudreaox, G.J. (1970). Isolation and structure of the biologically reduced aflatoxin B_1. *J.A.O.A.C.*, 18: 1090-1091.

Rogers, S.R., Pesti, G.M. & Wyatt, R.D. (1991). Effect of tryptophan supplementation on aflatoxicosis in laying hens. *Poult. Sci.*, 70: 307-312.

Sala, N. & Ueno, Y. (1997). Comparative toxicities of trichothecenes. In: Mycotoxins in Human and Animal Health. Eds: Rodrick, J.V., Hesseltine, C.A. and Mehlman, M.A., Prof. Conf. Hum. Anim. Health, Univ. of Maryland, USA, pp. 295-307.

Salwa, A. & Anwer, W. 2009. Effect of naturally contaminated feed with aflatoxin on performance of laying hens and the carry over of aflatoxin B_1 residues in table eggs. *Pak. J. Nutr.*, 8: 181-186.

Scott, P.M. (1989). The Natural occurrence of trichothecenes. In: *Trichothecene mycotoxicosis: pathophysiologic effects.* 1: 1-26.

Sharlin, J.S., Howarth, B., Thompson, JR. F.N. & Wyatt, R.D. (1981). Decreased reproductive potential and reduced consumption in mature WL males fed aflatoxin. *Poult. Sci.*, 60: 2071-2708.

Shetty, S.N., Asuzu, I.U. & Anika, S.M. (1987). Aflatoxin contamination of animal feedstuffs in Anambra State. *Trop. Vet.*, 5: 21-25.

Sims, W.M., Kellys, JR.D.C. and Sanford, P.E. (1970). A study of aflatoxicosis in laying hens. *Poult. Sci.*, 49:1082-1084.

Stanley, V.G., Ojo, R., Woldensenbet, S. & Hutchinson, D.H. (1993). The use of *Saccharmyces cerevisiae* to suppress the effects of aflatoxicosis in broiler chicks. *Poult. Sci.*, 72: 1867-1872.

Stanley, V.G., Winsman, M., Dunckley, C., Daley, M. Kruger, W.F., Sefton, A.E. & Hinton, A. (2004). Impact of yeast culture residue on the supplementation of dietary aflatoxin in performance of broiler breeder hens. *J. Appl. Poult. Res.*, 13: 533-539.

Stephen, R., Pesti, G.M. & Wyatt, D. (1991). Effect of tryptophan supplementation and aflatoxicosis in laying hens. *Poult. Sci.*, 70: 307-312.

Svetlana Grigorova (2005). Dry biomass of fresh water algae of chlorella genus in the combined forages for laying hens. *J. Cent. Euro. Agri.*, 6: 625-630.

Swamy, H.V.L.N., Smith, T.K., Karrow, N.A. & Boermans, H.J. (2004). Effects of feeding blends of grains naturally contaminated with Fusarium mycotoxins on growth and immunological parameters of broiler chickens. *Poult. Sci.,* 83: 533-543.

Sypecka, Z., Kelly, M. & Brereton, P. (2004). Deoxynivalenol and zearalenone residues in eggs of laying hens fed with a naturally contaminated diet: Effects on egg production and estimation of transmission rates from feed to eggs. *J. Agric. Food Chem.,* 52: 5463–5471.

Thapa, N.K. (2008). Pathological effects of aflatoxicosis in layer chicken with special emphasis on reproductive pathology. M.V.Sc. Thesis submitted to Tamil Nadu Veterinary and Animal Sciences University.

Tiwari, R.P., Virdi, J.S., Gupta, L.K., Saini, S.S. & Vadehra, D.V. (1989). Development of chicks exposed to aflatoxin B1 during embryogenesis. *Indian J. Anim. Sci.,* 59: 1473-1478.

Trenholm, H.L., Charmley, L.L. & Prelusky, D.B. (1996). Mycotoxin binding agents: An update on what we know. In: *Biotechnology in the Feed Industry.* Eds: Lyons, T.P. and Jacques, K.A., In: *Proc. Alltech's 17th Annual Symp., KY, USA,* pp. 327-349.

Trucksess, M.W., Stoloff, L. & Young, K. (1983). Aflaoxicol and aflatoxins B_1 and M_1 in eggs and tissues of laying hens consuming aflatoxin contaminated feed. *Poult. Sci.,* 62: 2176-2182.

Verma, J., Johri, T.S., Swan, B.K. & Ameena, S. (2004). Effect of graded levels of aflatoxin and their combination on the performance and immune response of broilers. *Br. Poult. Sci.,* 45: 512-518.

Wang, R., Miao, R., Chong, N., Sawatwiroj, K., Dawson, S., Fu & Feng, D., 2003. Feed and feed ingredient mycotoxin contamination in China: A survey. In: *Proc. of Alltech's 17th Annual Asia Pacific Lecture Tour.*

Washburn, K.W., Wyatt, R.D., Potts, P.L. & Lanze, G.M., 1985. Effects and mechanism of aflatoxin variation in shell strength. *Poult. Sci.,* 64: 1302-1305.

Weiss, W.P. (2002). Amntioxidants nutrients and milk quality. In: *Roche symposium, pacific Northwest nutrition conference,* Vancouver, British Columbia, Canada.

Wolzak, A., Pearson, A.M., Coleman, T.H., Pestka, J.J. & Gray, J.I. (1985). Aflatoxin deposition and clearance in the eggs of laying hens. *Food Chem. Toxixo.,* 32: 1057-1061.

Xue, C.Y., Wang, G.H., Chen, F., Zhang, X.B. & Cao, Y.C. (2010). Immunopathological Effects of Ochratoxin A and T-2 Toxin Combination on Broilers. *Poult. Sci.,* 89: 1162-1166.

Yegani, M., Smith, T. K., Leeson, S. & Boermans, H.J. (2006). Effects of Feeding Grains Naturally Contaminated with *Fusarium* Mycotoxins on Performance and Metabolism of Broiler Breeders. *Poult. Sci.,* 85: 1541-1549.

Zaghini, A., Martelli, G., Ronchada, P. & Rizzi, L. (2005). Mannanoligosaccharides and aflatoxin B_1 and M_1 residues in fed of laying hens. Effect on egg quality, aflatoxin B_1 and M_1 residue in eggs and aflatoxin B_1 levels in liver. *Poult. Sci.,* 84: 825-832.

12

Aflatoxins and Aflatoxicosis in Human and Animals

D. Dhanasekaran[1], S. Shanmugapriya[1],
N. Thajuddin[1] and A. Panneerselvam[2]
[1]*Department of Microbiology, School of Life Sciences,*
Bharathidasan University, Tiruchirappalli,
[2]*P.G. & Research Department of Botany & Microbiology,*
A.V.V.M. Sri Pushpam College, (Autonomous), Poondi, Thanjavur District, Tamil Nadu
India

1. Introduction

Moldy feed toxicosis was recognized as a serious livestock problem in the 1950's but it was only in 1960 during the investigations in the United Kingdom of moldy feed toxicosis which was called Turkey "x" disease, that *A. flavus* and *A. parasiticus* were identified as the organisms responsible for the elaboration of the toxin in the feed. The earliest symptoms of the disease are lithargy and muscular weakness followed by death. The term aflatoxin now refers to group of bisfuranocoumarin metabolites isolated from strains of *A. flavus* group of fungi. The toxic material derived from the fungus *A. flavus* was given the name "aflatoxin" in 1962 (Sargeant *et al.*, 1963).

Aflatoxins fluoresce strongly in ultra violet light. The major members are designated as B1, B2, G1 and G2. B1 and B2 fluoresces blue, while Gl and G2 fluoresces green. In some animal species in dairy cattle, aflatoxin B1 and B2 are partially metabolized to the hydroxylated derivates namely M1 and M2, respectively.

Aflatoxin P1 is a urinary metabolite of Bl in monkeys. All aflatoxins absorb UV light in the range of 362-363nm, a characteristic used in preliminary identification. The growth of toxigenic molds and elaboration of the toxin occurs if moisture conditions are ideal following harvest and storage.

Although initially aflatoxin was detected in the peanut meal it is now known that a variety of cereals, and other plant products are susceptible to fungal invasion and mycotoxin production. The occurrence of aflatoxins in agricultural commodities depends on such factors as region, season and the conditions under which a particular crop is grown, harvested or stored.

Because of the wide spread nature of fungi producing aflatoxins in food materials, international agencies have now permitted the presence of 20 ppb of aflatoxin in food materials as the maximum permissible level. In 1993 aflatoxin by the World Health Organization (WHO) for cancer research institutions designated as a Class 1 carcinogen, is a highly poisonous toxic substances. Aflatoxin is harmful to human and animal liver tissue has damaging effects, serious, can lead to liver cancer or even death. In the natural food contaminated with aflatoxin B1 is most common, is also its most toxic and carcinogenic.

2. Occurence of aflatoxin in food and feed

Aflatoxin found in soil, plants and animals, all kinds of nuts, especially peanuts and walnuts. In soybean, rice (Fouzia Begum & Samajpati, 2000), corn, pasta, condiments, milk, dairy products, edible oil products are also often found aflatoxin. Aflatoxins often occur in crops in the field prior to harvest. Post harvest contamination can occur if crop drying is delayed and during storage of the crop if water is allowed to exceed critical values for the mould growth. Insect or rodent infestations facilitate mould invasion of some stored commodities.

Aflatoxins are found occasionally in milk, cheese, peanuts, cottonseed (Fig. 1), nuts, almonds, figs, spices, and a variety of other foods and feeds. Milk, eggs, and meat products are sometimes contaminated because of the animal consumption of aflatoxin contaminated feed. Cottonseed, Brazil nuts, copra, various tree nuts and pistachio nuts are the other commodities quite susceptible to the invasion of aflatoxin producing fungi.

Fig. 1. Contamination of cotton seeds

2.1 Types of aflatoxins

Although 17 aflatoxins have been isolated (WHO, 1979), only 4 of them are well known and studied extensively from toxicological point of view. Being intensely fluorescent in ultraviolet light the four are designated by letters B1, B2, G1 and G2 representing their blue and green fluorescence in UV light. Two other familiar aflatoxins are M1 and M2. Because of their presence in milk of animals previously exposed to B1 and B2. Of all the above-named aflatoxins, aflatoxin B I (AFB1) is the most acutely toxic to various species.

Toxigenic *A. flavus* isolates generally produce only aflatoxins B1 and B2, whereas *A. parasiticus* isolates generally produce aflatoxins B1, B2, G1 and G2 (Davis and Diener, 1983). Other metabolites $B2_a$, aflatoxicol, aflatoxicol H1 and aflatoxins P1 and Q1 have been identified (FDA, 1979). Aflatoxin M1 is a metabolite of aflatoxin B1 in humans and animals. Aflatoxin M2 is a metabolite of aflatoxin B1 in milk of cattle fed on contaminated foods. Although aflatoxins B1, B2 and G1 are common in the same food sample, AFB1 predominates (60-80% of the total aflatoxin content). Generally AFB2, AFG1 and AFG2 do not occur in the absence of AFB1. In most cases AFG1 is found in higher concentrations than AFB2 and AFG2 (Weidenborner, 2001).

2.2 Favorable conditions for aflatoxin biosynthesis production

The formation of aflatoxins is influenced by physical, chemical and biological factors. The physical factors include temperature and moisture. The chemical factors include the composition of the air and the nature of the substrate. Biological factors are those associated with the host species (Hesseltine, 1983).

Specific nutrients, such as minerals (especially zinc), vitamins, fatty acids, amino acids and energy source (preferably in the form of starch) are required for aflatoxins formation (Wyatt, 1991). Large yield of aflatoxins are associated with high carbohydrate concentrations, such as wheat rice and to a lesser extent in oilseeds such as cottonseed, soyabean and peanuts (Diener and Davis, 1968).

The limiting temperatures for the production of aflatoxins by *A. flavus* and *A. parasiticus* are reported as 12 to 41°C, with optimum production occurring between 25 and 32°C (Lillehoj, 1983). Synthesis of aflatoxins in feeds are increased at temperatures above 27°C (80 F), humidity levels greater than 62% and moisture levels in the feed above 14% (Royes and Yanong, 2002).

2.3 Causes of aflatoxin

Crops grown under warm and moist weather in tropical or subtropical countries are especially more prone to aflatoxin contamination than those in temperate zones. Groundnuts and groundnut meal are by far the two agricultural commodities that seem to have the highest risk of aflatoxin contamination. Although these commodities are important as substrates, fungal growth and aflatoxin contamination are the consequence of interactions among the fungus, the host and the environment. The appropriate combination of these factors determines the infestation and colonization of the substrate and the type and amount of aflatoxin produced.

Water stress, high-temperature stress and insect damage of the host plant are major determining factors in mould infestation and toxin production. Similarly, specific crop growth stages, poor fertility, high crop densities and weed competition have been associated with increased mould growth and toxin production. The moisture content of the substrate and temperature are the main factors regulating the fungal growth and toxin formation. A moisture content of 18% for starchy cereal grains and 9-10% for oil-rich nuts and seeds has been established for maximum production of the toxin (WHO, 1979). On the other hand, the minimum, optimum and maximum temperatures for aflatoxin production have been reported to be 12^0 - 27^0C and 40^0-42^0C respectively. Frequent contamination of corn and other commodities with high levels of aflatoxins has been a serious problem all over the world resulting in significant economic losses to farmers and a health hazard to farm animals and humans as well.

2.4 Structure of aflatoxin

In 1963, Asao *et al.*; Van Dorp *et al.* and Van der Zijden characterized the chemical and physical nature of the aflatoxins B1, B2, G1 and G2. Chemically, aflatoxins are difurocoumarolactones (difurocoumarin derivatives). Their structure consists of a bifuran ring fused to a coumarin nucleus with a pentenone ring (in B and M aflatoxins) or a six-membered lactone ring in G aflatoxins (Fig. 2). The four compounds are separated by the color of their fluorescence under long wave (Devero, 1999) ultraviolet illumination (B=blue,

G= green). Two other aflatoxins M_1 and M_2 were isolated from urine and milk and identified as mammalian metabolites of AFB1 and AFB2 (Patterson *et al.*, 1978).

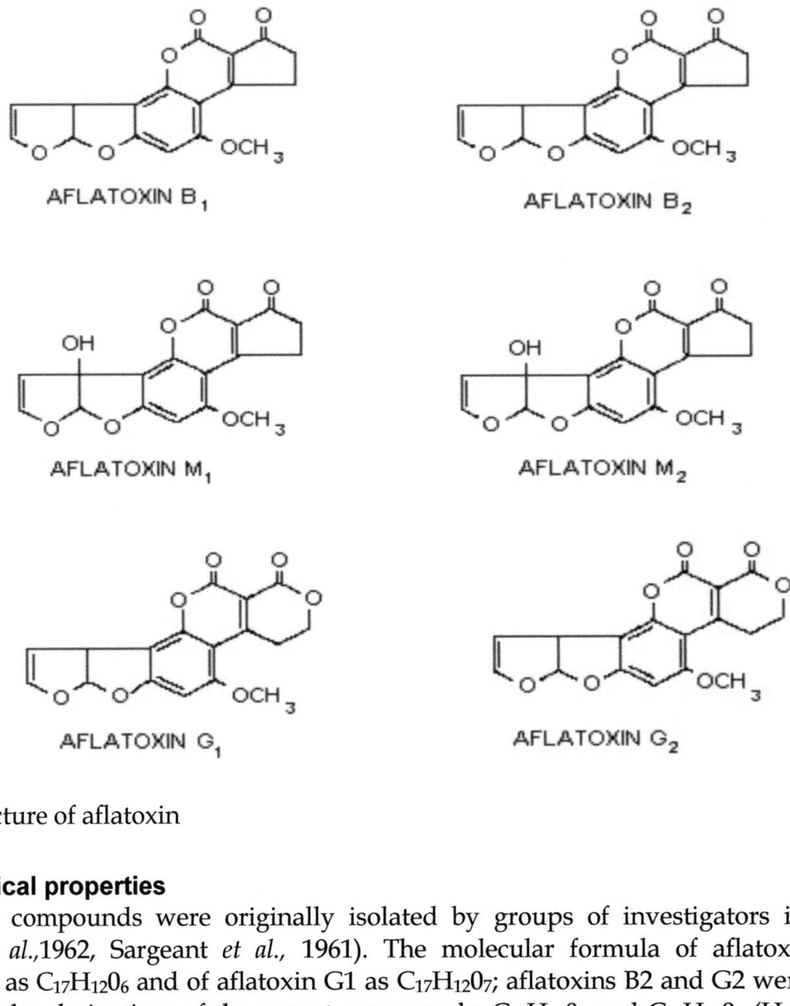

Fig. 2. Structure of aflatoxin

2.4.1 Physical properties

These four compounds were originally isolated by groups of investigators in England (Nesbitt *et al.*,1962, Sargeant *et al.*, 1961). The molecular formula of aflatoxin B1 was established as $C_{17}H_{12}0_6$ and of aflatoxin G1 as $C_{17}H_{12}0_7$; aflatoxins B2 and G2 were found to be the dihydro derivatives of the parent compounds, $C_{17}H_{14}0_6$ and $C_{17}H_{14}0_7$ (Hartley *et al.*, 1963). Some physical properties of the compounds are summarized in following Table 1.

Aflatoxin	Molecular Formula	Molecular Weight	Melting Point C	[α]D[23]
B1	$C_{16}H_{12}O_6$	312	268-269*	-559
B2	$C_{17}H_{14}O_6$	314	286-289*	-492
G1	$C_{17}H_{12}O_7$	328	244-246*	-533
G2	$C_{17}H_{14}O_7$	330	237-240*	-473

*Decomposes

Table 1. Physical properties of aflatoxin

The spectral characteristics of the aflatoxins have been determined by several investigators and are summarized in Table 2. The ultraviolet absorption spectra are very similar, each showing maxima at 223, 265 and 363 mμ. The molar extinction coefficients at the latter two peaks, however, demonstrate that B1 and G2 absorb more intensely than G1 and B2 at these two wavelengths. Because of the close similarities in structural configuration, the infrared absorption spectra of the four compounds are also very similar, as illustrated. The fluorescence emission maximum for B1 and B2 has been reported to be 425 mμ and that for G1 and G2 is 450 mμ. The intensity of light emission, however, varies greatly among the four compounds, a property of significance in the estimation of concentrations of the compounds by fluorescence techniques.

Aflatoxin	Ultraviolet absorption (ε)		Infrared absorption (cm^{-1}) $\nu_{max}^{CHCl_3}$					Fluorescence emission (mμ)
	265 mμ	363 mμ						
B1	13400	21800	1760	1684	1632	1598		425
B2	9200	14700	1760	1685	1625	1600		425
G1	10000	16100	1760	1695	1630	1595	1562	450
G2	11200	19300	1760	1694	1627	1597		450

Table 2. Spectral properties of aflatoxin

2.4.2 Chemical properties

The chemical reactivity and behavior of the aflatoxins has received relatively little systematic study beyond work associated with structure elucidation. However, it has been shown (Asao *et al.*, 1963) that catalytic hydrogenation of aflatoxin B1 to completion results in the uptake of 3 moles of hydrogen with the production of the tetrahydrodeoxy derivative. Interruption of the hydrogenation procedure after the uptake of 1 mole of hydrogen results in the production of aflatoxin B2 in quantitative yield.

Aflatoxin B1 has also been reported to react additively with a hydroxyl group under the catalytic influence of a strong acid (Andrellos *et al.*, 1964). Treatment with formic acid-thionyl chloride, acetic acid thionyl chloride or trifluroacetic acid results in addition products of greatly altered chromatographic properties, but relatively unchanged fluorescence characteristics. Ozonolysis results in fragmentation of aflatoxin B1 and the products of this reaction include levulinic, succinic, malonic and glutaric acids (Van Drop *et al.*, 1963). The presence of the lactone ring makes the compound labile to alkaline hydrolysis, and partial recyclization after acidification of the hydrolysis product has been reported (De Iongh *et al.*, 1962).

Although few systematic studies have been carried out on the stability of the aflatoxins, the general experience would seem to indicate that some degradation takes place under several conditions. The compounds appear partially to decompose, for example, upon standing in methanolic solution, and this process is greatly accelerated in the presence of light or heat. Substantial degradation also occurs on chromatograms exposed to air and ultraviolet or visible light. These processes may give rise to some of the nonaflatoxin fluorescent compounds typically seen in chromatograms of culture extracts. The nature of the decomposition products is still unknown, and the chemical reactions involved in their formation remain to be established (Wogan *et al.*, 1966).

2.5 Biological effects of the aflatoxins

The discovery of this group of compounds as contaminants of animal feeds and the potential public health hazards involved, have stimulated considerable research effort concerned with their effects in various biological assay systems. The toxic properties of the aflatoxins manifest themselves differently depending on the test system, dose and duration of exposure. Thus, they have been shown to be lethal to animals and animal cells in culture when administered acutely in sufficiently large doses and to cause histological changes in animals when smaller doses were administered subacutely. Chronic exposure for extended periods has resulted in tumor induction in several animal species (Wogan *et al.*, 1966).

2.5.1 Aflatoxin and animal diseases

Aflatoxin poisoning (Aflatoxicosis) mainly on animal liver injury, the injured individual species of animals, age, gender and nutritional status vary. The aflatoxin can cause liver dysfunction, reduced milk production and egg production and to reduced immunity of animals. Susceptible to the infection of harmful microorganisms. In addition, long-term consumption of food containing low concentrations of aflatoxin in feed can also result in embryo toxicity. Usually young animals are more sensitive to aflatoxin. The clinical manifestations of aflatoxin are digestive disorders, reduced fertility, reduced feed efficiency and anemia. Aflatoxins not only the decline in milk production, but also the transformation of the milk containing aflatoxin M1 and M2. According to the U.S. agricultural economy scientists to statistics, the consumption of aflatoxin contaminated feed, make at least 10% of the U.S. livestock industry suffered economic losses.

2.5.2 Metabolism and mechanisms of action of aflatoxin B1

The absorption from the gastrointestinal tract should be complete since very small doses, even in the presence of food, can cause toxicity. After the absorption, highest concentration of the toxin is found in the liver (Mintzlaff *et al.*, 1974). Once in liver, aflatoxin B1 is metabolized by microsomal enzymes to different metabolites through hydroxylation, hydration, demethylation and epoxidation (Fig. 3). Thus hydroxylation of AFB1 at C4 or C22 produces, AFM1 and AFQ1, respectively. Hydration of the C2 - C3 double bond results in the formation of AFB2a which is rapidly formed in certain avian species (Patterson and Roberts, 1970). AFP1 results from o-demethylation while the AFB1 - epoxide is formed by epoxidation at the 2,3 double bond. Aflatoxicol is the only metabolite of AFB1 produced by a soluble cytoplasmic reductase enzyme system.

The liver is the target organ for toxic effects of aflatoxin B1. As a result, metabolism of proteins, carbohydrates and lipids in liver is seriously impaired by AFB1. The toxin inhibits RNA polymerase and subsequent protein synthesis at a faster rate in ducks than in rats probably because of faster liver metabolism of AFB1 in ducks than in rats (Smith, 1965). In day-old chicks, AFB1 reduces the activity of liver UDP glucose-glycogen transglucosylase resulting in depletion of hepatic glycogen stores (Shankaran *et al.*, 1970). On the other hand, there is lipid accumulation in the liver of chickens and ducklings exposed to aflatoxin (Carnaghan *et al.*, 1966; Shank and Wogan, 1966). With regard to its toxic effects on liver microsomal enzymes, AFB1 is known to decrease microsomal glucose-6-phosphatase activity (Shankaran *et al.*, 1970) whereas stimulation of microsomal enzyme activity by inducers seems to be unaffected by AFB1 (Kato *et al.*, 1970). In fact, pretreatment with the toxin actually stimulates its own metabolism in the rat when this is assayed in vitro

(Schabort and Steyn, 1969). Since aflatoxin inhibits protein synthesis, it is conceivable why aflatoxin reduces resistance of poultry to infection with *Pasteurella multocida, salmonella sp.,* Marek disease virus, Coccidia and *Candida albicans* (Smith *et al.*, 1969; Hamilton and Harris, 1971). Another effect of aflatoxin is that it causes anticoagulation of blood. This is probably because AFB1 inhibits synthesis of factors II and VII involved in prothrombin synthesis and clotting mechanism (Bababunmi and Bassir, 1969).

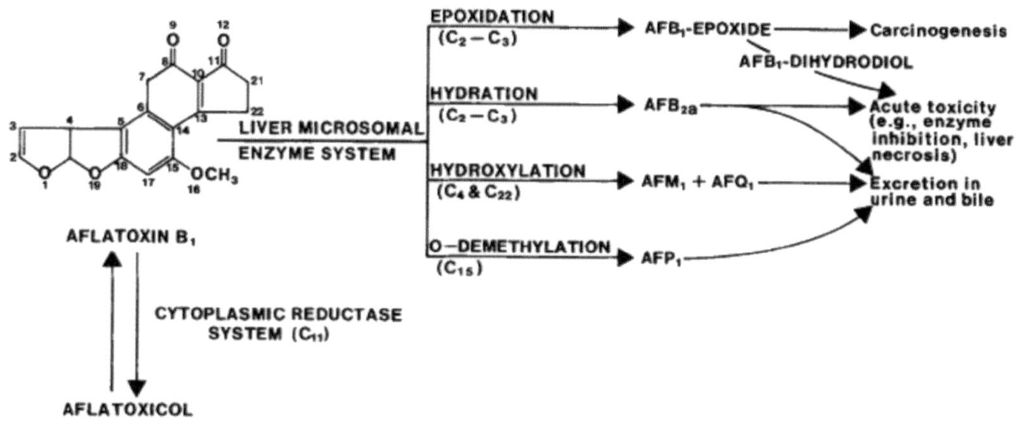

Fig. 3. Metabolism of aflatoxin in liver

Aflatoxin molecules in the double-furan ring structure, the structure is an important toxicity. Studies show that aflatoxin cytotoxic effects of toxins, is an interference RNA and DNA synthesis of information, thereby interfere with cell protein synthesis, resulting in systemic damage to animals (Nibbelink, 1988). Huangguang Qi *et al.*, (1993) research indicates that aflatoxin B1 to form with the tRNA adduct, aflatoxin-tRNA adduct can inhibit tRNA binding activity of some amino acids on protein synthesis in the essential amino acids such as lysine, leucine, arginine and glycine and tRNA binding, have different inhibitory effect, thereby $v_{\max}^{CHCl3}$ interfering with the translation level of protein biosynthesis, affect cell metabolism.

Aflatoxin B1 is excreted in urine and feces, and also in milk of lactating animals either unchanged or as various metabolites (Nabney *et al.*, 1967; Allcroft *et al.*, 1968). Only one milk metabolite, namely AFM1, appears to be the major metabolite of AFB1 that has shown appreciable oral toxicity (Holzapfel *et al.*, 1966). Its toxicity is considered to be nearly as potent as AFB1. Even so this metabolite may be detoxified by conjugation with taurocholic and glucuronic acids prior to excretion in the bile or brine (Delongh *et al.*, 1964; Bassir and Osiyemi, 1967). In this respect, two other metabolites of AFB1, namely, AFP1 and AFQ1 are similar in that they also undergo this type of detoxication (Dalezios *et al.*, 1971; Buchi *et al.*, 1973; Masri *et al.*, 1974). Both of these metabolites are several-fold less toxic than AFB1. For example, toxicity tests showed that AFP1 causes some mortality in newborn mice at 150 mg/kg as compared to the LD50 of 9.5 mg/kg for AFB1 under comparable conditions.

2.6 Aflatoxin and human health

Human health hazards by aflatoxin were mainly due to people eating aflatoxin contaminated food. For the prevention of this pollution is very difficult, the reason is due to fungi in the food or food materials in the presence of a very common. The state health department has been heavily polluted enterprises to use against the grain for food production and supervision of enterprises to develop the implementation of the relevant standards. But with lower concentrations of aflatoxin food and food cannot be controlled. In developing countries, consumption of aflatoxin contaminated food and incidence of cancer was positively correlated. Asian and African research institutions, disease research showed that aflatoxin in food and liver cells cancer (Liver Cell Cancer, LCC) showed positive correlation. For a long time with low concentrations of aflatoxin consumption of food was the leading cause liver cancer, stomach cancer, colon cancer and other diseases. In 1988, International Agency for Research on Cancer (IARC) classified the aflatoxin B1 as a human carcinogen.

The median lethal dose of aflatoxin Bl 0.36 mg / kg body weight is a special range of highly toxic poison (aflatoxin animal half of the lethal dose is found in the strongest carcinogens). Its carcinogenicity is 900 times more than dimethylnitrosamine induced liver cancer in the large capacity 75 times higher than the 3,4-benzopyrene, a large 4000-fold. It is mainly to induce liver cancer in animals, can also induce cancer, renal cancer, colorectal cancer and breast, ovary, small intestine and other sites of cancer.

2.7 Aflatoxicosis

The disease caused by the consumption of substances or foods contaminated with aflatoxin is called aflatoxicosis.

2.7.1 Aflatoxicosis in humans

2.7.2 Human exposure conditions

Two pathways of the dietary exposure have been identified:

a. Direct ingestion of aflatoxins (mainly B1) in contaminated foods of plant origin such as maize nuts and their products.
b. Ingestion of aflatoxins carried over from feed into milk and milk products including cheese and powdered milk, where they appear mainly as aflatoxin M1. In addition to the carryover into milk, residues of aflatoxins may be present in the tissues of animals that consume contaminated feed (WHO, 1979). Aflatoxin residues have been found in animal tissues, eggs and poultry following the experimental ingestion of aflatoxin-contaminated feed (Rodricks and stoloff, 1977). Contamination of milk, egg and meat can result from animal consumption of mycotoxin contaminated feed. Aflatoxins, ochratoxin and some trichothecences have been given considerable attention, because they are either carcinogenic or economic concern in animal health (CAST, 1989).

Aflatoxin M1 is believed to be associated with casein (protein) fraction of milk. Cream and butter contain lower concentration of M1 than the milk from which these products are made, while, cheese contains higher concentration of M1 about 3-5 times the M1 in the original milk (Kiemeier and Buchner, 1977; Stoloff, 1980; Brackett and Marth, 1982).

The expression of aflatoxin related diseases in humans may be influenced by factors such as age, sex, nutritional status, and concurrent exposure to other causative agents such as viral hepatitis (HBV) or parasite infestation.

2.8 Effect on human health

2.8.1 Acute toxicity

The disease was characterized by high fever, high colored urine, vomiting, and edema of feet, Jaundice, rapidly developing ascitis, portal hypertension and a high mortality rate. The disease was confirmed to the very poor, who were forced by economic circumstances to consume badly molded corn containing aflatoxins between 6.25 -15.6 ppm, an average daily intake per person of 2-6 mg of aflatoxins (Krishnamachari *et al.*, 1975a and 1975b; Keeler and Tu, 1983).

2.8.2 Chronic toxicity

Long exposure to aflatoxins in the diet increases risk with a synergistic effect from increased alcohol consumption. Aflatoxin B1 has also been implicated as a cause of human hepatic cell carcinoma (HCC). Aflatoxin B1 also chemically binds to DNA and caused structural DNA alterations with the result of genomic mutation (Groopman *et al*, 1985).
Ingestion of aflatoxin, viral diseases, and hereditary factors has been suggested as possible aetiological agents of childhood cirrhosis. There are evidences to indicate that children exposed to aflatoxin breast milk and dietary items such as unrefined groundnut oil, may develop cirrhosis. Malnourished children are also prone to childhood cirrhosis on consumption of contaminated food. Several investigators have suggested aflatoxin as an aetiological agent of Reye's syndrome in children in Thailand, New Zealand etc. Though there is no conclusive evidence as yet. Epidemiological studies have shown the involvement of aflatoxins in Kwashiorkor mainly in malnourished children. The diagnostic features of Kwashiorkor are edema, damage to liver etc. These out breaks of aflatoxicosis in man have been attributed to ingestion of contaminated food such as maize, groundnut etc. Hence it is very important to reduce the dietary intake of aflatoxins.

2.9 Aflatoxicosis in animals

Aflatoxin can cause oncogenesis, chronic toxicity or peracute signs depending on the species, age of animal, dose and duration of aflatoxin exposure (Smith, 2002). All animal species are susceptible to aflatoxicosis, but outbreak occurs mostly in pigs, sheep and cattle (Radostits *et al.*, 2000). Beef and dairy cattle are more susceptible to aflatoxicosis than sheep or horses. Young animals of all species are more susceptible than mature animals to the effects of aflatoxin. Pregnant and growing animals are less susceptible than young animals, but more susceptible than mature animals (Cassel *et al.*, 1988). Nursing animals may be affected by exposure to aflatoxin metabolites secreted in the milk (Jones *et al.*, 1994). For most species, oral LD50 values of aflatoin B1 vary from 0.03 to 18 mg/kg body weight.

2.10 Effects on animal health

There are differences in species with respect to their susceptibility to aflatoxins, but in general, most animals, including humans, are affected in the same manner.

2.10.1 Acute toxicity

Acute toxicity is less likely than chronic toxicity. Studies have shown that ducklings are the species most susceptible to acute poisoning by aflatoxins. The LD50 of a day old duckling is 0.3mg/kg bodyweight.

Species	Oral LD50/Lethal dose (mg/Kg)
Chick embryo	0.025
Duckling	0.3
Turkey poultry	0.5
Chicken, New Hampshire	2.0
Chicken, Rhode Island	6.3
Sheep	5.0
Rat(male)	7.2
Rat(female)	17.9
Rabbit	0.3
Cat	0.6
Pig	0.6
Guinea pig	1.4
Hamster	10.2
Mouse	9.0
Baboon	2.0

Table 3. Comparative LD50 or lethal values for Aflatoxin B1 (Edds, 1973 &WHO, 1979).

The principal target organ for aflatoxins is the liver. After the invasion of aflatoxins into the liver, lipids infiltrate hepatocytes and leads to necrosis or liver cell death. The main reason for this is that aflatoxin metabolites react negatively with different cell proteins, which leads to inhibition of carbohydrate and lipid metabolism and protein synthesis. In correlation with the decrease in liver function, there is a derangement of the blood clotting mechanism, icterus (jaundice), and a decrease in essential serum proteins synthesized by the liver. Other general signs of aflatoxicosis are edema of the lower extremities, abdominal pain, and vomiting.

2.10.2 Chronic toxicity

Animals which consume sub-lethal quantities of aflatoxin for several days or weeks develop a sub acute toxicity syndrome which commonly includes moderate to severe liver damage. Even with low levels of aflatoxins in the diet, there will be a decrease in growth rate, lowered milk or egg production and immunosupression. There is some observed carcinogenicity, mainly related to aflatoxin B1. Liver damage is apparent due to the yellow color that is characteristic of jaundice, and the gall bladder will become swollen. Immunosuppression is due to the reactivity of aflatoxins with T-cells, decrease in Vitamin K activities and a decrease in phagocytic activity in macrophages.

2.10.3 Cellular effects

Aflatoxins are inhibitors of nucleic acid synthesis because they have a high affinity for nucleic acids and polynucleotides. They attach to guanine residues and form nucleic acid adducts. Aflatoxins also have been shown to decrease protein synthesis, lipid metabolism, and mitochondrial respiration. They also cause an accumulation of lipids in the liver, causing a fatty liver. This is due to impaired transport of lipids out of the liver after they are synthesized. This leads to high fecal fat content. Carcinogenisis has been observed in rats, ducks, mice, trout, and subhuman primates and it is rarely seen in poultry or ruminants. Trout are the most susceptible. In fact, 1ppb of aflatoxin B1 will cause liver cancer in trout. Carcinogenisis occurs

due to the formation of 8,9-epoxide, which binds to DNA and alters gene expression. There is a correlation with the presence of aflatoxins and increased liver cancer in individuals that are hepatitis B carriers. Animals of different species vary in their susceptibility to acute aflatoxin poisoning with LD_{50} values ranging from 0.3 to 17.9 mg/kg (Table 4).
In fact duckling liver metabolized aflatoxin very rapidly in vitro (Patterson and Allcroft, 1970), although the species is sufficiently susceptible for day old birds to be used widely in a sensitive bioassay for the toxin (Patterson, 1973). Studies indicated that rabbit, duckling and guinea-pig constitute a "fast metabolizing group" being apparently capable of handling an LD_{50} dose in under 12 minutes. Chick, mouse, pig and sheep fall into an intermediate group, metabolizing an LD_{50} dose in a few hours. So far, the rat is the only example of a "slow metabolizing group" in which LD_{50} dose would probably disappear from the liver over a period of days (Patterson, 1973).
Factors that influence aflatoxin toxicity residue levels in animal species include: species and breeds of animals and poultry, levels and duration of exposure, nutrition and health of animals, age, sex and diseases, drugs and other mycotoxins (FDA, 1979).

Toxin	Animal	Age/Size	LD_{50}(mg/kg)
AFB1	Duckling	Day old	0.37
AFB2			1.69(84.8μg/50gm duckling)
AFG1			0.79
AFG2			2.5(172.5μg/duckling)
AFM1			0.8(16.6μg/duckling)
AFB1	Rabbit		0.3-0.5
	Cat		0.55
	Pig	6.0-7.0 kg	0.62
	Turkey		0.5-1.0
	Dog	Puppies	0.5-1.0
	Cattle	Young calves	0.5-1.0
	Guinea pig		1.4-2.0
	Horse		2.0
	Sheep		2.0
	Monkey		2.2
	Chickens	Young foals	6.5-16.5
	Mouse		9.0
	Hamster		10.2
	Rat, male,	21 days	5.5
	female		7.4
	Rat male	100 gm	17.9

Table 4. A comparison of single oral LD50 values for AFB1 in various species. (Agag, 2004)

2.10.4 Aflatoxicosis in ruminants

Aflatoxin ingested in the feed by cattle is physically bound to ruminal contents, and as little as 2-5% reach the intestine. Levels of AFB1 in excess of 100 μg/kg of feed are considered to be poisonous for cattle (Radostits *et al.*, 2000). The effects of aflatoxin fed to cattle depend on the level of aflatoxin in the ration, the length of feeding period and the age of animal (Jones *et al.*, 1994).

2.10.5 Calves

The LD_{50} dosage of AFB1 in calves has been estimated to be 0.5-1.5 mg/kg. Affected calves had anorexia, depression, jaundice, photosensitization of unpigmented skin, submandibular edema, severe keratocojunctivitis and diarrhea with dysentery. Collapse and death followed. Postmortem findings showed hemorrhages in subcutaneous tissues, skeletal muscles, lymph nodes, pericardium, beneath the epicardium and serosa of the alimentary tract. The liver was pale and carcass jaundiced. Histopathological examination of the liver revealed that hepatocytes were markedly enlarged, especially in the periportal areas and occasional hepatocyte nuclei were up to 5 times the diameter of their companions. Hepatocyte cytoplasm was finely vacuolated, many of these vacuoles containing fat. Serum enzymes of hepatic origin and bilirubin were elevated.

In calves who have consumed contaminated rations for several weeks, the onset of clinical signs is rapid. The most consistent features are blindness, circling and falling down, with twitching of the ears and grinding of the teeth. Severe tenesmus and erosion of the rectum are seen in most cases, and death of some cases (Humphreys, 1988).

2.10.6 Dairy and beef cattle

The signs most commonly reported with acute toxicosis in cattle include anorexia, depression, dramatic drop in milk production, weight loss, lethargy, ascitis, icterus, tenesmus, abdominal pain (animals may stretch or kick at their abdomen), bloody diarrhea, abortion, hepatoencephalopathy, photosensitization and bleeding (Colvin *et al.*, 1984; Cook *et al.*, 1986; Ray *et al.*, 1986; Eaton and Groopman, 1994; Reagor, 1996). Other signs associated with acute aflatoxicosis include blindness, walking in circles, ear twitching, frothy at the mouth, keratoconjunctivitis and rectal prolapse (Radostits *et al.*, 2000).

Hepatic damage is a constant finding in acute aflatoxicosis. Lesions include fatty degeneration, megalocytosis and single-cell necrosis with increasing fibrosis, biliary proliferation and veno occlusive lesions as the disease progresses (Burnside *et al.*, 1957; Morehouse, 1981; Colvin *et al.*, 1984).

In additions, chronic aflatoxicosis may impair reproductive efficiency including abnormal estrous cycle (too short and too long) and abortions, induce immunosuppression and increase susceptibility to disease (Cassel *et al.*, 1988). The immunotoxic effect of AFB1 was expressed via the cell-mediated immune system (Raisbeck *et al.*, 1991).

Other symptoms including decreased milk production, birth of smaller and healthy calves, diarrhea, acute mastitis, respiratory disorders, prolapsed rectum and hair loss are also observed in chronically exposed dairy cattle (Guthrie, 1979). High aflatoxin levels (4 ppm) can cause milk production to drop within one week while, lower levels (0.4 ppm) can cause production drop in 3 to 4 weeks (Hutjens, 1983).

Another character of aflatoxin exposure in dairy cattle is the conversion to AFM1 in milk (Price *et al.*, 1985). Experiments have shown that milk will be free of aflatoxin after 96 hours of feeding non-contaminated feed. The level of aflatoxin in the feed and milk at the stating point will influence clearance time (Lynch, 1972; Hutjens, 1983).

The concentration of AFM1 in milk seems to depend more on intake of AFB1 than on milk yield (Vander Linde *et al.*, 1965). However, the toxin content of milk appears to increase rapidly when milk yield is reduced as a result of high toxin intake (Masri *et al.*, 1969). Rate of metabolism by the liver and rate of excretion by other routes (urine and feces) are also likely to influence the toxin level in milk (Applebaum *et al.*, 1982).

Decreased performance (i.e. rate of gain, milk production) is one of the most sensitive indicators of aflatoxicosis (Richard *et al.*, 1983). The ultimate cause of this effect is probably multifactorial, involving not only nutritional interactions, but also the compounding influences of anorexia, deranged hepatic protein and lipid metabolism and disturbances in hormonal metabolism (Raisbeck *et al.*, 1991). Aflatoxins have shown to affect rumen motility (Cook *et al.*, 1986) and rumen function by decreasing cellulose digestion, volatile fatty acid production and proteolysis (Fehr and Delag, 1970; Bodine and Mertins, 1983).

2.10.7 Sheep and goats

Anorexia, depression and icterus were observed in sheep and goats exposed to aflatoxin. The goats also developed a nasal discharge and dark brown urine was noted in the sheep (Hatch *et al.*, 1971; Samarajeewa, *et al.*, 1975; Abdelsalam *et al.*, 1989).

Anorexia and diarrhea occurred in sheep given aflatoxin at a rate of 0.23mg/kg. These signs were accompanied by excessive salivation, tachypnea and pyrexia at dosages of 0.59 mg/kg or more. Heavy mucous diarrhea and dysentery were observed in sheep dosed at a rate of 1.28 to 2.0mg/kg. Sheep that died within 24 hours of dosing had marked centrilobular necrosis of the liver. Sheep that survived until the 7th day after dosing had periportal congestion of the liver, widely dilated sinusoids and necrosis of liver cells (Armbrecht *et al.*, 1970).

Acutely intoxicated sheep with 4mg/kg showed anorexia, diarrhea, excessive salivation, rumen atony, scour, rectal prolapse, fever and death (Wylie and Morehouse, 1978).

2.10.8 AFB1 toxicity in equine species

The existing information on aflatoxicosis in the horse is inconclusive, although a total dietary concentration of 500–1000 μg/kg has been shown to induce clinical changes and liver damage, depending on the duration of exposure (Meerdink, 2002). The target organ in horses, as in all affected animals, is the liver, where the toxin induces centrilobular necrosis (Stoloff and Trucksess, 1979). Horses suffering from aflatoxicosis exhibit non-specific clinical signs, such as inappetence, depression, fever, tremor, ataxia and cough (Larsson *et al.*, 2003). Necropsy findings include yellow-brown liver with centrilobular necrosis, icterus, haemorrhage, tracheal exudates and brown urine (Angsubhakorn *et al.*, 1981; Cysewski *et al.*, 1982; Bortell *et al.*, 1983; Vesonder *et al.*, 1991).

Clinical signs of toxicity were observed in adult male Shetland ponies given a daily oral dose of 0.075 (over 36 or 39 days), 0.15 (over 25 or 32 days) and 0.3 mg/kg (over 12 or 16 days) of partially purified aflatoxin (47% AFB1 activity) (Cysewski *et al.*, 1982). Daily clinical observations revealed the appearance of depression, inappetence and weakness 5-7 days before the onset of severe illness and death. The onset of signs was generally dose related and appeared first in the ponies given 0.3mg/kg. Prothrombin time, sulfabromophthalein clearance time, total plasma bilirubin and the icteric index increased markedly before death. An increase in the plasma activity of aspartate amino transferase (AST) was observed. The AST activity grew significantly and generally remained at elevated levels in all animals. At the time of necropsy, pathological changes, such as generalised icterus, haemorrhage, a brown to tan liver, dark reddish brown urine and dark brown kidneys, were consistently observed. Microscopic lesions, including centrilobular fatty change, hepatic cell necrosis and periportal fibrosis, were observed in all ponies (Cysewski *et al.*, 1982).

2.10.9 Aflatoxicosis in equine species: Case reports

The first case of probable equine aflatoxicosis in which a 15-year-old Arabian stallion died, was reported by Greene and Oehme (1976). The reported symptoms included anorexia, icterus and rapid weight loss immediately prior to death. On post mortem examination, the liver was described as being black, of firm consistency and enlarged.

Histopathological examination revealed marked centrilobular hepatic necrosis and necrotic areas were engorged with erythrocytes. Kupffer cells were prominent and many contained phagocytosed haemosiderin, which was the likely cause of the black coloured liver. Bile-duct hyperplasia, congestion of renal vessels and adrenal cortex were found. Samples of the feed revealed AFB1 levels of 58.4 μg/kg, which exceeded the limit recommended by the FDA of 20 μg/kg (Greene and Oehme, 1976).

Two outbreaks of aflatoxicosis in two separate horse farms in different geographical areas of the world were later reported by Angsubhakorn *et al.* (1981). The first episode occurred in a horse breeding farm outside Bangkok, Thailand and started a few days after the introduction of a new lot of feed prepared from stored ground corn and stored peanut meal. A number of young animals had clinical signs of illness and 12 yearling colts died. At post mortem examination, a swollen fatty liver, pale swollen kidneys and hemorrhagic enteritis were found. A few animals necropsied soon after death, had pale hearts, focal myocardial necrosis and epicardial petechiae. When the cranial cavity was opened, the brain was seen to be swollen and the gyri compressed.

The second episode occurred in a farm for riding mares in southern Georgia, USA, in the autumn of 1978. After the introduction of a new mixed ration, a number of horses reduced their feed intake over a period of 3–4 days and showed signs of clinical illness. Within 5 days of the onset of clinical signs, two 2 year old horses and one 7 year old mare died. Necropsy findings were similar to those observed in the episode in Thailand. The swollen, fatty liver, epi- and endo-cardial petechiae; pale, mottled myocardium; swollen kidneys (although the paleness was more pronounced in the Georgia horses), haemorrhagic enteritis and variable mesenteric oedema.

Hematological and serum chemical examinations revealed hypoglycaemia, hyperlipidaemia and depletion of lymphocytes. The increase in haematocrit and the number of red blood cells was probably due in part to dehydration. The diagnosis of aflatoxicosis in these two episodes was based on histological examinations and isolation of the toxin from the feed and animal tissues. Microscopically, the lesions of the Georgia horses appeared the same as those seen in the Thai horses. Assay of liver specimens from two of the Georgia horses confirmed the presence of AFB1 in both samples and the results showed that AFB1 was metabolized and rapidly cleared from the liver and other tissues.

Peanut meal and corn were found to be the sources of aflatoxins in Thailand where representative samples of the mixed ration contained approximately 0.2 mg of AFB1 and 0.2 mg of AFB2 per kg (Caloni and Cortinovis, 2011).

2.10.10 Aflatoxins and COPD

Larsson *et al.* (2003) suggested a possible link between chronic obstructive pulmonary disease (COPD) and inhaled mycotoxins. *A.fumigatus* and *Mycropolyspora faeni* are potential causes of COPD in horses (Halliwell *et al.*, 1993), which is characterised by asthma like symptoms, such as chronic cough, nasal discharge, expiratory dyspnoea and reduced exercise tolerance (Gillespie and Tyler, 1969; Cook, 1976). The olfactory and respiratory mucosa of horses may be exposed to mycotoxins and other xenobiotics via inhalation of contaminated feed-dust particles (Sorensson *et al.*, 1981; Burg and Shotwell, 1984). The inhaled aflatoxins and other xenobiotics may be activated by CYP-enzymes in the epithelial linings of the respiratory tract and contribute to the aetiology of COPD (Larsson *et al.*, 2003; Tydèn *et al.*, 2008).

2.10.11 Canine

Canine aflatoxicosis was first reported in 1952 by Seibold and Bailey who described a liver disease called hepatitis "x" which was observed in dogs fed moldy contaminated feed. Dogs and cats are extremely sensitive to aflatoxins. The LD_{50} of AFB1 in dogs is 0.5-1.5 mg/kg and in cats is 0.3-0.6 mg/kg (Rumbeiha, 2001). Feed containing AFB1 concentrations of 60 ppb or greater have caused outbreaks of aflatoxicosis in companion animals. As with other toxic compounds, sensitivity depends on individual susceptibility which in turn depends on age, hormonal status (pregnancy), nutritional status, among other factors (Rumbeiha, 2001).

Dogs exposed to aflatoxin developed the typical anorexia, depression, icterus, prostration and blood in the feces, but also may have hemorrhages, melena and pulmonary edema (FDA, 1979; Liggett *et al.*, 1986; Bastianello *et al.*, 1987; Thornburg and Raisbeck, 1988). Moreover, vomiting, increased water consumption, polyuria, polydipsia, jaundice and elevation of serum liver enzymes in acute aflatoxicosis in dogs and cats (Rumbeiha, 2001).

At neuopsy, the liver is swollen, petechial hemorrhages are observed on the gums, along the gastrointestinal tract, in the lungs, pleura, epicardium and urinary bladder. The hemorrhages are associated with a yellow, reddish-yellow, or orange discoloration of the liver, icterus of the conjunctiva, oral mucosa, serous membranes and in body fat (Chaffee *et al.*, 1969; FDA, 1979; Rumbeiha, 2001). Lymphoid depletion and necrosis of the thymus, spleen and lymph nodes, gross uterine and placental hemorrhage and congestion and hemorrhage in the adrenal cortex were also reported (Newberne *et al.*, 1966).

In subacute aflatoxicosis, affected dogs and cats will present with lethargy, anorexia, polyuria, polydipsia, elevated liver enzymes and jaundice. In chronic aflatoxicosis, dogs and cats will have clinical signs similar to subacute aflatoxicosis, with prominent jaundice. Chronic aflatoxicosis may cause also immunosuppression, followed by non-specific clinical signs, including increased susceptibility to viral, bacterial, fungal or parasitic infections (Rumbeiha, 2001).

Histologically, there is severe fatty degeneration with distinct vacuolation of hepatocytes, bile canaliculi are distended with bile, portal and central veins are congested with bile, and portal and central veins are congested in acute cases. In subacute cases, the distinct feature is bile duct proliferation and there is evidence of liver regeneration. In chronic cases, there is extensive liver fibrosis and bile duct proliferation (Rumbeiha, 2001).

2.10.12 Pigs

Young swine are extremely sensitive to aflatoxins but susceptibility decreased with age (Diekman *et al.*, 1992). The toxicity of aflatoxin is both-dose related and time related and age is an important factor in susceptibility (Lawlor and Lynch, 2001). Sows and boars normally tolerate levels > 0.5 ppm in the feed for short periods but, when fed for extended periods, contamination levels in the feed should not exceed 0.1 ppm (Blaney and Williams, 1991). Levels in excess of 0.5 ppm in the dites of lactating sows will depress growth rates in suckling pigs due to aflatoxin in milk. For growing and finishing pigs residues will build up in the liver at concentrations of even less than 0.1 ppm in the feed (Osweiler, 1992). The LD_{50} in young pigs dosage was determined to be 0.8 mg/kg (Jones and Jones, 1978). The clinical syndrome in pigs include rough coat, depression, anorexia, decreased feed conversion, decreased rate of gain, weight loss, muscular weakness and shivering, tremors, bloody rectal discharge and icterus (Sisk *et al.*, 1968; Jones and Jones, 1978; Hoerr and D' Andrea, 1983; Radostits *et al.*, 2000). Aflatoxins also suppress the immune system and thus make pigs more susceptible to bacterial viral or parasitic diseases (Diekman *et al.*, 1992).

At necropsy, the livers from swine receiving toxic levels of AFB1 in their ration vary in close from tan to pale yellow with atrophic gall bladders, the livers contain increased fibrous connective tissue with resistance to cutting. There is icterus and petechial hemorrhages on the heart and massive hemorrhage into the ileum or throughout the digestive tract. Microscopic lesions include irregular shaped cells, centrilobular congestion, karyorrhexis and pyknosis, vacuolation, disappearance of nuclei, bile duct proliferation and extensive connective tissue in the inter and intralobular areas (FDA, 1979).

2.11 Aflatoxicosis in poultry

Aflatoxicosis have produced severe economic losses in the poultry industry affecting ducklings, broilers, layers, turkeys and quail (CAST, 1989). Susceptibility of poultry to aflatoxins varies among species, breeds and genetic lines. Comparative toxicological studies in avian species have shown that ducklings and turkey poultry are the most sensitive species to aflatoxins. Goslings, quails and pheasants show intermediate sensitivity while chickens appear to be the most resistant (Leeson *et al.*, 1995). The susceptibility ranges from ducklings > turkey poults > goslings > pheasant chicks > chickens (Muller *et al.*, 1970).

Ducklings are 5 to 15 times more sensitive to the effects of aflatoxins than are laying hens, but when laying hen strains are compared, certain strains of hens may be as much as 3 times more sensitive than other strains (Jones *et al.*, 1994). In comparing sensitivity of different strains of leghorn chicks (Table 5), it was found there is up to a 2.5 difference in the LD_{50} dose at 6 weeks of age (FDA, 1979)

Strain	LD_{50} mg/kg
A	6.5
B	7.25
C	9.25
D	9.50
E	11.50
F	16.50

Table 5. Sensitivity in different leghorn strains of chicks

In poultry, aflatoxin impairs all important production parameters including weight gain, feed intake, feed conversion efficiency, pigmentation, processing yield, egg production, male and female reproductive performance. Some influences are direct effects of intoxication, while others are indirect, such as from reduced feed intake (Calnek *et al.*, 1997).

The direct and indirect effects of aflatoxicosis include increased mortality from heat stress (broiler breeders, Dafalla *et al.*, 1987a), decreased egg production in leghorns, (Bryden *et al.*, 1980), anemia, hemorrhages and liver condemnations (Lamont, 1979), paralysis and lameness (Okoye *et al.*, 1988), impaired performance in broilers, (Jones *et al.*, 1982), increased mortality rate in ducks, (Bryden *et al.*, 1980), impaired ambulation and paralysis in quail, (Wilson *et al.*, 1975), impaired immunization in turkeys, (Hegazy *et al.*, 1991), and increased susceptibility to infectious diseases (Bryden *et al.*, 1980 and Calnek *et al.*, 1997).

2.11.1 Chickens

Susceptibility of chickens to toxic effects of AFB1 varies with several factors such as breed, strain, age, nutritional status, amount of toxin intake and also the capacity of liver microsomal enzymes to detoxify AFB1 (Edds, 1973; Veltmann, 1984). Acute toxicity of aflatoxins in chickens may be characterized by hemorrhage in many tissues and liver necrosis with icterus.

Although number of field cases of aflatoxicosis in chickens has been diagnosed in various countries, the most severe spontaneous outbreak occurred in North Carolina, in which 50% of a flock of laying hens died within 48 hr of being fed highly toxic maize containing 100 ppm aflatoxin (Hamilton, 1971). The necropsy revealed that liver damage was the most important lesion showing paleness, occasional white pinhead-sized foci and petechial hemorrhages while gallbladder and bile ducts were distended.

Levels of aflatoxin B1 in moldy feed normally vary from 0 to 10 ppm. At low levels of feed contamination, exposed chickens show, in general, weakness, failure to gain weight with concomitant decline in feed efficiency and egg production (Smith and Hamilton, 1970; Doerr *et al.*,1983). Hepatic damage is manifested by enlarged and putty-colored liver, petechial hemorrhages, marked vacuolation of hepatic cells and bile duct proliferation. Feed levels of AFB1 as low as 250-500 ppb given to New Hampshire chickens have been reported to result in liver damage, decreased hemoglobin, and hypoproteinemia (Brown and Abrams, 1965).

Experimental trials with naturally contaminated feed containing aflatoxin levels ranging from 1-1.5 ppm have caused growth retardation in chickens. Mortality was low but marked hepatic damage was manifested by enlarged and hemorrhagic liver (Carnaghan *et al.*, 1966).

Relatively, high dietary levels of aflatoxin B1 (0-10 ppm) given to Rock type broiler chickens have been reported to cause substantial decrease in weight gain, feed efficiency and hepatic microsomal drug metabolizing enzymes with concomitant increase in serum glutamic oxalacetic transaminase activity reflecting liver damage (Dalvi and McGowan, 1984; Dalvi and Ademoyero, 1984).

Metabolic alterations caused by aflatoxins in chickens result in elevated lipid levels(Tung *et al.*, 1972; Donaldson *et al.*, 1972), disruptions in hepatic protein synthesis (Tung *et al.*,1975) which result in several blood coagulation disorders (Doerr *et al.*,1976; Bababunmi and Bassir, 1982), immunosuppression and decreased plasma amino acid concentrations (Voight *et al.*, 1980).

2.11.2 Ducks

Lethal aflatoxicosis in ducklings occurred as inappetance, reduced growth, abnormal vocalizations, feather picking, purple discoloration of legs and feet and lameness. Ataxia, convulsions and opisthotonus preceded death (Asplin and Carnaghan, 1961).

At necropsy, livers and kidneys were enlarged and pale. With chronicity, ascitis and hydropericardium developed accompanied by shrunken firm nodular liver, distention of the gall bladder and hemorrhages (Asplin and Carnagham, 1961; Calnek *et al.*, 1997), distended abdomen due to liver tumors and secondary ascitis (Hetzel *et al.*, 1984).

Microscopic lesions in the liver were fatty change in hepatocytes, proliferation of bile ductules and extensive fibrosis accompanied by vascular and degenerative lesions in pancreas and kidney (Asplin and Carnagham, 1961 and Calnek *et al.*, 1997). Bile duct hyperplasia and bile duct carcinoma are also reported (Hetzel *et al.*, 1984) in aflatoxicated Campbell ducks.

2.11.3 Turkeys

The initial clinical signs reported during the outbreak of Turkey "x" disease were anorexia and weight loss followed by depression, ataxia and recumbency. Affected birds died with in a week or two and at the time of death frequency had opisthotonus characterized by arched neck, head down back and legs extended backwards (Hamilton *et al.*, 1972).

Along with decreased feed conversion and weight gain, reduced spontaneous activity, unsteady gait, recumbency, anemia and death (Siller and Ostler, 1961; Wannop, 1961; Giambrone *et al.*, 1985 ; Richard *et al.*, 1987).

At necropsy, the body condition was generally good but there was generalized congestion and edema. The liver and kidney were congested, enlarged and firm, the gall bladder was full, and the duodenum was distended with catarrhal content (Siller and Ostler, 1961; Wannop, 1961; Calnek *et al.*, 1997).

2.11.4 Broilers

Decreased water and feed intake, weight loss, dullness, stunting, ruffled feathers, poor appearance and paleness, trembling, ataxia, lameness, paralysis of the legs and wings gasping, prostration and death are frequency seen in experimental and natural outbreak of aflatoxicosis in broilers (Asuzu and Shetty, 1986; Okoye *et al.*, 1988; Rao and Joshi, 1993 ; Leeson *et al.*, 1995).

The most characteristic gross lesions appeared in the livers which were enlarged, pale yellow to grayish brown and had a prominent reticular pattern. Petecheal hemorrhages were observed on the surface of some livers. Gall bladders were enlarged and bile duct distended and there were blood in the intestinal lumen (Archibald *et al.*, 1962; Azuzu and Shetty, 1986). The liver, spleen and kidney were increased in size, whereas the bursa of fabricius and thymus were decreased (Smith and Hamilton, 1970; Huff and Doerr, 1980).

Lethal aflatoxicosis can cause either dark red or yellow discoloration of the liver due to congestion or fat accumulation, respectively (Slowik *et al.*, 1985). At chronicity livers became shrunken, firm and nodular and gall bladder was distended (Asplin and Carnaghan, 1961). The kidneys of affected birds appeared enlarged and congested (Tung *et al.*, 1973) and the spleen will be enlarged and mottled in appearance (Tung *et al.*, 1975 a).

Histopathology of the liver revealed congestion of hepatic sinusoids, fecal hemorrhages, centro-lobular fatty cytoplasmic vacuolation and necrosis, biliary hyperplasia and nodular lymphoid infiltration. In the kidney, the epithelial cells of many tubules were vacuolated (Dafalla *et al.*, 1987 b). Azuza and Shetty (1986) and Okoye *et al.*, (1988) observed severe degeneration of hepatocytes, dilation of central veins, bile duct proliferation and lymphocytic depletion in lymphoid organs in field outbreaks of aflatoxicosis in broilers.

2.11.5 Laying hens

Reduced egg production and egg weight, enlarged liver and increased liver fat are the most prominent manifestations of experimental aflatoxicosis in layers (Nesheim and Lvy, 1971; Hamilton and Garlich, 1973; Leeson *et al.*, 1995). High mortality and dramatic reduction of egg production were reported to occur during a natural outbreak (Hamilton, 1971). Egg size, egg weight and yolk as percent of total egg size are decreased (Huff *et al.*, 1975). In Japanese quail, decreased feed conversion, egg production, egg weight, hatchability and exterior and interior egg quality were detected (Sawhney *et al.*, 1973a & b). Dhanasekaran *et al.*, (2009) reported that histopathological analysis of aflatoxin ingested hens reveals that lesions were observed in tissues of liver, kidney, intestine (Plate 1). Jayabharathi and Mohamudha parveen (2010) tested the aflatoxicosis in hens. Haematological analysis showed the decreased haemoglobin than the control group (Plate2).

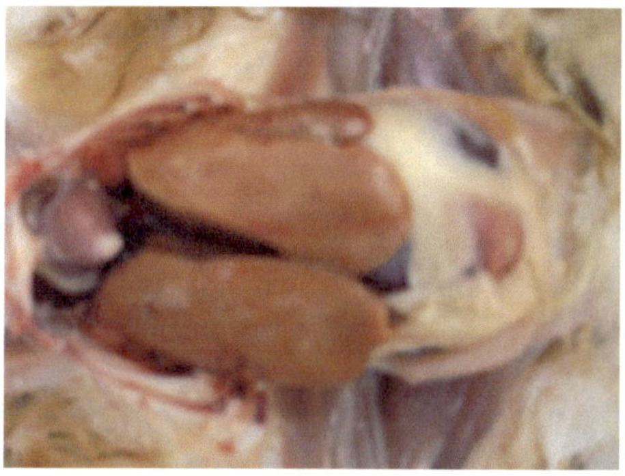

Control Hens with various organs

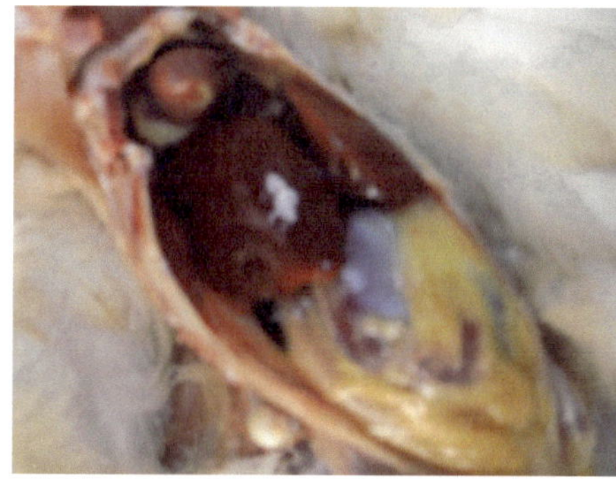

Test Hens with accumulation of fatty layer

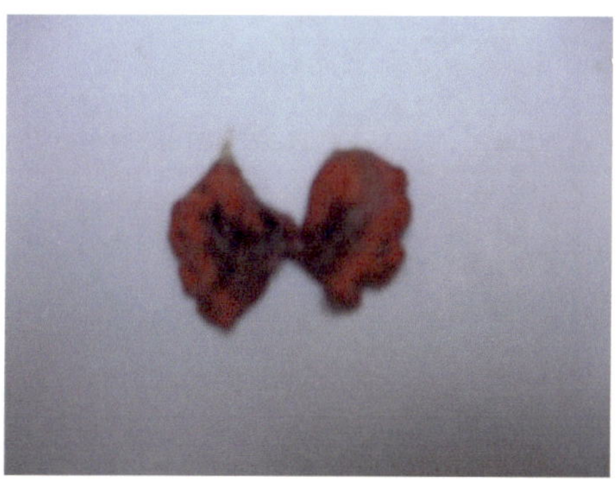

Lung (Control)

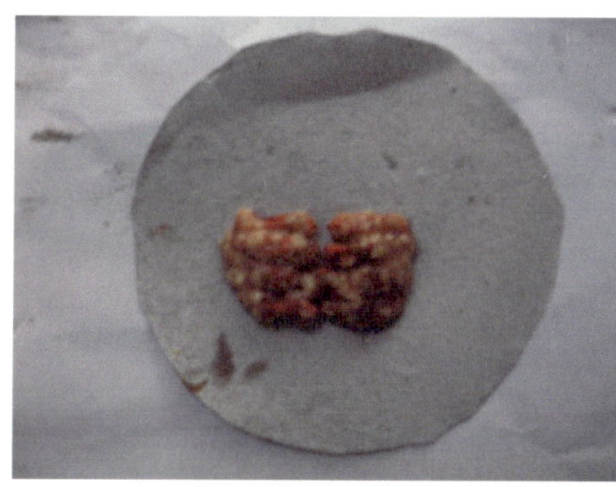

Lung (Test)

Plate 1. Organal view of Hens with Aflatoxicosis

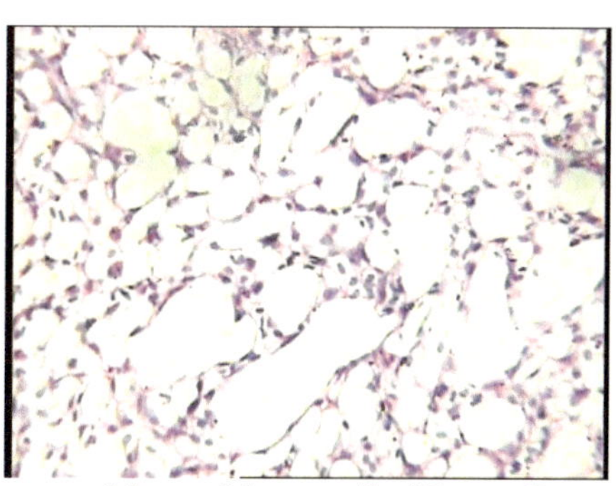

Normal lung cells (Control)

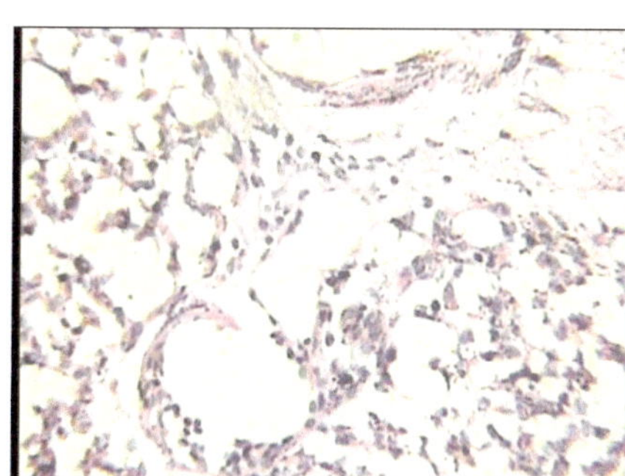

Mild infiltration by L and P (Test)

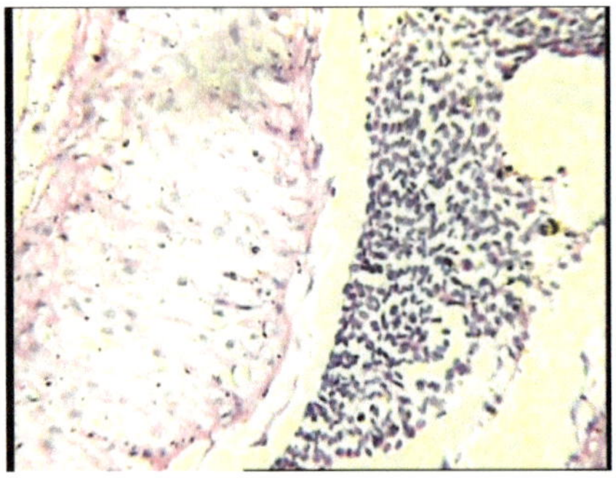

Dense infiltration by L and P (Test)

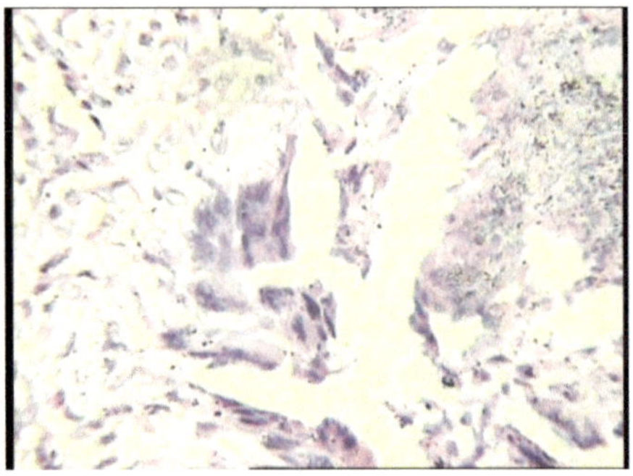

Infiltration by granular structures suspicious of carcinoma

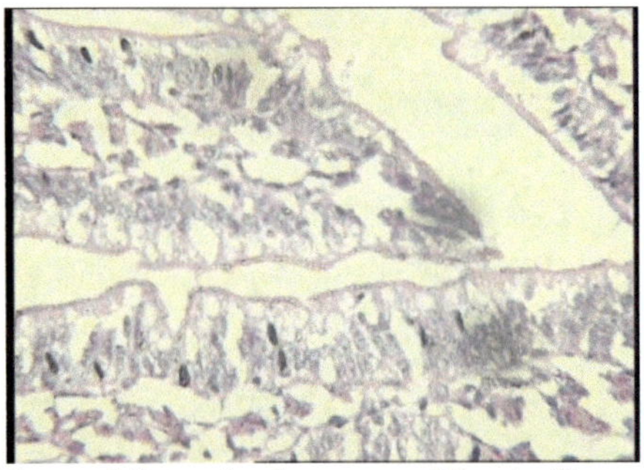

Normal Intestinal mucosa (Control)

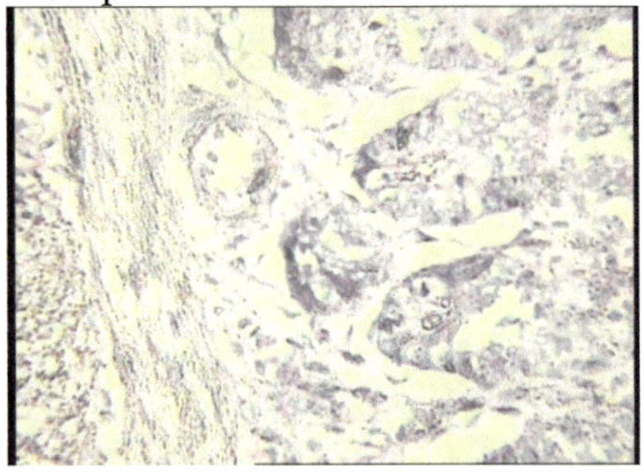

Atypical granular structures some with enlarged nuclei (Test)

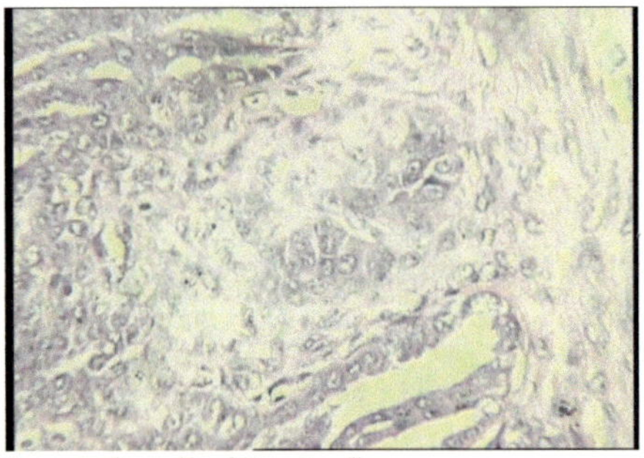

Stomach cells (Control)

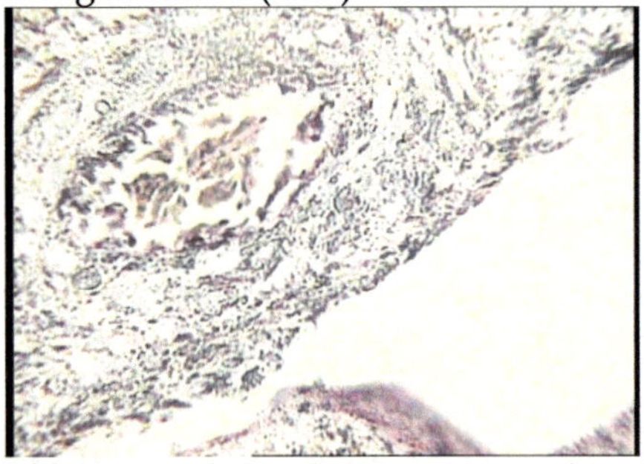

Infiltration of glandular structure (Test)

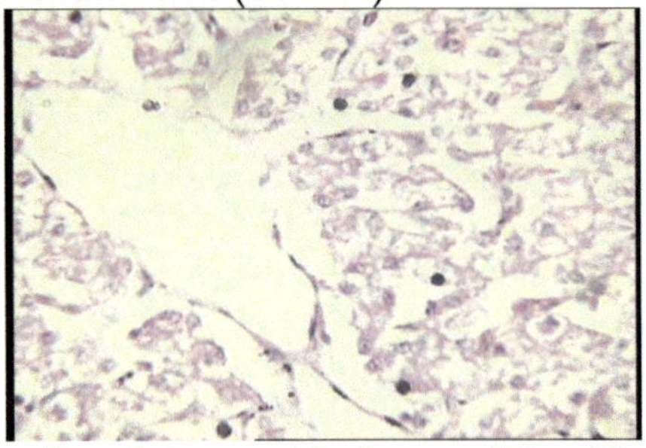

Renal tissue with abnormalities (Test)

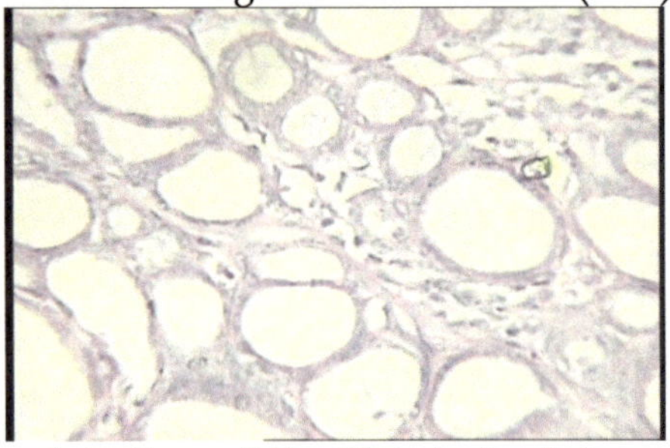

Intestine mildly dilated glandular structure (Test)

Plate 2. Histopathological analysis of various organs of hens with aflotoxicosis

2.12 Reproduction and hatchability

Aflatoxins causes delayed maturation of both males and females (Doerr, 1979; Doerr and Ottinger, 1980). Aflatoxicosis in white leghorn males resulted in decreased feed consumption, body weight, testes weight and semen volume (Sharlin *et al.*, 1980), and decreased plasma testosterone values (Sharlin *et al.*, 1980). While in broiler breeder males reduction in body weight and mild anemia with no alterations in semen characteristics were observed (Wyatt, 1991; Briggs *et al.*, 1974).

In mature laying hens experiencing aflatoxicosis, enlarged and fatty liver and marked decrease in egg production were observed (Hamilton and Garlich, 1972). Severe decline in hatchability was recorded in mature broiler breeder hens after consumption of aflatoxin (Howarth and Wyatt, 1976). Hatchability declines before egg production and is the most sensitive parameter of aflatoxicosis in broiler breeder hens (Howarth and Wyatt, 1976).

The immediate and severe decline in hatchability was found to arise from an increase in early embryonic mortality rather than infertility of the hens. The cause of the increased embryonic mortality is the transfer of toxic metabolites from the diet of the hen to the egg (Wyatt, 1991). The delayed response in egg production is thought to occur due to reducing synthesis and transport of yolk precursors in the liver (Huff *et al.*, 1975).

2.13 Immunosuppression

Aflatoxin induces immunosuppression and increases susceptibility of toxicated birds to bacterial, viral and parasitic infections. Immunosuppression caused by AFB1 has been demonstrated in chickens and turkeys as well as laboratory animals (Sharma, 1993).

Aflatoxin decreases the concentrations of immunoglobulins IgM, IgG and IgA in birds (Giambrone *et al.*, 1978a & b). The presence of low levels of AFB1 in the feed appears to decrease vaccinal immunity and may therefore lead to the occurrence of disease even in properly vaccinated flocks (Leeson *et al.*, 1995).

Thaxton *et al.*, (1974) recorded reduced antibody production following injection of sheep red blood cells in chickens experiencing aflatoxicosis. Batra *et al.*, (1991) found that chickens fed AFB1 and vaccinated against Marek's disease showed a significantly higher frequency of gross and microscopical lesions of Marek's disease than did chickens fed aflatoxin-free diet.

Cell-mediated immune response and effector cell function are also affected during aflatoxicosis (Leeson *et al.*,1995). Aflatoxin decrease complement activity in chickens (Campbell *et al.*, 1983 and Stewart *et al.*, 1985) and turkeys (Corrier, 1991). Since complement is required for normal phagocytosis, impairment in complement activity may partially explain impairment of phagocytosis in chickens experiencing aflatoxicosis (Gewurz and Suyehira, 1976; Wyatt, 1991).

Chang and Hamilton (1979a) demonstrated reduced chemotactic ability of leucocytes, impaired phagocytosis of heterophils and impaired cellular and serum factors required for optimal phayocytosis in aflatoxicated chickens. Although thrombocytic counts are depressed by dietary aflatoxin (Mohiuddin *et al.*, 1986) their phagocytic activity is not affected by aflatoxin (Chang and Hamilton, 1979b). However, other phagocytic cells (heterophils, macrophages and monocytes) were affected by dietary aflatoxin (Chang and Hamilton, 1979a).

Chickens receiving aflatoxin-contaminated diets showed higher susceptibility to Marek's disease (Edds and Bortell, 1983), infectious bursal disease virus (Giambrone *et al.*, 1978a & b), congenitally acquired salmonellosis (Wyatt and Hamilton, 1975) and duodenal and cecal coccidiosis (Edds *et al.*, 1973) than chickens receiving aflatoxin free diet.

From the aforementioned, it is postulated that aflatoxin interferes with normal function of B and T lymphocytes, rather than causing destruction of these cells (Wyatt, 1991). The impairment of protein synthesis caused by dietary aflatoxin could account for the lack of humoral immunity without the necessity of B and T cell destruction (Wyatt, 1991). Regardless the atrophy of the bursa of fabricius and thymus gland, the apparent alteration of splenic function is also of diagnostic significance and implies alteration in the immunocompetence of birds with aflatoxicosis (Richard *et al.*, 1975).

2.14 Hematological and biochemical alterations

Aflatoxin causes hematopoietic suppression and anemia observed as decreases in total erythrocytes, packed-cell volume and hemoglobin (Reddy *et al.*,1984; Huff *et al.*, 1986; Mohiuddin *et al.*, 1986). Total leucocytes are increased and differential leucocytic counts vary among studies with concurrent lymphopenia (Tung *et al.*, 1975a; Lanza *et al.*, 1980), monocytoses and heterophilia (Wannop, 1961).

Aflatoxin is known to produce hemolytic anemia by decreasing the circulating mature erythrocytes. Lysis of erythrocytes will result in above the normal levels of cellular debris in circulation (Tung *et al.*, 1975a) and consequently the spleen appear congested because of an unusually high concentration of inorganic iron and debris from the circulation (Wyatt, 1991).

Several biochemical parameters are affected by aflatoxin exposure. Aflatoxin decreases total serum proteins, alpha, beta and gamma globulins, with IgG being more sensitive than IgM (Tung *et al.*, 1975a). Total serum proteins contents are depressed due to reduced values of alpha and beta globulins and albumen, while gamma globulins are affected more variably (Pier, 1973).

Serum lipoproteins, cholesterol, triglycerides, uric acid and calcium are also decreased (Garlich *et al.*, 1973; Doerr *et al.*, 1983; Reddy *et al.*,1984; Huff *et al.*, 1986). The activity of serum or plasma enzymes has been extensively used as a measure of aflatoxin activity in chickens. Increased activities of sorbitol dehydrogenase, glutamic dehydrogenase, lactate dehydrogenase, alkaline phosphatase, acid phosphatase, aspartate aminotransferase and alanine aminotransferase were reported in aflatoxicated chickens (Dafalla *et al.*, 1987b; Rao and Joshi, 1993; Leeson *et al.*, 1995). The increase in the levels of serum enzymes measured was interpreted as a consequence of hepatocyte degeneration and subsequent leakage of enzymes (Leeson *et al.*, 1995).

Aflatoxin has also shown to alter both the extrinsic and common clotting pathways in chickens. Aflatoxins causes biochemical changes in thromboplastin clotting factors V, VII and X and reduces plasma prothrombin and fibrinogen (Doerr *et al.*, 1976), and consequently increases whole blood clotting and prothrombin times (Doerr *et al.*, 1974). The elevated prothrombin time was considered to be the result of impaired hepatic synthesis of clotting factors caused by the toxication of aflatoxin on the liver cells (Huff *et al.*, 1983).

The activity of some digestive enzymes, the absorption of carotenoid compounds from the gastrointestinal tract, and the metabolism of lipids can be altered by aflatoxin exposure (Leeson *et al.*, 1995). Dietary aflatoxin produced a malabsorption syndrome characterized by steatorrhea, hypocarotenoidemia and decreased concentrations of bile salts and pancreatic lipase, trypsin, amylase and Rnase (Osbrone *et al.*,1982). In another experiment, the specific activities of pancreatic chymotrypsin, amylase and lipase, but not trypsin were increased significantly by aflatoxin (Richardson and Hamilton, 1987).

The effect of aflatoxin on the renal function of broiler chickens was reported by Glahn (1993). Aflatoxin treated birds showed decreased fractional excretion of phosphate, total plasma calcium concentration, decreased total plasma proteins, plasma 25-hydroxyl vitamin D and plasma 1, 25-dihydroxy vitamin D.

2.15 Wild life

Birds, fishes and mammals vary among species in susceptibility to aflatoxins. Birds such as bobwhite quail and wild turkey appear to be more susceptible than mammals (Horn *et al.*, 1989). It is difficult to document the extent to which wildlife species are affected because wild animals are free roaming and elusive. In many cases, predators and scavengers may consume dead or dying animals before the dead animals are found by humans (Stewart and Larson, 2002).

Clinical signs of aflatoxicosis in wildlife vary according to the dose received, the time period of exposure, and species of animal. Toxic effects can be divided into acute, subacute and chronic exposures (Stewart and Larson, 2002). Acute effects reflect severe liver disease. Animals may be anemic and may exhibit difficulty in breathing. Sudden death with no clinical signs may occur. Subacute effects may allow animals to live for a longer period of time. These animals have yellow eyes, mucous membranes, or yellowed skin along with abnormalities in blood clotting. Bruising, nose bleeds and hemorrhaging may be observed. Chronic effects are generally related to impaired liver function. Long-term, low-level consumption of aflatoxins may result in reduced feed efficiency, weight loss, lack of as appetite and increased receptivity to secondary infectious diseases. Lesions may occur in the liver and other organs and fluid may accumulate in the body cavity.

2.15.1 Fish

Fish have been found susceptible to aflatoxin and trichothecenes. Aflatoxicosis is most prevalent among fishes. The extent of lesions caused by consumption of aflatoxins depends upon the age and species of the fish. Fry are more susceptible to aflatoxicosis than adults and some species of fish are more sensitive to aflatoxins than others (Royes and Yanong, 2002). Rainbow trout are the most sensitive species to aflatoxin. Feeding trout diets containing less than 1 ppb will cause liver tumors in 20 months. (Horn *et al.*, 1989). Diet containing AFB1 at 0.4 ppb for 15 months had a 14% chance of developing tumors. Feeding trout a diet containing 20 ppb for 8 months resulted in 58% occurrence of liver tumors and continued feeding for 12 months resulted in 83% incidence of tumors (Royes and Yanong, 2002).

Deaths quickly occur in 50% of stock if dietary levels of 500 to 1000 ppb are consumed. Warm water fishes such as channel catfish (*Ictalurs punctatus*) are much less sensitive than rainbow trout, and the level required to cause 50% mortality is approximately 30 times that of rainbow trout (Horn *et al.*, 1989). Channel catfish fed a diet containing purified AFB1 at 10.000 ppb for 10 weeks exhibited decreased growth rate and moderate internal lesions (Royes and Yanong, 2002).

Initial findings associated with aflatoxicosis in fishes include pale gills, impaired blood clotting, poor growth rates or lack of weight gain. Prolonged feeding of low concentrations of AFB1 causes liver tumors, which appear as pale yellow lesions and which can spread to the kidney. Increased in mortality may be observed (Royes and Yanong, 2002).

Aflatoxin can also depress the immune system indirect through their effect on enextial nutrients in the diet, making fish more susceptible to bacterial, viral and parasitic diseases. Moreover, aflatoxin can cause slow growth rate and reduced weight of the finished product of warm-water fish (Royes and Yanong, 2002).

3. Conclusion

This chapter describes the food sources of aflatoxin contamination and their diseases in human and animals such as cattles, poultry, fish and other wild animals. Quality of food and feed plays the most important role in the farming as its share is 70%. Good quality food and resistant strain of animals can lead to greater production and more profit for the poultry, dairy, fishery former. However, the acute shortage of chicken, mutton, fish meat has pushed its prices steeply upwards. It is suggested that use of chicks, cow, sheep, fish, dog, horses are resistant to aflatoxicosis, would help in minimizing problem of poor growth rate and poor feed conversion which perhaps are the two most important factors in animal management.

4. References

Abdelsalam, E. B.; Eltayeb, A.F.; Noreidin, A.A. & Abdulmagid, A. M.(1989). Aflatoxicosis in fattening sheep. *Vet. Rec*, Vol. 124, pp. 487-488.

Agag, B.I. (2004). Mycotoxins in foods and feeds 1-Aflatoxins. *Ass Univ. Bull. Environ. Res*, Vol.7, No.1, pp.173-205

Allcroft, R.; Roberts, B. A. & Lloyd, M. K. (1968). Excretion of aflatoxin in a lactating cow. *Food Cosmet. Toxicol*, Vol.6, pp.619- 625

Andrellos, P. J. & Reid, G.B (1964). Confirmatory tests for aflatoxin B. *J. Assoc. Offic. Ag, Chemists* Vol.47, pp.801-803.

Angsubhakorn, S.; Poomvises, P. Romruen, K. & Newbern, P.M. (1981). Aflatoxicosis in horses. *Am, Vet. Med. Asso,*. Vol.178, pp.274-278.

Applebaum, R.S.; Brackett, R.E., Wiseman, D.W., & Marth, E.H. (1982). Aflatoxin toxicity to dairy cattle and occurrence in milk and milk products. *J. Food. Pro,* Vol.45, No.8, pp.752-777.

Archibald, R.; Smith, H.J. & Smith, J.D. (1962). Brazilian groundnut toxicosis in Canadian broiler chickens. *Can. Vet. J*, Vol.3. pp.322-325.

Armbrecht, B.H.; Shalkop, W.T. Rollins, L.D. Pohland, A.E. & Stoloff, L. (1970). Acute toxicity of AFB1 in weathers. *Nature*, Lond. Vol.225, pp.1062-1063

Asao, T. G.; Buchi, M. M, Abdel kader, S. B, Chang, E. L, Wick, & Wogan G.N. (1963). Aflatoxins B and G. *J. Am. Chem. So,* Vol.85, pp.1706-1707

Asplin, F.D.& Carnaghan, R.B. (1961). The toxicity of certain groundnut meals for poultry with special reference to their effect on ducklings and chickens. *Vet. Res*, Vol.3, pp. 1215-1219.

Asuzu, I.U. & Shetty, S.N. (1986). Acute aflatoxicosis in broiler chicken in Nsuka, Nigeria. *Trop. Vet*, Vol.4, pp.79-80

Bababunmi, E. A.& Bassir, O. (1969). The effect of aflatoxin on blood clotting in the rat. *Brit. J. Pharmacol*, Vol.37, pp.497-500

Bassir, O. & Osiyemi, F. (1967). Biliary excretion of aflatoxin in the rat after a single dose. *Nature*, Vol.215, pp.882

Bastianello, S.S.; Nesbit, J.W. Willians, M.C. & Lange, A.L. (1987). Pathological findings in a natural outbreak of aflatoxicosis in dogs. Ondersteport *J. Vet. Res*, Vol.54, pp.635-640.

Batra, P.; Pruthi, A.K. & Sadana, J.R. (1991). Effect of AFB1 of the efficacy of turkey herpes virus vaccine against Marek's disease. *Res. Vet. Sc*, Vol.51, pp.115-119

Blaney, B.K. & Williams, K.C. (1991). Effective use in livestock feeds of mouldy and weather damaged grain containing mycotoxins. Case histories and economic assessments pertaining to pig and poultry industries of Queensland. *Aust. J. Agric. Res*,Vol.42, pp.993-1012

Bodine, A.B. & Mrtens, D.R. (1983). Toxicology, metabolism and physiological effects of aflatoxin in the bovine. pp.46-50. In: U.L. Diener; R.L. Asquith and J.W. Dickens Eds. Aflatoxin and *A. flavus* in corn. *AAES, Auburn Univ., Alabama.*

Bortell, R.; Asquith, R.L. Edds, G.T. Simpson, C.F. Aller, W.W. (1983). Acute experimentally induced aflatoxicosis in the weanling pony. *American Journal of Veterinary Research*, Vol.44, pp.2110–2114

Brackett, R.E. & Marth, E.H. (1982). Association of AFM1 with casein.*Z. Lebensm Unters Forsch*, Vol.174, pp.439

Briggs, D.M.; Wyatt, R.D.& Hamilton, P.B. (1974).The effect of dietary aflatoxin on semen characteristics of mature broiler breeder males. *Poult. Sci*, Vol.53, pp.2115-2119

Brown, J.M.M. & Abrams, L. (1965). Biochemical studies on aflatoxicosis. Onderstepoort, *J. Vet. Res.*, Vol. 32, pp. 119-146.

Bryden, W.L.; Lioyd, A.B. & Cumming, R.B. (1980). Aflatoxin contamination of Australian animal feeds and suspected cases of mycotoxicosis. *Aust. Vet. J*, Vol.56, pp.176-180

Buchi, G.; Spitzner, D. Paglialunga, S. & Wogan, G. N. (1973). Synthesis and toxicity evaluation of aflatoxin P1. *Life Sci*, Vol .3, pp.1143-1149

Burg, W.R.& Shotwell, O.L. (1984). Aflatoxin levels in airborne dust generated from contaminated corn during harvest and at an elevator in 1980. *Journal of the Association of Official Analytical Chemists* Vol.67, pp.309–312

Burnside, J.E.; Sipple, W.L. Forgacs, J, Carll, W.T, Atwood, M.B. & Doll, E.R. (1957). A disease of swine and cattle caused by eating moldy corn. II. Experimental production with pure cultures of molds. *Am. J. Vet. Res*, Vol.18, pp.817

Calnek, B.C.; Barnes, H.J. MCDougald, L.R. & Saif, Y.M. (1997). *Diseases of poultry*. 10th ed., pp.951-979. Mosby-Wolfe, Iowa state Univ. press, Ames, Iowa, USA.

Caloni, F. & Cortinovis.C (2011). Toxicological effects of aflatoxins in horses. *The veterinary journal*. Vol.188, pp. 270-273

Campbell, T.C. & Hayes, J.R. (1976). *Toxicol. Appl. Pharmacol*, Vol.35, pp.199-222. Cited by Hsieh (1983).

Carnaghan, R. B. A.; Lewis, G. Patterson, D. S. P. & Allcroft R. (1966). Biochemical and pathological aspects of groundnut poisoning in chickens. *Pathol. Vet*.3, pp.601-615

Cassel, E.K.; Barao, S.M. and Carmal, D.K. (1988). Aflatoxicosis and ruminants. *Texas Vet. Med. Diagnostic lab, Tesas college*. The national dairy database (1992) NDB, Health, Test, Hf100200. TxT

CAST, Council for Agric. Sci. and Technol. (1989). Mycotoxins economic and health risks. Task force report No.116

Chaffee, V.W.; Edds, G.T. Himes, J.A. & Neal, F.C. (1969). Aflatoxicosis in dogs. *Am. J. Vet. Re.,*Vol.30, pp.1737-1749

Chang, C.F. & Hamilton, P.B. (1979a). Impaired phagocytosis by heterophils from chickens during aflatoxicosis. *Toxicol. Appl. Pharmacol*, vol.6, pp.459

Chang, C.F. & Hamilton, P.B. (1979b). Refractory phagocytosis by chicken thrombocytes during aflatoxicosis. *Poult. Sci*, Vol.58, pp.559-561

Colvin, B.M.; Harrison, L.R. Grosser, H.S. & Hall, R.F. (1984). Aflatoxicosis in feeder cattle. *J. Am. Vet. Med. Assoc.*Vol.184. pp.956.

Cook, W.O.; Richard, J.L. Osweiller, G.D. & Trampel, D.W. (1986). Clinical and pathological changes in acute bovine aflatoxicosis. Rumen motility and tissue and fluid concentrations of aflatoxins B1 and M1. *Am. J. Vet. Res*, Vol.47, pp.1817-1825

Cook, W.R. (1976). Chronic bronchitis and alveolar emphysema in the horse. *Veterinary* Record,Vol.99, pp.448–451

Corrier, D.E. (1991). Mycotoxicosis: Mechanism of immunosuppression. *Vet. Immunol. Immuno Pathol*, Vol.30, pp.73-87

Cysewski, S.J.; Pier, A.C. Baetz, A.L.& Cheville, N.F. (1982). Experimental equine aflatoxicosis. *Toxicology and Applied Pharmacology*, Vol.65, pp. 354–365

Dafalla, R.; Hassan, Y.M. & Adam, S.E.I. (1987a). Fatty and hemorrhagic liver and kidney syndrome in breeding hens caused by AFB1 and heat stress in the Sudan. *Vet. Hum. Toxicol*, Vol.29, pp.222-226

Dafalla, R.; Yagi, A.I. & Adam, S.E.I. (1987b). Experimental aflatoxicosis in Hybro-type chicks: Sequential changes in growth and serum constituents and histopathological changes. *Vet. Human Toxicol*,Vol.29, pp.222-225

Dalezios, J. I.; Wogan, G. N. & Weinhreb, S. M. (1971). Aflatoxin P1: A new aflatoxin metabolite in monkeys. *Science* ,Vol.171, pp.584-585

Dalvi, R.R. & McGowan, C. (1984). Experimental induction of chronic aflatoxicosis in chickens by purified aflatoxins B1 and its reversal by activated charcoal, Phenobarbital and reduced glutathione. *Poult. Sci*, Vol. 63, PP. 485-491.

Dalvi, R.R. & Ademoyero,A.A. (1984). Toxic effects of aflatoxins B1 in chickens given feed contaminated with aspergillus flavus and reduction of the toxicity by activated charcoal and some chemical agents. *Avia. dis*, Vol. 28, PP. 61-69.

De Iongh, H.; Beerthuls, R. K. Vles, R. 0. Barrett, C. B. & Ord, W. 0. (1962). Investigation of the factor in groundnut meal responsible for Turkey "x" disease. *Biochim. Biophys. Acta*, Vol.65, pp.548-551

DeIongh, H., Vles, R. O. & VanPelt, J. G. (1964). Milk of mammals fed an aflatoxin containing diet. *Nature*, Vol. 202, pp.466-467

Devero, A. (1999). Aflatoxins: The effects on human and animal health. Biol, 4900; Fall, 1999

Dhanasekaran, D.; Panneerselvam, A. & Thajuddin,N. (2009). Evaluation of Aflatoxicosis in hens fed with commercial poultry feed. *Turk. J. Vet. Anim. Sci*, Vol.33, No.5. pp.385-391

Diekman, A.; Coffey, M.T. Purkhiser, E.D. Reeves, D.E. & Young, L.G. (1992). Mycotoxins and swine performance. CES, PTH-129, Purdue Univ., West Lafayette, Indian.

Diener, U.L.& Davis, N.D. (1968): "Effect of environment on aflatoxin production in peanuts. *Tropical science*, Vol.10, pp.22-25

Doerr, J.A. & Ottinger, M.A. (1980). Delayed reproductive development resulting from aflatoxicosis in Juvenile Japanese quail. *Poult. Sci*, Vol.59, pp.1995-2001

Doerr, J.A. & Ottinger, M.A. (1980). Delayed reproductive development resulting from aflatoxicosis in Juvenile Japanese quail. *Poult. Sci.*, Vol.59, 1995- 2001.

Doerr, J.A. (1979). *Mycotoxicosis and avian hematosis*. Diss. Abstr. B Sci. Eng., 4127.

Doerr, J.A.; Huff, W.E. Wabeck, C.J. Chaloupka, G.W. May, J.D. & Merkley, J.W. (1983). Effects of low level chronic aflatoxicosis in broiler chickens. *Poult. Sci*,Vol.62, pp.1971-1977

Doerr, J.A.; Huff, W.E. Wyatt, R.D. & Hamilton, P.B. (1974). Survey of T-2 toxin, ochratoxin and aflatoxin for their effects on the coagulation of blood in young broiler chickens. *Poult. Sci*, Vol.53, pp.1729

Doerr, J.A.; Wyatt, R.D. & Hamilton, P.B. (1976). Impairment of coagulation function during aflatoxicosis in young chickens. *Toxicol. Appl. Phormacol*, Vol.35, pp.437

Donaldson, W, E.; Tung, H.T. & Hamilton, P.B. (1072). Depression of fatty acid synthesis in chick liver (*Gallus domesticus*) by aflatoxins. *Comp. Biochem. Physiol*, Vol.41, pp. 843-847.

Eaton, D.L. and Groopman, J.D. (1994). The toxicology of aflatoxins. Human health, veterinary and agricultural significance. pp. 6-8 Academic press, San Diego, Ca.

Edds, G.T. & Bortell, R.A. (1983). Biological effects of aflatoxins-poultry. pp. 56-61. In:L U.L. Diener, R.L. Asquity and J.W. Dickens Eds. Aflatoxin and *A. flavus* in corn AAES, Auburn Univ, Alabama

Edds, G.T.; Nair, N.P. & Simpson, C.F. (1973). Effect of AFB1 on resistance in poultry against cecal coccidiosis and Marek's disease. *Am. J. Vet.* Res, Vol.34, pp.819-826

FDA, Food and Drug Administration (1979). Conference on mycotoxins in animal feeds and grains related to animal health. Rockville, Maryland

Fehr, P.M. & Delage, J. (1970). Effect de l'aflatoxine sur les fermentations du rumen. *Can. Nutr. Diet*, Vol.5, pp59-61

Fouzia Begum. & Samajpati.N(2000). Mycotoxins production on rice, pulses and oilseeds. *Naturwissenschaften*, Vol.87, pp.275-277

Garlich, J.D.; Tung, H.T. & Hamilton, P.B. (1973). The effects of short term feeding of aflatoxin on egg production and some plasma constituents of laying hens. *Poult. Sci*, Vol.52, pp.2206.

Gewurz, H. & Suyehira, L.A. (1976). *Manual of clinical immunology*. Rose, N.R. and Friedman, H.R. Eds., American Society of Microbiologists, Washongton, D.C., 36.

Giambrone, J.J.; Diener, U.I. Davis, N.D. Panangala, V.S. & Hoerr, F.J. (1985). Effects of aflatoxin on young turkeys and broiler chickens. *Poult. Sci*, Vol.64, pp.1678-1684

Giambrone, J.J.; Ewert, D.L. Wyatt, R.D. & Eidson, C.S. (1978a). Effect of aflatoxin on the humoral and cell-mediated immune systems of chicken. *Am. J. Vet. Res*,Vol.39, pp. 305

Giambrone, J.J.; Partadiredja, M. Eidson, C.S. Kleven, S.H. and Wyatt, R.D. (1978b). Interaction of aflatoxin with infectious bursal disease virus infection in young chickens. *Avian Dis*,Vol.22, pp.431

Gillespie, J.R.& Tyler, W.S. (1969). Chronic alveolar emphysema in the horse. *Advances in Veterinary Science and Comparative Medicine*, Vol.13, pp.59–99

Glahn, R.P. (1993). Mycotoxins and the avian kidney: Assessment of physiological function.*wld. Poult. Sci. J*, Vol.49, pp.242-250

Greene, H.J.& Oehme, F.W. (1976). A possible case of equine aflatoxicosis. *Clinical Toxicology*, Vol.9, pp.251–254

Groopman, J.D.; Donahue, P.R. Zhu, J. Chen, J. & Wogan, G.N. (1985). Aflatoxin metabolism in humans. Detection of metabolites and nucleic acid adducts in urine by affinity chromatography. *Proc. Natl. Acad. Sci.*, USA, Vol.82, pp.6492-6497

Guthrie, L.D. (1979). Effect of aflatoxin in corn on production and reproduction in dairy cattle. *J. Dairy Sci*, Vol.62, pp.134

Halliwell, R.E.W.; McGorum, B.C. Irving, P. & Dixon, P.M. (1993). Local and systemic antibody production in horses affected with chronic obstructive pulmonary disease. *Veterinary Immunology and Immunopatholog*, Vol.38, pp. 201–215

Hamilton, P. B.& Harris, J. R. (1971). Interaction of aflatoxicosis with *Candida albicans*: infections and other stresses in chickens. *Poult. Sci*, Vol. 50, pp.906-912

Hamilton, P.B. & Garlich, J.D. (1972). Failure of vitamin supplementation to alter the fatty liver syndrome by aflatoxin. *Poult. Sci*, Vol.51, pp.688

Hamilton, P.B. (1971). A natural and extremely severe occurrence of aflatoxicosis in laying hens. *Poult. Sci*, Vol.50, pp.1880-1882

Hartley, R. D.; Nesbitt, B. F. & O'Kelly, J. (1963). Toxic metabolites of *Aspergillus flavus*. *Nature*, Vol.198, pp.1056-1058

Hatch, R.C.; Clark, J.D. Jain, A.V. & Mahaffey, E.A. (1971). Experimentally induced acute aflatoxicosis in goats treated with ethyl maleate glutathione precursors or thiosulfate. *Am. J. Vet. Res*, Vol.40, pp.405-411

Hegazy, S.M.; Azzam, A. & Gabal, M.A. (1991). Interaction of naturally occurring aflatoxins in poultry feed and immunization against fowl cholera. *Poult. Sci*, Vol.70, pp. 2425-2428

Hesseltine, C.W. (1983). Introduction, definition and history of mycotoxins of importance to animal production. In: Interactions of mycotoxins in animal production. National Academy of Science, Washington, DC.

Hetzel, D.J.S.; Hoffmann, D. Van De Ven, J. & Soeripto, S. (1984). Mortality rate and liver histopathology in four breeds of ducks following long term exposure to low levels of aflatoxins. *Sing. Vet. J*, Vol.8, pp6-14

Holzapfel, C. W.; Steyn, P. S. & Purchase, I. F. H. (1966). Isolation and structure of aflatoxins M I and M 2. *Tetrahedron Lett*,Vol. 25, pp.2799-2803

Horn, C.W.; Boleman, L.L. Coffman, C,G. Deton, J.H. & Lawhorn, D.B. (1989). Mycotoxins in feed and food producing crops college state Texas. *Texas Vet. Med. Diagnostic*, The National Dairy Database (1992)

Horr, F.J. & D'Andrea, G.H. (1983). Biological effects of aflatoxin in swine. pp.51-55. In: U.L. Diener; R.L. Asquith and J.D. Dickens Eds. Aflatoxin and *A. flavus* in corn AAES, Aubura Univ. Alabama

Howarth, B.Jr. & Wyatt, R.D. (1976). Effect of dietary aflatoxin on fertility, hatchability and progeny performance of broiler breeder hens. *Appl. Environ. Microbiol*, Vol.31, pp.680-684

Huff, W.E. & Doerr, J.A. (1980). Synergism between aflatoxin and OA in broiler chickens. *Poult. Sci*, Vol.60, pp.55

Huff, W.E.; Doerr, J.A. Wabeck, C.J. Chaloupka, O.W. May, J.D. & Merkley, J.W. (1983). Individual and combined effects of aflatoxin and ochratoxin A on bruising in broiler chickens. *Poult. Sci*, Vol.62, pp.1764-1771

Huff, W.E.; Kubena, R.B. Harvey, R.H. Corrier, D.E. & Mollenhaur, H.H. (1986). Progression of aflatoxicosis in broiler chickens. *Poult. Sci*, Vol.65, pp.1891-1899

Huff, W.E.; Wyatt, R.D. & Hamilton, P.B. (1975). Effects of dietary aflatoxin on certain egg yolk parameters. *Poult. Sci*, Vol.54, pp.2014-2018

Humphreys, D.J. (1988). *Veterinary toxicology.*3rd ed. pp. 286, Bailiere Trindell, London

Hutjens, M. (1983). *Aflatoxin contaminated feed and dairy*. In: J. Jacobsen, Aflatoxin 83. Texas Vet. Med. Diagnostic lab., Texas collage.

Jayabarathi, P. & Mohamudha Parveen, R. (2010). Biochemical and histopathological analysis of aflatoxicosis in Growing hens fed with commercial poultry feed. *International Journal of Pharmaceutical Sciences Review and Research* , Vol.3, No.2, pp. 127-130

Jones, F.T.; Beth, M. Genter, M.M. Hagler, W.M. Hansen, J.A. Mowrewy, B.A. Poore, M.H. & Whitlow, L.W. (1994). Understanding and coping with effects of mycotoxins in livestock feed and forage. Electronic publication No. DRO-29, NCCES, North Carolina State Univ., Raleigh, North Carolina.

Jones, F.T.; Hagler, W.H. & Hamilton, P.B. (1982). Association of low levels of aflatoxin in feeds with productivity losses in commercial broiler operations. *Poult. Sci*, Vol.6, pp. 861-868

Jones, N.R. & Jones, B.D. (1978). *Aflatoxins*. Voeding, Vol.30, pp.330

Kato, R., Takaoka, A. Onoda, K. & Omori, Y. (1970). Different effect of aflatoxin on the induction of tryptophan oxygenase and of microsomal hydroxylase system. *J. Biochem*, (Tokyo). Vol.68, pp.589-592

Keeler, R.F. & Tu, A.T. (1983). *Handbook of natural toxins*. Vol. I. *Plant and fungal toxin*, pp.308, New York, Marcal Dekker, Ine

Kiemeier, F. & Buchner, M. (1977). Zur verteilung van aflatoxin M1 auf Molke and Bruch bei der Kasehersteilung. *Z. Lebensm. Unters. Forsch*, Vol.164, pp.87

Krishnamachari, K.A.; Bhat, R.V. Nagarajan, V. & Tilak, T.B. (1975a). Investigations into outbreak of hepatitis in parts of Western India. *Indian J. Med. Res*, Vol.63, pp.1036-1049

Krishnamachari, K.A.; Nagarajan, V.; Bhat, R.V. & Tilak, T.B. (1975b). Hepatitis due to aflatoxicosis. An outbreak in Western India. Lencet i. pp.1061-1062

Lamont, M.H. (1979). Cases of suspected mycotoxicosis as reported by veterinary investigation centers. *Proc. Mycotoxins Anim. Dis*, Vol.3, pp.38-39

Lanza, G.M.; Washburn, K.W. & Wyatt, R.D. (1980). Strain variation in hematological response of broilers to dietary aflatoxin. *Poult. Sci*, Vol.59, pp.2686-2691

Larsson, P., Persson, E., Tydén, E. & Tjälve, H. (2003). Cell-specific activation of aflatoxin B1 correlates with presence of some cytochrome P450 enzymes in olfactory and respiratory tissues in horse. *Research in Veterinary Science*, Vol.74, pp.227–233

Lawlor, P.G. & Lynch, P.B. (2001). Mycotoxins in pig feeds. 2: clinical aspects. *Irish Vet. J*, Vol.54, No.4, pp.172-176

Leeson, S.; Diaz, G.J. & Summers, J.D. (1995). *Poultry metabolic disorders and mycotixns*. pp. 249-298, University Books, Guelph, Ontario, Canada

Liggett, A.D.; Colvin, B.M. Beaver, R.W. & Wilson, D.M. (1986). Canine aflatoxicosis: A continuing problem. *Vet. Hum. Toxicol*, Vol.28, pp.428-430

Lillehoj, E.B. (1983). Effect of environmental and cultural factors of aflatoxin contamination of developing corn kernels. In: V.L. Diener, R.L. Asquith and J.W. Dickens (Eds). Aflatoxin and *A. flavus* in corn. Southern Coop Serv. Bull. 279, Craftmaster, Opelika, Ala, 112P.

Lynch, G.P. (1972). Mycotoxins in feedstuffs and their effect on dairy cattle. *J. Dairy Sci*, Vol.55, pp.1243-1255

Masri, M. S.; Booth, A. N. and Hsieh, D. P. H. (1974). Comparative metabolic conversion of aflas B1 to aflatoxin M1 and aflatoxin Q by monkey, rat, and chicken. *Life Sci*, Vol.15, pp.203-212

Masri, M.S.; Catcia, V.C. & Page, J.R. (1969). AFM content of milk from cows fed known amounts of aflatoxin. *Vet. Rec*, Vol.84, pp.146-147

Meerdink, G.L. (2002). Mycotoxins. *Clinical Techniques in Equine Practice 1*. pp.89–93

Mintzlaff, H. J., Lotzsch, R. Tauchmann, F. Meyer, W. & Leistner, L. (1974). Aflatoxin residues in the liver of broiler chicken given aflatoxin-containing feed. *Fleischwirtschaft*, Vol.54, pp.774-778

Mohiuddin, S.M.; Reddy, M.V. Reddy, M.M. & Ramakrishna, K. (1986). Studies on phagocytic activity and haematological changes in aflatoxicosis in poultry. *Indian Vet. J*, Vol.63, pp.442-445

Morchouse, L.G. (1981). Mycotoxins-their toxicology and principle lesions. *Proc. U.S. Anim. Health Assoc.* pp.232.

Muller, R.D.; Carlson, C.W. Semeniuk, G. & Harshfield, G.S. (1970). The response of chicks, duckling, goslings, pheasants and poults to graded levels of aflatoxins. *Poult. Sci*, Vol.49, pp.1346-1350

Nabney, J.; Burbage, M. B. Allcroft, R. and Lewis, G. (1967). Metabolism of aflatoxin in sheep. Excretion pattern in the lactating ewe. *Food Cosmet. Toxicol.*Vol.5. pp.11-17

Nesbitt, B.; O'Kelly, K.; Sargeant, K. & Sheridan, A. (1962). Toxic metabolites of *Aspergillus flavus*. *Nature*, Vol.195, pp.1062-1063

Newberne, P.M.; Russo, R. and Wogan, G.N. (1966). Acute toxicity of AFB1 in the dog. *Pathol. Vet.* Vol.3. pp.331-340.

Nibbelink, S.K. (1986). Aflatoxicosis in food animals. *A clinical review*. Iowa State Univ.Vet. Vol.48, pp. 28-31.

Nesheim, M.C. & Ivy, C.A. (1971). Effect of aflatoxin on egg production and liver fat in laying hens. Cornell Nutr. Conf. Buffalo, pp. 126-129.

Okoye, J.O.A.; Asuzu, I.U. & Gugnani, J.C. (1988). Paralysis and lameness associated with aflatoxicosis in broilers. *Avian Pathol*, Vol.17, pp.731-734

Osborne, D.J.; Huff, W.E. Hamilton, P.B. & Burtmeister, H.R. (1982). Comparison of ochratoxin, aflatoxin and T-2 toxin for their effects on selected parameters related to digestion and evidence for specific metabolism of carotenoids in chickens. *Poult. Sci*, Vol.61, pp.1646-1652.

Osweiller, G.D. (1992). Mycotoxins. In: *Diseases of swine.*7th ed. pp.735-743. Edited by A.D. Leman, B.E. Straw, W.L. Mengeling, S.D.; Allaire & D.J. Taylor Wolfe publishing, London

Patterson, D.S.P. & Roberts, B.A. (1970). The formation of aflatoxin $B2_a$ and $G2_a$ and their degradation products during the *in vitro* detoxification of aflatoxin by livers of certain avian and mammalian species. *Food cosmet. Toxicol*, Vol. 8, pp. 527-538.

Patterson, D.S.P. & Allcroft, R. (1970). Metabolism of aflatoxins in susceptible and resistant animal species. *Fd. Cosmet. Toxicol*, Vol.8, pp.43

Patterson, D.S.; Galaney, E.M. & Roberts, B.A.(1978). The estimation of AFM1 in milk using 2-dimensional TLC. *Fd.Cosmet toxicol*, Vol.16, pp.49-50

Pier, A.C. (1973). Effects of aflatoxin on immunity. *J. Amer. Vet. Med. Assoc*, Vol.163, pp.1268-1269

Pier, A.C.; Cysewski, S.J. Richard, J.L. Baetz, A.L. & Mitchell, L. (1976). *Proc. U.S., Anim. Hlth. Assoc*, Vol.80, pp.130. In: Mackenzie *et al.* (1981)

Price, R.L.; Paulson, J.H. Lough, O.E. Gingg, C. & Kurtz, A.G. (1985). Aflatoxin conversion by dairy cattle consuming naturally contaminated whole cottonseed. *J. Fd. Prot*, Vol.48, pp.11-15

Radostits, O.M.; Gay, C.C. Blood, D.C. & Hinchcliff, K.W. (2000). *Veterinary medicine*, pp.1684-1688, W.B. Saunders Co. Ltd., London.

Raisbeck, M.F.; Rottinghaus, G.E. & Kendall, J.D. (1991). Effects of naturally occurring mycotoxins on ruminants. pp. 647-677.

Rao, V.N. & Joshi, H.C. (1993). Effect of certain drugs on acute induced aflatoxicosis in chicken (4 mg AFB1/ kg b.wt.). *Ind. Vet. J*, Vol 70, pp.344-347

Ray, A.C.; Abbitt, B. Cotter, S.R. Murphy, M.J.; Reagor, J.C. Robinson, R.M. West, J.E. and Whitford, H.W. (1986). Bovine abortion and death associated with consumption of aflatoxin contaminated peanuts. *J. Am. Vet. Med. Assoc*, Vol.184, pp.956

Reagor, J.C. (1996). *Implications of mycotoxins in horses.* WEVR, 96, Cybersteed, Ine.

Reddy, D.N.; Rao, P.V. Reddy, V.R. and Yadgiri, B. (1984). Effect of selected levels of dietary aflatoxin on the performance of broiler chickens. *Indian, J. Anim. Sci*,Vol.54, pp.68-73

Richard, J.L.; Pier, A.C. Stubblefiels, R.D. Shotwell, O.L. Lyon, R.L. and Cutlip, R.C. (1983). Effect of feeding cornnaturally contaminated with aflatoxin on feed efficiency, on physiology, immunologic and pathologic changes and in tissue residues in steers. *Am. J. Vet. Re*, Vol.44, pp.1294

Richard, L.J.; Stubblefield, R.D. Lyon, R.L. Peden, W.M. Thurston, J.R. & Rimler, R.B. (1987). Distribution and clearance of aflatoxin B1 and M1 in turkeys fed diets containing 50 or 150 ppb aflatoxin from naturally contaminated corn. *Avian Dis*, Vol.30, pp.788-793

Richard, T.L.; Thurston, J.R. & Pier, A.C. (1975). *Mycotoxin-induced alterations in immunity.* In: Microbiology, Schlessinger, D. Ed., American Society of Microbiologist, Washington, D.C.

Richardson, K.E. & Hamilton, P.B. (1987). Enhanced production of pancreatic digestive enzymes during aflatoxicosis in egg-type chickens. *Poult. Sci*, Vol.66, pp.1470-1478

Rodricks, J.V. & Stoloff, L. (1977). Mycotoxins in human and animal health. Pathotox publishers, park forest, south, 1L, pp.67-69

Royes, J.B. & Yanong, R.P. (2002). Molds in fish feeds and aflatoxicosis. Copyright by the University of Florida, *Institute of Agric. Sci.* (UF/ IFAS)

Rumbeiha, W. (2001). *Mycotoxicosis in pets.* DCPAH-Newsletter.Vol.18, No. 2.

Samarajeewa, U.; Arseculeratne, S.N. & Tennekoon, G.E. (1975). Spontaneous and experimental aflatoxicosis in goats. *Res. Vet. Sci*, Vol.19, pp.269-277

Sargeant, K.; Carraghan, R.B. & Allcroft, R. (1963).Toxic products in groundnuts. Chemistry and origin. *Chem. And Ind*, pp.53-55

Sargeant, K.; Sheridan, J.; O'Kelly, J. & Carnaghan, R. B. A. (1961). Toxicity associated with certain samples of groundnuts. *Nature*, Vol.192, pp.1096-1097

Sawhney, D.S.; Vadehra, D.V. & Baker, R.C. (1973a). Aflatoxicosis in the laying Japanese quail. *Poult. Sci*, Vol.52, pp.485-493

Sawhney, D.S.; Vadehra, D.V. & Baker, R.C. (1973b). The metabolism of ^{14}C aflatoxins in laying hens. *Poult. Sci*, Vol.52, pp.1302-1309

Schabort, J. C. & Steyn, M. (1969). Substrate and Phenobarbital inducible aflatoxin-4-hydroxylation and aflatoxin metabolism by rat liver microsomes. *Biochem. Pharmacol*, Vol.1, pp.2241-2252

Shank, R. C. & Wogan, G. N. (1966). Acute effects of aflatoxin B1 on liver composition and metabolism in the rat and duckling. *Toxicol. Appl. Pharmacol*, Vol.9, pp.468-476.

Shank, R.C. (1977). *Environmental cancer.* pp.291-318. In: H.F. Kraybill and M.A. Mehlman (eds.), John Wiley sons Ltd New York.

Shankaran, R.; Raj H. G. & Venkatasubramanian T. A. (1970). Effect of aflatoxin on carbohydrate metabolism in chick liver. *Enzymlogia*, Vol.39, pp.371-378

Sharlin, J.S.; Howarth, B.Jr. Thompson, E.N. and Wyatt, R.D. (1981). Decreased reproductive potential and reduced feed consumption in nature white leghorn males fed aflatoxin. *Poult. Sci*, Vol.60, pp.2701

Sharma, R.P. (1993). Immunotoxicity of mycotoxins. *J. Dairy Sci*, Vol.76, pp.892-897

Siller, W.G. & Ostler, D.C. (1961). The histopathology of an entero-hepatic syndrome of turkey poults. *Vet. Rec*, Vol.73, pp.134-138

Sisk, D.B.; Carlton, W.W. and Curtin, T.M. (1968). Experimental aflatoxicosis in young swine. *Am. J. Vet. Res*, Vol.29, pp.207-215

Slowik, J.; Graczyk, S. and Madej, J.A. (1985). The effect of single dose of AFB1 on the value of nuclear index of blood lymphocytes and on histopathological changes in the liver, bursa of fabricious, suprarenal glands and spleen in ducklings. *Folia Histochem. Cytobiol* Vol.3. pp.71-80.

Smith, B.P. (2002). *Large animal internal medicine.* 3rd ed. pp.1627-1637. Mosby, Inc. USA.

Smith, J. W.; Prince, W. R. & Hamilton, P. B. (1969). Relationship of aflatoxicosis to *Salmonella gallirarem* infections of chickens. *Appl. Micobiol*, Vol.18, pp.946-947

Smith, J.E. & Hamilton, P.B. (1970). Aflatoxicosis in the broiler chicken. *Poult. Sci*, Vol.49, pp.207

Smith, R. H. (1965). The inhibition of amino acid activation in liver and *E. coli* preparations by aflatoxin in vivo. *Biochem. J*, Vol.95, pp.438-448

Sorensson, W.G.; Simpson, J.P. Peach, M.J. Thedell, T.D. & Olenchock, S.A. (1981). Aflatoxin in respirable dust particles. *Journal of Toxicology and Environmental Health*, Vol.7, pp.669–672

Stewart, D. & Larson, E. (2002). *Aflatoxicosis in wildlife*. Information Sheet 1582 Mississippi State Univ. Extension Service., Cooperating with U.S. Dept. of Agriculture.

Stewart, R.G.; Skeeles, J.K. Wyatt, R.D. Brown, J. Page, R.K. Russell, I.D. & Lukert, P.D. (1985). The effect of aflatoxin on complement activity in broiler chickens. *Poult. Sci*, Vol.64, pp.616-619

Stoloff, L. & Trucksess, M.W. (1979). Distribution of aflatoxins B1 and M1 in contaminated calf and pig livers. *Journal of the Association of Official Analytical Chemists*, Vol.62, pp1361–1362

Stoloff, L. (1980). Aflatoxin M1 in perspective. *J. Food Prot*, Vol.43, pp.226

Thaxton, J.P.; Tung, H.T. & Hamilton, P.B. (1974). Immunosuppression in chickens by aflatoxin. *Poult. Sci*, Vol.53, pp.721

Tung, H.T.; Cook, F.W. Wyatt, R.D. & Hamilton, P.B. (1975a). The anemia caused by aflatoxin. *Poult. Sci*, Vol.54, pp.1962-1969

Tung, H.T.; Wyatt, R.D. Thaxton, P. & Hamilton, P.B. (1973). Impairment of kidney function during aflatoxicosis. *Poult. Sci*, Vol.52, pp.873

Tung, H.T.; Wyatt, R.D. Thaxon, P. and Hamilton, P.B. (1975b). Concentrations of serum proteins during aflatoxicosis. *Toxicol. Appl. Pharmacol*, Vol.34, pp.320-326

Tydèn, E.; Olsèn, L. Tallkvist, J. Tjälve, H.& Larsson, P. (2008). Cytochrome P450 3A, NADPH cytochrome P450 reductase and cytochrome b in the upper airways in horse. *Research in Veterinary Science*, Vol.85, pp.80–85

Van der Linde, L.A.; Van der Frens, A.M. & Van Esch, G.J. (1965). *Experiments with cows fed groundnut meal containing aflatoxin*. pp. 247. In: G.N. Wogan, ed. *Mycotoxins in foodstuffs*. MIT Press, Cambridge, mass.

Van Der Zijden, A. S. M.; Koelensmid, W. A. A. Boldingh, J. Barrett, C. B. Ord, W. 0. W. 0. & Philip, J. (1962). Isolation in crystalline form of a toxin responsible for turkey X disease. *Nature*, Vol.195, pp.1060-1062

Van dorp, D. A.; Van Der Zijden, A. S. M. Beerthuis, R. K. Sparreboom, S. Ord, K. De Jong, & Keuning, R. (1963). Dihydroaflatoxin B, a metabolite of *Aspergillus flavus* Remarks on the structure of aflatoxin B. *Rec. Trav. Chim*, Vol.82, pp.587-592

Van dorp, D. A.; Van der zijden, A. S. M.; Beerthuis, R. K.; Sparreboom, S.; Ord, W. 0.; De jong, K.& Keuning, R..(1963) Dihydroaflatoxin B, a metabolite of *Aspergillus flavus*. Remarks on the structure of aflatoxin B. *Rec. Trav. Chim*. Vol. 82, pp.587-592

Veltmann, J. R. (1984). Reducing effects of mycotoxins through nutrition. Poult. Digest, pp. 190-194.

Vesonder, R.; Haliburton, J. Stubblefield, R. Gilmore, W. & Peterson, S. (1991). *Aspergillus flavus* and aflatoxin B1, B2 and M1 in corn associated with equine death. *Archives of Environmental Contamination and Toxicology*, Vol.20, pp.151–153

Voight, M. N.; Wyatt, R.D.; Ayers, J.C. and Koehler, P. (1980). Abnormal concentrations of B vitamin and amino acids in plasma, bile and liver of chicks with aflatoxicosis. Appl. Environ. Microbiol, Vol. 40, pp. 870-875

Wannop, C.C. (1961). The histopathology of turkey "x" disease in great Britain. *Avian Dis*, Vol.5, pp. 371-381

Weidenborner, M. (2001). *Encyclopedia of food mycotoxins.*"Spronger Publisher Berlin, New York, London.

WHO, World Health Organization (1979). Environmental Health Criteria, Safety evaluation of certain food additives. pp. 1-127

Wilson, H.R.; Douglas, C.R. Harms, R.H. & Edds, G.T. (1975). Reduction of aflatoxin effects on quail. *Poult. Sci*, Vol.54, pp.923-925

Wogan, N. (1966). Chemical nature and biological effects of the Aflatoxins. *Bacteriological Review*, Vol.30, No.2, pp.460-470

Wyatt, R.D. & Hamilton, P.B. (1975). Interaction between aflatoxicosis and a natural infection of chickens with *Salmonella. Appl. Microbiol*, Vol.30, pp.870-872

Wyatt, R.D. (1991). *Poultry*, pp.553-578, In: Smith, J.E. and Henderson, R.S. (1991).

Wylie, T.D. & Morehouse, L.G. (1978). *Mycotoxic fungi, mycotoxins and mycotoxicosis.* Vol.2, pp.414-427. Marcel Dekker, INC, USA

13

Aflatoxins and Their Impact on Human and Animal Health: An Emerging Problem

Eva G. Lizárraga-Paulín[1],
Ernesto Moreno-Martínez[2] and Susana P. Miranda-Castro[1]
[1]*Laboratorio de Biotecnología, Facultad de Estudios Superiores Cuautitlán, UNAM. Cuautitlán Izcalli, Estado de México,*
[2]*Unidad de Investigación en Granos y Semillas, Facultad de Estudios Superiores Cuautitlán, CAT, UNAM. Cuautitlán Izcalli, Estado de México, México*

1. Introduction

Aspergillus is a fungus that essentially belongs to grains storage flora. It grows optimally at 25 °C with a minimum necessary water activity of 0.75. It starts to produce secondary metabolites at 10-12 °C, but the most toxic ones are produced at 25°C with a water activity of 0.95 (Hesseltine 1976). Those toxic secondary metabolites named aflatoxins (AF) is a group of mycotoxins produced by a large number of Aspergillus species, basically by three phylogenetically distinct sections. The main producers are *A. flavus*, and *A. parasiticus*, but it has been demonstrated that *A. nomius, A. pseudotamarii, A. parvisclerotigenus,* and *A.bombycis* of section Flavi, *A. ochraceoroseus* and *A. rambellii* from section Ochraceorosei and *Emericella astellata* and *E. venezuelensis* from Nidulatans section also generate aflatoxins (IARC 2002; Frisvad et. al., 2005). All of them contaminate a large fraction of the world's food, including maize, rice, sorghum, barley, rye, wheat, peanut, groundnut, soya, cottonseed, and other derivative products made from these primary feedstuffs in low-income countries (Rizzi et al., 2003; Saleemullah et al., 2006; Strosnider et. al., 2006; Masoero et. al., 2007; Caloni, 2010).
Although aflatoxins have been a problem throughout history, until 1960 they have been recognized as significant contaminants within agriculture, because in this year they were initially isolated and identified as the causative toxins in "Turkey-X-disease" after 100,000 turkeys died in England from an acute necrosis of the liver and hyperplasia of the bile duct after consuming groundnuts infected with Aspergillus flavus (Asao et. al., 1965; D'Mello, 1997; Strosnider et. al., 2006).
Williams et al. estimated in 2004 that 4.5 billion of the world's population is exposed to aflatoxins because they are also everywhere. Some essential factors that affect aflatoxin contamination include the climate of the region, the genotype of the crop planted, the soil type, the minimum and maximum daily temperatures, and the daily net evaporation (Strosnider et. al., 2006). Moreover, aflatoxin contamination is also promoted by stress or damage to the crop due to drought before harvest, the insect activity, a poor timing of harvest, the heavy rains during and after harvest, and an inadequate drying of the crop before storage. Levels of humidity, temperature, and aeration during storage are also important factors that are

intimately related with the actual problems of climate changes and environmental warming around the whole world (Cotty & Jaime-García, 2007; Paterson & Lima, 2010).
There have been identified 18 types of aflatoxins, nevertheless, the naturally occurring and well-known ones are aflatoxin B1 (AFB1), aflatoxin B2 (AFB2), aflatoxin G1 (AFG1) and aflatoxin G2 (AFG2) (Gimeno, 2004; Saleemullah et. al., 2006; Strosnider et. al., 2006). These names were given due to their blue (B) or green (G) fluorescence properties under ultraviolet light and their migration patterns during chromatography (Wogan & Busby, 1980; Dikeman & Green, 1992). The International Agency for Research on Cancer (IARC, 2000) has classified aflatoxin B1 as a group 1 carcinogen (that means carcinogenic to humans) since 1987, and a group 1 carcinogenic agent since 1993 due to the exposure to hepatitis B virus (Castegnaro & McGregor, 1998). AFB1 is the most prevalent aflatoxin usually found in cases of aflatoxicosis, and is responsible for acute toxicity, chronic toxicity, carcinogenicity, teratogenicity, genotoxicity and immunotoxicity. AFM1 is a metabolic derivate of AFB1, and AFM2 is a metabolic derivate of AFB2; both come from the metabolism of some animals, and are normally found in milk and urine (Gimeno, 2004; Strosnider et. al., 2006).
The B-toxins are characterized by the fusion of a cyclopentenone ring to the lactone ring of the coumarin structure, while G- toxins contain an additional fused lactone ring. Aflatoxin B1 and to a lesser extent AFG1, are responsible for the biological potency of aflatoxin-contaminated feed. These two toxins possess an unsaturated bond at the 8,9 position on the terminal furan ring. Aflatoxin B2 and AFG2 are essentially biologically inactive unless these toxins are first metabolically oxidized to AFB1 and AFG1 in vivo (Verma, 2004). AFM1 and AFM2 are hydroxylated derivatives of AFB1 and AFB2 that may be found in milk, milk products or meat (hence the designation M1). They are formed by the metabolism of B1 and B2 in the body of the animals following absorption of contaminated feeds (Gimeno, 2004; Verma, 2004; Wild & Gong, 2010).
In animals, aflatoxins impair growth and are immunosuppressive. B aflatoxin has been reported to induce liver and kidney tumors in rodents, and there has been found a possible link to increased esophageal cancer. Aflatoxins have been recently considered as an important sanitary problem because it has been demonstrated that human exposure to mycotoxins may result from consumption of plant derived foods that are contaminated with toxins and their metabolites (which are present in animal products such as milk, meat, visceral organs and eggs) or exposure to air and dust containing toxins (Jarvis, 2002). It has been reported that aflatoxins, once ingested (because of their low molecular weight), are rapidly adsorbed in the gastro-intestinal tract through a non-described passive mechanism, and then quickly appear as metabolites in blood after just 15 minutes and in milk as soon as 12 hours post-feeding (Yiannikouris & Jouany, 2002; Moschini et. al., 2006). Aflatoxins are hepatocarcinogenic particularly in conjunction with chronic hepatitis B virus infection, and cause aflatoxicosis in episodic poisoning outbreaks. Recent studies also suggest that the B aflatoxins may cause neural tube defects in populations that consume maize as a staple food (Wild & Gong, 2010).
Due to this important global issue, some organizations and institutions have been purposing a great number of practical primary and secondary prevention strategies, especially for developing countries, in order to reduce the risks given by this public problem, but they could be beneficial if political wills and financial investments are applied to what remains a largely ignored worldwide health matter.

2. Aflatoxins in food products from contaminated grains

In many low-income countries, mycotoxins, and particularly aflatoxins, affect staple foods including cereals (maize, wheat and rice principally) and their derivates; oilseeds (cotton, peanut, rapeseed, coconut, sunflowers and others), cassava, groundnuts and other nuts, and a great variety of foods which are consumed by humans like dry fruits, delicatessen products, spices, wines, legumes, fruits, milk and milk derivates (Gimeno, 2004; Wild & Gong 2010). Maize and groundnuts are major sources of human exposure because of their greater susceptibility to contamination and frequent consumption throughout the world. Table 1 shows some of the most important commodities affected by aflatoxins producer species, according to a review made by Abdin and collaborators in 2010.

Type of aflatoxin	Producer fungal species	Affected commodities
B (B1, B2)	A. flavus, A. parasiticus, A. tamarii, A. pseudotamarii, A. bombycis, A. parvisclerotigenus, A. nomius, A. minisclerotigenes, A. oryzae, A. toxicarius, A. versicolor, A. rambellii, A. arachidicola, A. ochraceoroseus, Emericella astellata, E. venezuelensis.	Cotton seed, peanuts, peanut butter, pea, sorghum, rice, pistachio, maize, oilseed rape, maize flour, sunflower seed, figs, spices, meats, dairy products, fruit juices (apple, guava)
G (G1, G2)	A. parasiticus, A. nomius, A. bombycis, A. pseudotamarii, A. terreus, A. versicolor, A. arachidicola, A. toxicarius, A. minisclerotigenes.	Peanuts, cotton seed, sunflower seed, tree nuts, pistachio, peanut butter, maize flour, pea, cereals, corn, figs, meats, spices, dairy products, fruit juices (apple, guava)

Table 1. Major commodities affected by aflatoxins (Taken from Abdin et. al., 2010.)

Aflatoxins are most prevalent in latitudes between 40° N and 40° S of the equator, but the greatest health risk lies within developing countries in tropical regions, which rely on these commodities as their staple food source (Strosnider et. al., 2006). Even, in some processed typical food like Mexican pozol, there have been found large amounts of aflatoxins, being the AFB2 the more prevalent and abundant toxin, suggesting that AFB2 is more resistant than AFB1 to the alkaline conditions given during hard processes like nixtamalization (Kamimura, 1989; Méndez & Moreno, 2004).

In wealthy grain-producing countries of the world, economic resources exist to ensure that regulations to limit aflatoxin exposure in the food supply are implemented. Furthermore, in markets of grain commodities, the prices of corn and groundnuts are often dictated by aflatoxin content, which contributes to lower levels of exposure in wealthy countries. Thus, a result of these regulations and market forces is that people in economically developing countries are exposed to far higher levels of aflatoxins in the diet (Groopman et. al., 2008).

The presence of aflatoxins in food means a risk for both animals and human beings. This is because not only grains (generally consumed by people), but also whole plants and grasses from which they emerge, could be contaminated by mycotoxins. This is a serious threat for animals, particularly livestock, because the herbaceous food they consume (commonly known as ensilage or forage) contains a large amount of aflatoxins, particularly if field was contaminated. A potentially hazardous feed is ground high-moisture corn, unless it is treated with adequate preservatives (e.g., propionates); the moisture content promotes the growth of the toxigenic molds and grinding of the kernel destroys the natural barrier to infestation. Hay (unless it contains a large complement of cereal grain infested in the field) is rarely if ever a source of appreciable aflatoxin (however, hay and forage may be sources of other mycotoxins such as ergot alkaloids, sporodesmin, slaframine, etc.). The fungus must gain access to susceptible parts of the plant (e.g., the corn kernel, cotton seed, etc.) before it grows and elaborates aflatoxins. Seasonal peaks in aflatoxin content are seen in key years when drought-damaged plants or insect-damaged crops are rendered more susceptible to fungal invasion. Wet harvest seasons also may contribute to high levels of aflatoxin in certain crops. Aflatoxin sometimes develops in crops stored at levels of moisture content > 15% or properly dried crops stored in leaky bins (Pier, 1992).

Grains for animal feed in the United States are allowed 300 ppb aflatoxin, because this concentration not only provides protection against acute aflatoxicosis but also is low enough to allow most of the grain produced to be traded. In these animal feeding situations, the long-term risk of cancer is not a concern, except for the most susceptible species. Consequently, veterinary research has examined higher levels of exposure but for shorter time periods. This research provides most of the information on the toxicities of aflatoxin at intermediate rates of exposure (100–500 ppb) and is the most potentially relevant information that is appropriate for the human situation in developing countries where no control of aflatoxin is exercised. However, the differences between species in response to aflatoxin introduce a measure of speculation into the extension of farm animal–derived information to the human situation (Williams et. al., 2004).

3. Aflatoxins in food products from contaminated animals

Aflatoxins M1 and M2 (whose names are derived from milk aflatoxins, and then related to meat aflatoxins too), are thermo-resistant hydroxylated metabolites produced by lactating animals consuming aflatoxin contaminated feeds. The ingested AFB1 and AFB2 are metabolized by livestock into AFM1 and AFM2 respectively, with estimated conversion ratio of 1–3% between AFB1 and AFM1 (Barbieri et. al., 1994; Ali et. al., 1999; Herzallah, 2009). The accepted limits of AFB1 and total aflatoxins in foods are 5 and 10 μg/kg, respectively, in more than 75 countries around the world whilst they are 2 and 4 μg/kg in the European Union (López et. al., 2003; Van Egmond & Jonker, 2004).

The most alarming problem through time has been the presence of aflatoxin contaminated milk, because cows and goats (the major producers of drinking milk) are largely affected when eating contaminated forage all around the world (Helferich et. al., 1986; López et. al., 2003). By the way, it is important to consider that AFM1 concentrations in milk vary not only in the cow breed, but also in the concentration of AFB1 in the diet, the amount and duration of consumption of contaminated food and the animal health.

There have been found differences between the amounts of AFM´s produced by different bovine species. In a review, Gimeno (2004) reports that in dairy cows, the relationship between the concentration of AFB1 in the final consumed ration and AFM1 excreted in breast milk could be 300:1; nevertheless this relation is only an approximation because the range is from 34:1 to 1600:1. In Holstein dairy cows consuming final rations with 80, 86, 470, 557, 1493 and 1089 μg of AFB1/Kg (ppb) on dry substance, there were found in milk AFM1 concentrations of 1.5, 0.245, 13.7, 4.7, 12.4 and 20.2 mg/L (ppb) respectively. On the other hand, when diet of Brindle cows was contaminated with 540 ppb of AFB1, 0.92 ppb of AFM1 was produced. In other cows, the values of contamination in the diet ranged between 64 and 1799 ppb of AFB1 giving some residues in milk between 0.35 and 14.2 ppb of AFM1. With an intake of AFB1 for 2-60 mg / cow / day, AFM1 residues in milk could range between 1 and 50 pp.

It is known that cows can transform AFB1 into AFM1 within 12-24 hours after ingestion of contaminated food. Even at six hours after ingestion, AFM1 residues can appear in milk, and the highest levels are reached after a few days. When the intake of AFB1 is stopped, the AFM1 concentration in the milk decreases to an undetectable level after 72 hours (Gimeno, 2004; Özdemir, 2007).

Many studies have dealt with the transfer of AFB1 in milk as AFM1 when lactating animals ingested contaminated feed continuously, especially in cows. It has been suggested that an increase in AFM1 occurs due to *Staphylococcus aureus* infection and other bacterial infections related with somatic cells diseases (Veldman et. al., 1992; Masoero et. al., 2007). In contrast, little research has been conducted on the transfer of AFM1 into milk as a result of a single assumption of AFB1. From a practical standpoint, the use of highly contaminated feed by dairy farmers is unlikely; however, a single accidental feeding of contaminated feed may happen and can lead to milk AFM1 content above tolerance levels (Mazzete et. al., 2009).

As mentioned before, goats are a clue target of aflatoxins too, so they have been studied as a good model for understanding the generating toxins metabolism (Smith et. al., 1994; Mazzete et. al., 2009). Mazzette and collaborators found that AFB1 ingested by lactating goats is quickly transferred to milk as AFM1. The maximum concentration of AFM1 was reached at 3-6 hours after the single oral administration of pure AFB1. Nevertheless, it showed a negative exponential trend and the toxin was no longer detected after 72 hours from administration. Therefore, an occasional oral assumption of AFB1 can lead to a transient contamination of AFM1 in goat's milk.

Milk has derivates that are consumed principally by humans. Among them we can find cheeses, butter, yogurt, cream and butterfat. The AFM1 distribution in some dairy foods made from contaminated milk is approximately: 40-60% in cheese, 10% in butterfat and <2% in buttermilk. AFM1 is highly soluble in water, so it is incomprehensible why this toxin is deposited in the cheese but not in the milk whey (Yusef & Marth, 1989).

Aflatoxins are not only present in cow, goat and sheep milk and derivates even after pasteurizing processes, there have also been found in other food animal products like turkey and hen eggs. Residues of aflatoxins and their metabolites in foodstuff animal tissues (like beef and sheep meat) may be a source of aflatoxin contamination in human foods (Rodricks & Stoloff, 1977; Herzallah, 2009); nevertheless, milk has been the most studied food, because of its implication in human nutrition at all growing stages.

4. Major human diseases caused by aflatoxins consumption

Populations of developing countries are the most susceptible to aflatoxicosis illness. This is because security blankets in crops at pre-harvest and post-harvest level are not as strict as in developed countries. The same occurs with milk derivates, because developing countries have not accepted and assumed amenities as quick as developed countries. It has been estimated that more than 5 billion people in developing countries worldwide are at risk of chronic exposure to aflatoxins through contaminated foods (Shephard, 2003; Williams et. al., 2004). Because of being an alarming number, aflatoxins have been recently considered as an important public health issue. Adult humans usually have a high tolerance of aflatoxin, and, in the reported acute poisonings, there are usually children who die (Cullen & Newberne, 1994).

The adverse effects of aflatoxins in humans and animals have been categorized in two general forms:

a. Acute aflatoxicosis.

It is produced when moderate to high levels of aflatoxins are consumed. Specific, acute episodes of disease include hemorrhage, acute liver damage which manifests as severe hepatotoxicity with a case fatality rate of approximately 25%, edema, absorption and/or metabolism of nutrients and alteration in digestion. The early symptoms of hepatotoxicity from aflatoxicosis can include anorexia, malaise, and low-grade fever. Acute high-level exposure can progress to potentially lethal hepatitis with vomiting, abdominal pain, jaundice, fulminant hepatic failure and death (Walderhaug, 1992; Cullen & Newberne, 1994; Strosnider et. al., 2006).

b. Chronic aflatoxicosis.

It results from ingestion of low to moderate levels of aflatoxins. The effects are usually subclinical and difficult to recognize. Some of the common symptoms are impaired food conversion and slower rates of growth with or without the production of an overt aflatoxin syndrome (Walderhaug, 1992).

There have been memorable clinic cases that had woken up the interest of scientists on investigating deeply the mode of action of aflatoxins in humans and take matters into. One of the most peculiar cases occurred in 1976, with a 68 old british chemical engineer who worked for three months on a method of sterilizing Brazilian peanutmeal which was contaminated by *Aspergillus flavus*. Three months after finishing this work he became ill with high fever and began to expectorate thick, white sputum. X-ray examination showed cavitation in the left lower lobe of the lung. At first the process was considered to be due to tuberculosis, and later to mycotic disease. After two months further lesions developed in both lungs. The condition of the patient became worse and he died 11 months after the onset of his illness. Necropsy showed enlarged, heavy lungs diffusely infiltrated with firm yellow-white or reddish lesions, mitotic figures were rare, the picture was of pulmonary adenomatosis, no metastases or tumors in other organs were found and bacteriological examination was negative. A sample of lung tissue was taken for chemical investigation. Thin-layer chromatography of the extract showed a blue fluorescent spot in 365 nm UV light similar to that of a commercial sample of aflatoxin B1, the same color change as standard B1 when treated with 50% H_2SO_4, and an RF value identical to that of the commercial aflatoxin sample B1. A colleague of this patient who had been doing the same work died three years before of pulmonary adenomatosis, but no chemical investigations were done in his case. The conclusion was that his illness was caused by the inhalation of A. flavus spores that, together with another factors lead to chronic damage to the lung, determining that aflatoxicosis is an opportunistic disease. Also, there is evidence that air-borne particles of dust contaminated by aflatoxins contribute to the development of pulmonary cancer (Dvorackova, 1976).

Aflatoxicosis is not only caused by inhalation, but also, as mentioned before, is caused by aflatoxin ingestion. In places like Brazil and Abu Dhabi, there have been found lots of cases in which infants were exposed to aflatoxin M1 from mother's breast milk. Aflatoxins have also been found in infant formula (Aksit et. al., 1997; Saad et. al.,1995; Navas et. al., 2005). There are lots of earlier studies reporting the presence of aflatoxins and derivatives in human urine, blood, and human cord blood that apparently can enter the developing fetus in humans and animals (Denning et. al., 1990).

The 80's and 90's were globally fatal decades. In India, at least 400 people were affected by eating infected corn, and 104 of them died. In Kenya, 12 people were also killed by high consumption of aflatoxins (Mehan & Mc Donald, 1991). In Southeast Asia, 19 patients after eating rice and pasta became jaundiced and sick within hours; 17 of them presented symptoms of hepatitis, and in total, 14 died because of liver failure and 7 because of renal failure. In biopsies, there were found high concentrations of aflatoxin in liver, lungs and other organs (Hendrickse, 1999).

It has been well documented that chronic aflatoxin exposure causes Hepatocellular Carcinoma (HCC), generally in association with hepatitis B virus (HBV) or other risk factors. That's why the International Agency for Research on Cancer (IARC) recognized aflatoxins as carcinogenic in 1976 (Chen et. al., 2001; Henry et. al., 2002; Omer et. al., 2004; Qian et. al., 1994; Wang et. al., 1996). HCC is the sixth most prevalent cancer worldwide. Developing countries have a higher incidence rate, with approximately 82% of the 600,000 new cases each year occurring in developing countries (Parkin et. al., 2005).

Unsafe sex associated with aflatoxicosis has been identified as a risk factor largely because of the HIV epidemic. Whereas the risk is behavioral, the disease is viral, and the progress of the epidemic is determined by disease transmission, rate of disease progress, and opportunistic infections. The disease of HIV is complicated, and the ways in which the virus interacts with another immunocompromising agent is also likely to be complicated. The animal data on immune suppression and nutritional interference has shown aflatoxicosis symptoms to be similar to HIV infection symptoms, differing mainly in that the removal of aflatoxin from the diet reverses the symptoms. The animal data on immune suppression suggest that the parameters of the epidemiologic model are likely to be modulated by aflatoxin at some level of exposure, either directly or indirectly through the known toxicities of aflatoxin. Nutrition is also a general area in which aflatoxin exposure can be expected to modulate HIV (Williams et. al., 2004).

5. Major animal diseases caused by aflatoxins consumption

Effects of aflatoxin consumption are similar in all animals; however, the susceptibility varies by species, age, and individual variation. Symptoms of acute aflatoxicosis consist of depression, anorexia, weight loss, disease, gastrointestinal bleeding, pulmonary edema and liver damage. Signs of acute hepatic injury are seen as coagulopathy, increased capillary fragility, hemorrhage, and prolonged clotting times. Blood pigments may appear in the urine and mucous membranes are icteric. The liver shows gross changes caused by centralobular congestion and hemorrhage and fatty changes of surviving hepatocytes. Death of the animal may occur within hours or a few days. Symptoms of prolonged exposure to moderate to aflatoxins may be reflected in a decline in feed consumption and production (growth and production of eggs and milk). It can also affect the quality of milk and milk products, and represent a risk for the presence of AFM1 as derived from AFB1 consumed by lactating females. In chronic aflatoxin poisoning, most of the effects are still referable to hepatic injury, but on a milder scale. The most sensitive clinical sign of chronic aflatoxicosis is reduced rate of growth of young animals. Other signs include prolonged clotting time, increases in serum glutamic oxalacetic trans-aminase, ornithine carbamyl transferase, and cholic acid levels. Hepatic pathology includes a yellow to brassy color, enlarged gall bladder, dilute bile, histologic signs of fatty changes in the hepatocytes, and bile duct proliferation. Frequently the signs of chronic aflatoxins are so protean that the condition goes undiagnosed for long periods. Chronic aflatoxin poisoning, however, is the manner in which animals are most frequently affected and the economic consequences are often considerable (Pier, 1992; Denli & Pérez, 2006).

AFB1 is absorbed via the gastrointestinal tract into the portal blood system and is carried to the liver where it is metabolized. A portion of aflatoxin is activated and set in hepatic tissues. Some water-soluble conjugated metabolites of AFB1 are excreted into the bile and go to the stool. Other water-soluble conjugated metabolites, AFB1 degradation products and non conjugated metabolites are excreted into the blood circulatory system and distributed systemically. Eventually, these residues are referred to milk, eggs, muscle and edible tissues (Dennis & Hsieh, 1981). AFM1 is one of those metabolic derivatives that taint milk. Other metabolites are formed from AFB1, including aflatoxicol (18 times less toxic than AFB1) and aflatoxin B2a (not toxic). The animal organism usually produces those metabolic products as an autodetoxification system (Gimeno, 2004).

AFB1 mainly affects birds, pigs and other monogastric animals. Ruminants are less vulnerable to aflatoxin ingestion. In monogastric animals, clinical symptoms may occur after consumption of feed contaminated with concentrations above 50 ppb while the symptoms in cattle occurs at concentrations above 1.5 to 2.23 mg/kg. Depending on the presence of other concurrent factors, small amounts of AFB1 (greater than 20 ppb) can cause toxic effects. In these conditions, an aflatoxin level above 100 ppb may be also toxic in ruminants (Denli & Pérez, 2006).

Experimental animal evidence suggests that chronic exposure to aflatoxins may lead to impaired immunity and reduced uptake of nutrients from the diet too (Hall & Wild, 1994; Miller & Wilson, 1994). Furthermore, diseases caused by aflatoxins can cause subclinical losses in production, and increase the risk and incidence of other diseases (Denli & Pérez, 2006).

Below, we describe some of the most important diseases that some animal species develop when they eat aflatoxin contaminated food or, in some cases, inhale the fungal spores from the air. Those data are summarized in Table 2.

5.1 Horses

There have been reported some cases of aflatoxicosis on horses since 1976. The reported symptoms included anorexia, icterus and rapid weight loss immediately prior to death. On post mortem examination, the liver was described as being black, of firm consistency and enlarged. Histopathological examination revealed marked centrilobular hepatic necrosis and necrotic areas were engorged with erythrocytes. Kupffer cells were prominent and many contained phagocytosed haemosiderin, which was the likely cause of the black coloured liver. Bile-duct hyperplasia, congestion of renal vessels and adrenal cortex were found. Samples of the feed revealed AFB1 levels of 58.4 μg/kg, which exceeded the limit recommended by the FDA (20 μg/kg) (Greene & Oehme, 1976). Other reports mention that AFB1 content in horse feeds was also within tolerable limits (10 μg/kg), with an average AFB1 concentration of 1.98 ± 0.71 μg/ kg (Greene & Oehme, 1976; Basalan et. al., 2004; Caloni & Cortinoivis, 2010).

It is thought that a possible link between chronic obstructive pulmonary disease (COPD) and inhaled mycotoxins exist. A. fumigatus and Mycropolyspora faeni are potential causes of COPD in horses, which is characterized by asthma-like symptoms, such as chronic cough, nasal discharge, expiratory dyspnoea and reduced exercise tolerance. The olfactory and

respiratory mucosa of horses may be exposed to mycotoxins and other xenobiotics via inhalation of contaminated feed-dust particles (for a complete review see Caloni & Cortinovis, 2010).
The existing information on aflatoxicosis in horses is inconclusive, although a total dietary concentration of 500–1000 μg/kg has been shown to induce clinical changes and liver damage, depending on the duration of exposure (Meerdink, 2002). Horses suffering from aflatoxicosis exhibit non-specific clinical signs, such as inappetence, depression, fever, tremor, ataxia and cough. Necropsy findings include, as intoxication by feeding, yellow-brown liver with centrilobular necrosis, icterus, hemorrhage, tracheal exudates and brown urine (Caloni & Cortinoivis, 2010).

5.2 Chickens

Broiler-type chickens are considered to be more resistant to aflatoxin toxicity than are other poultry species (Arafa et. al., 1981). In the poultry industry, AFB1 is called "the silent murderer" because its chronic consumption at levels below 20 ppb does not induce evident clinical symptoms; however, it reduces the absorption of food and causes immunosuppression. The final result is a low productivity, because birds show a low growth and low stance. Additionally, due to induced immunosuppression, birds are much more susceptible to opportunistic infectious agents and respond poorly to vaccination programs (AgroBioTek, 2009).
Some studies conducted on the 1960's decade, showed that aflatoxins ingestions caused periportal fatty infiltrations, increase in connective tissue and hemorrhages in most of sick chickens (Newberne & Butler, 1969). Later, in 1984, Chen and collaborators made an study un which they fed some broiler chicken with aflatoxins contaminated food, and 3 hours after the withdrawal of the contaminated feed, measurable amounts of AFB1 and AFB2 were found in all of the tissues of the birds that had been fed the aflatoxin-contaminated ration. The highest levels were found in the gizzards, followed by the livers in second place, and kidney contained the third highest levels. The capacity of the liver and kidneys to concentrate aflatoxins is probably associated with their important role in the metabolism and elimination of xenobiotics. After four days on an aflatoxin-free diet, there were no detectable levels of aflatoxins in any of the tissues. This suggests that four days on an aflatoxin-free diet before slaughter is adequate to remove detectable levels of free aflatoxins and their metabolites from the tissues of chickens that had previously been fed a highly contaminated diet. These aflatoxin residues are rapidly cleared from the tissues alter removal of the contaminated food (Chen et. al., 1984).
It has been demonstrated that broiler chickens fed with a diet rich in aflatoxins record significantly lower performances, growth and survival rate than controls. That's why lowered growth rate and increase of mortality have been associated with contaminated feeding broiler diets (Oguz & Kutoglu, 2000). In 2010, Okiki and collaborators confirmed this, because they found that chickens fed with aflatoxin decreased their growth rate and showed a weight loss of up to 400g when compared with controls after having been fed for 56 days with contaminated food.

5.3 Pigs

In pigs, aflatoxin also induces a low growth rate and increases the expression of opportunistic infections to cause immunosuppression (Gimeno, 2004). Since old times, major

lesions have been identified in liver, which turns swollen, congested, and friable; liver surface shows occasional petechiae and animals surviving beyond 24 hours often had ascites and hydrothorax. The gall bladder seems edematous and the mucosa turns petechiated and ecchymotic (Newberne & Butler, 1969).
In some studies with guinea pigs, it was found that aflatoxins cause acute gastrointestinal effects too (Luzi et. al., 2002). In a study conducted in 1982, swine with a rich-aflatoxins diet presented a peracute toxicity that caused collapse and deaths within several hours while acute toxicity caused deaths within 12 hours; subacute toxicity deaths occurred after 3 weeks on a toxic ration. Anorexia and ill thrift affecting only growing animals were seen with chronic toxicity. Extensive centrilobular liver necrosis and haemorrhage occurred with peracute toxicity and in cases of acute poisoning there was hepatic centrilobular cellular infiltration, hepatocyte swelling and bile stasis. With subacute toxicity hepatocyte vacuolation together with bile stasis and bile ductule hyperplasia was seen (Ketterer et. al., 1982).
Although aflatoxicosis in pigs is a big health problem, it has not been considered as important as fumonisin toxicosis, which nowadays is the biggest swine threat in this specie (Mallman & Dilkin, 2007; D'Mello et. al., 1999; Placinta et. al., 1999; Straw et. al., 1999).

5.4 Cattle

The first case of poisoning in cattle by groundnut was reported in 1961. Calves (3-9months of age), had eaten for at least six weeks a compounded aflatoxin contaminated groundnut. Livers of animals exhibited areas of fibrosis with biliary proliferation and venocclusive disease. In other reported cases, it was found an increase in connective tissue too, and degeneration of centrilobular hepatic cells was described. Icterus, weight loss and dead were reported (For a review look for Newberne & Butler, 1969).
Milk from cattle is mainly affected because of the infection mechanism it suffers. Pathological, hematological and plasma enzymatic studies were made on milk cattle affected by chronic aflatoxicosis caused by the prolonged feeding of concentrate feed mixtures containing contaminated groundnut cake having aflatoxin B1 (110 μg/kg groundnut cake at the time of sampling), B2, G1 and G2. Clinical and necropsy observations on liver included proliferation of connective tissue along portal triads leaving small group of hepatocytes intact. Liver function tests showed liver damage in three of the four affected animals studied (Vaid et. al., 1981)
In dairy cattle, aflatoxin B1 in contaminated food consumed is metabolized and processed in approximately 5% of aflatoxin M1, which is secreted in milk. Although the transformation from B1 to aflatoxin M1 turns it about 1,000 times less toxic, M1 levels in milk are regulated to 0.5 ppb, because the milk is consumed primarily by children and is at the stage of development when immune system is more susceptible to the suppressive effects of aflatoxin. Therefore, milk with aflatoxin M1 levels above 0.5 ppb is not fit for human consumption (Gimeno, 2004).
Goats, because of being one of the major fonts of milk production as in cows, are very susceptible to present liver damages and milk contaminations.

5.5 Other species

Species described above have been the most studied ones because they are the basis of human feeding, and furthermore, they are the ones who imply the most economic gain worldwide. Nevertheless, there are other specific susceptible species to be affected by aflatoxins like turkeys (Richard et. al., 1986; Mckenzie et. al., 1998, Klein et. al., 2002) ducks

(Ostrowski-Meissner,, 1983; Cova et. al., 1990; Bintvihok, 2001), sheep, rats, mouse, frogs, dogs, cats, rabbits and monkeys to name a few (Newberne & Butler, 1969)

SPECIE	DISEASE	SYNTOMPHS	REFERENCES
Horses	When eating: Liver damage, centrilobular hepatic necrosis phagocytosed haemosiderin in Kupffer cells, bile-duct hyperplasia, congestion of renal vessels and adrenal cortex. When inhaling: Chronic obstructive pulmonary disease (COPD), yellow–brown liver with centrilobular necrosis, icterus hemorrhage, tracheal exudates and brown urine.	When eating: Anorexia, icterus, rapid weight loss and dead. When inhaling: Chronic cough, nasal discharge, expiratory dyspnoea reduced exercise tolerance, inappetence, depression, fever, tremor, ataxia, cough and dead.	Greene & Oehme, 1976; Meerdink, 2002; Basalan et al., 2004; Caloni & Cortinovis, 2010.
Chickens	Immunosuppression, liver and kidney damage, periportal fatty infiltrations, increase in connective tissue, hemorrhages, susceptibility to opportunistic infectious agents and poor response to vaccination programs.	Low productivity, low growth, low weight, low stance, but no evident clinical symptoms and death.	Newberne & Butler, 1969; Arafa et al. 1981; Chen et al., 1984; Oguz & Kutoglu, 2000; Okiki et al. 2010.
Swine	Immunosuppression, expression of opportunistic infections, liver swollen, liver congestion, hydrothorax, edematous gall bladder, petechiated and ecchymotic mucosa, extensive centrilobular liver necrosis, haemorrhage, hepatic centrilobular cellular infiltration, hepatocyte swelling, bile stasis, hepatocyte vacuolation and bile ductule hyperplasia.	Low growth rate, gastrointestinal problems, anoroexia, ill thrift and dead.	Newberne and & Butler, 1969; Ketterer et al, 1982; Luzi et al, 2002; Gimeno, 2004.
Cattle	Fibrosis with biliary proliferation in livers, venocclusive disease increase in connective tissue, degeneration of centrilobular hepatic cells, proliferation of connective tissue, generalized liver damage and immunosupression.	Icterus, rapid weight loss and dead.	Newberne and & Butler, 1969 Vaid et al. 1981) (Gimeno 2004.
Other Animals	Pulmonary edema, generalized liver damage, coagulopathy, capillary fragility, hemorrhage, prolonged clotting times, urine pigmentation, icterus and hepatic injury.	Depression, anorexia, weight loss, bleeding, decline in feed consumption and production, gastrointestinal damage and death.	Newberne & Butler, 1969; Ostrowski-Meissner, 1983; Richard et al, 1986; Cova et al. 1990; Mckenzie et al, 1998; Klein et al, 2002; Bintvihok 2001.

Table 2. Major diseases caused by aflatoxicosis in some animal species.

6. Aflatoxins metabolism and mutagenesis

Most of the research for understanding the metabolism and mutagenesis of aflatoxins inside the consumer organism have been done using different animals as models. Those investigations have let us know that aflatoxin B1 may not itself be a toxic molecule but is metabolized in the animal body in a complex network of reactions and it is the result of this metabolism which determines both the acute and chronic toxicity (Moss, 2002).

When AFB1 is ingested, once inside the body, it is absorbed by the intestine and carried to the liver. There, AFB1 is activated and metabolized by cytochromes p450 (CYP) of hepatocytes to AFB1-8,9-exo-epoxide and AFB1-8,9-endo-epoxide. CYP3A4, 3A5, 3A7 and 1A2 are the enzymes involved in aflatoxin metabolism. Aflatoxin undergoes enzymatic conversion by the microsomal mixed function oxidase (MFO) primarily present in the liver, but probably also present in the lungs, kidneys and elsewhere. The overall contribution of these enzymes to AFB1 metabolism in vivo will depend on affinity and expression; CYP3A4 appears to be the most important, with the relative contribution of CYP3A5 varying by individual. Polymorphisms identified in the CYP3A5 promoter region have been associated with different levels of aflatoxin biomarkers, suggesting that this interindividual variation could influence susceptibility to aflatoxin. Given the fact that aflatoxin is known to cross the placenta, it is also of interest that CYP3A7, a major CYP in human fetal liver, has the capacity to activate AFB1 to 8,9-epoxide (Hendrickse, 1991; Moss, 2002; Wild & Turner, 2002; Kamdem et. al., 2006).

AFB1-8,9-exo-epoxide is highly unstable when joining to the nitrogen of guanine, which binds to DNA to form the predominant 8,9-dihydro-8-(N7-guanyl)-9-hydroxy-AFB1 (AFB1–N7-Gua) adduct. AFB1–N7-Gua adduct confers the mutagenic properties of the compound. This may be the most important product from the carcinogenic point of view. The binding of the exo-epoxide to guanine reflects the geometry of intercalation between base pairs in the DNA helix; 5′ intercalation appears to facilitate adduct formation by positioning the epoxide for in-line nucleophilic reaction with the N7 guanine. The epoxide ring is positioned above the plane and in trans to the 5a and 9a protons in the endo-epoxide, hindering reaction, but in the exo-epoxide, the epoxide ring is positioned below the plane and in cis to the 5a and 9a protons, assisting reaction (Wild & Turner, 2002; Verma, 2004).

The 8,9-epoxide is not only known to react with DNA, but also to do so at the guanine residues of specific sites, one of these being the third base position of codon 249 of the p53 gene. A considerable insight into the nature of this reaction is provided by the chemical synthesis of aflatoxin B1 epoxide and its use in reactions with model oligodeoxynucleotides. Indeed, there is evidence that a dose-dependent relationship between dietary aflatoxin B1 intake and codon 249ser p53 mutations was observed in hepatocellular carcinoma from Asia, Africa and North America (Moss, 2002).

Cytochrome p450 3A4 can activate and detoxicate AFB1. Only the 8,9-exoepoxide appears to be mutagenic and others are detoxification products. The putative AFB1 epoxide is generally accepted as the active electrophilic form of AFB1 that may attack nucleophilic nitrogen, oxygen and sulphur heteroatoms in cellular constituents (Verma, 2004).

Both humans and animals possess enzymes systems, which are capable of reducing the damage to DNA and other cellular constituents caused by the 8,9-epoxide. For example, glutathione-S-transferase mediates the reaction (termed conjugation) of the 8,9-epoxide to the endogenous compound glutathione. This essentially neutralizes its toxic potential. The exo and endo-epoxides can also undergo rapid non-enzymatic hydrolysis to AFB1-8,9-dihydrodiol that in turn is subject to slow, base-catalysed ring opening to a dialdehyde phenolate ion. The dihydrodiol can react with the e-amino group of lysine in serum albumin resulting in aflatoxin–albumin adducts, used as biomarkers. A further metabolic step involves aflatoxin aldehyde reductase catalysing the nicotinamide adenine dinucleotide phosphate (NADPH) dependent reduction of the dialdehydic phenolate ion to a dialcohol. Animal species such as the mouse that are resistant to aflatoxin carcinogenesis have 3-5 times more glutathione-S-transferase activity than susceptible species such as the rat. Humans have less glutathione-S-transferase activity or 8,9-epoxide conjugation than rats or mice suggesting that humans are less capable of detoxifying this important metabolite (Guengerich, 1996; Verma, 2004; Johnson et. al., 2008).

In addition to errors in DNA transcription due to its binding to AFB1 exo-8-9- epoxide, it can be configured a similar adduct when binding to albumin or lysine; that's why this two compounds are used at clinical level to determine the consumption of AFB1. AFM1 also been detected in urine, indicating that this toxin is also capable of reacting with DNA and form adducts (Unusan, 2006). In circulation, aflatoxin binds with plasma proteins (especially albumin) to form an aflatoxin-albumin adduct. The protein adduct by binding with 8,9-epoxy aflatoxin, initially forms dihydrodiol with sequential oxidation to dialdehyde and condensation with the S-amino group of lysine. This protein adduct is a completely modified aflatoxin structure retaining only the coumarin and cyclopentenone rings of the parent compound. These adduct represent the cumulative dose of aflatoxin intake over previous weeks. The average half-life of albumin in people is about 20 days. Therefore, an accumulated dose of aflatoxin will be present in albumin long after the dietary exposure has ceased. This is a property not found for DNA adduct because the half-life of DNA adduct is about 12 hour and then rapidly excreted in urine (Verma, 2004).

In a next phase, the challenge is to stabilize and inactivate the epoxide, hydrolyzing and conjugating it with glutathione to form AFB1-Glutathione (AFB-SG) that will be excreted in urine. In this metabolic stage are also originated three major hydroxylated metabolites: AFQ1, AFP1 and AFM1, which begin to distribute systemically and can be found in milk, eggs and tissues from intoxicated animals. Another important derivative from AFB1 metabolism is aflatoxicol, which extends the presence of AFB1 in the organism; it comes from reducing AFB1, and it can be reoxidized back to AFB1 by NADHP (Arangurén & Argüelles, 2009).

In an extensive review made by Verma in 2004, he mentions that aflatoxin concentration recorded in the serum of human beings varies with the amount and duration of aflatoxin-ingested and the physiological state of the body. Unmetabolized (B1, B2, G1 and G2) and metabolized forms (aflatoxicol, M1 and M2) of aflatoxins are excreted in the urine, stool, milk and saliva. Aflatoxin excreted/secreted through saliva might be getting absorbed in gastrointestinal tract and passing again to the blood stream. This explains a sort of recycling of aflatoxin in the body. Aflatoxin (0.35-3.5 μg/ml) exposure to hepatocytes in vitro caused pronounced swelling, polymorphic condition, bleb formation and cell lysis. Aflatoxin B1 is reported to induce cytotoxicity and transformation in culture cells. The earliest effect of aflatoxin is to reduce protein biosynthesis by forming adducts with DNA, RNA and protein, to inhibit RNA synthesis and DNA dependent RNA polymerase activity and to cause degranulation of the endoplasmic reticulum. Some of this information can be well understood in Figure 1.

In summary, once the toxin has entered the liver cell, the agency causing tissue injury in particular animal species is dictated by the rate and pattern of aflatoxin metabolism. When it is metabolized slowly, the untransformed toxin activates the molecular species that cause chronic liver damage as the most probable result. When it is metabolized rapidly, metabolites are the ones involved in diseases. Acute liver damage may be caused by the intracellular formation of aflatoxin hemiacetal in many species (Patterson, 1973).

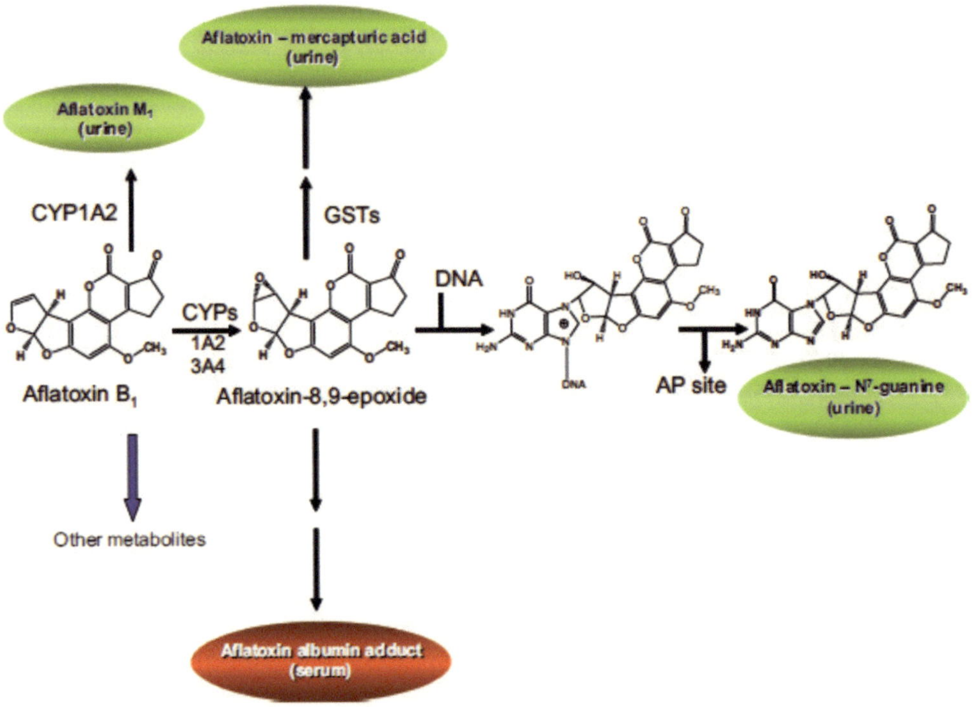

Fig. 1. Biomarkers of aflatoxin exposure in an internal dose and a biologically effective dose. Biomarkers of exposure include aflatoxin M1, the internal dose includes the aflatoxin-mercapturic acid and aflatoxin- albumin adduct, and the biologically effective dose is reflected by the excretion of the aflatoxin-N7-gua- nine adduct formed by depurination leading to an apurinic (AP) site in DNA (Taken from Groopman et. al., 2008).

7. Biomarkers and immunoassays

Since the chemical structures of the major aflatoxin-DNA and aflatoxin-protein adducts were identified, an extensive research has been conducted to validate these structures for biomarker applications. Groopman and collaborators, in their review published in 2008, mention that early experimental studies around 1980 demonstrated that the major aflatoxin-nucleic acid adduct, AFB1- N7-Gua, was excreted exclusively in the urine of exposed rats. The serum aflatoxin-albumin adduct was also examined as a biomarker of exposure because the longer half-life of albumin would be expected to integrate exposures over longer time periods, i.e., months instead of days. Studies in experimental models found that the formation of aflatoxin-DNA adducts in liver, urinary excretion of aflatoxin-nucleic acid adduct and formation of the serum albumin adduct were highly correlated events. These investigations provided the rationale for exploring the application of these biomarkers in human studies. An immunoaffinity clean-up/HPLC procedure was developed for aflatoxin metabolites in urine samples. With this approach, initial validation studies investigated the dose-dependent excretion of urinary aflatoxin biomarkers in rats after a single aflatoxin B1 (AFB1) exposure. Investigators found a linear relationship between AFB1 dose and excretion of the AFB1-N7-Gua adduct in urine over the initial 24 hours period. Subsequent studies in rodents that assessed the formation of aflatoxin macromolecular adducts after chronic administration also supported the use of DNA and protein adducts as molecular measures of exposure. For example, in rats treated with relatively low doses of AFB1 (3.5 μg) twice daily for 24 days there was an accumulation of aflatoxin binding to peripheral blood albumin followed by steady-state levels, which illustrated the potential for this biomarker (aflatoxin-albumin adduct) to integrate exposure over time. Many different analytical methods are now available for quantitation of chemical adducts in biological samples, each with unique specificity and sensitivity (Santella, 1999; Poirier, 2004; Wogan et. al., 2004; Scholl et. al., 2006).

Initial studies of aflatoxin biomarkers in human populations began in the Philippines, where investigators demonstrated that an oxidative metabolite of aflatoxin, AFM1, could be measured in urine as an internal dose marker. Subsequent works conducted in China and Gambia (areas with high incidences of HCC) determined that the levels of urinary aflatoxin biomarkers followed a dose-dependent relationship with aflatoxin intake. However, as in the earlier experimental studies, this relationship was dependent on the specific urinary marker under study; for example, AFB1-N7-Gua and AFM1 showed strong correlations with intake, whereas urinary AFP1, a different oxidative metabolite, showed no such link. In other studies, levels of aflatoxin-albumin adducts were measured and there was observed a highly significant association between intake of aflatoxin and level of adduct. This kind of studies, to measure dietary aflatoxin intake and biomarkers at the individual level, is crucial to validate a biomarker for exposure assessment and is often over-looked in molecular epidemiology. In Gambia, there was observed that urinary aflatoxin metabolites reflected day-to-day variations in aflatoxin intake, whereas the aflatoxin-albumin adducts integrated exposure over the week-long study. Data from these initial cross-sectional biomarker studies demonstrated short-term dose-response relationships for a number of the aflatoxin metabolites, including the major nucleic acid adduct, serum aflatoxin-albumin adduct, and AFM1. This supported the validity of these exposure biomarkers for use in epidemiological studies, including investigations of intervention strategies and studies of the mechanisms underlying susceptibility (Groopman et. al., 2008).

Thin layer chromatography (TLC), High Performance Liquid Chromatography (HPLC), minicolumns, immunoassays such as Enzyme-Linked Immunosorbent Assay (ELISA) and Immunoaffinity Columns (IAC) are employed in testing biological samples like blood, serum, plasma, urine, stool, breastmilk and other body exudates. Taking cost, speed of analysis, availability of personnel and facilities as well as the characteristics of the tests (sensitivity, specificity and reproducibility), TLC, HPLC, ELISA and other immunoassays have been identified as the preferred methods for aflatoxins detection. Since mycotoxins were added to the list of materials covered by international conventions relating to bio-terrorism, maintaining standards has become a major issue (WHO, 2005).

Nevertheless, information regarding the interpretation and application of AFB1 adducts and urine immunoassay is also limited. Aflatoxin metabolites or adducts in urine and serum indicate exposure, but do not necessarily equate to adverse health effects. Some studies have examined the correlation of aflatoxin intakes to biomarker levels (Reviewed by Strosinder et. al., 2006). Aflatoxin B1 adducts and urine immunoassay for epidemiologic studies, biomarkers in serum and urine provide a better estimate of aflatoxin exposure than food analysis. Aflatoxin metabolites in urine reflect recent exposure (i.e. 2-3 days) whereas the measurement of aflatoxin albumin adducts in blood reflects exposure over a longer period (i.e. 2-3 months); these analysis are labor-intensive and expensive (Groopman et. al., 1994; FAO, 2005).

More research is needed to further elucidate the correlation between aflatoxin levels in biologic specimens and adverse health effects. Research must also clarify the relationship between aflatoxin levels in biologic specimens and levels in food.

8. Permissible worldwide aflatoxin levels

Aflatoxins are considered unavoidable contaminants of food and feed, even where good manufacturing practices have been followed. The FDA has established specific guidelines on acceptable levels of aflatoxins in human food and animal feed by establishing action levels that allow for the removal of violative lots from commerce. However, it is very difficult to accurately estimate aflatoxins concentration in a large quantity of material because of the variability associated with testing procedures; hence, the true aflatoxin concentration in a lot cannot be determined with 100% certainty. Table 3 summarizes some FDA action levels for aflatoxins (Cornell University, 2009).

Aflatoxins are regulated quite differently than food additives and other chemical substances included in food. In developed countries, human populations are protected because regular surveillance keeps contaminated foods out of the food supply. Unfortunately, in countries where populations are facing starvation, or where regulations are either nonexistent or unenforced, routine ingestion of aflatoxin is very common (FAO, 1997).

However, not only the FDA in USA, but also some European countries have been establishing special committees and commissions to create and recommend guidelines, test standardized assay protocols, and maintain up-to-date information on regulatory statutes of aflatoxins and other mycotoxins. Those guidelines are developed from epidemiological data and extrapolations from animal models, taking into account the inherent uncertainties associated with both types of analysis. Estimates of "safe doses" are usually stated as a "tolerable daily intake". For example, in the United States, the Food and Drug Administration guideline is 20 ppb total aflatoxin in food destined for human consumption and 100 ppb is the limit for breeding cattle and mature poultry (FDA, 1998; Bennett et. al., 2007).

Commodity	Level (ng/g)
All products, except milk, designated for humans	20
Milk	0.5
Corn for immature animals and dairy cattle	20
Corn for breeding beef cattle, swine and mature poultry	100
Corn for finishing swine	200
Corn for finishing beef cattle	300
Cottonseed meal (as feed ingredient)	300
All feedstuff other than corn	20

Table 3. FDA action levels for aflatoxins (Taken from Cornell University,2009)

According to Tedesco et al, (2008), few countries regulate AFB1 in feedstuffs for dairy cattle. Limiting AFB1 in animal feeds is the most effective means of controlling aflatoxin M1 in milk. A limit of 5 µg AFB1/kg feed for dairy cow and a limit of 20 µg AFB1/kg in feed for cattle, sheep, goats, swine and poultry are applied in the European Union countries. This limit is applied by countries in the European Free Trade Association (EFTA), in many of the candidate EU countries and sporadically outside Europe. A limit of 20 μg AFB1/kg feed for dairy animals and a limit of 100 μg AFB1/kg intended for breeding beef cattle, breeding swine, or mature poultry is applied in the United States, Africa and Latin America. Regulations for AFM1 existed in 60 countries at the end of 2003, a more than threefold increase as compared to 1995 (FAO, 2005). EU, EFTA, candidate EU countries and some other countries in Africa, Asia and Latin America, apply a maximum level of 0.05 μg AFM1/kg in milk and a maximum level of 0.025 μg AFM1/kg in infant formula. A limit of 0.5 μg AFM1/kg in milk is applied in the United States, several Asian, European countries and in Latin America, where it is also established as a harmonized MERCOSUR (a trading block consisting of Argentina, Brazil, Paraguay and Uruguay) limit.

The Codex Committed on Food Additives and Contaminants (CCFAC) contain the result of discussions envisaged for the maximum level of AFM1 contamination. Given the public health concerns, the EU continues to maintain the maximum level of 0.05 ppb in milk AFM1 and 0.025 ppb in dairy foods for infants (CCFAC, 1999, 2000, 2001). This contradicts the regulations in America, where 0.5 ppb is an aflatoxin permissible level.

According to aflatoxins levels in human and animal health, Gimeno reviewed in 2004 that, after studies presented by the World Health Organization in 2005, it is known that the risk of liver cancer is almost null if concentrations of 0.05 ppb to 0.5 ppb are present; but exposure to any level of genotoxic carcinogens as AFM1, may pose a health risk to consumers, especially for children, so, the exposure level should be zero for a zero risk to liver cancer that may be caused by aflatoxins in general. Countries which defended an AFM1 maximum level of 0.5 ppb argue that those concentrations they could cause adverse economic consequences due to the difficulty of milk exports to countries that accept only a maximum level of 0.05 ppb. Delegates from some other countries argue that the level of 0.05 ppb is difficult to achieve in most regions of the world, so, a level of 0.5 ppb is enough to promote public health protection.

The Codex Alimentarius Comitee has reported some recommendations to institutions and consumers in general if AFB1 is detected. Some of the most important ones are:

1. In all cases, be sure that the level of aflatoxin B1 in the finished feed is suitable for its intended purpose (i.e., according to the maturity and animal species which are going to be fed) and if it conforms to codes and guidelines or qualified veterinary advices.
2. Consider the restriction of contaminated feed with aflatoxin B1 to a percentage of daily rations, so that the daily intake of AFB1 does not lead the presence of significant residues of AFM1 in milk.
3. If the feed restriction cannot be put into effect, the use of contaminated feed could be diverted to non-dairy animals.

Nevertheless, it is important to unify regulations of permissible aflatoxin levels in order to homogenize the consumption laws and amenities worldwide, in order to avoid risks and health problems derived from importing and exporting contaminated food.

9. Treatment and prevention of diseases caused by aflatoxins

As it has been mentioned before, most aflatoxicosis results from eating contaminated foods. Unfortunately, except for supportive therapy (e.g., diet and hydration) there are almost no treatments for aflatoxin exposure. However, there have been described few and specific methods for veterinary management of mycotoxicosis; for example, there is evidence that some strains of Lactobacillus effectively bind dietary mycotoxins. Similarly, clay-based enterosorbents have been used to bind aflatoxins in the gastrointestinal tract. It has been demonstrated that selenium supplementation modifies the negative effects of aflatoxin B1 in Japanese quail, while butylated hydroxytoluene gives some protection in turkeys. Oltipraz, a drug originally used to treat schistosomiasis, has been tested in human populations in China with some apparent success (Bennett et al, 2007). In Figure 2 we reproduce an overview for preventing acute aflatoxicosis in countries in development purposed in 2006.

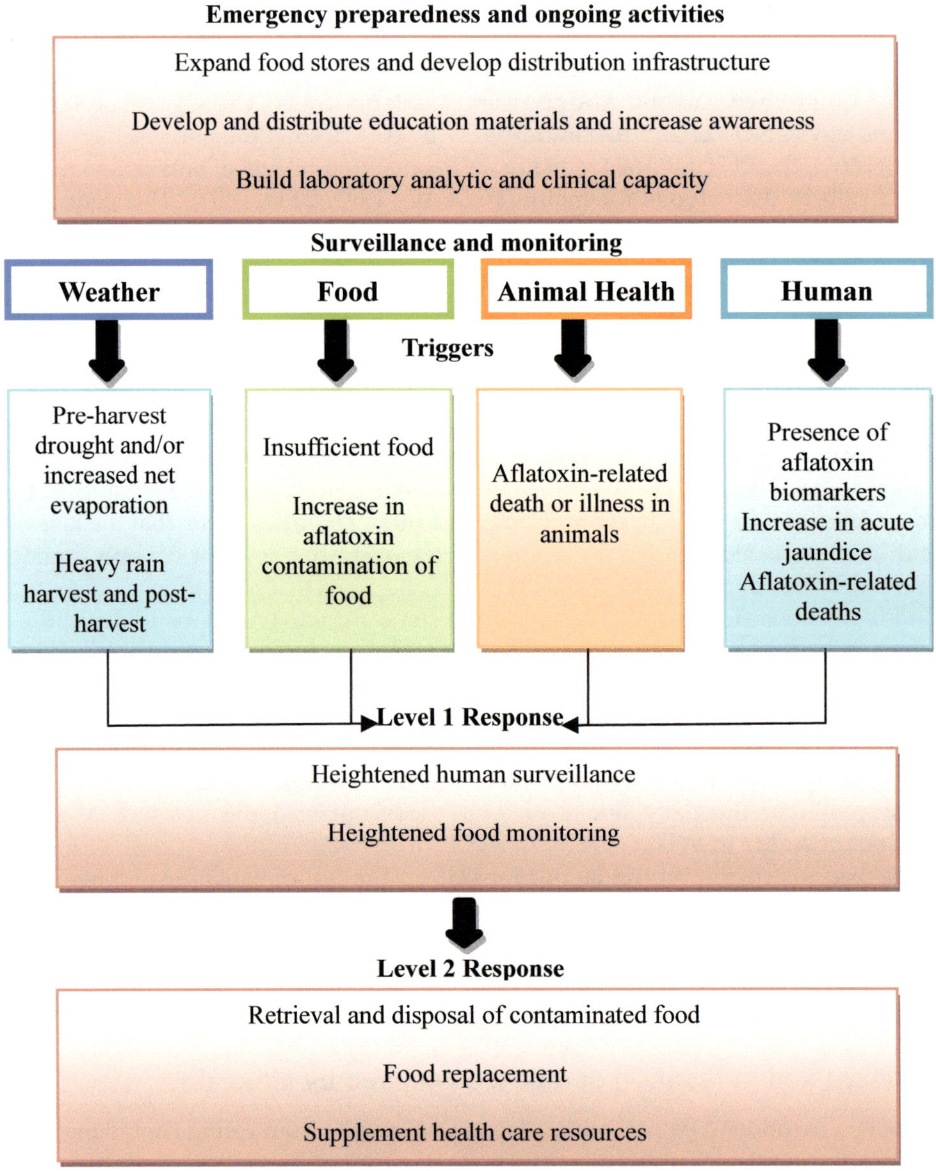

Fig. 2. Overview of preparedness, surveillance, and response activities for preventing acute aflatoxicosis in countries in development (Strosnider et. al.,2006).

Methods for controlling aflatoxin exposure are largely prophylactic. In a primary prevention trial, the goal is to reduce exposure to aflatoxins in the diet. A range of interventions includes planting pest-resistant varieties of staple crops, attempting to lower mold growth in harvested crops, improving storage methods following harvest, and using trapping agents that block the uptake of unavoidably ingested aflatoxins. In secondary prevention trials, one goal is to modulate the metabolism of ingested aflatoxin to enhance detoxification processes, thereby reducing internal dose and subsequent risk (Groopman, 2008). The aflatoxin problem sits at the interface of agriculture, health and economics, whose detailed explanation is presented below:

9.1 Agricultural sector

It consists principally in a good agricultural practice, including an appropriate drying of crops after harvest and avoidance of moisture during storage.

Pre-harvest interventions

Although the initial focus of research was on the prevention of contamination in storage, it was established in about 1970 that fungal contamination could start in the field before harvest. For peanuts, environmental conditions such as drought during the grain growth stage, insect damage in the field, variety and soil characteristics have proven to be determining factors in pre-harvest contamination. These conditions are now sufficiently well understood for computer simulation models to describe the risk of contamination of major crops (Williams et. al., 2004). Pre-harvest would be the most effective point of control because this is the point at which the crop is first infected by the toxin-producing fungus (Wild & Turner, 2002)

According to extensive reviews (Cotty & Bhatnagar, 1994; Williams et. al., 2004; Strosnider et. al., 2006; Bennett et. al., 2007; Wild & Gong, 2010), the presence and growth of Aspergillus on pre-harvested crops can be reduced through agricultural practices such as proper irrigation and pest management. Pre-harvest interventions include choosing crops with resistance to abiotic stresses (like drought, temperature and moisture content) and reducing crop stresses in general, developing host resistance through plant breeding, and choosing varieties that are genetically more resistant to fungal growth and aflatoxins production, diseases and pests. However, these processes may not be economically feasible in many high risk regions. The use of staple crops resistant to fungal colonization or genetically modified crops that inhibit fungal invasions (transgenic crops), joined to the elimination of inoculum sources (such as infected debris from the previous harvest) may prevent infection of the crop. Years before, the use of fungicides, pesticides and insecticides were a good way for controlling infections, but nowadays, the use of biocontrol agents is the most appropriated in order to avoid consumers chemical intoxications. For example, biopesticides consisting of a nonaflatoxigenic strain of Aspergillus may competitively exclude toxic strains from infecting crops, but the allergenic and human health aspects of the atoxigenic strain need still to be evaluated.

Post-harvest interventions

Post-harvest interventions can be practiced at three stages: drying level, storage level and in food preparation; nevertheless, the last mentioned is not practiced as commonly as the first ones, which are properly physicochemical methods practiced by grain producers.

Before storage, properly drying crops can prevent the development of aflatoxins. Sorting and disposing of visibly moldy or damaged kernels before storage is an effective method for reducing, but not eliminating, the development of aflatoxins (Fandohan et. al., 2005; Turner et. al., 2005). Moisture, insect and rodent control during storage can prevent damage to the crop, which would promote aflatoxin development.

Aflatoxins often accumulate during food storage and therefore post-harvest control at the subsistence farm aims to minimize fungal growth and aflatoxin production. The growth of Aspergillus is influenced most critically by temperature, moisture content and storage time. Studies conducted in Guinea, revealed a high HCC incidence and aflatoxin exposure mainly attributable to contamination of groundnuts following storage. A primary prevention study is underway where the intervention incorporates a package of post-harvest procedures, including improved sun drying prior to storage, drying on cloth rather than directly on the earth, removal of visibly mouldy nuts by hand sorting, storage in jute sacks rather than plastic, use of wooden pallets for storage to avoid contact with the earth and to improve ventilation and, finally, use of insecticides to control insect damage and spread of fungal spores. The outcomes of the study are being determined by measuring both food levels of the toxin and, more importantly, blood AF–albumin biomarker levels at three time points post-harvest. Primary intervention strategies to reduce mycotoxin exposures at the post-harvest level may have a significant impact in high exposure populations, but are unlikely to eliminate exposure. In addition, these approaches cannot be targeted specifically to high risk individuals. Therefore, intervention strategies also encompass chemoprevention, using compounds that interfere with the absorption or metabolism of aflatoxins once ingested (Reviewed by Wild & Turner, 2002). From here derives the health sector.

9.2 Health sector

It refers basically to those kinds of food we can eat and how hygienically does food is prepared.

Chemoprotection is one of the major used post-harvest techniques, and consists in the use of chemicals (e.g. oltipraz [4-methyl-5-(2- pyrazinyl)-1,2-dithiole-3-thione], chloro- phylin) or dietary intervention (e.g., eating broccoli sprouts, drinking green tea) to alter the susceptibility of humans to carcinogens, and has been considered as a strategy to reduce the risk of HCC in populations with high exposures to aflatoxins (Strosnider, 2006). The dietary intervention is maybe the easiest way to prevent cancer disease; however, for many communities in developing countries a change in diet is simply not feasible because they do not have the culture of eating a balanced diet, joined to a great skepticism about eating organic food, and moreover, that money isn´t enough to buy non-staple food.

Finally, is important to consider that simple food preparation methods such as sorting, washing, crushing, and grain dehulling, may reduce aflatoxin levels (Fandohan et. al., 2005; Park, 2002). In the case of maize, the fight against the fungal species has focused mainly through processes such as nixtamalization in which product aflatoxins are eliminated (Méndez & Moreno, 2009), or by the addition of low concentrations of Sodium Hydroxide which achieves the elimination of a large amount of aflatoxins (Carrillo 2003). Aflatoxin may be prevented by packing the dried products in polyethylene or propylene bags (Siriacha, et. al., 1990).

Most efforts to address the mycotoxin problem involve analytic detection, government regulation, and diversion of mycotoxin-contaminated commodities from the food supply. Basic research on the biosynthesis and molecular biology of aflatoxins has been a priority because a full understanding of the fundamental biological processes may yield new control strategies for the abolition of aflatoxin contamination of food crops.

9.3 Economical sector

This is maybe the post complicated treatment and preventive sector, because it includes the government security blankets to face this global problem. Because of the global threat that aflatoxicosis represents, the World Health Organization has started to respond and highlight the need for action (Strosnider et. al., 2006). However, aflatoxins and mycotoxins in general have not been widely prioritized from a public health perspective in low-income countries. This is because knowledge of mycotoxins and the full range and scale of their adverse health effects is incomplete and the known risks are poorly communicated to governments in regions where the contamination is greatest (Wild & Gong, 2010). Matters that have to be considered by government to avoid diseases from aflatoxicosis are: an opportune and non-expensive analytic detection, unifying worldwide government regulations, deviation of aflatoxin-contaminated commodities from the food supply, improving research on the biosynthesis and molecular biology of aflatoxins, and designing new control strategies for the abolition of aflatoxin contamination of food crops, inter alia.

10. Conclusions

Aflatoxins are not only a big problem at crop production level, but also it has become a global health issue because of the consequences that the consumption of this toxin generates in animals and human beings. Diverse worldwide established groups have the challenge of identifying public health strategies, which complement the agricultural ones in order to reduce aflatoxin exposure, especially in developing countries. Although there have been documented extensive researches about how to prevent and control aflatoxicosis, populations of developing countries know just a little about aflatoxin exposure and the resulting health effects.

It is known that acute aflatoxicosis is preventable, and chronic exposure can be reduced, even without a complete understanding of the public health problem caused by aflatoxins. Efforts to reduce aflatoxin exposure require the commitment of sufficient resources and the collaboration between the agriculture and public health communities as well as local, regional, national, and international governments.

Because of the recent investigations conducted in this area, it is important to take actions to prevent damage and diseases; that´s why, at first, governments supported by scientific research groups should report publicly the risks that aflatoxins consumption means by quantifying the human health impacts and the burden of disease due to the toxin exposure; then, they should compile inventory and worldwide statistics in order to evaluate the efficacy of the current intervention strategies. It is also important to increase disease surveillance, food monitoring, laboratory detection of mycotoxins and public health response capacity of affected regions. Public health services should offer immediate attention to aflatoxicosis diagnoses and opportunistic diseases caused by them in order to reduce mortality rates in humans and animals. Finally, it is important to develop response protocols to be used in an event of an outbreak of acute aflatoxicosis, which could become in an epidemic stage.

11. Acknowledgements

To PAPIIT program No. IT220411-3 "Estudio fitopatológico, bioquímico y molecular de la respuesta contra estreses bióticos y abióticos en plántulas de maíz", for financial support.

To CONACYT, Clave: 106133, "Mecanismos genéticos involucrados en la herencia y variación de la interacción patógeno y hospedero en la contaminación del maíz con aflatoxinas en el almacén y en el campo", for financial support.

12. References

Abdin MZ, Ahmad MM & Javed S. (2010) Advances in molecular detection of Aspergillus: an update. Archives of microbiology, 192(6):409-25.

AgroBioTek International. (2009). Las Micotoxinas en la Agroindustria - Aflatoxinas. AgroBioTek Dominicana. http://www.agrobiotek.com

Aksit S, Caglayan S, Yaprak I & Kansoy S. (1997) Aflatoxin: Is it neglected threat for formula-fed infants? Acta Pediátrica Japónica, 39:34-36

Ali N, Hashim NH & Yoshizawa T. (1999) Evaluation and application of a simple and rapid method for the analysis of aflatoxins in commercial foods from Malaysia and the Philipines. Food Additives and Contaminants, 16:273–280.

Arafa AS, Bloomer RJ, Wilson HR, Simpson CF & Harms RH. (1981) Susceptibility of various poultry species to dietary aflatoxin. British Poultry Science, 22:431-436.

Aranguréh EM, Argüelles MJ. (2009) Detección de aflatoxina M1 en quesos frescos comercializados en el municipio de Yopal, Casnare, mediante la técnica de Elisa. Thesis, Pontificia Universidad Javeriana, Bogotá, Colombia, 25pp.

Asao T, Buchi G, Abdel-Kader MM, Chang SB, Wick EL & Wogan GN. (1965) Aflatoxins B and G. Journal of American Chemical Society, 87:882-886

Barbieri G, Bergamini C, Ori E & Pesca P. (1994). Aflatoxin M1 in parmesan cheese: HPLC determination. Journal of Food Science, 59:1313–1331.

Basalan M, Hsmogullar SE, Hsmogullar AA & Flaz A. (2004) Fungi and aflatoxin B1 in horse and dog feeds in Western Turkey. Revue de Médecine Vétérinaire 156:248–252.

Bennett JW, Kale S & Yu J. (2007) Aflatoxins: Background, Toxicology, and Molecular Biology. In: Simjee S. Infectious Disease: Foodborne Diseases. Humana Press Inc, Totowa NJ, pp 355-373.

Bintvihok BA. (2001) Controlling aflatoxin danger to ducks and duck meat. Control, 17(11):1-2.

Caloni F & Cortinovis C. (2010) Toxicological effects of aflatoxins in horses. The Veterinary Journal, doi:10.1016/ j.tvjl.2010.06.002.

Carrillo, L. (2003) Microbiología agrícola. En: Carrillo, L. Los hongos de los alimentos y forrajes. Universidad Nacional de Salta. Cap. 6, p.1-7.

Castegnaro M & McGregor D. (1998) Carcinogenic risk assessment of mycotoxins. Revue de Médecine Vétérinaire, 149:671- 678.

CCFAC (Codex Committe on Food Additives and Contaminants) (1999). Micotoxinas presentes en Alimentos y Piensos. Observaciones sobre el proyecto de nivel máximo para la aflatoxina M1 en la leche (Tema 16 del programa). http://www.fao.org/docrep/meeting/005/x7137s/x7137s0n.htm

CCFAC (Codex Committe on Food Additives and Contaminants) (2000). Micotoxinas en los Alimentos y los Piensos. Observaciones sobre el proyecto de nivel máximo para la aflatoxina M1 en la leche (Tema 15 del programa). http://www.fao.org/docrep/meeting/005/y0474s/y0474s0m.htm

CCFAC (Codex Committe on Food Additives and Contaminants). (2001). La Comisión Europea; Seguridad Alimentaria. Observaciones de la Comunidad Europea para la

Comision del Codex Alimentarius, 24° Reunión. 2-7 de Julio de 2001, Ginebra, Suiza. "Proyecto del nivel máximo para la aflatoxina M1 en la leche" (ALINORM 01/12 A-apendice X)". http://europa.eu.int/comm/foods/fs/ifsi/eupositions/cac/archives/cac_item10a_es.html

Chen C, Pearson AM, Coleman TH, Gray JI, Pestka JJ & Aust SD. (1984) Tissue deposition and clearance of aflatoxins from broiler chickens fed a contaminated diet. Food and Chemical Toxicology, 22(6):447-51.

Chen SY, Chen CJ, Chou SR, Hsieh LL, Wang LY, Tsai WY, Ahsan H & Santella RM. (2001) Association of aflatoxin B(1)-albumin adduct levels with hepatitis B surface antigen status among adolescents in Taiwan. Cancer Epidemiology, Biomarkers & Prevention, 10(11):1223–1226.

Cornell University. (2009) Aflatoxins: Occurrence and Health Risks. In Plant Poisonous in Livestock. Cornell University, Department of Animal Science. http://www.ansci.cornell.edu/plants/toxicagents/aflatoxin/aflatoxin.htm

Cotty PJ & Bhatnagar D. (1994) Variability among atoxigenic Aspergillus flavusstrains in ability to prevent aflatoxin contamination and production of aflatoxin biosynthetic pathway enzymes. Applied and Environmental Microbiology, 60(7):2248–2251.

Cotty PJ & Jaime-Garcia, R. (2007) Influences of climate on aflatoxin producing fungi and aflatoxin contamination. International Journal of Food Microbiology, 119(1-2):109-115.

Cova L, Wild C, Mehorota R, Turusov V, Shirai T, Lambert V, Jacuqet C & Tomatis L. (1990) Contribution of Aflatoxin B1 and Hepatitis B Virus Infection in the Induction of Liver Tumors in Ducks. Cancer Research, 50:2156-2163.

Cullen JM & Newberne PM. (1994) Acute hepatotoxicity of aflatoxins. In: Eaton DL, Groopman JD, eds. The toxicology of aflatoxins: human health, veterinary, and agricultural significance, London, Academic Press, 1993:1–26.

D'Mello JPF, Macdonald AMC. (1997) Mycotoxins. Animal Feed Science Technology, 69:155-166.

D'Mello JPF, Placinta CM & Mcdonald AMC. (1999) Fusarium mycotoxins: a review of global implication for animal health, welfare and productivity. Animal Feed Science and Technology, 80:183-205.

Denli M, Pérez JF. (2006) Contaminación por Micotoxinas en los piensos: Efectos, tratamiento y prevención. XXII Curso de ESpecialización FEDNA, p.1-18.

Denning, DW, Allen R, Wilkinson AP & Morgan MRA. (1990) Transplacental transfer of aflatoxin in humans. Carcinogenesis, 11(6):1033-1035.

Dennis P & Hsieh H. (1981) International Symposium and Workshop on Mycotoxins, Cairo, Dokki, Egypt, Proceedings International Symposium on Mycotoxins, p.151-165.

Diekman MA & Green, ML. (1992) Mycotoxins and reproduction in domestic livestock. Journal of Animal Science, 70(5): 1615-27.

Dvorackova I. (1976) Aflatoxin inhalation and alveolar cell carcinoma. Case Report. British Medical Journal. March 20, pp 691.

Fandohan P, Gnonlonfin B, Hell K, Marasas WF & Wingfield MJ. (2005) Natural occurrence of Fusarium and subsequent fumonisin contamination in preharvest and stored maize in Benin, West Africa. International Journal of Food Microbiology, 99(2):173–183.

Fandohan P, Zoumenou D, Hounhouigan DJ, Marasas WF, Wingfield MJ, Hell K. (2005) Fate of aflatoxins and fumon- isins during the processing of maize into food products in Benin. International Journal of Food Microbiology 98(3):249–259.

FAO. (1997) Food and Agricultural Organization of the United Nations Food and Nutrition Paper 64. Worldwide Regulations for Mycotoxins. A Compendium, Rome.

FDA. (1988) Food and Drug Administration, USA. (1988) Action levels for added poisonous or deleterious substances in food. Notice Fed. Register 53, 5043–5044.

Frisvad JC, Smedsgaard J, Larsen TO & Samson RA. (2004) Mycotoxins, drugs and other extrolites produced by species in Penicillium subgenus Penicillium. Studies in Mycology, 49:201–242.

Gimeno A. (2004) Aflatoxina M1 no leite. Riscos para a saúde pública, prevenção e controlo. Alimentação Animal (Revista de la Associação Portuguesa dos Industriais de Alimentos Compostos para Animais (IACA)), 49:32-44.

Greene HJ & Oehme FW. (1976) A possible case of equine aflatoxicosis. Clinical Toxicology 9, 251–254.

Groopman JD, Wogan GN, Roebuck BD & Kensler TW. (1994) Molecular biomarkers for aflatoxins and their application to human cancer prevention. Cancer Research, 54,7-1907s-1911s.

Groopman, JD, Kensler, TW & Wild CP. (2008) Protective Interventions to Prevent Aflatoxin-Induced Carcinogenesis in Developing Countries. Annual Review of Public Health, 29(1):187-203.

Guengerich FP, Johnsen WW, Ueng YF, Yamazaki H, Shimada T. (1996) Involvement of cytochrome P450, glutathione S-transferase and epoxide hydrolase in the metabolism of aflatoxin B1 and relevance to risk of human liver cancer. Environmental Health Perspectives, 104: 557-562.

Hall AJ & Wild CP (1994). Epidemiology of Aflatoxin-Related Disease. In: The Toxicology of Aflatoxins: Human Health, Veterinary, and Agricultural Significance. Eaton DL and Groopman JD eds. San Diego, CA, Academic Press, Inc.: 233-258.

Helferich WG, Baldwin RL & Hsieh DPH. (1986) [14 C]-Aflatoxin B1 Metabolism in Lactating Goats and Rats. Journal of Animal Science, 62:697-705.

Hendrickse R. (1999) Of sickturkeys, kwashiorkor, malaria, perinatal mortality, heroin addicts and food poisoning: Research on the influence of aflatoxins on child health in the tropics. Annals of Tropical Pediatrics, 19:229-36.

Hendrickse RG. (1991) Clinical implications of food contaminated by aflatoxins. Annals Academy of Medicine Singapore, 20: 84-90.

Henry SH, Bosch FX & Bowers JC. (2002) Aflatoxin, hepatitis and worldwide liver cancer risks. Advances in Experimental Medicine and Biology, 504:229–233.

Herzallah, SM. (2009) Determination of aflatoxins in eggs, milk, meat and meat products using HPLC fluorescent and UV detectors. Food Chemistry, 114(3):1141-1146. .

Hesseltine CW. (1976) Conditions Leading to Mycotoxin Contamination of Foods Feeds. In: Mycotoxins, Other Fungal Related Food Problems. Joseph V. Rodricks (Ed), American Chemical Society, Washington DC. pp.1-22.

IARC. (2002) Some traditional herbal medicines, some mycotoxins, naphthalene and styrene. Summary of data reported and evaluation. IARC Monographs on the evaluation of the carcinogenic risk to humans. Vol. 82. International Agency for Research on Cancer, Lyon, France.

Jarvis BB. (2002) Chemistry and toxicology of molds isolated from water-damaged buildings, Mycotoxins and Food Safety. Advances in Experimental Medicine & Biology, 504:43–52

Johnson DN, Egner PA, O'Brien G, Glassbrook N, Roebuck BD, Sutter T, Payne GA, Kensler TW & Groopman JD. (2008). Quantification of Urinary Aflatoxin B1 Dialdehyde Metabolites Formed by Aflatoxin Aldehyde Reductase Using Isotope Dilution Tandem Mass Spectrometry. Chemical Research in Toxicology, 21:752-760.

Kamdem LK, Flockhart DA & Desta Z. (2006) Dominant contribution of P450 3A4 to the hepatic carcinogenic activation of aflatoxin B1. Chemical Research in Toxicology, 19:577–586.

Kamimura H. (1989) Removal of mycotoxins during food processing. In: Natori, S., Hashimoto, K., Ueno, Y. (eds.). Mycotoxins and Phycotoxins. Elsevier, Amsterdam, pp. 169–176.

Ketterer PJ, Blaney BJ, Moore CJ, McInnes IS & Cook PW. (1982). Field cases of aflatoxicosis in pigs. Australian Veterinary Journal, 59(4):113-117.

Klein PJ, Van Vleet TR, Hall JO & Coulombe RA (2002) Dietary Butylated Hydroxytoluene Protects against Aflatoxicosis in Turkeys. Toxicology and Applied Pharmacology, 182(1):11-19.

López CE, Ramos LL, Ramadan SS & Bulacio LC. (2003) Presence of aflatoxins M1 in milk for human consumption in Argentina. Food Control, 14:31–34.

Luzi A, Cometa MF & Palmery M. (2002) Acute effects of aflatoxins on guinea pig isolated ileum. Toxicology in Vitro, 16:525-529.

Mallman CA & Dilkin P. (2007) Fumonisinas. In: "Micotoxinas e Micotoxicoses em Suinos. Sociedade Vicente Pallotti- Editora, Santa María, Brasil, pp. 105-427.

Masoero F, Gallo A, Moschini M, Piva G & Díaz D. (2007) Carryover of aflatoxin from feed to milk in dairy cows with low or high somatic cell counts. Animal, 1:1344–1350.

Mazzette A, Decandia M, Acciaro M, Fenu, Francesconi AD & Battacone AH. (2009) Excretion of Aflatoxin M1 in milk of goats fed diet contaminated by Aflatoxin B1, Italian Journal of Animal Science, 8(2):631-633.

Mckenzie KS, Kubena LF, Denvir AJ, Rogers TD, Hitchens GD & Bailey RH. (1998). Aflatoxicosis in Turkey Poults is Prevented by Treatment of Naturally Contaminated Corn with Ozone Generated by Electrolysis. Environment and Health, 77(8):1094-1102.

Meerdink GL. (2002) Mycotoxins. Clinical Techniques in Equine Practice, 1:89–93.

Mehan V & Mc Donald D. (1991) The groundnut aflatoxin problem: Review and Literature Database. Patanchercu India: International Crops Research Institute for the semi-Arid Tropics, pp. 64-115

Méndez-Albores A & Moreno-Martínez E. (2009) Las Micotoxinas: Contaminantes naturales de los alimentos". Revista Ciencia, julio-septiembre 2009:1-7.

Méndez-Albores JA. (2004) Aflatoxins in pozol, a nixtamalized, maize-based food. International Journal of Food Microbiology, 94:211-215.

Miller DM & Wilson DM. (1994) Veterinary Diseases Related to Aflatoxins. In: The Toxicology of Aflatoxins: Human Health, Veterinary, and Agricultural Significance. DL. Eaton and JD Groopman (eds) San Diego, CA, Academic Press, Inc.: 347-364.

Moschini R, Sisterna M & Carmona M. (2006) Modelling of wheat black point incidence based on meteorological variables in the southern Argentinean Pampas region. Australian Journal of Agricultural Research 57:1151-1156.

Moss M. (2002) Risk assessment for aflatoxins in foodstuffs. International Biodeterioration & Biodegradation, 50(3-4):137-142.

Navas SA, Sabino M, Rodríguez-Amaya DB. (2005) Aflatoxin M1 and ochratoxin A in a human milk bank in the city of Sao Paulo, Brazil. Food Additives and Contaminants, 22(5):457–462.

Newberne and, PM & Butler WH. (1969) Acute and Chronic Effects of Aflatoxin on the Liver of Domestic and Animals: A Review. Cancer Research, 29(January):236-250.

Oguz H. & Kutoglu V. (2000) Effect of clinoptilolite on fattening performance of broiler chickens during experimental aflatoxicosis. British Poultry Science, 41:512- 517.

Okiki, PA, Ojeizeh TI & Ogbimi AO. (2010) Effects of Feeding Diet Rich in Mycotoxins on the Health and Growth Performances of Broiler Chicken. International Journal of Poultry Science, 9(12):1136-1139.

Omer RE, Kuijsten A, Kadaru AM, Kok FJ, Idris MO, El Khidir IM, et al, (2004) Population-attributable risk of dietary aflatoxins and hepatitis B virus infection with respect to hepatocellular carcinoma. Nutrition and Cancer, 48(1):15–21.

Ostrowski-Meissner HT. (1983) Effect of contamination of diets with aflatoxins on growing ducks and chickens. Tropical animal health and production, 15(3):161-8.

Özdemİr M. (2007) Determination of aflatoxin M 1 levels in goat milk consumed in Kilis province. Ankara Üniversitesi Veteriner Fakültesi Dergisi, 54:99-103.

Park DL. (2002) Effect of processing on aflatoxin. Advances in Experimental Medicine and Biology, 504:173–179.

Parkin, MD, Bray F, Ferlay J & Pisani P. (2005) Global cancer statistics, 2002. CA: A Cancer Journal for Clinicians, 55(2):74-108.

Paterson RRM & Lima N. (2010) How will climate change affect mycotoxins in food? Food Research International, 43(7):1902-1914.

Patterson DSP (1973). Metabolism as a Factor in Determining the Toxic Action of the Aflatoxins in Different Animal Species. Food and Cosmetics Toxicology 11:287-294.

Pier A. (1992) Major Biological Consequences of Aflatoxicosis in Animal Production. Journal of Animal Science, 70:3964-3967.

Placinta CM, D'Mello JPF & Macdonald AMC. (1999) A review of worldwide contamination of cereal grains and animal feed with mycotoxins. Animal Feed Science and Technology, 78:21-37.

Poirier MC. (2004) Chemical-induced DNA damage and human cancer risk. Nature, 4:630–37

Qian GS, Ross RK, Yu MC, Yuan JM, Gao YT, Henderson BE, et al. (1994) A follow-up study of urinary markers of aflatoxin exposure and liver cancer risk in Shanghai, People's Republic of China. Cancer Epidemiology, Biomarkers & Prevention, 3(1):3–10.

Richard JL, Stubblefield RD, Lyon BR, Peden WM, Thurston JR & Rimier RB. (1986) Distribution and clearance of aflatoxins B1 and M1 in turkeys fed diets containing 50 or 150 ppb aflatoxin from naturally contaminated corn. Avian diseases, 30(4):788-93.

Rizzi L, Simioli M, Roncada P & Zaghini A. (2003) Aflatoxin B1 and clinoptilolite in feed for laying hens. Effect on egg quality, mycotoxin residues in livers and hepatic mixed function oxidase activities. Journal of Food Protection, 66:860–865.

Rodricks JV & Stoloff L. (1977) Aflatoxin residues from contaminated feed in edible tissues of feed producing animals.]n: Mycotoxins in Human and Animal Health. Edited by J. V. Rodricks, C. W. Hesseltine & M. A. Mehlman. Pathotox Publishers Inc., Park Forest South, IL, pp. 67.

Saad AM, Abdelgadir AM & Moss MO. (1995) Exposure of infants to aflatoxin M1 from mothers' breast milk in Abu Dhabi, UAE. Food Additives and Contaminants 12:255–261.

Saleemullah AI, Khalil IA, Shah H. (2006) Aflatoxin contents of stored and artificially inoculated cereals and nuts. Food Chemistry, 98(2006):699-703.

Santella RM. (1999) Immunological methods for detection of carcinogen-DNA damage in humans. Cancer Epidemiology, Biomarkers & Prevention, 8:733–39

Scholl PF, Turner PC, Sutcliffe AE, Sylla A, Diallo MS, et al. (2006) Quantitative comparison of aflatoxinB1serum albumin adducts inhumansby isotope dilution mass spectrometry and ELISA. Cancer Epidemiology, Biomarkers & Prevention15:823–26.

Shephard GS. (2003) Aflatoxin and food safety: recent African perspectives. Journal of Toxicology, 22(2&3):267–286.

Siriacha P, Kawashima K, Saito M, Tonboon Ek P & Buangsuwon D. (1990) Prevention of Thai maize from the infection by Aspergillus flavus and aflatoxin contamination in various packages. Source Proceedings of the Japanese Association of Mycotoxicology. 32:41-46.

Smith EE, Phillips TD, Ellis JA, Harvey RB, Kubena LF, Thompson J & Newton G. (1994) Dairy goat milk and effects on milk production and components Dietary Hydrated Sodium Calcium Aluminosilicate Reduction of Aflatoxin M1 Residue in Dairy Goat Milk and Effects on Milk Production and Components. Journal of Animal Science, 72:677-682.

Straw BE, D'Allaire S, Mengeling W & Taylor DJ. (1999) Disseases of Swine, Iowa State University Press. AMES, Iowa, USA, 8th Edition, pp. 731-742

Strosnider H, Azziz-Baumgartner E, Banziger M, Bhat RV, Breiman R, et al. (2006). Public Health Strategies for Reducing Aflatoxin Exposure in Developing Countries: A Workgroup Report. Environmental Health Perspectives, 12:1898-1903.

Tedesco, D, Barbieri C, Lugano S & Garavaglia L. (2008) Aflatoxin contamination risk: Bioactive natural compounds for animal health and healthy food. In B. F. A. Y. Sinyavskiy, ed. Impact of pollution on Animal Products. pp. 177-184.

Turner PC, Sylla A, Gong YY, Diallo MS, Sutcliffe AE, Hall AJ, et al. (2005) Reduction in exposure to carcinogenic aflatoxins by postharvest intervention measures in West Africa: a community-based intervention study. Lancet 365(9475):1950–1956.

Unusan N. (2006) Ocurrence of aflatoxin M1 in UHT milk in Turkey. Food Chemical Toxicology, 44:1897-1900.

Vaid J, Dawra RK, Sharma OP & Negi SS. (1981) Chronic aflatoxicosis in cattle. Veterinary & Human Toxicology, 23(6):436-8

Van Egmond HB & Jonker MA. (2004) Current situation on regulation for mycotoxins. In: T. Yoshizawa, S. Kumagai, & T. Goto (Eds.), New horizon of mycotoxicology for assuring food safety. Tokyo: Japanese Association of Mycotoxicology, pp. 1-15

Veldman A, Meijs JAC, Borggreve GJ & Heeres-van der Tol J.J. (1992). Carry-over of aflatoxin from cows' food to milk. Journal of Animal Production, 55:163-168.

Verma RJ. (2004) Aflatoxin Cause DNA Damage. International Journal of Human Genetics, 4(4):231-236.

Walderhaug M. (1992). Ciguatera. Foodborne Pathogenic Microorganisms and Natural Toxins Handbook. Food and Drug Administration, U.S.

Wang LY, Hatch M, Chen CJ, Levin B, You SL, Lu SN, et al. (1996) Aflatoxin exposure and risk of hepatocellular carcinoma in Taiwan. International Journal of Cancer 67(5):620–625

WHO (World Health Organization). (2005) Public Health Strategies for Preventing Aflatoxin Exposure. pp.1-26.

Wild C & Gong Y. (2010) Mycotoxins and human disease: a largely ignored global health issue. Carcinogenesis, 31:71-82.

Wild CP & Turner PC. (2002) The toxicology of aflatoxins as a basis for public health decisions. Mutagenesis, 17(6):471-81.

Williams JH, Phillips TD, Jolly PE, Stiles JK, Jolly CM &Aggarwal D. (2004) Human aflatoxicosis in developing countries: a review of toxicology, exposure, potential health consequences, and interventions. American Journal of Clinical Nutrition, 80(5):1106–1122.

Wogan GN & Busby WF. (1980) Natural occurring carcinogens. In: Toxic Constituents in Plant Foodstuffs. Liener IE, ed. Academic Press NY, 502 pps.

Wogan GN, Hecht SS, Felton JS, Conney AH & Loeb LA. (2004) Environmental and chemical carcinogenesis. Seminars in Cancer Biology, 14,473(868):733–739.

Yiannikouris A & Jouany JP. (2002) Les mycotoxins dans les aliments des ruminants, leur devenir et leurs effets chez l'animal. INRA, Producción Animal, 15:3-16.

Yousef AE & Marth EH. (1989) Stability and Degradation of Aflatoxin M1. In: Mycotoxins in Dairy Products. Hans P.Van Egmond (Ed.) Elsevier Applied Science, London and New York. Chapter 5, pp.127-161.

Part 4

Miscellaneous

14

Aflatoxins: Mechanisms of Inhibition by Antagonistic Plants and Microorganisms

Mehdi Razzaghi-Abyaneh[1],
Masoomeh Shams-Ghahfarokhi[2] and Perng-Kuang Chang[3]
[1]*Department of Mycology, Pasteur Institute of Iran, Tehran,*
[2]*Department of Mycology, Faculty of Medical Sciences, Tarbiat Modares University, Tehran,*
[3]*Southern Regional Research Center, Agricultural Research Service, U.S. Department of Agriculture, New Orleans, LA,*
[1,2]*Iran*
[3]*USA*

1. Introduction

Aspergillus flavus and some other closely-related members of *Aspergillus* section *Flavi* especially *A. parasiticus* and *A. nomius* are common fungi that normally inhabit as saprobes in soils and on a wide variety of decaying organic matters (Samson, et al., 2000; Varga et al., 2010). Besides being an etiological agent of systemic aspergillosis and allergic reactions, *A. flavus* has received major attention due to its ability to produce the carcinogenic aflatoxins (AFs) (Hedayati et al., 2007). AFs are a group of structurally related compounds found worldwide in a wide array of food and feed crops including maize, peanuts, tree nuts and oilseeds. AF contamination of agricultural crops is a major concern due to economical losses resulting from inferior crop quality, reduced animal productivity and impacts on trade and public health. It has been estimated that about one-fourth of global food supply is contaminated annually with AFB_1, the most toxigenic among AFs. Hence, an action level of 20 ppb for total AFs in foods for human consumption has been imposed by Food and Drug Administration (FDA) in the United States and by regulatory agencies in many other countries. The biosynthetic pathway of AF is one of the best known pathways of secondary metabolism in microorganisms (Trail et al., 1995). Our current understanding about chemistry and molecular biology of AF biosynthesis is a direct consequence of bioconversion experiments using several blocked mutants as well as successful cloning and characterization of the majority of the AF biosynthetic pathway genes (Georgianna and Payne, 2009). About 30 genes and related intermediates are involved in the production of AFs. To ensure global safety on food and feed supplies, extensive researches have been carried out to effectively control and manage AF contamination of crops. Conventional procedures have been used to prevent contamination process in crops before and after harvest, the majority of which are expensive, time consuming and with limited success. These technologies include crop rotation, use of fungicides, and alteration in planting time. The rapid expansion in our knowledge about inhibition of AF biosynthesis by plants and microorganisms has enable us to utilize them as potential AF biocontrol agents (Alinezhad et al., 2011; Holmes et al., 2008; Razzaghi-Abyaneh et al., 2008, 2009, 2010). A large number of compounds and extracts from natural sources including plants, bacteria, microalgae, fungi and actinomycetes have now been screened for the ability to inhibit toxigenic fungal growth and/or AF production. Substantial efforts have been carried out in identifying organisms inhibitory to AF biosynthesis through co-culture with AF-producing fungi with the aim of finding potential biocontrol agents as well as novel inhibitory metabolites. Despite the positive prospective in combating AF contamination, control of AF contamination of food and feed has not yet been achieved. Understanding the mechanisms by which plants and microorganisms and their bioactive metabolites affect AF biosynthesis is a major focus in the molecular biological study of AF. This endeavor would also help us to advance the knowledge about host plant-toxigenic fungus interactions, one of the most important aspects of AF contamination of crops. Genomics, proteomics and metabolomics studies have revealed novel inhibitory mechanisms on AF production by bioactive compounds from natural sources (Bhatnagar et al., 2008; Brown et al, 2010; Kim et al., 2008). In this review, we describe AF inhibitors from plants and microorganisms with an emphasis on their potential mechanisms of action at cellular and molecular levels. We highlight direct inhibition of AF biosynthesis via interfering with the signal transduction regulatory networks involved in gene expression and also by blocking activities of AF biosynthetic enzymes. We also address indirect inhibition of AF production by plants and organisms that affect toxin synthesis by the mechanisms including (i) down-regulation of fungal genes responsible for oxidative stress defense system which combats metabolic and environmental stressors, (ii) enhancement of plant defense mechanisms through genetic engineering and (iii) disruption of mitochondrial respiration which is critical for providing the first substrate of AF biosynthesis, acetyl-CoA.

2. *Aspergillus* species producing G-type aflatoxins in addition to B-type aflatoxins

Aspergilli that produce only B-type aflatoxins, i.e., AFB_1 and AFB_2 have been reported from species of three groups. They include many isolates of *A. flavus* and infrequent isolates of *A. pseudotamarii* in the section *Flavi*, rare isolates of *Emericella astellata* and *E. venezuelensis* (producing AFB_1 only) from the genus *Emericella* (anamorph: *Aspergillus* section *Nidulantes*) and rare isolates of *A. rambellii*, and *A. ochraceoroseus* from a newly proposed section *Ochraceorosei* (Cary et al., 2005; Frisvad et al., 2005). Several aspergilli besides producing B-type aflatoxins produce the G-type aflatoxins, AFG_1 and AFG_2. Species currently recognized to produce both B- and G-type aflatoxins belong exclusively to section *Flavi* and include *A. parasiticus, A. nomius, A. bombycis, A. minisclerotigenes*, and *A. arachidicola* (Pildain et al., 2008; Varga et al., 2010). Isolates of *A. toxicarious* and *A. parvisclerotigenus* also have been mentioned in this category, but their species status needs further clarification.

3. Plants: Target sites of inhibitory action on AF production

The history of herbal medicine dates back thousands of years ago (Samuelsson 2004). Beneficial plants are widely distributed all over the world, and they are rich sources of useful secondary metabolites often as compounds with therapeutic roles in defense against a wide array of pathogens including viruses, bacteria and fungi, herbivores (both insects and mammals) and environmental stresses like UV light and ozone (Bakkali et al. 2008; Korkina 2007). Plant bioactive metabolites can be divided into major groups including terpens (terpenoids, isoterpenoids), phenylpropanoids (flavonoids, tannins, glycosides, and lignins), phenolics and nitrogen-containing compounds (alkaloids and heterocyclic aromatics). Search of natural sources for novel inhibitors of AF biosynthesis has been a subject of intense study and a variety of bioactive AF inhibitory compounds have been reported from medicinal plants (Review by Razzaghi-Abyaneh et al. 2010 and references therein).

3.1 Phenylpropanoids from *Anethum graveolens* and *Petroselinum crispum*

Anethum graveolens L. (dill) is a short-lived annual herb cultivated as a native plant in southwest and central Asia including Iran. *Petroselinum crispum* (parsley) is a bright green hairless biennial herbaceous plant in temperate climates, an annual herb in sub-tropical and tropical areas. It is native to the central Mediterranean region including Iran, southern Italy, Algeria and Tunisia and widely cultivated as a herb, a spice and a vegetable. Several biological activities of both *A. graveolens* and *P. crispum* (Apiaceae family) have been attributed to major constituents of the whole plants including monoterpenes, flavonoids, furanocumarins and phenylpropanoids (Crowden et al. 1969). Phenylpropanoids are a large class of plant phenols with a three-carbon side chain and a phenyl ring derived from phenylalanine, an initial precursor, through shikimic acid pathway (Korkina 2007). A large number of plant-derived phenolics including flavonoids, cumarins and lignins are by-products of phenylpropanoid metabolism (MacRae & Towers, 1984). Phenylpropanoids are involved in plant defense against pathogenic and symbiotic microorganisms through cell wall strengthening and repair, direct antimicrobial activity and coordinating signaling and chemotaxis pathways against naturally occurring stressors. They are known for a wide range of biological activities from antimicrobial to adaptogenic, neurotropic, immunostimulatory, antioxidant, antiulcer, anticancer and antiproliferative properties (Korkina 2007 and references therein). In a recent study, we reported the isolation of a phenylpropanoid compound named dillapiol, from leaf essential oil of *A. graveolens* as specific inhibitor of AFG_1 production by *A. parasiticus* with an IC_{50} (50% inhibitory concentration) equal to 0.15 μM without obvious effect on fungal growth and AFB_1 synthesis (Razzaghi-Abyaneh et al. 2007). Another phenylpropanoid, apiol, isolated from the seed essential oil of *P. crispum* in the same study showed similar effects to dillapiol with an IC_{50} value of 0.24 μM for AFG_1. It is proposed that these phenylpropanoids may inhibit AFG_1 biosynthesis via inhibition of CypA, a cytochrome P450 dependent monooxygenase involved in conversion of *O*-methylsterigmatocystin to AFG_1 in AF biosynthetic pathway. More than 20 enzymes are involved in the formation of AFB_1, AFB_2, AFG_1, and AFG_2. Among them, six are P450 monooxygenases, which include OrdA, CypA, AvnA, CypX, VerA and VerB. AvnA is responsible for the conversion of averantin to 5′-hydroxyaverantin (Yu et al., 1997) and CypX for the conversion of averufin to hydroxyversicolorone (Wen et al., 2005). VerA and VerB are both involved in the conversion of versicolorin A to demethylsterigmatocystin (Keller et al., 1994; Keller et al., 1995). Cytochrome P450 monooxygenases belong to the superfamily of proteins that contain a heme cofactor. The active site of a P450 is a heme-iron center. The iron is tethered to the P450 protein via a thiolate ligand derived from a cysteine residue. This cysteine heme-iron ligand signature, F[S/G/E] XGXRXCXG, is present at the N terminal portion of the six P450s involved in AF biosynthesis. It is believed that OrdA and CypA may correspond to the microsomal enzymes and NadA the cytosol enzyme that are involved in the formation of AFG_1 and AFG_2 (Yabe et al., 1999; Zeng et al., 2011). The CypA gene encodes a P450 monooxygenase, and its knockout in *A. parasiticus* abolished production of AFG_1 and AFG_2 but not AFB_1 and AFB_2. Fig. 1 shows that CypA likely catalyzes epoxidation of a closed ring intermediate of 370 Da (**2**) that is derived from HOMST epoxidized by OrdA followed by first cleavage of the A ring (Ehrlich et al., 2008). This separates AFG_1 formation from AFB_1 formation; the latter requires a second cleavage of the 370 Da intermediate to give an open-ring form (**3**, 388) followed by demethylation, reclosure, a decarboxylation/dehydration step (Udwary et al., 2002). Most recently, Zeng et al. (2011) confirmed in *A. parasiticus* that HOMST was converted to AFG_1 and AFB_1. Cai et al., 2008 showed that the *nadA* gene previously assigned to the adjacent sugar utilization gene cluster instead is required for the formation of AFG_1 and AFG_2. The *nadA* knockout mutants of *A. parasiticus* accumulate a new 360 Da precursor, NADA. As illustrated in the Fig. 1, OrdA performs two epoxidation reactions on the OMST aryl A-ring whereas CypA performs an epoxidation reaction on an alkene (non-aromatic) substrate. They likely have very different catalytic structures. The 370 Da (**2**) substrate has an *O*-methyl group situated next to the alkene where epoxidation by CypA occurs. This structure is similar to the moiety of dillapiol, apiol, and myristicin where *O*-methyl group(s) links to the planar methylenedioxyphenyl ring (Razzaghi-Abyaneh et al., 2007). These *O*-methyl containing phenylpropanoids like other polysubstrate monooxygenase inhibitors (Casida, 1970; MacRae & Towers, 1984) may be able to access and bind to the heme in the active site pocket of CypA but are unable to access or bind to the catalytic heme residue in OrdA, AvnA, CypX, VerA and VerB. This could explain why formation of AFG_1 and AFG_2 but not AFB_1 and AFB_2 was specifically inhibited by these types of compounds. The observed dosage effects may suggest that these inhibitory compounds compete for access with the 370 Da (**2**) substrate.

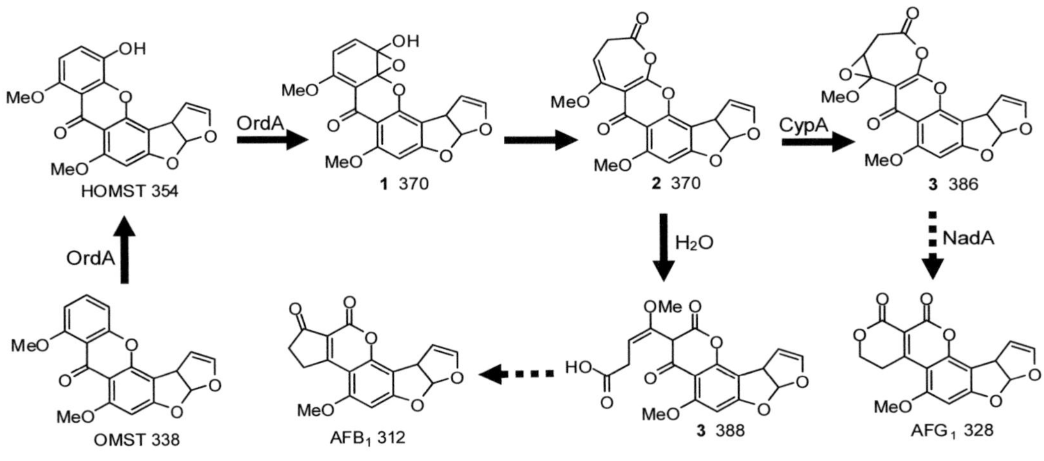

Fig. 1. Proposed pathway of aflatoxin G_1 formation from HOMST. It is believed that apiols exert their specific inhibition on formation of AFG toxins by affecting the pathway specific gene product i.e. CypA or that the inhibition is a result of structural similarities of apiols to some pathway intermediates, which enable apiols to inhibit CypA via binding to the heme in the active site pocket of the protein.

3.2 Spiroethers from *Matricaria recutita*

Matricaria recutita L. (syn: *M. chamomilla* L.; German chamomile) resides in the Asteraceae family and is one of the most widely used medicinal plants in the world (Salamon, 1992). It has a long history of application in herbal medicine which dates back ancient Greece and Rome periods where it was referenced by Hippocrates, Galen and Asclepius (Franke & Schilcher, 2005). The plant is an annual herb with erect branching and finely divided leaves growing to 50–90 cm tall. It has a stable natural monocyclic sesquiterpene alcohol named α-bisabolol as the main constituent (Tolouee et al., 2010). The plant has no reported toxic compounds and acute toxicity to human and animals. Consequently, it has been listed as generally recognized as safe (GRAS) by FDA (Bradley, 1993; Newall et al., 1996). A diverse range of pharmacological effects have been recognized for the plant including antimicrobial, anti-inflammatory, antioxidant, antispasmodic, antiviral, carminative, sedative and antiseptic properties. Other potentially active constituents are terpenoids, flavonoids, coumarins, and spiroethers which are believed to be responsible in part for the plant's wide range of biological activities (Newall et al., 1996). By screening 110 commercial essential oils from different plants using a microbioassay technique, Yoshinari et al. (2008) found a novel biological activity from *M. recutita* that completely inhibited AFG_1 production in *A. parasiticus* NRRL2999 at a concentration of 100 μg/ml. The components associated with the activity were identified as *(E)*- and *(Z)*-spiroethers (Martinez et al. 1987). Both compounds inhibited AFG_1 production with IC_{50} values of 2.8 and 20.6 μM without affecting fungal growth. However, they increased AFB_1 production in a concentration dependent manner. An *in vitro* feeding experiment with OMST as the substrate using *A. parasiticus* ATCC24690 (norsolorinic acid accumulating mutant) in the presence of spiroethers showed that these compounds specifically inhibited the pathway from OMST to AFG_1. In contrast, the mutant received only exogenous OMST produced both AFB_1 and AFG_1. Spiroethers have been implied to be able to inhibit human cytochrome P450 enzymes, CYP1A2 and CYP3A4 (Ganzera et al., 2005). Yoshinari et al. (2008) suggested that a possible inhibitory mechanism of spiroethers on AFG_1 production was on the activity of a cytochrome P450-dependent enzyme. In support of this notion, the authors further showed that spiroethers efficiently inhibited 3-acetyldeoxynivalenol by inhibiting TRI4, a key P450 monooxygenase involved in the early step of trichothecene biosynthesis. The target of spiroethers inhibition on AFG_1 formation likely is the aforementioned CypA, a P450 monooxygenase enzyme.

3.3 *Ageratum conyzoides*: Possible role for precocenes

Ageratum conyzoides is a species of invasive plants belonging to the family Asteraceae with high degree of environmental adaptability. It is native to tropical America, especially Brazil where it is commonly known as "Mentrasto". Several medicinal properties of this plant such as antimicrobial effects against different bacteria and fungi have been attributed to its main chemical constituents including terpenoids, flavonoids and phenolics (Kong et al., 2004). Nogueria et al. (2010) reported the complete inhibition of AFB_1 production by *A. flavus* IMI190 by the essential oil of the plant's aerial parts (concentrations ≥ 0.1 μg/ml) along with retarded fungal growth. Electron microscopic examination on oil-treated fungal structures revealed ultrastructural changes mainly in plasma membrane and memberanous organelles especially the mitochondria. In the excellent two-branch model presented for subcellular compartmentalization, translocation, AF gene expression and aflatoxisome biogenesis, Chanda et al. (2009) proposed that acetyl-CoA molecules necessary for early steps of AF biosynthesis are originated from β-oxidation of short chain fatty acids inside mitochondria.

Therefore, destruction of mitochondria resulting from exposure to *A. conyzoides* may account in part for its inhibitory effects on AFB_1 production. Two dimethylchromenes i.e. precocenes I and II were identified as the main constituents in the essential oil of *A. conyzoides* (Nogueria et al., 2010). These compounds purified also from *Matricaria recutita* essential oil have been recently reported as potent inhibitors of trichothecene biosynthesis by *Fusarium graminearum* (Yaghchi et al., 2009). Inhibition of 3-acetyldeoxynivalenol, a precursor of deoxynivalenol, by precocene II might be a consequence of decreased mRNA levels of *Tri4* and some other genes encoding proteins required for deoxynivalenol biosynthesis. Whether a direct inhibition of AFB_1 production by *A. conyzoides* is at the gene expression level by precocene II warrants further investigation.

3.4 Caffeic acid: An example of antioxidant-based inhibition of AF biosynthesis

Oxidative stress is an imbalanced state where excessive amounts of reactive oxygen species (ROS) overcome endogenous cell scavenging (antioxidant) capacity, leading to oxidation of a wide array of macromolecules such as enzymes, proteins, nucleic acids and lipids (Dai & Mumper, 2010). It has been shown that oxidative stress is a prerequisite and stimulatory factor for AF biosynthesis (Jayashree and Subramanyam, 2000; Reverberi et al., 2005). Oxidative stress is a result of exposure of fungal cells to ROS such as superoxide anion ($O_2^{\cdot -}$), hydrogen peroxide (H_2O_2), hydroxyl radicals ($HO^{\cdot}$) and lipoperoxides (LOOH), which are formed from unsaturated fatty acids. These molecules can be generated as by-products of cell respiration or as a consequence of fungal response to environmental stressors like infectious microorganisms. ROS production at physiological concentrations plays an important role in fungal developmental processes such as conidiation and differentiation (Reverberi et al., 2008 and references therein). If the ROS level exceeds the cell-scavenging capacity, it can damage cell membranes and cell metabolism. Under such conditions, fungal cells preserve themselves by activating free radical scavenging system composed of the antioxidant enzymes such as superoxide dismutase (SOD), catalase and glutathione peroxidase (GPX). It is believed that AF and its precursors produced by toxigenic fungi as a defense mechanism in response to abnormally elevated levels of ROS (Narasaiah et al., 2006; Reverberi et al., 2008). Reverberi et al. (2008) first described the role of a gene Ap*yapA* from *A. parasiticus* in regulating cell differentiation and aflatoxin biosynthesis following ROS formation and the activation of antioxidant defense mechanisms. This gene is an orthologue of *Yap1* from *Saccharomyces cereviciae* which modulates the expression of many antioxidant related genes. The authors proposed that consecutive events starting from *A. parasiticus* growth followed by oxidative burst, triggered the expression of Ap*yapA*, modulation of antioxidant defense and finally led to initiation of AF biosynthesis. Many plants have antioxidant properties. Hence, they may exert inhibitory effects on AF biosynthesis by affecting oxidative stress responses in toxigenic fungi. Caffeic acid is one of the most important naturally occurring plant secondary metabolites with well known antioxidant activity. It decreases lipid peroxidation and GSH depletion resulting in reduced cell death (Lima et al., 2006). Caffeic acid is found in all plants because it is a key intermediate in the lignin biosynthesis. Recently, it has been used as a marker for the elucidation of antioxidant-based inhibition of AF biosynthesis (Kim et al. 2008). Addition of caffeic acid at a final concentration of 12 mM to fat-riched growth media reduced by >95% of AF production by *A. flavus* NRRL3357 without any obvious effect on fungal growth. Microarray-based gene expression profiling showed that expression of all genes in the AF biosynthetic cluster except for *norB* and the AF pathway regulatory gene, *aflJ*, were downregulated in caffeic acid-treated *A. flavus* compared with non-treated controls. Further microarray analysis of a number of genes involved in lipid metabolism, cell wall integrity and transport, and oxidative/antioxidative activities suggested a combination of events for the caffeic acid-induced inhibition of AF production by *A. flavus* (Kim et al., 2008).

3.5 Plant-based naturally occurring phenolics

Phenolics are the most abundant secondary metabolites of plants with more than 8,000 known structures ranging from simple compounds such as phenolic acids to complex structures such as tannins (Dai & Mumper, 2010). Phenolic acids, flavonoids, tannins, stilbenes and lignans are the main classes of plant phenolics. Besides the role in defense against UV, pathogens, parasites and predators, plant polyphenols have received special attention regarding their potent antioxidant properties which make them promising in suppressing oxidative stress associated disorders such as cancer. Inhibition of fungal growth and AF production by phenolics has been a subject of many studies (Hua et al., 1999; Kim et al., 2005, 2006; Razzaghi-Abyaneh et al., 2008). Hua et al. (1999) showed that plant phenolics i.e. acetosyringone, syringaldehyde and sinapinic acid efficiently inhibited the biosynthesis of AFB_1 in *A. flavus*. Using a norsolorinic acid (NOR) accumulating mutant of *A. flavus*, they proposed that these phenolics exert their inhibitory effects on AFB_1 biosynthesis at one or more early steps in the AF biosynthetic pathway. Razzaghi-Abyaneh et al. (2008) showed a novel biological activity from leaf essential oil of *Satureja hortensis* L. as strong inhibition of AF biosynthesis by *A. parasiticus* NRRL2999. The active substances purified by column chromatography were identified as phenolics, thymol and carvacrol. Bioassay with HPLC purified fractions revealed that both carvacrol and thymol effectively inhibited fungal growth. Inhibition of AFB_1 and AFG_1 production by these phenolics exhibited a dose-dependent manner at concentrations of 0.041 to 1.32 mM throughout all two-fold dilutions. The IC_{50} values of the compounds for AFB_1 and AFG_1 was as 0.50 and 0.06 mM. Since these phenolics are potent antioxidants, they likely exert their inhibitory activities on AF production through mediation of oxidative stress levels in the fungus. Kim et al. (2006), using a *S. cerevisiae* model system, demonstrated an effective synergism of natural phenolics with known antifungal chemicals such as carboxin and strobilurin. They showed that growth inhibition of *A. flavus* by the phenolics salicylic acid, thymol, vanillyl acetone, vanillin and cinnamic acid is via targeting the mitochondrial oxidative stress defense system. Since mitochondria are responsible for providing acetyl-CoA which is a main precursor for AF biosynthesis, disruption of mitochondrial respiration chain may account in part for the inhibitory effects of antifungal phenolics on AF production.

3.6 *Azadirachta indica*: A global tree for global problems

Azadirachta indica A. Juss (syn. *Melia azadirachta* L., Neem, Margosa) is a subtropical tree native to the drier areas of Asia and Africa. The plant is known for its medicinal, spermicidal, antiviral, antibacterial, antiprotozoal, insecticidal, insect repellent, antifungal and antinematode properties (Allameh et al., 2001; Bhatnagar et al., 1988). It is indigenous to the Indian subcontinent where it has been used in agriculture, medicine and cosmetics. Several active substances from different parts of the plant have been identified. Extracts from different parts contain terpenoids, desactylimbin, quercetin and sitosterol. It has been shown that aqueous extracts of leaves and seeds inhibit AF production by *A. parasiticus* at concentrations higher than 10% (v/v) without affecting fungal growth. Studies suggested that the inhibitory components of these extracts are non-volatile substances that affect the synthesis of enzymes in the early steps of AF biosynthetic pathway (Bhatnagar et al., 1988; Zeringue and Bhatnagar, 1990). Allameh et al. (2001) did not find a positive correlation between AF production and the activity of fatty acid synthase, a key enzyme involved in AF production on neem-treated *A. parasiticus*. Razzaghi-Abyaneh et al. (2005) showed that AF production at 96 h in cultures containing 50% neem leaf and seed extracts was inhibited by 90 and 65%, respectively. Electron microscopy examination of treated fungus and non-treated controls revealed an association between decreased AF production and morphological changes suggesting that the integrity of cell barriers particularly cell wall is crucial in the regulation of AF biosynthesis and excretion.

3.7 Caffeine: An alkaloid from cocoa and coffee beans

Caffeine is a xanthine alkaloid which was isolated from coffee in 1820 by a German chemist, Friedlieb Ferdinand Runge. This compound also is found in different quantities in the beans, leaves and fruits of some plants, and acts as a natural pesticide against plant pathogens. Caffeine has been reported to inhibit fungal growth and mycotoxin (sterigmatocystin, citrinin, patulin and ochratoxin A) production by some *Aspergillus* and *Penicillium* species (Buchanan & Lewis, 1984 and references therein). Its mechanism of action was elucidated by Buchanan & Lewis (1984). They observed nearly complete inhibition of AF production along with a marked suppression (80-90%) in growth of *A. parasiticus* in submerged cultures containing 2 mg/ml caffeine. Based on the results of the feeding experiments with [U-C^{14}] glucose and enzymatic assays, Buchanan & Levis proposed that caffeine blocks AF production by affecting respiratory system of fungal cells and by inhibiting glucose uptake which is necessary for the production of acetyl-CoA, the building block of AFs. It seems that caffeine inhibits glucose uptake by directly affecting glucose transport system rather than altering the level or activity of enzymes associated with the glucose metabolism.

3.8 Gallic acid from walnuts

Gallic acid is a phenolic compound and a key component of hydrolysable tannins found in different plant species such as walnuts, oat bark and tea leaves. It is synthesized from an early intermediate named 5-dehydroshikimate in shikimate pathway. Among diverse biological activities reported for gallic acid, antimicrobial, antioxidant and antitumor properties are involved in plant defense against environmental stressors and pathogens. Inhibition of AF prduction by gallic acid without obvious effect on fungal growth was first described by Cary et al. (2003). Investigation on the mechanisms of action of gallic acid has shown that the compound affects AF biosynthesis by i) inhibition of the expression of AF biosynthetic pathway genes *nor1* and *ver1* without affecting transcription of the regulatory gene i.e. *aflR*, ii) disruption of the signal transduction pathway of oxidative stress system and iii) suppression of expression of regulatory genes of AF biosynthetic pathway such as *laeA*, whose expression is triggered by oxidative stress (Cary et al., 2003; Kim et al., 2005; Mahoney & Molyneux, 2004).

3.9 Salicylaldehyde: A volatile natural plant compound

Salicylaldehyde is an aroma compound of *Fagopyrum esculentum* and other buckwheat which acts as a key precursor of a variety of chelating agents with commercial importance. Little has been documented about physiological roles and biological properties of this volatile compound. Recently, Kim et al. (2010) showed that salicylaldehyde inhibits AF production in *A. flavus* and *A. parasiticus* by 13-45% at a concentration of 9.5 mM; it also caused retardation in fungal growth. Using the model of yeast gene deletion mutants, they suggested that the fungal antioxidant system is the molecular target of salicylaldehyde and that vacuolar detoxification plays an important role in fungal resistance to the inhibitory effects of salicylaldehyde.

3.10 Carotenoids from maize and other plants

A large number of plant carotenoids have been reported as inhibitors of AF biosynthesis (Norton, 1997 and references therein). Norton (1997) studied the effects of maize carotenoids on AF biosynthesis by *A. flavus* and found all 11 carotenoids tested except α-tocopherol markedly suppressed AFB_1 production. The compounds containing the α-ionone ring i.e. α-carotene, lutein and α-ionone were the most active carotenoids capable of inhibiting >90% AFB_1 production. Exposure of a norsolorinic acid (NOR) accumulating mutant of *A. parasiticus* SRRC162 to α-carotene resulted in production of low levels of both NOR and AF, indicating that a target site(s) of α-carotene likely are at early steps of AF pathway before NOR formation. Comparative analysis of chemical structures of tested carotenoids showed the conjugated tail and the double-bond arrangement of the ring to be the determinants of the AF inhibitory activities. Based on the structure/activity data, modification of cell membranes that indirectly affect cytosolic polyketide synthase and specific interaction with hydrophobic domains of AF pathway enzymes were postulated to result in the observed inhibition (Norton, 1997).

3.11 Resistance associated proteins from maize kernel (RAPs)

The information derived from genomics, proteomics and metabolomics has provided us a better understanding of how the AF producing fungi survive in the field and how they invade host plants and produce AF (Bhatnagar et al., 2008; Brown et al., 2010; Kim et al., 2007; Rajasekaran et al., 2006). Published studies revealed that plants respond to fungal invasion and infection through: i) producing inhibitors to fungal cell wall degrading enzymes, ii) producing specific inhibitors against fungal growth and/or AF production, iii) producing ROS and stress responsive proteins, iv) increasing lignification and cell wall cross-linking and v) triggering host cell death at the site of infection (Bhatnagar et al., 2008; Liang et al., 2006). Among crops susceptible to AF contamination, maize has been the subject of several studies because of its importance as human staple and as animal feeds worldwide. Natural resistance to AF contamination has been noticed in maize genotypes during field screening. Comparative proteomics studies have identified maize kernel resistance associated proteins (RAPs) as promising breeding markers (Bhatnagar et al., 2008; Brown et al., 2010). RAPs were classified in three major groups including antifungal, storage and stress-responsive proteins. A RAP from maize kernel, the 14 kDa trypsin inhibitor, in resistance to fungal invasion and AF contamination has been confirmed (Brown et al., 2010 and references therein). This trypsin inhibitor indirectly suppresses AF production by inhibiting α-amylase of *A. flavus*, a fungal pathogenesis factor (Chen et al., 1998; Fakhoury & Woloshuk, 1999). Extracellular hydrolases of *A. flavus* including α-amylase are responsible for degrading starch to glucose and maltose used for fungal growth. Fakhoury & Woloshuk (1999) first described *Amy1* as α-amylase gene of *A. flavus* and confirmed the role of α-amylase in AF biosynthesis. They showed that α-amylase produced by *A. flavus* generated sugar concentrations sufficient to induce AF biosynthesis. How different classes of RAPs contribute to resistance of maize to AF accumulation is a key question remains to be answered.

4. Microorganisms and their bioactive metabolites

Beneficial microorganisms especially bioactive fungi, bacteria and actinomycetes are cell factories that can produce a wide array of biologically active substances inhibitory to AF production. It has been reported that, on average, two or three antibiotics derived from microorganisms enter into the market each year (Clark, 1996). Hundreds of antifungal compounds also have been isolated from different fungi, bacteria and actinomycetes. Terrestrial actinomycetes especially those classified in the genus *Streptomyces* are rich sources of antifungal and AF inhibitory metabolites (Deshpande et al., 1988). Most recently, the roles of mycoviruses and RNA silencing in relation to AF control have gained special attention (Hammond et al., 2008; Schmidt 2009).

4.1 Bacteria and actinomycetes

4.1.1 Cyclo(L-Leucyl-L-Prolyl); A cyclic dipeptide from *Achromobacter xylosoxidans*

Achromobacter xylosoxidans is a non-fermentative, gram-negative bacillus belonging to the family Alcaligenaceae. It has been associated with a variety of clinical cases ranging from superficial sepsis to potentially fatal nosocomial infections. *A. xylosoxidans* is a newly emerging microorganism isolated with increased frequency from the lungs of patients with cystic fibrosis, but information about its clinical relevance is limited (Saiman & Siegel, 2004). Yan et al. (2004) reported a new biological activity for an environmental strain of *A. xylosoxidans* that inhibited AF production by *A. parasiticus*. The inhibitory metabolite was successfully isolated by a combination of chromatographic techniques and identified as a heat and chemical resistant cyclic dipeptide, "cyclo(L-Leucyl-L-Prolyl)". This compound has also been reported from different organisms including *Streptomyces* sp., an ascomycete (*Rosellinia necatrix*), a marine sponge (*Rhaphisia pallida*) and *Halobacillus litoralis*, a marine bacterium (Yan et al., 2004 and references therein). By using a tip culture method, the IC_{50} value of the compound on AFB_1 production was determined to be 200 μg/ml. It was inhibitory to *A. parasiticus* growth at a high concentration of 6000 μg/ml. RT-PCR analyses showed that cyclo(L-Leucyl-L-Prolyl) inhibited AF biosynthesis by repressing transcription of AF pathway genes *aflR*, *hexB*, *pksL1* and *dmtA*. The feeding experiment for conversion of sterigmatocystin (ST) to AF in the presence of the compound showed the loss of the most enzymes involved in the pathway from ST to AF. Cyclo(L-Leucyl-L-Prolyl) may direct or indirect affect the expression of the pathway regulatory gene, *aflR*. Further studies are needed in order to elucidate the underlying mechanisms of the inhibition.

4.1.2 Dioctatin A from *Streptomyces* sp. SA-2581

Dioctatin A was first isolated from *Streptomyces* sp. SA-2581 by Takeuchi et al. (1991) as an inhibitor of human dipeptidyl peptidase II, a property which accounts for immunosuppressive effects of the compound. Dioctatin A is a white powder with molecular formula of $C_{12}H_{39}N_3O_4$, molecular mass of 397.6 Da and a melting point of 263-265 °C. Yoshinari et al. (2007) demonstrated that Dioctatin A also is a strong inhibitor of AF production by *A. parasiticus*. AF production was inhibited with an IC_{50} value at 4.0 μM without any obvious effect on fungal growth. Using RT-PCR, they showed that dioctatin A inhibited the transcription of *pksA*, *ver1* and *omtA* and significantly repressed the pathway regulatory gene, *aflR*. Besides inhibition in AFB_1 and AFG_1 production, an efficient suppression of conidiation was observed on solid medium accompanied by a concomitant reduction in the mRNA level of *brlA*, which encodes a conidiation-specific transcription factor. Based on the data about inhibition of ST biosynthesis and conidia formation in *A. nidulans*, they proposed that dioctatin A may target the G protein signaling pathway and thus results in inhibition of AF biosynthesis. Dioctatin A may be a good bioactive agent for the control of AF contamination based on several proven benefits including simple structure, no toxicity to mammals, inhibiting AF production in a model infection system on raw peanuts, inhibition of AF and conidiogenesis without affecting fungal growth (lowering the chance for spread of resistance), and targeting only secondary but not primary metabolism.

4.1.3 Aflastatin A from *Streptomyces* sp. MRI142

Aflastatin A was first isolated from solvent extracts of the mycelial cake of a soil isolate of *Streptomyces* sp. MRI142 (Ono et al., 1997). Using NMR and chemical degradation analyses, it was revealed that the compound has a skeleton of tetramic acid derivative with a highly oxygenated long alkyl chain (Ono et al., 1998). It is active against different yeasts, mycelial fungi and gram-positive but not gram-negative bacteria (Ono et al., 1997). Aflastatin A completely inhibited AF production by *A. parasiticus* NRRL2999 in liquid and solid cultures at a concentration of 0.5 μg/ml without affecting fungal growth (Ono et al., 1997). Its inhibitory mechanism was studied by evaluating the effect on AF biosynthetic pathway and glucose metabolism in *A. parasiticus* (Kondo et al., 2001). Inhibition of NOR production was observed when *A. parasiticus* ATCC24690 was cultured on potato dextrose agar plates in the presence of 0.1% (v/v) of aflastatin A. Glucose metabolism and ethanol accumulation were accelerated with a marked suppression in transcription of related genes *aldA* and *facA*. RT-PCR of the AF biosynthesis genes showed a significant reduction in transcription of *pksA*, *ver1*, *omtA* and *aflR* when cells were exposed to this compound. Aflastatin A may suppress AF biosynthesis either directly via affecting *aflR* transcription or indirectly by causing a marked disturbance in the regulatory machinery of carbon metabolism.

4.1.4 Blasticidins A and S from *Streptomyces griseochromogenes*

Blasticidin A, a peptidyl nucleoside antibiotic, was first reported by Fukunaga et al. (1958) as an anti-phytopathogenic substance and its absolute configuration was determined by Sakuda et al. (2007). It is a potent inhibitor of AF biosynthesis by *A. parasiticus* (Sakuda et al., 2000). Using two-dimensional differential gel electrophoresis (2D-DIGE) Yoshinari et al. (2010) showed that blasticidin A inhibited AF (total of B_1 and G_1) production and fungal growth with IC_{50} values of 0.25 and 1.6 μM, respectively. MALDI-TOF MS analysis of protein spots on the 2D-DIGE gel of blasticidin A-treated *A. flavus* showed decreased amounts of AF biosynthetic enzymes including Vbs, OmtB, OmtA, NorA, Ver-1 and Nor-1 after 36 h treatment. Levels of other proteins with unknown functions were also decreased. It was suggested that protein synthesis in toxigenic fungi maybe the possible target site of blasticidin A. Blasticidin S, another peptidyl nucleoside antibiotic highly similar to blasticidin A, was reported in the same study as an inhibitor of AF production (IC_{50} = 28 μM) with weak inhibitory effect on fungal growth (IC_{50} >1000 μM).

4.2 Mushrooms and microfungi

4.2.1 β-glucans and culture filtrates from *Lentinula edodes* and *Trametes versicolor*

Mushrooms have received major attention with regard to their biological properties including healing effects against different diseases, antioxidant, anticancer, antiviral, and antibacterial properties and hepatoprotective effects against AF (Zjalic et al., 2006 and references therein). Recently, mushrooms have been explored as potential control agents for AF contamination (Reverberi et al., 2005; Zjalic et al., 2006). Reverberi et al. (2005) concluded that culture filtrate and purified β-glucans from *Lentinula edodes*, an edible basidiomycetous mushroom native to East Asia, significantly inhibited AF production by *A. parasiticus* without affecting fungal growth. Their RT-PCR analyses of treated *A. parasiticus* mycelia showed a delay in activation of AF biosynthetic pathway genes *aflR* and *norA* as well as a simultaneous activation of *hsf2*-like, a transcription factor involved in oxidative stress responses. They suggested that AF production inhibition by β-glucans and culture filtrate of *L. edodes* resulted from a stimulation of fungal anti-oxidant system that activates antioxidant enzymes such as SOD, catalase and glutathione peroxidase. As a consequence, a delay in AF gene transcription leads to a marked suppression of AF biosynthesis. Mushroom constituents have some advantages over chemicals and plants extracts including low toxicity, simple extraction procedures and easy production on waste materials. Therefore, *L. edodes* culture filtrate is a promising tool to control pre- and post-harvest AF contamination of crops. Zjalic et al. (2006) examined another industrially important mushroom, *Trametes versicolor*, for its antifungal activity against an AF producing *A. parasiticus*. They showed that lyophilized culture filtrates and purified exopolysaccharides of different strains of *T. versicolor* efficiently inhibited AF (B and G series) production in the range of 40-90% in submerged cultures and on maize and wheat seeds without affecting fungal growth. Antioxidant activity of *T. versicolor* culture filtrate possibly is associated with β-glucan, a free radical scavenging agent that suppresses AF biosynthesis. RT-PCR analyses of AF biosynthetic genes showed that *T. versicolor* filtrate also significantly inhibited expression of *norA* and markedly delayed *aflR* transcription.

4.2.2 Wortmannin from *Penicillium* and other microfungi

Wortmannin is a furanosteroid fungal metabolite produced by different mycelial fungi especially *Penicillium funiculosum* and *Talaromyces wortmannii* (Bräse et al., 2009). It is well known for its biological activity as a specific covalent inhibitor of phosphoinositide 3-kinases. These group of enzymes are responsible for regulating various cell survival signaling pathways including growth and proliferation, receptor mediated endocytosis, apoptosis and membrane trafficking in mammalian cells (Shepherd et al., 1988). Recently, Lee et al. (2007) showed that wortmannin at a concentration of 1 μM inhibited fungal growth, asexual sporulation, AF production and expression of AF pathway genes *ver1* and *nor1*. The inhibition on AF biosynthesis appears to interfere the phopsphatidyl inositol 3-kinase-mediated signaling pathway similar to that described for mammalian hepatocytes by Rondione et al. (2000). In the model proposed, a cascade of events including blocking of the phopsphatidylinositol 3-kinase activity, inhibition of phosphodiesterase activation, accumulation of cAMP levels to higher than the physiological state, reduction of *aflR* expression and activity, and reduction of promoter activity of *nor1* and *ver1* genes occur in wortmannin-treated *A. parasiticus* which concertedly led to strong inhibition of AF production.

4.3 dsRNA viruses; RNA silencing as a mechanism for AF suppression

Viruses, the fundamental component of life, are involved in modulating intracellular gene activities. They are unique among microorganisms in terms of adaptability, propagation of genomic materials and cellular metabolism. The history of interaction between viruses and aflatoxigenic fungi dates back more than 20 years ago when Schmidt and co-workers (1986) described the effects of viruses on AF production by *A. flavus*. Transfection experiments with naked and complete dsRNA virus from *Penicillium chrysogenum* (*PcV*) which shared similarities in structure and size to dsRNA materials of a non-toxigenic strain of *A. flavus* resulted in a stable suppression of AF biosynthesis by toxigenic *A. flavus*. The recent descriptions of RNA interference (Schmidt 2004) and the interaction of *Aspergillus* mycoviruses with their host via RNA interference (Hammond & Keller, 2008; Hammond et al., 2008) suggest that dsRNA virus from *P. chrysogenum* may degrades transcripts of AF genes by the RNA interference mechanism (Schmidt et al., 2009). However, further experiments using morphological traits revealed that the *PcV*-gene suppressing effect on AF biosynthesis is probably nonspecific because it also affected genes involved in both morphogenesis and secondary metabolism (Schmidt, 2004). Schmidt (2009) proposed that suppression of *veA* gene by *PcV*-induced siRNAs eventually led to the blocking of AF biosynthesis in the virus transfected *A. flavus*.

5. Concluding remarks and future prospective

Despite the rapid growth of our knowledge in genetics and molecular biology of AF biosynthesis in recent years, little has been documented on how we can practically combat the global problem of AF contamination of crops and agricultural commodities. The information derived from genomic resources such as whole genome sequence and expressed sequence tag (EST) of *A. flavus*, as well as from proteomics and metabolomics studies will provide us a better understanding of how AF-producing fungi survive in the field and how they invade host plants and produce the carcinogenic AFs. A large number of compounds originated from plants and microorganisms have been proven as strong inhibitors of AF biosynthesis. Recent advances in the identification of the target sites of these inhibitors have shown that they may act via i) interfering with the signal transduction regulatory networks involved in AF gene expression, ii) blocking activities of AF biosynthetic enzymes, iii) down-regulating fungal genes of the oxidative stress defense system that combats metabolic and environmental stressors, iv) inhibiting fungal pathogenesis factors and v) disrupting mitochondrial respiration, a critical process that provides acetyl-CoA for AF biosynthesis. Elucidation of the underlying mechanisms by which plants and microorganisms and their bioactive metabolites affect AF biosynthesis is a major focus in the current molecular biological studies of AF biosynthesis. This endeavor also will advance the knowledge on the complex host plant-toxigenic fungus interactions, which is one of the most important aspects in solving the AF contamination problem.

6. Acknowledgements

The authors wish to thank Dr. Kenneth Ehrlich for his kind contribution in preparing Fig. 1.

7. References

Alinezhad, S.; Kamalzadeh, A.; Shams-Ghahfarokhi, M.; Rezaee, M. B.; Jaimand, K.; Kawachi, M.; Zamani, Z.; Tolouei, R. & Razzaghi-Abyaneh, M. (2011). Search for novel antifungals from 49 indigenous medicinal plants: *Foeniculum vulgare* and *Platycladus orientalis* as strong inhibitors of aflatoxin production by *Aspergillus parasiticus*. *Annals of Microbiology*, doi: 10.1007/s13213-010-0194-1.

Allameh, A.; Razzaghi-Abyaneh, M.; Shams, M.; Rezaee, M. B. & Jaimand, K. (2001). Effects of neem leaf extract on production of aflatoxins and activities of fatty acid synthetase, isocitrate dehydrogenase and glutathione S-transferase in *Aspergillus parasiticus*. *Mycopathologia*, 154, 79-84.

Bakkali, F.; Averbeck, S.; Averbeck, D. & Idaomar, M. (2008). Biological effects of essential oils-A review. *Food and Chemical Toxicology*, 46, 446-475.

Bhatnagar, D. & McCormic, S. P. (1988). The inhibitory effect of neem (*Azadirachta indica*) leaf extracts on aflatoxin synthesis in *Aspergillus parasiticus*. *Journal of the American Oil Chemists Society*, 65, 1166–1168.

Bhatnagar, D.; Rajasekaran, K.; Payne, G. A.; Brown, R. L.; Yu, J. & Cleveland, T. E. (2008). The 'omics' tools: genomics, proteomics, metabolomics and their potential for solving the aflatoxin contamination problem. *World Mycotoxin Journal*, 1, 3-12.

Bradley, P. (1993). *The British Herbal Compendium: Vol. 1: A Handbook of Scientific Information on Widely Used Plant Drugs*. British Herbal Medicine Association, London, UK.

Bräse, S.; Encinas, A.; Keck, J. & Nising, C. F. (2009). Chemistry and biology of mycotoxins and related fungal metabolites. *Chemical Reviews*, 109, 3903-3990.

Brown, R. L.; Chen, Z. -Y.; Warburton, M.; Luo, M.; Menkir, A.; Fakhoury, A. & Bhatnagar, D. (2010). Discovery and characterization of proteins associated with aflatoxin-resistance: Evaluation their potential as breeding markers. *Toxins*, 2, 919-933.

Brown, R. L.; Cotty, P. J. & Cleveland, T. E. (1991). Reduction in aflatoxin content of maize by atoxigenic strains of *Aspergillus flavus*. *Journal of Food Protection*, 54, 623-626.

Buchanan, R. l. & Lewis, D. F. (1984). Caffeine inhibition of aflatoxin synthesis: Probable site of action. *Applied and Environmental Microbiology*, 47, 1216-1220.

Cai, J.; Zeng, H.; Shima, Y.; Hatabayashi, H.; Nakagawa, H.; Ito, Y.; Adachi, Y.; Nakajima, H. & Yabe, K. (2008). Involvement of the *nadA* gene in formation of G-group aflatoxins in *Aspergillus parasiticus*. *Fungal Genetics and Biology*, 45, 1081-1093.

Cary, J. F.; Harris, P. Y.; Molyneux, R. J. & Mahoney, N.E. (2003). Inhibition of aflatoxin biosynthesis by gallic acid. In: Proceedings of the 3rd Fungal Genomics, 4th Fumonisin, and 16th Aflatoxin Elimination Workshop. 13–15 October, 2003, Savannah, GA., p 51.

Cary, J. W.; Klich, M. A. & Beltz, S. B. (2005). Characterization of aflatoxin-producing fungi outside of *Aspergillus* section *Flavi*. *Mycologia* 97, 425-432.

Chanda, A.; Roze, L. V.; Kang, S.; Artymovich, K. A.; Hicks, G. R.; Raikhel, N. V.; Calvo, A. M. & Linz, J. E. (2009). A key role for vesicles in fungal secondary metabolism. *Proceedings of the National Academy of Sciences USA*, 106, 19533-19538.

Chang, P. -K. & Hua, S. -S. T. (2007). Nonaflatoxigenic *Aspergillus flavus* TX9-8 competitively prevents aflatoxin accumulation by *A. flavus* isolates of large and small sclerotial morphotypes. *International Journal of Food Microbiology*, 114. 275-279.

Chen, Z. -Y.; Brown, R. L.; Lax, A. R.; Guo, B. Z.; Cleveland, T. E.; & Russin, J. S. (1998). Resistance to *Aspergillus flavus* in corn kernels is associated with a 14-kDa protein. *Phytopathology*, 88, 276-281.

Clark, A. M. (1996). Natural products as a resource for new drugs, *Pharmaceutical Research*, 13, 1133-1141.

Crowden, R. K.; Harborne, J. B. & Heywood, V. H. (1969). Chemosystematics of the Umbelliferae–A general survey. *Phytochemistry*, 8, 1963-1984.

Dai, J. & Mumper, R. J. (2010). Plant phenolics: extraction, analysis and their antioxidant and anticancer properties. *Molecules*, 15, 7313-7352.

Deshpande, B. S.; Ambedkar, S. S. & Shewale, J. G. (1988). Biologically active secondary metabolites from *Streptomyces*. *Enzyme and Microbial Technology*, 10, 455-473.

Donner, M.; Atehnkeng, J.; Sikora, R. A.; Bandyopadhyay, R. & Cotty, P. J. (2010). Molecular characterization of atoxigenic strains for biological control of aflatoxins in Nigeria. *Food Additives and Contaminants*, 27, 576-590.

Dorner, J. W. (2004). Biological control of aflatoxin contamination of crops. *Journal of Toxicology and Toxin Reviews*, 23, 425–450.

Ehrlich, K. C.; Scharfenstein, L. L. Jr.; Montalbano, B. G. & Chang, P. K. (2008). Are the genes *nadA* and *norB* involved in formation of aflatoxin G_1? *International Journal of Molecular Sciences*, 9, 1717-1729.

Fakhoury A. M. & Woloshuk, C. P. (1999). *Amy1*, the α-amylase gene of *Aspergillus flavus*: involvement in aflatoxin biosynthesis in maize kernels. *Phytopathology*, 89, 908-914.

Franke, R. & Schilcher, H. (2005). Chamomile: Industrial Profiles (Medicinal and Aromatic Plants). CRC Press, Boca Raton, FL, USA.

Frisvad, J. C.; Skouboe, P. & Samson, R. A. (2005). Taxonomic comparison of three different groups of aflatoxin producers and a new efficient producer of aflatoxin B_1, sterigmatocystin and 3-*O*-methylsterigmatocystin, *Aspergillus rambellii* sp. nov. *Systematics and Applied Microbiology*, 28, 442-453.

Fukunaga, K.; Misato, T.; Ishii, I. & Asakawa, M. (1958). Blasticidin, a new anti-phytopathogenic fungal substance. *Part I. Bulletin of the Chemical Society of Japan*, 19, 181–188.

Ganzera, M.; Schneider, P. & Stuppner, H. (2005). Inhibitory effects of the essential oil of chamomile (*Matricaria recutita* L.) and its major constituents on human cytochrome P450 enzymes. *Life Sciences*, 78, 856–861.

Garber, R. K. & Cotty, P. J. (1997). Formation of sclerotia and aflatoxins in developing cotton bolls infected by the S strain of *Aspergillus flavus* and potential for biocontrol with an atoxigenic strain. *Phytopathology*, 87, 940–945.

Georgianna, D. R. & Payne, G. A. (2009). Genetic regulation of aflatoxin biosynthesis: From gene to genome. *Fungal Genetics and Biology*, 46, 113-125.

Hammond, T. M.; Andrewski, M.D.; Roossninck, M. J. & Keller, N. P. (2008). *Aspergillus* mycoviruses are targets and suppressors of RNA silencing. *Eukaryotic Cell*, 7, 350–357.

Hammond, T. M. & Keller, N. P. (2008). RNA silencing in the Aspergilli. *Mycology Series*, 26, 197–209.

Hedayati, M. T.; Pasqualetto, A. C.; Warn, P. A.; Bowyer, P. & Denning, D. W. (2007). *Aspergillus flavus*: human pathogen, allergen and mycotoxin producer. *Microbiology*, 153, 1677-1692.

Holmes, R. A.; Boston, R. S. & Payne, G.A. (2008). Diverse inhibitors of aflatoxin biosynthesis. *Applied Microbiology and Biotechnology*, 78, 559-572.

Hua, S. -S. T. (2004). Application of a yeast, *Pichia anomala* strain WRL-076 to control *Aspergillus flavus* for reducing aflatoxin in pistachio and almond. *Bulletin of International Organization for Biological and Integrated Control of Noxious Animals and Plants* 27, 291–294.

Hua, S. -S. T.; Grosjean, O.-K. & Baker, J. L. (1999). Inhibition of aflatoxin biosynthesis by phenolic compounds. *Letters in Applied Micobiology*, 29, 289-291.

Jayashree, T. & Subramanyam, C. (2000). Oxidative stress as a prerequisite for aflatoxin production by *Aspergillus parasiticus*. *Free Radical Biology & Medicine*, 29, 981-985.

Keller, N. P.; Kantz, N. J. & Adams, T. H. (1994). *Aspergillus nidulans verA* is required for production of the mycotoxin sterigmatocystin. *Applied and Environmental Microbiology*, 60, 1444-1450.

Keller, N. P.; Segner, S.; Bhatnagar, D. & Adams, T. H. (1995). stcS, a putative P-450 monooxygenase, is required for the conversion of versicolorin A to sterigmatocystin in *Aspergillus nidulans*. *Applied and Environmental Microbiology*, 61, 3628-3632.

Kim, J. H.; Campbell, B. C.; Yu, J.; Mahoney, N.; Chan, K. L.; Molyneux, R. J.; Bhatnagar, D. & Cleveland, T.E. (2005). Examination of fungal stress response genes using *Saccharomyces cerevisiae* as a model system: targeting genes affecting aflatoxin biosynthesis by *Aspergillus flavus* Link. *Applied Microbiology and Biotechnology*, 67, 807–815.

Kim, J. H.; Mahoney, N.; Chan, K. L.; Molyneux, R. & Campbell, b. C. (2006) Controlling Food-contaminating fungi by targeting antioxidant stress-response system with natural phenolic compounds. *Applied Microbiology and Biotechnology*, 70, 735-739.

Kim, J. H.; Yu, J.; Mahoney, N.; Chan, K. L.; Molyneux, R. J.; Varga, J.; Bhatnagar, D.; Cleveland, T. E.; Nierman, W. C. & Campbell, B. C. (2008). Elucidation of the functional genomics of antioxidant-based inhibition of aflatoxin biosynthesis. *International Journal of Food Microbiology*, 122, 49-60.

Kim, J. H.; Campbell, B. C.; Mahoney, N.; Chan, K. L. & Molyneux, R. J. (2010). Chemosensitization of aflatoxigenic fungi to antimycin A and strobilurin using salicylaldehyde, a volatile natural compound targeting cellular antioxidant system. *Mycopathologia*, doi: 10.1007/s11046-010-9356-8.

Kim, Y. H.; Woloshuk, C. P.; Cho, E. H.; Bae, J. M.; Song, Y.-S. & Huh, G. H. (2007) Cloning and functional expression of the gene encoding an inhibitor against *Aspergillus flavus* α-amylase, a novel seed lectin from *Lablab purpureus* (*Dolichos lablab*).*Plant Cell Reports*, 26, 395-405.

Kondo, T., Sakurada, M., Okamoto, S., Ono, M., Tsukigi, H., Suzuki, A., Nagasawa, H. & Sakuda, S. (2001). Effects of aflastatin A, an inhibitor of aflatoxin production, on aflatoxin biosynthetic pathway and glucose metabolism in *Aspergillus parasiticus*. *The Journal of Antibiotics (Tokyo)*, 54, 650–657.

Kong, C.; Hu, F.; Xu, X.; Liang, W. & Zhang, C. (2004). Allelopathic plants. XV. *Ageratum conyzoides* L. *Allelopathy Journal*, 14, 1–12.

Korkina, L. G. (2007). Phenylpropanoids as naturally occurring antioxidants: from plant defense to human health. *Cellular and Molecular Biology*, 53, 15-25.

Lee, J. -W.; Roze, L. V. & Linz, J. E. (2007). Evidence that a wortmannin-sensitive signal transduction pathway regulates aflatoxin biosynthesis. *Mycologia*, 99, 562-568.

Liang, X. Q.; Luo, M. & Guo, B. Z. (2006). Resistance mechanisms to *Aspergillus flavus* infection and aflatoxin contamination in peanut (*Arachis hypogaea*). *Plant Pathology Journal*, 5, 115-124.

Lima, C.F.; Fernandes-Ferreira, M. & Pereira-Wilson, C. (2006). Phenolic compounds protect HepG2 cells from oxidative damage: relevance of glutathione levels. *Life Sciences*, 79, 2056–2068.

MacRae, W. D. & Towers, G. H. N. (1984). Biological activities of lignans. *Phytochemistry* 23, 1207-1220.

Mahoney N, Molyneux RJ (2004) Phytochemical inhibition of aflatoxigenicity in *Aspergillus flavus* by constituents of walnut (*Juglans regia*). *Journal of Agricultural and Food Chemistry*, 52, 1882–1889.

Martinez, V.; Barbera, O.; Sanchez-Parareda, J. & Marco, J. A. (1987). Phenolic and acetylenic metabolites from *Artemisia assoana*. *Phytochemistry*, 26, 2619–2624.

Narasaiah, K. W.; Sashidhar, R. B. & Subramanyam, C. (2006) Biochemical analysis of oxidative stress in the production of aflatoxin and its precursor intermediates. *Mycopathologia*, 162, 179–189.

Newall, C. A.; Anderson, L. A. & Phillipson, J. D. (1996). *Herbal Medicines: a guide for healthcare professionals*. Pharmaceutical Press, London.

Nogueira, J. H. C.; Gonçalez, E.; Galleti, S. R.; Facanali, R.; Marques, M. O. M. & Felício, J. D. (2010) *Ageatum conyzoides* essential oil as aflatoxin suppressor of *Aspergillus flavus*. *International Journal of Food Microbiology*, 137, 55-60.

Norton, R. A. (1997). Effect of carotenoids on aflatoxin B_1 synthesis by *Aspergillus flavus*. *Phytopathology*, 87, 814-821.

Ono, M.; Sakuda, S.; Suzuki, A. & Isogai, A. (1997). Aflastatin A, a novel inhibitor of aflatoxin production by aflatoxigenic fungi. *The Journal of Antibiotics (Tokyo)*, 50, 111–118.

Ono, M.; Sakuda, S.; Ikeda, H.; Furihata, K.; Nakayama, J.; Suzuki, A. & Isogai, A. (1998). Structures and biosynthesis of aflastatins: novel inhibitors of aflatoxin production by *Aspergillus parasiticus*. *The Journal of Antibiotics (Tokyo)*, 51, 1019–1028.

Pildain, M. B.; Frisvad, J. C.; Vaamonde, G.; Cabral, D.; Varga, J. & Samson, R. A. (2008). Two novel aflatoxin-producing *Aspergillus* species from Argentinean peanuts. *International Journal of Systematic and Evolutionary Microbiology*, 58, 725-735.

Rajasekaran, K.; Cary, J. W. & Cleveland, T. E. (2006) Prevention of preharvest aflatoxin contamination through genetic engineering of crops. *Mycotoxin Research*, 22, 118-124.

Razzaghi-Abyaneh, M,; Allameh, A,; Tiraihi, T,; Shams-Ghahfarokhi, M. & Ghorbanian, M. (2005). Morphological alterations in toxigenic *Aspergillus parasiticus* exposed to neem (*Azadirachta indica*) leaf and seed aqueous extracts. *Mycopathologia*, 159, 565-570.

Razzaghi-Abyaneh, M.; Yoshinari, T.; Shams-Ghahfarokhi, M.; Rezaee, M. B.; Nagasawa, H. & Sakuda, S. (2007). Dillapiol and apiol as specific inhibitors of the biosynthesis of aflatoxin G_1 in *Aspergillus parasiticus*. *Bioscience, Biotechnology and Biochemistry*, 71, 2329-2332.

Razzaghi-Abyaneh, M.; Shams-Ghahfarokhi, M,; Yoshinari, T,; Rezaee, M. B,; Nagasawa, H. & Sakuda, S. (2008). Inhibitory effects of *Satureja hortensis* L. essential oil on growth and aflatoxin production by *Aspergillus parasiticus. International Journal of Food Microbiology*, 123, 228-233.

Razzaghi-Abyaneh, M.; Shams-Ghahfarokhi, M.; Rezaee, M. B.; Jaimand, S.; Alinezhad, S.; Saberi, R. & Yoshinari, T. (2009). Chemical composition and antiaflatoxigenic activity of *Carum carvi* L., *Thymus vulgaris* and *Citrus aurantifolia* essential oils. *Food Control*, 20, 1018-1024.

Razzaghi-Abyaneh, M.; Shams-Ghahfarokhi, M.; Rezaee, M. B. & Sakuda, S. (2010). Natural Aflatoxin Inhibitors From Medicinal Plants, In: *Mycotoxins in Food, Feed and Bioweapons*, Rai, M. & Varma, A. (ed.), 329-352, Springer-Verlag, Berlin, Heidelberg.

Reverberi, M.; Fabbri, A. A.; Zjalic, S.; Ricelli, A.; Punelli, F. & Fanelli, C. (2005). Antioxidant enzymes stimulation in *Aspergillus parasiticus* by *Lentinula edodes* inhibits aflatoxin production. *Applied Microbiology and Biotechnology*, 69, 207–215.

Reverberi, M.; Zjalic, s.; Ricelli, A.; Punelli, F.; Camera, E.; Fabbri, C.; Picardo, M.; Fanelli, C. & Fabbri, A. A. (2008). Modulation of antioxidant defence in *Aspergillus parasiticus* is involved in aflatoxin biosynthesis: a role for the Ap*yap*A gene. *Eukaryotic Cell*, 7, 988-1000.

Rondinone, C. M.; Carvalho, E.; Rahn, T.; Manganiello, V. C.; Degerman, E. & Smith, U. P. (2000). Phosphorylation of PDE3B by phosphatidylinositol 3-kinase associated with the insulin receptor. *Journal of Biological Chemistry*, 275, 10093–10098.

Saiman, L. & Siegel J. (2004). Infection control in cystic fibrosis. *Clinical Microbiology Reviews*, 17, 57–71.

Sakuda, S.; Ono, M.; Ikeda, H.; Nakamura, T.; Inagaki, Y.; Kawachi, R.; Nakayama, J.; Suzuki, A.; Isogai, A. & Nagasawa, H. (2000a). Blasticidin A as an inhibitor of aflatoxin production by *Aspergillus parasiticus. The Journal of Antibiotics (Tokyo)*, 53, 1265–1271.

Salamon, I. (1992). Chamomile, a medicinal plant. *The Herb, Spice, and Medicinal Plant Digest*, 10, 1–4.

Samson, A. R., Hoekstra, E. S.; Frisvad, J. C. & Filtenborg, O. (2000). *Introduction to Food and Airborne Fungi*, (Sixth edn.), CBS-Utrecht, The Netherlands.

Samuelsson, G. (2004). *Drugs of Natural Origin: a Textbook of Pharmacognosy*, 5th Swedish Pharmaceutical Press, Stockholm.

Schmidt, F. R.; Davis, N. D.; Diener, U. L. & Lemke, P. A. (1983). Cycloheximide induction of aflatoxin synthesis in a nontoxigenic strain of *Aspergillus flavus. Nature Biotechnology*, 1, 794–795.

Schmidt, F. R.; Lemke, P. A. & Esser, K. (1986). Viral influences on aflatoxin formation by *Aspergillus flavus. Applied Microbiology and Biotechnology*, 24, 248–258.

Schmidt, F. R. (2004). RNA interference detected 20 years ago? *Nature Biotechnology*, 22, 267–268.

Schmidt, F. R. (2009). The RNA interference-virus interplay: tools of nature for gene modulation, morphogenesis, evolution and a possible mean for aflatoxin control. *Applied Microbiology and Biotechnology*, 83, 611-615.

Shepherd, P. R.; Withers, D. J. & Siddle, K. (1988). Phosphoinositide 3-kinase: the key switch mechanism in insulin signaling. *Biochemical Journal*, 333, 471–490.

Takeuchi, T.; Aoyanagi, T.; Okami, Y.; Osanawa, H.; Iinuma, H. & Ogawa, K. (1991). Novel aminopeptidase-inhibiting dioctatins and their manufacture with *Streptomyces* species. Japanese Patent JP 03077857 A2 19910403 Heisei.

Tolouee, M.; Alinezhad, S.; Saberi, R.; Eslamifar, A.; Zad, S. J.; Jaimand, K.; Taeb, J.; Rezaee, M. B.; Kawachi, M.; Shams-Ghahfarokhi, M. & Razzaghi-Abyaneh, M. (2010). Effect of *Matricaria chamomilla* L. flower essential oil on the growth and ultrastructure of *Aspergillus niger* van Tieghem. *International Journal of Food Microbiology*, 139, 127-133.

Trail, F.; Mahanti, N. & Linz, J. (1995). Molecular biology of aflatoxin biosynthesis. *Microbiology*, 141, 755-765.

Udwary, D. W.; Casillas, L. K. & Townsend, C. A. (2002). Synthesis of 11-hydroxyl O-methylsterigmatocystin and the role of a cytochrome P-450 in the final step of aflatoxin biosynthesis. *Journal of the American Chemical Society*, 124, 5294-5303.

Varga, J.; Frisvad, J. C. & Samson, R. A. (2010). A reappraisal of fungi producing aflatoxins. *World Mycotoxin Journal*, 2, 263-277.

Wen, Y.; Hatabayashi, H.; Arai, H.; Kitamoto, H. K. & Yabe, K. (2005). Function of the *cypX* and *moxY* genes in aflatoxin biosynthesis in *Aspergillus parasiticus*. *Applied and Environmental Microbiology*, 71, 3192-3198.

Yabe, K.; Nakamura, M. & Hamasaki, T. (1999). Enzymatic formation of G-group aflatoxins and biosynthetic relationship between G- and B-group aflatoxins. *Applied and Environmental Microbiology*, 65, 3867-3872.

Yaguchi, A.; Yoshinari, T.; Tsuyuki, R.; Takahashi, H.; Nakajima, T.; Sugita-Konishi, Y.; Nagasawa, H. & Sakuda, S. (2009). Isolation and identification of precocenes and piperitone from essential oils as specific inhibitors of trichothecene production by *Fusarium graminearum*. *Journal of Agricultural and Food Chemistry*, 57, 846-851.

Yan, P. -S.; Song, Y.; Sakuno, E.; Nakajima, H.; Nakagawa, H. & Yabe, K. (2004). Cyclo(L-Leucyl-L-Prolyl) produced by *Achromobacter xylosoxidans* inhibits aflatoxin production by *Aspergillus parasiticus*. *Applied and Environmental Microbiology*, 70, 7466-7473.

Yoshinari, T.; Akiyama, T.; Nakamura, K.; Kondo, T.; Takahashi, Y.; Muraoka, Y.; Nonomura, Y.; Nagasawa, H. & Sakuda, S. (2007). Dioctatin A is a strong inhibitor of aflatoxin production by *Aspergillus parasiticus*. *Microbiology*, 153, 2774-2780.

Yoshinari, T.; Yaguchi, A.; Takahashi-Ando, N.; Kimura, M.; Takahashi, H.; Nakajima, T.; Sugita-Konichi, Y.; Nagasawa, H. & Sakuda, S. (2008). Spiroethers of German chamomile inhibit production of aflatoxin G and trichothecene mycotoxin by inhibiting cytochrome P450 monooxygenases involved in their biosynthesis. *FEMS Microbiology Letters*, 284, 184-190.

Yoshinari, T.; Noda, Y.; Yoda, K.; Sezaki, H.; Nagasawa, H. & Sakuda, A. (2010). Inhibitory activity of blasticidin A, a strong aflatoxin production inhibitor, on protein synthesis of yeast: selective inhibition of aflatoxin production by protein synthesis inhibitors. *The Journal of Antibiotics*, 63, 309-314.

Yu, J.; Chang, P. -K.; Cary, J. W.; Bhatnagar, D. & Cleveland, T. E. (1997). *avnA*, a gene encoding a cytochrome P-450 monooxygenase, is involved in the conversion of averantin to averufin in aflatoxin biosynthesis in *Aspergillus parasiticus*. *Applied and Environmental Microbiology*, 63, 1349-1356.

Zeng, H.; Hatabayashi, H.; Nakagawa, H.; Cai, J.; Suzuki, R.; Sakuno, E.; Tanaka, T.; Ito, Y.; Ehrlich, K. C.; Nakajima, H. & Yabe, K. (2011). Conversion of 11-hydroxy-O-methylsterigmatocystin to aflatoxin G_1 in *Aspergillus parasiticus*. *Applied Microbiology and Biotechnology*. Doi: 10.1007/s00253-010-2999-z.

Zeringue, H. J. Jr. & Bhatnagar, D. (1990). Inhibition of aflatoxin production in *Aspergillus flavus* infected cotton bolls after treatment with neem (*Azadirachta indica*) leaf extracts. *Journal of the Americal Oil Chemists Society*, 67, 215–216.

Zjalic, S.; Reverberi, M.; Ricelli, A.; Mario-Granito, V.; Fanelli, K. & Fabbri, A. A. (2006). *Trametes versicolor*: A possible tool for aflatoxin control. *International Journal of Food Microbiology*, 107, 243-249.

15

The Evolutionary Dynamics in the Research on Aflatoxins During the 2001-2010 Decade

Martín G. Theumer and Héctor R. Rubinstein
Centro de Investigaciones en Bioquímica Clínica e Inmunología (CIBICI), CONICET, Departamento de Bioquímica Clínica, Facultad de Ciencias Químicas, Universidad Nacional de Córdoba, Argentina

1. Introduction

The interest on aflatoxins began in the late fifties and the early sixties, after more than 100 000 young turkeys' deaths were registered in the course of a few months in 1960 on poultry farms in England. Since the deaths were caused by an apparently new disease, it was termed "Turkey X disease". Shortly after, it was observed that the disease also affected ducklings and young pheasants with high rates of mortalities. The disease was initially suspected to be induced by toxins of fungal origin, and later was proved to be caused by aflatoxins, fungal secondary metabolites synthesized by toxigenic stocks of *Aspergillus* spp. This discovery has led to a growing awareness of the potential hazards of these substances as potential inducers of illnesses and even death in humans and other mammals (Kensler et al., 2011). Thus, the "Turkey X disease" outbreak is widely considered as the initial step in the era of the aflatoxins research.

The genus *Aspergillus* is an extremely common contaminant in stored products in tropical and subtropical regions; mainly grains, nuts and spices, and several *Aspergillus* species are frequently involved in its decomposition. These molds, and consequently the food contamination with mycotoxins that they synthesize, are ubiquitous in warm regions. Despite this, because the cold-climate countries import grain of geographical areas with tropical and subtropical climates, aflatoxins have gained importance worldwide. Moreover, nowadays the technological processes of food production fail to completely eliminate aflatoxins and therefore are part of food for humans and farm animals, thus favoring the diet-mediated intake of aflatoxins (Park, 2002).

The scientific literature on aflatoxins includes more than 8000 research articles, and the toxicological data regarding these compounds led to the International Agency for Research on Cancer to classify 1993 the naturally occurring aflatoxins as Group 1 (carcinogenic to humans), and the aflatoxin M1 (AFM1, an oxidative metabolic product of AFB1) as Group 2B (possibly carcinogenic to humans) (IARC, 1993).

In this chapter it is attempted to describe somehow the information generated in the aflatoxins field in the past decade on the basis of an exhaustive categorization of the publications retrieved by entering the keyword "aflatoxin" in the PubMed search engine (http://www.ncbi.nlm.nih.gov/sites/entrez?db=pubmed) of the United States National Library of Medicine web page, a database of biomedical literature widely used around the world. The initial searching criteria included all the articles published (print or electronic) between 2001/01/01 and 2010/12/31.

2. Quantitative characterization of the research on aflatoxins through the 2001-2010 period

2.1 Number of publications

Two thousands two hundreds and six results were retrieved by inserting the searching criteria specified above. These results were then further classified according to the year of publication (print or electronic), in order to evaluate the behavior of the research on the aflatoxins area throughout the decade. The figure 1 shows the evolution of the number of annual contributions.

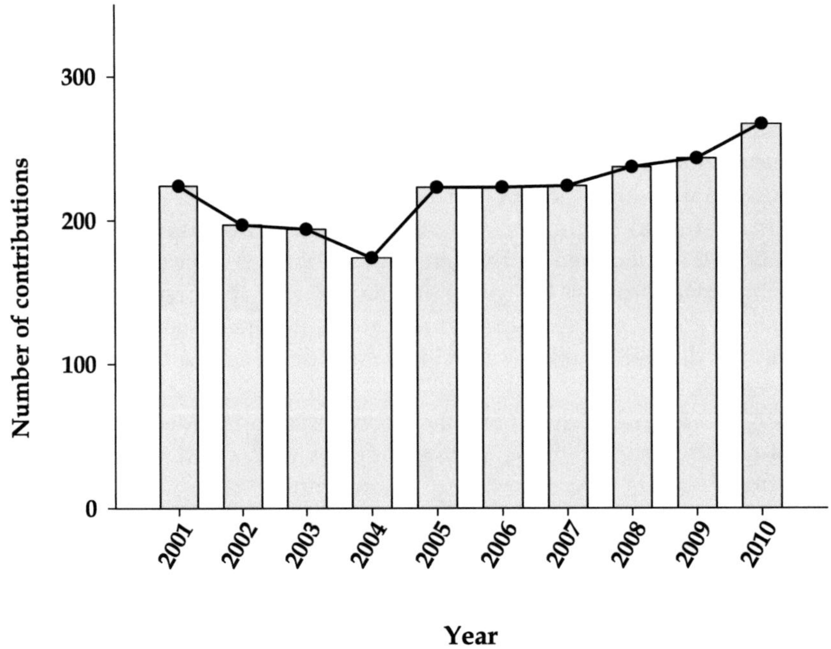

Fig. 1. The behavior in the research on aflatoxins in the 2001-2010 decade, assessed by the number of publications by year.

The average number of contributions in the full period was 220 articles by year. Assessed by the number of annual publications, the research on the aflatoxins field decreased annually from years 2001 through 2004; however this tendency was reverted according to the number of contributions registered in 2005, when the interest in the aflatoxins research increased till to the end of the decade, with 267 contributions registered in 2010.

2.2 Type of articles

The PubMed search engine also offers the possibility of filtering the requested information according to the type of contribution. Thus the searching results may include only *clinical trials (phases I to IV), letters, original or corrected and republished journal articles, reviews, case reports,* or *congresses,* for instance. This tool allows a fast characterization of the type of contribution on a specific issue. With regards to the specific field of the aflatoxins, the main publication type was original journal articles, as depicted in figure 2. The number of reviews was fluctuating in the 2001-2010 decade, and the number of journal articles published in the period showed a similar behavior to that of the total annual contributions.

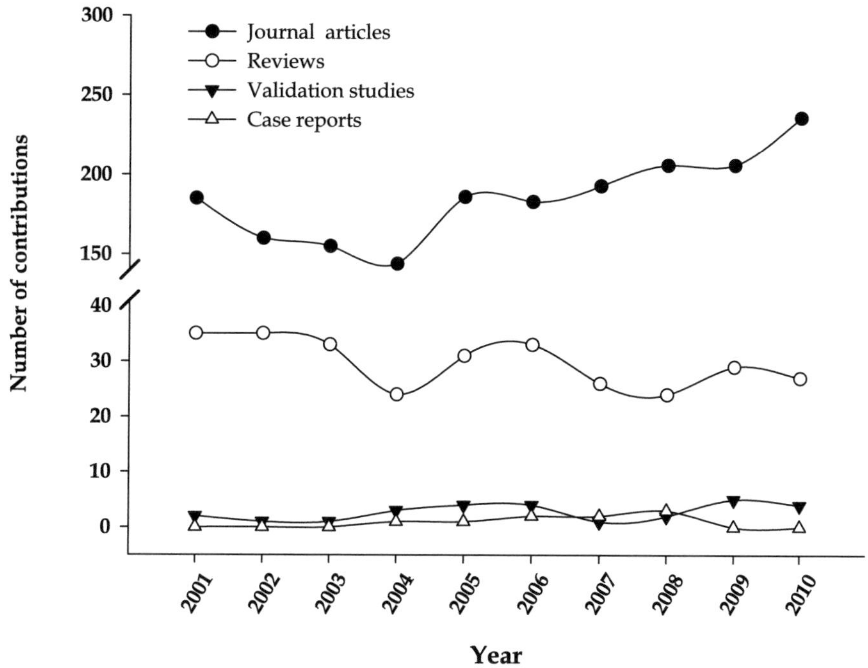

Fig. 2. The behavior in the research on aflatoxins in the 2001-2010 decade, assessed by the annual number of every main type of contribution.

The case reports are of special interest in mycotoxicology, since they may provide evidence for the linking of symptoms with diseases induced by fungal metabolites, or even may introduce new insights on the possible mechanisms of pathogenesis occurring in the natural poisonings with fungal toxins, for instance.

A total of 317 cases of acute aflatoxicosis were reported in Kenya by 20 July 2004, with a case fatality rate of 39%. The 2004 outbreak resulted from widespread aflatoxin contamination of locally grown maize, which occurred during storage of the maize under damp conditions (Probst et al., 2007).

Mwanda and co-workers (2005) reported a 17-year-old schoolboy case of aflatoxicosis, highlighting the salient clinical findings in order to increase the index of suspicion, enhance early diagnosis and improve management of such cases.

Case reports of probable aflatoxicosis in animals were also informed in 2004 and 2006. Osman et. al (2004) reported the death of twenty young female adult one-humped racing camels (*Camelus dromedarius*) kept in camps scattered outside Al Ain city and aged between 3- and 6-years-old, after a short clinical illness. On the other hand, in the USA more than 100 dogs apparently died as a result of the toxic effects of contaminated products, as it was suggested by medical history, clinical signs, progression of disease, and necropsy findings (Stenske et al., 2006).

Abstracts of Congresses such as the 3rd Fungal Genomics, 4th Fumonisin, and 16th Aflatoxin Elimination Workshops, 2003, Savannah, Georgia, USA; the Workshop on Aflatoxin and fumonisin elimination and fungal genomics, 2002, San Antonio, Texas, USA; and the Aflatoxin/Fumonisin Elimination and Fungal Genomics Workshops, 2001, Phoenix, Arizona, USA; were also published in the 2001-2010 decade.

2.3 Main contributors

A further analysis of journal articles and reviews retrieved by the PubMed search engine was carried out in order to determine the countries with a mayor contribution to the aflatoxins research field, and the results are depicted in figs. 3 and 4.

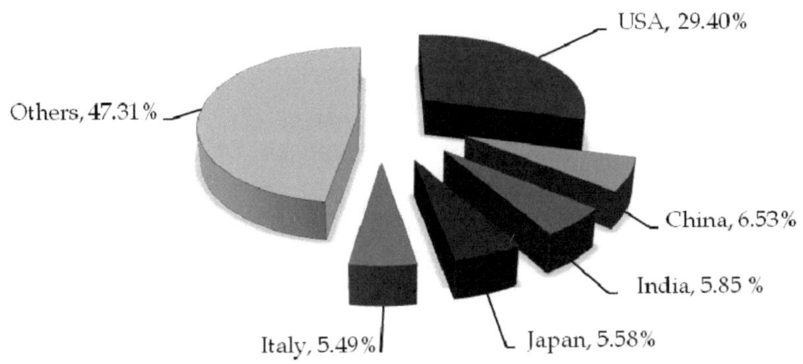

Fig. 3. Main contributor countries in the aflatoxins research, according to the number of articles published in the 2001-2010 decade.

According to the information provided by PubMed, the contributions in the research field of aflatoxins in the 2001-2010 decade were carried out in the following countries, alphabetically listed: Albania, Algeria, Argentina, Australia, Austria, Bangladesh, Belgium, Benin, Botswana, Brazil, Brunei, Bulgaria, Burundi, Cameroon, Canada, China, Colombia, Croatia, Cuba, Czech Republic, Denmark, Ecuador, Egypt, Ethiopia, Finland, France, The Gambia, Germany, Ghana, Greece, Hong Kong, Hungary, India, Indonesia, Iran, Iraq, Ireland, Israel, Italy, Japan, Kenya, North Korea, South Korea, Kuwait, Lebanon, Malawi, Malaysia, Mexico, Morocco, Nepal, Netherlands, New Zealand, Nigeria, Norway, Oman, Pakistan, Philippines, Poland, Portugal, Qatar, Romania, Russia, Samoa, Saudi Arabia, Serbia and Montenegro, Sierra Leone, Slovakia, Slovenia, South Africa, Spain, Sri Lanka, Sudan, Sweden, Switzerland, Taiwan, Tanzania, Thailand, Trinidad and Tobago, Tunisia, Turkey, Uganda, Ukraine, United Arab Emirates, United Kingdom, United States, Venezuela, Vietnam, Zambia.

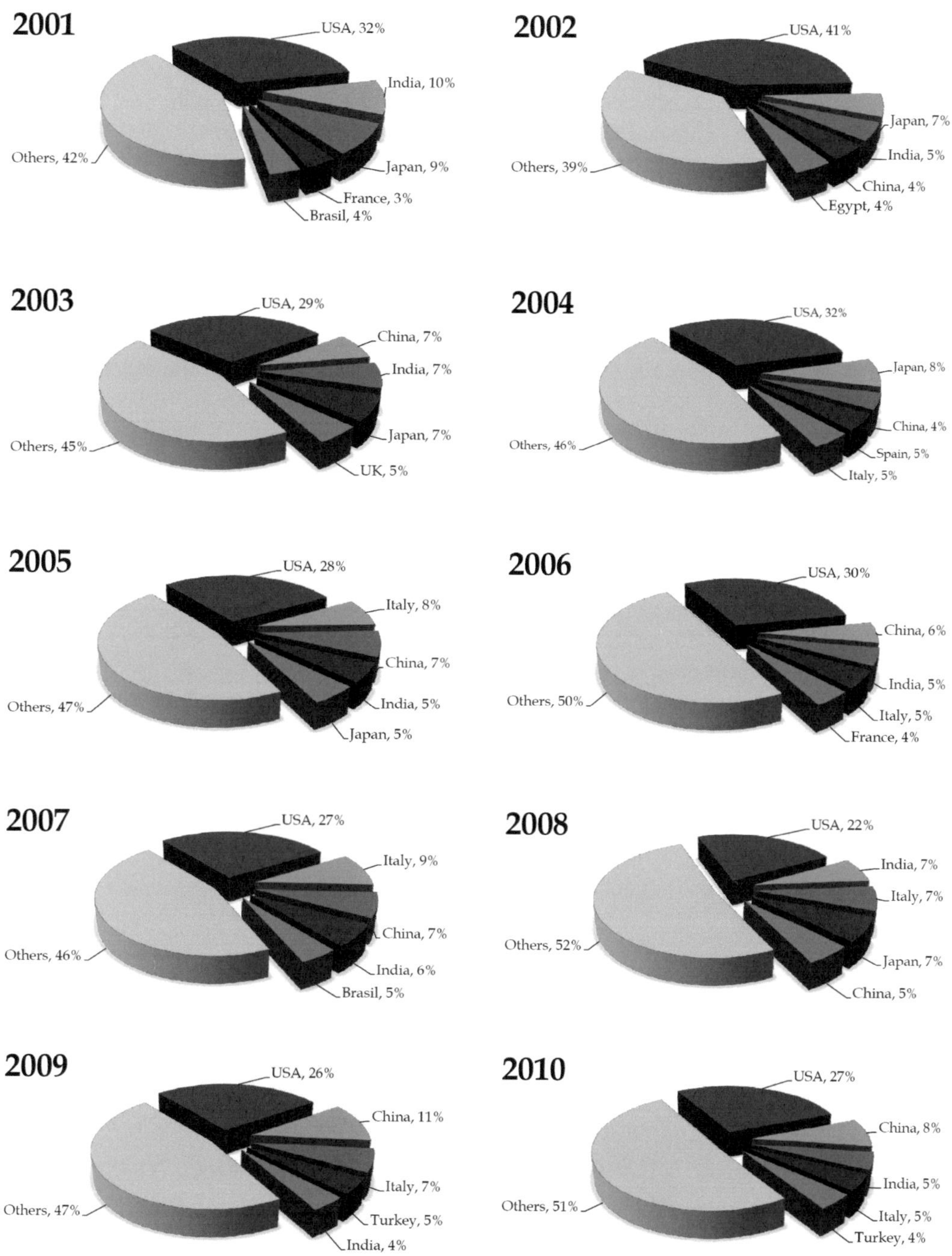

Fig. 4. The dynamic of the five main contributor countries in the aflatoxins research, according to the number of articles published by year in the 2001-2010 decade.

The United States of America did the main contribution of original journal articles and reviews published in the aflatoxins research field in the whole period (fig. 3), which was related to highest annual contributions of this country from 2001 to 2010 (fig. 4). China, India, Japan and Italy were also quantitatively important contributors in the full period, with 6.53, 5.85, 5.58 and 5.49% of the total publications, respectively. These five countries contributed with more than a half of the articles related to aflatoxins within the decade. Detailed information on the top-five countries with the highest annual number of publications is shown in fig. 4.

3. Qualitative characterization of the research on aflatoxins through the 2001-2010 period

3.1 General overview

The articles related to aflatoxins in the 2001-2010 decade were further classified qualitatively according to the focusing of the work into the following five main sub-areas of interest:

i. those oriented in the study of the plants commonly infected by aflatoxin-producing *Aspergillus* spp.;
ii. those where the main objective was to elucidate some issue related to the fungal physiology and its relationship with the environment;
iii. those where the interest was in the strategies for preservation and prevention of food contamination with aflatoxins;
iv. those focused in the elucidation of the physiological and biochemical disturbances associated with the exposure of cells/organisms to the aflatoxins, and the potential exogenous interventions to prevent them;
v. those concerning to the analytical methods commonly applied to detect and quantify aflatoxins in commodities or its by-products, for instance.

The fig. 5 shows the percentage of articles focused in every sub-area mentioned above.

More than a half of the publications in the 2001-2010 decade were oriented on topics related to the toxicology of aflatoxins. The same percentage of publications was focused in the fungal physiology and its relationship with the environment, and in the strategies for preservation and prevention of food contamination with aflatoxins. The minor proportion of the total contributions were concerning to the plants commonly infected by aflatoxin-producing *Aspergillus* spp.

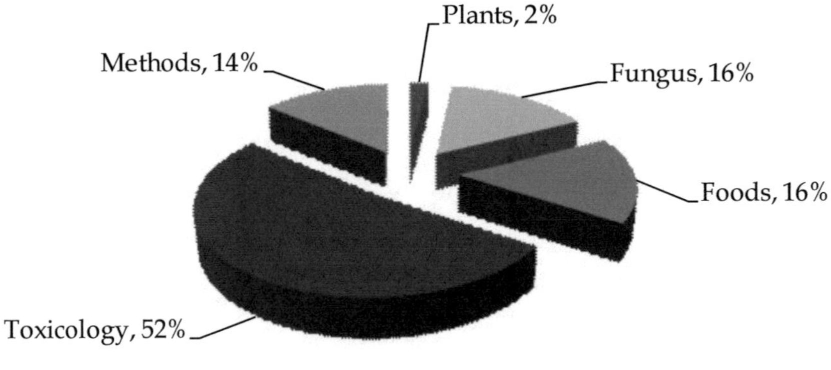

Fig. 5. Distribution of the total contributions in the decade into the main five focusing sub-areas in the research on aflatoxins.

The same classification of the articles into the five sub-areas of interest was carried out in order to identify possible inter-annuals variations in the percentages of articles in every sub-area. The results of this categorization are shown in the table 1.

Year	Sub-area of interest				
	Plants	Fungi	Food	Toxicology	Methods
2001	3,6	10,0	15,9	58,2	12,3
2002	2,5	13,2	18,3	57,9	8,1
2003	0,5	13,4	16,0	62,9	7,2
2004	1,7	19,1	12,7	56,1	10,4
2005	1,9	17,2	17,6	51,1	12,2
2006	3,3	14,0	17,6	52,0	13,1
2007	1,8	21,0	15,6	42,4	19,2
2008	1,3	18,6	17,7	48,9	13,5
2009	1,2	18,2	14,9	43,0	22,7
2010	1,9	12,5	17,4	50,8	17,4

Table 1. Annual distribution of the contributions focusing into the main five sub-areas of interest in the research on aflatoxins.
The data are expressed as percentages of the total number of publications in the specified year.

The outstanding finding of this classification was the increasing proportion of the publicationsin the second part of the decade focusing in different aspects of the analytical methods. This could be related at least in part, to the increasing demand of new methods to detect simultaneously co-contamination of foods with several micotoxins, which also should be ideally feasible of applying to different matrices.

3.2 Approaches in the articles regarding to plants

Most of the work carried out in this sub-area aimed to pinpoint the biochemical basis of the susceptibility or resistance of plants to their infection with aflatoxin-producing fungus, and its subsequent contamination with this mycotoxin. Some examples of the advances in this area are mentioned below.

Baker et al. (2009a) used a proteomic approach in order to identify proteins that may be associated with the resistance of maize to the fungus. The authors identified a higher expression of a protein (ZmCORp) in maize resistant to *A. flavus* infection. The fungistatic activity exhibited by this protein was related with its inhibitory effect on the conidia germination and mycelia growth, concluding that ZmCORp may play an important role in enhancing kernel resistance to *A. flavus* infection and aflatoxin accumulation.

Trypsin inhibitors may also play a role in the susceptibility/resistance of maize to *Aspergillus* spp. infection. Baker et al. (2009b) observed that an overexpressed trypsin inhibitor reduced the germination of conidia as well as the mycelia growth from several maize pathogens, although the effect of this trypsin inhibitor on fungal growth was weaker than the previously reported for other trypsin inhibitors.

A proteomic approach performed by Chen and co-workers (2006) led to the identification of a pathogenesis-related protein (PR-10) which is expressed fivefold higher in maize resistant lines. It was also observed that the expression of this protein was induced upon *A. flavus* infection in a resistant but not in a susceptible genotype, thus suggesting the participation of this protein in kernel resistance to the infection by *A. flavus*.

In a recent contribution, Gao and collaborators (2009) studied the potential interference of plant-derived oxylipins (a set of substances produced by the oxidation of polyunsaturated fatty acids by plant and fungal lipoxygenases) in the fungal development and in the secondary metabolism. The authors found contrasting results with different fungal pathogens, suggesting that the modulation of the plant-pathogen interactions are pathogen specific.

A novel PLD gene, encoding a putative phospholipase D, was identified in *Arachis hypogea* by Gu et al. (2006). The authors concluded that this PLD may be involved in drought sensitivity and tolerance response, and they also suggest the study of the PLD gene expression as a tool in germplasm screening for drought tolerance.

3.3 Approaches in the articles regarding to fungus

Several aspects were evaluated with regards to the fungi involved in the aflatoxins biosynthesis, including the molecular characterization of isolates in order to elucidate their phylogenetic relationships, the fungal biology and the genetic regulation leading to the mycotoxins production, the biochemical basis underlying the fungal reproduction, for instance. According to the publications registered in the 2001-2010 decade, much of the researchers' attention was on the identification of physical, chemical or biological conditions that ultimately reduce the fungal-induced food alterations and their contamination with aflatoxins. A wide spectrum of physical, chemical and biological agents was tested in attempts to solve such problems.

The essential oils from plants are complex mixtures of substances that in some cases exert strong biological activities against microorganisms. This property attracts the attention of researchers looking for new strategies of chemoprevention of the food contamination with mycotoxins.

Gandomi et al. (2010) tested the effects of *Zataria multiflora* Boiss. essential oil (EO) on *A. flavus*. The EO suppressed the size of colonies as well as the fungal esporulation. Electronic microscopy revealed morphological alterations ranging from loss of turgidity and uniformity of mycelia at low concentrations of EO to evident destruction of the hyphae at higher concentration of EO, mechanisms that could be involved in the fungal growth and in the biosynthesis of aflatoxins.

On the other hand, nine different oils were evaluated by Juglal and Govinden (2002) on the growth of *A. parasiticus* and *Fusarium moniliforme*. The highest growth inhibitory activity was found when clove oil (eugenol) was used, followed by cinnamon (cinnamic aldehyde), oregano (thymol and carvacol) and mace oils (myristin). Neem and eucalyptus oil (cineole) did not affect the fungal growth. Then, commonly occurring mycotoxigenic fungi could be controlled with clove oil (eugenol).

In other study, Kumar et al. (2007) observed that the EO extracted from the leaves of *Chenopodium ambrosioides* Linn. completely inhibited the mycelial growth and the aflatoxin synthesis, also exerting a broad fungitoxic spectrum against several toxigenic and non-toxigenic funguses.

Reverberi and co-workers (2005) studied the mechanisms underlying the aflatoxin-inhibiting effect of the *Lentinula edodes* culture filtrates. The authors reported that *L. edodes* lyophilised filtrates stimulate *A. parasiticus* anti-oxidant enzymes (superoxide dismutase, catalase, glutathione peroxidase) and aflatoxin inhibition was better correlated with beta-glucan content than with anti-oxidant activity of the filtrates.

Several studies were performed with chemical substances purified from different sources. For instance, Cleveland et al. (2009) studied the effects of volatile compunds (including

aldehydes, alcohols, ketones, and furans) produced upon exposure of soybean homogenates to lipase, on the *A. flavus* growth and aflatoxin production. They found that aldehydes inhibited up to 100% of the observed fungal growth and AFB1 production.
Goncalez and co-workers (2001) studied the effects of four biflavonoids isolated from *Ouratea* species on the *A. flavus* development and production of aflatoxins. These authors observed that the four biflavonoids showed inhibitory activity on aflatoxin B1 and B2 production, without alterations in the mycelial growth.
Analytical and industrial grade food additives effects on toxigenic *Aspergillus* were also tested, in order to identify a potential use as preservatives in foods. Passone et al. (2007) assayed the effects of three-food grade antioxidants: butylated hydroxyanisole (BHA), butylated hydroxytoluene (BHT) and propyl paraben (PP) to prove their fungitoxic effect on *Aspergillus* section Flavi strains. The authors concluded that idustrial grade antioxidant mixtures could be used for controlling the growth of mycotoxigenic species.
The biological control is a promising field to solve, at least in part, the problem of food contamination with aflatoxins. Several microorganisms could be use for this purpose, including bacteria and fungus.
Kong et al. (2010) evaluated a strain of marine *Bacillus* with regards to its activity in reducing postharvest decay of peanut kernels caused by *A. flavus*, and demonstrated that this marine bacterium could be used as a biocontrol agent against postharvest fungal disease.
Other bacteria such as *Bacillus subtilis* (Mohammadipour et al., 2009), isolations from almond flowers and mature nut fruits (Palumbo et al., 2006), and the *Lactobacillus plantarum* ATCC 8014 (Xu et al., 2003) were evaluated for their potential biocontrolling activities against aflatoxin-producing *Aspergillus*.
Studies with nontoxigenic strains of *A. flavus* and *A. parasiticus* separately and in combination were also conducted to determine the efficacy for reducing aflatoxin contamination in corn (Dorner, 2009) and peanuts (Dorner & Horn, 2007). Whereas the results of some of these experiments indicated that the treatment with nontoxigenic *A. flavus* strains could be more effective than with the *A. parasiticus* strains, others results were inconclusive.
Penna & Etcheverry (2006) observed that the interaction of *Kluyveromyces* spp. with *Aspergillus* aflatoxigenic strains induced changes on conidia germination, mycelial growth and aflatoxin B1 accumulation.
Since biological control can potentially have positive and negative effects on biodiversity, additional efforts are needed to clearly determine a minimum impact on nature by the introduction of such biological control strategies.

3.4 Approaches in the articles regarding to foods

Most of the articles classified within this category were surveys on the mycoflora and contamination of foods with aflatoxins. The fungal and aflatoxins contamination was reported in rice in Turkey (Aydin et al., 2010); in maize from south-western Nigeria (Bankole & Mabekoje, 2004), Brazil (Rocha et al., 2009) and northern Italy (Pietri et al., 2004); in domestic and imported beers in Canada (Mably et al., 2005); and in peanuts in the Cote d'Ivoire (Sangare-Tigori et al., 2006), between others.
Since contamination of grains with mycotoxins is not efficiently eliminated by most of the food processing operations such as sorting, trimming, cleaning, milling, brewing, cooking, baking, frying, roasting, canning, flaking, alkaline cooking, nixtamalization, and extrusion;

the fungal toxins are commonly found in grain by-products. Thus, the design of new technological food-processing methods that completely eliminate the mycotoxins is a field of growing interest for researchers worldwide.

Castells et al. (2008) studied the distribution of aflatoxins in various corn processed fractions during industrial cornflake processing. These authors observed that the application of dry milling of corn led to a heterogeneous distribution of the two groups of mycotoxins in the different parts of the grain, with increased levels in fractions processed from outer layers (animal feed flour and corn flour) and decreased levels in fractions processed from inner portions, such as corn meal and flaking grits.

The changes in concentration of aflatoxin M1 during manufacture of skim milk powder were assessed by Deveci & Sezgin (2006) in cow's milk contaminated artificially with aflatoxin M1 (AFM1) at two different levels. Pasteurization, concentration, and spray drying reduced the AFM1 contents with efficiencies dependant on the contamination levels.

Park & Kim (2006) studied the effect of pressure cooking on aflatoxin residues in polished rice. They found aflatoxin losses of 78-88% after pressure cooking, which correlated with the reduction of aflatoxin-induced mutagenic potential.

Interestingly, it was recently showed that raw material contamination with aflatoxins substantially affected the course of subsequent fermentation phases of maize mashes, influencing on characteristic factors such as alcohol concentration, productivity, yield and energy (Klosowski et al., 2010), and altering the composition of alcoholic fermentation volatile by-products in raw spirits (Klosowski & Mikulski, 2010).

Notwithstanding the development of new technological strategies tending to reduce the food contamination with aflatoxins; the possible formation of mycotoxin degradation products (even more toxic that the mycotoxin itself), during the application of such technologies should be considered in all cases. Most of the food processes have variable effects on mycotoxins, with those that utilize the highest temperatures having greatest effects.

Physical, chemical and biological methods of decontamination of aflatoxin-contaminated materials are another strategy feasible of application to avoid the toxicological effects of aflatoxins in human beings and in animals. Frequently, the physical reduction of the mycotoxin contents is achieved by the mechanical removal of highly contaminated fractions.

Aly & Hathout (2011) suggested that the manufacture of hydrolyzed vegetable protein is a suitable method for decontamination of aflatoxin in highly contaminated grains, especially gluten fractions. This was concluded by the application of hydrochloric acid to aflatoxin B1 contaminated corn gluten.

Enzymes of diverse origins may also be a useful tool to reduce the aflatoxin contamination of different matrices, as it was suggested by Wang et al. (2011). In line with these observations, microorganisms may have advantages with respect to the use of a purified enzyme, since they provide a subset of metabolic pathways that potentially could reduce the aflatoxins levels (Guan et al., 2010).

3.5 Approaches in the articles regarding to the toxicology of aflatoxins

The determination of quantitative or qualitative value of risk related to the exposure of population to aflatoxins is a major concern, and is currently under evaluation in different geographic areas. Sugita-Konishi and co-workers (2010) evaluated the risk of exposure to aflatoxin B1 in Japan on the basis of the contamination of 24 foods from a 3-year retail market survey and data available on the food consumption by the population studied.

Probabilistic approaches were used to estimate the aflatoxin B1 intake and the potential risk of cancer. The authors concluded that the current dietary intake of aflatoxin B1 in Japan has no appreciable effect on health.

The main etiology of hepatocellular carcinoma (HCC) is chronic infection with hepatitis B and hepatitis C viruses. However, other important factors such as obesity, diabetes, non-alcoholic steatohepatitis and dietary exposures to food toxins like aflatoxins, were linked with the development of this disease. Emerging evidence suggests that the etiology of many cases of HCC is in fact multifactorial. A clear relationship between the exposure to high levels of aflatoxins and the prevalence of HCC was identified in several geographic areas; however the precise aflatoxin-virus interaction in the induction of this disease remains unclear. The current knowledge concerning the participation of aflatoxins in the induction of HCC was recently reviewed by Sanyal et al. (2010) and Whittaker et al. (2010).

Biomarkers are currently used in the aflatoxins exposure assessment. They are classified as biomarkers of exposure (generally the aflatoxins itself or any of its metabolites, that can be measured in the body or after excretion from the body), effect (all the quantifiable changes that persists in an individual, which indicates the exposure to aflatoxins and may indicate a resulting health effect), and susceptibility (includes all natural characteristics of an organism that make it more susceptible to the effects of aflatoxins exposure).

Johnson et al. (2010) studied the levels of biomarkers of exposure (serum AFB1-lysine adduct and urinary AFM1) in order to assess the aflatoxins exposure in a community with elevated incidence of hepatocellular carcinoma. The results of this survey suggest that the incidence and level of AF exposure were less than those observed in a high-risk population.

Theumer et al. (2010) studied the changes in some biomarkers of effect by the co-exposure of spleen mononuclear cells to a mixture of aflatoxin B1 and fumonisins, and concluded that a possible protective effect of the fumonisins-AFB1 mixtures may exist with regards to the genotoxicity induced individually by the toxins.

It was suggested that the CYP3A5 polymorphism found in the population of The Gambia, wich is associated with increased levels of the mutagenic AFB1-exo-8,9-epoxide particularly in individuals with low CYP3A4, may modulate individual risk of HCC (Wojnowski et al., 2004), and thus could be potentially used as a biomarker of susceptibility to aflatoxins.

Metabonomics and proteomics are promising approaches to elucidate the biochemical perturbations of metabolism caused by aflatoxins. An overview of biochemical consequences of AFB1 exposure and comprehensive insights into the metabolic aspects of AFB1-induced hepatotoxicity in rats was reported by Zhang et al. (2011), who studied the AFB1-induced metabonomic changes in multiple biological matrices (plasma, urine, and liver) of rats to understand the mammalian systems responses to aflatoxin B1 exposure. Li and co-workers (2008) performed a proteome analysis of aflatoxin B1-induced hepatocarcinogenesis in tree shrew (*Tupaia belangeri chinensis*) In order to explore the proteins responsible for HCC. They found a protein that differentially expressed in hepatocellular carcinoma (Peroxiredoxin II), with a probable important role in hepatocarcinogenesis, possibly through its function in regulating peroxidation and hereby providing a favorable microenvironment for cancer cell surviving and progressing.

Several studies were conducted in the 2001-2010 decade to characterize the impact of aflatoxins in the production performance of animals. For instance, Pandey & Chauhan (2007) found alteration of several productive parameters, including weight gain, feed intake, feed gain ratio, age at sexual maturity, production and quality of eggs and retention of nutrients; in White Leghorn female chicks exposed to different levels of aflatoxins.

The toxicity of aflatoxins to the central nervous, digestive, renal, reproductive and immunological systems was evaluated in multiple experimental designs with microorganisms, cell lines and primary cells from laboratory animal exposed to aflatoxins. The immunotoxic effects of aflatoxins were studied by Theumer et al. (2003) in subchronic mycotoxicoses in Wistar rats poisoned with aflatoxin B1 alone and mixed with fumonisins. Essential oils from plants, a wide spectrum of chemical substances, and microorganisms were co-administered with the aflatoxins with the aim of identify possible protective effects against the toxicity of afatoxins. Sghaier et al. (2010) found antimutagenic effects of the essential oil extracted from the aerial parts of *Teucrium ramosissimum* against aflatoxin B1, in the bacterial reverse mutation assay in *Salmonella typhimurium* TA98, TA100, and TA1535. In a similar experimental system, Polivkova and co-workers (2010) observed antimutagenic effects of lycopene and tomato puree, wich was related with a significant reduction of DNA damage in mice cells assessed by the micronucleus test.
Probiotics microorganisms could be used in order to amelliorate the adverse effects of exposure to aflatoxins (Kabak et al., 2009), since it was suggested that whole microbiological cells or fractions of them could adsorb mycotoxins, including the aflatoxins.

3.6 Approaches in the articles regarding to methodologies

One of the leading advances in the 2001-2010 decade, concerning the analytical methodologies feasible of being applied in the field of aflatoxins, was the development of methods to quantify multiple mycotoxins including aflatoxins in several grains and byproducts. For instance, Zachariasova et al. (2010) developed and optimized a simple, high-throughput method for the control of 32 mycotoxins (*Fusarium* and *Alternaria* toxins, aflatoxins, ergot alkaloids, ochratoxins, and sterigmatocystin) in beer.
Studies aimed to compare the performance of analytical methods were conducted by diverse laboratories. Zhang & Chen (2004) compared the post-column derivatization technique for the determination of aflatoxins B1, B2, G1 and G2 and the rapid procedure with fluorometric analysis for the determination of total aflatoxins, concluding that the later was not the suitable method for the determination of total aflatoxins in medicinal herbs and plant extracts.
Rahamani et al. (2010) described the method validation for quantitative analysis of aflatoxins, ochratoxin A and zearalenone in cereals using HPLC with fluorescence detector; and concluded that it is suitable for the simultaneous determination of aflatoxins, ochratoxin A and zearalenone in cereals and is suitable for routine analysis.
The potential pre-analytical sources of error in the quantification of aflatoxins in foods were also considered in the 2001-2010 decade. Gallo et al. (2010) highlighted the possibility of underestimating the food contamination with aflatoxins by inefficient extraction procedures in cattle feed containing adsorbents. Brera and co-workers (2010) emphasized the use of proper sampling methods throughout the agri-food chain when it comes to effectively detecting contaminants in foods and feeds.

4. Acknowledgment

This work was partially supported by grants from Secretaría de Ciencia y Técnica de la Universidad Nacional de Córdoba (SeCyT-UNC), Agencia Nacional de Ciencia y Tecnología grants PICT 2005 N° 15-32256 and PICT 2010 N° 1232, and Ministerio de Ciencia y Tecnología de la provincia de Córdoba grant 0279-005429/2006. Martin G. Theumer is career investigator of the National Research Council of Argentina (CONICET).

5. References

Aly, S.E. & Hathout, A.S. (2011). Fate of aflatoxin B(1) in contaminated corn gluten during acid hydrolysis. *Journal of the Science of Food and Agriculture* Vol. 91, No. 3, (Feb), pp. 421-427, ISSN 1097-0010 (Electronic), 0022-5142 (Linking)

Aydin, A., Aksu, H. & Gunsen, U. (2010). Mycotoxin levels and incidence of mould in Turkish rice. *Environ Monit Assess*, (Sep 9), ISSN 1573-2959 (Electronic), 0167-6369 (Linking)

Baker, R.L., Brown, R.L., Chen, Z.Y., Cleveland, T.E. & Fakhoury, A.M. (2009a). A maize lectin-like protein with antifungal activity against Aspergillus flavus. *Journal of Food Protection* Vol. 72, No. 1, (Jan), pp. 120-127, ISSN 0362-028X (Print), 0362-028X (Linking)

Baker, R.L., Brown, R.L., Chen, Z.Y., Cleveland, T.E. & Fakhoury, A.M. (2009b). A maize trypsin inhibitor (ZmTIp) with limited activity against Aspergillus flavus. *Journal of Food Protection* Vol. 72, No. 1, (Jan), pp. 185-188, ISSN 0362-028X (Print), 0362-028X (Linking)

Bankole, S.A. & Mabekoje, O.O. (2004). Occurrence of aflatoxins and fumonisins in preharvest maize from south-western Nigeria. *Food Additives and Contaminants* Vol. 21, No. 3, (Mar), pp. 251-255, ISSN 0265-203X (Print), 0265-203X (Linking)

Brera, C., De Santis, B., Prantera, E., Debegnach, F., Pannunzi, E., Fasano, F., Berdini, C., Slate, A.B., Miraglia, M. & Whitaker, T.B. (2010). Effect of sample size in the evaluation of "in-field" sampling plans for aflatoxin B(1) determination in corn. *Journal of Agricultural and Food Chemis ry* Vol. 58, No. 15, (Aug 11), pp. 8481-8489, ISSN 1520-5118 (Electronic), 0021-8561 (Linking)

Castells, M., Marin, S., Sanchis, V. & Ramos, A.J. (2008). Distribution of fumonisins and aflatoxins in corn fractions during industrial cornflake processing. *International Journal of Food Microbiology* Vol. 123, No. 1-2, (Mar 31), pp. 81-87, ISSN 0168-1605 (Print), 0168-1605 (Linking)

Cleveland, T.E., Carter-Wientjes, C.H., De Lucca, A.J. & Boue, S.M. (2009). Effect of soybean volatile compounds on Aspergillus flavus growth and aflatoxin production. *J Food Sci* Vol. 74, No. 2, (Mar), pp. H83-87, ISSN 1750-3841 (Electronic), 0022-1147 (Linking)

Chen, Z.Y., Brown, R.L., Rajasekaran, K., Damann, K.E. & Cleveland, T.E. (2006). Identification of a Maize Kernel Pathogenesis-Related Protein and Evidence for Its Involvement in Resistance to Aspergillus flavus Infection and Aflatoxin Production. *Phytopathology* Vol. 96, No. 1, (Jan), pp. 87-95, ISSN 0031-949X (Print), 0031-949X (Linking)

Deveci, O. & Sezgin, E. (2006). Changes in concentration of aflatoxin M1 during manufacture and storage of skim milk powder. *Journal of Food Protection* Vol. 69, No. 3, (Mar), pp. 682-685, ISSN 0362-028X (Print), 0362-028X (Linking)

Dorner, J.W. (2009). Biological control of aflatoxin contamination in corn using a nontoxigenic strain of Aspergillus flavus. *Journal of Food Protection* Vol. 72, No. 4, (Apr), pp. 801-804, ISSN 0362-028X (Print), 0362-028X (Linking)

Dorner, J.W. & Horn, B.W. (2007). Separate and combined applications of nontoxigenic Aspergillus flavus and A. parasiticus for biocontrol of aflatoxin in peanuts.

Mycopathologia Vol. 163, No. 4, (Apr), pp. 215-223, ISSN 0301-486X (Print), 0301-486X (Linking)

Gallo, A., Masoero, F., Bertuzzi, T., Piva, G. & Pietri, A. (2010). Effect of the inclusion of adsorbents on aflatoxin B1 quantification in animal feedstuffs. *Food Addit Contam Part A Chem Anal Control Expo Risk Assess* Vol. 27, No. 1, (Jan), pp. 54-63, ISSN 1944-0057 (Electronic), 1944-0057 (Linking)

Gandomi, H., Misaghi, A., Akhondzadeh Basti, A., Hamedi, H. & Ramezani Shirvani, Z. (2010). Effect of Zataria multiflora Boiss. essential oil on colony morphology and ultrastructure of Aspergillus flavus. *Mycoses*, (Oct 11), ISSN 1439-0507 (Electronic), 0933-7407 (Linking)

Gao, X., Brodhagen, M., Isakeit, T., Brown, S.H., Gobel, C., Betran, J., Feussner, I., Keller, N.P. & Kolomiets, M.V. (2009). Inactivation of the lipoxygenase ZmLOX3 increases susceptibility of maize to Aspergillus spp. *Molecular Plant-Microbe Interactions* Vol. 22, No. 2, (Feb), pp. 222-231, ISSN 0894-0282 (Print), 0894-0282 (Linking)

Goncalez, E., Felicio, J.D. & Pinto, M.M. (2001). Biflavonoids inhibit the production of aflatoxin by Aspergillus flavus. *Brazilian Journal of Medical and Biological Research* Vol. 34, No. 11, (Nov), pp. 1453-1456, ISSN 0100-879X (Print), 0100-879X (Linking)

Guan, S., Zhao, L., Ma, Q., Zhou, T., Wang, N., Hu, X. & Ji, C. (2010). In Vitro Efficacy of Myxococcus fulvus ANSM068 to Biotransform Aflatoxin B(1). *Int J Mol Sci* Vol. 11, No. 10, pp. 4063-4079, ISSN 1422-0067 (Electronic), 1422-0067 (Linking)

Guo, B.Z., Xu, G., Cao, Y.G., Holbrook, C.C. & Lynch, R.E. (2006). Identification and characterization of phospholipase D and its association with drought susceptibilities in peanut (Arachis hypogaea). *Planta* Vol. 223, No. 3, (Feb), pp. 512-520, ISSN 0032-0935 (Print), 0032-0935 (Linking)

IARC, 1993. IARC Monographs on the evaluation of carcinogenic risks to humans., Some naturally occurring substances: Food items and constituents, heterocyclic aromatic amines and mycotoxins. Aflatoxins, International Agency for Research on Cancer, Lyon, pp. 249-395

Johnson, N.M., Qian, G., Xu, L., Tietze, D., Marroquin-Cardona, A., Robinson, A., Rodriguez, M., Kaufman, L., Cunningham, K., Wittmer, J., Guerra, F., Donnelly, K.C., Williams, J.H., Wang, J.S. & Phillips, T.D. (2010). Aflatoxin and PAH exposure biomarkers in a U.S. population with a high incidence of hepatocellular carcinoma. *Science of the Total Environment* Vol. 408, No. 23, (Nov 1), pp. 6027-6031, ISSN 1879-1026 (Electronic), 0048-9697 (Linking)

Juglal, S., Govinden, R. & Odhav, B. (2002). Spice oils for the control of co-occurring mycotoxin-producing fungi. *Journal of Food Protection* Vol. 65, No. 4, (Apr), pp. 683-687, ISSN 0362-028X (Print), 0362-028X (Linking)

Kabak, B., Brandon, E.F., Var, I., Blokland, M. & Sips, A.J. (2009). Effects of probiotic bacteria on the bioaccessibility of aflatoxin B(1) and ochratoxin A using an in vitro digestion model under fed conditions. *Journal of Environmental Science and Health. Part B: Pesticides, Food Contaminants, and Agricultural Wastes* Vol. 44, No. 5, (Jun), pp. 472-480, ISSN 1532-4109 (Electronic), 0360-1234 (Linking)

Kensler, T.W., Roebuck, B.D., Wogan, G.N. & Groopman, J.D. (2011). Aflatoxin: a 50-year odyssey of mechanistic and translational toxicology. *Toxicological Sciences* Vol. 120 Suppl 1, (Mar), pp. S28-48, ISSN 1096-0929 (Electronic), 1096-0929 (Linking)

Klosowski, G. & Mikulski, D. (2010). The effect of raw material contamination with mycotoxins on the composition of alcoholic fermentation volatile by-products in raw spirits. *Bioresour Technol* Vol. 101, No. 24, (Dec), pp. 9723-9727, ISSN 1873-2976 (Electronic), 0960-8524 (Linking)

Klosowski, G., Mikulski, D., Grajewski, J. & Blajet-Kosicka, A. (2010). The influence of raw material contamination with mycotoxins on alcoholic fermentation indicators. *Bioresour Technol* Vol. 101, No. 9, (May), pp. 3147-3152, ISSN 1873-2976 (Electronic), 0960-8524 (Linking)

Kong, Q., Shan, S., Liu, Q., Wang, X. & Yu, F. (2010). Biocontrol of Aspergillus flavus on peanut kernels by use of a strain of marine Bacillus megaterium. *International Journal of Food Microbiology* Vol. 139, No. 1-2, (Apr 30), pp. 31-35, ISSN 1879-3460 (Electronic), 0168-1605 (Linking)

Kumar, R., Mishra, A.K., Dubey, N.K. & Tripathi, Y.B. (2007). Evaluation of Chenopodium ambrosioides oil as a potential source of antifungal, antiaflatoxigenic and antioxidant activity. *International Journal of Food Microbiology* Vol. 115, No. 2, (Apr 10), pp. 159-164, ISSN 0168-1605 (Print), 0168-1605 (Linking)

Li, Y., Qin, X., Cui, J., Dai, Z., Kang, X., Yue, H., Zhang, Y., Su, J., Cao, J., Ou, C., Yang, C., Duan, X. & Liu, Y. (2008). Proteome analysis of aflatoxin B1-induced hepatocarcinogenesis in tree shrew (Tupaia belangeri chinensis) and functional identification of candidate protein peroxiredoxin II. *Proteomics* Vol. 8, No. 7, (Apr), pp. 1490-1501, ISSN 1615-9861 (Electronic), 1615-9853 (Linking)

Mably, M., Mankotia, M., Cavlovic, P., Tam, J., Wong, L., Pantazopoulos, P., Calway, P. & Scott, P.M. (2005). Survey of aflatoxins in beer sold in Canada. *Food Additives and Contaminants* Vol. 22, No. 12, (Dec), pp. 1252-1257, ISSN 0265-203X (Print), 0265-203X (Linking)

Mohammadipour, M., Mousivand, M., Salehi Jouzani, G. & Abbasalizadeh, S. (2009). Molecular and biochemical characterization of Iranian surfactin-producing Bacillus subtilis isolates and evaluation of their biocontrol potential against Aspergillus flavus and Colletotrichum gloeosporioides. *Canadian Journal of Microbiology* Vol. 55, No. 4, (Apr), pp. 395-404, ISSN 0008-4166 (Print), 0008-4166 (Linking)

Mwanda, O.W., Otieno, C.F. & Omonge, E. (2005). Acute aflatoxicosis: case report. *East African Medical Journal* Vol. 82, No. 6, (Jun), pp. 320-324, ISSN 0012-835X (Print), 0012-835X (Linking)

Osman, N., El-Sabban, F.F., Al Khawli, A. & Mensah-Brown, E.P. (2004). Effect of foodstuff contamination by aflatoxin on the one-humped camel (Camelus dromedarius) in Al Ain, United Arab Emirates. *Aust Vet J* Vol. 82, No. 12, (Dec), pp. 759-761, ISSN 0005-0423 (Print), 0005-0423 (Linking)

Palumbo, J.D., Baker, J.L. & Mahoney, N.E. (2006). Isolation of bacterial antagonists of Aspergillus flavus from almonds. *Microbial Ecology* Vol. 52, No. 1, (Jul), pp. 45-52, ISSN 0095-3628 (Print), 0095-3628 (Linking)

Pandey, I. & Chauhan, S.S. (2007). Studies on production performance and toxin residues in tissues and eggs of layer chickens fed on diets with various concentrations of aflatoxin AFB1. *British Poultry Science* Vol. 48, No. 6, (Dec), pp. 713-723, ISSN 1466-1799 (Electronic), 0007-1668 (Linking)

Park, D.L. (2002). Effect of processing on aflatoxin. *Advances in Experimental Medicine and Biology* Vol. 504, pp. 173-179, ISSN 0065-2598 (Print), 0065-2598 (Linking)

Park, J.W. & Kim, Y.B. (2006). Effect of pressure cooking on aflatoxin B1 in rice. *Journal of Agricultural and Food Chemis ry* Vol. 54, No. 6, (Mar 22), pp. 2431-2435, ISSN 0021-8561 (Print), 0021-8561 (Linking)

Passone, M.A., Resnik, S. & Etcheverry, M.G. (2007). Antiaflatoxigenic property of food grade antioxidants under different conditions of water activity in peanut grains. *International Journal of Food Microbiology* Vol. 118, No. 1, (Aug 15), pp. 8-14, ISSN 0168-1605 (Print), 0168-1605 (Linking)

Penna, M.L. & Etcheverry, M. (2006). Impact on growth and aflatoxin B1 accumulation by Kluyveromyces isolates at different water activity conditions. *Mycopathologia* Vol. 162, No. 5, (Nov), pp. 347-353, ISSN 0301-486X (Print), 0301-486X (Linking)

Pietri, A., Bertuzzi, T., Pallaroni, L. & Piva, G. (2004). Occurrence of mycotoxins and ergosterol in maize harvested over 5 years in Northern Italy. *Food Additives and Contaminants* Vol. 21, No. 5, (May), pp. 479-487, ISSN 0265-203X (Print), 0265-203X (Linking)

Polivkova, Z., Smerak, P., Demova, H. & Houska, M. (2010). Antimutagenic effects of lycopene and tomato puree. *J Med Food* Vol. 13, No. 6, (Dec), pp. 1443-1450, ISSN 1557-7600 (Electronic), 1096-620X (Linking)

Probst, C., Njapau, H. & Cotty, P.J. (2007). Outbreak of an acute aflatoxicosis in Kenya in 2004: identification of the causal agent. *Applied and Environmental Microbiology* Vol. 73, No. 8, (Apr), pp. 2762-2764, ISSN 0099-2240 (Print), 0099-2240 (Linking)

Rahmani, A., Jinap, S. & Soleimany, F. (2010). Validation of the procedure for the simultaneous determination of aflatoxins ochratoxin A and zearalenone in cereals using HPLC-FLD. *Food Addit Contam Part A Chem Anal Control Expo Risk Assess* Vol. 27, No. 12, (Dec), pp. 1683-1693, ISSN 1944-0057 (Electronic), 1944-0057 (Linking)

Reverberi, M., Fabbri, A.A., Zjalic, S., Ricelli, A., Punelli, F. & Fanelli, C. (2005). Antioxidant enzymes stimulation in Aspergillus parasiticus by Lentinula edodes inhibits aflatoxin production. *Applied Microbiology and Biotechnology* Vol. 69, No. 2, (Nov), pp. 207-215, ISSN 0175-7598 (Print), 0175-7598 (Linking)

Rocha, L.O., Nakai, V.K., Braghini, R., Reis, T.A., Kobashigawa, E. & Correa, B. (2009). Mycoflora and co-occurrence of fumonisins and aflatoxins in freshly harvested corn in different regions of Brazil. *Int J Mol Sci* Vol. 10, No. 11, (Nov), pp. 5090-5103, ISSN 1422-0067 (Electronic), 1422-0067 (Linking)

Sangare-Tigori, B., Moukha, S., Kouadio, H.J., Betbeder, A.M., Dano, D.S. & Creppy, E.E. (2006). Co-occurrence of aflatoxin B1, fumonisin B1, ochratoxin A and zearalenone in cereals and peanuts from Cote d'Ivoire. *Food Additives and Contaminants* Vol. 23, No. 10, (Oct), pp. 1000-1007, ISSN 0265-203X (Print), 0265-203X (Linking)

Sanyal, A.J., Yoon, S.K. & Lencioni, R. (2010). The etiology of hepatocellular carcinoma and consequences for treatment. *The Oncologist* Vol. 15 Suppl 4, pp. 14-22, ISSN 1549-490X (Electronic), 1083-7159 (Linking)

Sghaier, M.B., Boubaker, J., Neffati, A., Limem, I., Skandrani, I., Bhouri, W., Bouhlel, I., Kilani, S., Chekir-Ghedira, L. & Ghedira, K. (2010). Antimutagenic and antioxidant potentials of Teucrium ramosissimum essential oil. *Chem Biodivers* Vol. 7, No. 7, (Jul), pp. 1754-1763, ISSN 1612-1880 (Electronic), 1612-1872 (Linking)

Stenske, K.A., Smith, J.R., Newman, S.J., Newman, L.B. & Kirk, C.A. (2006). Aflatoxicosis in dogs and dealing with suspected contaminated commercial foods. *J Am Vet Med Assoc* Vol. 228, No. 11, (Jun 1), pp. 1686-1691, ISSN 0003-1488 (Print), 0003-1488 (Linking)

Sugita-Konishi, Y., Sato, T., Saito, S., Nakajima, M., Tabata, S., Tanaka, T., Norizuki, H., Itoh, Y., Kai, S., Sugiyama, K., Kamata, Y., Yoshiike, N. & Kumagai, S. (2010). Exposure to aflatoxins in Japan: risk assessment for aflatoxin B1. *Food Addit Contam Part A Chem Anal Control Expo Risk Assess* Vol. 27, No. 3, (Mar), pp. 365-372, ISSN 1944-0057 (Electronic), 1944-0057 (Linking)

Theumer, M.G., Canepa, M.C., Lopez, A.G., Mary, V.S., Dambolena, J.S. & Rubinstein, H.R. (2010). Subchronic mycotoxicoses in Wistar rats: assessment of the in vivo and in vitro genotoxicity induced by fumonisins and aflatoxin B(1), and oxidative stress biomarkers status. *Toxicology* Vol. 268, No. 1-2, (Jan 31), pp. 104-110, ISSN 1879-3185 (Electronic), 0300-483X (Linking)

Theumer, M.G., Lopez, A.G., Masih, D.T., Chulze, S.N. & Rubinstein, H.R. (2003). Immunobiological effects of AFB1 and AFB1-FB1 mixture in experimental subchronic mycotoxicoses in rats. *Toxicology* Vol. 186, No. 1-2, (Apr 15), pp. 159-170, ISSN 0300-483X (Print), 0300-483X (Linking)

Wang, J., Ogata, M., Hirai, H. & Kawagishi, H. (2011). Detoxification of aflatoxin B1 by manganese peroxidase from the white-rot fungus Phanerochaete sordida YK-624. *FEMS Microbiology Letters* Vol. 314, No. 2, (Jan), pp. 164-169, ISSN 1574-6968 (Electronic), 0378-1097 (Linking)

Whittaker, S., Marais, R. & Zhu, A.X. (2010). The role of signaling pathways in the development and treatment of hepatocellular carcinoma. *Oncogene* Vol. 29, No. 36, (Sep 9), pp. 4989-5005, ISSN 1476-5594 (Electronic), 0950-9232 (Linking)

Wojnowski, L., Turner, P.C., Pedersen, B., Hustert, E., Brockmoller, J., Mendy, M., Whittle, H.C., Kirk, G. & Wild, C.P. (2004). Increased levels of aflatoxin-albumin adducts are associated with CYP3A5 polymorphisms in The Gambia, West Africa. *Pharmacogenetics* Vol. 14, No. 10, (Oct), pp. 691-700, ISSN 0960-314X (Print), 0960-314X (Linking)

Xu, J., Wang, H., Ji, R. & Luo, X. (2003). [Study on the effect of the growth and aflatoxin production by Aspergillus flavus parasiticus NRRL 2999 in the present of Lactobacillus plantarum ATCC 8014]. *Wei Sheng Yan Jiu* Vol. 32, No. 4, (Jul), pp. 334-338, ISSN 1000-8020 (Print), 1000-8020 (Linking)

Zachariasova, M., Cajka, T., Godula, M., Malachova, A., Veprikova, Z. & Hajslova, J. (2010). Analysis of multiple mycotoxins in beer employing (ultra)-high-resolution mass spectrometry. *Rapid Communications in Mass Spectrometry* Vol. 24, No. 22, (Nov 30), pp. 3357-3367, ISSN 1097-0231 (Electronic), 0951-4198 (Linking)

Zhang, L., Ye, Y., An, Y., Tian, Y., Wang, Y. & Tang, H. (2011). Systems responses of rats to aflatoxin B1 exposure revealed with metabonomic changes in multiple biological matrices. *J Proteome Res* Vol. 10, No. 2, (Feb 4), pp. 614-623, ISSN 1535-3907 (Electronic), 1535-3893 (Linking)

Zhang, X.H. & Chen, J.M. (2004). [Comparison between the post-column derivatization with bromine by HPLC and the fluorometric analysis for determination of aflatoxins in medicinal herbs and plant extracts]. *Yao Xue Xue Bao* Vol. 39, No. 12, (Dec), pp. 997-1000, ISSN 0513-4870 (Print), 0513-4870 (Linking)

16

Binding of Aflatoxin B_1 to Lactic Acid Bacteria and *Saccharomyces cerevisiae* in vitro: A Useful Model to Determine the Most Efficient Microorganism

Romina P. Pizzolitto, Dante J. Bueno, María R. Armando, Lilia Cavaglieri, Ana M. Dalcero and Mario A. Salvano

Universidad Nacional de Río Cuarto
Argentina

1. Introduction

Mycotoxins are toxic fungal metabolites found as contaminants in many agricultural products. Feeds contaminated with mycotoxins have a health risk to animals and, as a consequence, may cause big economical losses due to the low efficacy of animal husbandry (Richard, 2007). In addition, directly or indirectly (animal by-products) contaminated foods may also have a health risk to humans (CAST, 2003; Hussein & Brasel, 2001; Wild, 2007).

Aflatoxins (AFs), a group of potent mycotoxins with mutagnic, carcinogenic, teratogenic, hepatotoxic and immunosupresive properties, are of particular importance because of their major occurrence and adverse effects on animal and human health, generalized as "aflatoxicosis" (CAST, 2003; Hussein & Brasel, 2001; Magnoli et al., 2011).

The AFs are produced by genus *Aspergillus*, mainly *A. flavus, A. parasiticus* and *A. nomius*, that grow on a variety of raw material during growth, harvest, storage and transportation of for example, the cereal used in the preparation of food and feed commodities (Ito et al., 2001; Kurtzman et al., 1987; Payne, 1998; Pereyra et al., 2010).

The investigation of strategies to prevent the presence of AFs in foods, as well as, to eliminate, inactivate or reduce the bio-availability of these mycotoxins in contaminated products include physical, chemical, and biological methods (Bueno et al., 2001; CAST, 2003; Kabak et al., 2006). Limitations such as the loss of nutritional and sensory qualities of the product, the expensive equipment required for these techniques and the impossibility to guarantee the desired results, have allowed us to consider the hipothesis that foods and feeds can always be potentially contaminated with aflatoxins. For instance, in the poultry industry aflatoxin B_1 (AFB_1) is almost an unavoidable feed contaminant and levels from 0-200 ng/g have been reported (Dalcero et al., 1997).

On the other hand, it is known that lactic acid bacteria (LAB) and some yeast, principally *Saccharomyces cerevisiae*, are capable to bind AFs in liquid media, apparently to cell wall components, polysaccharides and peptidoglycans of LAB (Haskard et al., 2001; Latinen et al., 2004) and glucomannans of yeast (Karaman et al., 2005; Raju & Devegowda, 2000) and therefore could be used as potential mycotoxin decontaminating (Armando et al., 2011; El-Nezami et al., 1998; Haskard et al., 2000, 2001; Hernandez-Mendoza et al., 2009; Lee et al., 2003; Peltonen et al., 2001; Shetty et al., 2007). The inclusion of appropriate microorganisms in the contaminated diet could prevent the absorption of mycotoxins during their passage in the gastrointestinal tract and eliminated in the faeces (Bueno et al., 2007; El-Nezami et al., 2000; Gratz et al., 2004, 2007). Moreover, Kankaanpää et al. (2000) showed that the binding of AFB_1 to the surface of LAB reduced their adhesive properties, and the accumulation of aflatoxins in the intestine may therefore be reduced via the increased excretion of an aflatoxin-bacteria complex.

These considerations encouraged the recent emphasis on biological methods, but mainly focused on preventing AFs absorption in the gastrointestinal tract of the consumers, including these microorganisms in the diet and so prevent the aflatoxicosis effects.

The first step in this direction is the selection of the most efficient microorganism for AFB_1 removing and while many researchers have assayed LAB and yeast with AFB_1 binding abilities (Ciegler et al., 1966; El-Nezami et al., 1998; Gourama & Bullerman, 1995; Haskard et al., 2001; Line et al., 1994; Oatley et al., 2000) no clear mechanism for this effect has been provided. Thus, this selection frequently is performed using a single concentration of AFB_1, but we demonstrated that the microorganism efficiency may change when the mycotoxin concentration is modified (Bueno et al., 2007; Pizzolitto, 2011), therefore the microorganism selected could not be the most competent.

In this context, we investigated the nature of the interaction between different microorganisms and AFB_1 molecule, which led us to develop a model to explain the binding of AFB_1 by LAB and *Saccharomyces cerevisiae* strains. This model allows an estimation of two important parameters related to a microorganism's capacity for dietary decontamination: the number of binding sites for AFB_1 in the surface microorganism (M) and the equilibrium constant of the process involved (K_{eq}), both of them are useful in the selection of the most suitable microorganism in a wide range of AFB_1 concentration (Bueno et al., 2007).

In adittion, studies of viability of the microorganisms in the salivary and gastrointestinal tract, cell adhesion, autoaggregation, coaggregation and antimicrobial activity against pathogen strains, were also evaluated as a way to research potential beneficial properties on the host (Armando et al., 2011).

Thus, in this chapter we describe the development and application of an *in vitro* methodology to evaluate the aflatoxin B_1 binding ability, gastrointestinal tolerance and potential beneficial properties of *Saccharomyces cerevisiae* strains, useful to select the more appropriated microorganism to be assayed in expensive, complicated but necessary *in vivo* studies.

2.1 Study of microorganism-aflatoxin B_1 interaction

To select accurately the most efficient microorganism to bind AFB_1, is very important so as to protect against aflatoxicosis by inclusion of microorganisms in the diet. Usually, the methodology assayed has been a selection of several candidates using a unique mycotoxin concentration (Haskard et al., 2001; Hernandez-Mendoza et al., 2009; Peltonen et al., 2001; Shetty & Jespersen, 2006). Table 1, developed in our laboratory, is a clear example of this methodology and its analysis shows that the efficiency of the microorganisms is strain dependent, so that toxin removal ranged from 13 to 42% for LAB strains and 16 to 40% for the yeast strains tested.

Microorganism	Source	% AFB_1 bound ± SD[a]
***Lactobacillus acidophilus* Po22**	Poultry cecum	42.8 ± 1.7
***L. acidophilus* Po7**	Poultry	34.6 ± 1.6
***L. acidophilus* 24**	Dairy	32.6 ± 2.0
***L. casei* 1**	Dairy	27.6 ± 1.5
***L. fermentum* 23**	Human	34.6 ± 3.2
***L. acidophilus* CRL 1014**	ATCC collection[b]	25.4 ± 1.7
***L. fermentum* subsp. *cellobiosus* 408**	Poultry	13.2 ± 9.8
***Saccharomyces cerevisiae* RC016**	Pig gut	15.8 ± 3.6
***S. cerevisiae* RC012**	Feedstuff	29.6 ± 2.4
***S. cerevisiae* RC008**	Feedstuff	20.6 ± 2.6
***S. cerevisiae* RC009**	Feedstuff	16.4 ± 1.2
***S. cerevisiae* 01**	Poultry faeces	28.6 ± 3.5
***S. cerevisiae* 03**	Poultry faeces	26.6 ± 2.9
***S. cerevisiae* 05**	Poultry faeces	33.4 ± 1.9
***S. cerevisiae* 08**	Poultry faeces	36.4 ± 2.7
***S. cerevisiae* CECT 1891**	STCC collection[c]	40.0 ± 2.5

[a] The percentage of AFB_1 bound to cells was calculated as the difference between the total AFB_1 (5 µg ml^{-1}) and the amount of free AFB_1 (supernatant). Values are means ± standard deviations for duplicate samples.
[b]ATCC, American Type Culture Collection, Manassas, VA, USA.
[c] STCC, Spanish Type Culture Collection, University of Valencia, Valencia, Spain.

Table 1. Percentage of AFB_1 bound to cells upon exposure to viable microorganisms Bacteria, 2 x 10^8 CFU ml^{-1} and yeast 1 x 10^7 CFU ml^{-1} were incubated with 1 ml of AFB_1 (5 µg ml^{-1}) in PBS for 30 min at 37°C. The microorganisms were then pelleted by centrifugation, and the supernatant was collected for free AFB_1 analysis by HPLC according to Bueno et al. (2007).

The same experiment, but using three different concentrations of aflatoxin B_1, shows that the microorganism assayed to one concentration could not be the most efficient when the latter is changed (Table 2 and 3). Thus when AFB_1 concentration was 50 ng ml^{-1}, *S. cerevisiae* RC 016 was the most effective strain, but *S. cerevisiae* 08 and *S. cerevisiae* CECT 1891 were the best when AFB_1 concentration was increased at 100 ng ml^{-1}, and with 500 ng ml^{-1} of AFB_1 *S. cerevisiae* RC 016 was again the most efficient strain. In addition, *S. cerevisiae* 01 and *S. cerevisiae* 03 removed AFB_1 with similar ability when their concentrations were 50 and 100 ng ml^{-1}; however *S. cerevisiae* 01 was more effective at 500 ng ml^{-1} (Table 2). Lactic acid bacteria strains also showed the same behaviour, because *L. rhamnosus* I, *L. acidophilus* 24 and *L. casei* subsp. *rhamnosus* were the best at 50, 100 and 500 ng ml^{-1} respectively (Table 3).

Strains	AFB$_1$ concentration (ng ml^{-1})					
	50		100		500	
	AFB$_1$ binding					
	(ng ml^{-1})	%	(ng ml^{-1})	%	(ng ml^{-1})	%
S. cerevisiae CECT 1891	10.0 ± 4.3	20.0	**57.6 ± 8.6**	**57.6**	255.0 ± 32.1	51.0
S. cerevisiae RC 008	33.8 ± 0.1	67.6	45.6 ± 7.1	45.6	197.9 ± 24.1	38.2
S. cerevisiae RC 012	15.3 ± 1.6	29.6	21.5 ± 3.1	21.5	103.7 ± 9.4	20.2
S. cerevisiae RC 009	8.4 ± 0.8	16.8	21.5 ± 0.8	21.5	159.3 ± 1.2	31.8
S. cerevisiae RC 016	**41.6 ± 1.9**	**82.0**	49.1 ± 1.4	49.1	**328.8 ± 5.2**	**65.7**
S. cerevisiae 01	19.3 ± 1.2	38.6	31.7 ± 1.2	31.7	164.0 ± 6.9	32.8
S. cerevisiae 03	23.3 ± 2.9	46.6	34.5 ± 1.8	34.5	128.7 ± 5.8	25.7
S. cerevisiae 05	16.7 ± 1.2	33.4	24.0 ± 1.7	24.0	92.0 ± 6.9	18.4
S. cerevisiae 08	23.2 ± 2.8	46.4	**58.9 ± 2.1**	**58.9**	187.0 ± 18.2	37.4

Cells (10^7 CFU ml^{-1}) were suspended in PBS in the presence of AFB$_1$ at the indicated concentration and incubated as described in Table 1. AFB$_1$ analysis by HPLC was performed according to Trucksess et al. (1994). Data are means ± standard deviations from three experiments in duplicate.

Table 2. Aflatoxin B$_1$ remotion by *Saccharomyces cerevisiae* strains at three different mycotoxin concentrations

Strains	AFB$_1$ concentration (ng ml^{-1})					
	50		100		500	
	AFB$_1$ binding					
	(ng ml^{-1})	%	(ng ml^{-1})	%	(ng ml^{-1})	%
Lactobacillus casei subsp. *rhamnosus*	18.0 ± 3.4	36.0	56.3 ± 5.4	56.3	**338.0 ± 7.2**	**67.6**
L. rhamnosus I	**35.8 ± 1.3**	**71.6**	61.8 ± 3.1	61.8	254.7 ± 4.2	50.9
L. fermentum 23	22.3 ± 1.2	44.6	41.0 ± 2.8	41.0	225.3 ± 10.8	45.1
L. acidophilus 24	26.3 ± 1.9	52.6	**82.5 ± 3.2**	**82.5**	254.0 ± 25.9	50.8
L. casei 1	13.8 ± 0.9	27.6	27.5 ± 1.3	27.5	59.0 ± 1.7	11.8

Cells (5 x 10^8 CFU ml^{-1}). AFB$_1$ binding to cells was performed as described in Table 2. Data are means ± standard deviations from three experiments in duplicate.

Table 3. Aflatoxin B$_1$ remotion by lactic acid bacteria strains at three different mycotoxin concentrations

Therefore, as AFB$_1$ concentration is highly variable in foods and feeds, the methods of selection using a unique aflatoxin B$_1$ concentration may lead to erroneous results. Another very important condition to consider is how to quantify the concentration of employed microorganisms, CFU ml^{-1} or total cells ml^{-1}, where to evaluate whether viable and nonviable cells remove AFB_1 with the same efficiency becomes necessary.

Thus, the solution to these problems will be to know the mechanism involved in cell-aflatoxin interaction, and probably will also allow us to find the microorganism able to protect against aflatoxicosis in a wide range of mycotoxin concentration. In this sense, one of the objectives of the present study was to develop a theoretical model able to explain the binding of AFB_1 by LAB and *Saccharomyces cerevisiae* strains. With this purpose we studied as influences on the process involved the following conditions: microorganism-AFB_1 time contact, aflatoxin B_1 and microorganism concentration, cell viability, release of AFB_1 bound by cells and importance of the microorganism cell wall.

2.1.1 Microorganism-AFB_1 time contact

Table 4 summarizes some representative results (three LAB and three yeast strains) of the studies done in our laboratory when contact time between AFB_1 and the microorganisms were tested. By varying the incubation time, no significant difference in the amount of AFB_1 removed for LAB and yeast strains were observed. Furthermore, the process was fast, since in 1 minute the microorganism was able to bind the same amount of mycotoxin as in 6 h.

Time (min)	AFB₁ Binding (ng ml⁻¹)					
	***S. cerevisiae* 08**	***S. cerevisiae* RC016**	***S. cerevisiae* CECT 1891**	***L. fermentum* subsp. *cellobiosus* 408**	***L. casei* 1**	***L. acidophilus* P_{22}**
1	182.8 ± 18.2	318.5 ± 5.2	258.0 ± 14.2	61.5 ± 4.3	57.5 ± 2.2	89.4 ± 4.2
5	194.0 ± 12.6	312.8 ± 6.7	245.5 ± 10.3	64.3 ± 5.1	54.3 ± 3.1	91.8 ± 3.3
60	178.9 ± 15.7	332.0 ± 10.2	252.7 ± 15.6	62.8 ± 4.6	59.0 ± 4.2	88.9 ± 5.1
360	193.2 ± 14.6	326.6 ± 12.1	267.2 ± 12.6	63.8 ± 3.5	56.8 ± 2.9	92.3 ± 4.7

Table 4. Effect of incubation time on aflatoxin B_1 binding by viable cells of yeasts and LAB AFB_1 concentration: 0.5 µg ml^{-1}. Cells number: yeasts 1 x 10^7 CFU ml^{-1}; LAB 5 x 10^8 CFU ml^{-1}. AFB_1 binding was performed as described in Table 1, except that the incubation time varied as indicated in column 1. Data are means ± standard deviations from three experiments in duplicate. There is not significant differences ($P<0.05$) in the means values of each column.

These results are in agreement with other authors (El-Nezami et al., 1998; Peltonen et al., 2001) who have found no significant differences in AFB_1 removal by *E. coli*, *Propionibacterium* and several LAB strains after 72 h of incubation with the toxin. As for yeast, our results are consistent with those reported by Shetty et al. (2007), who have not observed differences between 0.5 and 12 h of time contact.

If the process needs so little time (one minute), it could suggest that neither the entrance of AFB_1 into cell nor its metabolic conversion is necessary, therefore microorganism cell wall components may be involved in aflatoxin B_1 remotion, as was suggested by various authors (Haskard et al., 2001; Karaman et al., 2005; Lahtinen et al., 2004; Raju & Devegowda, 2000).

2.1.2 Mycotoxin and microorganisms concentration

Effects of different AFB_1 concentration on toxin removal by LAB and yeast strains are shown in figure 1. Regardless of the studied strain, mycotoxin binding was dependent of its solution concentration and was always lineal at low values of AFB_1 and showed the transition to a plateau with higher toxin concentrations. The amount of toxin removed increased with increasing AFB_1 concentration, but the percentage removed decreased with increasing toxin concentration, because the saturation started. This behaviour strongly suggests that the microorganisms have a limited number of sites to bind AFB_1 either as free or occupied sites.

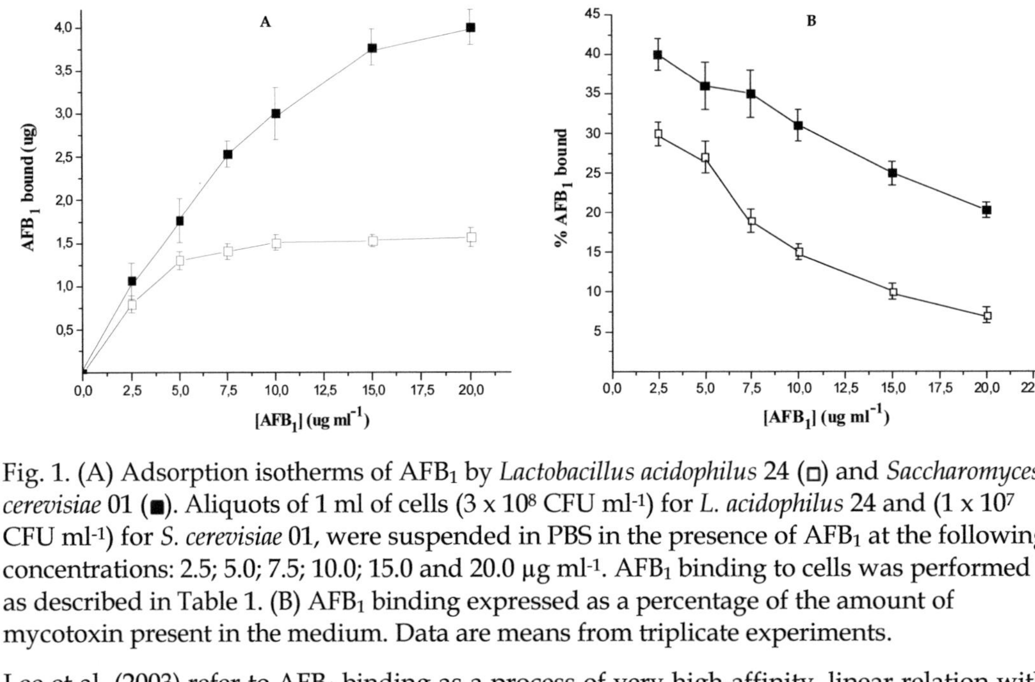

Fig. 1. (A) Adsorption isotherms of AFB_1 by *Lactobacillus acidophilus* 24 (□) and *Saccharomyces cerevisiae* 01 (■). Aliquots of 1 ml of cells (3 x 10^8 CFU ml^{-1}) for *L. acidophilus* 24 and (1 x 10^7 CFU ml^{-1}) for *S. cerevisiae* 01, were suspended in PBS in the presence of AFB_1 at the following concentrations: 2.5; 5.0; 7.5; 10.0; 15.0 and 20.0 μg ml^{-1}. AFB_1 binding to cells was performed as described in Table 1. (B) AFB_1 binding expressed as a percentage of the amount of mycotoxin present in the medium. Data are means from triplicate experiments.

Lee et al. (2003) refer to AFB_1 binding as a process of very high-affinity, linear relation with the toxin concentration used, and therefore, the amount of AFB_1 bound should be "limitless"; in other words they conclude that the bacterial surface does not have a defined number of binding sites. Our results do not support this idea. An important difference could be the number of microorganisms used in the experiments (10^{10} for Lee et al. and 10^8 for us), including more than a hundred times higher than ours for similar concentrations of AFB_1, so that the saturation phenomenon could not be observed.

When a growing number of microorganisms were suspended in PBS in the presence of a fixed AFB_1 concentration, we observed that the increase in bacterial or yeast concentration also reported an increase in AFB_1 binding, but it was never sufficient to bind all toxins present in the medium. Figure 2 shows the results with *Lactobacillus casei* subsp. *rhamnosus* which are similar to those obtained with all LAB and yeast strains we analysed.

Other authors, including El-Nezami et al. (1998), have also reported that removal of the toxin is cell number dependent and the bacterial increase was never sufficient to bind more than 90% of the toxin present in the environment. These results suggest that the process reached equilibrium between bound toxins (occupied sites) and unbound toxins (free sites) and therefore a reversible process could be involved in AFB_1 decontamination.

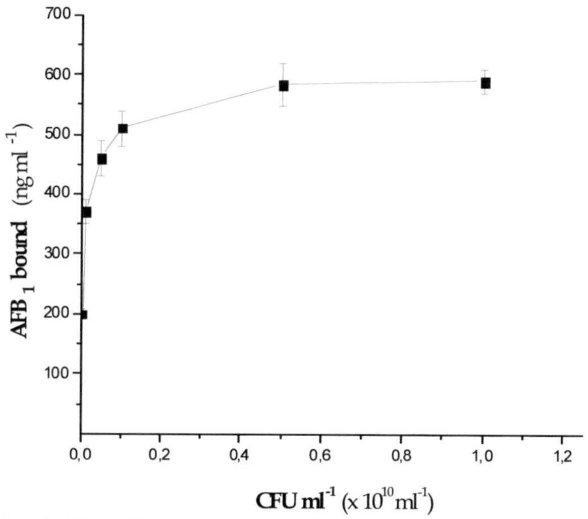

Cells, at the concentration indicated, were suspended in PBS in the presence of AFB_1 at a concentration of 750 ng ml^{-1} and incubated for 30 min at 37°C. AFB_1 binding to cells was performed as described in Table 2. Data are means ± standard deviations for triplicate samples.

Fig. 2. Effect of bacterial concentration on AFB_1 removal by *Lactobacillus casei* subsp. *rhamnous*

2.1.3 Cell viability

To further study the mechanism involved in AFB_1 removing, we examined whether cell viability affects the AFB_1 binding. Results in Table 5 did not show significant differences in remotion of the toxin by viable and nonviable cells (obtained by heat treatment).

Our results are similar to those obtained by El-Nezami et al. (1998) and some of the tested strains by Haskard et al. (2001), but different from other authors as Lee et al. (2003) and Shetty et al. (2007) who reported that heat treatment of cells enhanced their binding abilities among 20-50%. However, this effect was not observed in our study with none of the microorganisms tested (seven LAB and nine yeast strains, nine of them including in table 5). Therefore, the fact that non viable and viable microorganisms are able to eliminate AFB_1 with similar efficiency, suggests that the process involved does not require metabolic conversion of the toxin by cells and on the other hand, that the total microorganism number (cells ml^{-1}) should be employed in the estimation of cellular concentration, instead of CFU ml^{-1}. Additionally, these results are indicating that the inclusion of viable or nonviable microorganisms in the diet of animals would be equally effective against aflatoxicosis. This is very important because the possibility of using nonviable cells decreases the risks of their inclusion in the diet. Although the studies to ensure that these microorganisms are innocuous for animal health are not easy they become highly necessary.

Strains	AFB_1 binding (ng ml^{-1})	
	Viable	Nonviable
L. fermentum 23	41.8 ± 3.6	45.0 ± 1.1
L. acidophilus 24	83.4 ± 3.5	78.7 ± 2.9
L. casei subsp. *rhamnosus*	54.7 ± 6.5	56.3 ± 10.5
L. rhamnosus I	61.8 ± 3.1	63.0 ± 2.1
L. paracasei subsp. *paracasei*	72.4 ± 5.5	69.5 ± 4.5
L. casei 1	27.5 ± 3.8	23.5 ± 1.1
S. cerevisiae CECT 1891	55.2 ± 7.5	53.0 ± 5.4
S. cerevisiae 08	58.9 ± 2.1	55.8 ± 2.3
S. cerevisiae RC016	49.1 ± 1.4	52.4 ± 1.7

Cells (10^7-10^8 cell ml^{-1}) were suspended in PBS in the presence of AFB_1 at a concentration of 100 ng ml^{-1} and incubated for 30 min at 37°C. AFB_1 binding to cells was performed as described in Table 2. Non viable cells were obtained by heat treatment (autoclaving for 20 min at 121°C) from the same sample of viable cells. Data are means ± standard deviations from three experiments in duplicate. There is not significant differences ($P<0.05$) in the means values of each row.

Table 5. Aflatoxin B_1 binding upon exposure to viable and nonviable cells

2.1.4 Aflatoxin B_1 released

The stability of the complex aflatoxin B_1-microorganism was studied by the application of repeated washings with PBS buffer (aqueous solvent) or acetonitrile (organic solvent in which AFB_1 is very soluble) of the cellular pellets that previously bound the mycotoxin, additionally the effect of the variation in the washing time was included.

Table 6 shows some representative results from all LAB and a yeast strains assayed in our laboratory. After five washings with PBS or acetonitrile, the AFB_1 released was ≥ 90% for the organic solvent treatment and close to 50% for the PBS treatment; on the other hand, the washing time did not change the release percentages when it varied from 1 to 60 minutes.

These results suggest that the process involved is fast and reversible. Moreover they confirm that metabolic conversion of the toxin by cells did not take place, because the aflatoxin B_1 was released in the same chemical form from microorganisms.

Reversibility of the process has previously been reported by other authors, for example, twelve LAB strains in both viable and non viable forms, tested by Haskard et al. (2001) exhibited reversible binding of AFB_1 after five washings too. Moreover, they also noted that the release of bound toxin was dependent on the washing solution employed, because only 6% to 11% of the bound AFB_1 was released using water, but when the complex was suspended in methanol, acetonitrile, chloroform, or benzene 83% to 99% of the bound AFB_1 was released. Similar findings have been reported by Peltonen et al. (2001) and Hernandez-Mendoza et al. (2009), the latter authors showed that employing PBS around 20-30% of AFB_1 bound was released and suggested that the aflatoxin B_1 is attached to the bacteria by weak, non covalent interactions that could be at least partially reversible.

Time (min)	AFB$_1$ released (%)							
	PBS				Acetonitrile			
	L. acidophilus P$_{22}$	*L. fermentum* subsp. *cellobiosus* 408	*L. casei* 1	*S. cerevisiae* CECT 1891	*L. acidophilus* P$_{22}$	*L. fermentum* subsp. *cellobiosus* 408	*L. casei* 1	*S. cerevisiae* CECT 1891
1	57.9 ±7.0	50.4 ± 3.5	42.3 ± 1.8	40.8 ± 2.1	98.3 ± 9.8	95.7 ± 5.6	94.4 ± 4.3	90.6 ± 4.1
10	58.7 ± 8.8	51.9 ± 2.7	43.8 ± 2.1	39.7 ± 1.2	97.0 ± 4.5	89.6 ± 8.9	91.8 ± 4.8	92.4 ± 3.9
60	61.0 ± 3.5	53.8 ± 3.9	44.1 ± 3.4	41.2 ± 1.5	98.4 ± 5.9	95.4 ± 5.8	95.9 ± 4.8	94.1 ± 4.2

For the aflatoxin release assay, cells (10^7-10^8 cells ml^{-1}) were incubated with AFB$_1$ at a concentration of 0.5 μg ml^{-1} for 30 min at 37°C and then centrifuged. Pelleted cells with bound AFB$_1$ were suspended in 1 ml of PBS or acetonitrile and incubated at the indicated times at 37°C with shaking. After that the microorganisms were pelleted by centrifugation, the supernatant containing the released AFB$_1$ was collected by HPLC analysis as was performed in Table 2. This process was repeated five times. The AFB$_1$ released (total from five washing) was expressed as a percentage of the total AFB$_1$ bound. Data are means ± standard deviations from triplicate experiments.

Table 6. Effects of different solvents and washing time on AFB$_1$ released by microorganisms

2.1.5 Importance of the microorganism cell wall

Although bibliographic data suggest that structural components of the cell wall of yeasts and LAB, are responsible in AFB$_1$-microorganism interaction (Karaman et al., 2005; Lahtinen et al., 2004), we designed an experiment to evaluate if yeast without cell wall (spheroplasts) are able to remove AFB$_1$, in order to confirm this assertion.

Sample*	AFB$_1$ binding			
	0.25 μg ml^{-1}#		2.5 μg ml^{-1}#	
	(μg ml^{-1})	(%)	(μg ml^{-1})	(%)
Cells control	0.14 ± 0.01^a	56.0	1.25 ± 0.10^a	50.0
Spheroplasts	0.02 ± 0.01^b	8.0	0.05 ± 0.01^b	2.0
Supernatant of spheroplasts	0.02 ± 0.00^b	8.0	0.08 ± 0.01^c	3.2

*Cells and spheroplasts: 10^7 ml^{-1}. #AFB$_1$ concentration in liquid media. The samples were suspended in 1 ml PBS in the presence of AFB$_1$ at a concentration of 0.25 μg ml^{-1} or 2.5 μg ml^{-1} and incubated for 30 min at 37°C. AFB$_1$ binding was performed as described in Table 2. Data are means ± standard deviations from triplicate experiments. Means with different letters in the same column differ significantly ($P<0.05$).

Table 7. Aflatoxin B$_1$ binding by cells, spheroplasts and supernatant of spheroplasts of *Saccharomyces cerevisiae* CECT 1891

The spheroplasts were obtained by the treatment with Zymolyase containing a protease activity with affinity for mannoproteins, and a β1,3 glucanase activity. The actions of both enzymatic activities were required to lyses yeast cell wall (Ovalle et al., 1998) and allowed us to obtain spheroplasts and the released products of the enzymatic breakdown, separately.

Thus, to determine the AFB_1 binding with three different samples: i) whole cells of *S. cerevisiae* CECT 1891 (cells control), ii) spheroplasts from cell control and iii) a concentrate of the supernatant from spheroplasts corresponding to 10^7 cells, was possible.
As Table 7 shows, neither spheroplasts nor its supernatant were able to remove AFB_1 from liquid medium, since very low uptake rates not even changed when the concentration of aflatoxin B_1 in the medium was increased 10 times, suggesting that these binds were nonspecific.
These results confirm that the compounds involved in AFB_1 binding to yeast, are components of the cell wall and that it must keep its structure in order to remove AFB_1 effectively. Similar results were reported by Hernandez-Mendoza et al. (2009), who demonstrated that the data obtained in binding assays with bacterial cell wall indicated that these purified fragments effectively bind AFB_1 as reported previously by Lahtinen et al. (2004). Furthermore, loss of the bacterial cell wall in response to treatments with enzymes showed a reduction in AFB_1 binding capacity relative to that of whole cells. These results demonstrate the importance of cell wall integrity in binding AFB_1 by LAB strains, and confirm the role of a cell wall-related physical phenomenon as opposed to a metabolic degradation reaction.

2.1.6 Mechanism proposed for the interaction between Aflatoxin B_1 with yeast and lactic acid bacteria strains

According to an integrated synthesis of the results reported above, it is clear that: (*i*) the removal and release of toxins is a fast and reversible process, (*ii*) this process does not involve AFB_1 chemical modification, (*iii*) the amount of AFB_1 removed is toxin- and cell concentration-dependent, (*iv*) the same result is obtained with viable and nonviable (heat-treated) cells and (*v*) the cell wall of the microorganism integrity is necessary for the "binding" mechanism between AFB_1 and the cells. Briefly, the process involved is, by nature, reversible and fast kinetic. Accordingly, this process can be analyzed as a **PHISICAL ADSORPTION** (physisorption), and probably the binding forces involved are a weak Van der Waals type, hydrogen bonds, or hydrophobic interaction.
Following this an adsorption phenomenon to the external microorganism surface to explain AFB_1 binding is proposed (Figure 3). This model considers the attachment of AFB_1 molecules to the microorganism surface. The relationship between the amounts of the AFB_1 at the microorganism surface as a function of its solution concentration is described by an adsorption isotherm. The shape of the isotherm shows linearity at low values of AFB_1 and then shows the transition to a plateau (Figure 4A). This type of isotherm can be described by the following equation:
Adsorption = $M\,[AFB_1]_{eq} \times K_{eq} / 1 + [AFB_1]_{eq}\,K_{eq}$ (Figure 3)
where **M** is the maximum number of adsorption sites per microorganism, and K_{eq} (expressed in liters per mole) is equivalent to the affinity (or cohesion force) of AFB_1 molecules for the adsorption sites. The linearized form of the isotherm is the double-reciprocal plot from the saturation curve (**1/adsorption = $1/[AFB_1]_{eq}\ 1/M\ K_{eq} + 1/M$**), as shown in Figure 4B. From the slope and interception of the resulting line, factors **M** and K_{eq} can be determined. The most efficient microorganism would be that having maximal M and K_{eq} values. Note that this physisorption model does not discriminate between viable and nonviable cells, i.e., cell concentrations should be determined by a hemocytometer instead of CFU per milliliter.

$$AFB_1 + S \rightleftarrows S\text{-}AFB_1$$

$$K_{eq} = \frac{[S\text{-}AFB_1]}{[AFB_1]\ [S]} = \frac{Xso}{[AFB_1]\ Xsf}$$

$$Xso + Xsf = 1 \quad \text{and} \quad \varnothing = Xso$$

$$Xsf = 1 - \varnothing \quad \text{and} \quad K_{eq} = \varnothing /\ [AFB_1]\ (1 - \varnothing)$$

$$\varnothing = [AFB_1]\ K_{eq} \ /\ 1 + [AFB_1]\ K_{eq}$$

$$\text{Adsorption} = M \times \varnothing \quad \text{and} \quad \text{Adsorption} = \frac{M\,[AFB_1]\ K_{eq}}{1 + [AFB_1]\ K_{eq}}$$

$$1/\text{adsorption} = 1/[AFB_1]_{eq}\ \ 1/M\,K_{eq} + 1/M$$

Fig. 3. Theoretical model proposed to explain the adsorption process of AFB_1 by LAB and *S. cerevisiae*. The equations permit the calculation of the total binding sites per microorganism (M) and the equilibrium constant (K_{eq}) involved in the process. Sf is the amount of free sites in the surface cellular. So represents the occupied sites in the cell surface and is equivalent to S-AFB_1, determined as the AFB_1 bound to the cell. [AFB_1] is the AFB_1 concentration in the equilibrium condition of the system, determined as the free AFB_1 in the medium. Xso (Ø) is the molar fraction of occupied sites (mol So/mol total). Xsf (1 - Ø) is the molar fraction of free sites (mol Sf/mol total). Adsorption is the amount of molecules of AFB_1 bound per cell (M x Ø).

Figure 4A shows the saturation curve of *L. acidophilus* Po_{22} and *L. fermentum* subsp. *cellobiosus* 408 when the process is considered physisorption and indicates that the two strains have similar AFB_1 binding efficiency per bacterium, particularly for low concentrations of toxin. This result is different from the ones reported in Table 1 based only on viable bacteria (CFU ml^{-1}) and assayed with a unique mycotoxin concentration, since *L. acidophilus* Po_{22} was more efficient in AFB_1 binding (42.8%) than *L. fermentum* subsp. *cellobiosus* 408 (13.2%). However, we determined that in these assays Po_{22} strain had more dead cells than 408 strain, consequently Po_{22} had more total cells and showed higher percentage of AFB_1 binding. In terms of our proposed adsorption model, *L. fermentum* subsp. *cellobiosus* 408 has lower M values but higher K_{eq} values than *L. acidophilus* Po_{22} (Table 8), and these two factors balance to give toxin adsorption efficiencies of 9.37 and 6.25 x 10^{10} respectively.

According to adsorption model, a larger cell surface is associated with higher total sites per cell (M). To test this possibility, we measured AFB_1 adsorption in three yeast strains and the saturation curves are shown in Figure 5.

The adsorption of AFB_1 by *S. cerevisiae* RC016 (from pig gut), *S. cerevisiae* RC008 (from feed stuff) and *S. cerevisiae* CECT 1891 (from culture collection) was dependent on the toxin concentration in the medium, which is similar to the results showed in Figure 1 by

S. cerevisiae 01 and *L. acidophilus* 24. The data from Figure 5 were employed to construct the respective adsorption isotherms to obtain the M and K_{eq} values for these systems, and they are shown in Table 8. The M values were 25- to 1,000-fold higher for the yeast strains than for the bacteria, whereas K_{eq} values were similar, as differences never exceeded 3 times. Thus, the yeast strains respect to bacterial strains, showed a ~50-300-fold higher efficiency to AFB_1 removal from the medium, mainly for their high M values.

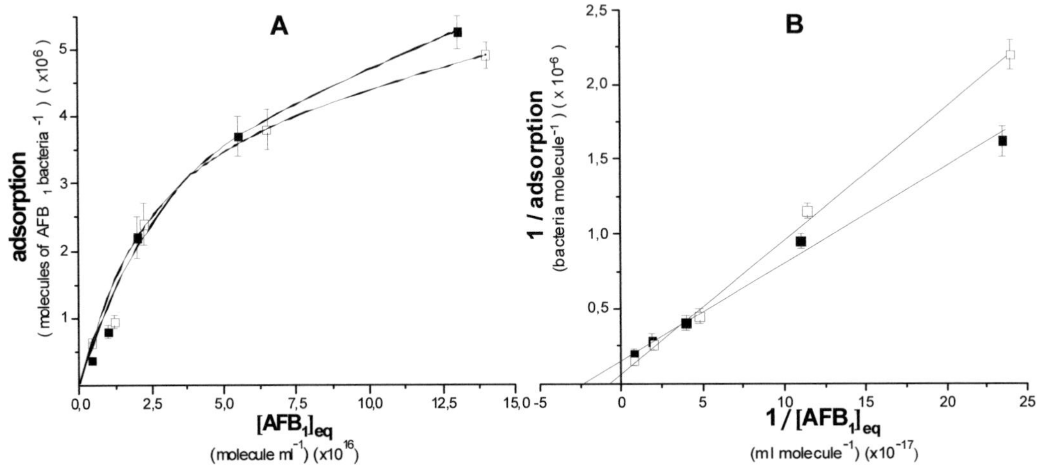

Fig. 4. Adsorption isotherms of AFB_1 by *L. fermentum* 408 (■) and *L. acidophilus* Po$_{22}$ (□). Aliquots of 1 ml of cells (0.89 x 10^9 cells for *L. fermentum* 408; 1.20 x 10^9 cells for *L. acidophilus* Po$_{22}$) were suspended in PBS in the presence of AFB_1 at the following concentrations: 0.5; 1.0; 2.5; 5.0 and 10.0 μg ml^{-1}. Then, the bacteria were incubated for 30 min at 37°C and pelleted by centrifugation. The supernatant was collected for free AFB_1 analysis by HPLC according to Bueno et al. (2007). AFB_1 bound to cells was calculated as the difference between the total AFB_1 and the amount of free AFB_1. The adsorption was calculated as the ratio between the molecules of AFB_1 bound to cells and the amount of cells in the incubation medium. The $[AFB_1]_{eq}$ was equivalent to the free AFB_1. (A) Saturation curve. (B) Inverse plot of the same data as (A). Data are means from triplicate experiments.

Lee et al. (2003) described an adsorption process by three different bacteria strains, in viable and nonviable forms, as a function of AFB_1 concentration. For comparative purposes, we applied our theoretical model to the data of Lee et al. (Table 8, lines 3 through 5), and the M and K_{eq} values for their three LAB strains were calculated. *L. rhamnosus* LC- 705 had the most efficient AFB_1 removal. However, all three strains were less efficient than P_{22} and 408, mainly because they had smaller M values. Strain LGG-V of Lee et al. was 10-fold less efficient than 408, since its M value was 10-fold smaller as well (Table 8).

We proposed a theoretical model of adsorption applicable to microorganisms that bind AFB_1 and this model allows an estimation of the number of aflatoxin B_1 binding sites per microorganism (M), the system equilibrium constant (K_{eq}), and the efficiency of cells to remove AFB_1 from liquid medium (M x K_{eq}). We analyzed three systems: two *Lactobacillus* strains (*L. acidophilus* Po$_{22}$ and *L. fermentum* subsp. *cellobiosus* 408) no tested before, three

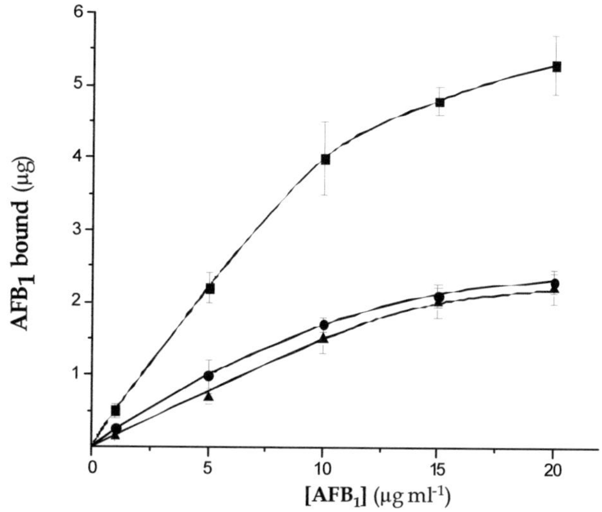

Cells, at the concentration indicated, were suspended in 1 ml of PBS with different AFB_1 concentrations (1; 5; 10; 15; 20 µg ml^{-1}) and incubated for 30 min at 37°C. AFB_1 binding to cells was performed as described in Fig. 1. Data are means ± standard deviations for triplicate samples.

Fig. 5. AFB_1 binding upon exposure to *S. cerevisiae* CECT 1891 (1 x 10^7 cells ml^{-1}) (■), *S. cerevisiae* RC008 (4.8 x 10^7 cells ml^{-1}) (●), *S. cerevisiae* RC016 (2.5 x 10^7 cells ml^{-1}) (▲).

Strains	M (1x10^6 sites $cell^{-1}$)	K_{eq} (1x10^4 M^{-1})	Efficiency (1x10^{10})
***L. acidophilus* Po$_{22}$**	8.33	0.75	6.25
L. *fermentum* subsp. *cellobiosus* 408	6.25	1.50	9.37
LC705-NV	1.48	3.12	4.62
PJS-NV	1.00	2.80	2.80
LGG-V	0.64	1.40	0.89
***S. cerevisiae* RC016**	580.00	0.80	460.00
***S. cerevisiae* RC008**	200.00	2.20	440.00
***S. cerevisiae* CECT 1891**	1,000.00	3.12	3,120.00

Table 8. Total binding sites per microorganism (M), equilibrium constant (K_{eq}) and efficiency (M x K_{eq}) for different strains. M, K_{eq} and M x K_{eq} for various microorganisms were calculated by linear regression by the following equation: 1/adsorption = (1/[AFB_1]) x 1/M x K_{eq} + 1/M, as described in Fig. 3. The data for linear regression construction were extracted from Fig. 4A for *L. acidophilus* Po$_{22}$ and L. *fermentum* subsp. *cellobiosus* 408 strains, from Fig. 5 for *S. cerevisiae* strains and from Fig. 1 from Lee et al. (2003) for LC705-NV (*Lactobacillus rhamnosus* strain, nonviable cells), PJS-NV (*Propionibacterium freudenreichii* subsp. *Shermanii* JS, nonviable cells), and LGG-V (*L. rhamnosus* GG, viable cells).

yeast strains (*S. cerevisiae* RC016, *S. cerevisiae* RC008 and *S. cerevisiae* CECT 1891) no studied before either, and LAB studied by another laboratory (Lee et al. 2003). The most efficient microorganism was *S. cerevisiae* CECT 1891, mainly because binds more AFB_1 per cell (Table 8). As we mentioned before, in AFB_1 binding to the yeast, the main components involved are cell wall glucomannans (Karaman et al., 2005), whereas cell wall peptidoglycans are responsible for AFB_1 removal by LAB (Lahtinen et al., 2004). Unexpectedly, bacteria and yeast had a K_{eq} value within the same order of magnitude, varying from 0.75 to 3.12 x 10^4 M^{-1}; in addition, *S. cerevisiae* CECT 1891 and LC705-NV had the same K_{eq} (Table 8). High efficiency resulted from high M value rather than from any variation in the K_{eq} value. These observations suggest that binding efficiency is a more quantitative phenomenon (large surface area) than a qualitative phenomenon (chemical structure involved in the interaction), and in this sense the analysis of the glucomannan and peptidoglycan chemical structures did not indicate major differences in the capacity to produce hydrophobic interactions or hydrogen bonds, which are presumably involved in the binding. The main application of the proposed model is its usefulness for the selection of the most efficient microorganism to remove AFB_1. *Sacharomyces cerevisiae* CECT 1891 was clearly the most efficient microorganism in the present study because it had high values of M and K_{eq}. When the election is between microorganisms with similar efficiencies (M x K_{eq})—for example, *L. acidophilus* P_{22} and LC705-NV or *S. cerevisiae* RC016 and *S. cerevisiae* RC008 (see Table 8)—the model suggests that the election should be the cell with the high K_{eq} factor (LC705- NV and *S. cerevisiae* RC008, respectively). In fact, when a microorganism-AFB_1 complex is exposed to consecutive washings, for example during gastrointestinal transit, the microorganism with the high value of K_{eq} will release few toxins. This is consistent with the results of Lee et al. (2003), who determined the constants of AFB_1 release for LC705-NV, PJS-NV, and LGG-V, which showed an inverse correlation with the respective K_{eq} values calculated by us (Table 8) and therefore the greatest constant of release of minor K_{eq} factors. Our results for *L. acidophilus* P_{22} and *S. cerevisiae* CECT 1891 also show the same correlation, i.e., that *L. acidophilus* P_{22} has a greater constant of release of minor K_{eq} factors than *S. cerevisiae* CECT 1891 (Table 8) and releases a greater percentage of AFB_1 than yeast strain does, when both are exposed to washings (Table 6). Therefore, it is clear that the microorganism with the greater K_{eq} factor has less toxin releases and this fact will be very important during passage through the gastrointestinal tract, where mainly aflatoxins are absorbed (Gratz et al., 2004).

When we analyzed the effect of mycotoxin and microorganism concentration, we remark the differences with the results obtained by Lee et al. (2003) who did not report saturability during the AFB_1 binding process, and we suggest that the high cell concentration employed by these authors was responsible of this non-saturation. For the same reason, our experiments with yeast strains were made with concentrations of 10^7 cells per ml, 100 times less than for bacteria; otherwise, the saturation was hidden.

Another important use of this adsorption model is its capacity to explain probable changes in the adsorption efficiency of the microorganisms (M x K_{eq}) after physical, chemical, or genetic treatments on the microorganisms. In fact, we can determine whether such changes are induced by modifications of M, K_{eq}, or both. This knowledge will help us to select the most efficient microorganisms for protection against AFB_1.

Although the results described come from in vitro experiments, some preliminary experiments in our laboratory (data not shown) suggest that the addition of *S. cerevisiae* CECT 1891 to poultry diet would help to prevent acute aflatoxicosis, chronic aflatoxicosis, or both.

In summary, when experiments *in vitro* made with a definite microorganism concentration and variable AFB_1 concentration show saturability, the adsorption model will be applicable and will allow a determination of M and K_{eq} values, both of which are useful when selecting the most efficient microorganism to remove AFB_1 in a wide range of mycotoxin concentration.

2.2 Evaluation of the gastrointestinal tolerance and anti-pathogenic bacterial influence of *Saccharomyces cerevisiae* strains

An interesting alternative to protect against aflatoxicosis, after the selection of the most efficient microorganism in AFB_1 remotion is to research its potential beneficial properties on

the host. While our proposed model above includes the possibility of using dead microorganisms and with the same aflatoxin B_1 binding efficiency, the inclusion in the diet of viable cells would add other beneficial effects to the animal we want to protect.
So, if the aim is to use living microorganisms, the first study to perform is to evaluate their tolerance to the gastrointestinal conditions, because acid in the stomach and bile salts in the gut are the first biological barriers to be overcome after ingestion (Gueimonde & Salminen, 2006).
In recent years, much attention has been paid to the design of functional foods that contain probiotic microbial strains responsible for health benefits in the host (Kumura et al., 2004). Several authors have demonstrated the safety of *Saccharomyces cerevisiae* for nutritional and pharmaceutical use in animal feed in the European Economic Community and in Japan (Nitta & Kobayashi, 1999). Moreover, the Food and Drug Administration has given level of insurance or GRAS (generally regarded as safe) microorganism degree to *Saccharomyces cerevisiae* strains (http://www.fda.gov/Food/FoodIngredientsPackaging/ucm078956.htm).
Microorganisms capable of withstanding the gastrointestinal transit (≥70% survival), have drawn attention to their possible use as probiotics (Lourens-Hattingh & Viljoen, 2001), and in this sense these microorganisms will be considered for further studies in order to evaluate their potential beneficial properties.
We employed this strategy to study four *Saccharomyces cerevisiae* strains isolated from pig environment, previously analyzed for their AFB_1 binding ability (Table 1 and 2), including assays for: *i*) gastrointestinal tolerance, *ii*) auto-aggregation, *iii*) cellular adhesion, *iv*) co-aggregation and *v*) antagonism to pig pathogens (Armando et al., 2011)

2.2.1 Selection of *Saccharomyces cerevisiae* strains for their tolerance to gastrointestinal conditions

The effect of simulated GI transit on viability of *S. cerevisiae* strains is presented in Table 9. All the yeast strains were able to survive under gastrointestinal conditions. In general, the strains retained viability through salivary (M2) conditions. *Saccharomyces cerevisiae* RC009 strain showed an increase in viability during transit to gastric (M3) and intestinal (M4) conditions; whereas *S. cerevisiae* RC008 strain did not increase the cell density but maintained the number of viable cells throughout the gastrointestinal transit. *Saccharomyces cerevisiae* RC012 showed a small decrease in viability after passage intestinal conditions (M4). In contrast, the RC016 strain was able to grow at these conditions.

Source	*S. cerevisiae* Strains	Viable count ($\log_{10}$ CFU ml^{-1}) during simulated GI transit tolerance			
		M_1	M_2	M_3	M_4
Feedstuff	**RC008**	6.4 ± 0.40 [a]	6.7 ± 0.17 [a]	6.5 ± 0.06 [a]	6.8 ± 0.11 [a]
Feedstuff	**RC009**	6.7 ± 0.23 [b]	6.4 ± 0.11 [b]	7.2 ± 0.17 [a]	7.5 ± 0.21 [a]
Feedstuff	**RC012**	6.4 ± 0.26 [b]	6.9 ± 0.15 [b]	6.4 ± 0.21 [b]	6.1 ± 0.26 [c]
Pig gut	**RC016**	7.3 ± 0.17 [a]	7.1 ± 0.15 [b]	5.8 ± 0.17 [c]	7.5 ± 0.06 [a]

Results are shown as mean ± SD (standard deviation). M_1: cell counts prior to assaying the GI transit tolerance. M_2: salivary conditions tolerance assay. M_3: gastric conditions tolerance assay. M_4: intestinal conditions tolerance assay. Values with the same letter are not significantly different according to Fisher's protected LSD test ($p<0.0001$). Statistical analysis compared means obtained from each yeast strain separately.

Table 9. Effect of simulated gastrointestinal (GI) transit on viability of *Saccharomyces cerevisiae* strains

The ability to survive gastrointestinal simulated conditions is an absolute need of probiotic microorganisms, and it is generally included among the criteria used to select potential probiotic strains (Morelli, 2000). In this work, all the assayed strains were able not only to resist gastrointestinal passage but also to grow under these conditions. Other authors have reported the same results with *S. cerevisiae* strains isolated from infant faeces and feta cheese (Psomas et al., 2001) and with *S. cerevisiae* var *boulardii* strain isolated from food (van der Aa Kuhle et al., 2005). However, the use of bacteria has demonstrated a very low recovery after being subjected to these gastrointestinal *in vitro* conditions (Gusils et al., 2002). Lin et al. (2007) reported a 2 or 3 log decrease in *Lactobacillus fermentum* strains isolated from swine and poultry. In agreement with our results, Pennacchia et al. (2008) reported that more than 50% of *S. cerevisiae* strains exposed to simulated passage through the human GI tract, showed 70% survival.

2.2.2 Auto-aggregation assay

Table 10 shows the auto-aggregation ability of the tested *S. cerevisiae* strains. Results showed that RC008, RC009 and RC016 strains exhibited a strong auto-aggregation score while RC012 strain exhibited weak auto-aggregation. This ability, or formation of multicellular clumps between microorganisms of the same strain, is a measure of the adhesion ability to epithelial cells and could be related to biofilm formation. The percentages of auto-aggregation obtained ranged from 85.3 to 97.9%, indicating that all strains showed an auto-aggregative phenotype.

S. cerevisiae Strains	OD_{600} (t_0)[1]	OD_{600} (t_1)[2]	[1- (OD_{600} t_1/ OD_{600} t_0)] x 100 %	Aggregation Score[3]
RC008	1.144	0.088	92.3	++
RC009	1.173	0.025	97.9	++
RC012	0.974	0.144	85.3	+
RC016	1.099	0.079	92.9	++

[1]Cells were harvested by centrifugation and suspended in PBS to optical density ~1 (O.D.) units at 600 nm. [2]Optical density of the upper suspension (O.D.) units at 600 nm after incubation at 37°C for 2 h. [3] (++) strong; (+) weak; (-) negative

Table 10. Auto-aggregation ability of *Saccharomyces cerevisiae* strains

2.2.3 Adhesion assay

The number of yeast cells adhesive to 100 Vero cells at different inoculums sizes of *S. cerevisiae* is shown in Table 11.

Results indicate that the adhesion ability of yeasts varied among strains according to the density of yeast cells. At 10^6 cells ml^{-1} the strains showed different adhesion ability: some of them had high adhesion capability while in others it was low. At 10^7 cells ml^{-1} all strains showed a similar strong adhesive ability to Vero cells. Adhesion ability of probiotcs to intestinal mucus and to enterocytes is an important prerequisite for transient colonization of the host intestinal tract. Adhesion to the mucosa is also considered to be an important factor for modulation of the immune system and for antagonistic activity against enteropathogens

(Ouwehand et al., 2002). Some authors reported a strong relationship between auto-aggregation and adhesiveness (Del Re et al., 2000), in line with the results obtained by us.

S. cerevisiae Strains	Cells ml^{-1}	Yeast cell number / 100 Vero cells	
		Mean ± SD[1]	LSD test[2]
RC008	10^6	18.5 ± 0.16	g
	10^7	154.0 ± 0.09	b
RC009	10^6	107.3 ± 0.09	d
	10^7	178.2 ± 0.02	a
RC012	10^6	109.5 ± 0.13	c
	10^7	154.8 ± 0.26	b
RC016	10^6	48.9 ± 0.04	f
	10^7	105.9 ± 0.15	e

Twenty randomized microscopic fields per coverslip were counted. Each adhesion assay was conducted in duplicate with cells from three successive passages. Adhesion was expressed as a mean of yeast adhering per 100 Vero cells. Controls without *S. cerevisiae* strains were included. [1]Mean ± Standard Deviation (SD). [2]Values corresponding to the same letter are not significantly different according to Fisher's protected LSD test ($P<0.05$).

Table 11. Adhesion of *Saccharomyces cerevisiae* strains to Vero cells

2.2.4 Co-aggregation assay

Co-aggregation among *S. cerevisiae* and pathogens is shown in Table 12. The results showed that the capacity of yeast to bind to a microorganism varies according to yeast strain and the microorganism involved. Some yeast strains varied from non co-aggregation to good co-aggregation ability when they interact with *E. coli, Salmonella* sp. or *Enterobacter* sp.. *Saccharomyces cerevisiae* RC016 strain was able to co-aggregate with all the pathogens tested, while the rest of the strains only co-aggregate with one of the pathogenic bacteria.

S. cerevisiae Strains	Co-aggregation score[1]		
	E. coli	*Enterobacter sp.*	*Salmonella sp.*
RC008	0	2	0
RC009	0	0	1
RC012	0	0	1
RC016	1	1	1

Yeast suspension (1 x 10^7 CFU ml^{-1} in PBS) was mixed with each pathogen strain (1 x 10^7 CFU ml^{-1} in PBS). Suspension was mixed and incubated for 4 h at 37ºC, under agitation at 200 rpm. Suspensions were observed by optic microscopy after Gram staining. [1]The score is based upon a scale described by Mastromarino et al. (2002), from 0 for no aggregation to 3 for maximum aggregation.

Table 12. Co-aggregation assay between *Saccharomyces cerevisiae* and pathogen strains

Co-aggregation and inhibition of pathogens are some of the most important beneficial health claims of probiotics. The co-aggregation ability of probiotic strains might enable them to form a barrier that prevents colonization by pathogenic bacteria (Collado et al., 2007).

2.2.5 Antagonism to pig pathogens

The antibacterial activity of yeast strains towards potential pathogenic bacteria was assayed (Table 13). Three of the tested *S. cerevisiae* were inhibitory against the pathogens studied. The RC008 and RC016 strains had the strongest antimicrobial activity; they were able to inhibit all the pathogens tested.

Most yeast strains are able to produce diffusible antimicrobial substances that inhibit the growth of pathogens. It was reported that certain *S. cerevisiae* strains secrete a protein toxin that kills not only sensitive strains of the same species but also other yeasts (Kitamoto et al., 1999). In our study, yeasts were able to inhibit pathogenic bacteria, but the nature of the antimicrobial substance is unknown.

S. cerevisiae Strains	Antimicrobial activity[1]		
	Escherichia coli	*Salmonella enterica*	*Enterobacter cloacae*
RC008	++	++	++
RC009	-	-	-
RC012	+	+	+
RC016	++	++	++

[1] +: inhibition zone ≥ 3 mm and ≤ 9 mm;
++: inhibition zone ≥ 10 mm and ≤ 15 mm;
-: inhibition zone ≤ 3 mm.
A central streak of each yeast strain was performed on Petri dishes containing YPD agar and incubated for 48 h at 37°C. Ten millilitres (10 ml) of additional liquid YPD agar were added to each plate and each pathogenic strain was streaked (perpendicularly) across the same agar plate. After 24 h incubation, antagonistic effect was determined by the appearance of clear zones surrounding the junctions of the streak lines.

Table 13. *Saccharomyces cerevisiae* antimicrobial activity against pathogen strains

In summary, the results let us predict that *S. cerevisiae* RC016 and *S. cerevisiae* strain RC008 may be regarded as the most promising beneficial yeast candidate for functional feed product development because they both had good and similar AFB_1 remotion efficiency (Table 8), higher than *S. cerevisiae* RC009 and *S. cerevisiae* RC012 (data not shown) and the strongest capacity of pathogen inhibition. They were also able to survive under gastrointestinal conditions and adhere to intestinal cells. Other functional and technological tests should be performed for the validation of these strains as suitable probiotics for animals, and future studies should be conducted to evaluate the influence of *S. cerevisiae* strains on non-pathogenic bacteria. But evidently *S. cerevisiae* RC008 and *S. cerevisiae* RC016 are the best candidates, from the four yeast strains analyzed, for further studies *in vivo* designed to prevent aflatoxicosis.

3. Conclusions

As we have seen, the potential presence of aflatoxins in animal diet is unavoidable, therefore a protection against aflatoxicosis is necessary, and the inclusion in the diet of microorganisms able to remove AFB_1 is the most suitable alternative. Moreover, if the microorganisms capable to avoid the aflatoxin absorption during its gastrointestinal transit have beneficial properties on the host, this way to prevent aflatoxicosis is highly promising. With the increasing interest in food safety throughout the world, the yeast and LAB cultures with high mycotoxin binding abilities and probiotic abilities are of immense value in reducing aflatoxin exposure.

The first section of this work presents *in vitro* experiments useful to the development of a model that considers the aflatoxin-microorganism interaction as a fast, reversible and strain specific process, concluding that it is a physical adsorption to the cell wall of the microorganism. The model is a tool for selecting the most efficient microorganism to remove AFB_1 in a wide range of mycotoxin concentration, since feed AFB_1 contamination is variable. Another important use of this model is its capacity to explain probable changes in the adsorption efficiency ($M \times K_{eq}$) of the microorganism after than physical, chemical or genetic treatments designed to obtain the ideal microorganism; i. e. if either the process equilibrium constant (K_{eq}) or the number of binding sites for AFB_1 in the surface microorganism (M) change, or both.

The application of the adsorption model to yeast and LAB strains showed that *Saccharomyces cerevisiae* CECT 1891 was the most efficient microorganism in AFB_1 remotion, but in general also showed that the yeasts had higher efficiencies than LAB, mainly because the M values were 20- to 1,000- fold higher for the yeasts than for the bacteria, whereas the K_{eq} values were similar.

The second part of this chapter shows the design of simple experiments *in vitro* to evaluate potential beneficial properties of the microorganisms under study, on the host. In this sense, the tolerance of gastrointestinal conditions, auto and co-aggregation, cell adhesion and antibacterial activity towards pathogenic bacteria, of four yeast strains from pig environmental, were assayed. The results showed that *Saccharomyces cerevisiae* RC016 and *Saccharomyces cerevisiae* RC008 were the best microorganisms, mainly because they showed high and similar AFB_1 remotion efficiency and the strongest antimicrobial activity against pathogen strains.

Thus, the combination of both strategies allows us to select, among all the microorganisms tested (LAB and yeasts), the best candidate for future *in vivo* studies. This selection is very important because, although *in vivo* studies are necessary, they are expensive, complicated and long lasting as well. So, in summary, this research represents an efficient *in vitro* strategy to select the correct microorganisms for future *in vivo* studies, useful to prevent aflatoxicosis in farm animals.

4. Acknowledgments

This work was carried out thanks to grants from SECYT-Universidad Nacional de Río Cuarto, Ministerio de Ciencia y Tecnología de Córdoba, ANPCYT-PICT and ANPCYT-PICT-CNPq.

5. References

Armando, M.R., Dogi, C.A., Pizzolitto, R.P., Escobar, F., Peirano, M.S., Salvano, M.A., Sabini, L.I., Combina, M., Dalcero, A.M. & Cavaglieri, L.R. (2011). *Saccharomyces cerevisiae* strains from animal environmental with aflatoxin B_1 detoxification ability and anti pathogenic bacteria influence *in vitro*. *World Mycotoxin Journal,* Vol. 4, No.1, (February 2011), pp. 59-68, ISSN 1875-0710

Bueno, D., Salvano, M.A., Silva, J.O., González, S.N. & Oliver, G. (2001). Mycotoxins: Diagnosis and prevention in poultry. Boletín micológico. Vol.16, pp. 23-36 ISSN 0716-114X

Bueno, D., Casale, C., Pizzolitto, R., Salvano, M. & Oliver, G. (2007). Physical adsorption of aflatoxin B_1 by lactic acid bacteria and *Saccharomyces cerevisiae*: A theoretical model. *Journal of Food Protection*, Vol.70, No.9, (September 2007) pp. 2148-2154, ISSN 0362-028X

CAST (Council for Agricultural Science and Technology). (2003). Mycotoxins: Risks in Plant, Animal and Human Systems, Task Force 478 Report N°139. Ames, Iowa, USA.

Ciegler, A., Lillehoj, E., Peterson, R. & Hall, H. (1966). Microbial detoxification of aflatoxin. *Journal of Applied Microbiology*, Vol.14, No.6, (November 1966) pp. 934–939, ISSN 1364-5072

Collado, M.C., Meriluoto, J. & Salminen, S. (2007). Measurement of aggregation properties between probiotics and pathogens: in vitro evaluation of different methods. *Journal of Microbiological Methods*, Vol.71, No.1, (October 2007), pp. 71-74, ISSN 0167-7012

Dalcero, A.M., Magnoli, C., Chiacchiera, S., Palacios, G. & Reynoso, M. (1997). Mycoflora and incidence of aflatoxin B_1, zearalenone and deoxynivalenol in poultry feeds in Argentina. *Mycopathologia*, Vol.137, No.3, (n.d.), pp. 179-184, ISSN 0301-486X

Del Re, B., Sgorbati, B., Miglioli, M. & Palenzona, D. (2000). Adhesion, autoaggregation and hydrophobicity of 13 strains of Bifidobacterium longum. *Letters Applied Microbiology,* Vol.31, No.6, (December 2000), pp. 438-442, ISSN 0266-8254

El-Nezami, H., Kankaanpää, P., Salminen, S. & Ahokas, J. (1998). Physicochemical alterations enhance the ability of dairy strains of lactic acid bacteria to remove aflatoxin from contaminated media. *Journal of Food Protection*, Vol.61, No.4, (April 1998), pp. 466-468, ISSN 0362-028X

El-Nezami, H., Mykkänen, H., Kankaanpää, P., Suomalainen, T., Salminen, S. & Ahokas, J. (2000). Ability of a mixture of *Lactobacillus* and *Propionibacterium* to influence the faecal aflatoxins contents in healthy Egyptian volunteers: a pilot clinical study. *Bioscience Microflora*, Vol.19, No.1 (n.d.), pp. 41-45, ISSN 1342-1441

Gourama, H. & Bullerman, L. (1995). Inhibition of growth and aflatoxin production of *Aspergillus flavus* by *Lactobacillus* species. *Journal of Food Protection*, Vol.58, No.11, (November 1995), pp. 1249–1256, ISSN 0362-028X

Gratz, S., Mykkänen, H., Ouwehand, A., Juvonen, R., Salminen, S. & El-Nezami, H. (2004). Intestinal mucus alters the ability of probiotic bacteria to bind aflatoxin B_1 *in vitro*. *Applied and Environmental Microbiology*, Vol.70, No.10, (October 2004), pp. 6306-6308, ISSN 0099-2240

Gratz, S., Wu, Q., El-Nezami, H., Juvonen, R., Mykkänen, H. & Turner, P.C. (2007). *Lactobacillus rhamnosus* strain GG reduces aflatoxin B_1 transport, metabolism and toxicity in caco-2 cells. *Applied and Environmental Microbiology,* Vol.73, No.12, (June 2007), pp. 3958-3964, ISSN 0099-2240

Gueimonde, M. & Salminen, S. (2006). New method for selecting and evaluating probiotics. *Digestive and Liver Disease,* Vol.38, No.2, (December 2006) pp. 242-247, ISSN 1590-8658

Gusils, C., Bujazha, M. & González, S. (2002). Preliminary studies to design a probiotic for use in swine feed. *Interciencia,* Vol.27, No.8, (August 2002), pp.409-413, ISSN 0378-1844

Haskard, C., Binnion, C. & Ahokas, J. (2000). Factors affecting the sequestration of aflatoxin by *Lactobacillus rhamnosus* strain GG. *Chemico-Biological Interactions.* Vol.128, No.1, (August 2000), pp. 39-49, ISSN 0009-2797

Haskard, C., El-Nezami, H., Kankaanpää, P., Salminen, S. & Ahokas, J. (2001). Surface binding of aflatoxin B_1 by lactic acid bacteria. *Applied and Environmental Microbiology,* Vol.67, No.7, (July 2001), pp. 3086-3091, ISSN 0099-2240

Hernandez-Mendoza, A., Garcia, H.S. & Steele, J.L. (2009). Screening of *Lactobacillus casei* strains for their ability to bind aflatoxin B_1. *Food and Chemical Toxicology,* Vol.47, No.n.d., (February 2009), pp. 1064-1068, ISSN 0278-6915

Hussein, H. & Brasel, J. (2001) Toxicity, metabolism, and impact on humans and animals. *Toxicology,* Vol.167, No.2, (October 2001), pp. 101-134, ISSN 0300-483X

Ito, Y., Peterson, S., Wicklow, D. & Goto, T. (2001). *Aspergillus pseudotamarii,* a new aflatoxin producing species in *Aspergillus* section Flavi. *Mycological Research,* Vol.105, No.2, (February 2001), pp. 233-239, ISSN 0953-7562

Kabak, B., Alan, D. & Var, I. (2006). Strategies to Prevent Mycotoxin Contamination of Food and Animal Feed: A Review. *Critical Reviews in Food Science and Nutrition,* Vol.46, No.8, (n.d.), pp. 593-619, ISSN 1040-8398

Kankaanpää, P., Tuomola, E., El-Nezami, H., Ahokas, J. & Salminen, S. (2000). Binding of aflatoxin B_1 alters the adhesion properties of *Lactobacillus rhamnosus* strain GG in a Caco-2 model. *Journal of Food Protection,* Vol.63, No.3, (n.d.), pp. 412-414, ISSN 0362-028X

Karaman, M., Basmacioglu, H., Ortatatli, M. & Oguz, H. (2005). Evaluation of the detoxifying effect of yeast glucomannan on aflatoxicosis in broilers as assessed by gross examination and histopathology. *British Poultry Science,* Vol.46, No.3, (June 2005), pp. 394-400, ISSN 1466-1799

Kitamoto, H.K., Hasebe, A., Ohmomo, S., Suto, E. G., Muraki, M. & Iimura, Y. (1999). Prevention of aerobic spoilage of maize silage by a genetically modified killer yeast, *Kluyveromyces lactis,* defective in the ability to grow on lactic acid. *Applied and Environmental Microbiology,* Vol.65, No.10, (October 1999), pp. 4697-4700, ISSN 0099-2240

Kumura, H., Tanoue, Y., Tsukahara, M., Tanaka, T. & Shimazaki, K. (2004). Screening of dairy yeast strains for probiotic applications. *Journal of Dairy Science,* Vol.87, No.12 , (December 2004), pp. 4050-4056, ISSN 0022-0302

Kurtzman, C., Horn, B. & Hesseltine, C. (1987). *Aspergillus nomius*, a new aflatoxin-producing species related to *Aspergillus flavus* and *Aspergillus tamarii*. *Antonie van Leeuwenhoek*, Vol.53, No.3, (n.d.), pp. 147-158, ISSN 0003-6072

Lahtinen, S. J., Haskard, C. A., Ouwehand, A. C., Salminen, S. J. & Ahokas, J. T. (2004). Binding of aflatoxin B_1 to cell wall components of *Lactobacillus rhamnosus* strain GG. *Food Additives and Contaminant*, Vol.21, No.2, (February 2004), pp. 158–164, ISSN 0265-203X

Lee, Y., El-Nezami, H., Haskard, C., Gratz, S., Puong, K., Salminen, S. & Mykkänen, H. (2003). Kinetics of adsorption and desorption of aflatoxin B_1 by viable and nonviable bacteria. *Journal of Food Protection*, Vol.66, No.3, (March 2003), pp. 426-430, ISSN 0362-028X

Lin, W.H., Yu, B., Jang, S.H. & Tsen, H.Y. (2007). Different probiotic properties for Lactobacillus fermentum strains isolated from swine and poultry. *Anaerobe*, Vol.13, No.3-4 , (August 2007), pp. 107-113 ISSN 1075-9964

Line, J., Brackett R. & Wilkinson, R. (1994). Evidence for degradation of aflatoxin B_1 by *Flavobacterium aurantiacum*. *Journal of Food Protection*, Vol.57, No.9, (September 1994), pp. 788–791, ISSN 0362-028X

Lourens-Hattingh, A. & Viljoen, B.C. (2001). Growth and survival of a yeast in dairy products. *Food Research International*, Vol.34, No.9, (November 2001), pp. 791–796, ISSN 0963-9969

Magnoli, A.P., Texeira, M., Rosa, C.A., Miazzo, R.D., Cavaglieri, L.R., Magnoli, C.E., Dalcero, A.M. & Chiacchiera, S.M. (2001). Sodium bentonite and monensin under chronic aflatoxicosis in broiler chickens. *Poultry Science*, Vol.90, No.2, (February 2011), pp.352-357, ISSN 0032-5791

Mastromarino, P., Brigidi, P., Macchia, S., Maggi, L., Pirovano, F., Trinchieri,V., Conte, U. & Matteuzzi, D. (2002). Characterization and selection of vaginal *Lactobacillus* strains for the preparation of vaginal tablets. *Journal Applied Microbiology*, Vol.93, No.5, (November 2002), pp.884-893, ISSN 1364-5072

Morelli, L. (2000). *In vitro* selection of probiotic lactobacilli: a critical appraisal. *Current Issues in Intestinal Microbiology*, Vol.1, No.2, (September 2000), pp. 59-67, ISSN 1466-531X

Nitta, K. & Kobayashi F. (1999). Brewer's yeast as health foodstuff. New Food Industry, Vol.41, pp. 17-23

Oatley J., Rarick M., Ji G. & Linz J. (2000). Binding of aflatoxin B_1 to bifidobacteria *in vitro*. *Journal of Food Protection*, Vol.63, No.8, (August 2000), pp. 1133–1136, ISSN 0362-028X

Ouwehand, A.C., Salminen, S. & Isolauri, E. (2002). Probiotics: an overview of beneficial effects. *Antonie Van Leeuwenhoek*, Vol.82, No.1-4, (August 2002), pp.279-289, ISSN 0003-6072

Ovalle, R., Lim, S.T., Holder, B., Jue, C.K., Moore, C.W. & Lipke, P.N. (1998). A spheroplast rate assay for determination of cell wall integrity in yeast. *Yeast*, Vol. 14, No.13, (September 1998), pp. 1159-1166. ISSN 0749-503X

Payne, G. (1998). *Process of contamination by aflatoxin-producing fungi and their impact on crops.* K. K. S. Sinha and D. Bhatnagar (Eds.). Mycotoxins in Agriculture and Food Safety Marcel Dekker, Inc., New York.

Peltonen, K., El-Nezami, H., Haskard, C., Ahokas, J. & Salminen, S. (2001). Aflatoxin B_1 binding by dairy strains of lactic acid bacteria and bifidobacteria. *Journal of Dairy Science*, Vol.84, No.10, (October 2001), pp. 2152-2156, ISSN 0022-0302

Pennacchia, C., Blaiotta, G., Pepe, O. & Villani, F. (2008). Isolation of *Saccharomyces cerevisiae* strains from different food matrices and their preliminary selection for a potencial use as probiotic. *Journal Applied Microbiology*, Vol.105, No.6, (December 2008), pp. 1919-1928, ISSN 1364-5072

Pereyra, C., Cavaglieri, L., Chiacchiera, S. & Dalcero, A. (2010) Fungi and Mycotoxins in feed intended for sows at different reproductive stages in Argentina. *Veterinary Medicine International*, Vol. 2010, (June 2010), pp.1-7, ISSN 2042-0048

Pizzolitto, R. (2011). Prevención de micotoxicosis en la industria avícola por microorganismos bioeliminadores de toxinas fúngicas. PhD. Thesis. Departamento de Biología Molecular, Universidad Nacional de Río Cuarto, Argentina

Psomas, E., Andrighetto, C., Litopoulou-Tzanetaki, E., Lombardi, A. & Tzanetakis, N. (2001). Some probiotic properties of yeast isolates from infant faeces and Feta cheese. *International Journal of Food Microbiology*, Vol.69, No.1-2, (September 2001), pp.125-133, ISSN 0168-1605

Raju, M. & Devegowda, G. (2000). Influence of esterified-glucomannan on performance and organ morphology, serum biochemistry and haematology in broilers exposed to individual and combined mycotoxicisis. *British Poultry Science*, Vol.41, No.5, (December 2000), pp. 640-650, ISSN 0007-1668

Richard J.L. (2007). Some major mycotoxins and their mycotoxicoses an overview. *International Journal of Food Microbiology*, Vol.119, No.1-2, (October 2007), pp. 3-10, ISSN 0168-1605

Shetty P., Hald B. & Jespersen L. (2007). Surface binding of aflatoxin B_1 by *Saccharomyces cerevisiae* strains with potential decontaminating abilities in indigenous fermented foods. *International Journal of Food Microbiology*, Vol.113, No.1, (January 2007), pp. 41-46, ISSN 0168-1605

Shetty, P. & Jespersen, L. (2006). *Saccharomyces cerevisiae* and lactic acid bacteria as potential mycotoxin decontaminating agents. *Trends in Food Science and Technology*, Vol.17, No.2, (February 2006), pp. 48-55, ISSN 0924-2244

Trucksess, M.W., Stack, M.E., Nesheim, S., Albert, R.H. & Romer, T.R. (1994). Multifunctional column coupled with liquid chromatography for determination of aflatoxins B_1, B_2, G_1 and G_2 in corn, almonds, Brazil nuts, peanuts and pistachio nuts: collaborative Study. *Journal of AOAC International*, Vol.77, No.6, (November 1994) pp. 1512-1521 ISSN 1060–3271

van der Aa Kühle, A., Skovgaard, K. & Jespersen, L. (2005). In vitro screening of probiotic properties of *Saccharomyces cerevisiae* var. *boulardii* and food-borne *Saccharomyces cerevisiae* strains. *Internal Journal of Food Microbiology* Vol.101, No.1, (May 2005), pp. 29-39, ISSN 0168-1605

Wild, C. (2007). Aflatoxin exposure in developing countries: the critical interface of agruculture and health. *Food and Nutrition Bulletin*, Vol.28, No.2, (June 2007), pp. 5372-5380, ISSN 0379-5721

17

The Population Dynamics of Aflatoxigenic Aspergilli

Geromy G. Moore[1], Shannon B. Beltz[1], Ignazio Carbone[2], Kenneth C. Ehrlich[1] and Bruce W. Horn[3]

[1]*United States Department of Agriculture, Agricultural Research Service, Southern Regional Research Center, New Orleans, Louisiana*

[2]*Center for Integrated Fungal Research, Department of Plant Pathology, North Carolina State University, Raleigh, North Carolina*

[3]*United States Department of Agriculture, Agricultural Research Service, National Peanut Research Laboratory, Dawson, Georgia*

United States of America

1. Introduction

A. flavus is a saprophyte that possesses the characteristics of an opportunistic pathogen with a wide host range (plants and animals) (Wilson, Mubatanhema et al. 2002). This fungus has substantial impact on agriculture because it produces aflatoxins (AFs), the most toxic and carcinogenic of all mycotoxins (Cary, Bhatnagar et al. 2000). Contamination of susceptible crops with AFs leads to serious health risks in developing countries and significant economic losses in the U.S. and other developed countries (Yu, Payne et al. 2008). Major commodities affected by AFs include corn, cotton, peanuts, tree nuts, rice, peppers, spices, and figs. AF contamination of crops usually results from opportunistic invasion due to some type of injury to the plant. In most infestations, the fungus causes little overall damage to the plant or its fruiting structures. Because *A. flavus* does not produce true plant pathogenicity factors, infection depends on both the plant's internal defenses and the ability of *A. flavus* to invade plant tissues. *A. flavus* is one of several known species that produces aflatoxins, including, *A. parasiticus, A. nomius, A. bombycis, A. pseudotamarii, A. ochraceoroseus,* and others. *A. flavus* and *A. parasiticus*, in particular, are disseminated widely in agricultural regions.

2. *A. flavus* diversity

2.1 Sclerotial size

A. flavus as a species contains two morphotypes that differ in sclerotial size and in their ability to produce AFs. Large (L) and small (S) sclerotial strains are often found in soils from both agricultural fields (Cotty, Bayman et al. 1994; Horn 2007) and non-agricultural areas (Ehrlich, Kobbeman et al. 2007) throughout the world. Recently, it was suggested that the S- strain, which is capable of producing much higher concentrations of aflatoxins than the L- strain, may be a more important source of aflatoxin contamination in corn and cotton in some regions (Zhang and Cotty 2007). Beyond sclerotial size, another difference between the L- and S- strain is colony morphology, since S- strain isolates produce many more sclerotia and, in the dark, fewer conidia (Cotty, Bayman et al. 1994). Furthermore, S- strain isolates have a higher virulence toward cotton as measured by production of pectinase (Mellon and Cotty 2004). The specific ability of *A. flavus* morphotypes to adapt to agricultural environments is poorly understood.

2.2 Vegetative compatibility

A. flavus populations are composed of sub-clades consisting of different vegetative compatibility groups (VCGs). Fungal isolates from different VCGs are presumed to have limited genetic exchange and recombination in the parasexual cycle since vegetative compatibility is determined by a series of heterokaryon incompatibility loci whose alleles must all be identical for stable hyphal fusions to occur (Leslie 1993). Populations of *A. flavus* and other section *Flavi* species are complex and consist of many VCGs (Bayman and Cotty 1991; Horn, Greene et al. 1995). The number of VCGs in a field may vary based on soil type, climate, and type of crop, and may even vary from year to year. No particular VCG has been consistently associated with a specific niche, and it is not clear if the VCG trait is important for adaptation although characters associated with VCG may be important. VCG analyses have shown that genetically distinct *A. flavus* strains frequently interact during dispersal, growth, and crop infection (Bayman and Cotty 1991; Cotty, Bayman et al. 1994). Isolates from different VCGs often vary in their ability to produce AF (Bayman and Cotty 1991; Horn, Greene et al. 1995). How strong a barrier vegetative compatibility is to recombination within a fungal species, due to parasexuality, is unknown. A comparison of three genes in *A. flavus* isolates of the same VCG performed by Ehrlich et al. (Ehrlich, Montalbano et al. 2007), and array comparative genome hybridization assays by Federova et al. (Federova, Harris et al. 2009) showed that isolates from the same VCG have identical genomes. These results suggest that isolates in the same VCG represent a clonal assembly. A study by Grushiba and Cotty (2009) estimated divergence time between examined VCGs to range from 50,000 to 200,000 years.

2.3 *A. flavus* non-aflatoxigenicity

Non-aflatoxigenic *A. flavus* strains are represented by both domesticated and wild-type forms. Domesticated *A. oryzae* is genetically indistinct from *A. flavus. A. oryzae* is used as a starter culture for the preparation of fermented foods and alcoholic beverages, and is an important source of many enzymes, such as glucoamylase, alpha-amylases and proteases used for starch processing, baking, and brewing worldwide (Machida, Yamada et al. 2008). Wild-type non-aflatoxigenic *A. flavus* is a common inhabitant of soil and crops from agricultural areas worldwide and co-exists with its toxigenic brethren. In agricultural fields, up to 40% of the isolates lack the ability to produce aflatoxins due to deletions or mutations within the aflatoxin cluster (Cotty and Bhatnagar 1994; Chang, Horn et al. 2005). Non-aflatoxigenic (AF-) *A. flavus* strains are currently being introduced into agricultural fields as biological control agents to lower crop aflatoxin contamination. Within the non-aflatoxigenic group are isolates incapable of producing the mycotoxin cyclopiazonic acid, while other isolates produce this mycotoxin. Although AF- domesticated forms of *A. parasiticus* are known (e.g. *A. sojae*), AF- isolates of *A. parasiticus* are rarely found in natural environments.

3. *A. flavus* and *A. parasiticus* exhibit evidence of recombination

P.A. Micheli characterized the genus *Aspergillus* as an asexually-reproducing group of fungi in 1729. Nearly a century later the first teleomorph was discovered and named *Eurotium herbariorum* by J.H.F. Link, which Anton DeBary later identified as the sexual state of *A. glaucus* (Raper and Fennell 1965). Though other teleomorphs have been identified over the years, the genus *Aspergillus* is still considered to be predominantly asexual due to lack of morphological evidence of sex in nature.

Populations of aflatoxigenic fungi were originally considered to result from the clonal amplification of individuals with the same chemotype profile, in contrast to the possibility of creating novel chemotype profiles, or restoring aflatoxin production in individuals that are AF-, through genetic exchange and recombination (Geiser, Pitt et al. 1998; Pitt and Hocking 2006). The belief that the aflatoxigenic Aspergilli reproduce asexually was the foundation for trusting in the stability of non-aflatoxigenic biocontrol strains in competitively excluding indigenous aflatoxigenic Aspergilli in fields (Geiser, Pitt et al. 1998; Moore, Singh et al. 2009). Currently there are two U.S. Environmental Protection Agency-approved biocontrol strains, both species of *A. flavus*, who are unable to synthesize aflatoxins due to either mutations in aflatoxin-synthesis pathway genes, or the loss of those genes altogether. A cryptic sexual state was reported in aflatoxigenic (AF+) *A. flavus* suggesting a history of recombination (Geiser, Pitt et al. 1998); therefore, the possibility of genetic exchange and recombination within populations of AF+ fungi may necessitate a re-evaluation of biocontrol selection strategies.

As molecular research to understand, and subsequently prevent, contamination by aflatoxigenic fungi has evolved, the *Aspergillus* research community has begun to explore population structure of AF+ Aspergilli. The tools to observe and interpret population dynamics are also evolving, beyond comparisons of mutational differences, allowing for stronger inferences regarding the mechanisms that maintain functional and genetic diversity in AF+ fungi. This chapter will review population genetics methods employed to obtain evidence of recombination within populations of two of the more common agents of aflatoxin contamination in agricultural commodities, *A. flavus* and *A. parasiticus*; the subsequent (current) research developments that support inferences made regarding recombination in these fungi; and where the investigations are leading researchers who continue to seek protection for the global food supply.

3.1 How is recombination detected in a population?

Recombination can be defined as the exchange of genetic material between individuals at the chromosomal level, oftentimes resulting in novel genomic compositions in offspring, since chromosomes from both "parents" are being broken apart and shuffled around during meiosis (Milgroom 1996). Detecting recombination depends on the amount of allele shuffling that is occurring in a population. If there is high genetic diversity and frequent shuffling, then detection is easier; conversely, low genetic diversity accompanied by little to no shuffling of alleles, might indicate no evidence of recombination (Nordborg and Tavaré 2002). Figure 1 illustrates how recombination may be detected in two simplified populations. Both populations are composed of the same sample size, and each population shares the same sequence breakpoints (X), with uniform recombination rates, between three contiguous loci. The nucleotide bases are color-coded blue or red with random mutations colored black. Population 1 has an approximately equal distribution of blue and red alleles that are shuffled around sequence breakpoints. Population genetics tools would easily detect recombination within this population. Population 2 appears clonal because of a predominance of the red allele, perhaps because it confers some selective advantage; conversely, the blue allele is found in one individual which may be a recent migrant, the survivor of a recent bottleneck, or a more fit individual under certain environmental conditions and therefore maintained, at a low frequency, in the population via balancing selection. The power of population genetic tools to detect evidence of recombination for population 2 is limited since there is not as much shuffling (or mixing) of both alleles. Increasing the sample size of population 2 may or may not alter the results. One benefit to studying a non-recombining population is the potential to detect phenotypic targets of balancing selection in the genomes of its individuals (Carbone, Jakobek et al. 2007; Moore, Singh et al. 2009). Though selection may be acting on all three loci, our ability to observe evidence of the phenotypic targets of selection depends on finding a region that is not subject to frequent shuffling which would obscure genotype-phenotype associations.

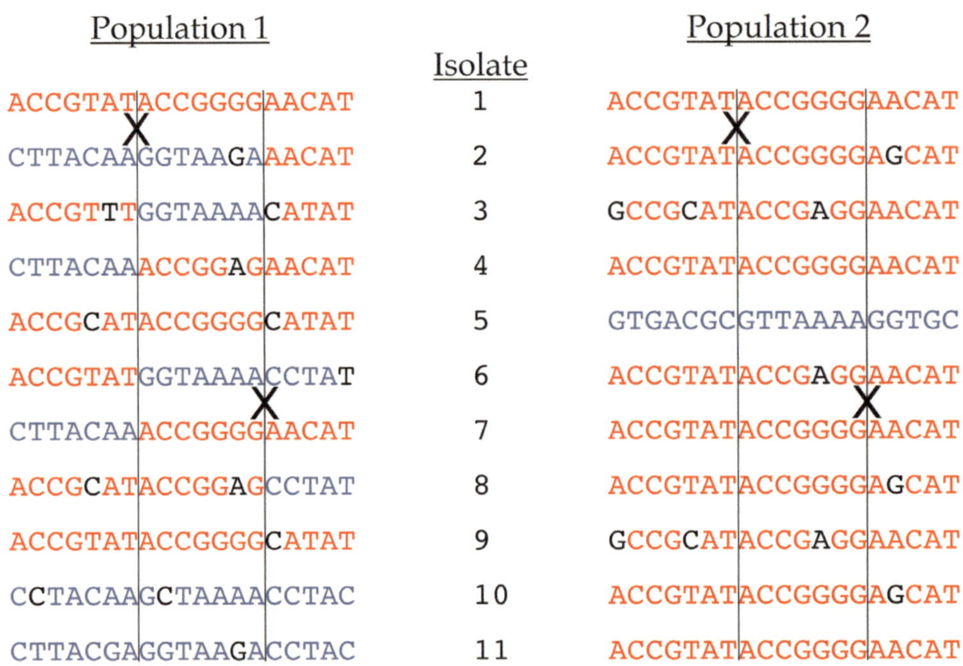

Fig. 1. Patterns of recombination among three concatenated linked loci in two isolated populations; loci boundaries are delineated using vertical lines.

3.2 Population genetics tools

A plethora of genetics tools are available for population analysis. These can be grouped into three broad categories: 1) non-parametric and summary statistic methods, 2) genealogical estimators of population parameters, and 3) coalescent and Bayesian approaches for estimating the direction and magnitude of evolutionary forces.

Non-parametric and summary statistic methods are frequency-based. These non-genealogical methods follow a simple model known as Hardy-Weinberg equilibrium which assumes only Mendelian inheritance and no effect on populations from evolutionary forces; so as long as the assumptions are not violated, allele frequencies will persist in a constant state with each generation (Hartl and Clark 1989; Hey and Machado 2003). Methods such as

Classical Wright's FST statistics, neutrality tests (Tajima 1989; Fu 1996), population subdivision (Hudson 2000), population mutation rate estimators (e.g. Watterson's θ), and compatibility methods (Quesne 1982) are frequently examined under a simple model of evolution.

Genealogical estimators of population parameters involve simple populations exhibiting the Wright-Fisher model of random genetic drift (Hey and Machado 2003). This means that every generation has a constant population size, and genetic drift is the only evolutionary force altering allele frequencies. Genealogical approaches can enhance the power of neutrality tests (Fu and Li 1993), haplotype phenotype association testing (Templeton, Routman et al. 1995), and demographic inference.

Coalescent and Bayesian approaches are considered parametric (involve parameters) and follow a complex genealogical model. As previously mentioned, the methods implemented here are for estimating the direction and magnitude of evolutionary forces. Coalescent approaches estimate: θ, recombination, and migration rates (Beerli and Felsenstein 1999). Bayesian inference uses probability distributions to quantify uncertainty (Nielsen and Wakeley 2001; Grünwald and Goss 2011). This includes equilibrium and non-equilibrium models, which relate to continuous versus sporadic processes, respectively.

Several population software tools are available that implement the various methods described above; for a comprehensive review see Grunwald and Goss (2011). Some useful analysis packages are Arlequin (Excoffier, Laval et al. 2005), GENEPOP (Raymond and Rousset 1995) and POPGENE (Yeh and Boyle 1997). Price and Carbone assembled a collection of population analysis programs and named it SNAP Workbench (2005). Imported sequence alignments are collapsed into haplotype groupings as "individuals" using SNAP Map (Aylor, Price et al. 2006) followed by various analyses to detect recombination. Within SNAP, analyses such as linkage disequilibrium (LD), the ancestral recombination graph (ARG), and allele coalescence have been used to detect recombination in populations of *A. flavus* and *A. parasiticus*.

3.2.1 Linkage disequilibrium (LD)

LD is a means to infer distinct evolutionary histories in populations (Nordborg and Tavaré 2002). The power of detecting LD depends on distinct SNP-patterns found in either a single allele or across a group of alleles. An LD plot (Figure 2), is generated based on pairwise measures of LD. Three commonly used LD measures with slightly different properties are D', r^2, and d^2. D' is a normalized value (between 0 and 1) regardless of allele frequencies, r^2 is the squared association in allelic state between two loci existing in haplotypes, and d^2 measures association between alleles for each locus (Nordborg and Tavaré 2002). For example, a typical LD plot is divided into two triangular portions showing all pairwise comparisons in the data. Significant LD is assessed using a simple Fisher's Exact test for a 2 x 2 contingency table or a permutation method if more than two alleles are present. LD can be examined for both single-locus and multi-locus data.

In Figure 2, population 1 (from Fig. 1) harbors three distinct LD blocks, which may translate to distinct evolutionary histories in the data. LD is significant with at least two individuals sharing red and blue alleles; correlation is strong with many polymorphisms segregating for the red and blue alleles. The LD pattern for population 2 shows extensive LD with essentially one large haplotype block. Table 1 lists identities for haplotypes in Figs. 2 and 3, in which we observe higher haplotype diversity in population 1 than in population 2.

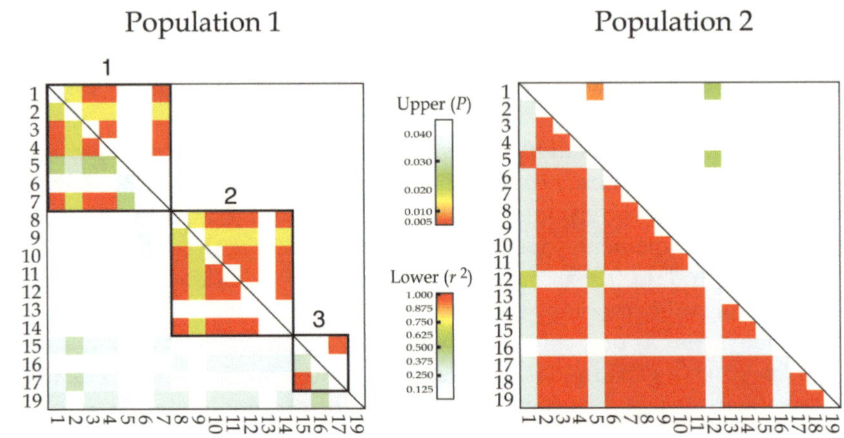

Fig. 2. LD plots based on combined loci for populations in Fig. 1.

Population 1		Population 2	
Haplotype	Isolate(s)	Haplotype	Isolate(s)
H1	1	H1	1, 4, 7, 11
H2	2	H2	3, 9
H3	10	H3	5
H4	4	H4	6
H5	7	H5	2, 8, 10
H6	11		
H7	5		
H8	8		
H9	3		
H10	6		
H11	9		

Table 1. Haplotype composition for hypothetical analyses shown in Figs. 2 and 3, based on collapsed sequences from Fig. 1.

3.2.2 Ancestral Recombination Graph (ARG)

The analysis software RecMin is often used to determine the minimum number of recombination events in a population (Myers and Griffiths 2003), whereas the ARG illustrates the relationships of homologous sequences undergoing recombination by coupling allele coalescence with recombination events (Lyngsø, Song et al. 2005). Both analyses are based on the four-gamete test that looks for patterns of incompatibility among pairs of sites exhibiting all four gametic types. Lyngsø et al. (2005) consider the ARG to be an improvement on RecMin, especially for large population samples. There are two different parsimony-based algorithms for inference of an ARG: Beagle and Kwarg. Each will attempt to infer the ancestral state from a group of nucleotide sequences whether or not a root is known. The four-gamete test is performed on sequence alignments as sections, reconstructing a history with each pass. Beagle searches for the most parsimonious reconstruction using branch and bound, so it may be limited by complex datasets with a great deal of recombination; therefore, Lyngsø later created a heuristic implementation that is less exact to handle more complex population datasets and named it Kwarg (http://www.stats.ox.ac.uk/~lyngsoe/section26/). In an ARG, mutations are shown as numbers along its branches (see Fig. 3), assuming a standard infinite sites model. An ARG shows population haplotypes/individuals encircled in red (present time), and the population's history is traced back to the most recent common ancestor at the top (in the past). Recombination events are encircled in blue and show the position of the variable site in the DNA sequence alignment where the breakpoint occurs. Preceding each recombination event are branches labeled with "P" (prefix) or "S" (suffix) that correspond to the 5′ and 3′-most segments, respectively, that are exchanged by a crossover between putative parental lineages. Graph nodes that correspond to coalescence/divergence events are shown as green dots in an ARG; a yellow dot denotes the inferred ancestral (root) sequence.

Using Beagle to infer ARGs, we observe evidence of recombination in the histories of populations 1 and 2 (Fig. 3). Population 1 has experienced recombination throughout its history, and haplotypes H2 and H9 were formed by recent recombination events between two parental haplotypes (H1+H6 and H10+H7, respectively). Population 2 shows only a single recombination event in its history whereby H3 is the putative offspring of H2 and H5. Clearly recombination has given rise to a greater number of haplotypes at these three loci in population 1 since there are eleven haplotypes compared to the five haplotypes for population 2.

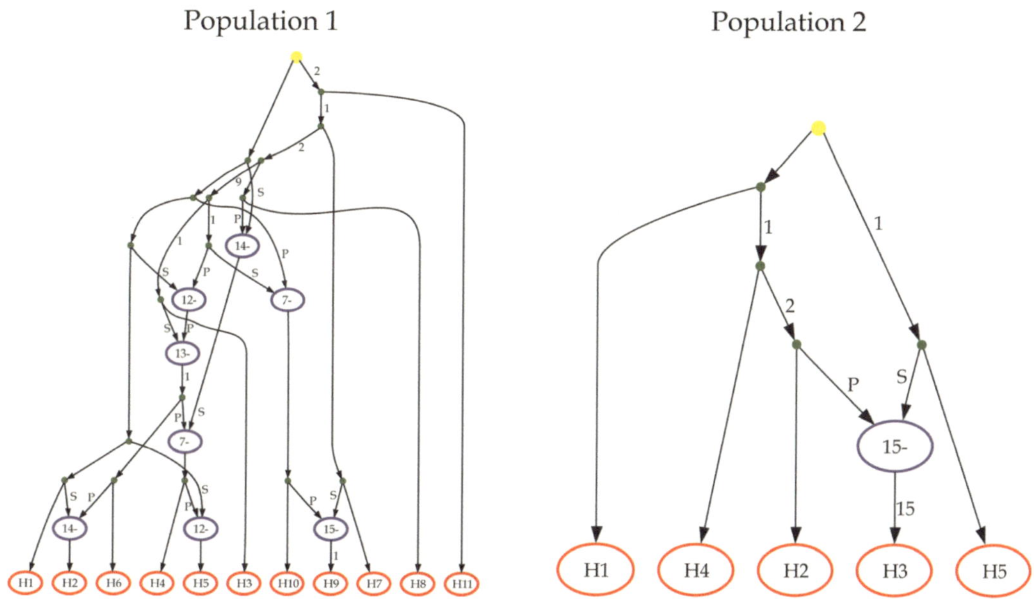

Fig. 3. ARGs based on combined loci for populations in Fig. 1.

3.2.3 Allele coalescence

Allele coalescence is rooted in the coalescent theory, which states that all alleles in a population are inherited from a single ancestor and are shared throughout the population. Coalescence is the reverse of divergence, so two lineages are traced back to their common ancestor. Coalescence may be inferred for a single locus or for multiple linked loci using GENETREE version 9.0 (http://www.stats.ox.ac.uk/~griff/software.html). Gene genealogies inferred using the coalescent reconstruct mutational relationships either within a "panmictic" population undergoing random mating, or among "subdivided" populations experiencing limited gene flow. Coalescent simulations can be performed assuming constant population sizes or exponential growth (Griffiths and Tavaré 1994). Similar in structure to an ARG, the coalescent will have the haplotypes at the tips of branches (present time) that coalesce to the most recent common ancestor (MRCA) in the past. Mutations are shown as dots along the branches. An interesting feature of the coalescent model is that coalescent time (shown as a scale above the genealogy in Fig. 4) can be converted to real time, giving approximate divergence times, in millions of years. Figure 4 shows examples of gene genealogies inferred for the three combined loci. The numbers in parentheses adjacent each haplotype refer to the isolate numbers from Figure 1.

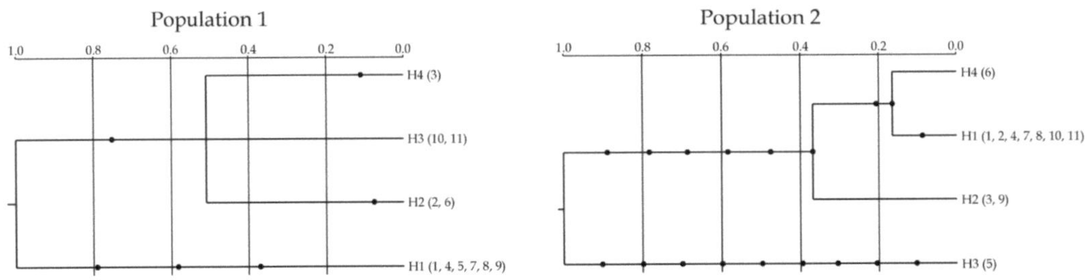

Fig. 4. Coalescent-based gene genealogies based on combined loci for populations in Fig. 1.

An important consideration when inferring gene genealogies, using a coalescent model that assumes no recombination, is that there are no incompatible sites in data. Frequent recombination in population 1 creates site incompatibilities since there may be different ancestral histories for each locus; as a result, only the largest non-recombining partition was examined. This affects inference of haplotype diversity and subsequently we see a reduction in haplotype number from eleven to four. Population 2 has only a single recombination event in its history, meaning fewer incompatible sites are present, so the haplotype number of four is much closer to initial population diversity (five haplotypes).

3.3 A comparison of recombination in *A. flavus* and *A. parasiticus*

3.3.1 Linkage disequilibrium

Carbone, Jakobek et al. (2007) used LD to investigate recombination in a single *A. parasiticus* population from Georgia. Starting with a sample of 76 single-spore isolates, an experimental sample of 24 isolates, each representing a separate VCG, was created (Carbone, Jakobek et al. 2007). Moore and co-workers investigated an *A. flavus* population from the same Georgia field from which they began with 92 single-spore isolates that were consolidated to a subset of 44 based on VCG testing (Moore, Singh et al. 2009). Both studies sequenced most of the intergenic regions across the aflatoxin gene cluster and imported those sequences into the SNAP Workbench (Price and Carbone 2005) for various statistical analyses. For *A. parasiticus*, LD analysis revealed evidence of recombination throughout the cluster by formation of five blocks of recombination (Fig. 1 in Carbone et al. 2007). The *A. flavus* LD, shown in Figure S1 of Moore et al. 2009, revealed six blocks of recombination, spanning contiguous late-pathway genes of the cluster (*aflE-aflO*). Recombination was not detected in the early pathway genes (*aflT-aflE*); however, Moore and co-workers did observe increased levels of recombination in a "hotspot" around *aflP*. Given that some of the blocks span multiple loci, it was inferred that shuffling a group of genes is as likely as shuffling only one gene. Also, since some of the sequences included exons, it is possible to infer that block junctions are in either the intergenic region or the exon. In comparing rates of recombination based solely on LD patterns, it would appear that *A. flavus* is experiencing similar rates of recombination as *A. parasiticus*. The *A. flavus* LD plot exhibited a more significant block structure across contiguous loci than the *A. parasiticus* LD which showed a less organized block structure intermittently throughout the cluster. Both *A. flavus* and *A. parasiticus* LD analyses showed evidence of balancing selection in a non-recombining region of the cluster (*aflW* and *hypE*, respectively). Both studies provided estimates of the minimum number of recombination events (RecMin) for their respective populations. *A. flavus* appeared to have more than four times as many recombination events as *A. parasiticus* in the same field. The higher rate of recombination was particularly evident in the ARG.

3.3.2 Ancestral recombination graphs

The Beagle algorithm was used for the *A. parasiticus* ARG analysis, but attempts to use Beagle for the *A. flavus* sample proved difficult; therefore, the Kwarg algorithm was used because it can handle more complex datasets. ARG analyses revealed that both *A. parasiticus* (Fig. 2B from Carbone et al. 2007) and *A. flavus* (Fig. 2 from Moore et al. 2009) populations from the same field show a history of recombination, but the *A. flavus* population exhibits a more extensive history of recombination. This is not surprising when one compares both the number of VCGs found for each population and the number of distinct multi-locus cluster haplotypes. *A. parasiticus* had fewer VCGs and distinct haplotypes (24 and 10, respectively) than *A. flavus* (44 and 27, respectively), even though the overall numbers of isolates sampled were comparable. It is well known that increased recombination will contribute to greater population diversity (Milgroom 1996).

3.3.3 Allele coalescence

Carbone, Jakobek et al. (2007) inferred a gene genealogy for a non-recombining region in the aflatoxin cluster of *A. parasiticus*, known as *hypE*. They found evidence of trans-speciation (balancing selection), which was supported by the separation of two groups of distinct chemotype classes (see Figure 3 in Carbone et al. 2007). Those distinct chemotype classes were reportedly related to G- and B- aflatoxin production versus *O*-methylsterigmatocystin (OMST) production. For the *A. flavus* study, Moore, Singh et al. (2009) performed coalescent analysis for a non-recombining cluster region known as *aflW* (*moxY*). They observed evidence of balancing selection for this region, and once again the genealogy (Fig. 5 in Moore et al. 2009) supported separation of the population into two groups based on distinct chemotype classes. In *A. flavus*, balancing selection appears to be between AF+ and AF- individuals because most full-cluster, and all partial-cluster, AF- isolates grouped as a separate clade/lineage from that of the AF+ isolates.

Both non-recombining loci offer glimpses into the ancestral states for each species, and perhaps the basis for balancing selection to maintain the ancestral states is cohesiveness of functional units. A plausible explanation for these results is that the gene *hypE*, and its cluster neighbor *aflN* (*verA*), encode proteins that catalytically function as a unit. Balancing selection would favor keeping these genes together in an evolutionary shuffling process. This suggests that, by a process not yet fully understood, the protein encoded by *hypE* encodes a helper protein that is required for the proper function of the cytochrome P450 monooxygenase, *aflN*, the first enzyme needed for conversion of versicolorin A to sterigmatocystin. Similarly, both *aflV* (*cypX*) and *aflW* (*moxY*), the oxidative enzymes required for conversion of averufin to versicolorin B, also function as a unit so balancing selection maintains these as non-recombining loci (Kenneth C. Ehrlich, unpublished).

3.4 Additional inferences resulting from recombination analyses

Moore, Singh et al. (2009) explored the role of G+C content in *A. flavus* recombination, on chromosome III, from the telomere to the centromeric end of the aflatoxin gene cluster. For Fig. 3 in Moore et al. 2009, a *z'* curve (Zhang and Zhang 2004) is shown that illustrates a steady increase of G+C content, from the telomere to the *aflF* (*norB*) gene, seen as a negative slope. The curve then rises sharply, as G+C content decreases, until the *aflE* (*norA*) gene where it plateaus to *aflQ* (*ordA*). G+C content begins to increase toward the end of the cluster, seen again as a negative slope of the curve. When one considers the lesser effort it takes to break apart an A+T [double] bond, then shuffling of genetic material might be easier to accomplish in genomic regions where the G+C content is lowest (Zhang and Zhang 2004). Perhaps the G+C content in *A. parasiticus* exists at higher levels than in *A. flavus*. Earlier it was mentioned that sequence breakpoints may exist in either the gene or the intergenic region. Moore et al. explored this concept by performing LD analysis on four complete cluster sequences. In Figure S2 from Moore et al. 2009, there was evidence of sequence breakpoints existing in both gene and intergenic region for *A. flavus* (AF13, AF36, and NRRL 3357) and *A. oryzae* (NRRL 5590). Therefore, even a single gene might have multiple histories if it has been experiencing frequent shuffling. Another finding worth noting pertains to the relative ages of partial-cluster to full-cluster AF- *A. flavus* strains. Referring once again to the gene genealogy from Fig. 5 in Moore et al. 2009, it appears the partial-cluster strains are recently derived from full-cluster ancestors, offering evidence of gene loss instead of gain. This suggests that partial-cluster strains are an evolutionary dead-end and that such strains may offer a more stable foundation as biocontrol agents.

3.5 Evolution of the aflatoxin gene cluster

Gene clusters that contain syntenic AF/sterigmatocystin (ST) biosynthesis genes have been found in very diverse fungi, including *Mycosphaerella pini*, the causative agent of needle blight in pine, (Bradshaw and Zhang 2006); *A. terreus*, a species commonly found in agricultural fields (Carbone, Ramirez-Prado et al. 2007); and the human pathogens *A. fumigatus* (Carbone, Ramirez-Prado et al. 2007) and *Coccidioides immitis* (Sharpton, Stajich et al. 2009; Ehrlich and Yu 2010). In *M. pini*, the genes encoding proteins necessary for production of the starter fatty acid hexanoyl CoA, HexA and HexB, and the decorating enzymes, OrdB, AvnA, HypC and Vbs, are in a separate cluster from the genes encoding polyketide synthase (PksA) and the decorating enzymes CypX, AvfA, EpoA (not found in AF or ST gene clusters) and MoxA (Cary and Ehrlich 2006). The latter two enzymes are involved in versicolorin A metabolism in AF/ST-producing *Aspergillus* species and are related to genes involved in melanin production in fungi. The regulatory proteins AflR and AflJ are also in a separate locus on the chromosome. Phylogenetic analysis revealed that the PksA types known to accept hexanoyl CoA as the starter unit have a different evolutionary history from other types of polyketide synthases (Ehrlich and Yu 2010).

Aspergillus species that produce both B and G aflatoxins, such as *A. parasiticus*, diverged prior to *A. flavus* (Cary and Ehrlich 2006). Loss of the ability to produce G-type AFs involves different mutations of the genes encoding enzymes required for G-toxin formation. There is evidence that loss of G AF production occurred at least three separate times among extant *Aspergillus* lineages. Phylogenetic evidence suggests that species with the *Flavus*-type gene cluster diverged from species with the *Nidulans*-type cluster about 75 Mya (Berbee and Taylor 2001; Heckman, Geiser et al. 2001; Kasuga, White et al. 2002). High identity between *Nidulans*- and *Flavus*-type cluster genes reflects purifying selection during divergence as measured by the ratio of rates of non-synonymous (Ka, nucleotide changes that result in amino acid changes) to synonymous nucleotide changes (Ks, nucleotide changes that result in conservation of amino acids) (Foster 2000). The Ka to Ks ratio was much lower for *ver-1* (ratio = 0.05) than for any of the 25 other *Flavus*-type cluster genes (ratio range = 0.2 to 1.04), suggesting that the most intense purifying selection was at this locus. Significantly higher Ka/Ks values in section *Flavi* compared to non-section *Flavi* species are evidence of adaptation and increased positive selection acting on genes in *Flavus*-type clusters with only partial gene sets. The gene *ver-1* encodes a reductase that is remarkably similar to the reductase required for melanin production. This reductase and *ordB* encode enzymes similar to the enzymes required for appressorium formation, a hardened mycelial structure needed for initial infection of plants and insects (Henson, Butler et al. 1999; Inagaki, Takano et al. 2000). This relationship suggests that AF production and other processes geared to offensive strategies for food acquisition by fungi arose from similar genetic roots. Therefore, while extant species of AF-producing Aspergilli are entirely saprophytic, they must have had, at one time, a more aggressively invasive lifestyle.

In the *Flavus*-type cluster there is considerable evidence that gene duplication is involved in cluster evolution (Carbone, Ramirez-Prado et al. 2007). Possible duplicated genes encode the following proteins (amino acid identity is in parentheses): the cytochrome P450 monooxygenases, VerB and AvnA (39%), NorA and NorB (50%), HypC and HypB (42%), and OrdB and AvfA (30%). Additional evidence for gene duplication is that after knockout of certain cluster genes, the resulting mutants still produce AF in addition to precursor intermediates. This "leaky" phenotype is found with the following knockout mutants: *nor-1, adhA, hypB1* or *hypB2, norA, norB,* and *ordB*. This suggests that gene homologs inside or outside of the cluster encode proteins that are able to substitute for the catalytic function of these *Nidulans/Flavus*-type cluster proteins (Ehrlich 2009).

Why are gene clusters maintained? Most fungal gene clusters represent regulatory islands within a region of chromatin that is at least partially heterochromatic (inactive and not able to be transcribed). It has been shown that heterologous genes inserted into the *Flavus*-type cluster are expressed under the same induction conditions as genes that are components of the cluster, whereas cluster genes inserted at loci removed from the cluster are severely down-regulated (Chiou, Miller et al. 2002; Keller, Bok et al. 2006). Many secondary metabolite clusters are in subtelomeric regions of the chromosome (regions generally rich in heterochromatin) (Galagan, Calvo et al. 2005; Wong and Wolfe 2005). Subtelomeric regions are particularly prone to chromatin modification. They tend to have an abundance of duplicated genes as well as retrotransposons and retrotransposon remnants. Such elements indicate a past history of DNA deposition from other locations in the genome (Robyr, Suka et al. 2002). Subtelomeric regions contain genes that permit growth under stress conditions and on unusual nitrogen and carbon sources, but which normally are silenced (Wong and Wolfe 2005). The small *aflR/aflJ* cluster in C. *immitis/posadasii* is associated with a retrotransposon.

After the ancestral cluster formed, gene losses occurred. The *Flavus*-type cluster is missing several *Nidulans*-type cluster genes (Ehrlich and Yu 2010). These genes, *stcC, stcD,* and *stcT,* lack promoter AflR-binding sites, and therefore, are probably not regulated by AflR during ST biosynthesis. They encode proteins with unknown functions and probably are not necessary for ST biosynthesis. A homolog to *stcC* is found in the dothistromin biosynthetic cluster (Bradshaw, Bhatnagar et al. 2002). Another gene loss appears to have occurred relatively recently, perhaps within the last 1 million years. This involves a deletion of the promoter region of the cytochrome P450 monooxygenase, *cypA,* a gene necessary for formation of AFG_1 (Ehrlich, Chang et al. 2004). Two types of deletion are found in *A. flavus*: a 1.2-kb deletion in *A. oryzae* and *A. flavus* S- strain, and a 0.8-kb deletion in most L- strains of *A. flavus*. It is possible that loss of G-toxin production was an adaptive change that permitted variant *A. flavus* to grow more readily on living plant tissues as opposed to being predominantly a resident in the soil.

The genes for AF/ST/dothistromin synthesis have been retained for greater than 125 million years and it is probable that the polyketides encoded by these clusters enabled the fungi to survive and prosper as free-living organisms in diverse environments (Bradshaw, Bhatnagar et al. 2002; Gomez and Nosanchuk 2003). We assume that the reason for maintaining such gene clusters is that the precursor metabolites are critical for fungal adaptation. Primordial fungi probably had a basal polyketide biosynthesis cluster that consisted of genes encoding only a PKS and a few enzymes capable of stabilizing the nascent polyketide. Such a basal cluster could have allowed fungi to synthesize aromatic, colorful molecules that, like plant pigments, may have aided spore dispersal (Cary and Ehrlich 2006). The colorful AF precursors, naphthoquinones and anthraquinones, are found in many types of fungi, including genera that preceded divergence of the Aspergilli. At this stage of cluster evolution, rather than being toxins, these metabolites may have been attractants (Cary and Ehrlich 2006). As the fungi dispersed into less hospitable niches, a need may have arisen for them to be competitive with associated microflora or they may have benefited by acquiring the ability to become virulent to plants (Cary and Ehrlich 2006). Such competitive conditions may have selected for the duplication and adaptation of genes to allow elaboration of an even greater variety of metabolites. There is increasing evidence that AF/ST biosynthesis and conidiospore development are regulated by a shared signal transduction mechanism (Hicks, Yu et al. 1997; Calvo, Wilson et al. 2002; Wilkinson, Ramaswamy et al. 2004).

4. *A. flavus* and *A. parasiticus* have sexual (teleomorphic) states

4.1 Sexuality in fungi

Many fungi have the ability to live a stable existence in either the haploid or diploid state. In yeast, the question was addressed concerning whether haploidy or diploidy confers an evolutionary advantage (Zeyl, Vanderford et al. 2003; Zeyl 2004). Conventional wisdom suggested that having an extra gene copy would prevent harmful mutations from accumulating in the genome whereas in haploid species, harmful mutations would be detrimental to the organism. However, adaptation depends on the rate that the mutation becomes "fixed" in the population. In haploid organisms, harmful mutations are rapidly lost and beneficial mutations are retained, while in diploid organisms, harmful and beneficial mutations take longer to become fixed in the population. Evolution in large populations is dominated by the accumulation of multiple mutations of moderate effect (Desai, Weissman et al. 2007). In wild populations of yeast, maximum genetic diversity was found in diploids (Ezov, Boger-Nadjar et al. 2006). Genetic diversity however, is only useful if selective pressure requires adaptation. Within an agricultural field, the demand for diversity may be less because a rich source of nutrients is readily available.

4.2 Identification of *MAT* loci in section *Flavi*

Until 2008, the only species in Section *Flavi* that had a known sexual state was homothallic *A. alliaceus* (teleomorph: *Petromyces alliaceus*). Little was known about the genetic structure of the mating-type locus in *A. alliaceus* or whether a mating-type locus existed for AF+ species like *A. flavus* and *A. parasiticus*. With evidence that both of these agriculturally important Aspergilli have histories of recombination, the Carbone lab began exploring the genomes for a *MAT* locus in all three species. Ramirez-Prado, Moore et al. (2008) published evidence of a heterothallic mating system in *A. flavus* and *A. parasiticus* by the amplification of a single idiomorph locus (chromosome VI) that was either *MAT1-1* (alpha-box) or *MAT1-2* (HMG). The *A. alliaceus* mating-type locus contained both idiomorphs on the same chromosome as shown in Fig. 1A from Ramirez-Prado, Moore et al. 2008. A diagnostic test was created to identify the mating-types of the isolates from the Georgia *A. flavus* and *A. parasiticus* populations previously studied.

4.3 *Aspergillus* mating studies

Once the *MAT* loci had been determined for populations of *A. flavus* and *A. parasiticus*, "parental" isolates were selected and paired for evidence of sexual reproduction (Horn, Moore et al. 2009; Horn, Ramirez-Prado et al. 2009). Parents were selected as belonging to both an opposite mating type and a different VCG. Resulting crosses offered morphological evidence of sex through the production of cleistothecia and ascospores. The teleomorphs within section *Flavi* were assigned to the genus *Petromyces* based on the morphology of the sexual state in which cleistothecia are borne within the matrix of stromata, which resemble the sclerotia produced by many section *Flavi* species. Since 2008, the teleomorphs of three AF+ species in section *Flavi* have been characterized: *P. parasiticus* (Horn, Ramirez-Prado et al. 2009), *P. flavus* (Horn, Moore et al. 2009), and *P. nomius* (Horn, Moore et al. 2011).

4.4 Potential impact of sex on population diversity

The greatest potential impact of sex on populations of AF+ fungi concerns the stability of biocontrol strains (Geiser, Pitt et al. 1998; Moore, Singh et al. 2009). What would we expect if an indigenous AF+ strain has sex, and exchanges genetic material, with a biocontrol strain? Tests are underway to determine if any possible offspring from such crosses differ from parents in the production of AFs. Mating in fungi allows for genetic modification to circumvent unfavorable conditions (Lee, Ni et al. 2010). When fields are inundated with a biocontrol strain, AF- genomes may become "rescued" by genetic material from an AF+ parent. Also, population chemotype diversity may increase when sex occurs in fields. One might expect, in an actively recombining population, a gradient of chemotype profiles. For instance, in *A. parasiticus* populations, there may be less distinction between G- or B-dominant isolates, i.e. more of the sample population exhibiting G1=B1. In *A. flavus* populations, there may be fewer AF- isolates and more AF producers. And if, by some chance, *A. flavus* and *A. parasiticus* are capable of mating, then there is the potential for even more chemotype diversity in a field.

5. Current research into *A. flavus* and *A. parasiticus* populations

5.1 Is recombination a global phenomenon?

Investigations are underway that compare global populations of AF+ and AF- Aspergilli. The main goal is to ascertain whether or not recombination observed in the previously examined Georgia populations is a global phenomenon. Other insights of interest relate to the rates of recombination in geographically-isolated fields, and if the inferred recombination breakpoints are conserved among species across the globe. Finally, research is underway to determine if the factors that influence chemotype diversity are the same no matter where the population is found.

5.2 What influences chemotype diversity?

Now that recombination has been shown for representative *A. flavus* and *A. parasiticus* populations in the United States, with the potential for recombination to exist for global populations of AF+ fungi, investigations should explore the factors that generate chemotype diversity in these populations. Is there a correlation between recombination and the different chemotype profiles in a sample population? Will a clonal population have lower overall aflatoxin concentrations compared to an actively recombining, more diverse population? By examining aflatoxin concentration and *MAT* designation for each individual, it may be possible to prove that not only does fungal sex increase population diversity, but it also increases overall aflatoxin production. And what of the potential for niche adaptation? Is there a connection between chemotype diversity and the environment in which the population is found? Ecological factors such as precipitation, soil type and temperature may be important to population diversity.

5.3 What are the boundaries to gene flow among populations?

If geographically isolated populations of AF+ fungi all have similar rates of recombination and aflatoxin chemotype diversity, does this offer evidence of gene flow? Future research should explore the barriers to gene flow among populations. Obvious factors such as species composition and geography are expected to prevent the mixing of genetic material. Would it then be feasible to seek a non-native biocontrol strain to prevent potential genetic exchange and recombination? One might expect global, geographically isolated populations of AF+ fungi to be evolving at different rates, sharing little recent evolutionary history with their distant relatives on the other side of the world. Continued interest in how these populations are evolving will be integral to control of AF+ fungi that may contaminate the global food supply.

6. Biocontrol of aflatoxin contamination using non-toxigenic *A. flavus* strains

6.1 Testing stability and longevity of current biocontrol strains

Research is being conducted that tests the stability of the AF- phenotype for the biocontrol strains when crossed with an AF+ strain. If mating and recombination occur, and any of the offspring are AF+, then we must re-evaluate how biocontrol strains are selected. It might be necessary to adopt a next-generation biocontrol strain that is more resistant to recombination in the field. Future research will also require exploration of the longevity of the current biocontrol strains. If we cannot find evidence of the persistence of the biocontrol strain in the field, does that mean the strain dies off, or does that mean the strain has recombined with indigenous AF+ strains and the AF- phenotype has been lost?

Researchers have also begun to explore how a biocontrol strain moves/persists in nature by creating a GFP-labeled transformant. Future endeavors could include releasing this labeled strain into a field and conducting a time-course study for the fluorescent strain. If an AF+ strain fluoresces then there will be indisputable evidence that the current biocontrol strains are recombining with indigenous strains as a result of successful out-crossing. Laboratory studies might include testing the aggressiveness of GFP-labeled strains when competing with AF+ strains, particularly in relation to various ecological conditions such as light, soil, and temperature.

6.2 Mating-type distribution in fields to be treated

Another important consideration in the selection of future biocontrol strains is the distribution of *MAT* idiomorphs in the field to be treated. If there is a predominance of one mating-type allele over another then it makes intuitive sense to release a biocontrol strain that shares the same *MAT* as the census population. Inundating a field in which the indigenous population is *MAT1-2* with a *MAT1-1* biocontrol agent may increase the incidence of recombination.

6.3 Genomic content of potential biocontrol strain

As mentioned earlier, it may be more beneficial to utilize a biocontrol strain that belongs to a lineage on an evolutionary dead-end. There has been no report of evidence that an AF- isolate which lacks the AF cluster can acquire a functional cluster configuration during recombination; therefore, an isolate lacking the aflatoxin gene cluster might be more difficult to "repair" than a point mutation in a gene subject to frequent shuffling. Another possible mechanism involves meiotic silencing, whereby unpaired homologs during prophase I of meiosis are silenced due to chromosomal misalignment (Shiu, Raju et al. 2001; Smith, Woloshuk et al. 2007).

7. Conclusion

A. flavus populations comprise a diverse assemblage of strains that are adapted to agricultural fields. Unlike other aflatoxigenic *Aspergillus* species, a portion of *A. flavus* populations has lost the ability to produce AFs. There is evidence that gene loss in the AF gene cluster of AF- isolates is irreversible and that balancing selection maintains non-aflatoxigenicity and lineage-specific gene loss in *A. flavus* populations (Donner, Atehnkeng et al.; Chang, Ehrlich et al. 2006; Ehrlich, Montalbano et al. 2007; Moore, Singh et al. 2009). Such AF- strains are being used for biocontrol of crop contaminations. Concerns have arisen about the possibility that introduced AF- strains of *A. flavus* used for biocontrol could re-acquire the ability to make AFs (Moore, Singh et al. 2009). If the introduced non-aflatoxigenic strain is more aggressive than native AF+ strains, a recombinant might have both increased AF-producing ability and increased ability to invade the plant. These concerns are based on the findings that *A. flavus, A. parasiticus* and *A. nomius*, aflatoxigenic fungi long considered to be incapable of forming a sexual state, can be induced to mate by crossing isolates from different mating types (Horn, Moore et al. 2009; Horn, Ramirez-Prado et al. 2009; Horn, Moore et al. 2011). The importance of understanding population structure is integral to preventing aflatoxin contamination of an increasingly precious food supply.

8. References

Aylor, D. L., E. W. Price, et al. (2006). "SNAP: Combine and Map modules for multilocus population genetic analysis." Bioinformatics 22(11): 1399-401.

Bayman, P. and P. J. Cotty (1991). "Vegetative compatibility and genetic diversity in the *Aspergillus flavus* population of a single field." Canadian Journal of Botany 69(8): 1707-1711.

Beerli, P. and J. Felsenstein (1999). "Maximum-likelihood estimation of migration rates and effective population numbers in two populations using a coalescent approach." Genetics 152(2): 763-73.

Berbee, M. L. and J. W. Taylor (2001). Fungal molecular evolution: gene trees and geologic time. In the Mycota. D. J. McLaughlin, E. McLaughlin and P. A. Lemke. Berlin, Springer: 229-246.

Bradshaw, R. and S. Zhang (2006). "Biosynthesis of dothistromin." Mycopathologia 162(3): 201-213.

Bradshaw, R. E., D. Bhatnagar, et al. (2002). "*Dothistroma pini*, a forest pathogen, contains homologs of aflatoxin biosynthetic pathway genes." Appl Environ Microbiol 68(6): 2885-92.

Calvo, A. M., R. A. Wilson, et al. (2002). "Relationship between secondary metabolism and fungal development." Microbiol Mol Biol Rev 66(3): 447-59, table of contents.

Carbone, I., J. L. Jakobek, et al. (2007). "Recombination, balancing selection and adaptive evolution in the aflatoxin gene cluster of *Aspergillus parasiticus*." Mol Ecol 16(20): 4401-17.

Carbone, I., J. H. Ramirez-Prado, et al. (2007). "Gene duplication, modularity and adaptation in the evolution of the aflatoxin gene cluster." BMC Evol Biol 7: 111.

Cary, J. W., D. Bhatnagar, et al. (2000). Aflatoxins: biological significance and regulation of biosynthesis. *Microbial Foodborne Diseases: Mechanisms of pathogenesis and toxin synthesis*. J. W. Cary, J. E. Linz and D. Bhatnagar. Lancaster, PA, Technomic Publishing Company, Inc.: 317-361.

Cary, J. W. and K. C. Ehrlich (2006). "Aflatoxigenicity in Aspergillus: molecular genetics, phylogenetic relationships and evolutionary implications." Mycopathologia 162(3): 167-77.

Chang, P. K., K. C. Ehrlich, et al. (2006). "Cladal relatedness among Aspergillus oryzae isolates and Aspergillus flavus S and L morphotype isolates." Int J Food Microbiol 108(2): 172-7.

Chang, P. K., B. W. Horn, et al. (2005). "Sequence breakpoints in the aflatoxin biosynthesis gene cluster and flanking regions in nonaflatoxigenic *Aspergillus flavus* isolates." Fungal Genet Biol 42(11): 914-23.

Chiou, C. H., M. Miller, et al. (2002). "Chromosomal location plays a role in regulation of aflatoxin gene expression in *Aspergillus parasiticus*." Appl Environ Microbiol 68(1): 306-15.

Cotty, P. J., P. Bayman, et al. (1994). Agriculture, aflatoxins and *Aspergillus*. *The Genus Aspergillus*. K. Powell. New York, Plenum Press: 1-27.

Cotty, P. J. and D. Bhatnagar (1994). "Variability among atoxigenic *Aspergillus flavus* strains in ability to prevent aflatoxin contamination and production of aflatoxin biosynthetic pathway enzymes." Appl Environ Microbiol 60(7): 2248-51.

Desai, M. M., D. B. Weissman, et al. (2007). "Evolution Can Favor Antagonistic Epistasis." Genetics: genetics.107.075812.

Donner, M., J. Atehnkeng, et al. "Molecular characterization of atoxigenic strains for biological control of aflatoxins in Nigeria." Food Addit Contam Part A Chem Anal Control Expo Risk Assess 27(5): 576-90.

Ehrlich, K. C. (2009). "Predicted Roles of the Uncharacterized Clustered Genes in Aflatoxin Biosynthesis." Toxins 1: 37-58.

Ehrlich, K. C., P. K. Chang, et al. (2004). "Aflatoxin biosynthesis cluster gene *cypA* is required for G aflatoxin formation." Appl Environ Microbiol 70(11): 6518-24.

Ehrlich, K. C., K. Kobbeman, et al. (2007). "Aflatoxin-producing *Aspergillus* species from Thailand." Int J Food Microbiol 114(2): 153-9.

Ehrlich, K. C., B. G. Montalbano, et al. (2007). "Analysis of single nucleotide polymorphisms in three genes shows evidence for genetic isolation of certain *Aspergillus flavus* vegetative compatibility groups." FEMS Microbiol Lett 268(2): 231-6.

Ehrlich, K. C. and J. Yu (2010). Aflatoxin-like Gene Clusters and How They Evolved. Mycotoxins in Food, Feed and Bioweapons. M. Rai and A. Varma, Springer Berlin Heidelberg. 1: 65-75.

Excoffier, L., G. Laval, et al. (2005). "Arlequin (version 3.0): an integrated software package for population genetics data analysis." Evol Bioinform Online 1: 47-50.

Ezov, T. K., E. Boger-Nadjar, et al. (2006). "Molecular-Genetic Biodiversity in a Natural Population of the Yeast *Saccharomyces cerevisiae* From "Evolution Canyon": Microsatellite Polymorphism, Ploidy and Controversial Sexual Status." Genetics 174(3): 1455-1468.

Federova, N. D., S. Harris, et al. (2009). "Using aCGH to study intraspecific genetic variability in two pathogenic molds, *Aspergillus fumigatus* and *Aspergillus flavus*." Medical Mycology 47: 1-8.

Foster, P. L. (2000). "Adaptive mutation: implications for evolution." Bioessays 22(12): 1067-74.

Fu, Y. X. (1996). "New statistical tests of neutrality for DNA samples from a population." Genetics 143(1): 557-70.

Fu, Y. X. and W. H. Li (1993). "Statistical tests of neutrality of mutations." Genetics 133(3): 693-709.

Galagan, J. E., S. E. Calvo, et al. (2005). "Sequencing of *Aspergillus nidulans* and comparative analysis with *A. fumigatus* and *A. oryzae*." Nature 438(7071): 1105-15.

Geiser, D. M., J. I. Pitt, et al. (1998). "Cryptic speciation and recombination in the aflatoxin-producing fungus *Aspergillus flavus*." Proc Natl Acad Sci U S A 95(1): 388-93.

Gomez, B. L. and J. D. Nosanchuk (2003). "Melanin and fungi." Curr Opin Infect Dis 16(2): 91-6.

Griffiths, R. C. and S. Tavaré (1994). "Ancestral inference in population genetics." Statistical Science 9: 307-319.

Grünwald, N. J. and E. M. Goss (2011). "Evolution and Population Genetics of Exotic and Re-Emerging Pathogens: Novel Tools and Approaches." Annu Rev Phyopathol 49: 5.1-5.19.

Grushiba, L. C. and P. J. Cotty (2009). "Parasexual recombination and migration maintain genotypic diversity in the aflatoxin-producing fungal plant pathogen *Aspergillus flavus*." Phytopathology 99: S47-S47.

Hartl, D. L. and A. G. Clark (1989). *Principles of Population Genetics*. Sunderland, MA, Sinauer Associates, Inc.

Heckman, D. S., D. M. Geiser, et al. (2001). "Molecular evidence for the early colonization of land by fungi and plants." Science 293(5532): 1129-33.

Henson, J. M., M. J. Butler, et al. (1999). "THE DARK SIDE OF THE MYCELIUM: Melanins of Phytopathogenic Fungi." Annu Rev Phytopathol 37: 447-471.

Hey, J. and C. A. Machado (2003). "The study of structured populations--new hope for a difficult and divided science." Nat Rev Genet 4(7): 535-43.

Hicks, J. K., J. H. Yu, et al. (1997). "*Aspergillus* sporulation and mycotoxin production both require inactivation of the FadA G alpha protein-dependent signaling pathway." Embo J 16(16): 4916-23.

Horn, B. W. (2007). "Biodiversity of *Aspergillus* section *Flavi* in the United States: a review." Food Addit Contam 24(10): 1088-101.

Horn, B. W., R. L. Greene, et al. (1995). "Effect of corn and peanut cultivation on soil populations of *Aspergillus flavus* and *A. parasiticus* in southwestern Georgia." Appl Environ Microbiol 61(7): 2472-5.

Horn, B. W., G. G. Moore, et al. (2009). "Sexual reproduction in *Aspergillus flavus*." Mycologia 101(3): 423-9.

Horn, B. W., G. G. Moore, et al. (2011). "Sexual reproduction in aflatoxin-producing *Aspergillus nomius*." Mycologia 103(1): 174-83.

Horn, B. W., J. H. Ramirez-Prado, et al. (2009). "The sexual state of *Aspergillus parasiticus*." Mycologia 101(2): 275-80.

Hudson, R. R. (2000). "A new statistic for detecting genetic differentiation." Genetics 155(4): 2011-4.

Inagaki, A., Y. Takano, et al. (2000). "Construction of an equalized cDNA library from *Colletotrichum lagenarium* and its application to the isolation of differentially expressed genes." Can J Microbiol 46(2): 150-8.

Kasuga, T., T. J. White, et al. (2002). "Estimation of nucleotide substitution rates in Eurotiomycete fungi." Mol Biol Evol 19(12): 2318-24.

Keller, N., J. Bok, et al. (2006). "*LaeA*, a global regulator of *Aspergillus* toxins." Medical Mycology 44(s1): 83-85.

Lee, S. C., M. Ni, et al. (2010). "The evolution of sex: a perspective from the fungal kingdom." Microbiol Mol Biol Rev 74(2): 298-340.

Leslie, J. F. (1993). "Fungal vegetative compatibility." Annu Rev Phytopathol 31: 127-50.

Lyngsø, R. B., Y. S. Song, et al. (2005). Minimum recombination histories by branch and bound. Proceedings of the 5th International Workshop on Algorithms in Bioinformatics.

Machida, M., O. Yamada, et al. (2008). "Genomics of *Aspergillus oryzae*: learning from the history of koji mold and exploration of its future." DNA Research 15(4): 173-183.

Mellon, J. E. and P. J. Cotty (2004). "Expression of pectinase activity among *Aspergillus flavus* isolates from southwestern and southeastern United States." Mycopathologia 157(3): 333-8.

Milgroom, M. G. (1996). "Recombination and the multilocus structure of fungal populations." Annu Rev Phytopathol 34: 457-77.

Moore, G. G., R. Singh, et al. (2009). "Recombination and lineage-specific gene loss in the aflatoxin gene cluster of *Aspergillus flavus*." Mol Ecol 18(23): 4870-87.

Myers, S. R. and R. C. Griffiths (2003). "Bounds on the minimum number of recombination events in a sample history." Genetics 163(1): 375-94.

Nielsen, R. and J. Wakeley (2001). "Distinguishing migration from isolation: a Markov chain Monte Carlo approach." Genetics 158(2): 885-96.

Nordborg, M. and S. Tavaré (2002). "Linkage disequilibrium: what history has to tell us." Trends in Genetics 18(2): 83-90.

Pitt, J. I. and A. D. Hocking (2006). "Mycotoxins in Australia: biocontrol of aflatoxin in peanuts." Mycopathologia 162(3): 233-43.

Price, E. W. and I. Carbone (2005). "SNAP: workbench management tool for evolutionary population genetic analysis." Bioinformatics 21(3): 402-4.

Quesne, W. J. L. (1982). "Compatibility analysis and its applications." Zoological Journal of the Linnean Society 74(3): 267-275.

Ramirez-Prado, J. H., G. G. Moore, et al. (2008). "Characterization and population analysis of the mating-type genes in Aspergillus flavus and Aspergillus parasiticus." Fungal Genet Biol 45(9): 1292-9.

Raper, K. B. and D. I. Fennell (1965). The genus *Aspergillus*. Baltimore, Williams & Wilkins.

Raymond, M. and F. Rousset (1995). "GENEPOP (Version 1.2): Population Genetics Software for Exact Tests and Ecumenicism." Journal of Heredity 86(3): 248-249.

Robyr, D., Y. Suka, et al. (2002). "Microarray deacetylation maps determine genome-wide functions for yeast histone deacetylases." Cell 109(4): 437-46.

Sharpton, T. J., J. E. Stajich, et al. (2009). "Comparative genomic analyses of the human fungal pathogens *Coccidioides* and their relatives." Genome Res 19(10): 1722-31.

Shiu, P. K., N. B. Raju, et al. (2001). "Meiotic silencing by unpaired DNA." Cell 107(7): 905-16.

Smith, C. A., C. P. Woloshuk, et al. (2007). "Silencing of the aflatoxin gene cluster in a diploid strain of *Aspergillus flavus* is suppressed by ectopic *aflR* expression." Genetics 176(4): 2077-86.

Tajima, F. (1989). "Statistical method for testing the neutral mutation hypothesis by DNA polymorphism." Genetics 123(3): 585-95.

Templeton, A. R., E. Routman, et al. (1995). "Separating population structure from population history: a cladistic analysis of the geographical distribution of mitochondrial DNA haplotypes in the tiger salamander, *Ambystoma tigrinum*." Genetics 140(2): 767-82.

Wilkinson, H. H., A. Ramaswamy, et al. (2004). "Increased conidiation associated with progression along the sterigmatocystin biosynthetic pathway." Mycologia 96(6): 1190-8.

Wilson, D. M., W. Mubatanhema, et al. (2002). "Biology and ecology of mycotoxigenic *Aspergillus* species as related to economic and health concerns." Adv Exp Med Biol 504: 3-17.

Wong, S. and K. H. Wolfe (2005). "Birth of a metabolic gene cluster in yeast by adaptive gene relocation." Nat Genet 37(7): 777-82.

Yeh, F. C. and T. J. B. Boyle (1997). "Population genetic analysis of co-dominant and dominant markers and quantitative traits." Belgian Journal of Botany 129: 157.

Yu, J., G. A. Payne, et al. (2008). Mycotoxin production and prevention of aflatoxin contamination in food and feed. *The Aspergilli Genomics, Medical Aspects, Biotechnology, and Research Methods*. S. A. Osmani and G. H. Goldman. Boca Raton, FL, CRC Press. 26: 457-472.

Zeyl, C. (2004). "Experimental studies of ploidy evolution in yeast." FEMS Microbiol Lett 233(2): 187-192.

Zeyl, C., T. Vanderford, et al. (2003). "An Evolutionary Advantage of Haploidy in Large Yeast Populations." Science 299(5606): 555-558.

Zhang, J. X. and P. J. Cotty (2007). "Diversity within communities of L and S strains of *Aspergillus flavus* in cotton fields in Texas and Arizona." Phytopathology 97: S172-S172.

Zhang, R. and C. T. Zhang (2004). "A systematic method to identify genomic islands and its applications in analyzing the genomes of *Corynebacterium glutamicum* and *Vibrio vulnificus* CMCP6 chromosome I." Bioinformatics 20(5): 612-22.

18

Aflatoxin in Agricultural Commodities and Herbal Medicine

Mehrdad Tajkarimi[1], Mohammad Hossein Shojaee[2], Hassan Yazdanpanah[3] and Salam A. Ibrahim[1]
[1]*North Carolina A&T State University,*
[2]*Faroogh Life Sciences Research Laboratory,*
[3]*School of Pharmacy, Shahid Beheshti Medical University*
[1]*USA*
[2,3]*Iran*

1. Introduction

Aflatoxins are a group of mycotoxins produced by *Aspergillus* species, including *A. flavus*, *A. parasiticus*, and *A. nomius*. A quarter of the world's food crops are estimated to be affected by mycotoxins; creating a large economical loss in the developed and developing countries (Kumar, Basu, & Rajendran, 2008; Reddy, Reddy, Abbas, Abel, & Muralidharan, 2008; Wagacha & Muthomi, 2008; Xu, Han, Huang, Li, & Jiang, 2008). Other reports indicate even higher contamination rate of aflatoxin (Njobeh, et al., 2009) . Exposure to higher levels of aflatoxin contamination increases cancer incidence, including risk of hepato-cellular carcinoma especially in 6- to 9-year-old girls and neural tube defects (Peng & Chen, 2009; Sun, et al., 2011a; Umoh, et al., 2011; Woo, et al., 2011).

One of the reason which makes aflatoxins one of the most challenging mycotoxin is the fact that it could be produced by the responsible fungi not only at pre-harvest time but also at post harvest stages including storage. However, later on, lack of regulations or poor enforcement, which make the use of such contaminated commodities inevitable, could lead to severe human and animal diseases too. Aflatoxin B1, B2, G1 and G2 are the most important members of the aflatoxin group, which chemically are coumarin derivatives with a fused dihydrofurofuran moiety. Presence of aflatoxin B1, B2, G1 and G2 may naturally occur in different ratios depending on different matrices. However, it was concluded that when aflatoxins are limited only to AFB1 and AFB2, such ratio is 1.0 to 0.1, while when all four aflatoxins occur (AFB1, AFB2, AFG1 and AFG2), they may be found in a ratio of 1.0:0.1:0.3:0.03 (Abbas, et al., 2010; Kensler, Roebuck, Wogan, & Groopman, 2011). Cereals notably corn, nuts such as peanuts, pistachio and Brazil nuts, oil seeds such as cottonseed, as well as copra, the dried meat of coconut, are some of the commodities with greater risk of aflatoxin contamination (Cornea, Ciuca, Voaides, Gagiu, & Pop, 2011; Head, Swetman, & Nagler, 1999; Idris, Mariod, Elnour, & Mohamed, 2010; Jelinek, Pohland, & Wood, 1989; Liang, Wang, Zhang, Chen, & Li, 2010; Moghadam & Hokmabadi, 2010; Pacheco & Scussel, 2009; Sun, et al., 2011b; Yassin, El-Samawaty, Moslem, Bahkali, & Abd-Elsalam, 2011).

Because peanuts, cottonseed and copra constitute the most important source of edible oils, they are of particular interest (Idris, Mariod, Elnour, & Mohamed, 2010). Commodities which are resistant or only moderately susceptible to aflatoxin contamination in the field include wheat, oats, millet, barley, rice, cassava, soybeans, beans, pulses and sorghum. However, when any of these commodities are stored under high moisture and temperature conditions, aflatoxin contamination may occur (Smith & Moss, 1985). Other commodities such as cocoa beans, linseeds, melon seeds and sunflower seeds have been infrequently contaminated with mycotoxins with lower importance rate compared to other commodities (Bankole, Adenusi, Lawal, & Adesanya, 2010; de Magalhaes, Sodre, Viscogliosi, & Grenier-Loustalot, 2011; Mngadi, Govinden, & Odhav, 2008; Pittet, 1998; Sanchez-Hervas, Gil, Bisbal, Ramon, & Martinez-Culebras, 2008). Aflatoxin is the single most important contaminant on The Rapid Alert System for Food and Feed (RASFF) of the European Union in a way that in 2008, aflatoxins alone were responsible for almost 30% of all the notifications to the RASFF system (902 notifications)(Energy., 2009). With increasing knowledge and awareness of aflatoxins as a potent source of health hazard to both human and animals, a great deal of effort has been made to completely eliminate the toxin or reduce its content in foods and feedstuffs to significantly lower levels. Although prevention is the most effective intervention, chemical, biological and physical methods have been investigated to inactivate aflatoxins or reduce their content in foodstuffs (Rustom, 1997). We will discuss the natural occurance and recent approaches on the fate and decontamination of aflatoxins in foods, herbs and feeds.

2. Natural occurrence of aflatoxin contamination in raw agricultural products

Natural occurrence of aflatoxins in raw agricultural products poses severe health and economic risks worldwide. The Food and Agriculture Organization (FAO) estimates that many basic foods could be contaminated with mycotoxin producing fungi, contributing to huge global losses of foodstuffs, about 1000 million metric tons each year (Bhat, Rai, & Karim, 2010). Contamination of feed materials with mycotoxins is an important issue for farmers due to both acute and chronic intoxication in animals. The economic impact of feed contamination with mycotoxins include productivity reduction and organ damage (Upadhaya, Park, & Ha, 2010). Aflatoxins, zearalenone, trichothecenes, fumonisins and ochratoxin A are the most frequently investigated toxins, although there are more than 300 recognized mycotoxins in animal feed (Griessler, Rodrigues, Handl, & Hofstetter, 2010; Rustemeyer, et al., 2010). Mycotoxin contamination reports in animal feed indicate a variety of contamination levels (de Oliveira, Sebastiao, Fagundes, Rosim, & Fernandes, 2010; Monbaliu, et al., 2010). Fungi which produce mycotoxin belong to *Aspergillus*, *Penicillium* and *Fusarium* species (Rustemeyer, et al., 2010). *Aspergillus* and *penicillium* constitute a major part of the fermented feed microbiota (Roige, et al., 2009). Intrinsic and extrinsic factors during storage and at field condition may interact with mycotoxin contamination (Griessler, Rodrigues, Handl, & Hofstetter, 2010). Animal feeds, such as hay and straw, might be contaminated during pre-harvest or drying stages (Bhat, Rai, & Karim, 2010). Mycotoxin contaminated animal feed causes serious effect on monogastric animals. However, the ruminants may be more resistant to mycotoxins due to biotransformation ability of the rumen microbiota. Other factors such as age, aflatoxin concentration and duration of exposure might also have some effect (Rustemeyer, et al., 2010; Upadhaya, Park, & Ha, 2010). The affected commodities by aflatoxins are corn, peanuts, cottonseed, millet, sorghum and other feed grains. In ruminants, a part of aflatoxin B_1 is degraded into aflatoxicol and the remaining is hydroxylated in the liver into aflatoxin M_1 (Upadhaya, Park, & Ha, 2010). Aflatoxin B_1 is considered as a group 1 carcinogen for humans by International Agency for

Research on Cancer (IARC) (Seo, Min, Kweon, Park, & Park, 2011). Aflatoxicosis may cause death in ruminants (Pierezan, et al., 2010). Despite extensive research done during the last few decades, which helped authorities around the world to establish control measures, still aflatoxin contamination in food and agricultural commodities remains as one of the most challenging and serious food safety problem.

Close study of the annual reports in the last decade (2003-2009) of the Rapid Alert System for Food and Feed (RASFF) showed four aforementioned groups contributed to the most aflatoxin contamination. Although, one should be careful in jumping to a bigger conclusion as these data also depends on the policy of EU countries, on products that go on a 100% check and those checked randomly. The detailed results are included in Table 1.

Literature review by Vinod Kumar, M.S. Basu, T.P. Rajendran (2008) on the incidence of mycotoxins in some commercially important agricultural commodities concluded that high-risk commodities for mycotoxin contamination were corn and groundnut (Kumar, Basu, & Rajendran, 2008).

Year	Total		nuts, nut products & seeds		fruit & vegetables		cereal products		herbs & spices		feed for food producing animals		pet food	
	No	%	No	%	No	%	No	%	No	%	No	%	No	%
2003	763	95	695	91	33	4	6	1	5	1	-	NA	-	NA
2004	844	95	699	83	42	5	12	1	7	1	-	NA	-	NA
2005	943	95	827	88	81	9	9	1	57	6	2	0	-	NA
2006	800	91	684	86	69	9	5	1	37	5	4	1	-	NA
2007	705	93	568	81	70	10	21	3	35	5	6	1	4	1
2008	902	97	710	79	103	11	46	5	26	3	11	1	3	0
2009	638	95	517	81	64	10	13	2	23	4	9	1	11	2

Summarized by the author's from the RASFF published reports(RASFF, 2011).

Table 1. Comparison of number of aflatoxin allert notifications according to product category reports in the RASFF system in years 2003-2009.

2.1 Nuts, nut products and seeds

As it is clear from Table 1 based on RASFF reports; nuts, nut products & seeds were the most rejected lots, and thus the most contaminated products in general too. These serve as very good substrates due to their high fat content. Also, aflatoxin producing fungi can cause toxin production in all steps including pre-harvest, drying process as well as storage.

Environmental conditions such as prolonged drought stress, play a major role in increasing the risk of aflatoxin contamination (Kumar, Basu, & Rajendran, 2008). Similar conclusion was also drawn by Wagacha and Muthomi (2008) from the African perspective too, in which aflatoxins widely occured in groundnuts (Wagacha & Muthomi, 2008). A close study of all mycotoxins rejected lots (1000 reports of 5979 at the time) based on online available information of RASFF from 16/12/2009 till 02/05/2011 revealed that highest aflatoxin levels were found again in this group (Table 2).

A Korean survey of different nuts and their products marketed in South Korea showed that 9 out of 85 samples including peanuts, peanut butter, and pistachios were contaminated with aflatoxins (10.6% of incidence). The most contaminated nut was peanut (roasted) with values ranging from 2.00–28.24 μg/kg and a mean of 10.67 ± 12.30 μg/kg for total aflatoxins

[7.97 ± 7.75 μg/kg for aflatoxin B1]. Similar data at slightly lower levels was found in one assorted nuts and 2 peanut butter samples (Chun, Kim, Ok, Hwang, & Chung, 2007). A Turkish study conducted from September 2008 to February 2009, detected aflatoxin B1 contamination in almost 49.2% (59/120) of unpacked and packed pistachio nut samples at levels lower than action limits of 5 μg/kg (Set & Erkmen, 2010). A.H.W. Abdulkadar et al (2004) found aflatoxin B1 contamination in different nuts in the range of not detected (ND)–81.64 μg/kg (Abdulkadar, Al-Ali, Al-Kildi, & Al-Jedah, 2004). In a study by Ismail et al 2010, from about 196 nuts and their products in Malaysia, 16.3% of the products showed contamination between 17.2 to 350 μg/kg (Ismail, Leong, Latif, & Ahmad, 2010). Forty-eight samples out of 95 were contaminated within a range of 0.007 to 7.72 μg/kg in pistachio nuts (Set & Erkmen, 2010).

Raw	Origin	Commodity	Maximum Contamination Levels (μg/kg)		Date of case	References
			B1	Total		
1	United States	Almonds	43800*	47800*	13/04/2010	2010.ARG
2	Italy, with raw material from Afghanistan	Shelled roasted pistachios	973	------	22/07/2010	2010.0996
3	Georgia	Hazelnut kernels	638	713	24/03/2011	2011.0396
4	Ghana	Groundnut paste	622	810	25/10/2010	2010.BVI
5	Egypt	Groundnuts in shell	614.0	646.4	07/12/2010	2010.CES
6	Turkey	Pistachios in shell	594	665	25/11/2010	2010.CCV
7	Iran	Pistachios	562	597.7	30/04/2010	2010.ATY
8	Syria	Pistachio kernels	333	369	29/03/2011	2011.ASB
9	Algeria	Dried Apricot kernels	333	342.5	16/12/2009	2009.CDC
10	Iran	Pistachio nuts	210	230	21/03/2011	2011.APX
11	China	Peanuts in shell	192	214	24/03/2010	2010.AOA
12	Italy	Dried sweet chestnut flakes	184		26/11/2010	2010.1615
13	Nigeria	Ground melon seeds	136.3	154.1	03/02/2011	2011.AGG
14	United States	Raw pistachios	120	134	18/03/2010	2010.AMX
15	India	Groundnut kernels	118.0	281.0	15/04/2011	2011.AWB
16	United States	Salted almonds	95.1	127.3	04/02/2010	2010.0130
17	United States	Almonds	61.5	69.2	12/11/2010	2010.BZI
18	Ukraine	Hulled walnuts	38		08/02/2011	2011.AHH

Table 2. Some of the highest values of aflatoxin contamination in the rejected lots of Nuts, nut products & seeds, based on The Rapid Alert System for Food and Feed (RASFF) Database*

* Both are very high value but it was reported as B1 = 43.8; Tot. = 47.8 mg/kg - ppm, Only changes has been made was modification of units from mg/kg (ppm) to μg/kg (ppb) Retrieved by the author's from the RASFF online Data Base(RASFF, 2011).

2.2 Fruits and vegetables

Close study of all mycotoxin rejected lots (277 reports of 672 at the time) from 01/01/2008 till 19/04/2011, based on online information from RASFF, revealed that highest aflatoxin levels were found in dried figs from Turkey followed by dried figs from Greece (Table 3). Natural occurrence of aflatoxin in fruits came to light more in the recent years.
Reports indicated that figs, dates and citrus fruits grown in susceptible regions (the high temperature conditions) could get contaminated with aflatoxins (Rivka Barkai-Golan, 2008), of which fig is most vulnerable to aflatoxin contamination. The reason for such high susceptibilities apart from their chemical properties is based on the fact that A. *Flavus* is able to enter and colonize in the internal cavity of the fruit (Doster, Michailides, & Morgan, 1996; Rivka Barkai-Golan, 2008). Although some surveys found only trace amount of aflatoxins in fig (Blesa, Soriano, Molto, & Manes, 2004), others found quite high levels and the contamination levels might go as high as 77,200 ng/g (Doster, Michailides, & Morgan, 1996). Aflatoxins were also reported, but in lesser extent, in other fruits such as dates, citrus fruits, raisins and olives (El Adlouni, Tozlovanu, Naman, Faid, & Pfohl-Leszkowicz, 2006; Ferracane, et al., 2007; Shenasi, Aidoo, & Candlish, 2002) . In case of citrus fruits, at least there is sound evidence of potential contamination risk (Bamba & Sumbali, 2005).

Raw	Origin	Commodity	Maximum Contamination Levels (μg/kg)		Date of case	References
1	Turkey	Dried figs	91.1	133.4	4/7/2008	2008.BAZ
2	Turkey	Dried figs	76	84	18/01/2008	2008.ADD
3	Greece	Dried figs	70.6	76.4	22/12/2010	2010.1742
4	Turkey	Dried figs	63.5	117.5	3/1/2008	2008.AAI
5	HUNGARY raw material from TURKEY	Dried figs	62.2	104.2	22/12/2008	2008.1672
6	Turkey	Dried figs	55	113	26/11/2010	2010.CDE
7	Turkey	Dried figs	54.2	58.3	20/10/2010	2010.BUC
8	Greece	Sun dried figs	47.9	86.7	3/12/2009	2009.1674
9	Turkey	Dried figs	41.80	51.23	19/11/2008	2008.BUT
10	Turkey	Dried figs	25.0	36.25	17/01/2008	2008.ACT

Table 3. Some of the highest values of aflatoxin contamination in the rejected lots of fruit & vegetables, based on The Rapid Alert System for Food and Feed (RASFF)*

2.3 Cereal products

Aflatoxin contamination of foodstuffs in Iran has been reviewed by Yazdanpanah (Yazdanpanah, 2006). Fifty-one maize samples, intended for animal feed and human

* Retrieved by the author's from the RASFF online Data Base(RASFF, 2011)

consumption, were collected from the four main maize production provinces in Iran and analyzed by high-performance liquid chromatography for contamination by four naturally occurring aflatoxin analogues (AFB1, AFB2, AFG1, and AFG2). AFB1 was detected in 58.3%, and 80% of the maize samples obtained from Kermanshah and Mazandaran provinces, respectively(Ghiasian, Shephard, & Yazdanpanah, 2011).

High levels of aflatoxin B1 contamination in rain-affected maize and rice at a level of 15600 and 1130 μg/kg respectively, was reported. Also, high levels of aflatoxin was found in parboiled rice (max 130 μg/kg). However, relatively lower values were reported in normal crops (Vasanthi S, 1998).

The crops with higher risk of aflatoxin contamination were groundnuts, maize and chilies. In one study, 21% and 26% of groundnuts and maize samples respectively, exceeded their national maximum limit of 30 μg/kg of aflatoxin contamination (Vasanthi S, 1998).

Vargas (2001) reported that 38.3% of maize samples were contaminated with aflatoxin B1 with a mean of 9.4 μg/kg and a maximum of 129 μg/kg. The investigators have reported that only 3.7% showed levels above 20 μg/kg. They found the co-occurrence of aflatoxin B1 and fumonisin B1 in all of the 82 aflatoxin-contaminated samples. Co-occurrence of these 2 mycotoxins with zearalenone was observed only in 18 samples (Vargas, Preis, Castro, & Silva, 2001).

Maize and groundnuts were reported to be a major source of aflatoxin contamination around the globe particularly in India, South America and the Far East in the late 90's. Other commodities which raised concerns with regard to high susceptibility to aflatoxin contamination were tropical and subtropical cereals, oilseeds, and tree nuts as well as cotton-seed meal.

The largest and the most severe documented aflatoxin poisoning has been reported at a level as high as 8,000 μg/kg in Kenya in 2004, causing 125 deaths out of 317 case-patients (Wagacha & Muthomi, 2008).

According to a study conducted by Sugita-Konishi et al (2006) about the contamination in various Japanese retail foods with aflatoxin B1, B2, G1, and G2, and other mycotoxins, between 2004 and 2005, aflatoxins were detected only in almost half of the peanut butter samples with the highest concentration of aflatoxin B1 at about 2.59 μg/kg. While in other products such as corn products, corn, peanuts, buckwheat flour, dried buckwheat noodles, rice, or sesame oil, aflatoxin contamination was not detected (Sugita-Konishi, et al., 2006).

Aflatoxin was also detected in the majority of dried yam chips samples surveyed in Benin with levels as high as 220 μg/kg, although the average was much lower (14 μg/kg).

More than 54% of dried yam chips in Nigeria were found positive for aflatoxin contamination, while high levels of aflatoxins ranging from 10–120 μg/kg was detected in slightly more than one third of the tiger nut (Cyperus esculentus) samples in the same country (Bankole & Mabekoje, 2004).

High aflatoxin levels in maize, in some other African countries, notably Benin and Togo have been reported and one third of the household grain, contained aflatoxins in the range of five-fold the safe limit (Wagacha & Muthomi, 2008).

Maize (Zea mays L.) grain was shown to be a good substrate for mould infection including A. *flavus*, A. *parasiticus* and production of aflatoxins. Indian scientists have reported several cases of aflatoxin epidemic in humans over the last decade mainly due to the consumption of heavily contaminated maize, that nominates maize as a high risk crop. Rice is another member of the cereal family which shows high level of aflatoxin contamination, as high as 2830 μg/kg, which according to some reports was even higher than levels compared to wheat and maize. Aflatoxin contamination in rice occurs in the preharvest stage. Delayed drying as well as high moisture content and crop storage can cause postharvest contamination. Although both white rice and parboiled rice could be contaminated with aflatoxin, parboiled rice (boiled rice in the husk), despite improvement in its nutritional profile especially its vitamin-B content (Beri-beri disease is common among the white rice-eating people), is more suitable for the storage fungi to enter if later drying is not adequate (Kumar, Basu, & Rajendran, 2008). Minh Tri Nguyen et al (2007) investigated the possible coexistence of aflatoxin B1, citrinin and ochratoxin in Vietnam. From 100 rice samples collected countrywide, 35 samples showed values higher than the limit of quantification (LOQ) of 0.22 μg/kg, with a mean of 3.31 μg/kg and a highest value of 29.8 μg/kg, for aflatoxin B1. The results also indicated a high percentage in co-occurrence of aflatoxin B1 and ochratoxin A in rice. Their findings showed significant effect of monsoons that increased the average of quantifiable samples of AFB1 and the ratio of detectable samples in rice, compared to those in the dry season. In some provinces, these were 5 times higher [mean of 10.08 μg/kg compared to 1.77 μg/kg] or even more [mean of 4.5 μg/kg compared to less than LOQ of 0.22 μg/kg]. Given the average daily intake of rice by aVietnamese adult to be 500 g, there is a cause for concern (Nguyen, Tozovanu, Tran, & Pfohl-Leszkowicz, 2007). Reports raised concern over the presence of citrinin in red yeast rice (Monascus fermented rice), a traditional natural food colorant in Asia, while no reports on aflatoxin was obtained (Lin, Wang, Lee, & Su, 2008). A study on Turkish wheat samples published in 2008 revealed 60% contamination level in a very low range indeed (maximum of 0.644 μg/kg)(Giray, Girgin, Engin, Aydin, & Sahin, 2007).

No aflatoxin was found in the 60 samples of corn meal and flour obtained from Sao Paulo Market in 2000 (Bittencourt, Oliveira, Dilkin, & Correa, 2005). A market research of various food products (cereal and cereal products, nuts and nut products, spices, dry fruits and beverages) in Qatar in 2002, revealed no detected levels of aflatoxin contamination in rice and wheat (Abdulkadar, Al-Ali, Al-Kildi, & Al-Jedah, 2004).

The highest aflatoxin levels were found in stone ground corn meal from India followed by mixed snacks from India, and rice from Thailand. Aflatoxin contamination in raw and processed food can be monitored using chromatography or antibody platforms (Seo, Min, Kweon, Park, & Park, 2011). Aflatoxin B1 was detected at the following levels in all samples of Nigerian grains : 17.01-20.53 μg/kg in wheat, 34.00-40.30 μg/kg in millet, 27.22-36.13 μg/kg in guinea corn, and 40.06-48.59 μg/kg in bread fruit (Odoemelam & Osu, 2009).

Close study of all mycotoxins rejected lots (249 reports of 249 at the time) from 14/02/2000 till 28/04/2011, based on online information available from RASFF, revealed that the third highest aflatoxin levels were found in this group (Table 4).

2.4 Herbs and spices

Medicinal plants are various plants with medicinal properties, which were the core of traditional therapy for the most of human history. Although the toxic effect of some were known for centuries, only in the recent modern time, the safety of these plants from the contamination point of view come to light.

Raw	Origin	Commodity	Maximum Contamination Levels (μg/kg)		Date of case	References
			B1	Total		
1	India	Stone ground corn meal	410	430	08/08/2008	2008.0970
2	Ghana	Dried roasted corn	336	383.6	15/10/2007	2007.CJI
3	India	Mixed snacks	184.07	188	12/12/2007	2007.CXJ
4	United Kingdom with raw material from Ghana	Kenkey (maize based product)	134	153	28/04/2011	2011.0553
5	Ghana	Fermented banku flour	57	127	03/09/2010	2010.BNA
6	Ghana	Maize flour	56	67	04/07/2008	2008.BAX
7	Thailand	Black rice	52.2	72.2	01/07/2004	2004.BMS
8	India	Corn meal in retail packs	47	51	6/02/2009	2009.AHE
9	India	Unpolished basmati rice	46.2	50.7	07/12/2007	2007.CVW
10	Hong Kong	Egg cake	45	54	01/12/2006	2006.CTX
11	-------------	Rice - red	35,0	43,6	14/08/2001	2001.JB
12	Malaysia	Glutinous rice balls with peanut butter	35*		19/03/2008	2008.AMU
13	Canada	Roasted red rice flour	32	37	18/09/2009	2009.1229
14	Pakistan	Broken rice	28	32.3	15/05/2006	2006.0315
15	Pakistan	Brown basmati rice	27		01/03/2007	2007.ANN
16	Pakistan	Brown basmati rice	22.1	23.7	13/03/2008	2008.ALH
17	Pakistan	Long grain white rice	18.9	25.6	03/03/2008	2008.AJX
18	Poland	Long grain white rice	16.7	18.4	23/03/2007	2007.0227
19	Bangladesh	Rice flakes	12.7	16.8	23/12/2008	2008.CBG
20	Pakistan	Basmati rice	12	14	27/11/2009	2009.1650
21	Pakistan	Broken rice	11.5	13	25/02/2009	2009.AKY

Table 4. Some of the highest values of aflatoxin contamination in the rejected lots of cereals and cereals products, based on The Rapid Alert System for Food and Feed (RASFF)**

One of the safety concerns in herbal medicine now a days is the presence of mycotoxins, notably aflatoxins, as their use have been increasing in the recent years after a decline in their use for almost a century. It has been reported that spices and herbs that was used for the improvement of some forms of liver disorder might be contaminated with high concentrations of aflatoxins, with aflatoxin B1 at an alarming level of 2230 μg/kg (Moss, 1998). Abdulkadar et al (2004) found aflatoxin B1 contamination in mixed spices powder in the range of 0.16–5.12 μg/kg, while chilli powder showed a higher range of 5.60–69.28 μg/kg (Abdulkadar, Al-Ali, Al-Kildi, & Al-Jedah, 2004).

* It might be because of presence of peanut butter

** Retrieved by the author's from the RASFF online Data Base(RASFF, 2011)

A Turkish study conducted from September 2008 to February 2009, detected aflatoxin B_1 contamination in 80% (48/60) of unpacked and packed ground red pepper samples within the range of 5-55.9 µg/kg (Set & Erkmen, 2010). Zinedine et al (2006) reported relatively low contamination levels in spice samples including paprika; ginger, cumin, and pepper. The highest level of aflatoxin was found in red paprika (9.68 µg/kg)(Zinedine, et al., 2006). Close study of all mycotoxin rejected lots (211 reports of 432 at the time) from 06/12/2007 till 19/04/2011, based on online information available from RASFF, revealed that the highest aflatoxin levels were found in curry powder from Nigeria, whole nutmeg from Indonesia, dried paprika from Peru and suya pepper from Ghana, followed by paprika powder from UK (Table 5). Contrary to the long history and the wide use of herbal medicines, there are only a few publications in regard to their mycotoxin contamination compared to the large number of publications on the contamination of cereals and oil seeds (Trucksess & Scott, 2008). The European Pharmacopeia has set limits for aflatoxin B1 and total aflatoxins at 2 and 4 µg/kg respectively, for some medicinal herbs (Pharmacopeia, 2007). Although in one study in South Africa, no aflatoxin contamination was found in some medicinal plants (Sewram, Shephard, van der Merwe, & Jacobs, 2006), while others reported levels ranging from 2.90–32.18 µg/kg (Yang, Chen, & Zhang, 2005). Roy et al. (1988) reported both high incidence (>93%) and high levels ranging from 90–1200 µg/kg in some common drug plants. *Piper nigrum* with a concentration of 1200 µg/kg was the highest contamination level reported. The second highest reported value was in the seeds of *Mucuna prurita* at a level of 1160 µg/kg. The third highest value was 1130 µg/kg, which found in the roots of *Plumbago zeylanica* (Roy, Sinha, & Chourasia, 1988). Aflatoxins were only found in 1 out of 5 Aerra lanata medicinal plant samples from Sri Lanka at 500 µg/kg (Abeywickrama & Bean, 1991). In another survey in India, 60% samples of medicinal plant seeds were contaminated with AFB1, ranging from 20 to 1180 µg/kg (Trucksess & Scott, 2008). In Thailand, five out of 28 herbal medicinal products were found to be contaminated with aflatoxins at 1.7–14.3 µg/kg using an immunoaffinity column (IAC) and high performance liquid chromatography (HPLC) method (Tassaneeyakul et al. 2004). None of the samples contained aflatoxins at levels above 20 ng/g (Tassaneeyakul, Razzazi-Fazeli, Porasuphatana, & Bohm, 2004). In Malaysia and Indonesia, 16 of the 23 commercial traditional herbal medicines, jamu and makjun, analyzed using IAC/LC method contained a low level of total aflatoxins (0.36 µg/kg)(Ali, et al., 2005). Romagnoli et al (2007) analyzed aflatoxins in 27 aromatic herbs, 48 herbal infusions and medicinal plants using LC with post-column derivatization and fluorescence detection. They found no contamination with aflatoxins (Romagnoli, Menna, Gruppioni, & Bergamini, 2007). In a study by Hitokoto et al., aflatoxins were not detected in the 49 powdered herbal drugs (Hitokoto, Morozumi, Wauke, Sakai, & Kurata, 1978). Ten percent of the tablets of Cascara sagrada dried bark were contaminated with aflatoxins in Argentina (Trucksess & Scott, 2008).

In a study on garlic samples, no aflatoxins were found at levels >0.1 µg/kg. However, aflatoxin levels between 4.2-13.5 µg/kg were detected in ginger (Patel, Hazel, Winterton, & Mortby, 1996).

A detailed UK study of aflatoxin contamination in some herbs and spices including curry powder, pepper, cayenne pepper, chilli, paprika, ginger, cinnamon and coriander showed 95% contamination below 10 µg/kg of total aflatoxins, while only 9 out of the 157 retail samples had higher levels (Macdonald & Castle, 1996). Study of ginseng root samples, both simulated wild and cultivated ones by D'Ovidio et al. (2006), showed approximately 15 µg/kg of total aflatoxins in only 2 of the simulated wild roots while none of the cultivated roots were contaminated with aflatoxins. Similar results (16 µg/kg) were found in just one mouldy ginseng root purchased from a grocery store (D'Ovidio, et al., 2006). Trucksess and Scott (2008) found that 30% of the ginseng products purchased in USA were contaminated with AFB1 at levels of about 0.1 µg/kg (Trucksess & Scott, 2008). In an aflatoxin survey done in Turkey, 17.1% and 23.1% of unpacked and packed ground red peppers respectively, were contaminated with total aflatoxins and aflatoxin B_1, with one out of the 82 samples over the legal limit (Set & Erkmen, 2010).

Raw	Origin	Commodity	Maximum Contamination Levels (µg/kg)		Date of case	References
1	India	Ground turmeric and whole nutmeg	700	1200	18/10/2010	2010.1405
2	Nigeria	Curry powder	570	1100	03/09/2008	2008.BJQ
3	Indonesia	Whole nutmeg	384.5	455.3	20/12/2007	2007.0950
4	India	Ground and broken nutmeg	230	249	03/09/2008	2008.BJM
5	Peru	Dried paprika	216	221	23/12/2009	2009.CER
6	Ghana	Suya pepper	169	215.9	04/01/2008	2008.AAZ
7	Spain	Paprika powder	145.3	160.8	10/08/2010	2010.1102
8	United Kingdom with raw material from Spain	Paprika powder	120.3	135.	11/2010	2010.1495
9	Indonesia	Nutmeg	120	140	03/12/2009	2009.CBP
10	Spain	Nutmeg	98	105	04/04/2011	2011.0444
11	India	Nutmeg powder	79+/-24	97+/-29	21/01/2010	2010.ACO
12	Indonesia	Nutmeg shrivels	57		26/10/2010	2010.BVK
13	Indonesia	Ground nutmeg	56	70.5	19/08/2010	2010.1143
14	India	Ground nutmeg	50	58.2	27/04/2010	2010.0515
15	India	Turmeric powder	48	53	24/12/2010	2010.CIJ
16	India	Turmeric powder	48	52	29/04/2009	2009.AWC
17	India	Chili powder	47.2	48.7	05/11/2010	2010.BXR
18	India	Organic ground nutmeg	41.1		28/05/2010	2010.0671
19	India	Crushed chillies	38	40	27/08/2010	2010.BMA
20	Pakistan	Chilli powder	30.3	32.1	16/08/2010	2010.BJO
21	India	Clove powder		29	17/02/2009	2009.AIU
22	India	Curry powder	26.4	27.4	14/07/2010	2010.BFZ
23	India	Chilli powder	24	25	31/08/2010	2010.BMK
24	India	Dried red chilli	23	25	17/12/2010	2010.CGS
25	China	Red pepper powder	22	26	09/07/2010	2010.0926
26	India	Dry whole chillies	20	21	24/11/2010	2010.CCP
27	India	Ginger	13.2	24	19/04/2011	2011.AXB

Table 5. Some of the highest values of aflatoxin contamination in the rejected lots of herbs & spices, based on The Rapid Alert System for Food and Feed (RASFF)*

* Retrieved by the author's from the RASFF online Data Base(RASFF, 2011)

2.5 Other foods

Tajkarimi et al. reported the contamination levels in milk within 0.057 μg/kg and between 0.041–0.065 μg/kg in another study in Iran (Tajkarimi, Aliabadi-Sh, et al., 2008; Tajkarimi, et al., 2007). Sixty two percent of the samples in North Western Iranian state were contaminated with values higher than 50 ng/l of aflatoxin M1 (Ghazani, 2009). Aflatoxin levels were 3.12- 3.65 fold more in whey, during an experimental Cheese production (Kamkar, Karim, Aliabadi, & Khaksar, 2008).

2.6 Feed

Aflatoxin contamination was detected in more than 60% of animal feed samples with an average concentration of 130.63 μg/kg. The most common contaminant was aflatoxin B_1 (Elzupir, Younis, Fadul, & Elhussein, 2009) , that was also similar with another study (Aksoy, Yavuz, Das, Guvenc, & Muglali, 2009). In a study conducted by Diaz et al., 2009, more than 50% of the feed samples were contaminated with *Aspergillus* spp. Maize (100%), cottonseed meal (80%), sorghum (60%) and wheat middlings (60%), showed the highest contamination level. Aflatoxin contamination range in this study was detected between 0.2 and 240.4 μg/g (Diaz, Lozano, & Acuna, 2009). Feed samples showed about 80% contamination in imported feed lots in Kuwait (Dashti, et al., 2009). In Lebanon, 4% of the commercial corn shipments were found to contain between 6 and 30 μg/kg of AFB_1 (Barbour, Farran, Usayran, & Daghir, 2008). Maize contamination in animal feed showed lower concentration compared to maize samples used for human consumption (Trung, et al., 2008). In Brazil, aflatoxin B_1 was found in 42% of the feed samples at levels of 1.0-26.4 μg/kg, with a mean of 7.1 +/- 7.2 μg/kg (Oliveira, Sebastiao, Fagundes, Rosim, & Fernandes, 2008). However, reported aflatoxin levels in South Africa was at a lower level of 0.8 +/- 0.2 μg/kg (Odhav, Mngadi, & Govinden, 2008). Goat feed and barely in Brazil were contaminated with aflatoxin B_1 at about 44% and 47% respectively, with a contamination range of 2.4 - 8.7 ng/g (Keller, et al., 2008). Aflatoxin B_1 contamination in maize in south Ethihopia was 22.72 μg/kg (Alemu, Birhanu, Azerefgne, & Skinnes, 2008). A study done on the composition and nutritional adequacy of six complete commercial feeds for pet rabbits, showed an aflatoxin B1 content of 11.36 ppb, which was slightly higher than the European recommended maximum amount of 10 μg/kg (Ricci, Sartori, Palagiano, & Zotte, 2010).

Seasonal variation in aflatoxin contamination may vary based on the feed type, processing and storage conditions. For example, Tajkarimi et al. (2008) demonstrated that the aflatoxin contamination was higher in winter (Tajkarimi, Aliabadi-Sh, et al., 2008; Tajkarimi, Faghih, et al., 2008). Higher rate for contamination during winter has also been indicated by Sugita-Konishi et al (2008), in corn samples (Sugita-Konishi, Sugiyama, & Hiraokai, 2008). However, in another study by Elzupir et al (2010), the level of aflatoxin contamination in summer was higher compared to winter (Elzupir & Elhussein, 2010). Early harvesting, proper drying, sanitation, proper storage and insect management are some other methods to control aflatoxin contamination (Wagacha & Muthomi, 2008).

3. Fate of aflatoxins during processing

Aflatoxins, like most of the mycotoxins, are stable compounds. Therefore, most of the processing steps during food production such as temperatures below 250 °C have little or no effect on their content, which may lead to contaminated finished cereal based products. However, there are some other processing steps such as alkaline cooking, nixtamalization (tortilla process), extrusion, roasting, flaking and modified processing methods that may reduce the aflatoxin content, but cannot eliminate the aflatoxin completely(Arzandeh & Jinap, 2011; Bullerman & Bianchini, 2007; Park, 2002; Perez-Flores, Moreno-Martinez, & Mendez-Albores, 2011; Yazdanpanah, Mohammadi, Abouhossain, & Cheraghali, 2005). Physical sorting is also another effective measure in the reduction of aflatoxins, as high as 40–80%(Bullerman & Bianchini, 2007). Marginal losses are considerable only if they are beyond the uncertainty of measurements at the given concentration. Some reports indicated a total destruction, at 1600 μg/kg of aflatoxin, in yellow dent contaminated corn by frying process (Magan, 2004). Using aflatoxin degradation enzyme named myxobacteria aflatoxin degradation enzyme (MADE), obtained from the extracellular enzyme of *Myxococcus fulvus,* was proposed to be an effective decontamination material with wide temperature range, pH tolerance and reasonable cost (Ji, et al., 2011). Application of different microorganisms to degrade aflatoxin was started in 1960 with a positive demonstration of removing aflatoxin by *Flavobacterium aurantiacum* in milk, vegetable oil, corn, peanut, peanut butter and peanut milk. It has been shown that pH and temperature influenced the uptake of toxin by the cells. However, the bright orange pigmentation caused by *Flavobacterium aurantiacum* ,limits its application in food (Smiley & Draughon, 2000). Other microorganisms also have aflatoxin degradation such as *Rhodococci* spp., *Lactobacillus rhamnosus and Enterococcus faecium* (Markov, Frece, Cvek, Lovric, & Delas, 2010; Topcu, Bulat, Wishah, & Boyaci, 2010). *Myxococcus fulvus,* with high activity and wide temperature and pH range, showed successful degradation activity against different aflatoxins (Ji, et al., 2011). Genetically modified plants could have aflatoxin lowering potential and also other applications (Davison, 2010; Halasz, Lasztity, Abonyi, & Bata, 2009; Montes, Reyes, Montes, & Cantu, 2009). Aqueous and organic extracts of plant materials such as viz. *Tagetes minuta, Lippia javanica, Amaranthus spinosus* and *Vigna unguiculata* have been successfully used against *Aspergillus flavus* and *A. parasiticus* (Houssou, et al., 2009; Katerere, Thembo, Vismer, Nyazema, & Gelderblom, 2010).

3.1 Cereal grains

Fandohana et al. studied the fate of aflatoxins through the traditional processing of naturally contaminated maize-based foods in West Africa. Aflatoxin levels were reduced by 7%, 8% and 60% during the preparation of makume, akassa and owo, respectively. The unit operations that resulted in marked mycotoxin removal included sorting, winnowing, washing and crushing, combined with dehulling of maize grains (Fandohan, et al., 2005). Stability of aflatoxins was more affected under alkaline, which led to partial degradation in cereals under heat based process. Reports indicated that fermentation process could destroy almost half of the aflatoxin B1 and G1 in wheat dough. Most of the aflatoxins remain intact during the baking of bread from contaminated wheat or corn flour with nil to maximum a quarter loss (Cheng, et al., 2010; Gumus, Arici, Daglioglu, & Velioglu, 2009; Magan, 2004). Reduction of aflatoxin B1 content in wheat through various cooking treatments such as washing, heating and steaming have been investigated by Hwang et al. (2006). Although the aflatoxin reductions were increased by increasing washing time (Jalili, Jinap, & Son, 2011), the most effective element was the temperature, irrespective of the origin of the wheat (J. H. Hwang & K. G. Lee, 2006).

Heating aflatoxin-contaminated corn grains at 160–180 °C, resulted in aflatoxin reduction from 383 to 60 μg/kg (Magan, 2004). The level of AFB1, during ordinary and pressure cooking of rice, reduced by 34% and 78–88%, respectively (Bullerman & Bianchini, 2007). The steam and aqueous treatment processes such as boiling may affect the aflatoxin content of the cereal matrix by degradation or extraction into the cooking liquid. In contrast, aflatoxins are relatively stable under dry conditions, which is affected at variable degrees in the presence of moisture. Reduction of aflatoxins in cooked rice was reported at variable ranges between 6–88%, depending on the ratio of water to rice used or the cooking condition. Similar range of aflatoxin reduction was reported for pasta, boiled buckwheat and for corn flour and corn grits with aflatoxin contamination. However, no substantial reduction in aflatoxin was reported in the preparation of 'Nshima' by boiling a thick paste. It might be because of the presence of other ingredients in cooking process (Magan, 2004). Dehulling and further pre-milling and soaking (eg. for 24 h) of corn kernels during corn flour processing at a village, reported by Njapau *et al.* (1998), resulted in 85–90% loss in their aflatoxin contents.

Different strategies have been applied for the elimination or inactivation of aflatoxins. However, problems still remain with the efficacy, safety and cost requirements for these methods (Ji, et al., 2011).Fermentation process is a very important step in reducing and controlling aflatoxins in storage and silage (Uegaki, Tsukiboshi, & Cai, 2010).

Tortillas production by alkaline cooking and steeping of the corn, also resulted in the reduction of aflatoxin content, which vary from almost 52% (tortillas) to 84% (tortilla chips) (Bullerman & Bianchini, 2007). Other tortilla production which involved the use of calcium hydroxide showed only a limited effect on Aflatoxin content (Magan, 2004). Generally, similar results were obtained by Elias-Orozco et al. (2002) while they used the extrusion process in the production of corn tortillas. They found that the use of 0.3% lime and 1.5% hydrogen peroxide was the most effective reducing process (Elias-Orozco, Castellanos-Nava, Gaytan-Martinez, Figueroa-Cardenas, & Loarca-Pina, 2002). Acidic conditions such as the use of 1N aqueous citric acid reported to inactivate aflatoxins *in-vitro* and in maize almost close to 100% (Mendez-Albores, Arambula-Villa, Loarca-Pina, Castano-Tostado, & Moreno-Martinez, 2005; Mendez-Albores, Del Rio-Garcia, & Moreno-Martinez, 2007).

3.2 Nuts

Effect of roasting on the aflatoxin content of pistachios have been recently investigated (Yazdanpanah, Mohammadi, Abouhossain, & Cheraghali, 2005). At 200 °C, most of the aflatoxins were destroyed (Jun-Ho Hwang & Kwang-Geun Lee, 2006). Microwave field (500 MHz–10 GHz) exposure could also result in aflatoxin destruction. Both microwave power and exposure time play a major role in the extent of such destruction. Reports indicated that only 16 min exposure of contaminated peanuts to a microwave power level of 1.6 kW resulted in a loss of almost 95% in its aflatoxin content. Similar results were obtained by higher power and lower exposure time (5 min treatment at a power level of 3.2 kW)(Magan, 2004). The presence of other substances, especially those alkaline in nature, also could considerably vary their destruction (Magan, 2004). Report by Hameed (1993) showed that addition of ammonia, either as hydroxide (0.7 and 1.0%) or as bicarbonate (0.4%) could increase the aflatoxin loss from 50–80% to 95%. Similar results were found through extrusion cooking of peanut meal, from 23–66% to 87% reduction, in the presence of ammonium hydroxide (2–2.5%) (Bullerman & Bianchini, 2007).

Yazdanpanah et al. (2005) studied the effect of roasting on the reduction of AF content in pistachio nuts. Although all treatment protocols showed some degree of AF degradation (ranging from 17% to 63%), roasting spiked samples at 120 °C for 120 min and 150 °C for 30–120 min, caused substantial reduction of aflatoxin in samples. Treatment of naturally contaminated whole pistachio kernels at 150 °C for 30 min, significantly reduced the level of aflatoxin contamination in samples. Degradation of aflatoxin was both time and temperature dependent. Roasting at 150 °C and 120 min condition, degraded more than 95% of aflatoxin B1 in pistachio(Yazdanpanah, Mohammadi, Abouhossain, & Cheraghali, 2005).

The efficiency of ozone on the degradation of aflatoxins in pistachio kernels and ground pistachios was evaluated. The efficiency of ozone on aflatoxin degradation in pistachios increased with increasing exposure time and ozone concentration. When pistachio kernels were ozonated at 9.0 mgL^{-1} ozone concentration for 420 min, the level of AFB1 and total aflatoxins reduced by 23 and 24%, respectively. While for ground pistachio nuts, under the same conditions, only a 5% reduction in AFB1 and total aflatoxin levels were obtained (Akbas & Ozdemir, 2006). The effectiveness of ozonation and mild heat in the degradation of aflatoxins in peanut kernels and flour were assessed. Degradation of aflatoxins were evaluated in peanut samples subjected to gaseous ozonation under various temperatures (25, 50, 75 °C) and exposure times (5, 10, 15 min). Higher temperatures and longer treatment times showed synergic effect on ozonation aflatoxin reduction effect. Among all aflatoxins, AFB1 and AFG1 showed the highest degradation levels. Greater efficiency in aflatoxin destruction was achieved in peanut kernels compared to flour. It was concluded that ozonation at room temperature for 10-15 min could be both economical as well as effective (Proctor, Ahmedna, Kumar, & Goktepe, 2004).

3.3 Other foods

Extrusion cooking is a technique that cooks the food product by heating under high pressure, while passing through continuous processing machinery, considerably reducing the food moisture content. Low transient time within the extruder can lead to limited aflatoxin loss despite high temperature and pressure.

Zorlugenc et al (2008) evaluated the effectiveness of the use of ozone and ozonated water on aflatoxin B_1 content of dried figs. Treatment of spiked dried fig samples with aflatoxins showed higher degradation of AFB_1 as ozonation time was increased in favour of the gaseous ozone compared with ozonated water (Zorlugenc, Kiroglu Zorlugenc, Oztekin, & Evliya, 2008). High temperature roasting of green coffee beans at 200 °C for 12 min reported 79% aflatoxin loss, which increased to 94% as the exposure time increased to 15 minutes (Magan, 2004). The aflatoxin reduction in coffee bean during roasting was also found to be dependent on the type and temperature of roasting with moderate reductions of approximately 42 to 56%(Bullerman & Bianchini, 2007). Gamma irradiation was also reported to decrease the total aflatoxins and aflatoxin B_1 levels gradually, with increase in gamma irradiation dose from 0 to 10 kGy (Ghanem, Orfi, & Shamma, 2008; Gupta, Bajpai, Mishra, Saxena, & Singh, 2009; Kumari, et al., 2009). However, in another study a 24-43% reduction in aflatoxin contamination was observed with irradiation at 60 kGY (Jalili, Jinap, & Noranizan, 2010). Gaseos ozone was effectively used against aflatoxin B1 at 13.8 mg/L (Zorlugenc, Kiroglu Zorlugenc, Oztekin, & Evliya, 2008).

3.4 Animal feed

Contaminated feed poses health risk, causes outbreaks in animals and leads to significant economic losses (Griessler, Rodrigues, Handl, & Hofstetter, 2010; Pierezan, et al., 2010). To protect severe loss and control the contamination of aflatoxin, the levels in many countries for food and feed is 20 μg/kg or less (Dorner, 2008; van Egmond, Schothorst, & Jonker, 2007). Contaminated feed with mycotoxins pose a health risk to animals and indirectly affects humans as well.

Principally, there are three possibilities to avoid the harmful effect of contamination of food and feed caused by mycotoxins: (1) prevention of contamination, (2) decontamination of mycotoxin-containing food and feed, and (3) inhibition of mycotoxin absorption into the digestive tract (Halasz, Lasztity, Abonyi, & Bata, 2009).

Direct and indirect effect of ammonia in feed can reduce aflatoxin levels (Safamehr, 2008; Tajkarimi, Riemann, et al., 2008). It has been demonstrated that vitamins such as A and pro vitamins such as carotene, carotenoids, phenolic compounds, curcuminoids, and sulfur containing compounds such as glutathione and glucomannan are capable of delivering antioxidant activity against aflatoxin B_1 toxicosis as well as mycotoxicosis (Donmez & Keskin, 2008; Gowda & Ledoux, 2008)

Citric acid was successfully used to reduce the aflatoxin level in sorghum from 17-92% with acceptable product color, viscosity, functional and textural properties (Mendez-Albores, Veles-Medina, Urbina-Alvarez, Martinez-Bustos, & Moreno-Martinez, 2009). High-grade sodium bentonite (HGSB) in broilers and dairy feed demonstrated reduction in aflatoxin toxicity (Manafi, Umakantha, Swamy, & Mohan, 2009; Schatzmayr, Fruhauf, & Vekiru, 2007). However, in another study, clinoptilolite and bentonite showed lower absorption index compared to hectorite (Dakovic, et al., 2008). Sodium and calcium aluminum silicate on silver catfish (*Rhamdia quelen*) fingerlings did not effectively reduce the aflatoxin levels (Lopes, et al., 2009).

3.4.1 Chicken feed

Poultry are highly susceptible to aflatoxin contamination (Diaz, Calabrese, & Blain, 2008; Souza, Vilar, Stamford, Bastos, & Filho, 2008). Aflatoxins can have severe effect on body weight, poultry organism, particularly liver and kidney and renal malfunction of laying hens (Khan, et al., 2010; Khan, Khan, Khan, & Hussain, 2010; Valdivia, et al., 2010). Several methods have been applied for the control and reduction of aflatoxins in feed including the application of different supplemental diet , *Canarium schweinfurthii* Engl. seed or maize cob (Kana, Teguia, Mungfu, & Tchoumboue, 2011). Sodium bentonite (0.5%), yeast (Saccharomyces cervisiae) 0.20%, hydrated sodium calcium aluminosilicate (HSCAS) (0.5%), ammonia (0.5%), formycine (0.1%), and toxiban (0.1%) demonstrated effective activity against aflatoxin B_1 as feed additives (Abousadi, Rowghani, & Honarmand, 2007). 0.5% sodium bentonite in another study demonstrated better results than 1% concentration of the chemical (Pasha, Farooq, Khattak, Jabbar, & Khan, 2007). Application of hydrated sodium calcium aluminosilicates (HSCAS) showed promising anti-aflatoxin effects compared to a combination of clay and yeast cell wall in preventing aflatoxicosis in broilers (Li, Suo, & Su, 2010; Zhao, et al., 2010). However, non selective nature of some of the anti-aflatoxin food additives may cause some nutritional materials unavailable due to the binding interaction. Sodium bentonite has been used as an effective anti-aflatoxin at 55 ppm in diet

(Chiacchiera, Magnoli, Monge, et al., 2011; Chiacchiera, Magnoli, Texeira, et al., 2011). Application of a combination of sodium treatments in itself and with the addition of 0.5% or 1.0% acetic acid, positively affected broiler performance (Magnoli, et al., 2008; Pasha, Mahmood, Khattak, Jabbar, & Khan, 2008). Application of hydrated aluminosilicate positively influenced some physiological parameters in broiler chicken meat (Prvulovic, Kojic, Grubor-Lajsic, & Kosarcic, 2008; Sehu, et al., 2007). *Bacillus subtilis* and *Bacillus lichenformis* are two bacterial strains that showed successful detoxification rate against mycotoxins and could be applied for aflatoxin (Wei, et al., 2010). Hydrated sodium calcium aluminosilicates and turmeric *Curcuma longa* powder (TMP) was effective against aflatoxin B_1 on growth performance and improving liver damage (Gowda, Ledoux, Rottinghaus, Bermudez, & Chen, 2008; Zhao, et al., 2010). Application of diatomaceous earth as a type of tectosilicate at levels of 30 ppm was suggested as an alternative treatment to other mycotoxin binders (Modirsanei, et al., 2008). Ribeiro et al (2009), reported that gamma irradiation can eliminate all Aspergillus spp. and other fungi strains at 8 kGy (Ribeiro, Cavaglieri, Vital, Kruger, & Rosa, 2009). Reduction of aflatoxin by turmeric (Curcuma longa) powder (TMP) supplementation in diets has also been demonstrated (Yarru, et al., 2009). Chlorophylin showed to reduce and prevent aflatoxin contamination by scavenging the free radicals and restoring the antioxidant defense mechanism activity (Thakur, Kumar, Reddy, Reddy, & Reddy, 2008).

3.4.2 Other feed

Despite the fact that mycotoxin contamination could be reduced in rumen flora due to various biotransformations (Fink-Gremmels, 2008; Upadhaya, Park, & Ha, 2010; Zain, 2011), there is still a need to develop different methods to control and reduce contamination levels in ruminants because the rumen barrier malfunction may increase absorption rates (Fink-Gremmels, 2008). In a study, application of freeze dried citrus peel (FDCP) showed reduction on aflatoxin contamination level without disrupting rumen fermentation (Ahn, Nam, & Garnsworthy, 2009). However, *Nocardia corynebacteroides* was successfully used against chicken feed contaminated with aflatoxin (Tejada-Castaneda, et al., 2008).

4. Conclusion

Temperature, food substrate, strain of the mould and other environmental factors are some parameters that effect mycotoxin production. Preventing mycotoxin production at farm level is the best way to control mycotoxin contamination (Sengun, Yaman, & Gonul, 2008).

Advances in molecular techniques and other decontamination methods such as gamma-irradiation and microwave heating could help to deal with these issues (Herzallah, Alshawabkeh, & Al Fataftah, 2008). Mycotoxins could be used as an energy source for a group of aerobic microorganisms, which are suitable to mycotoxin biodegradation. Several protocols have been provided to biodegrade mycotoxins in food and feed using potential bacteria such as Lactobacillus and Bifidobacterium (Awad, Ghareeb, Bohm, & Zentek, 2010; Fuchs, et al., 2008; Halasz, Lasztity, Abonyi, & Bata, 2009; Kabak & Var, 2008; Wei, et al., 2010).

However, there are varieties of responses between different microorganisms against mycotoxins. For example, *Bacillus brevis* were not affected by high concentrations of trichothecene.

Application of microorganisms needs to be evaluated from a safety point of view. Application of microorganisms on mycotoxin degradation, food and feed materials also need to be investigated (Halasz, Lasztity, Abonyi, & Bata, 2009). Further studies need to be conducted to address the seasonal variation of aflatoxin contamination in food and feed. Understanding the seasonal variation could help demonstrate and develop more effective decontamination methods. For example, it is postulated that mycotoxin issues due to monsoons in Hungary could possibly be concluded to technical difficulties in pre- and post-harvest operations. Application of advanced methods such as DNA biosensors and infrared spectroscopy for rapid and accurate detection of mycotoxin and related fungi is increasing dramatically (Fernandez-Ibanez, Soldado, Martinez-Fernandez, & de la Roza-Delgado, 2009; Maragos, 2009; Mascini, Tombelli, Scherm, Battacone, & Migheli, 2009; Tombelli, Mascini, Scherm, Battacone, & Migheli, 2009). Application of new and advanced detection techniques could enable the agricultural industry to deal more effectively with the occurrence of aflatoxin contamination.

5. Acknowledgement

Sangeetha Viswanathan is appreciated for her excellent editing in this chapter.
Professor Dean O. Cliver, recently deceased scholar is also gratefully acknowledged.

6. References

Abbas, H. K., Reddy, K. R. N., Salleh, B., Saad, B., Abel, C. A., & Shier, W. T. (2010). An overview of mycotoxin contamination in foods and its implications for human health. *Toxin Reviews,* Vol. 29, Mar, pp. 3-26,1556-9543

Abdulkadar, A. H. W., Al-Ali, A. A., Al-Kildi, A. M., & Al-Jedah, J. H. (2004). Mycotoxins in food products available in Qatar. *Food Control,* Vol. 15, Oct, pp. 543-548,0956-7135

Abeywickrama, K., & Bean, G. A. (1991). Toxigenic Aspergillus-Flavus and Aflatoxins in Sri-Lankan Medicinal Plant-Material. *Mycopathologia,* Vol. 113, Mar, pp. 187-190,0301-486X

Abousadi, M. A., Rowghani, E., & Honarmand, M. E. (2007). The efficacy of various additives to reduce the toxicity of aflatoxin B-1 in broiler chicks. *Iranian Journal of Veterinary Research,* Vol. 8, pp. 144-150,1728-1997

Ahn, J. H., Nam, I. S., & Garnsworthy, P. C. (2009). Effects of Freeze-dried Citrus Peel on Feed Preservation, Aflatoxin Contamination and In vitro Ruminal Fermentation. *Asian-Australasian Journal of Animal Sciences,* Vol. 22, May, pp. 674-680,1011-2367

Akbas, M. Y., & Ozdemir, M. (2006). Effect of different ozone treatments on aflatoxin degradation and physicochemical properties of pistachios. *Journal of the Science of Food and Agriculture,* Vol. 86, Oct, pp. 2099-2104,0022-5142

Aksoy, A., Yavuz, O., Das, Y. K., Guvenc, D., & Muglali, O. H. (2009). Occurrence of Aflatoxin B1, T-2 Toxin and Zearalenone in Compound Animal Feed. *Journal of Animal and Veterinary Advances,* Vol. 8, Mar, pp. 403-407,1680-5593

Alemu, T., Birhanu, G., Azerefgne, F., & Skinnes, H. (2008). Evidence for mycotoxin contamination of maize in Southern Ethiopia: The need for further multidisciplinary research. *Cereal Research Communications,* Vol. 36, pp. 337-339,0133-3720

Ali, N., Hashim, N. H., Saad, B., Safan, K., Nakajima, M., & Yoshizawa, T. (2005). Evaluation of a method to determine the natural occurrence of aflatoxins in commercial traditional herbal medicines from Malaysia and Indonesia. *Food and Chemical Toxicology,* Vol. 43, Dec, pp. 1763-1772,0278-6915

Arzandeh, S., & Jinap, S. (2011). Effect of initial aflatoxin concentration, heating time and roasting temperature on aflatoxin reduction in contaminated peanuts and process optimisation using response surface modelling. *International Journal of Food Science and Technology,* Vol. 46, Mar, pp. 485-491,0950-5423

Awad, W. A., Ghareeb, K., Bohm, J., & Zentek, J. (2010). Decontamination and detoxification strategies for the Fusarium mycotoxin deoxynivalenol in animal feed and the effectiveness of microbial biodegradation. *Food Additives and Contaminants Part a-Chemistry Analysis Control Exposure & Risk Assessment,* Vol. 27, pp. 510-520,1944-0049

Bamba, R., & Sumbali, G. (2005). Co-occurrence of aflatoxin B(1) and cyclopiazonic acid in sour lime (Citrus aurantifolia Swingle) during post-harvest pathogenesis by Aspergillus flavus. *Mycopathologia,* Vol. 159, Apr, pp. 407-411,0301-486X (Print)0301-486X (Linking)

Bankole, S. A., Adenusi, A. A., Lawal, O. S., & Adesanya, O. O. (2010). Occurrence of aflatoxin B1 in food products derivable from 'egusi' melon seeds consumed in southwestern Nigeria. *Food Control,* Vol. 21, Jul, pp. 974-976,0956-7135

Bankole, S. A., & Mabekoje, O. O. (2004). Mycoflora and occurrence of aflatoxin B-1 in dried yam chips from markets in Ogun and Oyo States, Nigeria. *Mycopathologia,* Vol. 157, Jan, pp. 111-115,0301-486X

Barbour, G. W., Farran, M. T., Usayran, N., & Daghir, N. J. (2008). Review of poultry production and the physical and chemical characteristics of imported corn and soybean meal in major feed operations in Lebanon. *Worlds Poultry Science Journal,* Vol. 64, Jun, pp. 177-185,0043-9339

Bhat, R., Rai, R. V., & Karim, A. A. (2010). Mycotoxins in Food and Feed: Present Status and Future Concerns. *Comprehensive Reviews in Food Science and Food Safety,* Vol. 9, Jan, pp. 57-81,1541-4337

Bittencourt, A. B. F., Oliveira, C. A. F., Dilkin, P., & Correa, B. (2005). Mycotoxin occurrence in corn meal and flour traded in Sao Paulo, Brazil. *Food Control,* Vol. 16, Feb, pp. 117-120,0956-7135

Blesa, J., Soriano, J. M., Molto, J. C., & Manes, J. (2004). Limited survey for the presence of aflatoxins in foods from local markets and supermarkets in Valencia, Spain. *Food Additives and Contaminants,* Vol. 21, Feb, pp. 165-171,0265-203X

Bullerman, L. B., & Bianchini, A. (2007). Stability of mycotoxins during food processing. *International Journal of Food Microbiology,* Vol. 119, Oct 20, pp. 140-146,0168-1605

Cheng, B. C., Wan, C. X., Yang, S. L., Xu, H. Y., Wei, H., Liu, J. S., Tian, W. H., & Zeng, M. (2010). Detoxification of Deoxynivalenol by Bacillus Strains. *Journal of Food Safety,* Vol. 30, Aug, pp. 599-614,0149-6085

Chiacchiera, S. M., Magnoli, A. P., Monge, M. P., Miazzo, R. D., Cavaglieri, L. R., Magnoli, C. E., Merkis, C. I., Cristofolini, A. L., & Dalcero, A. M. (2011). Effect of low levels of aflatoxin B(1) on performance, biochemical parameters, and aflatoxin B(1) in broiler liver tissues in the presence of monensin and sodium bentonite. *Poultry Science,* Vol. 90, Jan 1, pp. 48-58,0032-5791

Chiacchiera, S. M., Magnoli, A. P., Texeira, M., Rosa, C. A. R., Miazzo, R. D., Cavaglieri, L. R., Magnoli, C. E., & Dalcero, A. M. (2011). Sodium bentonite and monensin under chronic aflatoxicosis in broiler chickens. *Poultry Science,* Vol. 90, Feb 1, pp. 352-357,0032-5791

Chun, H. S., Kim, H. J., Ok, H. E., Hwang, J. B., & Chung, D. H. (2007). Determination of aflatoxin levels in nuts and their products consumed in South Korea. *Food Chemistry,* Vol. 102, pp. 385-391,0308-8146

Cornea, C. P., Ciuca, M., Voaides, C., Gagiu, V., & Pop, A. (2011). Incidence of fungal contamination in a Romanian bakery: a molecular approach. *Romanian Biotechnological Letters,* Vol. 16, Jan-Feb, pp. 5863-5871,1224-5984

D'Ovidio, K., Trucksess, M., Weaver, C., Horn, E., McIntosh, M., & Bean, G. (2006). Aflatoxins in ginseng roots. *Food Additives and Contaminants,* Vol. 23, Feb, pp. 174-180,0265-203X

Dakovic, A. S., Stojanovict, A. I., Matijasevic, S. D., Rottinghaus, G. E., Sekulic, Z. T., & Stanic, T. T. (2008). Adsorption of T-2 toxin by natural mineral adsorbents. *Hemijska Industrija,* Vol. 62, Mar-Apr, pp. 59-63,0367-598X

Dashti, B., Al-Hamli, S., Alomirah, H., Al-Zenki, S., Abbas, A. B., & Sawaya, W. (2009). Levels of aflatoxin M1 in milk, cheese consumed in Kuwait and occurrence of total aflatoxin in local and imported animal feed. *Food Control,* Vol. 20, Jul, pp. 686-690,0956-7135

Davison, J. (2010). GM plants: Science, politics and EC regulations. *Plant Science,* Vol. 178, Feb, pp. 94-98,0168-9452

de Magalhaes, J. T., Sodre, G. A., Viscogliosi, H., & Grenier-Loustalot, M. F. (2011). Occurrence of Ochratoxin A in Brazilian cocoa beans. *Food Control,* Vol. 22, May, pp. 744-748,0956-7135

de Oliveira, C. A. F., Sebastiao, L. S., Fagundes, H., Rosim, R. E., & Fernandes, A. M. (2010). Determination of aflatoxin B-1 in animal feed and aflatoxin M-1 in milk in dairy farms of Sao Paulo State. *Ciencia E Tecnologia De Alimentos,* Vol. 30, May, pp. 221-225,0101-2061

Diaz, G. J., Calabrese, E., & Blain, R. (2008). Aflatoxicosis in chickens (Gallus gallus): An example of hormesis? *Poultry Science,* Vol. 87, Apr, pp. 727-732,0032-5791

Diaz, G. J., Lozano, M. C., & Acuna, A. (2009). Prevalence of Aspergillus species on selected Colombian animal feedstuffs and ability of Aspergillus section Flavi to produce aflatoxins. *World Mycotoxin Journal,* Vol. 2, Feb, pp. 31-34,1875-0710

Donmez, N., & Keskin, E. (2008). The Effects of Aflatoxin and Glucomannan on Some Antioxidants and Biochemical Parameters in Rabbits. *Acta Veterinaria-Beograd,* Vol. 58, pp. 307-313,0567-8315

Dorner, J. W. (2008). Management and prevention of mycotoxins in peanuts. *Food Additives and Contaminants,* Vol. 25, pp. 203-208,0265-203X

Doster, M. A., Michailides, T. J., & Morgan, D. P. (1996). Aspergillus species and mycotoxins in figs from California orchards. *Plant Disease,* Vol. 80, May, pp. 484-489,0191-2917

El Adlouni, C., Tozlovanu, M., Naman, F., Faid, M., & Pfohl-Leszkowicz, A. (2006). Preliminary data on the presence of mycotoxins (ochratoxin A, citrinin and aflatoxin B1) in black table olives "Greek style" of Moroccan origin. *Mol Nutr Food Res,* Vol. 50, May, pp. 507-512,1613-4125 (Print)1613-4125 (Linking)

Elias-Orozco, R., Castellanos-Nava, A., Gaytan-Martinez, M., Figueroa-Cardenas, J. D., & Loarca-Pina, G. (2002). Comparison of nixtamalization and extrusion processes for a reduction in aflatoxin content. *Food Additives and Contaminants,* Vol. 19, Sep, pp. 878-885,0265-203X

Elzupir, A. O., & Elhussein, A. M. (2010). Determination of aflatoxin M1 in dairy cattle milk in Khartoum State, Sudan. *Food Control,* Vol. 21, Jun, pp. 945-946,0956-7135

Elzupir, A. O., Younis, Y. M. H., Fadul, M. H., & Elhussein, A. M. (2009). Determination of Aflatoxins in Animal Feed in Khartoum State, Sudan. *Journal of Animal and Veterinary Advances,* Vol. 8, pp. 1000-1003,1680-5593

Energy., E. C. D.-G. f. (2009). The rapid alert system for food and feed (RASFF) Annual report 2008. *European Communities,*

Fandohan, P., Zoumenou, D., Hounhouigan, D. J., Marasas, W. F. O., Wingfield, M. J., & Hell, K. (2005). Fate of aflatoxins and fumonisins during the processing of maize into food products in Benin. *International Journal of Food Microbiology,* Vol. 98, Feb 15, pp. 249-259,0168-1605

Fernandez-Ibanez, V., Soldado, A., Martinez-Fernandez, A., & de la Roza-Delgado, B. (2009). Application of near infrared spectroscopy for rapid detection of aflatoxin B1 in maize and barley as analytical quality assessment. *Food Chemistry,* Vol. 113, Mar 15, pp. 629-634,0308-8146

Ferracane, R., Tafuri, A., Logieco, A., Galvano, F., Balzano, D., & Ritieni, A. (2007). Simultaneous determination of aflatoxin B₁ and ochratoxin A and their natural occurrence in Mediterranean virgin olive oil. *Food Additives and Contaminants,* Vol. 24, pp. 173 - 180,0265-203X

Fink-Gremmels, J. (2008). Mycotoxins in cattle feeds and carry-over to dairy milk: A review. *Food Additives and Contaminants,* Vol. 25, pp. 172-180,0265-203X

Fuchs, S., Sontag, G., Stidl, R., Ehrlich, V., Kundi, M., & Knasmuller, S. (2008). Detoxification of patulin and ochratoxin A, two abundant mycotoxins, by lactic acid bacteria. *Food and Chemical Toxicology,* Vol. 46, Apr, pp. 1398-1407,0278-6915

Ghanem, I., Orfi, M., & Shamma, M. (2008). Effect of Gamma Radiation on the Inactivation of Aflatoxin B1 in Food and Feed Crops. *Brazilian Journal of Microbiology,* Vol. 39, Oct-Dec, pp. 787-791,1517-8382

Ghazani, M. H. M. (2009). Aflatoxin M1 contamination in pasteurized milk in Tabriz (northwest of Iran). *Food and Chemical Toxicology,* Vol. 47, Jul, pp. 1624-1625,0278-6915

Ghiasian, S. A., Shephard, G. S., & Yazdanpanah, H. (2011). Natural Occurrence of Aflatoxins from Maize in Iran. *Mycopathologia,* Mar 4,1573-0832 (Electronic)0301-486X (Linking)

Giray, B., Girgin, G., Engin, A. B., Aydin, S., & Sahin, G. (2007). Aflatoxin levels in wheat samples consumed in some regions of Turkey. *Food Control,* Vol. 18, Jan, pp. 23-29,0956-7135

Gowda, N. K. S., & Ledoux, D. R. (2008). Use of antioxidants in amelioration of mycotoxin toxicity: A review. *Animal Nutrition and Feed Technology,* Vol. 8, Jan, pp. 1-11,0972-2963

Gowda, N. K. S., Ledoux, D. R., Rottinghaus, G. E., Bermudez, A. J., & Chen, Y. C. (2008). Efficacy of turmeric (Curcuma longa), containing a known level of curcumin, and a hydrated sodium calcium aluminosilicate to ameliorate the adverse effects of aflatoxin in broiler chicks. *Poultry Science,* Vol. 87, Jun, pp. 1125-1130,0032-5791

Griessler, K., Rodrigues, I., Handl, J., & Hofstetter, U. (2010). Occurrence of mycotoxins in Southern Europe. *World Mycotoxin Journal,* Vol. 3, Aug, pp. 301-309,1875-0710

Gumus, T., Arici, M., Daglioglu, O., & Velioglu, M. (2009). Fate of aflatoxins during bread making. *Journal of Food Science and Technology-Mysore,* Vol. 46, Jul-Aug, pp. 384-387,0022-1155

Gupta, P. C., Bajpai, V., Mishra, V., Saxena, R. K., & Singh, S. (2009). Effect of gamma irradiation on microbial load, aflatoxins and phytochemicals present in Trigonella foenum-graecum. *World Journal of Microbiology & Biotechnology,* Vol. 25, Dec, pp. 2267-2271,0959-3993

Halasz, A., Lasztity, R., Abonyi, T., & Bata, A. (2009). Decontamination of Mycotoxin-Containing Food and Feed by Biodegradation. *Food Reviews International,* Vol. 25, pp. 284-298,8755-9129

Head, S. W., Swetman, T. A., & Nagler, M. J. (1999). Studies on deterioration and aflatoxin contamination in copra during storage. *Ocl-Oleagineux Corps Gras Lipides,* Vol. 6, Jul-Aug, pp. 349-359,1258-8210

Herzallah, S., Alshawabkeh, K., & Al Fataftah, A. (2008). Aflatoxin Decontamination of Artificially Contaminated Feeds by Sunlight, gamma-Radiation, and Microwave Heating. *Journal of Applied Poultry Research,* Vol. 17, Win, pp. 515-521,1056-6171

Hitokoto, H., Morozumi, S., Wauke, T., Sakai, S., & Kurata, H. (1978). Fungal Contamination and Mycotoxin Detection of Powdered Herbal Drugs. *Applied and Environmental Microbiology,* Vol. 36, pp. 252-256,0099-2240

Houssou, P. A., Ahohuendo, B. C., Fandohan, P., Kpodo, K., Hounhouigan, D. J., & Jakobsen, M. (2009). Natural infection of cowpea (Vigna unguiculata (L.) Walp.) by toxigenic fungi and mycotoxin contamination in Benin, West Africa. *Journal of Stored Products Research,* Vol. 45, pp. 40-44,0022-474X

Hwang, J.-H., & Lee, K.-G. (2006). Reduction of aflatoxin B1 contamination in wheat by various cooking treatments. *Food Chemistry,* Vol. 98, pp. 71-75,0308-8146

Hwang, J. H., & Lee, K. G. (2006). Reduction of aflatoxin B-1 contamination in wheat by various cooking treatments. *Food Chemistry,* Vol. 98, pp. 71-75,0308-8146

Idris, Y. M. A., Mariod, A. A., Elnour, I. A., & Mohamed, A. A. (2010). Determination of aflatoxin levels in Sudanese edible oils. *Food and Chemical Toxicology,* Vol. 48, Aug-Sep, pp. 2539-2541,0278-6915

Ismail, N., Leong, Y. H., Latif, A. A., & Ahmad, R. (2010). Aflatoxin occurrence in nuts and commercial nutty products in Malaysia. *Food Control,* Vol. 21, Mar, pp. 334-338,0956-7135

Jalili, M., Jinap, S., & Noranizan, A. (2010). Effect of gamma radiation on reduction of mycotoxins in black pepper. *Food Control,* Vol. 21, Oct, pp. 1388-1393,0956-7135

Jalili, M., Jinap, S., & Son, R. (2011). The effect of chemical treatment on reduction of aflatoxins and ochratoxin A in black and white pepper during washing. *Food Additives and Contaminants Part a-Chemistry Analysis Control Exposure & Risk Assessment,* Vol. 28, pp. 485-493,1944-0049

Jelinek, C. F., Pohland, A. E., & Wood, G. E. (1989). Worldwide occurrence of mycotoxins in foods and feeds--an update. *J Assoc Off Anal Chem,* Vol. 72, Mar-Apr, pp. 223-230,0004-5756 (Print)0004-5756 (Linking)

Ji, C., Zhao, L. H., Guan, S., Gao, X., Ma, Q. G., Lei, Y. P., & Bai, X. M. (2011). Preparation, purification and characteristics of an aflatoxin degradation enzyme from Myxococcus fulvus ANSM068. *Journal of Applied Microbiology,* Vol. 110, Jan, pp. 147-155,1364-5072

Kabak, B., & Var, I. (2008). Factors affecting the removal of aflatoxin M-1 from food model by Lactobacillus and Bifidobacterium strains. *Journal of Environmental Science and Health Part B-Pesticides Food Contaminants and Agricultural Wastes,* Vol. 43, pp. 617-624,0360-1234

Kamkar, A., Karim, G., Aliabadi, F. S., & Khaksar, R. (2008). Fate of aflatoxin M-1 in Iranian white cheese processing. *Food and Chemical Toxicology,* Vol. 46, Jun, pp. 2236-2238,0278-6915

Kana, J. R., Teguia, A., Mungfu, B. M., & Tchoumboue, J. (2011). Growth performance and carcass characteristics of broiler chickens fed diets supplemented with graded levels of charcoal from maize cob or seed of Canarium schweinfurthii Engl. *Tropical Animal Health and Production,* Vol. 43, Jan, pp. 51-56,0049-4747

Katerere, D. R., Thembo, K. M., Vismer, H. F., Nyazema, N. Z., & Gelderblom, W. C. A. (2010). Antifungal activity of four weedy plant extracts against selected mycotoxigenic fungi. *Journal of Applied Microbiology,* Vol. 109, Oct, pp. 1479-1486,1364-5072

Keller, K. M., Keller, L. A. M., de Oliveira, A. A., Almeida, T. X., Garcia, R. D., & Rosa, C. A. D. (2008). Mycotoxicological Evaluation of Feedstuffs Intended for Dairy Goats in Teresopolis, Rj, Brazil. *Revista Brasileira De Medicina Veterinaria,* Vol. 30, Apr-Jun, pp. 91-96,0100-2430

Kensler, T. W., Roebuck, B. D., Wogan, G. N., & Groopman, J. D. (2011). Aflatoxin: a 50-year odyssey of mechanistic and translational toxicology. *Toxicological Sciences,* Vol. 120 Suppl 1, Mar, pp. S28-48,1096-0929 (Electronic)1096-0929 (Linking)

Khan, M. Z., Hussain, Z., Khan, A., Javed, I., Saleemi, M. K., Mahmood, S., & Asi, M. R. (2010). Residues of aflatoxin B1 in broiler meat: Effect of age and dietary aflatoxin B1 levels. *Food and Chemical Toxicology,* Vol. 48, Dec, pp. 3304-3307,0278-6915

Khan, W. A., Khan, M. Z., Khan, A., & Hussain, I. (2010). Pathological Effects of Aflatoxin and Their Amelioration by Vitamin E in White Leghorn Layers. *Pakistan Veterinary Journal,* Vol. 30, pp. 155-162,0253-8318

Kumar, V., Basu, M. S., & Rajendran, T. P. (2008). Mycotoxin research and mycoflora in some commercially important agricultural commodities. *Crop Protection,* Vol. 27, Jun, pp. 891-905,0261-2194

Kumari, N., Kumar, P., Mitra, D., Prasad, B., Tiwary, B. N., & Varshney, L. (2009). Effects of Ionizing Radiation on Microbial Decontamination, Phenolic Contents, and Antioxidant Properties of Triphala. *JOURNAL OF FOOD SCIENCE,* Vol. 74, Apr, pp. M109-M113,0022-1147

Li, J. J., Suo, D. C., & Su, X. O. (2010). Binding Capacity for Aflatoxin B-1 by Different Adsorbents. *Agricultural Sciences in China,* Vol. 9, Mar, pp. 449-456,1671-2927

Liang, X. Q., Wang, T., Zhang, E. H., Chen, X. P., & Li, L. (2010). Identification of seed proteins associated with resistance to pre-harvested aflatoxin contamination in peanut (Arachis hypogaea L). *Bmc Plant Biology,* Vol. 10, Nov 30,1471-2229

Lin, Y. L., Wang, T. H., Lee, M. H., & Su, N. W. (2008). Biologically active components and nutraceuticals in the Monascus-fermented rice: a review. *Applied Microbiology and Biotechnology,* Vol. 77, Jan, pp. 965-973,0175-7598

Lopes, P. R. S., Pouey, J. L. O. F., Enke, D. B. S., Mallmann, C. A., Kich, H. A., & Soquetta, M. B. (2009). Use of adsorbent in diets containing aflatoxin for silver catfish fingerlings. *Revista Brasileira De Zootecnia-Brazilian Journal of Animal Science,* Vol. 38, Apr, pp. 589-595,1516-3598

Macdonald, S., & Castle, L. (1996). A UK retail survey of aflatoxins in herbs and spices and their fate during cooking. *Food Additives and Contaminants,* Vol. 13, pp. 121 - 128,0265-203X

Magan, N., Olsen, M. (2004). Mycotoxins in Food - Detection and Control. In. Cambridge Woodhead Publishing 978-1-85573-733-4

Magnoli, A. P., Cavaglieri, L. R., Magnoli, C. E., Monge, J. C., Miazzo, R. D., Peralta, M. F., Salvano, M. A., Rosa, C. A. D., Dalcero, A. M., & Chiacchiera, S. M. (2008). Bentonite Performance on Broiler Chickens Fed with Diets Containing Natural Levels of Aflatoxin B-1. *Revista Brasileira De Medicina Veterinaria,* Vol. 30, Jan-Mar, pp. 55-60,0100-2430

Manafi, M., Umakantha, B., Swamy, H. D. N., & Mohan, K. (2009). Evaluation of high-grade sodium bentonite on performance and immune status of broilers, fed ochratoxin and aflatoxin. *World Mycotoxin Journal,* Vol. 2, Nov, pp. 435-440,1875-0710

Maragos, C. M. (2009). Biosensors for mycotoxin analysis: recent developments and future prospects. *World Mycotoxin Journal,* Vol. 2, May, pp. 221-238,1875-0710

Markov, K., Frece, J., Cvek, D., Lovric, N., & Delas, F. (2010). Aflatoxin M-1 in raw milk and binding of aflatoxin by lactic acid bacteria. *Mljekarstvo,* Vol. 60, Oct-Dec, pp. 244-251,0026-704X

Mascini, M., Tombelli, S., Scherm, B., Battacone, G., & Migheli, Q. (2009). DNA biosensors for the detection of aflatoxin producing Aspergillus flavus and A-parasiticus. *Monatshefte Fur Chemie,* Vol. 140, Aug, pp. 901-907,0026-9247

Mendez-Albores, A., Arambula-Villa, G., Loarca-Pina, M. G., Castano-Tostado, E., & Moreno-Martinez, E. (2005). Safety and efficacy evaluation of aqueous citric acid to degrade B-aflatoxins in maize. *Food Chem Toxicol,* Vol. 43, Feb, pp. 233-238,0278-6915 (Print)0278-6915 (Linking)

Mendez-Albores, A., Del Rio-Garcia, J. C., & Moreno-Martinez, E. (2007). Decontamination of aflatoxin duckling feed with aqueous citric acid treatment. *Animal Feed Science and Technology,* Vol. 135, Jun 15, pp. 249-262,0377-8401

Mendez-Albores, A., Veles-Medina, J., Urbina-Alvarez, E., Martinez-Bustos, F., & Moreno-Martinez, E. (2009). Effect of citric acid on aflatoxin degradation and on functional and textural properties of extruded sorghum. *Animal Feed Science and Technology,* Vol. 150, Apr 14, pp. 316-329,0377-8401

Mngadi, P. T., Govinden, R., & Odhav, B. (2008). Co-occurring mycotoxins in animal feeds. *African Journal of Biotechnology,* Vol. 7, Jul 4, pp. 2239-2243,1684-5315

Modirsanei, M., Mansoori, B., Khosravi, A. R., Kiaei, M. M., Khazraeinia, P., Farkhoy, M., & Masoumi, Z. (2008). Effect of diatomaceous earth on the performance and blood variables of broiler chicks during experimental aflatoxicosis. *Journal of the Science of Food and Agriculture,* Vol. 88, Mar, pp. 626-632,0022-5142

Moghadam, M. M., & Hokmabadi, H. (2010). Study on the effect of pistachio testa on the reduction of Aspergillus flavus growth and aflatoxin B-1 production in kernels of different pistachio cultivars. *Australian Journal of Crop Science,* Vol. 4, Nov, pp. 744-749,1835-2693

Monbaliu, S., Van Poucke, C., Detavernier, C., Dumoulin, F., Van De Velde, M., Schoeters, E., Van Dyck, S., Averkieva, O., Van Peteghem, C., & De Saeger, S. (2010). Occurrence of Mycotoxins in Feed as Analyzed by a Multi-Mycotoxin LC-MS/MS Method. *Journal of Agricultural and Food Chemistry,* Vol. 58, Jan 13, pp. 66-71,0021-8561

Montes, G. N., Reyes, M. C. A., Montes, R. N., & Cantu, A. M. A. (2009). Incidence of potentially toxigenic fungi in maize (Zea mays L.) grain used as food and animal feed. *Cyta-Journal of Food,* Vol. 7, pp. 119-125,1947-6345

Moss, M. O. (1998). Recent studies of mycotoxins. *Symp Ser Soc Appl Microbiol,* Vol. 27, pp. 62S-76S,1467-4734 (Print)1467-4734 (Linking)

Nguyen, M. T., Tozovanu, M., Tran, T. L., & Pfohl-Leszkowicz, A. (2007). Occurrence of aflatoxin B1, citrinin and ochratoxin A in rice in five provinces of the central region of Vietnam. *Food Chemistry,* Vol. 105, pp. 42-47,0308-8146

Njobeh, P. B., Dutton, M. F., Koch, S. H., Chuturgoon, A., Stoev, S., & Seifert, K. (2009). Contamination with storage fungi of human food from Cameroon. *International Journal of Food Microbiology,* Vol. 135, Nov 15, pp. 193-198,0168-1605

Odhav, B., Mngadi, P. T., & Govinden, R. (2008). Co-occurring mycotoxins in animal feeds. *African Journal of Biotechnology,* Vol. 7, Jul 4, pp. 2239-2243,1684-5315

Odoemelam, S. A., & Osu, C. I. (2009). Aflatoxin B(1) Contamination of Some Edible Grains Marketed in Nigeria. *E-Journal of Chemistry,* Vol. 6, Apr-Jun, pp. 308-314,0973-4945

Oliveira, C. A. F., Sebastiao, L. S., Fagundes, H., Rosim, R. E., & Fernandes, A. M. (2008). Aflatoxins and cyclopiazonic acid in feed and milk from dairy farms in Sao Paulo, Brazil. *Food Additives & Contaminants Part B-Surveillance,* Vol. 1, Dec, pp. 147-152,1939-3210

Pacheco, A. M., & Scussel, V. M. (2009). Aflatoxins evaluation on in-shell and shelled dry Brazil nuts for export analysed by LC-MS/MS-2006 and 2007 harvests. *World Mycotoxin Journal,* Vol. 2, Aug, pp. 295-304,1875-0710

Park, D. L. (2002). Effect of processing on aflatoxin. *Mycotoxins and Food Safety,* Vol. 504, pp. 173-179,0065-2598

Pasha, T. N., Farooq, M. U., Khattak, F. M., Jabbar, M. A., & Khan, A. D. (2007). Effectiveness of sodium bentonite and two commercial products as aflatoxin absorbents in diets for broiler chickens. *Animal Feed Science and Technology,* Vol. 132, Jan 1, pp. 103-110,0377-8401

Pasha, T. N., Mahmood, A., Khattak, F. M., Jabbar, M. A., & Khan, A. D. (2008). The effect of feed supplemented with different sodium bentonite treatments on broiler performance. *Turkish Journal of Veterinary & Animal Sciences,* Vol. 32, pp. 245-248,1300-0128

Patel, S., Hazel, C. M., Winterton, A. G. M., & Mortby, E. (1996). Survey of ethnic foods for mycotoxins. *Food Additives and Contaminants,* Vol. 13, Oct, pp. 833-841,0265-203X

Peng, K. Y., & Chen, C. Y. (2009). Prevalence of Aflatoxin M-1 in Milk and Its Potential Liver Cancer Risk in Taiwan. *Journal of Food Protection,* Vol. 72, May, pp. 1025-1029,0362-028X

Perez-Flores, G. C., Moreno-Martinez, E., & Mendez-Albores, A. (2011). Effect of Microwave Heating during Alkaline-Cooking of Aflatoxin Contaminated Maize. *J Food Sci,* Vol. 76, Mar, pp. T48-52,1750-3841 (Electronic)0022-1147 (Linking)

Pharmacopeia, E. (2007). Determination of aflatoxin B1 in herbal drugs. In *European Pharmacopeia* (6 ed.). Strasbourg Directorate for the Quality of Medicines and HealthCare of the Council of Europe (EDQM).

Pierezan, F., Oliveira, J. C., Carmo, P. M., Lucena, R. B., Rissi, D. R., Togni, M., & Barros, C. S. L. (2010). Outbreak of aflatoxicosis in calves in southern Brazil. *Pesquisa Veterinaria Brasileira,* Vol. 30, May, pp. 418-422,0100-736X

Pittet, A. (1998). Natural occurrence of mycotoxins in foods and feeds - an update review. *Revue De Medecine Veterinaire,* Vol. 149, pp. 479–492

Proctor, A. D., Ahmedna, M., Kumar, J. V., & Goktepe, I. (2004). Degradation of aflatoxins in peanut kernels/flour by gaseous ozonation and mild heat treatment. *Food Additives and Contaminants,* Vol. 21, Aug, pp. 786-793,0265-203X

Prvulovic, D., Kojic, D., Grubor-Lajsic, G., & Kosarcic, S. (2008). The effects of dietary inclusion of hydrated aluminosilicate on performance and biochemical parameters of broiler chickens. *Turkish Journal of Veterinary & Animal Sciences,* Vol. 32, pp. 183-189,1300-0128

RASFF. (2011). RASFF portal. In DG SANCO https://webgate.ec.europa.eu/rasff-window/portal/

Reddy, K. R. N., Reddy, C. S., Abbas, H. K., Abel, C. A., & Muralidharan, K. (2008). Mycotoxigenic Fungi, Mycotoxins, and Management of Rice Grains. *Toxin Reviews,* Vol. 27, pp. 287-317,1556-9543

Ribeiro, J. M. M., Cavaglieri, L. R., Vital, H. D., Kruger, C. D., & Rosa, C. A. D. (2009). Gamma radiation on the mycoflora of poultry feed and Aspergillus species. *Ciencia Rural,* Vol. 39, Aug, pp. 1452-1458,0103-8478

Ricci, R., Sartori, A., Palagiano, C., & Zotte, A. D. (2010). Study on the Nutrient Adequacy of Feeds for Pet Rabbits Available in the Italian Market. *World Rabbit Science,* Vol. 18, Jul-Sep, pp. 131-137,1257-5011

Rivka Barkai-Golan, N. P. (2008). Mycotoxins in fruits and vegetables. In. Burlington Academic Press,978-0-12-374126-4

Roige, M. B., Aranguren, S. M., Riccio, M. B., Pereyra, S., Soraci, A. L., & Tapia, M. O. (2009). Mota and mycotoxins in fermented feed, wheat grains and corn grains in Southeastern Buenos Aires Province, Argentinaycobi. *Revista Iberoamericana De Micologia,* Vol. 26, Dec 31, pp. 233-237,1130-1406

Romagnoli, B., Menna, V., Gruppioni, N., & Bergamini, C. (2007). Aflatoxins in spices, aromatic herbs, herb-teas and medicinal plants marketed in Italy. *Food Control,* Vol. 18, pp. 697-701,0956-7135

Roy, A. K., Sinha, K. K., & Chourasia, H. K. (1988). Aflatoxin Contamination of Some Common-Drug Plants. *Applied and Environmental Microbiology,* Vol. 54, Mar, pp. 842-843,0099-2240

Rustemeyer, S. M., Lamberson, W. R., Ledoux, D. R., Rottinghaus, G. E., Shaw, D. P., Cockrum, R. R., Kessler, K. L., Austin, K. J., & Cammack, K. M. (2010). Effects of dietary aflatoxin on the health and performance of growing barrows. *Journal of Animal Science,* Vol. 88, Nov, pp. 3624-3630,0021-8812

Rustom, I. Y. S. (1997). Aflatoxin in food and feed: Occurrence, legislation and inactivation by physical methods. *Food Chemistry,* Vol. 59, May, pp. 57-67,0308-8146

Safamehr, A. (2008). The Performance and Heamatological Characters in Broiler Chicks Fed Ammonia-Treated Aflatoxin Contaminated Feed. *Journal of Animal and Veterinary Advances,* Vol. 7, Mar, pp. 331-336,1680-5593

Sanchez-Hervas, M., Gil, J. V., Bisbal, F., Ramon, D., & Martinez-Culebras, P. V. (2008). Mycobiota and mycotoxin producing fungi from cocoa beans. *Int J Food Microbiol,* Vol. 125, Jul 31, pp. 336-340,0168-1605 (Print)0168-1605 (Linking)

Schatzmayr, G., Fruhauf, S., & Vekiru, E. (2007). Deactivation of aflatoxin B1 in animal feed by using a selected bentonite. *Journal of Dairy Science,* Vol. 90, pp. 176-176,0022-0302

Sehu, A., Ergun, L., Cakir, S., Ergun, E., Cantekin, Z., Sahin, T., Essiz, D., Sareyyupoglu, B., Gurel, Y., & Yigit, Y. (2007). Hydrated sodium calcium aluminosilicate for reduction of aflatoxin in quails (Coturnix coturnix japonica). *Deutsche Tierarztliche Wochenschrift,* Vol. 114, Jul, pp. 252-259,0341-6593

Sengun, I. Y., Yaman, D. B., & Gonul, S. A. (2008). Mycotoxins and mould contamination in cheese: a review. *World Mycotoxin Journal,* Vol. 1, Aug, pp. 291-298,1875-0710

Seo, J. H., Min, W. K., Kweon, D. H., Park, K., & Park, Y. C. (2011). Characterisation of monoclonal antibody against aflatoxin B(1) produced in hybridoma 2C12 and its single-chain variable fragment expressed in recombinant Escherichia coli. *Food Chemistry,* Vol. 126, Jun 1, pp. 1316-1323,0308-8146

Set, E., & Erkmen, O. (2010). The aflatoxin contamination of ground red pepper and pistachio nuts sold in Turkey. *Food Chem Toxicol,* Vol. 48, Aug-Sep, pp. 2532-2537,1873-6351 (Electronic)0278-6915 (Linking)

Sewram, V., Shephard, G. S., van der Merwe, L., & Jacobs, T. V. (2006). Mycotoxin contamination of dietary and medicinal wild plants in the Eastern Cape Province of South Africa. *Journal of Agricultural and Food Chemistry,* Vol. 54, Jul 26, pp. 5688-5693,0021-8561

Shenasi, M., Aidoo, K. E., & Candlish, A. A. G. (2002). Microflora of date fruits and production of aflatoxins at various stages of maturation. *International Journal of Food Microbiology,* Vol. 79, pp. 113-119,0168-1605

Smiley, R. D., & Draughon, F. A. (2000). Preliminary evidence that degradation of aflatoxin B1 by Flavobacterium aurantiacum is enzymatic. *J Food Prot,* Vol. 63, Mar, pp. 415-418,0362-028X (Print)362-028X (Linking)

Smith, J. E., & Moss, M. O. (1985). *Mycotoxins. Formation, analysis and significance.* John Wiley & Sons Ltd.,ISBN 0-471-90671-9, Chichester

Souza, E. L., Vilar, E. A., Stamford, T. L. M., Bastos, S. T. G., & Filho, E. V. C. (2008). Short communication. Total aflatoxin and ochratoxin A in the liver, kidneys and plasma of experimentally contaminated chickens. *Spanish Journal of Agricultural Research,* Vol. 6, Sep, pp. 373-377,1695-971X

Sugita-Konishi, Y., Nakajima, M., Tabata, S., Ishikuro, E., Tanaka, T., Norizuki, H., Itoh, Y., Aoyama, K., Fujita, K., Kai, S., & Kumagai, S. (2006). Occurrence of aflatoxins, ochratoxin A, and fumonisins in retail foods in Japan. *Journal of Food Protection,* Vol. 69, Jun, pp. 1365-1370,0362-028X

Sugita-Konishi, Y., Sugiyama, K. I., & Hiraokai, H. (2008). Aflatoxin M-1 Contamination in Raw Bulk Milk and the Presence of Aflatoxin B-1 in Corn Supplied to Dairy Cattle in Japan. *Journal of the Food Hygienic Society of Japan,* Vol. 49, Oct, pp. 352-355,0015-6426

Sun, G., Wang, S., Hu, X., Su, J., Zhang, Y., Xie, Y., Zhang, H., Tang, L., & Wang, J. S. (2011a). Co-contamination of aflatoxin B1 and fumonisin B1 in food and human dietary exposure in three areas of China. *Food Addit Contam Part A Chem Anal Control Expo Risk Assess,* Vol. 28, Apr, pp. 461-470,1944-0057 (Electronic)1944-0057 (Linking)

Sun, G., Wang, S., Hu, X., Su, J., Zhang, Y., Xie, Y., Zhang, H., Tang, L., & Wang, J. S. (2011b). Co-contamination of aflatoxin B-1 and fumonisin B-1 in food and human dietary exposure in three areas of China. *Food Additives and Contaminants Part a-Chemistry Analysis Control Exposure & Risk Assessment,* Vol. 28, pp. 461-470,1944-0049

Tajkarimi, M., Aliabadi-Sh, F., Nejad, A. S., Poursoltani, H., Motallebi, A. A., & Mahdavi, H. (2008). Aflatoxin M-1 contamination in winter and summer milk in 14 states in Iran. *Food Control,* Vol. 19, Nov, pp. 1033-1036,0956-7135

Tajkarimi, M., Faghih, M. A., Poursoltani, H., Nejad, A. S., Motallebi, A. A., & Mahdavi, H. (2008). Lead residue levels in raw milk from different regions of Iran. *Food Control,* Vol. 19, May, pp. 495-498,0956-7135

Tajkarimi, M., Riemann, H. P., Hajmeer, M. N., Gomez, E. L., Razavilar, V., & Cliver, D. O. (2008). Ammonia disinfection of animal feeds --laboratory study. *Int J Food Microbiol,* Vol. 122, Feb 29, pp. 23-28,0168-1605 (Print)0168-1605 (Linking)

Tajkarimi, M., Shojaee Aliabadi, F., Salah Nejad, M., Pursoltani, H., Motallebi, A. A., & Mahdavi, H. (2007). Seasonal study of aflatoxin M1 contamination in milk in five regions in Iran. *Int J Food Microbiol,* Vol. 116, May 30, pp. 346-349,0168-1605 (Print) 0168-1605 (Linking)

Tassaneeyakul, W., Razzazi-Fazeli, E., Porasuphatana, S., & Bohm, J. (2004). Contamination of aflatoxins in herbal medicinal products in Thailand. *Mycopathologia,* Vol. 158, Jul, pp. 239-244,0301-486X

Tejada-Castaneda, Z. I., Avila-Gonzalez, E., Casaubon-Huguenin, M. T., Cervantes-Olivares, R. A., Vasquez-Pelaez, C., Hernandez-Baumgarten, E. M., & Moreno-Martinez, E. (2008). Biodetoxification of aflatoxin-contaminated chick feed. *Poultry Science,* Vol. 87, Aug, pp. 1569-1576,0032-5791

Thakur, C., Kumar, B. K., Reddy, A. G., Reddy, A. R., & Reddy, Y. N. (2008). Antioxidant and immunomodulatory effect of chlorophyllin on induced aflatoxicosis in broilers. *Indian Journal of Animal Sciences,* Vol. 78, Jul, pp. 696-699,0367-8318

Tombelli, S., Mascini, M., Scherm, B., Battacone, G., & Migheli, Q. (2009). DNA biosensors for the detection of aflatoxin producing Aspergillus flavus and A-parasiticus. *Monatshefte Fur Chemie,* Vol. 140, Aug, pp. 901-907,0026-9247

Topcu, A., Bulat, T., Wishah, R., & Boyaci, I. H. (2010). Detoxification of aflatoxin B(1) and patulin by Enterococcus faecium strains. *International Journal of Food Microbiology,* Vol. 139, May 15, pp. 202-205,0168-1605

Trucksess, M. W., & Scott, P. M. (2008). Mycotoxins in botanicals and dried fruits: A review. *Food Additives and Contaminants,* Vol. 25, pp. 181-192,0265-203X

Trung, T. S., Tabuc, C., Bailly, S., Querin, A., Guerre, P., & Bailly, J. D. (2008). Fungal mycoflora and contamination of maize from Vietnam with aflatoxin B(1) and fumonisin B(1). *World Mycotoxin Journal,* Vol. 1, Feb, pp. 87-94,1875-0710

Uegaki, R., Tsukiboshi, T., & Cai, Y. M. (2010). Aspergillus flavus Producing Aflatoxins Isolated from Materials of Commercial Feed in Japan. *Jarq-Japan Agricultural Research Quarterly,* Vol. 44, Oct, pp. 421-427,0021-3551

Umoh, N. J., Lesi, O. A., Mendy, M., Bah, E., Akano, A., Whittle, H., Hainaut, P., & Kirk, G. D. (2011). Aetiological differences in demographical, clinical and pathological characteristics of hepatocellular carcinoma in The Gambia. *Liver International,* Vol. 31, Feb, pp. 215-221,1478-3231 (Electronic)1478-3223 (Linking)

Upadhaya, S. D., Park, M. A., & Ha, J. K. (2010). Mycotoxins and Their Biotransformation in the Rumen: A Review. *Asian-Australasian Journal of Animal Sciences,* Vol. 23, Sep, pp. 1250-1260,1011-2367

Valdivia, A. G., Martinez-de-Anda, A., Jaramillo-Juarez, F., Reyes, J. L., Ortiz, R., Quezada, T., de Luna, M. C., & Rodriguez, M. L. (2010). Effects of aflatoxin chronic intoxication in renal function of laying hens. *Poultry Science,* Vol. 89, Aug 1, pp. 1622-1628,0032-5791

van Egmond, H. P., Schothorst, R. C., & Jonker, M. A. (2007). Regulations relating to mycotoxins in food. *Analytical and Bioanalytical Chemistry,* Vol. 389, Sep, pp. 147-157,1618-2642

Vargas, E. A., Preis, R. A., Castro, L., & Silva, C. M. (2001). Co-occurrence of aflatoxins B1, B2, G1, G2, zearalenone and fumonisin B1 in Brazilian corn. *Food Addit Contam,* Vol. 18, Nov, pp. 981-986,0265-203X (Print)0265-203X (Linking)

Vasanthi S, B. R. (1998). Mycotoxins in foods--occurrence, health & economic significance & food control measures. *Indian Journal of Medical Researches,* Vol. 108, Nov, pp. 212-224

Wagacha, J. M., & Muthomi, J. W. (2008). Mycotoxin problem in Africa: Current status, implications to food safety and health and possible management strategies. *International Journal of Food Microbiology,* Vol. 124, May 10, pp. 1-12,0168-1605

Wei, H., Cheng, B. C., Wan, C. X., Yang, S. L., Xu, H. Y., Liu, J. S., Tian, W. H., & Zeng, M. (2010). Detoxification of Deoxynivalenol by Bacillus Strains. *Journal of Food Safety,* Vol. 30, Aug, pp. 599-614,0149-6085

Woo, L. L., Egner, P. A., Belanger, C. R., Wattanawaraporn, R., Trudel, L. J., Croy, R. G., Groopman, J. D., Essigmann, J. M., & Wogan, G. N. (2011). Aflatoxin B1-DNA adduct formation and mutagenicity in livers of neonatal male and female B6C3F1 mice. *Toxicological Sciences,* Apr 19,1096-0929 (Electronic)096-0929 (Linking)

Xu, Z. R., Han, X. Y., Huang, Q. C., Li, W. F., & Jiang, J. F. (2008). Changes in growth performance, digestive enzyme activities and nutrient digestibility of cherry valley ducks in response to aflatoxin B(1) levels. *Livestock Science,* Vol. 119, Dec, pp. 216-220,1871-1413

Yang, M.-H., Chen, J.-M., & Zhang, X.-H. (2005). Immunoaffinity Column Clean-Up and Liquid Chromatography with Post-Column Derivatization for Analysis of Aflatoxins in Traditional Chinese Medicine. *Chromatographia,* Vol. 62, pp. 499-504,0009-5893

Yarru, L. P., Settivari, R. S., Gowda, N. K. S., Antoniou, E., Ledoux, D. R., & Rottinghaus, G. E. (2009). Effects of turmeric (Curcuma longa) on the expression of hepatic genes associated with biotransformation, antioxidant, and immune systems in broiler chicks fed aflatoxin. *Poultry Science,* Vol. 88, Dec 1, pp. 2620-2627,0032-5791

Yassin, M. A., El-Samawaty, A. M. A., Moslem, M., Bahkali, A., & Abd-Elsalam, K. (2011). Fungal Biota and Occurrence of Aflatoxigenic Aspergillus in Postharvest Corn Grains. *Fresenius Environmental Bulletin,* Vol. 20, pp. 903-909,1018-4619

Yazdanpanah, H. (2006). Mycotoxin Contamination of Foodstuffs and Feedstuffs in Iran. *Iranian Journal of Pharmaceutical Research,* Vol. 1, pp. 9-16

Yazdanpanah, H., Mohammadi, T., Abouhossain, G., & Cheraghali, A. M. (2005). Effect of roasting on degradation of Aflatoxins in contaminated pistachio nuts. *Food and Chemical Toxicology,* Vol. 43, Jul, pp. 1135-1139,0278-6915

Zain, M. E. (2011). Impact of mycotoxins on humans and animals. *Journal of Saudi Chemical Society,* Vol. 15, Apr, pp. 129-144,1319-6103

Zhao, J., Shirley, R. B., Dibner, J. D., Uraizee, F., Officer, M., Kitchell, M., Vazquez-Anon, M., & Knight, C. D. (2010). Comparison of hydrated sodium calcium aluminosilicate and yeast cell wall on counteracting aflatoxicosis in broiler chicks. *Poultry Science,* Vol. 89, Oct 1, pp. 2147-2156,0032-5791

Zinedine, A., Brera, C., Elakhdari, S., Catano, C., Debegnach, F., Angelini, S., De Santis, B., Faid, M., Benlemlih, M., Minardi, V., & Miraglia, M. (2006). Natural occurrence of mycotoxins in cereals and spices commercialized in Morocco. *Food Control,* Vol. 17, Nov, pp. 868-874,0956-7135

Zorlugenc, B., Kiroglu Zorlugenc, F., Oztekin, S., & Evliya, I. B. (2008). The influence of gaseous ozone and ozonated water on microbial flora and degradation of aflatoxin B(1) in dried figs. *Food Chem Toxicol,* Vol. 46, Dec, pp. 3593-3597,0278-6915 (Print) 0278-6915 (Linking)

19

A Review of Aflatoxin M_1, Milk, and Milk Products

Hamid Mohammadi
Kurdistan General Department of Standards and Industrial Research
Iran

1. Introduction

Aflatoxins are a group of closely related heterocyclic compounds produced predominantly by two filamentous fungi, *Aspergillus flavus* and *Aspergillus parasiticus*. Recent studies have shown that some *A. nominus* and *A. tamarii* strains are also aflatoxin producing, of which *A. nominus* is phenotypically similar to *A. flavus* (Kurtzman et al, 1987; Goto et al., 1997). Ito et al. (2001) isolated one more strain, *A. pseudotamarii*, which can produce aflatoxin. These fungi belong to the class Hyphomycetes, subdivision Deuteromycotina and family Aspergillaceae. They contaminate a vast array of food and agricultural commodities. Aspergillus species are capable of growing on a variety of substrates and under a variety of environmental conditions. Therefore, most foods are susceptible to aflatoxigenic fungi at some stage of production, processing, transportation, and storage. The outbreak of aflatoxicosis (famous as Turkey "X" disease) in England in 1960 caused the death of a large population of livestock (Blount, 1961) and led to the discovery of aflatoxin in groundnut meal contaminated by *A. flavus* (Hesseltine, 1979). Subsequently, aflatoxins were found in other feeds, especially maize (Chakrabarty, 1981) and cottonseed meal (Sharma et al., 1994). Aflatoxin M_1 (AFM_1) in milk and milk products is considered to pose certain hygienic risks for human health. Mammals that ingest aflatoxin B_1 contaminated diets eliminate into milk amounts of the principal 4-hydroxylated metabolite known as "milk toxin" or aflatoxin M_1 (Figure. 1). The economic impacts attributed to aflatoxin are incurred directly by loss in crops, livestock, and dairy and indirectly by a recurring expenditure in quality-control programs, research and education, lower foreign exchange earnings, and increased storage and packaging costs of vulnerable commodities. The potential hazard of aflatoxins to human health has led to worldwide monitoring programs for the toxin in various commodities as well as regulatory actions by nearly all countries.

2. Formation, toxicity, and regulation of aflatoxin M_1

Aflatoxin B_1 (AFB_1) is metabolized by the hepatic microsomal mixed-function oxidase system, but it also can undergo several metabolic conversations depending upon species (Marsi et al., 1974). The amounts of aflatoxin M_1 (AFM_1) excreted in milk as a percentage of AFB_1 averages 1-2%, varying from animal to animal, from day to day and from one milking to the other. The AFM_1 could be detected in milk 12-24 h after the first AFB_1 ingestion, reaching a high level after a few days. When the intake of AFB_1 is finished, the AFM_1

concentration in the milk decreases to an undetectable level after 72 h (van Egmond 1989). Battocone et al. (2003) observed that there was a linear relationship between AFB_1 dose and excretion of AFM_1 into ewes' milk.
International Agency for Research on Cancer (IARC, 1993) classified AFB_1 and AFM_1 as class 1 and 2B (or probable) human carcinogens, respectively. Lafont et al. (1989) observed a high genotoxic activity of AFM_1, although it was lower than that of AFB_1. The aflatoxins show both acute and chronic toxicity, and one of the outstanding features in the toxicology of the aflatoxins is the wide variation in response amongst different species of animals and even between the male and female of the same species. It can be seen that for some animals, such as the rat, aflatoxins are very carcinogenic, and yet in other species, it is difficult to demonstrate carcinogenicity. This considerable variation in biological response arises from the requirement that the mold metabolite itself has to be metabolized in order that a toxic response occurs, and the metabolites responsible for acute toxicity differ from those reponsible for carcinogenicity. There are several metabolic reactions (Fig. 1), such as the demethylation to aflatoxin P_1 and hydration to aflatoxin B_{2a}, which may lead to a decrease in toxicity.

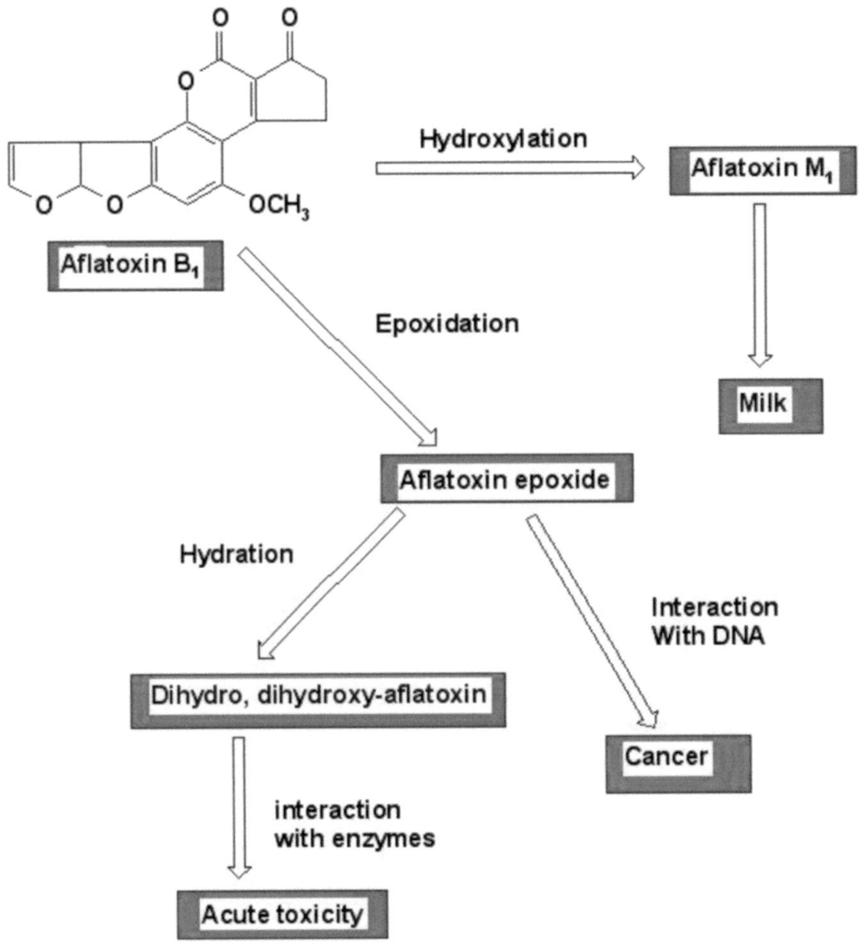

Fig. 1. Some metabolic products from AFB_1

It seems that milk has the greatest demonstrated potential for introducing aflatoxin residues from edible animal tissues into human diet. Aflatoxins are one of the major etiological factors in the development of hepatocellular carcinoma (IARC, 2002), and more recently associations between childhood aflatoxin exposure and both growth faltering (Gong et al., 2004) have been reported. Moreover, as milk is the main nutrient for growing young, whose vulnerability is notable and potentially more sensitive than that of adults, the occurrence of AFM_1 in human breast milk, commercially available milk, and milk products is one of the most serious problems of food hygiene.

In order to decrease aflatoxins risk nearly all developed countries are of the maximum permissible levels of AFB_1 in foods and feeds as well as the levels of AFM_1 in milk and milk products. Currently the limits of AFM_1 are highly variable (Table 1), depending upon the degree of development and economic standing of the countries. Some European Community and Codex Alimentarius prescribe that the maximum level of AFM_1 in liquid milk and dried or processed milk products should not exceed 50 ng/kg (Codex Alimentarius Commissions, 2001). However, according to US regulations the level of AFM_1 in milk should not be higher than 500 ng/kg (Stoloff et al., 1991). There are thus differences in maximum permissible limit of AFM_1 in various countries.

Region	Maximum Acceptabe Level (ng/l)	Type
European Union	50	Milk
Austria	50	Milk
Argentina	50	Milk
Bulgaria	500	Milk
Germany	50	Milk
Australia	20	Children's milk
Sweden	50	Liquid milk products
Netherlands	20	Butter
Switzerland	50	Milk and milk products
	250	Cheese
Belgium	50	Milk
USA	50	Milk
Czech Republic	100	Children's milk
	500	Adult's milk
Serbia	500	Milk
Iran	50	Raw, Pasteurized, and UHT milk
	200	Cheese
	20	Butter
France	30	Children's milk< 3 years
	50	Adult's milk
Turkey	50	Milk and milk products
	250	Cheese
Brazil	500	Milk

Table 1. Maximum acceptable level of AFM_1 in some regions

Regulatory limits seem to be a practical compromise between the need to have carcinogen-free commodities and the economic consequences of setting regulatory limits (Pohland and Yess, 1992). However, Stoloff et al. (1991) observed that as to aflatoxins there was little scientific basis, or the existing scientific information was not used in setting legal limits in most countries. Thus, even the low regulatory limits set by countries could not prevent chronic effects of aflatoxins, due to continued exposure to subacute levels of aflatoxins.
Because of the following reasons, it seems that monitoring and preventive program are the most effective strategies to decrease the risk of exposure to both human and animals:

1. Evaluation of human exposure levels and health risk based on animal toxicological research
2. Difficulties in assessing dietary intake
3. Decontamination and remove mycotoxins from human and animal diets

2.1 AFM_1 in milk

Milk, as a liquid, is a highly variable product that rapidly loses its quality and spoils if not to be treated. Since milk may be processed in numerous ways, the effects of storage and processing on stability and distribution of AFM_1 are of great concern.
Kiermier and meshaley (1977) reported the effect of cold treatments. They observed that detectable AFM_1 decreased by 11 to 25% after 3 days at 5ºC, 40% after 4 days at 0ºC, and 80% after 6 days at 0ºC. Whereas, McKinney et al. (1973) revealed that freezing at -18ºC for 30 days resulted in an apparent loss of 14%, with 85% lost after 53 days. Stoloff et al. (1981) suggested less degradation of AFM_1 at -18ºC with insignificant loss after 53 days. As to the effect of heating contradictory data have been reported. Kiermeier and Mashaley (1977), reported that various heat–time treatments caused reductions in the AFM_1 concentrations of milks between 12% and 40%. Choudhary et al. (1998) studied the effect of various heat-treatments on AFM_1 content of cow's milk and reported that sterilization of milk at 121 °C for 15 min caused 12.21% degradation of AFM_1, whereas boiling decreased AFM_1 by 14.50%. They concluded that destruction of AFM_1 depends on time and temperature combination of the heat treatment applied. In an investigation Conducted by Bakirci (2001), it was observed that pasteurization caused a decrease in the level of AFM_1 at the rate of 7.62%. Deveci (2007) showed that pasteurization can partially reduce the amount of AFM_1 in milk. However, some reports showing that aflatoxins are stable during heat-treatments such as pasteurization and sterilization (Van-Egmond et al. 1977; Wiseman and Marth, 1983; Yousef and Marth, 1989; Govaris et al. 2001) were also published. Fluctuation in data reported in literature could be attributed to the wide range of temperature, different analytical methods, and employment of both naturally and artificially contaminated milk.
AFM_1 distribution in milk is not homogeneous. Cream separation can affect AFM_1 distribution, since 80% is partitioned in the skim milk portion (Grant and Carlson, 1971) because of AFM_1 binding to casein (Brackett, 19982a). An amount of 30% of AFM_1 is indeed estimated to be associated with the nonfat milk solids and in particular with casein. According to Van Egmond and Paulsch (1986) the behavior of AFM_1 in processes which involve fat separation may be explained by its semipolar character, leading to predominance in the nonfat fraction.
Contradictory data have been reported on the influence of milk concentration on AFM_1. Kiermeier (1973) reported no losses of AFM_1, whereas some authors observed losses ranging from 60 to 75% following milk concentration (Moreau, 1976; Purchase, I. F. H., 1973). Data from the studies on the occurrence of AFM_1 in milk since the 1990s are reported in (Table 2.)

Year	Region	Milk type	Samples	+ samples	Range(ng/liter)	Reference
1991	Kazakhstan	C	*	0	NA	Nikov et al.
1992	Cuba	C	85	22	>500	Margolles et al.
1993	Japan	C	37	0	NA	Tabata et al.
1994	USA	D	10	4	95	Kawamura et al.
	China	D	28	21	102.8	
	Italy	D	14	0	NA	
	New Zealand	D	3	0	NA	
	Poland	D	3	1	85	
1995	India	R	504	89	100-3.500	Rajan et al.
1996	Italy	UHT	161	125	<1- 23.5	Galvano et al.
		D	92	49	<1- 79.6	
1998	Kuwait	C	9	5	*	Srivastava et al
		R	7	5	*	
1999	Portugal	R	31	25	*	Martins
1999	Portugal	UHT	70	60	*	Martins
1999	Argentina	R	56	6	12-30	Lopez et al.
		PW	5	4	10-14	
		P	16	8	10-17	
1999-2000	Iran	R	186	119	≤10-410	Ghiasian et al.
2001	Turkey	R	90	79	12.5-123.6	Bakirci
2001	Iran	R	111	85	15-280	Kamkar
2002-2003	Brazil	R	22	13	*	Shundo and Sabino
2002-2003	Brazil	P	43	32	*	Shundo and Sabino
2002-2003	Brazil	UHT	34	34	*	Shundo and Sabino
2002	Greece	R	54	*	*	Kaniou-Grigoriadou et al.
2003	Iran	P	624	624	*	Alborzi et al.
2004	Turkey	P	3	2	*	Gurbay et al.
		UHT	24	14	*	
2004	Iran	R	319	172	15.4[a]	Tajkarimi et al
2003-2004	Iran	R	98	*	53[a]	Tajkarimi et al.
2003-2004	Italy	R	208	36	5-36.1	Virdis et al.
*	Turkey	P	85	75	*	Celik et al.
2004-2005	Italy	R	344	5	*	Decastelli et al
2004-2005	Brazil	P	12	7	11-161	Oliveira and Ferraz
		UHT	12	10	11-161	
		PW	12	8	11-161	

Year	Region	Milk type	Samples	+ samples	Range(ng/liter)	Reference
2005	Pakistan	R	168	168	10-700	Hussain and Anwar
2005	Iran	P	128	128	31-113	Oveisi et al.
2005-2007	Kuwait	R	177	176	4.9-67.8	Dashti et al.
2006	Iran	R	*	*	43-59	Mohammadi et al.
2006	Iran	P	110	110	8-89	Karimi et al.
2006	Iran	PW	42	42	51-914	Kamkar
2006	Iran	P	*	*	178.8-253.5	Sefidgar et al.
2006-2007	Iran	R	240	226	12.56[a]	Mohammadian et al.
		P	32	31	12.43[a]	
2007	Iran	P	*	*	23.22[a]	Mohammadi et al.
2007	Iran	UHT	*	*	19.53[a]	Mohammadi et al.
2007	Pakistan	B	55	19	13[a]	Hussain et al.
		C	40	15	14[a]	
		G	30	6	2[a]	
		S	24	4	2[a]	
		Ca	20	0	0	
2008	Iran	P	50	50	*	Movassagh Ghazani
		UHT	49	49	*	
2007-2008	Serbia	C	3	*	10-50	Polovinski-Horvatović et al.
2007-2008	Iran	UHT	210	116	8-249	Heshmati and Milani
2007-2008	Iran	C	75	59	60.1[a]	Rahimi et al.
		B	75	29	31.9[a]	
		Ca	40	5	19.0[a]	
		S	51	19	28.1[a]	
		G	60	19	30.1[a]	
2008	Spain	R	72	68	9.69[a]	Cano-Sancho et al.
2008	Iran	C	88	74	13-394	Fallah et al.
		G	65	28	13-55	
		S	72	43	15-102	
2009	Iran	P	91	66	13-250	Fallah
2009	Croatia	R	61	*	0.6-58.7	Bilandzic et al.
2009	Sudan	R	44	42	220-6900	Elzupir and Elhussein

P: pasteurized; D: Dry milk; PW: Powdered Milk; R: Raw Milk; C: Cow milk; B: Buffalo milk; S: Sheep milk; G: Goat milk; Ca: Camel milk; NA: Not applicable
* Not reported
[a] Average of contamination

Table 2. Occurrence and content of AFM_1 in milk samples

Many authors showed that Seasonal effect influences concentration of aflatoxin M_1. They reported higher concentration of AFM_1 in cold seasons as compared to hot seasons (Applebaum et al., 1982; Blanco et al., 1988b; Hussain and Anwar, 2008; Tajkarimi et al. 2008; Fallah, 2010, Bilandzic et al., 2010), the reason being in winters mostly milking animals are fed with compound feeds and thus concentration of aflatoxin B1 increases which in turn enhances AFM_1 concentration in milk. Moreover, temperature and moisture contents also affect the presence of aflatoxin B_1 in feeds. *A. flavus* and *A. parasiticus* can easily grow in feeds having moisture between 13% and 18% and environmental moisture between 50% and 60%, furthermore, they can produce toxin (Jay, 1992). Another reason of low AFM_1 level in summer may be attributed to out-pasturing of milking cattle.

It is too difficult to compare the data from the literature due to wide differences between and within the countries related to feeding, animal and environmental factors, extraction and analysis procedures, and regulatory limits for aflatoxins in feeds and milk. However, in recent years the incident of AFM_1 contamination seems to have been balanced on the one hand by increasing precision of extraction and analysis procedures and on the other hand the setting of stricter regulatory limits for alatoxins in feeds and milk (Galvano et al., 1996). Today the high efficiency of immuno-enzymatic extraction and the accuracy of analytical methodology and equipment, such as high pressure liquid chromatography and fluorescence detectors, allow detection limits to decrease, improving the percentage of positive samples. Furthermore, in recent years attention to the concern of aflatoxins in feeds as well as in milk has increased in most of the developed countries.

2.2 AFM_1 in cheese

Occurrence of aflatoxin in cheese can be owing to three possible causes:

1. AFM_1 present in raw milk as a consequence of carry over of AFB_1 from contaminated animal feed to milk
2. Synthesis of aflatoxin (B1, B2, G1, and G2) by fungi that grow on
3. cheese (although the low level of carbohydrate does not make it a very suitable substrate)
4. The use of powdered milk contaminated with AFM1 for cheese production

Contrasting data have been reported on the influence of cheese preparation on AFM_1 recovery. Studies performed in the early years showed variable losses of AFM_1 during cheese production: 65%, 47%, <20% and <15% according to Purchase et al. (1972), McKinney et al. (1973), Grant and Carlson (1971) and Stubblefield and Shannon (1974), respectively. In contrast, later investigations of several authors (Brackett and Marth, 1982b; Brackett and Marth, 1982c; Munksgaard et al., 1987; Van Egmond and Paulsch, 1986; Bakirci, 2001, Govaris et al. 2001; Deveci, 2007) reported increases in AFM_1 concentration in cheese as a function of cheese type, technologies, and the amount of water eliminated during processing. For example, Mohammadi et al. (2008) investigated some factors, which are involved in the process of making Iranian white brine cheese. They reported that some factors such as renneting temperature, press time, and saturated brine pH affected the amount of water eliminated and in turn the content of AFM_1 in the cheese curds. However, many results have been drawn from experiments in which the processed milk contained the toxin at high levels, which seldom appear in the practice. Therefore, additional investigations should verify the influence of cheese making on AFM_1 occurrence to avoid uncertainty in actual practice when the concentration of the toxin in the processed milk is at around the maximum permissible level of 0.05 mg/kg that is frequently recorded in monitoring programmes.

The increase in AFM_1 concentration in cheese has been ascribed to the affinity of AFM_1 for casein (Allcroft and Carnagham, 1363; Applebaum et al., 1982; Brackett and Marth, 1982a; Grant and Carlson, 1971). Brackett and Marth (1982a) suggested that since is possible to extract AFM_1, it must not be covalently bound but linked by hydrophilic interactions hydrophobic areas of the casein. According to Dosako et al. (1980), AFM_1 is a water-soluble component and due to the hydrophobic sides of the casein molecule, AFM_1 has affinity to casein of milk. Therefore, they defined a factor named "Enrichment Factor" (EF) for cheeses. Further surveys should be done to find as for cheese manufacture influences on AFM_1 distribution.

Some tests heve been carried out on several kinds of cheeses as to overall stability of AFM_1 during ripening and storage. Fremy et al. (1990) and Dragacci et al. (1995) reported that the concentrations of AFM_1 in Camembert cheese were higher at the beginning than at the later time of ripening. These results were in agreement with studies by Govaris et al. (2001). Such results however, conflict with reports of earlier studies that indicate different behaviour of AFM_1 in various other types of cheeses. Thus, in Camembert and Tilsit (Kiermeier and Buchner 1977), Cheddar (Brackett and Marth 1982b) and Brick (Brackett et al., 1982) cheeses stored for 3, 14 and 6.5 months, respectively, the concentration of the toxin increased during the early stage of their ripening to decrease thereafter to reach about its initial concentration at the beginning of ripening. On the other hand, the concentration of AFM_1 in Parmesan cheese started high at the beginning of the ripening period, decreased until about the fifth month and then slowly increased up to the tenth month of storage (Brackett and Marth 1982d). In contrast, the AFM_1 content of Mozzarella remained almost constant during storage of 4.5 months (Brackett and Marth, 1982d). Additionally, studies by Deveci (2007), and Huseyin Oruc et al. (2007) showed that the amount of AFM_1 in white pickled and Kashar cheeses did not significantly affect over the storage. Kaniou-Grigoriadou et al. (2005) found that the final ripened cheese was free of aflatoxin M_1.

These different profiles of AFM_1 in various cheese products may be the result of several factors such as heat treatment (Brackett and Marth 1982b, Yousef and Marth 1989), proteolysis (Brackett and Marth 1982b, d, Brackett et al. 1982, Yousef and Marth 1989), exposure of contaminated milk to light (Yousef and Marth 1989), and especially to an inadequate method of analysis (Yousef and Marth 1989). Some results of studies on the behaviour of AFM_1 during cheese ripening seem to represent changes in the recovery of toxin by the method during the different phases of the study rather than real changes in the level of AFM_1 in cheese (Brackett and Marth 1982a).

Several investigations on the partitioning of AFM_1 during cheese production staring with different milk contamination levels reported a wide range of distribution of AFM_1 between whey and curd. Some authors observed that half or more of the AFM_1 was in the way: 50%, 50%, 66%, 100%, 60%, and 53-58% according to Grant and Carlson (1971), Stubblefield and Shannon (1974), Blanco et al. (1988a), Purchase et al. (1972), Lopez. et al. (2000), and Huseyin Oruc et al. (2007), respectively. In contrast, others reported that most of AFM_1 was with curd: ranging from 66% to 72%, from 73% to 77%, 80%, 100%, and 59.1% according to Marshaley et al. (1986), El Deeb et al. (1992), Mckinney et al. (1973), Allcroft and Carnaghan (1963), and Deveci (2007), respectively. Kaniou-Grigoriadou et al. (2005) observed that enrichment factor in the production of Feta cheese made from naturally contaminated milk ranged between 4.3 and 5.6. Kamakar et al. (2008) showed that the mean concentration of toxin in curd and cheese was 3.12 and 3.65-fold more than that in whey and 1.68 and 1.80 fold more than that in cheese milk, respectively.

It is thought that since AFM_1 is a semi-polar component, it has less affinity to serum protein (Applebaum et al., 1982). Regarding the affinity of AFM_1 with proteins, Recently, Barbiroli et al. (2007) indicated that there is no simple physical method to remove AFM_1 from ovine and caprine milk. Neither ultrafiltration, nor acidic or enzymatic treatments were able to influence the toxin's interaction with casein or whey proteins. Only the combined action of heat and low pH (as used in ricotta cheese production) was able to denature whey proteins to a point where they lost their AFM_1-binding capacity.
According to Blanco et al. (1988a) these contrasting results can be attributed to different factors such as extraction techniques, methodology, type and degree of milk contamination, differences in milk quality, expression of the results, the presence of a small portion of curd in whey which could influence AFM_1 concentration, and the cheese manufacture process.

Year	Region	Samples	+ samples	Range(ng/kg)	Refernce
1990	Syria	*	0	NA	Haydar et al.
1991	Kazakhestan	*	*	*	Nikov et al.
1993	Japan	37	0	NA	Tabata et al.
1995	Japan	41	0	NA	Taguchi et al.
1995	Spain	35	16	20-200	Jose Barios et al.
2001-2002	Turkey	600	30	100-800	Yaroglu et al.
2003-2004	Iran	80	66	150-2410	Kamkar
2004	Italy	41	4	79.5-389	Virdis et al.
2005	Turkey	100	99	0-4100	Tekinsen and Ucar
2005-2007	Kuwait	40	32	23.8-452	Dashti et al.
2008	Spain	72	0	-	Cano-Sancho et al
2008	Iran	75	49	30-313	Fallah et al.
2009	Iran	72	59	30-1200	Fallah

NA: Not Applicable
*: Not Reported

Table 3. Occurrence and content of AFM_1 in cheese samples

The incidence of positive cheese samples for AFM_1 (Table 3) seem to be widely variable. Taguchi et al. (1995) found no positive samples in imported cheese in Japan. Virdis et al. (2008) detected few positive samples, whereas Tekinsen and Ucar (2008) observed a high incidence of positive samples. As regards the contamination level, several authors (Kamkar, 2006; Fallah, 2010) found a maximum contamination level over 1000 ng of AFM_1 per kg. This latter contamination level could be hazardous.

2.3 AFM_1 in yogurt

Several studies have been conducted regarding the effect of yogurt manufacturing on AFM_1 content. Some authors reported no influence on aflatoxin M_1 content (Blanco et al. 1993; Stoloff, 1980; Stoloff et al. 1981; Van Egmond, 1983; Van Egmond and Paulsch, 1986). In contrast, Munksgaard et al. (1987) and Bakirci (2001) detected variable increases of AFM_1 content in yogurt related to the milk. The effect of fermentation was assessed by Govaris et al. (2002). They reported that AFM_1 levels in all yoghurt samples showed a significant decrease from those initially present in milk. This decrease in AFM_1 was attributed to factors such as low pH, formation of organic acids or other fermentation by-products, or even to the

presence of lactic acid bacteria. The low pH during fermentation alters the structure of milk proteins such as the caseins leading to formation of yoghurt coagulum. The change in casein structure during yoghurt production may affect the association of AFM_1 with this protein (Brackett and Marth 1982) causing adsorption or occlusion of the toxin in the precipitate.
As to AFM_1 stability over storage of yogurt, Van Egmond et al. (1977) observed no reduction of AFM_1 in yogurt stored for 7 days at 7 °C. Megalla and Hafez (1982) observed complete transformation AFB_1 in its hydroxy derivative AFB_2A caused by the acids present in yogurt. Whereas, Rasic et al. (1991) revealed a high reduction (up to 97%) of AFM_1 in yogurt and acidified milk. El Deeb et al. (1992) observed that enzymatic, microbial, and particularly acid coagulation caused degradation of AFM_1 in buffalo milk. Maryamma et al. (1990) reported a high reduction of AFM_1 in fermented goat milk. As a result of Study by Govaris et al. (2002), during refrigerated storage, AFM_1 was rather more stable in the yoghurts with pH 4.6 than with pH 4.0. The percentage loss of the initial amount of AFM_1 in milk was estimated at about 13 and 22% by the end of the fermentation, and 16 and 34% by the end of storage for yoghurts with pHs 4.6 and 4.0, respectively.
Since it is known that exposure of the aflatoxin molecule to strong acid, such as trifluoracetic acid, can cause its acid-catalyzed hydration, leading, for example, from AFB_1 to AFB_2A (Cohen and Lapointe, 1981), but not its degradation or neutralization, the effect of the weak acidity of yogurt on aflatoxin should be more investigated.
Some investigations have been conducted related to the effect of aflatoxin on nutritive properties of yogurt. El Deeb et al. (1989) observed some negative effects of AFM_1 on Lactobacillus bulgaricus (cell wall thickening and shortening of cell chain lenghth) and *Staphylococcus thermophilus* (cell wall thickening and cell shape changing from coccoid to oval). Rasic et al. (1991) found that *S. thermophilus* was affected by the presence of in milk during fermentation of yoghurt, exhibiting longer cell chains in the contaminated than in the uncontaminated yoghurt samples. Similarly, Govaris et al. (2002) observed that the growth rate of *S. Thermophilus* and curdling time were affected by the higher level and not by the lower level of AFM_1. Unlike cheese and milk samples, the presence of AFM_1 in yogurt has not frequently been studied. Thus, more investigations are needed because:

1. currently, human consumption of yogurt has greatly increased
2. there are contradictory data on AFM_1 stability over manufacture and storage in the literature
3. The presence of aflatoxin in yogurt could reduce the nutritional values of its consumption.

2.4 AFM_1 in other milk products

Many other milk products such as cream, butter, ice cream may contain AFM_1. The presence of AFM_1 in these products has rarely been investigated and could be of interesting aspects for study. Some surveys conducted on the occurrence of AFM_1 in milk products are reported in (Table. 4). In a study by Bakirci (2001), the levels of AFM_1 in the products made from contaminated milk namely butter, butter milk, cream, skim milk was investigated. The mean AFM_1 level found in cream samples was 64.4% of AFM_1 concentration of bulk-tank milk. Whereas, mean AFM_1 level of skim milks was 3% higher than those of bulk-tank milk. These values were close to the results given by Van Egmond et al. (1977), and lower than the values given by Wiseman et al. (1983). Levels of AFM_1 in butter samples in the study were less, and they were as 33.80% of AFM_1 amounts of bulk-tank milk. Mean AFM_1 levels obtained from buttermilk samples were similar to those of bulk-tank milk (mean 83% of it).

The same results were reported by Grant and Carlson (1971). During butter processing, protein membrane around fat globules is broken down and serum phase is separated. Due to the chemical structure of AFM_1 and its affinity to casein, it adsorbs on this fraction of protein (Yousef & Marth, 1989), therefore, cream contained less AFM_1 than milk, and butter contained less amount of AFM_1 than cream. As a result of the associate effects of these factors, AFM_1 concentration occurs in lipid phase (like butter and cream) less than serum phase and protein fraction (Grant & Carlson, 1971).

Year	Region	type	Samples	+ samples	Range(ng/kg)	Refernce
2005	Turkey	Butter	92	92	10-7000	Tekinsen and Ucar
2005	Turkey	Butter	27	25	0-100	Aycicek et al.
2005	Iran	Infant formula	120	116	1-14	Oveisi et al.
		Milk-based cereal weaning food	80	72	3-35	
2005	Egypt	Beast milk	443	248	4.2- 889	Polychronaki et al.
2005-2007	Kuwait	Breast milk	12	5	8.83-15.2	Dashti et al.
2008		Kashk	125	53	28-291	Fallah et al.
		Doogh	136	25	13-53	
2009	Iran	Butter	31	8	13-26	Fallah
		Ice Cream	36	25	15-132	

Table 4. Occurrence of AFM_1 reported in some regions

AFM_1 is frequently observed in the aflatoxin exposed individuals and in the breast milk. AFM_1 toxicity in this respect is important as it is known that within aflatoxin exposed nursing mothers it can provide a source of aflatoxin exposure to the infant (El-Nezami, 1995). The occurrence of AFM_1 in breast milk has been investigated in some regions. There is increased awareness of the link between growth and health of the fetus and infant, and disease risk in later life. Long term pre and postnatal exposure to aflatoxins could be one of the factors contributing to growth faltering and/or the early onset of hepatocellular carcinoma (HCC) in countries with a high incidence of the disease. Additionally, the presence of other aflatoxins, B_1, B_2, G1, G_2 and M_2, has also been reported in breast milk (IARC, 1993). The identification and understanding of factors determining the presence of toxicants in human milk is important and may provide a strong basis for controlling the transfer of chemicals to the infants through breast milk.

3. Conclusion

At present, since it considers that there is not enough information to establish a reasonable exposure level, The World Health Organisation (WHO) recommends the reduction of AFM_1 consumption to a minimum so as to minimize AFM_1 potential risks. The regulatory limits are widely variable and there has been little scientific basis in their setting. Efforts should be made in attempting to provide further and extensive scientific information on human health hazards related to low-level long term aflatoxin exposure and to standardize the already existing regulatory limits for aflatoxins.

Future studies should verify the effect of milk storage and processing on AFM_1 occurrence to avoid actual uncertainty. However, since it is generally assumed that neither storage nor processing determine reduction of AFM_1 content, further information on possible AFM_1 concentration following milk processing should be furnished.

The occurrence of AFM_1 in cow milk and milk products is widespread. For this reason, milk and milk products have to be controlled continuously by accurate and reliable analytical techniques for presence of AFM_1 contamination. It is also extremely important to maintain low levels of AFM_1 in the feeds of dairy animals. In order to achieve this, dairy cow feds should be kept away from contamination as much as possible. Therefore, animal feeds should be checked regularly for aflatoxin and, particularly important, storage conditions of feeds must be strictly controlled.

The occurrence of aflatoxin and their metabolites in human breast milk is of great concern. Since serious health hazards to the mother, fetus, and infant could occur. Therefore extensive and periodic surveys should be performed. Additionally, the incidence and occurrence of AFM_1 in dried milk infant formula should be more investigated.

4. References

Allcroft, R.; & Carnaghan, R. B. A. (1963). Groundnut toxicity: an examination for toxin in human food productsfrom animal fed toxin groundnut meal. *Veterinary Record*, No. 75, pp. (259-263)

Applebaum, R. S.; Brackett, R. E.; Wiseman, D. W.; & Marth, E. H. (1982). Aflatoxins: toxicity to dairy cattle and occurrence in milk and milk products. *Journal of Food Protection*, No. 45, pp. (752-777)

Aycicek, H.; Aksoy, A.; & Saygi, S. (2005). Determination of aflatoxin levels in some dairy and food products which consumed in Ankara, Turkey. *Food Control*, No. 39, pp. (263–266)

Bakirci, I. (2001). A study on the occurrence of aflatoxin M_1 in milk and milk products produced in Van province of Turkey. *Food Control*, No. 12, pp. (45-51)

Barbiroli, A.; Bonomi, F.; Benedetti, S.; Mannino, S.; Monti, L.; Cattaneo, T.; & Iametti, S. (2007). Binding of Aflatoxin M_1 to Different Protein Fractions in Ovine and Caprine Milk. *Journal of Dairy Science*, No. 90, pp. (532–540)

Battacone, G.; Nudda, A.; Cannas, A.; Cappio Borlino, A.; Bomboi, G.; & Pulina, G. (2003). Excretion of aflatoxin M_1 in milk of dairy ewes treated with different doses of aflatoxin B_1. *Journal of Dairy Science*, No. 86, pp. (2667-2675)

Bilandzic, N.; Varenina, I.; & Solomun, B. (2010). Aflatoxin M1 in raw milk in Croatia. *Food Control*, No. 21, pp. (1279–1281)

Blanco, J. L.; Domingues, L.; Gomes-Lucia, E.; Garayzabal, J. F. F.; Goyache, J.; and Suarez, J. (1988a). Behavior of aflatoxin during the manufacture, ripening and storage of Manchego-type cheese. *Journal of Food Science*, No. 53, pp. (1373-1376)

Blanco, J. L.; Dominguezs, L.; Gomez-Lucia, E.; Garayzabal, J. F. F.; Gracia, J. A.; & Suarez, G. (1988b). Presence of aflatoxin M_1 in commercial ultra-high temperature treated milk. *Applied Environmental Microbiology*, No. 56, pp. (1622–1623)

Blanco, J. L.; Carrion, B. A.; Liria, N.; diaz, S.;Garcia, M. E.; Domingues, L.; & Suarez, G. (1993). Behavior of aflatoxin during manufacture and storage yogurt. *Milchwissenchaft,* No. 48, pp. (385-387)

Blount, W.P.(1961). Turkey "X" disease. *Turkeys,* Vol. 9, No. 2, pp. (52, 55-58, 61,77)

Brackett, R. E.; & Marth, E. H. (1982a). Association of aflatoxin M1 with casein. Zeitschrift fur Lebensmittel untersuchung und forschung, No. 174, pp. (439-441)

Brackett, R. E.; & Marth, E. H. (1982b). Fate of aflatoxin M_1 in Cheddar cheese and in process cheese spread. *Journal of Food Protection,* No. 174, pp. (549-552)

Brackett, R. E.; & Marth, E. H. (1982c). Fate of aflatoxin M_1 in Cottage cheese. *Journal of Food Protection,* No. 45, pp. (903)

Brackett, R. E.; and Marth, E. H. (1982d). Fate of aflatoxin M_1 in parmesan and mozzarella cheese. *Journal of Food Protection,* No. 45, pp. (597-600)

Brackett, R. E.; Applebaum, R. S.; Wiseman, D. W.; & Marth, E. H. (1982). Fate of aflatoxin M1 in brick and limburger-like cheese. *Journal of Food Protection,* 45, pp. (553-556)

Cano-Sancho, G.; Marin, S.; Ramos, A. J.; Peris-Vicente, J.; & Sanchis, V. (2010). Occurrence ofaflatoxin M1 and exposure assessment in Catalonia (Spain). *Revista Iberoamericana de Micologia,* No. 3, pp. (130-135)

Celik, T. H.; Sarimehmetoglu, B.; Kuplulu, O. (2005). Aflatoxin M1 contamination in pasteurised milk. *Veterinarski Archiv,* Vol. 75, No. 1, pp. (57-65)

Chakrabarty, A. B. (1981). Detoxification of aflatoxin in Corn. *Journal of Food Protection,* No. 44, pp (173-176)

Choudhary, P. L.; Sharma, R. S.; & Borkartria, V. N. (1998). Effect of chilling and heating on aflatoxin M_1 content of contaminated Indian cow's milk. *Egyptian Journal of Dairy Science,* No. 26, pp. (223–229)

Codex Alimentarius Commissions .(2001). *Comments submitted on the draft maximum level for Aflatoxin M_1 in milk.* Codex committee on food additives and contaminants 33rd sessions, Hauge, The Netherlands.

Cohen, H.; Lapointe, M. (1981). High pressure liquid chromatography determination and fluorescence detection of aflatoxins in corn and dairy feeds. *Journal of the Association of Official Analytical Chemists,* No. 64, pp. (1372-1376)

Creppy, E. E. (2002): Update of survey, regulation and toxic effects of mycotoxins in Europe. *Toxicology Letters,* No.127, pp. (19-28)

Dashti, B.; Al-Hamli, S.; Alomirah, H.; Al-Zenki, S.; Bu Abbas, A.; Sawaya, W. (2009). Levels of aflatoxin M1 in milk, cheese consumed in Kuwait and occurrence of total aflatoxin in local and imported animal feed. *Food Control,* No. 20, pp. (686-690)

Decastelli, L.; Lai, J.; Gramaglia, M.; Monaco, A.; Nachtmann, C.; Oldano, F.; Ruffer, M.; Sezian, A.; Bandirola, C. (2007). Aflatoxins occurrence in milk and feed in Northern Italy during 2004–2005. *Food Control,* No. 18, pp. (1263–1266)

Deveci, O. (2007). Changes in the concentration of aflatoxin M1 during manufacture and storage of White Pickled cheese. *Food Control,* No. 9, pp. (1103-1107)

El Deeb, S. A. (1989). Interaction of aflatoxin M_1 on the physiological and morphological properties of *Lactobacillus bulgaricus* and *Streptococcus thermophilus. Alexandria Journal of Agriculture Research,* No. 34, pp. (61-72)

El Deeb, S. A.; Zaki, N.; Shoukry, Y. M. R.; & Kheadr, E. E. (1992). Effect of some technological processes on stability and distribution of aflatoxin M_1 in milk. *Egyptian Journal of Food Science,* No. 20, pp. (29-42)

El-Nezami, H.; Nicoletti, G.; Neal, G. E.; Donohue, D. C.; & Ahokas, J. T. (1995). Aflatoxin M1 in human breast milk samples from Victoria, Australia and Thailand. *Food andvChemical Toxicology,* No. 33, pp. (173–179)

Elzupir, A. O.; & Elhussein, A. M. (2010). Determination of aflatoxin M1 in dairy cattle milk in Khartoum State, Sudan. *Food Control,* No. 21, pp. (945-946)

Fallah, A. (2010). Aflatoxin M1 contamination in dairy products marketed in Iran during winter and summer. *Food Control,* No. 21, pp. (1478–1481)

Fallah, A. A.; Rahnama, M.; Jafari, T.; & Saei-Dehkordi, S. S. (2011). Seasonal variation of aflatoxin M_1 contamination in industrial and traditional Iranian dairy products. *Food Control,* In Press

Food and Agriculture Organization (1997). Worldwide regulations for mycotoxins, 1995. A compendium. FAO, Food and Nutrition Paper 64, Rome.

Fremy, J. M.; Roiland, J. C.; & Gaymard, A. (1990). Behavior of aXatoxin M1 during camembert cheese making. *Journal of Environmental Pathology, Toxicology and Oncology,* No. 10, pp. (95–98)

Galvano, F.; Galofaro, V.; & Galvano, G. (1996). Occurrrence and stability of aflatoxin M1 in milk and milk products: a worldwide review. *Journal of Food Protection,* No. 10, pp. (1079-1090)

Galvano, F.; Galofaro, V.; Ritieni, R.; Bognanno, M.; De Angelis, A.; & Galvano, G. (2001). Survey of the occurrence of aflatoxin M1 in dairy products marketed in Italy: second year of observation. *Food Additives and Contaminants,* No. 7, pp. (644-646)

Ghiasian, S. A.; Maghsood, A. H.; Neyestani, T. R.; Mirhendi, S. H. (2007). Occurrence of aflatoxin M_1 during the summer and winter seasons in Hamedan, Iran. *Journal of Food Safety,* No. 27, pp. (188-198)

Gong, Y.; Hounsa, A.; Egal, S.; Turner, P. C.; Sutcliffe, A. E.; Hall, A. J.; Cardwell, K.; & Wild, C. P. (2004). Postweaning exposure to aflatoxin results in impaired child growth: a longitudinal study in Benin, West Africa. *Environmental Health Perspectives,* No. 112, pp. (1334–1338)

Goto, T.; Peterson, S. W.; Ito, Y.; & *Wicklaw,* D. T.(1997). Mycotoxin producing ability of A. tamarii. *Mycotoxins,* No. 44, pp. (17-20)

Govaris, A.; Roussi, V.; Koidis, A.; & Botsoglou, N. A. (2001). Distribution and stability of aflatoxin M_1 during processing, ripening and storage of Telemes cheese. *Food Additives and Contaminants,* Vol. 18, No. 5, pp. (437-443)

Govaris, A.; Roussi, V.; Koudis, P. A.; & Botsoglou, N. A. (2002). Distribution and stability of aflatoxin M_1 during production and storage of yoghurt. *Food Additives and Contaminants,* No. 11, pp. (1043–1050)

Grant, D. W.; & Carlson, F. W. (1971). Partitioning behavior of aflatoxin M1 in dairy products. Bulletin of the Environmental Contamination and Toxicology, No. 6, pp. (521-524)

Gurbay, A.; Aydın, S.; Girgin, G.; Engin, A. B., & Sahin, G. (2006). Assessment of aflatoxin M1 levels in milk in Ankara, Turkey. *Food Controlo,* No. 17, pp. (1-4)

Haydar, M.; Benelli, L.; & Brera, C. (1990). Occurrence of aflatoxins in Syrian foods and foodstuffs: a preliminary study. Food Chemistry, No. 4, pp. (261-268)

Heshmati, A.; & Milani, J. M. (2010). Contamination of UHT milk by aflatoxin M1 in Iran. *Food Control,* No. 21, pp. (19-22)

Huseyin Oruc, H.; Cibik, R.; & Guns, E. (2007). Fate of aflatoxin M_1 in Kashar cheese. *Journal of Food Safety,* No. 27, pp. (82-90)

Hussain, I.; & Anwar, J. (2008). A study on contamination of aflatoxin M1 in raw milk in the Punjab province of Pakistan. *Food Control,* No. 19, pp. (293-295)

Hussain, I.; Anwar, J.; Asi, M. R.; Munawar, M. A.; & Kashif, M. (2010). Aflatoxin M1 contamination in milk from five dairy species in Pakistan. *Food Control,* No. 21, pp. (122-124)

IARC. (1993). Some naturally occurring substances - food items and constituents, heterocyclic aromatic amines and mycotoxins. *IARC Monogr Eval Carcinog Risk Hum,* Lyon, France, 56, pp. (245–391)

IARC. (2002). Some traditional herbal medicines, some mycotoxins, naphthalene and styrene. *IARC Monogr Eval Carcinog Risk Hum,*Lyon, France, pp. 1–556.

ISIRI, Institute of Standards and Industrial Research of Iran .(2002). Food and Feed-Mycotoxins Maximum Tolerated Level, Karaj, Iran. URL http://www.isiri.org/std/ 5925.pdf Accessed 23/12/2009.

Ito, Y.; Peterson, S. W.; Wicklaw, D. T.; & Goto T. (2001). Aspergillus pseudotamarii, a new aflatoxin producing sp. *Mycology Research,* No.105, pp. (233-239)

Jay, J. M. (1992). *Modern food microbiology,* pp. (1–701), Chapman and Hall, New York

Jose Barrios, M. ; Jesus Gualda, M. ; Cabanas, J. M. ; Medina, L. M. ; & Jordano, R. (1996). Occurrence of aflatoxin M_1 in cheeses from the South of Spain, *Journal of Food Protection,* No. 8, pp. (898-900)

Kamkar, A. (2005). A study on the occurrence of aflatoxin M1 in raw milk produced in Sarab city of Iran. *Food Control,* No. 16, pp. (593–599)

Kamkar, A. (2006). A study on the occurrence of aflatoxin M1 in Iranian Feta cheese. *Food Control,* No. 17, pp. (768–775)

Kamkar, A. (2007). Detection of aflatoxin M_1 in powdered milk samples by ELISA. Pajouhesh and Sazandegi, No .79, pp: (174-180)

Kamkar, A.; Karim, G.; Shojaee Aliabadi, F.; & Khaksar, R. (2008). Fate of aflatoxin M1 in Iranian white cheese processing. *Food and Chemical Toxicology,* No. 46, pp. (2236-2238)

Kaniou-Grigoriadou, I.; Eleftheriadou, A.; Mouratidou, T.; & Katikou, P. (2005). Determination of aflatoxin M1 in ewe's milk samples and the produced curd and Feta cheese. *Food Control,* No. 16, pp. (257–261)

Karaioannoglu, P. G. ; Mantia, A. ; Koufidia, D. ; Koidia, P. ; & Triantafillou, J. (1989). Occurrence of aflatoxin M_1 in raw and pasteurized milk and in Feta and Teleme cheese samples. *Milchwissenschaft,* No. 44, pp. (746-748)

Karimi, Gh.; Hassanzadeh, M.; Teimuri, M; Naziri, F.; & Nili, A. (2007). Aflatoxin M_1 Contamination in Pasteurized Milk in Mashhad, Iran. *Iranian Journal of Pharmaceutical Sciences,* Vol. 3, No. 3, pp. (153-156)

Kawamura, O.; Wang, D. S.; Liang, Y. X.; Hasegawa, A.; Saga, C.; Visconti, A.; & Ueno, Y. (1994). Further survey of aflatoxin M_1 in milk powders by ELISA. *Food and Agricultural Immunology,* No. 6, pp. (465-467)

Kiermeier, F. (1973). Aflatoxins residues in fluid milk. *Pure Applied Microbiology,* No. 35, pp. (271-273)

Kiermier, F. ; & Buchner, M. (1977). Verhalten von aflatoxin M1 wahrend der reifung and lugerung von kase. Zeitschrift fur Lebensmittel untersuchung und forschung, No. 164, pp. (87-91)

Kiermier, F.; & Meshaley, R. (1977). Einfluss der molkereitechnischen behanlung der rohmilich auf des aflatoxin M_1. Gehalt der daraus hergestellten produkte. Zeitschrift fur Lebensmittel untersuchung und forschung, No. 164, pp. (183-187)

Kurtzman, C. P.; Horn, B. W. & Hesseltine, C. (1987). *A nominus*, a new aflatoxin producing species related to *A. flavus* and *A. Tamarii. Antonnie Van Leewenhoek,* No. 53, pp. (147-158)

Lafont, P.; Siriwardana, M.; & Lafont, J. (1989). Genotoxicity of hydroxy-aflatoxins M1 and M4. *Microbiology Alimentarius Nutrition,* NO. 7, pp (1-8)

Lieu, F. Y.; Bullerman, L. B. (1977). Production and stability of aflatoxin, penicillic acid and patulin in several substrates. *Journal of Food Science,* No. 42, pp. (1222-1224)

Lopez, C.; Ramos, L.; Ramadan, S.; Bulacio, L.; & Perez, J. (2001). Distribution of aflatoxin M1 in cheese obtained from milk artificially contaminated. *International Journal of Food Microbiology,* No. 64, pp. (211–215)

Lopez, C. E. L.; Ramos, L. L.; Ramadan, S. S.; & Bulacio, L. C. (2003). Presence of aflatoxin M_1 in milk for human consumption in Argentina. *Food Control,* No. 14, pp. (31-34)

Margolles, E.; Escobar, A.; & Acosta, A. (1992). Aflatoxi B_1 residuality determination directly in milk by ELISA. *Revista de Salud Animal,* No. 12, pp. (35-38)

Marshaly, R. I.; Deeb, S. A.; & Safwat, N. M. (1986). Distribution and stability of aflatoxin M_1 during processing and storage of Karish cheese. *Alexandria Journal of Agriculture Research,* No. 31, pp. (219-228)

Marsi, M. S.; Booth, A. N.; & Hsieh, D. P. H. (1974). Comparative metabolic conversation of aflatoxin B1 in aflatoxin M1 and Q1. Life Science, No. 15, pp (203-209)

Martins, M. L.; & Martins, H. M. (2000). Aflatoxin M1 in raw and ultra high temperature-treated milk commercialized in Portugal. *Food Additives and Contaminants,* Vol. 17, No. 10, pp. (871-874)

Maryamma, K. L.; Rajan, A.; Gangadharan, B.; Ismail, P. K.; Valsalad, & Manomohan, C. B. (1990). Reduction of aflatoxin in milk by fermentation into curd. *Journal of Veterinary and Animal Sciences,* No. 21, pp. (102-107)

McKinney, J. D.; Cavanaugh, G. C.; Bell, G. T.; Bell, A. S.; Hoversland, A. S.; Nelson, D. M.; Pearson, J.; & Silkirk R. J. (1973). Effects of ammoniation on aflatoxins in rations fed lactating cows. *Journal of American Oil Chemical Society,* No. 50, pp. (79-84)

Megalla, S. E.; & Hafez, A. H. (1982). Detoxification of aflatoxin M_1 by acidogenous yogurt. *Mycopathologia,* No. 77, pp. (89-91)

Mohammadi, H.; Alizadeh, M.; R Bari, M.; Tajik, H. (2008). Minimization of aflatoxin M_1 content in Iranian white brine cheese. *International Journal of Dairy Technology,* No. 2, pp. (141-145)

Mohamadi, H.; Alizadeh, M.; Rahimi, J.;& Qasri, Sh. (2010). Assessment of aflatoxin M1 levels in selected dairy products in north-western Iran. *International Journal of Dairy Technology,* No. 2, pp.(262-265)

Mohammadian, B.; Khezri, M.; Ghasemipour, N.; Mafakheri, Sh.; & Poorghafour Langroudi, P. (2010). Aflatoxin M_1 contamination of raw and pasteurized milk produced in Sanandaj, Iran. *Archives of Razi Institute,* Vol. 65, No. 2, pp. (99-104)

Moreau, C. (1976). Les mycotoxines dans les produits laitiers. Lait, No. 55, pp. (286-303)

Movassagh Ghazani, M. H. (2009). Aflatoxin M_1 contamination in pasteurized milk in Tabriz (northwest of Iran). *Food and Chemical Toxicology,* No. 7, pp. (1624-1625)

Movassagh Ghazani, M. H. (2011). Presence of aflatoxin M1 in UHT milk in Tabriz (Northwest of Iran). *Journal of Food Safety,* No. 2, pp. (238-241)

Munksgaard, L.; Larsen, L.; Werner, H.; Andersen, P. E.; & Viuf, B. T. (1987). Carryover of aflatoxin from cows' feed to milk and milk products. *Milchwissenschaft,* No. 42, pp. (165-167)

Nikov, P. S.; Bukharbaeva, A. S.; Baimbetova, A. M.; & Amireeva, N. T. (1991). Contamination with aflatoxins of milk and milk products in the South-East of Kazakhstan. *Voprosy Pitaniya,* No. 5, pp. (72-74)

Oliveira, C. A. F.; & Ferraz, J.C.O. (2007). Occurrence of aflatoxin M1 in pasteurised, UHT milk and milk powder from goat origin. *Food Control,* No. 4, pp. 375-378

Oveisi, M.; Jannat, B.; Sadeghi, N.; Hajimahmoodi, M.; & Nikzad, A. (2007). Presence of aflatoxin M1 in milk and infant milk products in Tehran, Iran. *Food Control,* No. 18, pp. (1216 - 1218)

Pohland, A. E.; & Yess, N. J. (1992). Food Contaminants: sientific and public health implications. *Proceedings of the Nutrition Society of Australia,* No. 17, pp. (1-12)

Polychronaki, N.; West, R. W.; Turner, P. C.; Amra, A.; Abdel-Wahhab, M; Mykkannen, H.; & El-Nezami, H. (2007). A longitudinal assessment of aflatoxin M1 excretion in breast milk of selected Egyptian mothers. *Food and Chemical Toxicology,* No. 45, pp. (1210–1215)

Polovinski-Horvatović, M.; Juric, V.; Glamocic, D. (2009). Tow year study of incidence of aflatoxin M1 in milk in the region of Serbia. *Biotechnology in Animal Husbandry,* No. 5-6, pp. (713-718)

Purchase, I. F. H., & Steyn, M.; Rinsma, R.; & Tustin, R. C. (1972). Reduction of the aflatoxin M_1 in milk by processing. *Food Cosmetics Toxicolology,* No. 10, pp. (383-387)

Purchase, I. F. H. (1973). The control of aflatoxin residues in food of animal origion. *Pure and Applied Chemistry,* No. 35, pp. (283-289)

Rahimi, E.; Bonyadian, M.; Rafei, M.; & Kazemeini, H. R. (2010). Occurrence of aflatoxin M1 in raw milk of five dairy species in Ahvaz, Iran. *Food and Chemical Toxicology,* No. 48, pp. (129–131)

Rajan, A.; Ismail, P. K.; & Radhakrishnan, V. (1995). Survey of milk samples for aflatoxin M_1 in Thrissur, Kerala. *Indian Journal of Dairy Science,* No. 48, pp. (302-305)

Rasic, J. L; Skrinjar, M.; & Markov, S. (1991). Decrease of aflatoxin M_1 in yogurt and acidified milks. *Mycopathologia,* No. 113, pp. (117-119)

Sefidgar, S. A. A. ; Gholampour Azizi, V.; Mirzai, M.; Azimi, H.; & Norozi AbolHashemi, M. (2006). Survey of the Occurrence of Aflatoxin M_1 in Pasteurized Milk Consumption, Babol University of Medical Sciences, Iran

Sharma, R. S.; Trivedi, K. R.; Wadodkar, U. R.; Murthy, T. N.; & Punjarath J. S. (1994). Aflatoxin B1 content in deoiled cakes, cattle feeds and damaged grains during different seasons in India. *Journal of Food Science and Technology,* Vol. 31, No. 3, pp (244-246)

Shundo, L.; & Sabino, S. (2006). Aflatoxin M_1 in Milk by Immunoaffinity Column Cleanup with TLC/HPLC determination. *Brazilian Journal of Microbiology,* No. 37, pp. (164-167)

Srivastava, V. P.; Bu-Abbas, A., Basuny, A.; Al-Johar, W.; Al-Mufti, S.; & Sidiqui, M. K. J. (2001). Aflatoxin M_1 contamination in commercial samples of milk and dairy products in Kuwait. *Food Additives and Contaminants,* Vol. 18, No. 11, pp. (993-997)

Stoloff, L. (1980). Aflatoxin M_1 in perspective. *Journal of Food Protection,* No. 43, pp. (226-230)

Stoloff, L.; Park, D. L.; & Van Egmond, H. P. (1991). Rationales for the establishments of limits and regulations for mycotoxins. *Food Additives and Contaminants,* No. 8, pp. (213-221)

Stoloff, L.; Wood, G.; & Carter, G. (1981). Aflatoxin M_1 in manufactured dairy products produced in the United States in 1979. *Journal of Dairy Science,* No. 64, pp. (2426-2430)

Stubblefield, R. D.; & Shannon, G. M. (1974). Aflatoxin M1: analysis in dairy products and distribution in dairy foods made from artificially contaminated milk. Journal of the Association of Official Analytical Chemists, No. 57, pp. (847-851)

Tabata, S. ; Kamimura, H. ; Ibe, A.; Hashimoto, H. ; Idia, M.; Tamura, Y.; & Nishima, T. (1993). Aflatoxin contamination in foods and foodstuffs in Tokyo: 1986-1990. *Journal of AOAC International,* No. 76, pp. (32-35)

Taguchi, S. ; Fukushia, S. ; Sumimoto, T.; Yoshida, S. ; & Nishimune, T. (1995). Aflatoxins in food collected in Osaka, Japan, from 1988 to 1992. *Journal of AOAC International,* No. 78, pp. (325-327)

Tajkarimi, M.; Shojaee Aliabadi, F.; Salah Nejad, M.; Pursoltani, H.; Motallebi, A. A.; Mahdavi, H. (2007). Seasonal study of aflatoxin M_1 contamination in milk in five regions in Iran. *International Journal of Food Microbiology,* No. 116, pp. (346 - 349)

Tajkarimi, M.; Aliabadi-Sh, F.; Salah Nejad, A.; H. Poursoltani, H.; Motallebi, A. A.; Mahdavi, H. (2008). Aflatoxin M_1 contamination in winter and summer milk in 14 states in Iran. *Food Control,* No. 19, pp. (1033-1036)

Tekinsen, K. K.; & Ucar, G. (2008). Aflatoxin M_1 levels in butter and cream cheese consumed in Turkey. *Food Control,* No. 1, pp. (27-30)

Van Egmond, H. P. (1983). Mycotoxins in dairy products. *Food Chemistry,* No. 11, pp. (289-307)

Van Egmond, H. P. (1989). Aflatoxin M_1 : occurrence, toxicity, regulation, In: *Mycotoxins in Dairy Products,* Van Egmond, H. P., pp. (11- 55), Elsevier Applied Science, New York

Van Egmond, H. P.; & Paulsch, W. E. (1986). Mycotoxins in milk and milk products. *Nethelands Milk Dairy Journal,* No. 40, pp. (175-188)

Van Egmond, H. P.; & Paulsch, W. E.; Veringa, H. A.; & Schuller, P. E. (1977). The effect of processing on the aflatoxin M_1 content of milk and milk products. *Archive Institue, Pasteur Tunis,* No. 54, pp. (381-390)

Virdis, S.; Corgiolu, G.; Scarano, C.; Pilo, A. L.; & De Santis, E.P.L. (2008). Occurrence of aflatoxin M_1 in tank bulk goat milk and ripened goat cheese, Food Control, No. 1, pp. (44-49)

Wiseman, D. W.; & marth, E. H. (1983). Heat and acid stability of aflatoxin M_1 in naturally and artifcially contaminated milk. *Milchwissenschaft,* No. 38, pp. (464-466)

Wiseman, D. W. ; Applebaum, R. S. ; Brackett, R. E. ; & Marth, E. H. (1983). Distribution and resistance to pasteurisation of aflatoxin M1 in naturally contaminated whole milk, cream and skim milk. *Journal of Food Protection,* Vol. 46, No. 6, pp. (530-532)

Yaroglu, T.; Oruc, H. H., Tayar, M. (2005). Aflatoxin M1 levels in cheese samples from some provinces of Turkey. *Food Control,* No. 16, pp. (883–885)

Yousef, A. E., & Marth, E. H. (1989). Stability and degradation of aflatoxin M1. In H.P. Van Egmond, *Mycotoxins in dairy products,* pp. (127-161), London: Elsevier.

20

Aflatoxins: Contamination, Analysis and Control

Giniani Carla Dors et al.*
Instituto Federal Catarinense (IFC)
Brazil

1. Introduction

Aflatoxins are toxic metabolites produced by different species of toxigenic fungi, called mycotoxins. Humans can be exposed to aflatoxins by the periodic consumption of contaminated food, contributing to an increase in nutritional deficiencies, immunosuppresion and hepatocellular carcinoma (Wagacha & Muthomi, 2008).

Aflatoxins (AFs) have a wide occurrence in different kind of matrices, such as spices, cereals, oils, fruits, vegetables, milk, meat, etc. Among the 18 different types of aflatoxins identified, the major members are aflatoxin B1 (AFB1), B2 (AFB2), G1 (AFG1), G2 (AFG2), M1 (AFM1) and M2 (AFM2) which are produced by *Aspergillus flavus* and/or *Aspergillus parasiticus*. Strains of *A. flavus* can vary from non-toxic to highly toxigenic and are more likely to produce AFB1 than AFG1. Strains of *A. parasiticus* generally have less variation in toxigenicity and produce AFB1 and varying amounts of AFB2, AFG1 and AFG2 (Coppock & Christian, 2007).

Other fungi have been described in the literature as aflatoxins' producers such as *A. bombycis, A. ochraceoroseus* and *A. pseudotamari* (Klich et al, 2000; Mishra & Das, 2003). *A. flavus* and *A. fumigatus* have also been identified as pathogens for animals and humans (Zain, 2011).

The order of acute and chronic toxicity is AFB1 > AFG1 > AFB2 > AFG2, reflecting the role played by epoxidation of the 8,9-double bond and also the greater potency associated with the cyclopentenone ring of the B series, when compared with the six-membered lactone ring of the G series. AFM1 and AFM2 are hydroxylated forms of AFB1 and AFB2 (Mclean & Dutton, 1995; Wogan, 1966).

In the primary fungi metabolism a lot of interrelated reactions catalyzed by enzymes occur, with the objective of promoting energy and primary metabolites (synthetic intermediates and macromolecules), ensuring the growth and reproduction of fungi. Secondary

* Sergiane Souza Caldas[2], Vivian Feddern[3], Renata Heidtmann Bemvenuti[2], Helen Cristina dos Santos Hackbart[2], Michele Moraes de Souza[2], Melissa dos Santos Oliveira[4], Jaqueline Garda-Buffon[2], Ednei Gilberto Primel[2] and Eliana Badiale-Furlong[2]
1 Instituto Federal Catarinense – Campus Concórdia (IFC), Brazil
2 Federal University of Rio Grande (FURG), Brazil
3 Embrapa Swine and Poultry, Brazil
4 Instituto Federal Farroupilha – Campus Santo Augusto (IFF), Brazil

metabolites are synthesized by a variety of routes from primary metabolites (Obrian et al., 2003; Ueno, 1986; Wild & Montesano, 2009). The biosynthesis of aflatoxins, as all secondary metabolites, is strongly dependent on growth conditions such as substrate composition or physical factors such as pH, water activity, temperature or modified atmospheres. Depending on the particular combination of external growth parameters the biosynthesis of aflatoxin can either be completely inhibited, albeit normal growth is still possible or the biosynthesis pathway can be fully activated. Knowledge about these relationships enables an assessment of which parameter combinations can control aflatoxin biosynthesis. The biochemical correlation between aflatoxin production and oxidative stress suggest that the latter is a prerequisite for aflatoxin synthesis (Ellis et al., 1993; Giorni et al., 2008; Luchese & Harrigan, 1993; Molina & Giannuzzi, 2002; Ribeiro et al., 2006; Schmidt-Heydt et al., 2009).
The chapter gives a section on aflatoxin analysis, its occurrence in food and feed as well as its control, once aflatoxin is the major mycotoxin studied and thus is of great concern for human and animal's health due to its carcinogenic, mutagenic, teratogenic and immunosupressive effects.

2. Factors promoting contamination in aflatoxins and occurrence

Fungi which grow and produce toxins in grains during storage are influenced by factors related to inadequate moisture and temperature, combined with long residence time in warehouses, which are stressful situations and originate toxigenic potential outbreak (Dilkin, 2002).
The most important factors that help predict the occurrence of aflatoxins in food include weather conditions (temperature and atmospheric humidity), agronomical practices (crop rotation and soil cultivation) and internal factors of the food chain (drying and storage conditions). A comprehensive approach is needed to identify and control risks related to food production system that could present a potential hazard to human health, being necessary to identify emerging risks which may include "newly" identified risks, not previously observed risks in human or animal food chain as well as known risks. The emerging risks need to be identified as early as possible in order to take appropriate preventive measures. Thus, the specific risk can be prevented from becoming a danger (Van der Fels-Klers et al., 2008).
Several groups of researchers from the European Union reached a consensus on the most important indicators, based on three stages in food production chain. For cultivation stage the selected indicators were: relative humidity, temperature, crop rotation, tillage practices and water activity of seeds. For transportation and storage the following factors were included: water activity, relative humidity, ventilation, temperature, storage capacity and logistics. For processing the indicators were: data quality, the fraction of grain used, the water activity of seeds, implanted traceability and system quality (Park & Bos, 2007; Van der Fels-Klers et al., 2010).
According to Park & Bos (2007) and Marvin et al. (2009a) to anticipate emerging risks models are developed to assess the risk from the indicators identified. The next step is to identify the sources of information for these indicators, such as climate change (changes in temperature and rainfall), market and consumer trends (crop demand, price and production) and market research (economics, as inflation and taxes) global trade (import and export data and trade barriers), transportation (strikes and transport company records), technology (covers of scientific journals), prevalence of pests, changes in legislation (registration of pesticides). The risk categories within each of the selected indicators should be defined for each specific food.

Among the models currently available in the literature to predict the occurrence of fungi and mycotoxins, are the meteorological indicators in combination with agricultural information. With respect to management strategies, monitoring and prevention are the main indicators derived from the food chain (Dekkers et al., 2008). It is important to stand out the potential interactions among indicators which should be taken into account, for example: between relative humidity and temperature during cultivation; among storage conditions and drying and finally, between crop rotation and management policies (Van der Fels-Klers, 2010; Marvin et al., 2009b).

Due to the great health concern in relation to mycotoxin contaminated food ingestion, studies are being conducted worldwide to verify the occurrence of aflatoxins. The main food products susceptible to fungal growth and consequently to mycotoxins' production, include peanuts (raw, roasted, sweet and infrosted), corn (popcorn, hominy and grains), wheat, rice, nut, walnuts, hazelnuts, cashews, almonds, dried fruits, spices, cotton seed, cassava, vegetable oils, cocoa and others that are normally used in the composition of foods and feeds. Thus, animals are also subjected to aflatoxin contamination, and when meat and milk from these animals are ingested, human contamination may also occur (Kwiatkowski & Alves, 2007). Mycotoxins importance relies on harm caused to human and animal health, besides economical losses in agriculture (Amaral et al., 2006).

Rubert et al. (2010) evaluated a total of 22 samples obtained from a local supermarket (10 samples of malt, 7 samples of coffee and 5 samples of instant-based cereal-breakfast beverage). Four samples of the total malt samples were positive for AFG2 and AFG1, and traces of AFB1 and AFB2 were detected. Khayoon et al. (2010) verified the occurrence of AFB1, B2, G1 and G2 in 42 animal feeds, comprising corn (16), soybean meal (8), mixed meal (13), sunflower, wheat, canola, palm kernel, copra meals (1 each). The results showed that eight samples (19%) were contaminated with aflatoxins, ranging from 6.5 to 101.9 ng g^{-1}. Ibáñez-Vea et al. (2011) evaluated AFG2, AFG1 and ZEA mycotoxins in 20 barley samples. All of the samples analyzed presented levels of AFB1 above its LOD, but only 5 (25%) presented quantifiable levels (>LOQ), with 0.173 µg kg^{-1} and 0.185 µg kg^{-1} being the mean of the positive values and the maximum level found, respectively. Reiter et al. (2010) evaluated eighty-one rice samples purchased from different markets. The results revealed that AFB1 (0.45 to 9.86 µg kg^{-1}) could be quantified in 15 samples and AFB2 (1.5 µg kg^{-1}) in one sample. Matumba et al. (2010) investigated aflatoxins in sorghum grain and malt samples, traditional opaque sweet beverage (*thobwa*) and beer prepared from sorghum malts. All malt and beer samples, 15% and 43% of the sorghum and *thobwa* samples, respectively, were contaminated. The sorghum malt prepared for beer brewing, had a significantly ($p < 0.01$) total aflatoxin content (average 408 ± 68 µg kg^{-1}) than any other type of sample. Dors et al. (2011) conducted a survey of mycotoxins in parboiled and whole rice. From the samples analyzed, 9% were contaminated with AFB1 in levels ranging from 11 to 74 µg kg^{-1}. Coelho et al. (1999) studied aflatoxin and ochratoxin A migration during rice parboiling process under different conditions of soaking, autoclaving and drying. It was noted that there was mycotoxin migration from the husk to the starchy endosperm in the following proportions: 32% AFB1, 44% AFB2, 36% AFG1 and 22% AFG2. Dors et al. (2009) assessed mycotoxin migration to the starchy endosperm during the parboiling process and the results showed a lower trend of migration from AFB1 in 6 h soaking and 30 min autoclaving.

Amaral et al. (2006) examined 123 samples of food products based on corn and corn grain, of which 16 were positive with levels of 0.78 µg kg^{-1}. Ramos et al. (2008) detected the presence of *Aspergillus* spp. and aflatoxin contamination grain samples (12) analyzed and this result was correlated with the greatest amount of rain during harvest. Levels of contamination ranged from "not detected" (nd) to 277.8 μg kg^{-1}, for AFB1; from 0.7 to 14 μg kg^{-1} for AFB2; and from nd to 34.1 μg kg^{-1} for AFG2. Oliveira et al. (2010) found aflatoxin contamination in 70% of maize samples from criollo varieties, which have not undergone genetic intervention, at levels ranging from 1 to 2.6 μg kg^{-1}.

Almeida et al. (2009) collected 80 samples of maize for poultry feed in two feed mills, from these samples 10% were contaminated with levels varying from 1 to 5 mg kg^{-1}. Marques (2007) analyzed 47 samples of corn grits for animal consumption and 46 were positive for aflatoxin with a maximum of 50 μg kg^{-1}. D'Angelo et al. (2007) reported injury in calves for veal production that had a corn-based diet. The toxicological analysis of corn-based feed revealed contamination in the following levels: 1400 μg kg^{-1} AFB1, 120 μg kg^{-1} AFB2, 80 μg kg^{-1} AFG1 and 70 μg kg^{-1} AFG2. In the liver of three animals were found levels of total aflatoxins of 0.1, 0.3 and 0.6 μg kg^{-1}. Velazquez et al. (2009) analyzed 40 samples of feed for dairy cattle and 92% of them were contaminated with aflatoxins at levels between 4.82 a 2.89 μg Kg^{-1}.

Most of AFB1 and AFB2 ingested by mammals is eliminated through urine and faeces, however a fraction is biotransformed in the liver and excreted together with milk in the form of aflatoxins AFM1 and AFM2, respectively. AFM1 could be detected in milk 12-24 h after the first AFB1 ingestion, reaching a high level after a few days. The ratio between AFB1 ingested and AFM1 excreted has been estimated to be 1-3%. One of the most used treatments for milk processing is heating, however, AFM1 is resistant to any thermal treatment (Carvajal et al., 2003; Park, 2002; Van Egmond, 1989).

Rahimi et al. (2010) analyzed 311 samples of raw milk from cow, water buffalo, camel, sheep, and goat. AFM1 was found in 42.1% of the samples by average concentration of 43.3 ± 43.8 ng kg^{-1}. The incidence rates of AFM1 in raw cow, water buffalo, camel, sheep, and goat milks were, 78.7%, 38.7%, 12.5%, 37.3%, and 27.1%, respectively. Fallah (2010) investigated the occurrence of AFM1 in 225 commercial liquid milk samples composed of pasteurized milk (116 samples) and UHT milk (109 samples). AFM1 was detected in 151 (67.1%) samples, consisted of 83 (71.5%) pasteurized milk samples (mean: 52.8 ng L^{-1} ; range: 5.8–528.5 ng L^{-1}) and 68 (62.3%) UHT milk samples (mean: 46.4 ng L^{-1}; range: 5.6–515.9 ng L^{-1}).

Heshmati and Milani (2010) verified the levels of AFM1 in UHT milk samples. Two hundred and ten UHT milk samples were obtained from supermarkets in Tehran, Iran. AFM1 was found in 116 (55.2%) of 210 UHT milk samples examined. The levels of AFM1 in 70 (33.3%) samples were higher than the maximum tolerance limit (0.05 μg L^{-1}) accepted by some European countries while none of the samples exceeded the prescribed limit of US regulations. The same authors also studied AFM1 levels of 61 milk samples delivered from small milking farms. The maximum mean concentrations of AFM1 recorded in winter-spring season were in the range of 35.8–58.6 ng L^{-1} and in summer–autumn season in the range of 11.6–14.9 ng L^{-1}.

Cano-Sancho et al. (2010) found AFM1 occurrence in the main dairy products consumed, that is 94.4% (68/72) of whole UHT milk samples, in 2.8% (2/72) of yoghurt samples, but was not detected in cheese. The maximum level was detected in one yoghurt sample with 51.58 ng kg^{-1}. Martins & Martins (2004) determined the occurrence of AFM1 in 96 yoghurt samples, being 48 of them natural and 48 added by strawberry pieces. The results showed that 18.8% of the samples were contaminated with AFM1, being 2 samples of natural yoghurt (0.043 and 0.045 ug L^{-1}) and 16 from fruit added yoghurt (0.019 and 0.098 ug L^{-1}). Khoury et al. (2011) investigated the presence and levels of AFM1 in 138 dairy samples (milk and yogurt). Results obtained showed that AFM1 was found in 40.62 % and 32.81 % of milk and yogurt samples respectively. Fallah et al. (2009) studied 210 cheese samples composed of white cheese (116 samples) and cream cheese (94 samples). AFM1 at measurable level (50 ng kg^{-1}) was detected in 161 (76.6%) samples, consisting of 93 (80.1%) white and 68 (72.3%) cream cheese samples.

Dashti et al. (2009) evaluated a total of 321 milk samples (177 fresh, 105 long-life, 27 powdered milk and 12 human milk), 40 cheese samples and 84 feed samples were analyzed for AFM1. Results showed that all fresh milk samples except one were contaminated with AFM1 ranging from 4.9 to 68.7 ng kg^{-1}, for the long-life milk samples were below the detection limit to (88.8 ng kg^{-1}) while in powdered milk samples ranged from 2.04 to 4.14 ng kg^{-1}. From human milk samples, only five were contaminated, with levels ranging from 8.83 to 15.2 ng kg^{-1}. The cheese samples recorded 80% contamination with AFM1 with a range of 23.8–452 ng kg^{-1}. Manetta et al. (2009) investigated samples of whey, curd and a typical hard and long maturing cheese such as Grana Padano produced with naturally contaminated milk in a range of 30–98 ng kg^{-1}. Experimental results showed that, in comparison to milk, AFM1 concentration levels increased both in curd (3-fold) and in long maturing cheese (4.5-fold), while AFM1 occurrence in whey decreased by 40%. Under review done by Montagna et al. (2008), there is an increase in aflatoxin M1 concentration as cheese ripening stage progresses, due to water loss and the consequent concentration of substances present.

Sassahara et al. (2005) collected 98 feed and 42 raw milk samples and the results showed that there was contamination by AFM1 in 26% commercial feed samples, besides 53% of feed samples prepared at the farm and in 100% of corn samples used in animal nutrition. As a result of this aflatoxin incidence in animal diet, milk showed 24% contamination in the collected samples. Romero et al. (2010) evaluated the presence of AFM1 in human urine samples from a specific Brazilian population, as well as in corn, peanut, and milk consumption measured by two types of food inquiry. A total of 69 samples were analyzed and 45 of them (65%) presented contaminations ≥1.8 pg mL^{-1}, which was the limit of quantification (LOQ). Seventy eight percent ($n = 54$) of the samples presented detectable concentrations of AFM1 (>0.6 pg mL^{-1}). The AFM1 concentration among samples above LOQ ranged from 1.8 to 39.9 pg mL^{-1}. There were differences in food consumption profile among donors, although no association was found between food consumption and AFM1 concentration in urine. The high frequency of positive samples suggests exposure to aflatoxins by the studied population.

Aflatoxins are found in maize and peanuts, as well as in tree nuts and dried fruits (Zain, 2011). Nakai et al. (2008) evaluated the mycoflora and occurrence of aflatoxins in stored peanut samples (hulls and kernels). Analysis of hulls showed that 6.7% of the samples were contaminated with AFB1 and AFB2; in kernels, 33.3% of the samples were contaminated with AFB1 and 28.3% with AFB2. Analysis of the toxigenic potential revealed that 93.8% of the *A. flavus* strains isolated were producers of AFB1 and AFB2. Shenasi et al. (2002) detected aflatoxins in 12% of the samples at twenty-five varieties of dates (*Phoenix dactylifera*) although aflatoxigenic Aspergillus were detected in 40% of the varieties examined. Bircan (2009) tested aflatoxin contamination in 98 dried figs analyzed for OTA to determine the co-occurrence of both toxins. Seven samples were confirmed aflatoxin positive, in the range of 0.23–4.28 ng g^{-1} and only 2 samples contained both toxins, with a maximum concentration of 24.37 ng g^{-1} for OTA and 1.02 ng g^{-1} for AFB1.

More recently, Herzallah et al. (2009) studied aflatoxin contamination in meat products collected in 5 different months. The AFB1, AFB2, AFG1 and AFG2 contents in the analysed food products ranged from 1.10 to 8.32 $\mu g.L^{-1}$ and 0.15 to 6.36 $\mu g.L^{-1}$ in imported and fresh meat samples collected during March, respectively.
Fruits and vegetables do not appear to be of major concern as possible sources of mycotoxin contamination in food and feeds because they were only listed as minor sources in a statement of the Institute of Food Science and Technology Trust Fund (2006). Major sources on the list included mold damaged foodstuffs, specifically cereals and oilseeds.
FAO has done a lot of work on mycotoxins in developing countries, although economic dimensions are rarely observed. In horticultural crops, mycotoxins are primarily associated with dried fruits (figs and prunes), certain processed products (apple and grape juice) and are probably in apples and grapes (Dombrink-Kurtzman, 2008).
Although a large number of different mycotoxins exist, there are only a few of them that are regularly found in foods. Most reports concerning aflatoxin formation on fruits refer to figs or citrus fruits (Drusch & Ragab, 2003). Aflatoxins constitute a problem that is already present in the orchard. Little contamination occurs when firm, ripe fruits are dried immediately (Steiner et al., 1988). From a practical point of view, the best approach for eliminating mycotoxins from foods is to prevent mold growth at all levels of production, including harvesting, transport, and storage (Boutrif, 1998).
Thus, the occurrence of fungi and mycotoxins can be controlled by applying a number of preventive measures both before and after harvest, including insect control, good harvesting, drying, and storage practices. If mycotoxin contamination has occurred, the levels of toxins can be reduced by physical, chemical or biological decontamination. Milling, food processing, and regulatory control of toxins to safety levels can also have a positive impact on food safety (Trucksess & Diaz-Amigo, 2011).

3. Sampling, measurement and analysis

3.1 Sample preparation

Since AFs are inhomogeneous distributed in food and feed, high-contaminated hotspots can occur. Thus, sampling is an important step in the analysis of contaminated food and feed (Reiter et al., 2009).
Relating to the sample preparation techniques used in the last years, liquid-solid extraction has been widely employed. Usually the procedure consists of weighing a mass of the homogenized sample, add the extractor solvent and agitate in a shaker. Commonly, after these steps, filtration is carried out. In these extractions different volumes and solvent kinds were employed. Solvent volumes ranging from 20 to 250 mL and composed mainly of methanol/water or acetonitrile/water have been used. The choice for the best extraction solvent is directly related to the extraction efficiency and the number of co-extractives that this solvent extracts. In the work developed by Capriotti et al. (2010) the authors compared the use of methanol, acetonitrile and acetone for mycotoxins' extraction from cereals, being observed the highest recovery for the analytes in the acetone solution.
Another tool that has been employed during extraction is the ultrasound assisted extraction (Amate et al., 2010; Bacaloni et al., 2008; Capriotti et al., 2010; Quinto et al., 2009). Ultrasound is a simple and versatile method because it aggressively agitates the solution system improving transfer from the cell into the solvent. Bacaloni et al. (2008) employed ultrasound extraction and compared the technique with matrix solid-phase dispersion (MSPD) and homogenization. Recoveries comparable to those obtained with the

homogenization method were achieved with a sonication time of 10 min. The authors concluded that the employment of ultrasound is time-saving because it is easy to handle and many samples can be treated at the same time. Besides, ultrasonic extraction may be an efficient, safe and reliable alternative to homogenization and MSPD extractions.

MSPD technique has been employed for aflatoxins' extraction in food samples (Cavaliere et al., 2007; Rubert et al., 2010; Sebastià et al., 2010). MSPD involves the homogenization of the sample together with a suitable sorbent (usually octadecylsilica) using a pestle and mortar. The solid mixture is transferred to a cartridge and after, the aflatoxins are eluted and determined. Rubert et al. (2010) extracted the aflatoxins AFB1, AFB2, AFG1 and AFG2 from cereal using 1 g sample, 1 g C18 and 10 mL acetonitrile for the elution from the cartridges. Recoveries were reported to be between 64 and 91%, and limits of quantification of 1 $\mu g\ kg^{-1}$ were reached.

Pressurized fluid extraction (PFE), with trade name of accelerated solvent extraction (ASE) has also been employed for aflatoxins' extraction (Sheibani & Ghaziaskar, 2009; Desmarchelier et al., 2010). This technique employs solvents at elevated pressures and temperatures to achieve complete extraction of analytes from solid and semi-solid samples with lower solvent volumes and shorter extraction times (Sheibani & Ghaziaskar, 2009).

The accelerated extraction solvent was compared to QuEChERS procedure (acronym name for Quick, Easy, Cheap, Effective, Rugged, and Safe) for extraction of mycotoxins including aflatoxins from food samples in the study developed by Desmarchelier et al. (2010). Both methods showed high extraction efficiency in a broad range of cereal-based products and with a comparable sensitivity. Nevertheless, the easiness-to-handle of these extraction methods was definitely in favor of the QuEChERS-like procedure, avoiding any tedious preparation of extraction cells, requiring less reagents and glassware and involving less intermediate steps. Consequently, a higher sample throughput was possible, with up to 40 individual samples extracted over one working day as compared to the 24 individual samples processed over one and a half working days by the ASE procedure. On a routine basis, the QuEChERS-like method constitutes undeniably the best option.

Solid-phase extractions have been used for mycotoxins' extraction from different kinds of samples. Solid-phase microextraction (SPME) was used by Nonaka et al. (2009). The authors optimized the on-line in-tube SPME-LC-MS and concluded that using this approach it's possible to continuously extract aflatoxins from samples extracts with no requirement of any other pretreatments, which can then be analyzed by LC–MS. This method is automatic, simple, rapid, selective, and sensitive, and may be easily applied to the analysis of various food samples.

Solid-phase extraction (SPE) has also been applied for many years to mycotoxins analysis, once this technique enables the extraction, preconcentration and purification in one step (Alcaide-Molina et al., 2009).

3.2 Clean-up

Due to the large number of co-extractives that are present in the sample extracts, most matrices are unsuitable for direct chromatographic analysis, needing a clean-up step.

Some studies, according to the detection technique that will be employed only uses the dilution approach to reduce the matrix interferences, as we could observe in the work developed by Acharya & Dhar (2008). The authors describe a simple approach for performing broad-specific noncompetitive immunoassays for the determination of total aflatoxins (AFB1 +AFB2 +AFG1 +AFG2). Twenty grams sample were extracted with 100 mL MeOH:H_2O (70:30, v/v) and stirred for 0.5 h. Extracts were filtered through a filter paper.

The matrix interferences were eliminated by diluting the sample 10-fold with the assay buffer.

The most employed clean-up methods in some laboratories are the solid-phase extraction, multifunctional columns or immunoaffinity columns (IACs) (Bacaloni et al., 2008; Huang et al., 2010; Piermarini et al., 2009; Reiter et al., 2010). IACs in combination with HPLC are increasingly used nowadays as reference methods and allow a sufficient elimination of matrix interferences, due to their high selectivity. The immunoaffinity is based on the binding of the immobilized specific antibodies on the surface of a column (Shepard, 2009). Clean-up only with solvents is rarely found nowadays (Sheibani & Ghaziaskar, 2009). The advantages of IACs are the effective and specific extract purification provided, the economic use of organic solvents and the improved chromatographic performance achieved with cleaner samples (Shepard, 2009).

The clean-up step has an important role in the quantification techniques, avoiding false positives, allowing better recoveries and helping with the time-life of the equipments.

3.3 Separation and detection

Different techniques have been found for the determination of aflatoxins in the last years. Techniques based on ELISA detection (Li et al., 2009), electrochemical sensor (Tan et al., 2009), immunoassays (Saha et al., 2007), Liquid Chromatography tandem Mass Spectrometry (LC-MS) (Kokkonen & Jestoi, 2009; Rubert et al., 2010), Liquid Chromatography with Fluorescence Detection (LC-FLD) (Ibáñez-Vea et al., 2011), Liquid chromatography with ultraviolet detection (Fu et al., 2008) and adsorptive stripping voltammetry (Hajian & Ensafi, 2009) are found in the literature.

Aflatoxins separation has been performed for many years by HPLC, using mainly reversed-phase columns, with mobile phases composed of water, methanol and acetonitrile mixtures. Chromatographic performance has improved with column technology, particularly with reduced size of the column packing material (Shepard, 2009). Researches employing the Ultra-Performance Liquid Chromatography (UPLC) have brought lower run times and better peak shapes. Huang et al. (2010) employed the UHPLC-MS/MS for the separation and detection of aflatoxins after an extraction with acetonitrile and water and a clean-up with SPE, reaching limits of quantification between 0.012 and 0.073 µg kg^{-1}. The total run time for the separation of AFB1, AFB2, AFG1, AFG2, AFM1 and AFM2 was less than 9 min.

The AFs are named due to their properties under UV-irradiation, where AFB1 and AFB2 emit blue fluorescence (350 nm), AFG1 and AFG2 green fluorescence (350 nm). These important features can be used for rapid identification and detection (Reiter et al., 2009). So, although aflatoxins are naturally strongly fluorescent compounds, making them ideal subjects for fluorescence detection, various analogues exhibit solvent-dependent quenching in HPLC solvent systems. In the aqueous mixtures used for reversed-phase chromatography, the fluorescence of AFB1 and AFG1 are significantly quenched (Shepard, 2009). This is generally overcome by some derivatization procedure. In the last years works employing post-column derivatization have been found. Ariño et al. (2009) determined AFB1, AFB2, AFG1 and AFG2 with liquid chromatography using post-column photochemical derivatization for improved sensitivity and selectivity. This technique allowed a fluorescence enhancement about 30 times for aflatoxin B1 and G1. Results showed that post-column photochemical derivatization of aflatoxins increased detectability and selectivity of responses for the LC–FLD system. The average recovery was between 84 and 91%, and LOQ was 0.1 µg kg^{-1}.

The coupling of HPLC to mass spectrometry is the more commonly employed detection technique in the last years. The ionization sources employed based on atmospheric pressure ionization techniques such as electrospray ionization (ESI) or atmospheric pressure chemical ionization (APCI) has resulted in a range of new methods (Beltrán et al., 2011; Cavaliere et al., 2007; Kokkonen & Jestoi, 2009; Sulyok et al., 2007). The advantages of LC-MS techniques lie in the improved detection limits, the confirmation provided by mass spectral fragmentation and the ability to filter out by mass any impurities that interfere in spectrophotometric detectors. For the determination of 32 mycotoxins, including aflatoxins, in beer, Zachariasova et al. (2010), developed a study with the aim of optimize a simple and high-throughput method. For determination of analytes, ultra-high-performance liquid chromatography hyphenated with high-resolution mass spectrometry utilizing an orbitrap (U-HPLC–orbitrapMS) or time-of-flight (TOFMS) technology was used. Because of significantly better detection capabilities of the orbitrap technology, the U-HPLC–orbitrapMS method was chosen. The U-HPLC–orbitrapMS technology represents a progressive alternative equivalent to MS/MS. The U-HPLC–orbitrapMS system used within this study operates in APCI mode enabled rapid determination of trace levels of multiple mycotoxins potentially occurring in beer samples.

Relating to the source of ionization, for aflatoxin determination we have found more studies employing the ESI as source of ionization. Atmospheric pressure photoionization (APPI) is the latest interface introduced in the field of soft ionization techniques, and it was employed in the study developed by Capriotti et al. (2010). Using APPI, detection limits for the investigated compounds were lower than by using ESI, due to a much lower noise and matrix effect. For aflatoxins, LOQs between 0.1 and 0.5 $\mu g\ kg^{-1}$ were reached.

The application of aflatoxin-specific antibodies has produced a range of immunoassay analytical methods (Acharya & Dhar, 2008; Li et al., 2009; Saha et al., 2007). A number of commercial enzyme-linked immunosorbent assays (ELISAs) are well established and available. The essential principle of these assays is the immobilization on a suitable surface of antibody or antigen and the establishment of a competitive process involving this resource and components of the analytical solution (Shepard, 2009). Piermarini et al. (2009) developed a method, called ELIME-array (Enzyme-Linked-Immuno-Magnetic-Electrochemical-array) for the determination of AFB1 in corn samples. In order to determine AFB1 at a level of regulatory relevance, a sample treatment that employs extraction, clean-up and concentration steps, was selected. The recovery of the ELIME-array was calculated by analyzing replicates of four certificate reference materials (CRMs). The method showed recoveries between 95 and 114% with a LOQ of 1.5 $ng\ mL^{-1}$.

3.3.1 Matrix effect

Another special issue about the determination of contaminants, such as aflatoxins in a variety of samples is the matrix effect. Mainly related to the mass spectrometric techniques, the matrix effect is known as the change of ionization efficiency for the studied analytes in the presence of other compounds (Kruve et al., 2008).

Relating to this topic some procedures could be done to guarantee the trueness of the results, avoiding false positives. For aflatoxins' determination the approaches observed were: dilution, matrix-matched calibration, standard addition and use of internal standard. Some studies employ the AFM1 as I.S, and in others the use of a deuterated one ($^{13}C_{17}$-AFB1) was observed. The sample clean-up, many times is enough to avoid the matrix effects, but in other cases not.

3.4 Analytical criteria

Some performance criteria are important for obtaining reliable results for aflatoxins' determination. Table 1 shows a summary of some manuscripts published after 2007, showing which aflatoxins were determined, kind of sample, sample preparation, clean-up, matrix effect, detection, limit of quantification and recoveries.

Aflatoxins	Matrix	Sample preparation (sample mass, type and volume of extractor solvent)	Clean-up	Matrix Effect	Detection	LOQ	R%	Reference
total aflatoxins AFB1, AFB2, AFG1, AFG2	corn	20 g 100 mL MeOH:H_2O (70:30, v/v)	-	Dilution 10-fold to eliminate the matrix interferents	broad-specific noncompetitive immunoassay	5 μg kg^{-1} (LOD)	86-100	Acharya & Dhar, 2008
AFB1, AFB2 AFG1, AFG2	Olive leaves and drupes	Automatic SPE 5 g 25 mL MeOH:H_2O (70:30, v/v)	Automatic SPE	Matrix-matched calibration	LC-ESI-MS/MS	0.03 – 0.11 μg kg^{-1}	96-102	Alcaide-Molina et al., 2009
AFB1, AFB2 AFG1, FG2	spices	1 g 10 mL ACN ultrasonic bath (30 min)	-	Matrix-matched calibration	LC-ESI-MS/MS	1-20 μg kg^{-1}	100-139	Amate et al., 2010
AFB1, AFB2 AFG1, AFG2	pistachios	10 g 1 g NaCl 40 mL MeOH:H_2O (8:2, v/v) 20 mL hexane	Immunoafinity column		LC-FLD post-column photochemical derivatization	0.1 μg kg^{-1}	84-91	Ariño et al., 2009
AFB1, AFB2 AFG1, AFG2 AFM1 (IS)	hazelnuts	1 g 20 mL ACN: H_2O (80:20, v/v). Ultrasonic bath (10 min)	SPE (Carbograph-4)	Matrix-matched calibration and internal standard	LC-ESI-MS/MS	0.04 – 0.07 μg kg^{-1}	91-102	Bacaloni et al., 2008
AFB1, AFB2 AFG1, AFG2 AFM1	Baby food and milk	Cereals infant formula - 5 g 20 mL ACN:H_2O (80:20, v/v) Liquid samples - 8 g 32 mL ACN	immunoaffinity column	Cleanup eliminated the matrix effect	UHPLC-ESI-MS/MS	0.003 - 0.025 μg kg^{-1}	79 - 112	Beltrán et al., 2011
AFB1, AFB2 AFG1, AFG2	Baby food and paprika	Baby food - 50 g 5 g NaCl 250 mL MeOH:H_2O (80:20, v/v). Paprika - 25 g 2.5 g NaCl 100 mL MeOH:H_2O (80:20, v/v)	Immunoaffinity column	-	HPLC-FLD	0.02 - 0.2 μg kg^{-1}	86-96	Brera et al., 2011
AFB1, AFB2 AFG1, AFG2 AFM1	Cereals Wheat and maize samples	1 g 6 mL CH_3COCH_3:H_2O:CH_3COOH (80:19:1, v/v/v) ultrasonic bath (20 min)	-		LC-APPI-MS/MS	0.1 – 0.5 μg kg^{-1}	86-104	Capriotti et al., 2010
AFB1, AFB2 AFG1, AFG2 AFM1 (I.S)	Olive oil	MSPD (C18) 0.32 g 6 mL MeOH:H_2O (80:20, v/v)	-	Matrix-matched calibration	LC-ESI-MS/MS	0.04. -0.12 μg kg^{-1}	92-107%	Cavaliere et al., 2007
AFB1, AFB2 AFG1, AFG2	Curry Red pepper paste Ginger product Red pepper flour Black pepper Cinnamon powder	25 g 100 mL MeOH:H_2O (70:30, v/v) 1% NaCl	Immunoafinity column		HPLC-FLD	0.03-0.45 μg kg^{-1}	68.1-103.9	Cho et al., 2008

Aflatoxins	Matrix	Sample preparation (sample mass, type and volume of extractor solvent)	Clean-up	Matrix Effect	Detection	LOQ	R%	Reference
AFB1, AFB2 AFG1, AFG2	maize, wheat, rye, rice, oat, barley, soya, and infant cereals	QuEChERS - 5 g 10 mL ACN 0.5% CH_3COOH ASE - 5 g $ACN:H_2O:CH_3COOH$ (80:19:0.5, v/v/v)	defatting step with n-hexane	Standard addition	LC-ESI-MS/MS	1.0 – 2.0 μg kg^{-1}	QuEChERS 89-116 ASE 67-107	Desmarchelier et al., 2010
AFB1, AFB2 AFG1, AFG2 AFM1	Maize Walnuts Biscuits Breakfasts cereals	5 g 10 mL $ACN:H_2O$ (80:20, v/v) biscuit samples - 20 mL $ACN:H_2O$ (80:20, v/v) -	-	UHPLC-MS/MS	Matrix-marched calibration	0.03-3.5 μg kg^{-1}	71.3-104-7	Frenich et al., 2009
AFB1, AFB2 AFG1, AFG2	Corn peanuts	25 g 80 mL $ACN:H_2O$ (84:16, v/v)	Immunoafinnity column	-	UPLC-UV	0.63-1.07 μg kg^{-1}	83.4-94.7	Fu et al., 2008
AFB1, AFB2 AFG1, AFG2	Sorghum pistachios	10 g 40 mL MeOH: H_2O (80:20, v/v) 1 g NaCl 20 mL n-hexane	immunoaffinity column	-	HPLC-FLD	0.08-0.16 μg kg^{-1}	68.3-87.7	Ghali et al., 2009
AFB1, AFB2 AFG1, AFG2	Cassava flour	10 g 1 g NaCl 25 mL $MeOH:H_20$ (80:20, v/v)	immunoaffinity column	-	HPLC-FLD Post columns PHRED	5.0 μg kg^{-1}	52-89	Gnonlonfin et al., 2010
AFB1, AFB2 AFG1, AFG2	Peanuts	25 g 5 g NaCl 125 mL $MeOH:H_20$ (7:3 v/v)	immunoaffinity columns	-	HPLC-UV-FLD	0.1-3.5 ng mL^{-1}	65-90	Gonçalez et al., 2008
AFB1, AFB2	Groundnut	5 g 10 mL $MeOH:H_20$ (80:20, v/v) 5 mL hexane	-	Diluted 10-fold to avoid interferences	Adsorptive stripping voltametry	0.1-0.115 ng mL^{-1} (LOD)	-	Hajian & Ensafi, 2009
AFB1, AFB2 AFG1, AFG2 AFM1, AFM2	Traditional Chinese medicines	2 g 10 mL ACN:H2O (84:16, v/v)	SPE	Internal standard [$^{13}C_{17}$]-AFB1	UHPLC-ESI-MS/MS	0.1-0.39 μg kg^{-1}	85.6-117.6	Han et al., 2010
AFB1, AFB2 AFG1, AFG2 AFM, AFM2	Peanuts and their derivative products	2.5 g 10 mL $ACN:H_2O$ (84:16, v/v)	SPE	Matrix-matched calibration	UHPLC-ESI-MS/MS	0.012-0.273 μg kg^{-1}	74.7-86.8	Huang et al., 2010
AFB1, AFB2 AFG1, AFG2	barley	10 g 50 mL $ACN:H_2O$ (60:40, v/v)	immunoaffinity column		UHPLC-FLD	0.038 - 0.15 μg kg^{-1}	71.7-99.6	Ibáñez-Vea et al., 2011
AFB1, AFB2 AFG1, AFG2	Cereals Wheat Barley Oats	Automatic ASE 10 g sample Extraction with acetonitrile	-	Matrix-matched calibration	LC-ESI-MS/MS	20 – 65 μg kg^{-1}	61-94	Kokkonen & Jestoi, 2009
AFB1, AFB2 AFG1, AFG2	nuts, cereals, dried fruits, and spices	0.5 g 1 mL $MeOH:H_20$ (80:20, v/v) SPME		AFM1 (I.S.)	LC-ESI-MS	2.1 - 2.8 pg mL^{-1} (LOD)	80.8-109.1	Nonaka et al., 2009
AFB1	Corn	25 g 100 mL $ACN:H_2O$ (84:16, v/v)	Mycosep columns	Matrix-matched calibration	ELIME-array	1.5 ng mL^{-1}	95-114%	Piermarini et al., 2009
AFB1, AFB2 AFG1, AFG2	Cereal flours	2 g 10 mL of MeOH:PB1 (80:20, v/v) Ultrasonic bath (20 min) SPME	Immunoafinity column			0.1-0.63 μg kg^{-1}	49-59	Quinto et al., 2009
AFB1, AFB2 AFG1, AFG2	rice	50 g 100 mL $MeOH:H_20$ (80:20, v/v)	immunoaffinity columns		HPLC-FLD	0.44-0.6 μg kg^{-1}	83-102	Reiter et al., 2010

Aflatoxins	Matrix	Sample preparation (sample mass, type and volume of extractor solvent)	Clean-up	Matrix Effect	Detection	LOQ	R%	Reference
AFB1, AFB2 AFG1, AFG2	cereals	MSPD (1 g C18) 1 g 10 mL ACN		Matrix-matched calibration	LC-ESI-MS/MS	1 μg kg^{-1}	64-91	Rubert et al., 2010
AFB1	Chili	2 g 5 mL $MeOH:H_20$ (80:20, v/v)			Membrane-based immunoassay	2 μg kg^{-1}	88-101	Saha et al., 2007
AFB1, AFB2 AFG1, AFG2	Tigernuts and Their Beverages	MSPD (2 g C18) 1 g or 1 mL 10 mL hexane 10 mL ACN			LC-FLD	0.21-1.49 μg kg^{-1} tigernuts 0.13-0.57 μg L^{-1} beverages	72.3-82.1 (tigernuts) 74.0-86.3 (beverages)	Sebastià et al., 2010
AFB1, AFB2	pistachio	PFE 7 g 5 mL n-hexane $MeOH:H_20$ (80:20, v/v)	purified with chloroform		HPLC-FLD		100	Sheibani & Ghasiaskar, 2009
AFB1, AFB2, AFG1, AFG2	paprika	25 g 100 mL of $MeOH:H_20$ (60:40, v/v)	immunoaffinity column	-	HPLC-FLD	0.23-0.45 μg kg^{-1}	75.6-108	Shundo et al., 2009
AFB1, AFB2 AFG1, AFG2 AFM1	bread, fruits, vegetables, jam, cheese, chestnuts red wine	0.5 g 2 mL $ACN:H_2O:CH_3COOH$ (79:20:1, v/v/v)	-	Matrix matched calibration	HPLC/ ESI-MS/MS	0.7-1.5 μg kg^{-1} (LOD)	97-100	Sulyok et al., 2007
AFB1	Rice	1 g 5 mL $MeOH:H_20$ (80:20, v/v)			Electrochemical sensor	0.1 μg L^{-1} (LOD)	88.5-112	Tan et al., 2009
AFB1, AFB2 AFG1, AFG2	beer	4 mL beer 16 mL ACN		Matrix matched calibration	U-HPLC-orbitrapMS	0.5 - 3.0 μg L^{-1}	90-117	Zachariasova et al., 2010
AFB1, AFB2 AFG1, AFG2	wheat flour, corn flour, poultry feeds	50 g 250 mL $MeOH:H_20$ (80:20, v/v)	immunoaffinity column	-	LC- FLD	0.01 - 0.01 μg kg^{-1}	>65%	Zinedine et al., 2007

Table 1. Main parameters about extraction and determination of aflatoxins from 2007 to the present.

3.5 Conclusions and analysis tools of tomorrow

Determination of aflatoxins has been carried out using TLC, HPLC, LC-MS, LC-MS-MS, and immunological methods. Each one of the techniques has advantages and disadvantages. TLC provides an economical screening method. HPLC methods coupled with fluorescence detection are sensitive and the most widely used methods, but most require a derivatization step. Immunoassays provide rapid screening for total aflatoxin, but they may not be sufficiently reliable as quantitative methods for individual aflatoxins. LC-MS methods are specific and sensitive, and their use is becoming increasingly widespread. However, due to the low levels and the number of interferences from the matrices, usually, a sample preparation step is required to allow the extraction, preconcentration, and clean-up, enhancing the sensitivity and selectivity.

The advance in the extraction and determination of aflatoxins will continue increasing together with the improvement of analytical science. The search for sample preparation methods that allow fast extraction, good accuracy and precision, low extraction of interferences, low consumption of solvents will continue together with the increase in detection techniques with higher accuracy and sensibility. So, the determination of aflatoxins in foods will continue to be developed and improved.

4. Legislation, desintoxication and control

Concern about the potential hazards posed by dietary aflatoxins started in the 1960s after some 100000 turkey poults in Great Britain died as a result of aflatoxin exposure from their feed. When it became evident that aflatoxin exposure caused cancer in many species, most countries, established various regulations for aflatoxin levels (either total aflatoxins or for AFB1) in food and/or feed in order to limit exposure to this group of mycotoxins (Van-Egmond et al., 2007). These initial regulations on aflatoxins were not based on the derivation of a TDI (estimated tolerable daily intake), but rather on a desire to keep levels as low as technologically feasible (basis for regulations in some countries), or 'free' of aflatoxins by not allowing residues above the analytical detection limit (basis for regulations in some other countries). The early prudent actions regarding aflatoxins by governments have been justified, since AFB1 has been found to be a potent genotoxic agent and carcinogen in many test systems and animal species (Kuiper-Goodman, 1995; Wogan, 1974).

Worldwide, aflatoxins because of their prevalence and toxicity are important in public health. Public health concerns center on both primary poisoning from aflatoxins in commodities, food and feed stuffs, and relay poisoning from aflatoxins in milk. The allowable levels of aflatoxins in animal feedstuff and human foods vary with governmental jurisdictions (Coppock & Christian, 2007).

Aflatoxins are of great concern because of their detrimental effects on the health of humans and animals, including carcinogenic, mutagenic, teratogenic and immunosuppressive effects. AFB1 is the most potent hepatocarcinogen known in mammals and is classified by the International Agency of Research on Cancer as Group 1 carcinogen (Eaton & Gallagher, 1994 as cited in Zinedine, 2009).

The hazardous nature of aflatoxin to humans and animals has necessitated the need for establishment of control measures and tolerance levels by national and international authorities. Different countries have different regulations for aflatoxin. The general trend is that industrialized countries usually set lower tolerance levels than the developing countries, where most of the susceptible commodities are produced. However, such lack of harmony may give rise to difficulties in the trade of some commodities (Aibara & Maeda, 1989; Ismail, 1997).

The first legislative act was undertaken in 1965 by the Food and Drug Administration (FDA) of the USA, which proposed a tolerance level of 30 pg kg^{-1} of total aflatoxins (Bl + Gl + B2 + G2). With increasing awareness of aflatoxins as potent toxic substances, the proposed level was lowered to 20 pg kg^{-1} in 1969. The FDA has action levels for aflatoxins regulating the levels and species to which contaminated feeds may be fed (Table 2). In 1973, the European Economic Community (EEC) established legislation on maximum permitted levels of AFBl in different types of feedstuffs. The legislation has been frequently amended since then (EEC, 1974; FDA, 1977; Ismail 1997).

The European Community levels are more restrictive (Tables 3 and 4), 4 μg kg^{-1} total aflatoxin in food for human consumption are the maximum acceptable limits in the EU, the strictest in standard worldwide. Human foods are allowed 4–30 ppb aflatoxin, depending on the country involved (John, 2007).

Commodity	Concentration ($\mu g\ kg^{-1}$)
Cottonseed meal as a feed ingredient	300
Corn and peanut products for finishing beef cattle	300
Corn and peanut products for finishing swine	200
Corn and peanut products for breeding beef cattle, swine and mature poultry	100
Corn for immature animals and dairy cattle	20
All products, except milk, designated for humans	20
All other feedstuffs	20
Milk	0.5

Table 2. U.S. Food and Drug Administration action levels for total aflatoxins in food and feed ($\mu g\ kg^{-1}$).

Human food	AFB1 ($\mu g\ kg^{-1}$)	AFB1, B2, G1, G2 ($\mu g\ kg^{-1}$)	M1 ($\mu g\ kg^{-1}$)
Groundnuts, dried fruit and processed products thereof	2	4	-
Groundnuts subjected to sorting or physic treating	8	15	-
As above but for nuts and dried fruits	5	10	-
Cereals (including maize) and processed products thereof	2	4	-
Milk	-	-	0,05

Table 3. European Union for aflatoxins in human food ($\mu g\ kg^{-1}$).

The Brazilian National Agency for Sanitary Vigilance established the Resolution (RDC) nº 7 of February 2011 which provides for the maximum permissible (LMT) for aflatoxins (Table 5) and other mycotoxins in food.

Feed	AFB1 ($\mu g\ kg^{-1}$)	Feed	AFB1 ($\mu g\ kg^{-1}$)
Feed (exceptions below)	50	Complete feedstuff for pigs and poultry	20
Groundnuts, copra, palm kernel, cottonseed, babasu, maize and products derived from processing thereof	20	Other complete feedstuffs	10
Complete dairy feed	5	Complementary feedstuffs for cattle, sheep, goats (except dairy, calves and lambs)	50
Complete feed for lambs and calves	10	Complementary feedstuffs for pigs and poultry (except for young animals)	30

Table 4. European Union regulations for aflatoxins in feeds ($\mu g\ kg^{-1}$).

It is estimated that about 35% of human cancers are directly related to diet, and the presence of aflatoxins in foods is considered an important factor in the formation of liver cancer, mainly in tropical countries. The reduction of population exposure to aflatoxin, and the consequent reduction of health risks will only be possible with a job with the food producers and efficient actions of sanitary vigilance (Doll & Peto, 1981).

Mycotoxin	Commodity	Maximum limit tolerated ($\mu g\ kg^{-1}$)
AFB1, B2, G1, G2	Cereals and cereal products, except corn and derivatives, including malted barley	5
	Beans	5
	Chestnuts except Brazil-nut, including walnuts, pistachios, hazelnuts and almonds	10
	Dried and dehydrated fruits	10
	Brazil-nut shell for direct consumption	20
	Brazil-nut shelled for direct consumption	10
	Brazil-nut shelled for further processing	15
	Cereal-based foods for infant feeding (infants and toddlers)	1
	Infant formulas and follow-up formula for infants and toddlers	1
	Cocoa beans	10
	Cocoa and chocolate	5
	Spices: *Capsicum* spp. (dried fruits, whole or ground, including peppers, chili powder, cayenne and paprika), *Piper* spp. (the fruit, including white pepper and black pepper) *Myristica fragrans* (nutmeg) *Zingiber officinale* (ginger) *Curcuma longa* (turmeric). Spice mixtures that containing one or more of the spices listed above.	20
	Groundnut (in shell), (peeled, raw or roasted), peanut butter or peanut butter.	20
	Corn, grain (whole, broken, crushed, ground), flour or corn meal	20
Aflatoxin M1	Fluid milk	0,5
	Milk powder	5
	Cheese	2,5

Table 5. Maximum permitted (LMT) for aflatoxin in Brazil.

Aflatoxins can be detoxified or removed from contaminated food and nutrients by physical, chemical or biological methods. The inactivation of these compounds by physical and chemical methods have not proved to be effective and economically viable (Mishra & Das, 2003). However, biological degradation offers an attractive alternative to eliminate these toxins retaining food nutritional value. In the last decade it became clear that fungi are among the microorganisms that play a major role in mycotoxin degradation in particular AFB1 (Zucchi et. al., 2008).

Aflatoxins are thermostable, so the physical treatment by heat results in only small changes in their levels (Tripathi & Mishra, 2010). Chemical treatments using solvents are able to extract these compounds causing minimal effect on nutritional quality, however, this technology is still impractical and expensive, besides inducing odors and flavors. Ammoniation is also used as an effective and practical application for decontamination of agricultural products containing aflatoxins (Allameh et al., 2005). Ozonation is the chemical method that has been most studied for the decontamination of aflatoxins in foods, once ozone has been recognized as safe by the Food and Drug Administration in 2001 (Zorlugenç et al., 2008).

Currently, several studies have shown that aflatoxins are susceptible to some microorganisms such as fungi, bacteria and yeasts, being for this reason studied as a form of biological degradation. Taylor et al. (2010) studied some enzymes belonging to the group of actinomicetales specifically *Mycobacterium smegmatis* which is capable of catalyzing the ester group of aflatoxins by activating the molecules for the spontaneous hydrolysis and subsequent decontamination. Niu et al. (2008) studied several microorganisms from microbial sources that have coumarin as a carbon source. The results indicated that degradation was performed enzymatically by protease. Cacciamani et al. (2007) evaluated AFB1 and ochratoxin A degradation by solid fermentation using *A. oryzae* and *Rhizopus sp.* The first showed higher AFB1 decontamination (80%). There are several alternatives for detoxification of aflatoxins in foods, such as the use of acids and bases in the industry, being replaced by processes that involve components such as ozone GRAS and the use of fungi, bacteria or yeasts.

5. References

Acharya, D. & Dhar, T. (2008). A novel broad-specific noncompetitive immunoassay and its application in the determination of total aflatoxins. *Analytica Chimica Acta,* Vol.630, No.1, pp. 82–90, ISSN 0003-2670

Aibara, K.; Maeda, K. (1989). The regulation for aflatoxin in Japan and the current situation of aflatoxin contamination of food imported from throughout the world. In: *Mycotoxin and Phycotoxins,* S. Natori, K. Hashimoto; Y. Ueno, Elsevier, 978-90-8686-007-4, Amsterdam

Alcaide-Molina, M.; Ruiz-Jiménez, J.; Mata-Granados, J. & Luque de Castro, M. (2009). High through-put aflatoxin determination in plant material by automated solid-phase extraction on-line coupled to laser-induced fluorescence screening and determination by liquid chromatography–triple quadrupole mass spectrometry. *Journal of Chromatography A,* Vol.1216, No.7, pp. 1115–1125, ISSN 0021-9673

Allameh, A.; Safamehr, A.; Mirhadi, S.A.M.; Shivazad, M.; Razzaghi-Abyaneh, M. & Afshar-Naderi, A. (2005). Evaluation of biochemical and production parameters of broiler chicks fed ammonia treated aflatoxin contaminated maize grains. *Animal Feed Science Technology* Vol.122, No.3, pp 289-301, INSS 0377-8401

Almeida, A. V. A. F.; Botura, M. B.; Abreu, R. D.; Bittencourt, T. C. C. & Batatinha, M. J. M. (2009). Ocorrência de aflatoxinas em milho destinado à alimentação de aves no estado da Bahia. *Arquivos Instituto Biológico*, São Paulo, Vol.76, No.3, pp. 353-358, ISSN 0020-3653

Amaral, K. A. S.; Nascimento, G. B.; Sekiyama, B. L.; Janeiro, V.; Machinski Jr., M. (2006). Aflatoxinas em produtos à base de milho comercializados no Brasil e riscos para a saúde humana. *Ciência e Tecnologia de Alimentos*. Vol.26, No.2, pp. 336-342, ISSN 0101-2061

Amate, C.; Unterluggauer, H.; Fischer, R.; Fernández-Alba, A. & Masselter, S. (2010). Development and validation of a LC-MS/MS method for the simultaneous determination of aflatoxins, dyes and pesticides in spices. *Analytical and Bioanalytical Chemistry*, Vol.397, No.1, pp. 93–107, ISSN 1618-2650

Ariño, A.; Herrera, M.; Estopañan, G.; Rota, M.; Carramiñana, J.; Juan, T. & Herrera, A. (2009). Aflatoxins in bulk and pre-packed pistachios sold in Spain and effect of roasting. *Food Control*, Vol.20, No.9, pp. 811–814, ISSN 0956-7135

Bacaloni, A.; Cavaliere, C.; Cucci, F.; Foglia, P.; Samperi, R. & Laganà, A. (2008) Determination of aflatoxins in hazelnuts by various sample preparation methods and liquid chromatography–tandem mass spectrometry. *Journal of Chromatography A*, Vol.1179, No.2, pp. 182–189, ISSN 0021-9673

Beltrán, E.; Ibáñez, M.; Sancho, J.; Cortés, M.; Yusà, V. & Hernández, F. (2011). UHPLC–MS/MS highly sensitive determination of aflatoxins, the aflatoxin metabolite M1 and ochratoxin A in baby food and milk. *Food Chemistry*, Vol.126, No.2, pp. 737–744. ISSN 0308-8146

Bircan, C. (2009). Incidence of ochratoxin A in dried fruits and co-occurrence with aflatoxins in dried figs. *Food and Chemical Toxicology*, Vol.47, No.8, pp.1996–2001, ISSN 028-6915

Boutrif, E. (1998). Prevention of Aflatoxin in Pistachios, 17.01.2011, Available from <http://www.fao.org/docrep/W9474T/w9474t06.htm>

Brera, C.; Debegnach, F.; Santis, B.; Pannunzi, E.; Berdini, C.; Prantera, E.; Gregori, E. & Miraglia, M. (2011). Simultaneous determination of aflatoxins and ochratoxin A in baby foods and paprika by HPLC with fluorescence detection: A single-laboratory validation study. *Talanta*, Vol.83, No.5, pp. 1442–1446, ISSN 0039-9140

Cacciamani J. M.; Peres G. L.; Garda-Buffon J. & Badiale-Furlong E. (2007). Efeito dos tratamentos térmicos seco e úmido nos níveis de aflatoxina B1 e ocratoxina A presentes em farelo e farinhas cereais. *Boletim do Centro de Pesquisa de Processamento de Alimentos*, Vol.25, No.1, pp. 157-164, ISSN 19839774

Cano-Sancho, G.; Marin, S.; Ramos, A. J.; Peris-Vicente, J. & Sanchis, V. (2010). Occurrence of aflatoxin M1 and exposure assessment in Catalonia (Spain). *Revista Iberoamericana de Micología*, Vol.27, No.3, pp. 130-135, ISSN 1130-1406

Capriotti, A.; Foglia, P.; Gubbiotti, R.; Roccia, C.; Samperi, R.; & Laganà, A. (2010) Development and validation of a liquid chromatography/atmospheric pressure photoionization-tandem mass spectrometric method for the analysis of mycotoxins subjected to commission regulation (EC) No. 1881/2006 In cereals. *Journal of Chromatography A*, Vol.1217, No.39, pp. 6044–6051, ISSN 0021-9673

Carvajal, M.; Bolonos, A.; Rojo, F. & Mendez, I. (2003). Aflatoxin in pasteurized and ultrapasteurized milk with different fat content in Mexico. *Journal of Food Protection*. Ames, Vol.66, No.10, pp. 1885-1892, ISSN 0362-028X

Cavaliere, C.; Foglia, P.; Guarino, C.; Nazzari, M.; Samperi, R. & Laganà, A. (2007). Determination of aflatoxins in olive oil by liquid chromatography–tandem mass spectrometry, *Analytica Chimica Acta*, Vol.596, No.1, pp. 141–148, ISSN 0003-2670

Cho, S.; Lee, C.; Jang, M.; Son, Y.; Lee, S.; Choi, I.; Kim, S. & Kim, D. (2008). Aflatoxins contamination in spices and processed spice products commercialized in Korea. *Food Chemistry*, Vol.107, No.7, pp. 1283-1288, ISSN 0308-8146

Coelho, C.S.; Almeida, T.L. & Badiale-Furlong, E. (1999). Migração de micotoxinas durante o processo de parboilização do arroz. *Brazilian Journal Food Technology*, Vol.2, No.2, pp. 43-50, ISSN 1516-7275

Coppock, W.R. & Christian, R.G. (2007). Aflatoxins, In: *Veterinary Toxicology – Basic and Clinical Principles*, R. C. Gupta; pp. 939-950, Academic Press, ISBN 0123704677, San Diego.

D'Angelo, A.; Bellino, C.; Alborali, G. L.; Biancardi, A.; Borelli, A.; Capucchio, M. T.; Maurella, C. & Cagnasso, A. (2007). Neurological signs associated with aflatoxicosis in Piedmontese calves. *Veterinary Record*, Vol.160, No.20, pp. 698-700, ISSN 0042-4900

Dashti, B.; Al-Hamli, S.; Alomirah, H.; Al-Zenki, S.; Abbas, A.B. & Sawaya, W. (2009). Levels of aflatoxin M1 in milk, cheese consumed in Kuwait and occurrence of total aflatoxin in local and imported animal feed. *Food Control*, Vol.20, No.7, pp. 686-690, ISSN 0956-7135

Dekkers, S.; Van Der Fels-Klerx, H.J.; Jeurissen, S.M.F.; Kandhai, M. C.; Booij, C.J.H. & Bos, P.M.J. (2008). Development of a model to assess the occurrence of mycotoxins in wheat, maize and nuts, A holistic approach, Report 320111002, National Institute for Public Health and the Environment, Bilthoven, The Netherlands, 2008, 87 pp., Available from http://www.rivm.nl/bibliotheek/rapporten/320111002.pdf

Desmarchelier, A.; Oberson, J.; Tella, P.; Gremaud, E.; Seefelder, W. & Mottier, P. (2010). Development and comparison of two multiresidue methods for the analysis of 17 mycotoxins in cereals by liquid chromatography electrospray ionization tandem mass spectrometry. *Journal of Agricultural and Food Chemistry*, Vol.58, No.13, pp. 7510-7519, ISSN 0021-8561

Dilkin, P. (2002). Micotoxicose suína: aspectos preventivos, clínicos e patológicos. *Biológico*. Vol. 64 , No. 2, pp. 187-191, ISSN 1980 - 6221

Doll, R. & Peto, R. (1981). The causes of câncer: quantitative estimates of avoidable risks of câncer in the United States today. *Journal of the National Cancer Institute*. Vol.66, No.6, pp. 291-308, ISSN 0027-8874

Dombrink-Kurtzman, M.A. (2008). Economic Aspects of Mycotoxins in Fruits and Vegetables, In: *Mycotoxins in Fruits and Vegetables*, Barkai-Golan, R. & Nachman, P., pp. 27-44, Academic Press, ISBN 978-0-12-374126-4, New York

Dors, G.C.; Pinto, L.A.A. & Badiale-Furlong, E. (2009). Migration of mycotoxins into rice starchy endosperm during the parboiling process. *LWT - Food Science and Technology*, Vol.42, No.1, pp. 433-437, ISSN 0023-6438

Dors, G.C.; Bierhals, V.S. & Badiale-Furlong, E. (2011). Parboiled rice: chemical composition and the occurrence of mycotoxins. *Ciência e Tecnologia de Alimentos*, Vol.31, No.1, pp. 172-177, ISSN 0101-2061

Drusch, S. & Ragab, W. (2003) Mycotoxins in fruits, fruit juices, and dried fruit. *Journal of Food Protection*, Vol.66, No.8, pp. 1514–1527, ISSN 0362-028X

EEC (1974). Council Directive 74/63/EEC on the fixing of maximum permitted levels for undesirable substances and products in feedingstuffs. *Official Journal of the European Community*, No. L 38

Ellis, W.O.; Smith, P.J.; Simpson, B.K.; Khanizadeh, S. & Oldham, J. H. (1993). Control of Growth and Aflatoxin Production of Aspergillus flavus under Modified Atmosphere Packaging Conditions. *Food Microbiology*, Vol.10, No.1, pp. 9-21, ISSN 0740-0020

Fallah, A.A.; , Jafarib, T., Fallahc A. & Rahnamad M. (2009). Determination of aflatoxin M1 levels in Iranian white and cream cheese. *Food and Chemical Toxicology*. Vol.47, No.8, pp. 1872-1875, ISSN 0278-6915

Fallah, A.A. (2010). Levels of aflatoxin M1 in milk, cheese consumed in Kuwait and occurrence of total aflatoxin in local and imported animal feed. *Food and Chemical Toxicology*. Vol.48, No.3, pp. 988-991, ISSN 0278-6915

FDA (1977). FDA Proposed Guidelines for Aflatoxin. Cereal Foods World, 22, 532-533, 541.

Frenich, A.; Vidal, J.; Romero-González, R. & Aguilera-Luiz, M. (2009). Simple and high-throughput method for the multimycotoxin analysis in cereals and related foods by ultra-high performance liquid chromatography/tandem mass spectrometry. *Food Chemistry*, Vol.117, No.4, pp. 705–712, ISSN 0308-8146

Fu, Z.; Huang, X. & Min, S. (2008). Rapid determination of aflatoxins in corn and peanuts, *Journal of Chromatography A*, Vol.1209, No.1, pp. 271–274, ISSN 0021-9673

Ghali, R.; Belouaer, I.; Hdiri, S.; Ghorbel, H.; Maaroufi, K. & Hedilli, A. (2009). Simultaneous HPLC determination of aflatoxins B1, B2, G1 and G2 in Tunisian sorghum and pistachios. *Journal of Food Composition and Analysis*, Vol.22, No.7/8, pp. 751–755. ISSN 0889-1575

Giorni, P.; Battilani, P.; Pietri, A. & Magan, N. (2008). Effect of Aw and CO_2 level of Aspergillus flavus growth and aflatoxin production in high moisture maize post-harvest. *International Journal of Food Microbiology*, Vol.122, No.1-2, pp. 109-113, ISSN 0168-1605

Gnonlonfin, G.; Katerere, D.; Adjovi, Y.; Brimer L.; Shephard, G. & Sanni, A. (2010). Determination of aflatoxin in processed dried cassava root: validation of a new analytical method for cassava flour. *Journal of AOAC International*, Vol.93, No.6, pp. 1882-1887, ISSN 1944-7922

Gonçalez, E.; Nogueira, J.; Fonseca, H.; Felicio, J.; Pino, F. & Corrêa, B. (2008). Mycobiota and mycotoxins in Brazilian peanut kernels from sowing to harvest. *International Journal of Food Microbiology*, Vol.123, No.3, pp. 184–190, ISSN 0168-1605

Hajian, R. & Ensafi, A. (2009). Determination of aflatoxins B1 and B2 by adsorptive cathodic stripping voltammetry in groundnut. *Food Chemistry*, Vol.115, No.3, pp. 1034–1037. ISSN 0308-8146

Han, Z.; Zheng, Y.; Luan, L. Cai, Z.; Ren, Y. & Wu, Y. (2010). An ultra-high-performance liquid chromatography-tandem mass spectrometry method for simultaneous determination of aflatoxins B1, B2, G1, G2, M1 and M2 in traditional Chinese medicines. *Analytica Chimica Acta*, Vol.664, No.2, pp. 165–171. ISSN 0003-2670

Herzallah, S.M. (2009). Determination of aflatoxins in eggs, milk, meat and meat products using HPLC fluorescent and UV detectors. *Food Chemistry*, Vol.114, No.3, pp. 1141–1146, ISSN 0308-8146

Heshmati, A. & Milani, J. M. (2010). Contamination of UHT milk by aflatoxin M1 in Iran. *Food Control,* Vol.21, No.1, pp. 19-22, ISSN 0956-7135

Huang, B.; Han, Z.; Cai, Z.; Wub, Y. & Ren, Y. (2010). Simultaneous determination of aflatoxins B1, B2, G1, G2, M1 and M2 in peanuts and their derivative products by ultra-high-performance liquid chromatography–tandem mass spectrometry. *Analytica Chimica Acta,* Vol.662, No.1, pp. 62–68, ISSN 0003-2670

Ibáñez-Vea, M.; Corcuera, L.; Remiro, R.; Murillo-Arbizu, M.; González-Peñas, E. & Lizarraga, E. (2011). Validation of a UHPLC-FLD method for the simultaneous quantification of aflatoxins, ochratoxin A and zearalenone in barley. *Food Chemistry,* Vol.127, No.1, pp. 351–358, ISSN 0308-8146

Institute of Food Science and Technology Trust Fund. (2006). Mycotoxins, Information Statement, 02.03.2011, Available from <http://www.ifst.org/uploadedfiles/cms/store/ATTACHMENTS/Mycotoxins.pdf>

Ismail Y.S.R. (1997). Aflatoxin in food and feed: occurrence, legislation and inactivation by physical methods. *Food Chemistry,* Vol.59, No.1, pp. 57-67, ISSN 0308-8146

John, L.R. (2007). Some major mycotoxins and their mycotoxicoses - an overview. *International Journal of Food Microbiology,* Vol.119, No.1-2, pp. 3-10, ISSN 0168-1605

Khayoon, W.S.; Saad B.;, Yan C. B.; Hashim N. H.; Ali A. S. M., Salleh M. I. & Salleh B. (2010). Determination of aflatoxins in animal feeds by HPLC with multifunctional column clean-up . *Food Chemistry,* Vol.118, No.3, pp. 882-886, ISSN 0308-8146

Klich, M. Mendoza, C. Mullaney, E. Keller, N. & Bennett, J.M. (2001). A new sterigmatocystin-producing Emericella Variant from agricultural desert soils. *Systematic and Applied Microbiology,* Vol.24, No.1, pp. 131-138, ISSN 0723-2020

Kokkonen, M. & Jestoi, M. (2009). A Multi-compound LC-MS/MS Method for the Screening of Mycotoxins in Grains. *Food Analytical Methods,* Vol.2, No.2, pp. 128–140, ISSN 1936- 9751

Kruve, A.; Künnapas, A.; Herodes, K. & Leito, I. (2008) Matrix effects in pesticide multi-residue analysis by liquid chromatography–mass spectrometry. *Journal of Chromatography A,* Vol.1187, No.1/2, pp. 58–66, ISSN 0021-9673

Kuiper-Goodman, T. Mycotoxins: risk assessment and legislation. (1995) *Toxicology Letters,* Vol.82-83, (Dec, 1995), pp. 853-859, ISSN 0378-4274

KWIATKOWSKI, A. & ALVES, A.P.F. (2007). Importância da detecção e do controle de aflatoxinas em alimentos. *SaBios- Revista Saúde e Biologia,* Vol.2, No.2 pp. 45-54, ISSN 1980-0002

Li, P.; Zhang, Q.; Zhang, W.; Zhang, J.; Chen, X.; Jiang, J.; Xie, L. & Zhang, D. (2009). Development of a class-specific monoclonal antibody-based ELISA for aflatoxins in peanut. *Food Chemistry,* Vol.115, No.1, pp. 313–317, ISSN 0308-8146

Luchese, R.H. & Harrigan, W.F. (1993). Biosynthesis of Aflatoxin - the Role of Nutritional Factors. *Journal of Applied Bacteriology,* Vol.74, No.1, pp. 5–14, ISSN 0021-8847

Manetta, A.C.; Giammarco, M.; Di Giuseppe, L.; Fusaro, I.; Gramenzi, A.; Formigoni, A.; Vignola, G.; Lambertini, L. (2009) Distribution of aflatoxin M1 during Grana Padano cheese production from naturally contaminated milk. *Food Chemistry,* Vol.113, No.2, pp. 595-599, ISSN 0308-8146

Martins, M.L. & Martins, H.M. (2004). Aflatoxin M1 in yoghurts in Portugal. *International Journal of Food Microbiology,* Vol. 91, No.3, pp. 315-317, ISSN 0168-1605

Marques, P.J.L. (2007). Avaliação de aflatoxina e zearalenona em quirera (canjiquinha de milho) na região dos Campos Gerais. Dissertação apresentada para obtenção do título de mestre em Ciência e Tecnologia de Alimentos, Universidade Estadual de Ponta Grossa - PR

Marvin, H.J.P.; Kleter, G.A.; Prandini, A.; Dekkers, S. & Bolton, D.J. (2009a). Early identification systems for emerging foodborne hazards, *Food and Chemical Toxicology*, Vol.47, No.5, pp. 915–926, ISSN 0278-6915

Marvin, H.J.P.; Kleter, G.A.; Frewer, L.J. S.; Cope, M.T.A. & Wentholt, G. (2009b). Rowe, Aworking procedure for identifying emerging food safety issues at an early stage: implications for European and international risk management practices, *Food Control*, Vol. 20, No. 2, pp. 345–356, ISSN 0956-7135

Matumba, L.; Monjerezi, M.; Khonga, E.B. & Lakudzala D.D. (2010). Aflatoxins in sorghum, sorghum malt and traditional opaque beer in southern Malawi. *Food Control*, Vol.22, No.2, pp. 266-268, ISSN 0956-7135

Mclean, M. & Dutton, M.F. (1995). Cellular Interations and Metabolism of Aflatoxin: an Update. *Farmacology Therapeutics*, Vol.65, No.2, pp. 163-192, ISSN 0163-7258

D'Mello, J.P.F. (2003). Mycotoxins in Cereal Grains, Nuts and Other Plant Products. International. Food Safety: *Contaminants and Toxins*, pp. 65-90, CABI, ISBN 0-85199-607-8, Wallingford

Mishra, H.N. & Das, C. A review on biological control and metabolism of aflatoxin. (2003) *Critical Reviews in Food Science and Nutrition*, Vol.43, No.3, pp. 245–264, INSS 1040-8398

Molina, M. & Giannuzzi, L. (2002). Modelling of Aflatoxin Production by Aspergillus parasiticus in a Solid Medium at Different Temperatures, pH and Propionic Acid Concentrations. *Food Research International*, Vol.35, No.6, pp. 585-594, ISSN 0963-9969

Montagna, M.T; Napoli, C.; De Giglio, O.; Latta, R. & Barbutti, G. (2008). Occurrence of aflatoxin M1 in Dairy products in Southem Italy. *International Journal Molecular Sciences*. Basel, Vol.9, No.12, pp. 2614-2621, ISSN 1422-0067

Nakai, V.K.; Rocha, L.O.; Gonçalez, E.; Fonseca, H.; Ortega, E.M.M. & Corrêa, B. (2008) Distribution of fungi and aflatoxins in a stored peanut variety. *Food Chemistry* Vol.106, No.1, pp. 285–290, ISSN 0308-8146

Niu, T.; Guan, S.; Ji, C.; Zhou, T.; Li, J. & Ma, Q. (2008). Aflatoxin B1 degradation by Stenotrophomonas maltophilia and other microbes selected using coumarin medium. *International Journal of Molecular Sciences*. Vol.9, No.8, pp. 1489-1503, INSS 1422-0067

Nonaka, Y.; Saito, K.; Hanioka, N.; Narimatsu, S. & Kataoka, H. (2009). Determination of aflatoxins in food samples by automated on-line in-tube solid-phase microextraction coupled with liquid chromatography–mass spectrometry. *Journal of Chromatography A*, Vol.1216, No.20, pp. 4416–4422, ISSN 0021-9673.

OBrian, G.R.; Fakhoury, A.M.; Payne, G.A. (2003). Identification of genes differentially expressed during aflatoxin biosynthesis in *Aspergillus flavus* and *Aspergillus parasiticus* . *Fungal Genetics and Biology*, Vol.39, No.2, pp. 118-127, ISSN 1087-1845

Oliveira, T.R.; Barana, A.C.; Jaccoud-Filho, D.S. & Neto, F.F. (2010). Avaliação da contaminação por aflatoxinas totais e zearalenona em variedades de milho crioulo

(*Zea Mays L.*) através do método imunoenzimático Elisa. *Revista Brasileira de Tecnologia Agroindustrial*, Vol.4, No.2, pp. 179-185, ISSN 1981-3686

Park, D.L. (2002). Effect of processing on aflatoxin. Advances in experimental medicine and biology. New York, v. 504, pp. 173-179.

Park, M.V.D.Z. & Bos, P.M.J. (2007). Information sources for the detection of emerging mycotoxin risks, Report 320111001, National Institute for Public Health and the Environment, Bilthoven, 2007, 56 pp., Available from http://www.gencat.cat/salut/acsa/320111001.pdf

Piermarini, S.; Volpe, G.; Micheli, L.; Moscone, D. & Palleschi, G. (2009). An ELIME-array for detection of aflatoxin B1 in corn samples. *Food Control*, Vol.20, No.4, pp. 371–375. ISSN 0956-7135

Quinto, M.; Spadaccino, G.; Palermo, C. & Centonze, D. (2009). Determination of aflatoxins in cereal flours by solid-phase microextraction coupled with liquid chromatography and post-column photochemical derivatization-fluorescence detection. *Journal of Chromatography A*, Vol.1216, No.49, pp. 8636–8641, ISSN 0021-9673

Ramos, C.R.B.A.; Brasil, E.M.; Geraldine, R.M. (2008). Contaminação por aflatoxinas em híbridos de milho cultivados em três regiões do estado de Goiás. *Pesquisa Agropecuária Tropical*, Vol.38, No.2, pp. 95-102, ISSN 0278-6915

Reiter, E.; Zentek, J. & Razzazi, E. (2009). Review on sample preparation strategies and methods used for the analysis of aflatoxins in food and feed. *Molecular Nutrition & Food Research*. Vol.53, No.4, pp. 508-524, ISSN 1613-4133

Reiter, E.; Vouk, F.; Böhma, J. & Razzazi-Fazeli, E. (2010). Aflatoxins in rice - A limited survey of products marketed in Austria. *Food Control*, Vol.21, No.7, pp. 988–991, ISSN 0956-7135

Ribeiro, J.M.M.; Cavaglieri, J.R.; Fraga, M.E.; Direito, G.M.; Dalcero, A.M. & Rosa, C.A.R. (2006). Influence of Water Activity, Temperature and Time on Mycotoxins Procuction on Barley Rootlets. *Letters of Applied Microbiology*, Vol.42, No.2, pp. 179-184, ISSN 02268254

Romero, A.C.; Ferreira, T.R.B.; Dias, C.T.S; Calori-Domingues, M.A. & Gloria, E.M. (2010). Occurrence of AFM1 in urine samples of a Brazilian population and association with food consumption. *Food Control*, Vol.21, No.4, pp. 554-558, ISSN 0956-7135

Rubert, J.; Soler, C. & Mañes, J. (2010). Optimization of Matrix Solid-Phase Dispersion method for simultaneous extraction of aflatoxins and OTA in cereals and its application to commercial samples. *Talanta*, Vol.82, No.1, pp. 567–574, ISSN 0039-9140

Saha, D.; Acharya, D.; Roy, D.; Shrestha, D. & Dhar, T. (2007). Simultaneous enzyme immunoassay for the screening of aflatoxin B1 and ochratoxin A in chili samples. *Analytica Chimica Acta*, Vol.584, No.2, pp. 343–349, ISSN 0003-2670

Sassahara, M.; Pontes Neto, D. & Yanaka, E.K. (2005). Aflatoxin occurrence in foodstuff supplied to dairy cattle and aflatoxin M1 in raw milk in the North of Parana State, *Food and Chemical Toxicology*, Vol.43, No.6, pp. 981-984, ISSN 0278-6915

Sebastia, N.; Soler, C.; Soriano, J. & Mañes, J. (2010) Occurrence of Aflatoxins in Tigernuts and Their Beverages Commercialized in Spain. *Journal of Agricultural and Food Chemistry*, Vol.58, No.4, pp. 2609–2612, ISSN 0021-8561

Sheibani, A. & Ghaziaskar, H. (2009). Pressurized fluid extraction for quantitative recovery of aflatoxins B1 and B2 from pistachio. *Food Control*, Vol.20, No.2, pp. 124–128, ISSN 0956-7135

Shephard, G. (2009). Aflatoxin analysis at the beginning of the twenty-first century. *Analytical and Bioanalytical Chemistry*, Vol.395, No.5, pp. 1215–1224, ISSN 1618-2650

Schmidt-Heydt, M.; Abdel-Hadi, A.; Naresh, M. & Geisen, R. (2009). Complex Regulation of the Aflatoxin Biosynthesis Gene Cluster of Aspergillus flavus in Relation to Various Combinations of Water Activity and Temperature. *International Journal of Food Microbiology*. Vol.135, pp. 231-237, ISSN 0168-1605

Shundo, L.; Almeida, A.; Alaburda, J.; Lamardo, L.; Navas, S.; Ruvieri, V. & Sabino, M. (2009). Aflatoxins and ochratoxin A in Brazilian paprika. *Food Control*, Vol.20, No.12, pp. 1099–1102, ISSN 0956-7135

Steiner, W.E.; Rieker, R.H. & Battaglia, R. (1988). Aflatoxin contamination in dried figs: Distribution and association with fluorescence. *Journal of Agricultural and Food Chemistry*, Vol.36, No.1, pp. 88–91, ISSN 0021-8561

Sulyok, M.; Krska, R. & Schuhmacher, R. (2007). A liquid chromatography/tandem mass spectrometric multi-mycotoxin method for the quantification of 87 analytes and its application to semi-quantitative screening of moldy food samples. *Analytical and Bioanalytical Chemistry*, Vol.389, No.5, pp. 1505–1523. ISSN 1618-2650

Tan, Y.; Chu, X.; Shen, G. & Yu, R. (2009). A signal-amplified electrochemical immunosensor for aflatoxin B1 determination in rice. *Analytical Biochemistry*, Vol.387, No.1, pp. 82–86. ISSN 1096-0309

Taylor, M.C.; Jackson, C.J.; Tattersall, D.B.; French, N.; Peat, T.S.; Newman, J.; Briggs, L.J. & Oakeshott, J.G. (2010). Identification and characterization of two families of F420H2 dependent reductases from Mycobacteria that catalyse aflatoxin degradation. *Molecular Microbiology*. Vol.78, No.3, pp. 561-575, ISSN 1365-2958

Tripathi, S. & Mishra, H.N. (2010). Enzymatic coupled with UV degradation of aflatoxina B1 in red chili powder. *Journal of Food Quality*. Vol.33, SUPPL. s1, pp. 186-203, ISSN 1745-4557

Trucksess, M.W. & Diaz-Amigo C. (2011). Mycotoxins in Foods Encyclopedia of Environmental Health, 888-897.

Ueno, Y. (1986). Trichothecene as environmental toxicants. In: Reviews in Environmental Toxicology 2. Ernest Hodgson (ed). Elsevier, Amsterdan, N.York, Oxford

Van der Felx-Klers, H.J.; Dekkers, S.; Kandhai, M. C.; Jeurissen, S.M.F. & Booij, C.J.H. (2010). Indicators for early identification of re-emerging mycotoxin. NJAS- Wageningen *Journal of Life Sciences*, Vol.57, pp. 133-139, ISSN 1573-5214

Van Egmond, H.P. (1989). Aflatoxin M1: occurrence, toxicity, regulation. In Hans P. van Egmond (Ed.), Mycotoxins in dairy products (pp. 11–55). New York: Elsevier Applied Science.

Van-Egmond, H.P.; Schothorst, R.C. & Jonker, M.A. (2007). Regulations Relating to Mycotoxins in Food: Perspectives in a Global and European Context. *Analitical and Bioanalytical Chemistry*, Vol.389, No.1, pp. 147-157, ISSN 1618-2650

Velázquez, W.R.; Martínez, S.P.; Espinosa, V.H.I.; Vera, M.A.N.; Palacios, E.L. & Rojo, F. (2009). Aflatoxinas totales em raciones de bovinos y AFM 1 em leche cruda obtenida en establos del estado de Jalisco, México. *Técnica Pecuaria México*, Vol.47, No.2, pp. 223-230, ISSN 0040-1889

Wagacha, J.M. & Muthomi, J.W. (2008). Mycotoxin problem in Africa: current status, implications to food safety and health and possible management strategies. *International Journal of Food Microbioogy* . Vol.124, No.1, pp. 1–12, ISSN 0168-1606

Wild, C.P. & Montesano, R. (2009). A model of interaction: Aflatoxins and hepatitis viruses in liver cancer aetiology and prevention. *Cancer Letters,* Vol.286, No.1, pp. 22-28, ISSN 0304-3835

Wogan, G.N. (1966). Chemical Nature and Biological Effects of the Aflatoxins. *Bacteriological Reviews, Baltimore,* Vol.30, No.2, pp. 460-470, ISSN 0102-3829

Wogan, G.N.; Paglialunga, S. & Newberne, P. M. (1974) Carcinogenic Effects of Low Dietary Levels of Aflatoxin B1 in Rats. *Food and Cosmetics Toxicology,* Vol.12, No.5-6, pp. 681-685, ISSN 0015-6264

Zachariasova, M.; Cajka, T.; Godula, M.; Malachova, A.; Veprikova, Z. & Hajslova, J. (2010). Analysis of multiple mycotoxins in beer employing (ultra)-high-resolution mass spectrometry. *Rapid Communications in Mass Spectrometry,* Vol.24, No.22, pp. 3357-3367, ISSN 0951-4198

Zain, M.E. (2011). Impact of mycotoxins on humans and animals. *Journal of Saudi Chemical Society,* Vol.15, No.2, pp. 129–144, ISSN 1319-6103

Zinedine, A.; Juan, C.; Soriano, J.; Moltó, J.; Idrissi, L. & Mañes, J. (2007). Limited survey for the occurrence of aflatoxins in cereals and poultry feeds from Rabat, Morocco. *International Journal of Food Microbiology,* Vol.115, No.1, pp. 124–127, ISSN 0168-1605

Zinedine, A. & Mañes, J. (2009). Occurrence and legislation of mycotoxins in food and feed from Morocco. *Food Control,* Vol. 20, No 4, pp. 334-344, ISSN 0956-7135.

Zorlugenç, B.; Zorlugenç, F.K.; Öztekin, S. & Evliya, I.B. (2008). The influence of gaseous ozone and ozonated water on microbial flora and degradation of aflatoxina B1 in dried figs. *Food and Chemical Toxicology.* Vol.46, No.12, pp. 3593-3597, ISSN 0278-6915

Zucchi, T.D.; Moraes, L.A. & Melo, I.S. (2008). *Streptomyces* sp. ASBV-1 reduces aflatoxin accumulation by *Aspergillus parasiticus* in peanut grains. *Journal of Applied Microbiology,* Vol. 105, No.6, (November 2008), pp. 2153–2160, INSS 1364-5072

21

Estimated Daily Intake of Aflatoxin M_1 in Thailand

Nongluck Ruangwises[1], Piyawat Saipan[2] and Suthep Ruangwises[3]
[1]*Department of Pharmaceutical Chemistry, Faculty of Pharmacy, Mahidol University, Bangkok*
[2]*Department of Veterinary Public Health, Faculty of Veterinary Medicine, Khon Kaen University, Khon Kaen*
[3]*Department of Veterinary Public Health, Faculty of Veterinary Science, Chulalongkorn University, Bangkok*
Thailand

1. Introduction

Aflatoxins are a group of mycotoxins produced by certain species of *Aspergillus*. These molds grow on a variety of food and feed commodities and produce aflatoxins under appropriate temperature and humidity (Jay et al., 2005). Aflatoxin B_1 (AFB_1) is the most potent hepatocarcinogen of this group of mycotoxins. Aflatoxin M_1 (AFM_1) is a hydroxylated metabolite of AFB_1 and is secreted in the milk of mammals that have eaten contaminated foods. AFM_1 is also a hepatocarcinogen and is classified in Group 1 (carcinogenic to humans) by the International Agency for Research on Cancer (IARC, 2002). Exposure to AFM_1 through milk products is considered to be a serious public health problem. Several countries have established regulatory limits for AFM_1 in raw milk and milk products, which vary from country to country. According to the Food and Agriculture Organization of the United Nations (FAO), there are 60 countries that have established regulatory limits for AFM_1; the values vary from ND (not detectable) to 15 μg/L (FAO, 2004). The two most prevalent limits are 0.05 μg/L (34 countries) and 0.5 μg/L (22 countries). The European Community has set the maximum permitted level for AFM_1 in infant formulae and follow-on formulae, including infant milk and follow-on milk, at 0.025 μg/kg, and in raw milk and heat-treated milk at 0.05 μg/kg (European Commission, 2006). The U.S. regulatory limit for AFM_1 is 0.5 μg/L (FAO, 2004). However, several countries, including Thailand, have not yet established regulatory limits for AFM_1. The Notification of the Ministry of Public Health No. 265 - the law that regulates the quality of milk products in Thailand - only states that "*...milk products may be contaminated with aflatoxins at a level that is not harmful to human health*" (Ministry of Public Health, 2003).

A national food consumption survey was conducted in Thailand during the years 2002–2004; 18,746 participants were divided into five age groups (Groups 1–5): 0–3, >3–9, >9–19, >19–65, and >65 yr (National Bureau of Agricultural Commodity and Food Standards, 2006). The survey showed that the consumption amounts of four types of milk products - milk powder, school milk, commercial pasteurized milk, and UHT (ultra-high-temperature) milk - comprised approximately 93% of all milk products consumed by the Thai population

(National Bureau of Agricultural Commodity and Food Standards, 2006). In Thailand, in the morning on every school day, students in Grades 1–6 (age range, 6–12 yr) are served pasteurized milk (200 ml) provided by the School Milk Project (Ruangwises & Ruangwises, 2009). Thailand is administratively divided into 76 provinces; the 45 provinces with the highest population were selected for milk sample collection. The purposes of this study were to investigate whether the concentrations of AFM_1 in milk powder, school milk, commercial pasteurized milk, and UHT milk products consumed in Thailand are within the acceptable level for consumption, and to estimate the daily intake of AFM_1 for the Thai population.

2. Materials and methods

2.1 Chemicals

AFM_1 reference standard (from *Aspergillus flavus*), trifluoroacetic acid, methylene chloride, *n*-hexane, and silica gel 60 were purchased from Sigma-Aldrich (St. Louis, MO, USA). $AflaM_1^{TM}$ immunoaffinity columns were purchased from Vicam (Nixa, MO, USA). Spherisorb ODS-2 HPLC columns (5 μm, 4.6 × 250 mm) and C_{18} Sep-Pak columns were obtained from Waters Corporation (Milford, MA, USA). Solvents (HPLC grade) - acetonitrile, methanol, isopropyl alcohol, and water - were obtained from Merck (Darmstadt, Germany).

2.2 Milk sample collection and sample preparation

Milk powder samples were purchased from supermarkets (2 samples/province), while school milk samples were collected from 180 elementary schools (4 schools/province). Commercial pasteurized and UHT milk samples were purchased from supermarkets (2 samples each of pasteurized and UHT milk/province). All milk samples were collected between January 2007 and January 2008. Pasteurized milk samples were frozen at –20 °C until analysis (within one month from the manufacturing date). A total of 450 milk samples were analyzed in this study.

Milk powder sample (31.25 g) was reconstituted in 200 ml of distilled water in a 250-ml volumetric flask, mixed well, and adjusted to 250 ml with distilled water (dilution 1: 8). Concentrations of fat, protein, and solid-not-fat (SNF) in reconstituted milk powder samples, analyzed using a MilkoScan 133B (Foss Electric, Hillerød, Denmark), were 4.12 ± 0.36, 3.21 ± 0.08, and 8.59 ± 0.09 g/100 ml (n = 30), respectively. These milk compositions conformed with the Notification of the Ministry of Public Health No. 265, which states that fluid milk with full butter fat must contain fat, protein, and SNF of at least 3.2, 2.8, and 8.25 g/100 ml, respectively (Ministry of Public Health, 2003). The densities of reconstituted and liquid milk samples were determined using 50 ml of milk sample.

2.3 Extraction and determination of Aflatoxin M_1

AFM_1 was extracted from milk samples using an $AflaM_1^{TM}$ immunoaffinity column. The extraction procedure was according to the manufacturer's recommendations, as previously described by Ruangwises & Ruangwises (2009). In brief, an aliquot of 50 ml of reconstituted milk powder or liquid milk sample was transferred to a 50-ml plastic centrifuge tube and defatted by centrifugation at 3,500 *g* for 20 min. After the fat was separated, the resulting skimmed milk was transferred into a 50-ml plastic syringe which was attached to an immunoaffinity column. The skimmed milk was allowed to flow into the column by gravity at a rate of 1 ml/min. The column was then washed with 20 ml of water. AFM_1 was eluted with 1.25 ml of acetonitrile: methanol (3: 2), followed by 1.25 ml of HPLC water. A total volume of 2.5 ml of eluate was filtered through a nylon filter (0.45 µm) and used for analysis of AFM_1 using HPLC. All milk samples were analyzed in duplicate.

The complete chromatographic system (Class-LC10; Shimadzu, Kyoto, Japan) consisted of a HPLC pump (model LC-10AD), an auto injector (model SIL-10A), a column oven (model CTO-10A), and a fluorescence detector (model RF-10AXL). The HPLC conditions for analysis of AFM_1 were as follows: column – Spherisorb ODS-2; column temperature – 40°C; mobile phase – water: methanol: acetonitrile (57: 23: 20); flow rate – 1 ml/min; detector – fluorescence spectrophotometer (excitation 360 nm; emission 440 nm).

The AOAC Official Method 986.16 for detection of aflatoxins M_1 and M_2 in fluid milk (Cunniff, 1995) was performed to confirm the AFM_1 analysis using the immunoaffinity column. In brief, an aliquot of fluid milk or reconstituted milk powder sample (20 ml) was mixed with 20 ml of hot water (80°C), and AFM_1 was extracted using a C_{18} Sep-Pak column. AFM_1 was eluted from the column with ether. The ether phase was then cleaned up using a silica gel 60 mini-column; AFM_1 was eluted with a mixture of methylene chloride: ethanol (95: 5). The eluate was evaporated to dryness under nitrogen gas; the residue was then dissolved in 0.2 ml of *n*-hexane. To convert AFM_1 to AFM_{2a}, which has a higher extinction coefficient, the solution was derivatized with 0.2 ml of trifluoroacetic acid at 40°C and then evaporated to dryness under nitrogen gas. The residue was dissolved in 2 ml of water: acetonitrile (3: 1). The final solution was filtered through a nylon filter (0.45 µm) and used for analysis of AFM_1 by HPLC. The HPLC conditions for analysis of AFM_{2a} were as follows: column – Spherisorb ODS-2 (5 µm, 4.6 × 250 mm); mobile phase – water: acetonitrile: isopropanol (80: 12: 8); flow rate – 1 ml/min; detector – fluorescence spectrophotometer (excitation 365 nm; emission 455 nm).

A combination of the two methods was used to confirm the quantification of AFM_1 extracted using the AflaM$_1$™ immunoaffinity column. The eluate (2.5 ml) from the immunoaffinity column was evaporated under nitrogen gas to dryness. The residue was dissolved in 0.2 ml of *n*-hexane and was quantified as described in the AOAC Official Method 986.16, with the addition of 0.2 ml of trifluoroacetic acid.

2.4 Determination of limit of quantification

The limit of quantification (LOQ) for AFM_1 was determined using the Q2B method of the U.S. Food and Drug Administration (U.S. FDA, 1996). Milk samples (50 ml) were fortified with standard AFM_1 at concentrations of 0.0125, 0.025, 0.05, 0.25, and 0.5 µg/L; blank samples were not fortified with standard AFM_1. Concentrations of AFM_1 in fortified milk samples and blank samples were quantified as described in Extraction and Determintion of AFM_1 using the AflaM$_1$™ immunoaffinity column. Calibration curves (n = 12) were obtained by least-square linear regression analysis of the residual peak heights versus fortified AFM_1 concentrations. The calculation for LOQ was based on the standard deviation of *y*-intercepts of linear regression analysis (σ) and the slope (S) using the equation $LOQ = 10\ \sigma/S$. The LOQ of the method was 0.01 µg/L and the overall recovery across the five concentrations of fortified AFM_1 was 85.6%. The precision of the method, expressed as %CV (coefficient of variation), ranged from 2.8 to 5.6%, as previously discussed by Ruangwises & Ruangwises (2009).

2.5 Statistical analysis

A randomized block experiment was used to assess the differences between AFM_1 concentrations. Duncan's multiple comparison test was applied to obtain significance levels between the four types of milk samples ($P < 0.05$). SPSS Statistics version 17.0 for Windows was used for statistical analysis.

3. Results and discussion

Concentrations of AFM_1 extracted using an immunoaffinity column with HPLC quantification were comparable to those analyzed using AOAC Official Method 986.16, and also comparable to those obtained from a combination of the two methods (extraction with an immunoaffinity column, derivatization with trifluoroacetic acid, and quantification with HPLC). Comparative results for AFM_1 concentrations in liquid milk and milk powder samples obtained from the three procedures are presented in Table 1. In this study, the average density of reconstituted milk powder and liquid milk samples was 1.03 ± 0.027 g/ml (n = 90), which was used for unit conversion. Concentrations and incidence of AFM_1 in 450 milk samples are presented in Table 2. Of the 450 samples, 288 (64.0%) were found to be contaminated with AFM_1 equal to or above the LOQ of 0.01 µg/L. The incidence of AFM_1 in milk powder, school milk, commercial pasteurized milk, and UHT milk samples was 21.1% (19/90), 71.1% (128/180), 78.9% (71/90), and 77.8% (70/90), respectively. Average concentrations of AFM_1 found in the four types of milk samples were 0.004 ± 0.009, 0.035 ± 0.028, 0.048 ± 0.034, and 0.045 ± 0.034 µg/L, respectively. In this study, statistical analysis showed that the average concentration of AFM_1 in milk powder samples was significant lower than those found in the other three milk products. The average concentration of AFM_1 in school milk samples was significantly lower than those found in the commercial pasteurized and UHT milk samples. Of the 19 positive milk powder samples, only 2 samples were contaminated with AFM_1 above the EU limit for infant milk products of 0.025 µg/kg. For school milk, commercial pasteurized milk, and UHT milk samples, 68/180 (37.8%), 25/90 (27.8%), and 29/90 (32.2%) samples, respectively, were contaminated with AFM_1 within the EU limit of 0.05 µg/kg. Concentrations of AFM_1 found in all milk samples were within the U.S. regulatory limit of 0.5 µg/kg.

Sample	AFM_1 concentration (µg/L)*		
	AOAC Official Method 986.16	Immunoaffinity column + TFA	Immunoaffinity column
Milk Powder (dilution 1:8)			
Sample A	0.021	0.024	0.026
Sample A + 0.05 µg/L AFM_1	0.027	0.028	0.033
Sample A + 0.1 µg/L AFM_1	0.119	0.120	0.125
Sample A + 0.25 µg/L AFM_1	0.291	0.302	0.299
Sample B	0.046	0.047	0.042
Sample C	0.031	0.040	0.038
Sample D	0.022	0.028	0.023
Liquid Milk			
Sample 1	0.035	0.032	0.033
Sample 1 + 0.05 µg/L AFM_1	0.079	0.078	0.076
Sample 1 + 0.1 µg/L AFM_1	0.127	0.122	0.129
Sample 1 + 0.25 µg/L AFM_1	0.246	0.247	0.244
Sample 2	0.062	0.060	0.065
Sample 3	0.040	0.043	0.039
Sample 4	0.089	0.085	0.091

*All samples were analyzed in duplicate.

Table 1. AFM_1 concentrations in milk powder and liquid milk samples obtained from three procedures

Sample	Samples Analyzed	Positive (%)	AFM_1 concentration (μg/kg)		AFM_1 incidence[1]			
			Mean ± SD	Range[2]	0.010- 0.050 μg/kg	0.051- 0.075 μg/kg	0.076- 0.100 μg/kg	≥ 0.101 μg/kg
Milk powder	90	19 (21.1)[3]	0.004± 0.009[4] A[5]	0.011-0.048	19 (21.1)	-	-	-
Liquid milk								
Pasteurized school milk	180	128 (71.1)	0.035±0.028 B	0.012-0.104	68 (37.8)	50 (27.8)	7 (3.9)	3 (1.7)
Pasteurized commercial	90	71 (78.9)	0.048± 0.034 C	0.014-0.109	25 (27.8)	29 (32.2)	14 (15.6)	3 (3.3)
UHT commercial	90	70 (77.8)	0.045± 0.034 C	0.012-0.112	29 (32.2)	26 (28.9)	11 (12.2)	4 (4.4)
Overall	450	288 (64.0)	0.034± 0.031	0.011-0.112	141 (31.3)	105 (23.3)	32 (7.1)	10 (2.2)

[1]AFM_1 incidence of the positive samples
[2]Ranges of AFM_1 concentrations of the positive samples
[3]Numbers in parentheses are percentages of each milk product
[4]Concentrations of AFM_1 in reconstituted milk powder samples (1: 8)
[5]Different letters (A, B, C) denote significant differences between means of each milk product ($P < 0.05$)

Table 2. Concentrations and incidence of AFM_1 in milk powder and liquid milk samples

Since each of the five Thai population groups consumed different types of milk products, calculation of the daily AFM1 intake for each population group was based on the mean AFM1 concentrations in the corresponding milk products. The mean daily AFM1 intake by each group was calculated as follows: mean AFM1 intake (ng AFM1/day) = (mean daily milk intake, kg/day) x (mean AFM1 concentration in the corresponding milk products, μg/kg) x 1,000 (a unit conversion factor, ng/μg). The mean consumption amounts of milk products for the five Thai population groups were 0.429, 0.220, 0.138, 0.059, and 0.036 kg/day, respectively, while the average body weights were 10.1, 20.0, 46.2, 59.4, and 54.5 kg, respectively (National Bureau of Agricultural Commodity and Food Standards, 2006).

For Group 1 (age range, 0–3 yr), a mean AFM1 concentration in milk powder samples of 0.004 μg/kg (n = 90) was used for the calculation of daily AFM1 intake. A mean AFM1 concentration of 0.024 μg/kg in milk powder and school milk samples (n = 270) was used for the calculation of daily AFM1 intake for Group 2 (>3–9 yr), while a mean AFM1 concentration of 0.040 μg/kg in school milk, commercial pasteurized milk, and UHT milk samples (n = 360) was used for Group 3 (>9–19 yr). As school milk is consumed by students in Grades 1–6 (6–12 yr), concentrations of AFM1 in school milk samples were used for both Groups 2 and 3. For Group 4 (>19–65 yr) and Group 5 (>65 yr), a mean AFM1 concentration of 0.045 μg/kg in commercial pasteurized and UHT milk samples (n = 180) was used for the calculation of daily AFM1 intake. The estimated daily intakes of AFM1 for the five Thai population groups were 1.63, 5.29, 5.50, 2.63, and 1.62 ng/day, respectively. The estimated daily intakes of AFM1 with respect to body weight for the five population groups were 0.16, 0.26, 0.12, 0.04, and 0.03 ng/kg BW/day, respectively.

Daily intakes of AFM_1 in various regions and countries are presented in Table 3. The Joint FAO/WHO Expert Committee on Food Additives (JECFA) calculated that the daily AFM_1 intakes in Africa, the Middle East, Latin America, Europe, and the Far East (using the mean AFM_1 concentrations in four different milk products) were 0.002, 0.005, 0.022, 0.023, and 0.36 μg/kg, respectively (JECFA, 2001). The amounts of milk products consumed in the five regions were 0.042, 0.12, 0.16, 0.29, and 0.032 kg/day, respectively. The estimated intakes of AFM_1 in the five regions were 0.1, 0.6, 3.5, 6.8, and 12 ng/day, respectively. When AFM_1 intakes were calculated with respect to body weight (assuming 60 kg), the estimated daily intakes of AFM_1 were 0.002, 0.1, 0.058, 0.11, and 0.20 ng/kg BW/day, respectively (JECFA, 2001).

Region/Country		Milk consumption[1] (kg/day)	AFM_1 concentration (μg/kg)	Daily AFM_1 intake ng/day	ng/kg BW/day	Reference
Africa		0.042	0.002	0.1	0.002	JECFA (2001)
Middle East		0.12	0.005	0.6	0.1	JECFA (2001)
Latin America		0.16	0.022	3.5	0.058	JECFA (2001)
Europe		0.29	0.023	6.8	0.11	JECFA (2001)
Far East		0.032	0.36	12	0.20	JECFA (2001)
France	age 3-14 yr	0.312	0.005-0.05[2]	NA[3]	0.22	Leblanc et al. (2005)
	age >15 yr	0.229	0.005-0.05[2]	NA[3]	0.09	
Brazil	children	0.360[4]	0.061	23.92	1.04[5]	Shundo et al. (2009)
	adults	0.412[4]	0.031	11.28	0.188[5]	
Spain	age 4-9 yr	0.532	0.00969[6]	2.63	0.21	Cano-Sancho et al. (2010)
	age 10-19 yr	0.404	0.00969	2.01	0.07	
	age 20-65 yr	0.305	0.00969	1.44	0.04	
	age >65yr	0.407	0.00969	1.94	0.05	
Thailand	age 0-3 yr	0.429	0.004[7]	1.63	0.16	This study
	age >3- 9yr	0.220	0.024[7]	5.29	0.26	
	age >9-19 yr	0.138	0.040[7]	5.50	0.12	
	age >19-65 yr	0.059	0.045[7]	2.63	0.04	
	age >65yr	0.036	0.045[7]	1.62	0.03	
	Overall	0.176	0.032	3.33	0.12	

[1]Average milk consumption for each region or country

[2]AFM_1 concentrations in five milk products which were used for the calculation of AFM_1 intake for each population group (see text for details)

[3]NA = Data not available

[4]The values are average milk consumption amounts for both sexes calculated from milk consumption amounts by males and females in each of four population groups presented in the report

[5]Daily AFM_1 intakes were a summation of individual AFM_1 intakes from five milk products

[6]A single value of average AFM_1 concentration in three milk products was used for the calculation of daily AFM_1 intake for all four Spanish population groups

[7]Average AFM_1 concentrations in corresponding milk products for each Thai population group were used for the calculation of daily AFM_1 intakes (see text for explanation)

Table 3. Daily AFM_1 intakes in various regions/countries

Leblanc et al. (2005) used individual means of AFM_1 concentrations of five milk products - butter (0.05 μg/kg), desserts (0.05 μg/kg), cheeses (0.05 μg/kg), milk (0.005 μg/kg), and ultra-fresh dairy products (0.047 μg/kg) - and daily milk consumption to calculate daily AFM_1 intakes for two French population groups: children (3–14 yr) and adults (≥ 15 yr). Daily intakes of AFM_1 for each population group were the summation of daily AFM_1 intakes from the five milk products; the estimated AFM_1 intakes by French children and adults were 0.22 and 0.09 ng/kg BW/day, respectively. Shundo et al. (2009) used AFM_1 concentrations in different milk products to estimate daily AFM_1 intake by Brazilian children and adults in the city of Sao Paulo. Based on the mean AFM_1 concentration in powder milk samples (61 ng/kg) collected from municipal day-care centers and elementary schools, an average milk consumption of 0.412 kg/day, and a body weight of 23 kg, the estimated daily AFM_1 intake for children was 1.04 ng/kg BW. For adults, the estimated AFM_1 intake was 0.188 ng/kg BW/day, which was calculated using the mean AFM_1 concentrations in milk powder and fluid milk samples purchased from supermarkets (31 ng/kg), a daily milk consumption of 0.361 kg, and a body weight of 60 kg.

Cano-Sancho et al. (2010) used an average AFM_1 concentration of 9.69 ± 2.07 ng/kg found in three milk products (UHT milk, cheese, and yogurt samples) to estimate daily AFM_1 intake by four population groups in Catalonia, Spain. Average milk consumption for each of the four population groups - children (4–9 yr), teenagers (10–19 yr), adults (20–65 yr), and elderly (>65 yr) - was 0.532, 0.404, 0.305, and 0.407 kg/day, respectively; while the body weights were 26.2, 54.1, 73.7, and 73.3 kg, respectively. The estimated AFM_1 intakes for the four Spanish population groups were 0.21, 0.07, 0.04, and 0.05 ng/kg BW/day, respectively.

This study showed that 288 (64.0%) of the 450 milk samples collected from 45 provinces of Thailand were contaminated with AFM_1 equal to or more than the LOQ of 0.01 μg/L. Daily intakes of AFM_1 in five Thai population groups, calculated using corresponding AFM_1 concentrations in milk products consumed by each population group, were comparable to those of other regions and countries. Thai children (3–9 yr) had the highest AFM_1 intake, 0.26 ng/kg BW/day, which was comparable to those in the Far East (0.20 ng/kg BW/day), France (3–14 yr; 0.22 ng/kg BW/day), Brazil (adults, 0.188 ng/kg BW/day), and Spain (4–9 yr, 0.21 ng/kg BW/day). Thailand is one of several countries that have not yet established regulatory limits for AFM_1 in raw milk and milk products. The present study and our two previous reports (Ruangwises & Ruangwises, 2009, 2010) suggest regular monitoring of raw milk and milk products, and regulatory limits for AFM_1 to ensure the quality of raw milk and milk products in Thailand.

4. Acknowledgments

This study was financially supported in part by a 2005–2006 Fiscal Research Grant. The authors thank Mrs. Chailai Kuwattananukul and Mr. Sumate Thiangtham for their technical help, and Mr. Christopher Salisbury, Chiang Mai University, Thailand, for reviewing the manuscript.

5. References

Cano-Sancho, G., Marin, S., Ramos, A.J., Peris-Vicente, J., & Sanchis, V. (2010). Occurrence of aflatoxin M_1 and exposure assessment in Catalonia (Spain). *Revista Iberoamericana de Micología,* Vol. 27, No. 3, (July-September 2010), pp. 130-135, ISSN 1130-1406

Cunniff, P.A. (Ed.). (1995). *Official Methods of Analysis* (16th edition), AOAC International, ISBN 0-935584-54-4, Gaithersburg, Maryland [Section: AOAC Official Method 986.16 - Aflatoxins M_1 and M_2 in fluid milk]

European Commission (2006). Commission Regulation (EC) No. 1881/2006 of 19 December 2006 setting maximum levels for certain contaminants in foodstuffs. *Official Journal of the European Union*, 20.12.2006, pp. L 364/5 - L 364/24, ISSN 1725-2555

FAO (Food and Agriculture Organization of the United Nations) (2004). *Worldwide Regulations for Mycotoxins in Food and Feed in 2003*, Food and Nutrition Paper 81, FAO, ISBN 92-5-105162-3, Rome

IARC (International Agency for Research on Cancer) (2002). *IARC Monographs on the Evaluation of Carcinogenic Risks to Humans, Vol. 82: Some Traditional Herbal Medicines, Some Mycotoxins, Naphthalene and Styrene*, World Health Organization (WHO), IARC Press, ISBN 92-832-1282-7, Lyon, France

Jay, J.M., Loessner, M.J., & Golden, D.A. (2005). *Modern Food Microbiology* (7th edition), Springer Science and Business Media, ISBN 0-387-23180-3, New York

JECFA (Joint FAO/WHO Expert Committee on Food Additives) (2001). *Safety Evaluation of Certain Mycotoxins in Food: Aflatoxin M_1*, WHO Food Additives Series No. 47, ISBN 92-4-120940-2, Geneva, Switzerland

Leblanc, J.C., Tard, A., Volatier, J.L., & Verger, P. (2005). Estimated dietary exposure to principal food mycotoxins from the First French Total Diet Study. *Food Additives and Contaminants, Part A*, Vol. 22, No. 7, (July 2005), pp. 652–672, ISSN 1944-0049

Ministry of Public Health (2003). Notification of the Ministry of Public Health No. 265 (B.E. 2545): Cow's milk. *The Royal Thai Government Gazette*, Vol. 120, Special Issue 4D, (January 2003), Bangkok, Thailand (In Thai)

National Bureau of Agricultural Commodity and Food Standards (2006). *Food Consumption Data of Thailand*, Ministry of Agriculture and Co-operatives, ISBN 974-403-423-8, Bangkok, Thailand (In Thai)

Ruangwises, S., & Ruangwises, N. (2009). Occurrence of aflatoxin M_1 in pasteurized milk of the School Milk Project in Thailand. *Journal of Food Protection*, Vol. 72, No. 8, (August 2009), pp. 1761–1763, ISSN 0362-028X

Ruangwises, N., & Ruangwises, S. (2010). Aflatoxin M_1 contamination in raw milk within the central region of Thailand. *Bulletin of Environmental Contamination and Toxicology*, Vol. 85, No. 2, (August 2010), pp. 195–198, ISSN 0007-4861

Shundo, L., Navas, S.A., Lamardo, L.C.A., Ruvieri V., & Sabino, M. (2009). Estimate of aflatoxin M_1 exposure in milk and occurrence in Brazil. *Food Control*, Vol. 20, No. 7, (July 2009), pp. 655–657, ISSN 0956-7135

U.S. FDA (United States Food and Drug Administration) (1996). *Guidance for Industry – Q2B Validation of Analytical Procedures: Methodology*, Department of Health and Human Services, Center for Drug Evaluation and Research, Rockville, Maryland, Available from: http://www.fda.gov/downloads/Drugs/GuidanceComplianceRegulatory Information/Guidances/ucm073384. pdf

22

Influence of Soluble Feed Proteins and Clay Additive Charge Density on Aflatoxin Binding in Ingested Feeds

William F. Jaynes and Richard E. Zartman
Texas Tech University
USA

1. Introduction

Hartley et al. (1963) isolated and identified toxic metabolites of *Aspergillus flavus* as aflatoxins B1, B2, G1, and G2; named from the blue and green fluorescence of the compounds under ultraviolet light. The aflatoxins are a group of mycotoxins produced by *Aspergillus* fungi that are both toxic and carcinogenic to animals and humans (Murphy et al., 2006). Aflatoxin B1 (AfB1) and mixtures of B1, G1, and M1 are proven human carcinogens (IARC, 1993). Aflatoxin B1 is the most toxic (Figure 1), most abundant, and the most potent natural carcinogen known (Squire, 1981). An estimated 4.5 billion people living in developing countries are chronically exposed to uncontrolled amounts of aflatoxins (Williams et al., 2004). Iraq produced aflatoxins for use in biological warfare between 1985 and 1991, but the weapons had little military value (Zilinskas, 1997). After ingestion, aflatoxins are converted to the reactive 8,9-epoxide form that can bind to DNA and proteins. Aflatoxin consumption results in diseases that are loosely called aflatoxicoses. Chemically, aflatoxins are derivatives of difuranocoumarin (Bennett & Klich, 2003).

Various methods have been used to reduce the toxicity of aflatoxin-contaminated grains. Cleaning to remove damaged corn kernels is sometimes effective in reducing aflatoxin concentrations, but undamaged kernels can also contain high aflatoxin concentrations (Vincelli et al., 1995). Treatment with anhydrous ammonia can be used to detoxify grain that is to be used on the farm (Vincelli et al., 1995). Brekke et al. (1977) ammoniated trout feed contaminated with 180 μg/kg aflatoxins, which inactivated the aflatoxins and reduced the carcinogenicity to a level not significantly different than the control. Grove et al. (1981) examined the ammoniation products of aflatoxin model coumarins and determined that the keto group in the cyclopentene ring is required for ammonia-induced decomposition. Nixtamalization is an Aztec word that means lime-cooked corn (Herrera et al., 1986) and is an ancient method used to soften grain before it is used in foods. Nixtamalization also increases protein quality and niacin bioavailability (Sefa-Dedah et al., 2004). The strong alkalinity imparted by lime (CaO, $Ca(OH)_2$) might have a similar effect on aflatoxins as ammonia. Arrriola et al., (1988) examined the effect of nixtamalization on aflatoxin fate during tortilla preparation using 2-10% CaO. Nixtamalization decreased aflatoxin concentrations at even the lowest CaO concentrations. However, nixtamalization did not reduce the 1360-1896 μg/kg initial aflatoxin concentrations down to the allowable value of 20 μg/kg. Ammoniation and other detoxifying methods, however, are not approved or sanctioned by the U.S. Food and Drug Administration (Sweets & Wrather, 2009).

Clays are colloidal or near-colloidal hydrous aluminum silicates that are more or less plastic when moist (Bates, 1969). Bentonites are natural materials that dominantly consist of clay minerals in the smectite group (Hosterman & Patterson, 1992). Smectites are a group of phyllosilicate minerals that include montmorillonite, beidellite, nontronite, saponite, and hectorite (Odom, 1984). Bentonite was the name given by Wilbur G. Knight in 1898 to deposits in the Benton Shale near Rock River, Wyoming (Hosterman & Pattterson, 1992). Bentonite deposits contain altered volcanic ash glass shards and field evidence suggests that bentonites formed from ash that fell into shallow lakes or seas (Bates, 1969). Smectites, vermiculites, talc, and pyrophyllite are structurally-related 2:1 clay minerals, but talc and pyrophyllite have zero layer charge and do not expand in water. Smectites and vermiculites characteristically expand in water along the crystallographic c-axis to form an interlayer region. Structurally, the 2:1 clay minerals consist of an octahedral aluminum or magnesium oxide sheet sandwiched between two tetrahedral silica sheets. Unlike talc and pyrophyllite, smectites and vermiculites have isomorphic chemical substitutions of Al^{3+} for Si^{4+} (tetrahedral charge) and Mg^{2+} for Al^{3+} (octahedral charge) that impart a negative charge to the mineral surface and a cation exchange capacity (CEC). Inorganic exchange cations, such as Na^{+} and Ca^{2+}, compensate for the negative charge on smectite and vermiculite surfaces. Smectite CECs range from 50 to 129 cmol/kg and vermiculite CECs range from 130 to 210 cmol/kg (Mermut & Lagaly, 2001; van Olphen & Fripiat, 1979). Bentonites are relatively pure, commercial deposits of smectites found throughout the world that can be mined, but smectites, vermiculites, and other clay minerals commonly also occur in soils and sedimentary deposits. In soils, clays retain exchangeable cations, such as Ca^{2+}, Mg^{2+}, and NH_4^{+}, which are essential plant nutrients. Vermiculite expansion (Ca-saturated) in water is limited to ~1.5 nm, but smectites (Ca-saturated) expand to 1.9 nm or more (McEwan & Wilson, 1984). Free expansion of smectites in water is almost unlimited. Completely dispersed smectite particles consist of single unit cells with no c-axis direction repeat distance (Eberl et al., 1998). Interlayer expansion of air-dried (32% relative humidity) samples of Na-smectite and Ca-smectite is illustrated in Figure 1 with characteristic basal spacings after McEwan & Wilson (1984). The single water layer and Na^{+} cations in Na-smectite interlayers (Figure 1) is ~0.25 nm and the water bilayer and Ca^{2+} cations in Ca-smectite interlayers is ~0.52 nm. Calcium- and magnesium-saturated smectites yield similar basal spacings. The replacement of inorganic exchange cations in smectites and vermiculites with organic cations can result in interlayer expansion. Jaynes & Boyd (1991) exchanged the organic cation, hexadecyltrimethylammonium (HDTMA), for the inorganic cations in a low-charge smectite, a high-charge smectite, and a vermiculite, which produced expanded interlayer basal spacings of 1.8, 2.3, and 2.8 nm, respectively. Polymer adsorption to clays can also produce interlayer expansion. Polyvinylpyrrolidone (PVP) expanded SAz high-charge montmorillonite to ~2.3 nm (Blum & Eberl, 2004). Smectite layer charge ranges from 0.2 to 0.6 per $O_{10}(OH)_4$ unit , whereas, vermiculite layer charge ranges from 0.6 to 0.9 (Bailey, 1980). Vermiculites are hydrous minerals that form by the weathering of micas and have a platey mica-like morphology (Newman & Brown, 1987). The name "vermiculite" is more commonly used for macroscopic heat-expanded (800 - 1100 °C) vermiculite particles that are used as a packing material, plant media, insulation, and construction material. This heated vermiculite should not be be confused with the natural mineral because heat-treatment greatly alters the properties. Sepiolite, palygorskite, and the zeolites are structurally much different than smectites and vermiculites and do not have interlayers. Sepiolite and palygorskite are fibrous, non-expandable, hydrous magnesium aluminosilicates. There are some health concerns about the possible effects of inhaled fibrous minerals. Bellman et al. (1997) used intratracheal instillation studies in rats to evaluate the carcinogenic potential of sepiolites. A short-fiber sepiolite from Spain showed no evidence of carcingenic potential, but a long-fiber sepiolite from China had a more pronounced fibrotic response (Bell et al., 1997). Sepiolite (CEC = 20–40 cmol/kg) and palygorskite (CEC = 5-30 cmol/kg) contain internal channels with exchangeable cations and water (Singer, 2002). Zeolites (e.g. clinoptilolite, erionite, analcime, mordenite) are framework-structure, three-dimensional aluminosilicate minerals with interconnected channels and cages that contain exchangeable cations (CEC = 220-570 cmol/kg) and adsorbed water (Boettinger & Ming, 2002). The internal channels in sepiolite (0.37 x 1.06 nm), palygorskite (0.37 x 0.64 nm), and zeolites (0.26 x 0.26 to 0.74 x 0.74 nm) are too small to accomodate aflatoxins. Hence, aflatoxins can only adsorb to external sites on these minerals.

Commercial clay additives have been used to prevent caking and improve the physical properties of animal feeds. The decreased toxicity of aflatoxins observed for contaminated animal feed mixed with clay feed additives has stimulated research on clay additives to prevent mycotoxicosis. The commercial clay feed additives, Novasil, Novasil plus, Astra-Ben 20, and Astra-Ben 20A, are bentonites that primarily consist of the smectite group mineral, montmorillonite. Animal feeding studies have demonstrated that Novasil, Novasil plus, Astra-Ben 20, Astra-Ben 20A, Na-bentonite, zeolite, and sepiolite feed additives can effectively reduce or prevent the toxicity caused by feed contaminated with *Aspergillus* mycotoxins, such as AfB1 (Phillips et al., 1988, 1995; Scheideler 1993; Schell et al., 1993a, 1993b; Edrington et al., 1996; Abdel-Wahhab et al., 1999; Miazzo et al., 2000; Diaz et al., 2004; Pimpukdee et al. 2004; Bailey et al., 2006; Fairchild et al., 2008; Magnoli et al., 2008). Ruminant animals, such as cattle and sheep, can tolerate higher aflatoxin levels and longer low-level intake periods than simple-stomached animals (Vincelli et al., 1995). The adsorption of aflatoxins to ingested soil minerals might partly explain the greater aflatoxin tolerance of ruminants. Soil ingested by cattle averaged 14% of the dry weight of fecal matter and increased as forage availability decreased (Mayland et al., 1975). Soil ingestion by grazing sheep in March exceeded 30% of dry matter intake at 2 of the 11 sites in mid-Wales (Abrahams & Steigmajer, 2003). Winfree and Allred (1992) measured significant aflatoxin adsorption to bentonite from methanol/water, which is commonly used in the extraction and measurement of aflatoxins in contaminated feed. Gallo et al. (2010) developed a more aggressive extraction procedure using acetone rather than methanol to more accurately measure aflatoxins in feeds that contain feed additives. Deng et al. (2010) measured smectite interlayer expansion of >1.2 nm that was stable to 400 °C after AfB1 treatment, which demonstrated that AfB1 adsorbs to interlayer clay surfaces. Interlayer clay surfaces account for most of the ~800 m^2/g surface area of smectites, such as montmorillonite. From infrared spectroscopy, Deng et al. (2010) concluded that hydrogen bonds between AfB1 carbonyl groups and the hydration water of exchangeable cations in clays is the dominant bonding force under humid conditions. Aflatoxin adsorption from aqueous corn and peanut meal to feed additives was consistent with animal feeding studies that used the feed additives, Novasil, Novasil Plus, Astra-Ben 20, Astra-Ben 20A, sepiolite, and activated (Norit-A) carbon (Jaynes et al., 2007; Seifert et al., 2010). Feed additives that effectively reduced or prevented aflatoxin toxicity in feeding studies adsorbed more AfB1 from aqueous corn and peanut meal.

Animal feedings studies (*in vivo*) are the surest way to identify effective feed additives, but are much too expensive for routine use. Hence, various approaches have been used for the *in vitro* evaluation of potential feed additives. Feed additive *in vitro* test methods should produce results that are consistent with animal feeding studies. Unfortunately, most animal feeding studies evaluated only one or a few feed additives, which makes comparisons of relative effectiveness difficult. This also frustrates efforts to identify feed additive properties related to effective aflatoxin toxicity reduction. In contrast, Schell et al. (1992a) evaluated a sodium calcium bentonite (Novasil), a calcium bentonite (Astra-Ben 20), a sodium bentonite (FD-181), a zeolite (Zeobrite), a palygorskite (Min-U-Gel), and a sepiolite (Sepiolgel UF). Phillips et al. (1988) measured AfB1 adsorption from water to aluminas, zeolites, silicas, phyllosilicates, a Mn-exchanged phyllosilicate, and an acid-activated phyllosilicate. Winfree & Allred (1992) measured a 70% reduction in AfB1 concentrations in methanol/water extracts of trout feed 1 hour after 10% bentonite was added to moistened feed. Grant & Phillips (1998) fitted AfB1 adsorption from water to Novasil data to the Langmuir and other isotherm equations to calculate adsorption capacities. Jaynes et al. (2007) measured AfB1 adsorption from aqueous corn meal after extraction with 60% methanol. Siefert et al. (2010) measured AfB1 adsorption to clays in aqueous peanut meal and total extractable aflatoxins in peanut meal/clay water extracts. Vekiru et al. (2007) measured AfB1 adsorption to bentonites and charcoal from acetate buffer, artificial gastric fluid, and from gastric fluid. Thieu and Pettersson (2008) measured AfB1 adsorption to zeolite and bentonite in simulated gastrointestinal fluids. Dixon et al. (2008) used X-ray diffraction, infrared spectroscopy, cation exchange capacity, and particle size to measure smectite purity and the Langmuir isotherm to measure aflatoxin adsorption from water. To identify feed additives that can effectively bind aflatoxins in ingested feed, *in vitro* tests should model the environment of ingested feed/feed additive as accurately as practical. Proteins are soluble ionic polymers of amino acids that can adsorb to clay surfaces. Living organisms, animal feed, and human food contain proteins. Lipson & Stotzky (1984) showed that the proteins chymotrypsin, ovalalbumin, and lysozyme adsorbed to montmorillonite and reduced adsorption of the Reovirus. Perez-Castells et al. (1985) puried collagen protein from calf skin and adsorbed 0.4 mg of collagen to 1 mg of sepiolite. Similarly, Garwood adsorbed 97 g of the enzymatic protein, glucose oxidase, to 100 g of Na-montmorillonite. Ralla et al. (2010) used a smectite clay to adsorb and separate proteins. Aflatoxin-contaminated peanuts are unsuitable as food or feed and are used to produce peanut oil (Siefert et al., 2010) and defatted peanut meal is a byproduct of peanut oil production. Siefert et al. (2010) used Astra-Ben 20A to bind aflatoxins in defatted peanut meal in the insoluble residue and produce an aflatoxin-free water-soluble protein extract. Addition of 0.2% Astra-Ben 20A to peanut meal decreased total aflatoxins in the soluble protein extracts from 50 to 4.8 µg/kg. The addition of 2% Astra-Ben 20A decreased total aflatoxins in the soluble fraction to 0 µg/kg, but decreased protein recovery by 37%. Soluble protein adsorption to clay additives in ingested feed might adsorb to clay feed additives and block aflatoxin adsorption. Other soluble compounds in feed, such as polysaccharides, might also adsorb to clay feed additives and block potential aflatoxin binding sites. Chenu et al. (1985) adsorbed ~400 mg of the polysaccharide, scleroglucan, to 1 g of Na-montmorillonite.

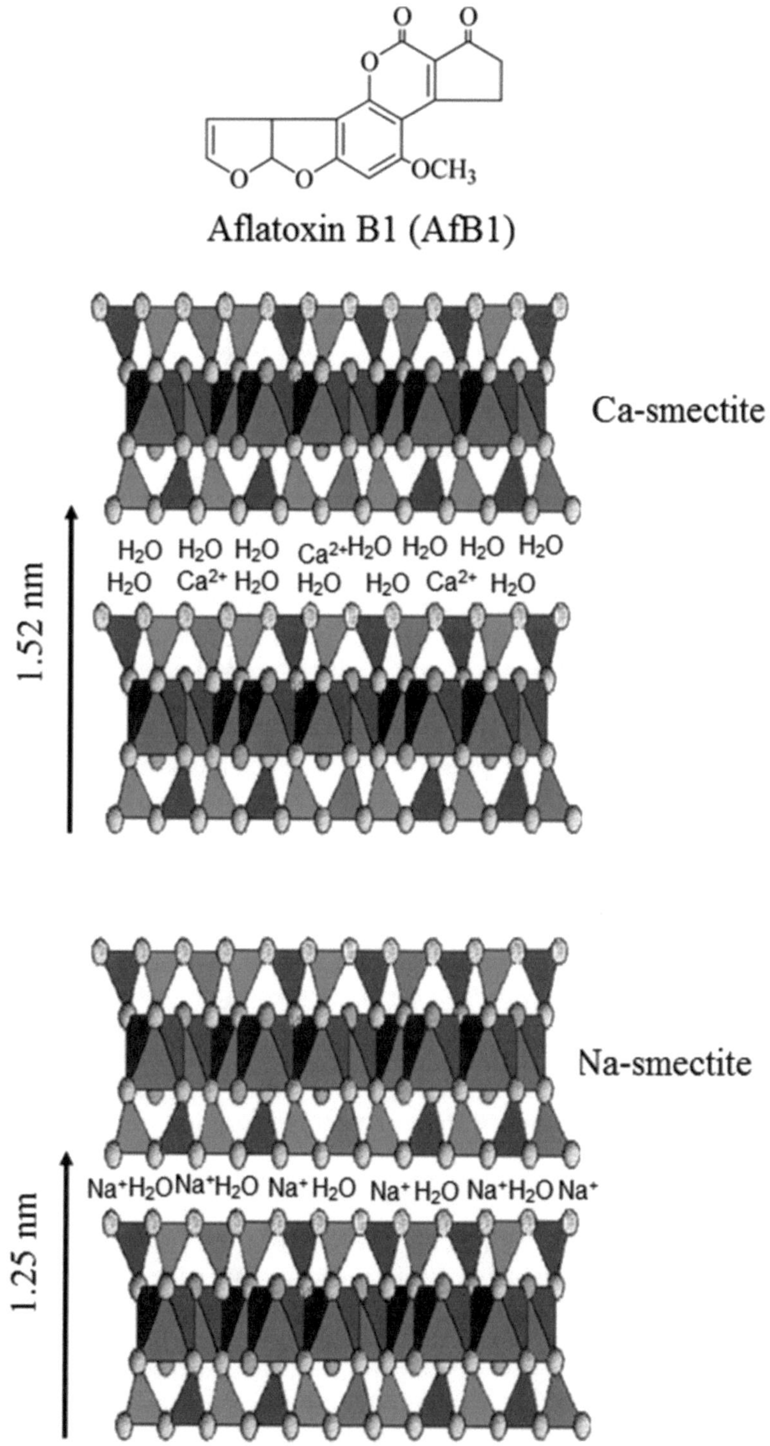

Fig. 1. Chemical structure of aflatoxin B1 and illustration of interlayer expansion of air-dried calcium-saturated and sodium-saturated smectites.

Decker (1980) measured the adsorption of 1 mg AfB1 to 100 mg of activated carbon (Norit-A) from aqueous media at pH 7 and suggested activated carbon might be used to prevent animal and human absorption of aflatoxins from contaminated foodstuffs. Early animal feeding studies by Hatch et al. (1982), Dalvi & Ademoyero (1984), and Dalvi & McGowan (1984) indicated that activated carbon can reduce aflatoxicosis. Similarly, bentonites and activated carbon reduced excretion of aflatoxin M1 in milk cows, turkey poults, and goats in feeding studies by Veldman (1992), Edrington et al. (1996), and Rao & Chopra (2001). However, animal feeding studies by Kubena et al. (1990), Bonna et al. (1991), Edrington et al. (1996), and Diaz et al. (2004) concluded that activated carbon does not effectively reduce aflatoxin toxicity to fed animals or is not as effective as clay additives. Diaz & Smith (2005) did not recommend routine inclusion of activated charcoal in diets after reviewing activated charcoal use in animal feeding studies. Vekiru et al. (2007) measured AfB1 binding by various sorbents and identified another drawback to activated carbon use as a feed additive; unlike the bentonites, activated carbon adsorbed the essential nutrients, vitamin B12 and biotin.

Clay feed additives are marketed and sold in the United States as anti-caking agents to improve the physical properties of feed because U.S. Food and Drug Administration (FDA) regulations do not permit feed additive companies to claim that feed additives can bind mycotoxins and reduce mycotoxicoses. Therefore, feed additive companies have little financial incentive to develop additives with improved mycotoxin binding. Feed additives are mixed with dry feeds and, hence, mycotoxin binding to clays must occur after ingestion. During digestion, pH, feed composition, and other factors can affect mycotoxin binding to feed additives. Mycotoxin adsorption to feed additives can sequester the toxins and limit absorption by animals or humans. However, feed additives that effectively remove mycotoxins from water might not prevent toxicity to animals from contaminated feed because the adsorption of soluble feed or digestive compounds might block mycotoxin adsorption to feed additives.

In this study, the effects of soluble feed compounds and clay layer charge on the adsorption of AfB1 to commercial feed additive clays, reference clays, and activated carbon will be examined. Adsorption of AfB1 to a variety of commercial feed additives and reference clays were measured from water and aqueous corn meal. Clays and activated carbon were treated with aqueous corn and peanut meal extracts to simulate the adsorption of soluble feed compounds to feed additives ingested with feed. The physical and chemical properties of materials treated with corn- and peanut-meal water extracts will be measured and related to aflatoxin adsorption. The adsorption of AfB1 to clays with a wide range in layer charge will be measured to determine layer charge effects on AfB1 adsorption.

2. Materials and methods

The reference clay samples, SWy-2 (SWy), SAz-1 (SAz), SepSp-1 (SepSp), and SHCa-1 (SHCa) were obtained from the Source Clay Repository of the Clay Minerals Society located at Purdue University (West Lafayette, IN). The SWy sample is a low-charge sodium montmorillonite from Wyoming and SAz is a high-charge calcium montmorillonite from Arizona. The SepSep sample is a sepiolite from Spain and SHCa is a hectorite from California. Sepiolite is a fibrous magnesium silicate clay mineral and hectorite is a low-charge magnesium layer silicate. Both montmorillonite and hectorite are expanding smectite-group minerals. Novasil and Novasil plus (low-charge montmorillonite) are products of Trouw Nutrition, which is a division of Englehard Corporation, Chemical

Catalysts Group (600 East McDowell Road, Jackson, MS). A vermiculite sample, VSC, from a mine in South Carolina was obtained from the W.R. Grace Company. Vermiculites have a higher CEC/layer charge than smectites. Clay mineral samples were Na-saturated by treatment with NaCl, <2 μm clay fractions were separated by centrifugation, and the <2 μm clay fractions were freeze-dried. Powdered activated carbon (alkaline Norit-A decolorizing carbon) was obtained from Fisher Scientific. Dispersions of the <2μm clays and activated carbon were prepared using an ultrasonic probe and a vortex mixer. Corn meal was purchased at a local grocery and defatted peanut meal was prepared by grinding and acetone-extraction of raw peanuts. Bovin serum albumin (BSA) protein, Bradford Reagent (Bradford, 1976) for total protein measurements, castor seed protein, and peanut protein (peanut agglutinin, PNA) were obtained from Sigma-Aldrich. Aflatoxin B1, rabbit anti-aflatoxin B1 antibody, aflatoxin B1-BSA conjugate, goat anti-rabbit horse radish peroxidase (HRP) conjugate, phosphate buffered saline with 0.05% Tween 20 (PBST), and o-phenylenediamine dihydrochloride (OPD) substrate tablets were obtained from Sigma-Aldrich. Stable AfB1 stock solutions were prepared in 95% toluene/5% acetonitrile and stored in a freezer (AOCS, 1999a). The aflatoxin stock solutions were calibrated by measuring the UV absorbance of AfB1 dissolved in methanol at 360 nm (AOCS, 1999a). A Shimadzu UV-1800 spectrometer and quartz glass cuvets were used to collect spectra of AfB1 solutions from 190 to 500 nm. Enzyme-linked immunoassay microplates were washed and read using a Bio-Tek ELx50 plate washer and a ELx800uv plate reader.

2.1 Preparation of reduced-charge clays

The cation exchange capacity (CEC) and layer charge of lithium-saturated montmorillonites decrease after heat treatment. Layer charge is a measure of the charge density of specific mineral particles, whereas, CEC is a bulk measure of the average charge density of all the mineral particles in a sample. Layer charge can be measured using the n-alkylammonium method of Lagaly & Weiss (1976). The CEC is directly related to layer charge in pure monomineralic samples, but not in samples with a mixed mineralogy. The decrease in montmorillonite CEC/layer charge by lithium/thermal treatment is termed the Hofmann-Klemen effect after Hofmann & Klemen (1950). Lithium charge reduction makes it possible to vary the CEC/layer charge of a particular clay sample and examine the effect of layer charge on other properties. Samples of <2μm SAz-1 montmorillonite with 35% Li/65% Na and 50% Li/50% Na on the cation exchange sites were prepared according to the method used by Jaynes & Bigham (1987). The 0.35Li,0.65Na-SAz and 0.50Li,0.50Na-SAz samples were heated at 250 °C in quartz glass crucibles to produce about 35% and 50% decreases in CEC. The reduced-charge SAz clays formed were designated 0.35Li-SAz and 0.50Li-SAz. The 0.35Li-SAz sample should have a CEC/layer charge comparable to SWy and the 0.50Li-SAz sample should have a CEC/layer charge about 20% less than SWy.

2.2 AfB1 adsorption from water

Batch adsorption isotherms (6 points in triplicate with 4 blanks) from water were prepared with an initial concentration of 1 μg AfB1/mL in 5-mL aqueous dispersions. The aqueous dispersions contained 10 to 180 μg of clay or activated carbon in 15-mL polypropylene centrifuge tubes. A stock solution containing 100 μg AfB1/mL in acetonitrile was prepared and aliquots of the stock solution were diluted with water and used to deliver aflatoxin to isotherm solutions (50 μL stock + 0.95 mL of water = 5 μg AfB1/mL). Blanks containing only AfB1 and water were prepared and samples and blanks were thoroughly mixed using a vortex mixer. After overnight agitation on a reciprocating shaker, the tubes were centrifuged, and the supernatants were passed through 0.2-μm filters and collected in 20-mL plastic vials. The AfB1 adsorption data were fitted to the Langmuir equation and monolayer adsorption capacities (X_m) were calculated (Hiemenz, 1986). Grant & Phillips (1998) similarly used the Langmuir equation and other model equations to fit data from AfB1 adsorption to Novasil from water isotherms.

2.3 AfB1 retention from aqueous corn meal with 60% methanol extraction

Aflatoxin retention from aqueous corn meal dispersions after 60% methanol extraction was used as a more applied and more conservative measure of aflatoxin binding to feed additives. Batch adsorption isotherms from aquous corn meal (6 points in triplicate with 4 blanks) were prepared with an initial concentration of 1.5 μg AfB1/mL (3 μg AfB1/g corn meal) in 2-mL aqueous dispersions containing 1 to 20 mg (0.1 to 2%) clay or 5 to 100 mg (0.5 to 10%) of activated carbon. Blanks were prepared similarly using only AfB1, corn meal, and water. The 3 μg AfB1/g corn meal is comparable to highly-contaminated (3000 μg aflatoxins/kg) grain. The samples were thoroughly mixed using a vortex mixer. After overnight agitation on a reciprocating shaker, 8 mL of a 60% methanol/40% 2M NaCl extracting solution were mixed with the samples using a vortex mixer. The tubes were centrifuged and the supernatants passed through 0.2-μm filters and collected in 20-mL plastic vials. Aflatoxins are more soluble in methanol than in water. The 60% methanol extraction was modified from the Asis et al. (2002) procedure to extract and measure aflatoxins in peanuts. The AOCS method for aflatoxins in corn, cottonseed, peanuts, and peanut butter similarly uses an 80% methanol/20% water extraction (AOCS, 1999b). The modified procedure first equilibrates AfB1/corn meal/clay in water (more like the environment of ingested aflatoxins) prior to extraction with 60% methanol. This contrasts with the immediate 60% methanol extraction used by Asis et al. (2002). The AfB1 adsorption data were fitted to a line using least squares linear regression and the slopes or adsorption coefficients (K_d) were calculated to compare relative adsorption capacities. Three-point isotherms were used to measure the effect of layer charge on AfB1 retention by natural and reduced-charge clays.

2.4 AfB1 retention from water with 60% methanol extraction

Clay and activated carbon samples (2 mg) in 15-mL polypropylene centrifuge tubes were equilibrated in 2 mL of water containing 60 μg AfB1 and later extracted with 8 mL of 60% methanol to distinguish between the effects of aqueous corn meal and 60% methanol extraction on AfB1 adsorption. A control sample with 60 μg AfB1 in 2 mL of water was similarly extracted with 8 mL of 60% methanol. Blanks containing 2 mg of clay or activated carbon without AfB1 were used to compensate for minor UV absorbance at 360 nm caused by trace amounts of suspended clay. Samples, blanks, and control were centrifuged to sediment suspended clay/activated carbon after overnight equilibration on a shaker and the addition of 8 mL of 60% methanol. The amounts of AfB1 retained by the clays and activated carbon were measured using UV/visible spectroscopy based on differences in the absorbance of AfB1 at 360 nm between samples and the control.

2.5 ELISA AfB1 measurement

A modification of the Asis et al. (2002) enzyme-linked immunoassay (ELISA) method was used for measuring aflatoxins. The modified method is thoroughly described in Jaynes et al.

(2007). Briefly, 96-well polystyrene microplates were coated with AfB1-BSA conjugate, washed, and saved for later use. In the first step, aflatoxin standards, sample solutions, and anti-AfB1 antibody were added to AfB1-BSA coated plates. The method is a competitive ELISA technique because AfB1 in solution (from standards or samples) competes with AfB1-BSA bound to the microplates for rabbit anti-AfB1 antibody. The AfB1 in high AfB1-concentration samples or standards binds to most of the rabbit anti-AfB1 antibody, which is subsequently washed out of the microplates. The AfB1 in low AfB1-concentration samples or standards does not bind as much of the anti-AfB1 antibody as the AfB1-BSA attached to the plates and most of the anti-AfB1 antibody is retained in the microplate. Goat anti-rabbit-HRP antibody is then added to the microplates and binds to any rabbit anti-AfB1 antibody that is attached to the microplates. Any unattached goat anti-rabbit-HRP is subsequently washed from the microplates. In the final step, OPD substrate is added to the microplates and the horse radish peroxidase enzyme in the goat anti-rabbit-HRP that is attached to the plates catalyzes color development. The optical densities of the colored solutions that are produced are inversely proportional to AfB1 concentration. The high AfB1-concentration standards and samples are lightly-colored and the blanks are darkly-colored. The optical densities are measured using a microplate reader and AfB1 concentrations are calculated based on a plot of standard concentration versus optical density.

2.6 Treatment of clays and activated carbon with corn and peanut meal water extracts

Corn and peanut meal extracts from 100-g samples were used to treat 2-g clay and activated carbon samples to model feed samples ingested with 2% feed additive. Corn and peanut meal samples (100 g) were suspended in 800 mL of deionized water and agitated on a shaker for 1 hour. The corn and peanut meal suspensions were centrifuged and filtered to remove solids and mixed with samples (2 g) of Novasil plus, SWy, SAz, Astra Ben 20A, SepSp, and activated carbon. The clay and activated carbon samples that were treated with corn and peanut meal extracts were stirred for 1 hour, centrifuged, and washed with deionized water 3 times to remove unadsorbed material, frozen, and freeze-dried.

2.7 Basal spacings, organic carbon contents, and specific surface areas

Oriented clay samples were prepared by air-drying suspensions on glass slides. The SWy-PNA sample was prepared by treating 30 mg of SWy dispersed in 5 mL of deionized water with 10 mg of Sigma-Aldrich peanut agglutinin dissolved in pH 5 acetate buffer. The SWy-PNA sample suspension was washed three times with deionized water before drying on a glass slide. The glass slides were further dried in a vacuum dessicator over anhydrous $CaCl_2$ and stored until collection of X-ray diffraction patterns. X-ray diffraction patterns were collected for the oriented clays by scanning from 2 to 15 °2θ using CuKα radiation and a Philips or Rigaku diffractometer interfaced to a computer. Organic carbon contents were measured using a LECO carbon analyzer. Clay and activated carbon samples were outgassed at 120 °C and nitrogen specific surface areas were measured by the single-point method using a Micromeritics Flowsorb II model 2300 surface area meter using a 30% N_2 / 70% He carrier gas and liquid N_2. The N_2 surface areas of expandable clays, such as montmorillonite, are attributed to external surfaces because the interlayers collapse under the dry conditions required for measurement. Specific surface areas were also measured by ethylene glycol monoethyl ether (EGME) adsorption using the method of Carter et al. (1986). Samples were dried over P_2O_5 in a vacuum dessicator, weighed, and subsequently transferred to another vacuum dessicator with $CaCl_2$-EGME solvate. Three mL of EGME were added to the (~1g) samples and excess unadsorbed EGME was removed under vacuum until constant weight was achieved. Surface areas were calculated using the weight of adsorbed EGME. Specific surface areas measured using EGME are attributed to total surface area because EGME is adsorbed into and expands the interlayers.

3. Results and discussion

3.1 AfB1 adsorption from water

The adsorption of AfB1 from water by activated carbon (X_m=179 g/kg) and the montmorillonites Novasil (X_m=248 g/kg), SWy (X_m=239 g/kg), and SAz (X_m=166 g/kg) were comparable, but SepSp (X_m=87 g/kg, sepiolite) adsorbed much less AfB1 (Figure 2). Monolayer AfB1 adsorption capacities (X_m) calculated from a fit of the data to the Langmuir model indicate relative AfB1 adsorption by the materials. The relative amounts of AfB1 adsorbed from water do not correlate with feeding study results. Animal feeding studies have shown that Novasil and sepiolite both effectively reduce or prevent aflatoxin toxicity, but activated carbon was not effective. Although Novasil and sepiolite were equally effective in a feeding study by Schell et al. (1992a), the sepiolite (SepSp) adsorbed far less AfB1 from water than Novasil. Simple water adsorption isotherms are clearly relevant to environmental contaminants, but not necessarily to ingested toxins. The relative amounts of aflatoxins that evidently must bind to feed additives in animal feeding studies seem unrelated to maximum adsorption from water.

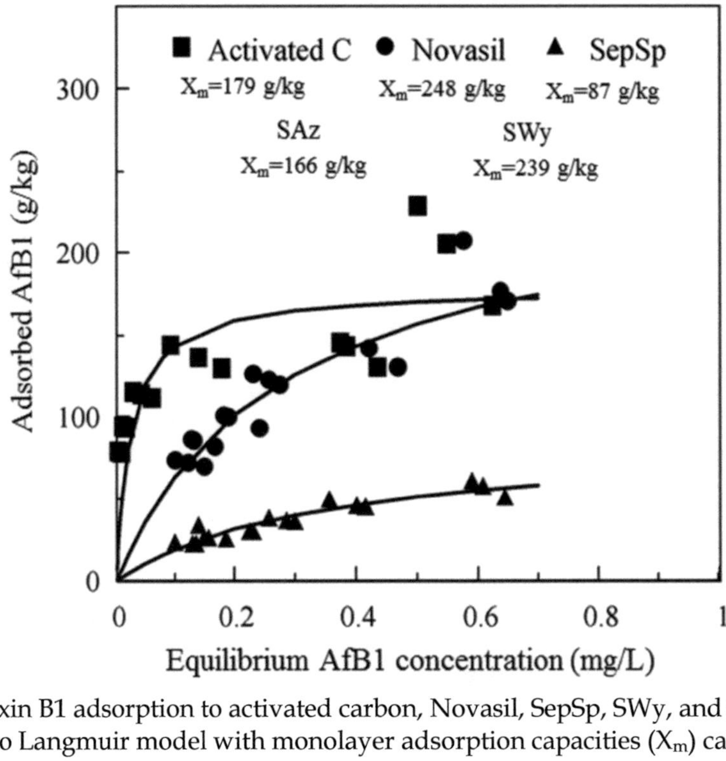

Fig. 2. Aflatoxin B1 adsorption to activated carbon, Novasil, SepSp, SWy, and SAz from water fitted to Langmuir model with monolayer adsorption capacities (X_m) calculated from Langmuir fit parameters.

3.2 AfB1 retention from aqueous corn meal after 60% methanol extraction

The amounts of AfB1 retained by the clays and activated carbon was 100 times less from aqueous corn meal extracted with 60% methanol (Figure 3) than from water (Figure 2); data were fitted to a linear model with adsorption coefficients (K_d) calculated from the linear fit.

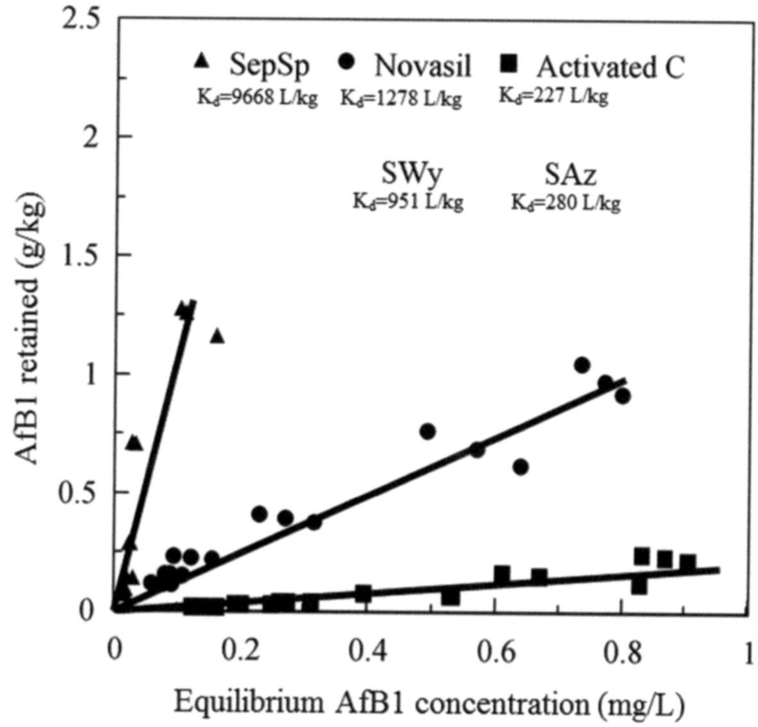

Fig. 3. Aflatoxin B1 retention from aqueous corn meal with 60% methanol extraction.

However, the relative amounts of AfB1 retained by Novasil (K_d=1278 L/kg), SepSp(K_d=9668 L/kg), and activated carbon (K_d=227 L/kg) are more consistent with animal feeding studies. Novasil and sepiolite effectively reduced aflatoxin toxicity in feeding studies and retained much more AfB1 from corn meal than activated carbon which was not effective. Aflatoxin adsorption from aqueous corn meal appears to model ingested aflatoxins more effectively than aflatoxin adsorption from water. The reference low- and high-charge montmorillonite samples, SWy (K_d=951 L/kg) and SAz (K_d=280 L/kg), retained significantly different amounts of AfB1. The SWy sample has a layer charge similar to Novasil and retained comparable amounts of AfB1. The high-charge SAz sample retained much less AfB1 that is comparable to the activated carbon. This suggests that high-charge montmorillonites would not reduce aflatoxin toxicity in ingested feed. However, no animal feeding studies have been identified that directly examined the effect of feed additive layer charge on toxicity reduction.

3.3 AfB1 retention from water after 60% methanol extraction

Both the soluble compounds in corn meal and extraction with 60% methanol probably contributed to the lower amounts of AfB1 retained by the clays and activated carbon in the aqueous corn meal adsorption isotherms (Figure 3) relative to the water isotherms (Figure 2). Aflatoxins are more soluble in methanol than in water and methanol might desorb weakly-adsorbed AfB1. To separate the effects of corn meal and methanol extraction, AfB1 adsorption from water to Novasil plus, SepSp, and activated carbon with 60% methanol extraction were measured (Figure 4). The Control, Novasil plus, SepSp, and activated carbon samples initially contained ~60 μg AfB1 in 2 mL of water. After overnight equilibration and the addition of 8 mL of 60% methanol, the control AfB1 concentration was 6.07 μg/mL. Adsorption of AfB1 by Novasil plus, SepSp, and activated carbon decreased the AfB1 concentration and the absorbance at 360 nm. Activated carbon (26.1 g/kg) retained far more AfB1 than Novasil plus (6.7 g/kg) and SepSp (4.0 g/kg), which indicates that 60% methanol extraction affected AfB1 retention by activated carbon less than the clays. Therefore, aqueous corn meal must reduce AfB1 retention to activated carbon more than methanol extraction. Activated carbon is generally used late in wastewater treatment after most organic materials (e.g. proteins, natural organic matter) have been removed by secondary wastewater treatment (Manahan, 2000). Activated carbon can be fouled or preloaded by the organic matter in untreated wastewater, which can limit uptake of pollutants, such as herbicides. Aqueous corn meal or feed contains a variety of soluble compounds, such as proteins, that can adsorb to the surface of activated carbon and limit access to internal adsorption sites. Soluble feed compounds can also adsorb to clays and block potential aflatoxin adsorption sites, but might not reduce aflatoxin adsorption to some clays as much as to activated carbon.

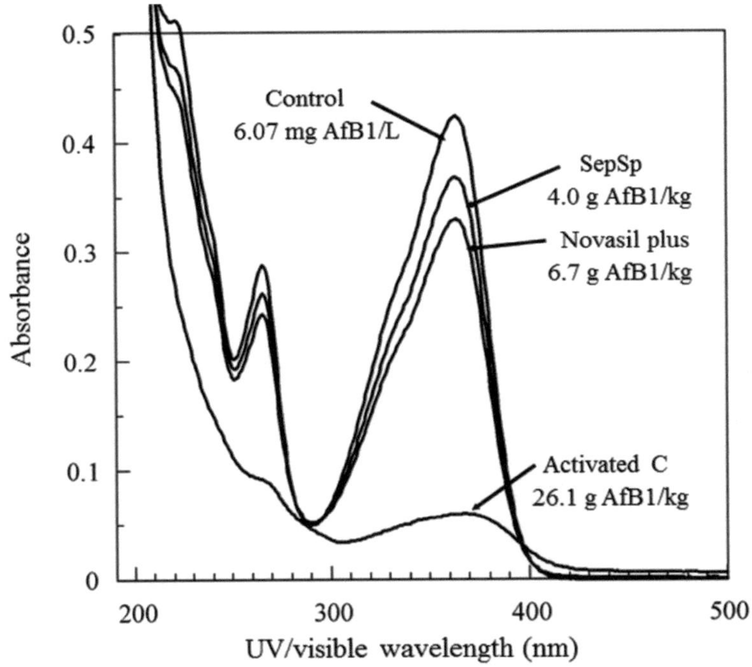

Fig. 4. Spectra of aqueous Control, Novasil plus, SepSp, and activated carbon samples (2 mg) after adsorption of aflatoxin B1 and extraction with 60% methanol. Maximum absorbance of aflatoxin B1 is at 360 nm. The amounts of aflatoxin adsorbed were calculated from the sample weights and absorbance differences from the control.

3.4 Corn and peanut meal water-extract-treated clay and activated carbon samples

The increased %C and %N contents after treatment with corn and peanut meal extracts indicate that soluble compounds were adsorbed to the clays and activated carbon (Table 1). The increased N contents suggest proteins were adsorbed. The higher protein content of defatted peanut meal (45 - 60%) relative to corn meal (8%) is reflected by the greater C and N contents of Novasil plus treated with peanut meal extract. Peanut agglutinin protein (PNA) is a 110,000 molecular weight lectin that contains four identical subunits. The pure PNA protein is readily available and is used to distinguish between normal and tumor tissues and for other medical uses (Vector Labs, 2011). X-ray diffraction patterns of SWy-Na, SWy-Mg, and SWy treated with PNA protein in Figure 5a are shown for comparison with the X-ray patterns of Novasil plus treated with corn meal and peanut meal extracts. Magnesium-and sodium-saturated SWy had basal spacings of 1.57 nm and 1.26 nm. Treatment with PNA protein greatly expanded the basal spacing to 2.65 nm, which indicates a ~1.65 nm interlayer thickness. Novasil plus (Figure 5b) has a basal spacing intermediate between SWy-Mg and SWy-Na, which indicates a mixture of Ca/Mg and Na. The Novasil plus sample was washed with NaCl, but apparently not enough to replace all of the Ca/Mg. Treatment with corn meal extract broadened, but did not appreciably shift the basal spacing of Novasil plus.

Sample	d001	C	N
	nm	%	%
Novasil plus	1.38	0.04	0.00
Novasil plus + corn extract	1.33	6.07	1.42
Novasil plus + peanut extract	2.56	22.10	5.83
SWy	nd	0.06	0.00
SWy + corn extract	nd	5.76	0.90
SWy-Na	1.26	nd	nd
SWy-Mg	1.57	nd	nd
SWy-HDTMA*	1.77	17.46	nd
SWy-PNA	2.65	nd	nd
SWy-ricin*	3.53	nd	nd
SAz	nd	0.07	0.00
SAz-HDTMA*	2.29	23.00	nd
SAz-PVP*	2.30	nd	nd
SAz + corn extract	nd	4.64	1.08
Astra Ben 20A	nd	0.04	0.01
Astra Ben 20A	nd	6.99	1.42
SepSp	na	0.12	0.00
SepSp + corn extract	na	4.27	0.83
Activated Carbon	na	78.40	0.68
Activated Carbon + corn extract	na	79.10	1.51

*na = not applicable, nd = not determined; SWy-ricin data from Jaynes et al., 2005; SWy-HDTMA and SAz-HDTMA data from Jaynes & Boyd 1991; and SAz-PVP data from Blum & Eberl 2004.

Table 1. Carbon and nitrogen contents from chemical analysis and d001 basal spacings from X-ray diffraction.

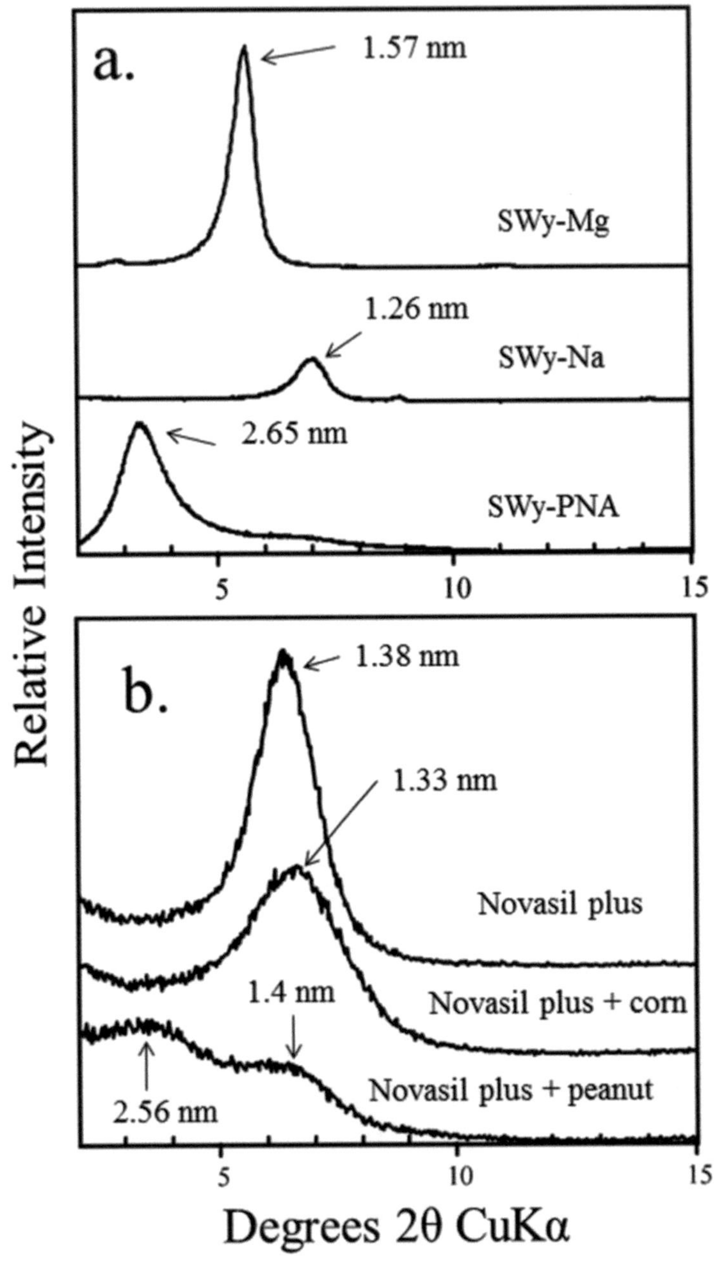

Fig. 5. X-ray diffraction patterns of a) SWy-Mg, SWy-Na, SWy-PNA; and b) Novasil plus, Novasil plus treated with corn meal extract, and Novasil plus treated with peanut meal extract.

Treatment with peanut meal extract yielded a broad peak centered at 2.56 nm similar to the 2.65-nm SWy-PNA peak. The peanut meal extract contains PNA protein, but also other proteins and soluble compounds. Adsorption of pure PNA protein yielded a sharp and intense 2.65 nm peak (Figure 5a), but adsorption of the mixture of soluble compounds in peanut meal extract yielded relatively weak, broad peaks. Proteins are polymers of amino acids and amino acid side-chain properties differ greatly. The 20 common amino acids have acidic, basic, aromatic, and aliphatic side chains (Stryer, 1975). Clays, such as montmorillonite, have negative charge sites compensated by inorganic cations. Basic amino acid side chains in proteins are protonated at pHs below the isoelectric point and can impart a net positive charge to the protein molecule. At pHs above the isoelectric point, proteins have a net negative charge. Proteins that have a large proportion of basic amino acid residues might adsorb more strongly to clays. The toxic castor seed globular protein (ricin) has an isoelectric point of 7.1 and contains 529 amino acid residues of which 43 are the basic amino acids histidine, lysine, and arginine (Merck, 2001; Katzin et al., 1991; Rutenber et al., 1991). Jaynes et al. (2005) examined castor seed protein adsorption to clay minerals and showed that adsorption was pH dependent. Castor protein adsorption at pHs below the isoelectric point (pH 5 and 7) was much greater than at pH 10 (Figure 6). Similarly, Garwood et al. (1983) adsorbed ~970 g/kg of the protein, glucose oxidase, to a Na-montmorillonite at pHs below the isoelectric point, which expanded the basal spacing to ~4.4 nm. The adsorption of purified peanut PNA protein and castor seed protein at pH 5 increased the basal spacing of SWy from 1.2 nm to 2.6 and from 1.2 nm to 3.5 nm (Table 1), respectively. Aflatoxin adsorption to clays does not show much pH dependence. However, protein adsorption to clays is pH dependent and might secondarily result in pH dependent aflatoxin adsorption.

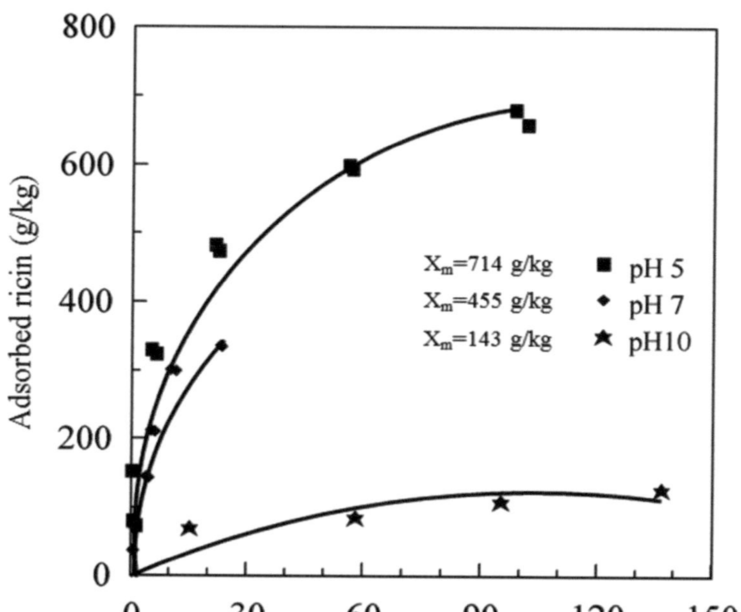

Fig. 6. Effect of solution pH on adsorption of castor protein (ricin) to SWy montmorillonite. Monolayer adsorption capacities (X_m) calculated from Langmuir model fit parameters.

The SepSp and activated C samples were not examined using X-ray diffraction because these samples are structurally different than montmorillonites and do not have basal spacings. The N_2 and EGME surface areas of the clays and activated carbon decreased after treatment with corn and peanut meal water extracts (Table 2). The adsorption of BSA protein and AfB1 generally decreased after treatment with corn and peanut meal extracts, which suggests that potential BSA and AfB1 adsorption sites were blocked by compounds that were adsorbed from the water extracts.

3.5 Layer charge and AfB1 retention from aqueous corn meal

Greater AfB1 retention by low-charge SWy than by high-charge SAz (Figure 3) suggests that aflatoxin adsorption to clays might be related to layer charge. The natural clays SHCa, SWy, SAz, and VSC range from low-charge to very high charge. The retention of AfB1 from aqueous corn meal decreased as layer charge increased (Figure 7). Aflatoxin retention by the high-charge vermiculite (VSC) was negligible compared to the low-charge hectorite (SHCa). The layer charge of hectorite (0.34 eq/$O_{10}(OH)_2$ formula unit) is about one-half of the South Carolina vermiculite (0.66 eq) layer charge (Laird et al., 1989). The layer charge of SWy (0.32 eq) is about 2/3 of the SAz (0.47 eq) layer charge (Laird et al., 1989). The effect of layer charge on AfB1 adsorption can also be determined using reduced-charge samples of one clay sample. The retention of AfB1 by SAz and reduced charge SAz was inversely proportional to layer charge (Figure 8). The 0.50LiSAz sample retained much more AfB1 than SAz. Low-charge clays more effectively adsorb AfB1 than high-charge clays. Lithium charge reduction would not be practical for clay feed additives, but layer charge might be used to identify natural clays that can effectively bind aflatoxins in feed. Other clay mineral properties, such as exchangeable cation and charge location (octahedral vs. tetrahedral), might also affect AfB1 adsorption.

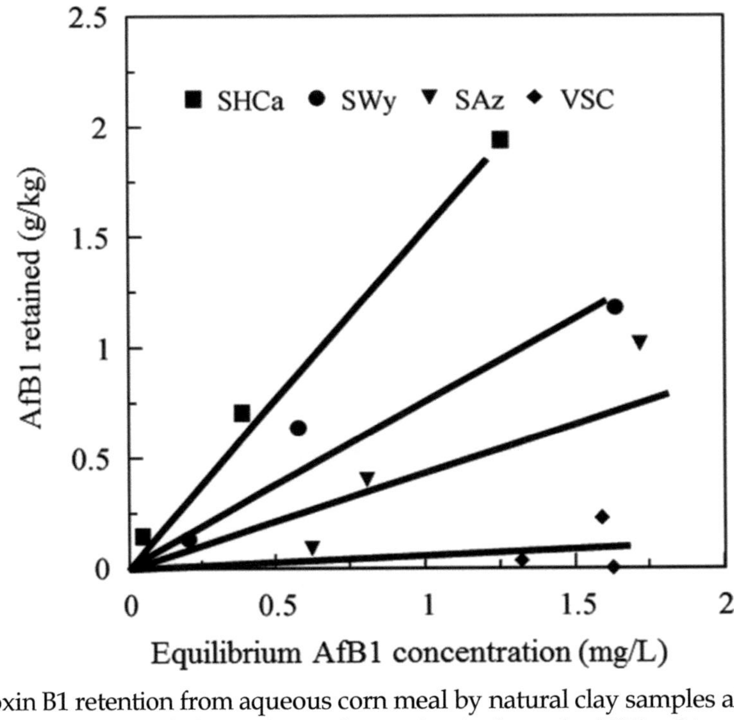

Fig. 7. Aflatoxin B1 retention from aqueous corn meal by natural clay samples after 60%methanol extraction with layer charge decreasing in the order VSC > SAz > SWy > SHCa.

Sample	N_2 surface area	EGME surface area	BSA adsorption	AfB1 adsorption
	m_2/g	m_2/g	g/kg	g/kg
Novasil plus	70	721	192	28
Novasil plus + corn extract	16	573	34	16
Novasil plus + peanut extract	8	167	11	14
SWy	29	691	192	29
SWy + corn extract	12	366	77	22
Astra Ben 20A	93	735	213	nd
Astra Ben 20A + corn extract	4	333	141	nd
SAz	63	715	212	nd
SAz + corn extract	55	112	152	nd
SepSp	250	460	53	6
SepSp + corn extract	92	385	77	3
Activated C	658	881	21	40
Activated C + corn extract	272	573	5	16
Activated C + peanut extract	180	494	11	10

Table 2. Nitrogen (N_2) and ethylene glycol monoethyl ether (EGME) surface areas, bovine serum serum (BSA) protein adsorption and aflatoxin B1 adsorption to clays and activated carbon. *nd = not determined.

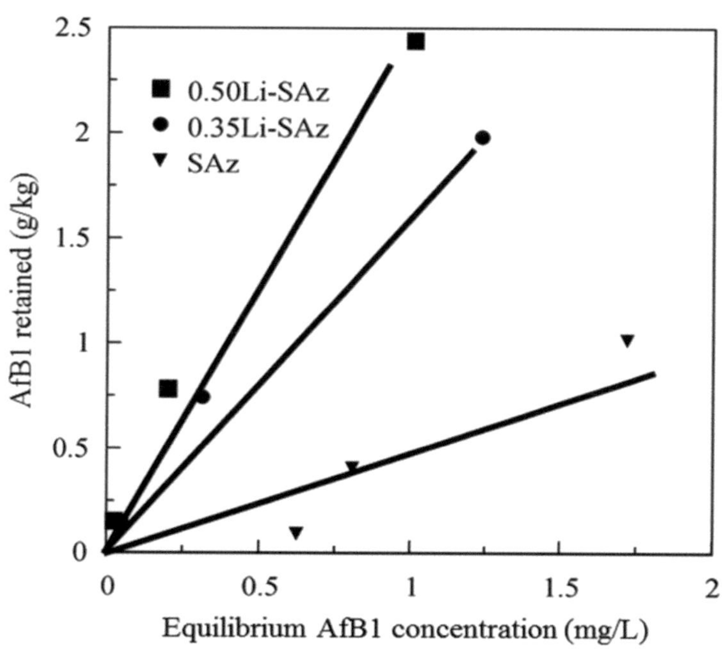

Fig. 8. Aflatoxin B1 retention from aqueous corn meal by natural (SAz) and reduced-charge SAz samples after 60% methanol extraction. Layer charge decreases in the order SAz > 0.35Li-SAz > 0.50Li-SAz.

4. Conclusions

The relative amounts of aflatoxin B1 retained from aqueous corn meal by clays and activated carbon after 60% methanol extraction were consistent with animal feeding study results, but animal feeding study results were not consistent with adsorption from water. Aflatoxin retention from aqueous feed seems to model aflatoxin adsorption to ingested feed additives more effectively than simple adsorption from water. Clay and activated carbon samples that were treated with aqueous extracts of corn and peanut meal had increased C and N contents and decreased surface areas. Smectite interlayer basal spacing was increased after treatment with the peanut extracts. The adsorption of bovine serum albumin protein and aflatoxin B1 also decreased after corn/peanut meal extract treatment. These data indicate that soluble compounds in corn and peanut meal extracts adsorbed to clay and activated carbon surfaces and adversely affected aflatoxin binding. Low-charge smectites, such as Novasil plus and reduced-charge SAz, retained more aflatoxin B1 from aqueous corn meal than higher charge minerals, such as vermiculite (VSC) and high-charge smectite (SAz). The selection of low-charge smectite feed additives might assure greater aflatoxin binding and toxicity reduction. The use of *in vitro* aflatoxin adsorption tests from aqueous feed might be used to identify feed additives that can effectively bind aflatoxins in ingested feed.

The correlation between animal feeding study results and *in vitro* aflatoxin adsorption from feed tests might be improved by use of gastrointestinal fluids or other factors to make the chemical environment of feed additives in adsorption tests more like ingested toxins. However, adsorption of soluble feed compounds to feed additives appears to have a strongly adverse effect on aflatoxin binding to ingested feed additives.

5. References

Abdel-Wahhab, M.A., Nada, S.A., & Amra, H.A. 1999. Effect of aluminosilicates and bentonite on aflatoxin-induced developmental toxicity in rat. *J. Appl. Toxicol.* 19, 199-204.

Abrahams, P.W., & Steigmajer, J. 2003. Soil ingestion by sheep grazing the metal enriched floodplain soils of mid-Wales. *Environ. Geochem. Health* 25, 17-24.

AOCS, 1999a. Aflatoxin standards. Official method Aj 0-88. Sampling and Analysis of Vegetable Oil Source Materials. American Oil Chemists Society.

AOCS, 1999b. Total aflatoxins (B1, B2, and G1) in corn, cottonseed, peanuts, and peanut butter. Official method Aj 6-95. Sampling and Analysis of Vegetable Oil Source Materials. American Oil Chemists Society.

Arriola, M.C., Porres, E., Cabrera, S., Zepeda, M., & Rolz, C. 1988. Aflatoxin fate during alkaline cooking of corn for tortilla preparation. *J. Agric. Food Chem.* 36, 530-533.

Asis, R., Di Paola, R.D., & Aldao, M.A.J. 2002. Determination of aflatoxin B1 in highly contaminated peanut samples using HPLC and ELISA. *Food and Agricultural Immunology* 14, 201-208.

Bailey, C.A., Latimer,G.W., Barr, A.C., Wigle, W.L., Haq, A.U., Balthrop, J.E. & Kubena, L.F. 2006. Efficacy of montmorillonite clay (Novasil plus) for protecting full-term broilers from aflatoxicosis. *J. Appl. Poult. Res.* 15, 198-206.

Bailey, S.W. 1980. Summary of recommendations of AIPEA nomenclature committee. *Clays Clay Miner.* 28, 73-78.

Bates, R.L. 1969. Clay. pp. 117-155. In: *Geology of The Industrial Rocks and Minerals*. Dover, New York.

Bell, B., Muhle, H., & Ernst, H. Investigations on health-related properties of two sepiolite samples. *Environ. Health Perspect.* 105 (Suppl 5) 1049-1052.

Bennett, J.W., & Klich, M. 2003. Mycotoxins. *Clinical Microbiology Reviews* 16, 497-516.

Blum, A.E., & Eberl, D.D. 2004. Measurement of clay surface areas by polyvinylpyrrolidone (PVP) sorption and its use for quantifying illite and smectite abundance. *Clays Clay Miner.* 52, 589-602.

Boettinger, J.L., & Ming, D.W. 2002. Zeolites. pp. 585-610. In Dixon, J.B., & Schulze, D.G. (eds.) *Soil Mineralogy with Environmental Applications*. SSSA Book Series, no. 7. Soil Science Society of America, Madison, WI.

Bonna, R.J., Aulerich, R.J., Bursian, S.J., Poppenga, R.H., Braselton, W.E., & Watson, G.L. 1991. Efficacy of hydrated sodium aluminosilicate and activated charcoal in reducing the toxicity of dietary aflatoxin to mink. *Arch. Environ. Contam. Toxicol.* 20, 441-447.

Bradford, M.M. 1976. A refined and sensitive method for the quantitation of microgram quantities of protein utilizing the principle of protein-dye binding. *Analytical Biochemistry*, 72, 248-254.

Brekke, O.L., Sinnhuber, R.O., Peplinski, A.J., Wales, J.H., Putnam, G.B., Lee, D.J., & Ciegler, A. 1977. Aflatoxin in corn: ammonia inactivation and bioassay with rainbow trout. *Appl. Environ. Microbiol.* 34, 34-37.

Carter, D.L., Mortland, M.M., & Kemper, W.D. 1986. Specific surface. pp. 413-423 In: *Methods of Soil Analysis: Part I. Physical and Mineralogical Methods*, 2nd edtiion (A. Klute, editor). Soil Science Society of America, Madison, WI.

Chenu, C., Pons, C.H., & Robert, M. 1985. Interaction of kaolinite and montmorillonite with neutral polysaccharides. In: Schultz, L.G., van Olphen, H., & Mumpton, F.A. (eds.) *Proceedings of the International Clay Conference 1985, Denver, CO.* pp. 375-381, The Clay Minerals Society, Bloomington, IN.

Dalvi, R.R. & Ademoyero, A.A. 1984. Toxic effect of aflatoxin B1 in chickens given feed contaminated with *Aspergillus flavus* and reduction of the toxicity by activated charcoal and some chemical agents. *Avin Dis.* 28, 61-69.

Dalvi, R.R. & McGowan, C. 1984. Experimental induction of chronic aflatoxicosis in chickens by purified aflatoxin B1 and its reversal by activated charcoal, phenobarbitol and reduced glutathione. *Poult. Sci.* 63, 485-491.

Decker, W.J. 1980. Activated charcoal adsorbs aflatoxin B1. *Vet. Hum. Toxicol.* 22, 388-389.

Deng, Y., Barrientos Velazquez, A.L., Billes, F., & Dixon, J.B. 2010. Bonding mechanisms between aflatoxin B1 and smectite. *Appl. Clay Sci.* 50, 92-98.

Diaz, D.E. & Smith, T.K. 2005. Mycotoxin sequestering agents: Practical tools for the neutralisation of mycotoxins. In Diaz, D.E. (editor) *The Mycotoxin Blue Book*. pp 323-340. Nottingham University Press.

Diaz, D.E., Hagler, W.M., Blackwelder, J.T., Eve, J.A., Hopkins, B.A., Anderson, K.L., Jones, F.T., & Whitlow, L.W. 2004. Aflatoxin binders II: reduction of aflatoxin M1 in milk by sequestering agents of cows consuming aflatoxin in feed. *Mycopathologia*, 157, 233-241.

Dixon, J.B., Kannewischer, Tenorio Arvide, M.G., & Barrientos Velazquez, A.L. 2008. Aflatoxin sequestration in animal feeds by quality-labeled smectite clays: an introductory plan. *Appl. Clay Sci.* 40, 201-208.

Eberl, D.D., Nüesch, R, Ŝuchá, V., & Tsipursky, S. 1998. Measurement of fundamental illite particle thicknesses by X-ray diffraction using PVP-10 intercalation. *Clays Clay Miner.* 46, 89-97.

Edrington, T.S., Sarr, A.B., Kubena, L.F., Harvey, R.B., & Phillips, T.D. 1996. Hydrated sodium calcium aluminosilicate (HSCAS), acidic HSCAS, and activated charcoal reduce urinary excretion of aflatoxin M1 in turkey poults. Lack of effect by activated charcoal on aflatoxicosis. *Toxicol. Lett.* 89, 115-122.

Fairchild, A.S., Croom, J., Grimes, J.L., & Hagler, W.M. 2008. Effect of Astra-Ben 20A on broiler chicks exposed to aflatoxin B1 or T2 toxin. *Int. J. Poult. Sci.* 7, 1147-1151.

Gallo, A., Masoero, F., Bertuzzi, T., Piva, G., & Pietra, A. 2010. Effect of the inclusion of adsorbents on aflatoxin B1 quantification in animal feedstuffs. *Food Additives and Contaminants* 27, 54-63.

Garwood, G.A., Mortland, M.M., & Pinnavaia, T.J. 1983. Immobilization of glucose oxidase on montmorillonite clay: hydrophobic and ionic modes of binding. *Journal of Molecular Catalysis*, 22, 153-163.

Grant, P.G. & Phillips, T.D. 1998. Isothermal adsorption of aflatoxin B1 on HSCAS clay. *J. Agric. Food Chem.* 46, 599-605.

Grove, M.D., Plattner, R.D., & Weisleder, D. 1981. Ammoniation products of an aflatoxin model coumarin. *J. Agric. Food Chem.* 29,1161-1164.

Hartley, R.D., Nesbitt, B.F., & O'Kelly, J. 1963. Toxic metabolites of *Aspergillus flavus. Nature* 198,1056-1058.

Hatch, R.C., Clark, J.D., Jain, A.V., & Weiss, R. 1982. Induced acute aflatoxicosis in goats: treatment with activated charcoal or dual combinations of oxytetracycline, stanozol and activated charcoal. *Am. J. Vet. Res.* 43, 644-648.

Herrera, M., Zyman, J., Pena, J.G., Segurajauregui, J.S., & Vernon, J. 1986. Kinetic studies on the alkaline treatment of corn (Zea mays) for tortilla preparation. *J. Food Sci.* 51, 1486-1490.

Hiemenz, P.C. 1986. *Principles of Colloid and Surface Chemistry.* Marcel Dekker, New York, 398-407.

Hofmann, U. & Klemen, R. 1950. Verlust der Austauschfahigkeit von Lithiumen an Bentonit durch Erhitzung: *Z. Anorg. Allg. Chem.* 262, 95-99.

Hosterman, J.W., & Patterson, S.H. 1992. Bentonite and fuller's earth resources of the United States: *U.S. Geological Survey Professional Paper* 1522. U.S. Government Printing Office.

IARC. 1993. Aflatoxins: Naturally occurring aflatoxins (Group 1), aflatoxins M1 . *Int. Agency Res. Cancer* 56, 245.

Jaynes, W.F. & Bigham, J.M. 1987. Charge reduction, octahedral charge, and lithium retention in heated, Li-saturated smectites. *Clays Clay Miner.* 35, 440-448.

Jaynes, W.F., & Boyd, S.A. 1991. Clay mineral type and organic compound sorption by hexadecyltrimethylammonium-exchanged clays. *Soil Sci. Soc. Am. J.* 55, 43-48.

Jaynes, W.F., Zartman, R.E., & Hudnall, W.H. 2007. Aflatoxin B1 adsorption by clays from water and corn meal. *Appl. Clay Sci.* 36, 197-205.

Jaynes, W.F., Zartman, R.E., Green, C.J., San Francisco, M.J., & Zak, J.C. 2005. Castor toxin adsorption to clay minerals. *Clays Clay Miner.* 53, 268-277.

Katzin, B.J., Collins, E.J., & Robertus, J.D. 1991. The structure of ricin A chain at 2.5 Å. *Proteins: Structure, Function, and Genetics,* 10, 251-259.

Kubena, L.F., Harvey, R.B., Phillips, T.D., Corrier, D.E., & Huff, W.E. 1990. Diminution of aflatoxicosis in growing chickens by the dietary addition of hydrated sodium calcium aluminosilicate. *Poult. Sci.* 69, 727-735.

Lagaly, G. & Weiss, A. 1976. The layer charge of smectitic layer silicates: In *Proc. Int. Clay Conf., Mexico City, 1975,* S.W. Bailey, ed., Applied Publishing, Wilmette, IL, 157-172.

Laird, D.A., Scott, A.D., & Fenton, T.E. 1989. Evaluation of the alkylammonium method of determining layer charge. *Clays Clay Miner.* 37, 41-46.

Lipson, S.M., & Stotzky, G. 1984. Effect of proteins on Reovirus adsorption to clay minerals. *Appl. Environ. Micro.* 48, 525-530.

MacEwan, D.M.C., & Wilson, M.J. 1984. Interlayer and intercalation complexes of clay minerals. pp. 197-248. In Brindley, G.W., & Brown, G. (eds.) *Crystal Structures of Clay Minerals and their X-ray Identification*. Mineralogical Society Monograph No. 5. Mineralogical Society, London.

Magnoli, A.P., Tallone, L, Rosa, C.A.R., Dalcero, A.M., Chiacchiera, S.M., & Torres Sanchez, R.M. 2008. Commercial bentontites as detoxifier of broiler feed contaminated with aflatoxin. *Appl. Clay Sci.* 40, 63-71.

Manahan, S.E. 2000. *Environmental Chemistry*. 7th edition. Lewis Publishers, New York. p. 237, 247.

Mayland, H.F.,, Florence, A.R., Rosenau, R.C., Lazar, V.A., & Turner, H.A. 1975. Soil ingestion by cattle on semiarid range as reflected by titanium analysis of feces. *J. Range Manage.* 28, 448-452.

Merck. 2001. *The Merck Index, An Encyclopedia of Chemicals, Drugs and Biologicals.* Merck Research Laboratories Division of Merck & Co., Inc. Whitehouse Station, NJ. p. 8290.

Mermut, A.R., & Lagaly, G. 2001. Baseline studies of the Clay Minerals Society Source Clays: Layer-charge determination and characteristics of those minerals containing 2:1 layers. *Clays Clay Miner.* 49, 393-397.

Miazzo, R., Rosa, C.A.R., De Queiroz Carvalho, E.C., Magnoli, C., Chiacchiera, S.M., Palacio, G., Saenz, M., Kikot, A., Basaldella, E., & Dalcero, A. 2000. Efficacy of synthetic zeolite to reduce the toxicity of aflatoxin in broiler chicks. *Poult. Sci.* 79, 1-6.

Murphy, P.A., Hendrich, S., Landgren, C., & Bryant, C. M. (eds.). 2006. Food mycotoxins: An update. *J. Food Sci.* 71, R51-R65.

Newman, A.C.D., & Brown, G. 1987. The chemical constitution of clays. In: Newman, A.C.D. (ed.) *Chemistry of Clays and Clay Minerals*. p. 56. Mineralogical Society Monograph No. 6, Longman Scientific & Technical, Essex, England.

Odom, I.E. 1984. Smectite clay minerals: properties and uses. *Phil. Trans. R. Soc. Lond. A* 311, 391-409.

Perez-Castells, R., Alvarez, A., Gavilanes, J., Lizarbe, M.A., Martinez Del Pozo, A., Olmo, N., & Santaren, J. 1985. Adsorption of collagen by sepiolite. In: Schultz, L.G., van Olphen, H., & Mumpton, F.A. (eds.) *Proceedings of the International Clay Conference, Denver, CO 1985*, pp. 359-362, The Clay Minerals Society, Bloomington, IN.

Phillips, T.D., Kubena, L.F., Harvey, R.B., Taylor, D.R., & Heidelbaugh, N.D. 1988. Hydrated sodium aluminosilicate: a high affinity sorbent for aflatoxin. *Poult. Sci.* 67, 243-247.

Phillips, T.D., Sarr, A.B., & Grant, P.G. 1995. Selective chemisorption and detoxification of aflatoxins by phyllosilicate clay. *Natural Toxins* 3, 204-213.

Pimpukdee, K., Kubena, L.F., Bailey, C.A., Huebner, H.J., Afriyie-Gyawu, E., & Phillips, T.D. 2004. Aflatoxin-induced toxicity and depletion of hepatic vitamin A in young broiler chicks: Protection of chicks in the presence of low levels of Novasil Plus in the diet. *Poult. Sci.* 83, 737-744.

Ralla, K., Sohling, U., Riechers, D., Kasper, C., Ruf, F., & Scheper, T. 2010. Adsorption and separation of proteins by a smectitic clay mineral. *Bioprocess Biosyst. Eng.* 33, 847-861.

Rao, S.B.N. & Chopra, R.C. 2001. Influence of sodium bentonite and activated charcoal on aflatoxin M1 excretion in milk of goats. *Small Ruminant Research* 41, 203-213.

Rutenber, E., Katzin, B.J., Ernst, S., Collins, E.J., Mlsna, D., Ready, M.P., & Robertus, J.D. 1991. Crystallographic refinement of ricin to 2.5 Å. *Proteins: Structure, Function, and Genetics*, 10, 240-250.

Scheideler, S.E. 1993. Effects of various types of aluminosilicates and aflatoxin B1 on aflatoxin toxicity, chick performance, and mineral status. *Poult. Sci.* 72, 282-288.

Schell, T.C., Lindemann, M.D., Kornegay, E.T, Blodgett, D.J., & Doerr, J.A. 1993a. Effectiveness of different types of clay for reducing the detrimental effects of aflatoxin-contaminated diets on performance and serum profiles of weanling pigs. *J. Anim. Sci.* 71, 1226-1231.

Schell, T.C., Lindemann, M.D., Kornegay, E.T., & Blodgett, D.J. 1993b. Effects of feeding aflatoxin-contaminated diets with and without clay to weanling and growing pigs on performance, liver function, and mineral metabolism. *J. Anim. Sci.* 71, 1209-1218.

Sefa-Dedeh, S., Cornelius, B., Sakyi-Dawson, E., & Afoakwa, E. O. 2004. Effect of nixtamalization on the chemical and functional properties of maize. *Food Chem.* 86, 317-324.

Seifert, L.E., Davis, J.P., Dorner, J.W., Jaynes, W.F., Zartman, R.E., & Sanders, T.H. 2010. Value-added processing of peanut meal: aflatoxin sequestration during protein extraction. *J. Agric. Food Chem.* 58, 5625-5632.

Singer, A. 2002. Palygorskite and sepiolite. pp. 555-586. In Dixon, J.B., & Schulze, D.G. (eds.) *Soil Mineralogy with Environmental Applications*. SSSA Book Series, no. 7. Soil Science Society of America, Madison, WI.

Squire, R.A. 1981. Ranking animal carcinogens: a proposed regulatory approach. *Science* 194, 877-880.

Stryer, R. 1975. Introduction to protein structure and function. Chapter 2 In: *Biochemistry*. W.H. Freeman & Company. San Francisco, CA.

Sweets, L.E., & Wrather, J.A. 2009. Aflatoxin in corn. University of Missouri, Delta Research Center. http://aes.missouri.edu/delta/croppest/aflacorn.stm.

Thieu, N.Q., & Pettersson, H. 2008. *In vitro* evaluation of the capacity of zeolite and bentonite to adsorb aflatoxin B1 in simulated gastrointestinal fluids. *Mycotoxin Research* 24, 124-129.

van Olphen, H., & Fripiat, J.J. 1979. Data Handbook for Clay Materials and other Non-metallic Minerals. p. 198. Pergamon Press, New York.

Vector Labs. 2011. Peanut agglutinin. www.vectorlabs.com/catalog.aspx? prodID=265.

Vekiru, E., Fruhauf, S., Sahin, M., Ottner, F., Schatzmayr, G., & Krska, R. 2007. Investigation of various adsorbents for their ability to bind aflatoxin B1. *Mycotoxin Research* 23, 27-33.

Veldman, A. 1992. Effect of sorbentia on carry-over of aflatoxin from cow feed to milk. *Michwissenschaft* 47, 777-780.

Vincelli, P., Parker, G., & McNeill. 1995. Aflatoxins in corn. Cooperative Extension Service, University of Kentucky. www.ca.uky.edu/agc/pubs/id/id59/id59.pdf.

Williams, J.H., Phillips, T.D., Jolly, P.E., Stiles, J.K., Jolly, C.M., & Aggarwal, D. 2004. Human aflatoxicosis in developing countries: a review of toxicology, exposure, potential health consequences, and interventions. *Am. J. Clin. Nutr.* 80, 1106-1122.

Winfree, R.A., & Allred, A. 1992. Bentonite reduces measurable aflatoxin B1 in fish feed. *The Progressive Fish-Culturist* 54, 157-162.

Zilinskas, R.A. Iraq's biological weapons: The past as future? *JAMA* 278, 418-424.